Argonne National Laboratory, 1946–96

Argonne National Laboratory, 1946–96

JACK M. HOLL

With the assistance of
Richard G. Hewlett and Ruth R. Harris

Foreword by
Alan Schriesheim

UNIVERSITY OF ILLINOIS PRESS

URBANA AND CHICAGO

© 1997 by the Board of Trustees of the University of Illinois
Manufactured in the United States of America
C 5 4 3 2 1

This book is printed on acid-free paper.

Unless otherwise noted, the photographs reproduced in this book were supplied by the Argonne National Laboratory.

Library of Congress Cataloging-in-Publication Data
Holl, Jack M.
Argonne National Laboratory, 1946–96 / Jack M. Holl ; with the assistance of Richard G. Hewlett and Ruth R. Harris ; foreword by Alan Schriesheim.
p.   cm.
Includes bibliographical references and index.
ISBN 0-252-02341-2 (cloth : alk. paper)
1. Argonne National Laboratory—History.
I. Hewlett, Richard G.
II. Harris, Ruth R.
III. Title.
QC789.2.U62A754   1997
621.042'07'277324—dc21
96-50031   CIP

Contents

Foreword, by Alan Schriesheim	ix
Acknowledgments	xv
Introduction	xix
1. Met Lab Days, 1940–46	1
2. A Troublesome Transition, 1946–48	47
3. Living in a Political World, 1949–52	81
4. The Reactor Laboratory, 1953–58	107
5. Accelerators and Basic Research, 1953–57	152
6. What Sort of Laboratory? 1957–61	175
7. Targeting Midwestern Science, 1961–67	207
8. New Research Priorities, 1967–72	242
9. Years of Crisis, Years of Challenge, 1973–77	278
10. The Multiprogram Laboratory, 1977–81	333
11. On the Brink, 1981–84	377
12. Challenging the Nuclear Option, 1984–94	430
13. New Paths to the Future, 1984–94	459
Appendixes	
1. Argonne National Laboratory Funding and Employment, 1979–94	505
2. Budgets and Staffing of DOE Multipurpose Laboratories, 1995	506
3. Argonne National Laboratory Key Personnel	507
4. Selected Argonne Projects and Technology Highlights, 1946–96	509
5. Argonne National Laboratory Nuclear Reactor Program	514
6. Multiprogram National Laboratories	517
Notes	519
Index	611

Illustrations

1. Albert Michelson and Arthur Compton — 4
2. Stagg Field stands — 11
3. 3:22 P.M., December 2, 1942 — 18
4. Chianti bottle — 19
5. CP-2 assembled in Argonne Forest Preserve — 23
6. Ernest O. Lawrence, Enrico Fermi, and Isidor Isaac Rabi — 30
7. Glenn T. Seaborg in his Met Lab office — 38
8. W. H. Zinn and Farrington Daniels — 41
9. Fourth anniversary reunion of CP-1 scientists — 45
10. Argonne National Laboratory site with farms — 52
11. Argonne National Laboratory, Palos Park Unit — 56
12. Quonset huts — 67
13. EBR-I lights four bulbs — 108
14. CP-2, CP-3, and CP-5 reactors — 111–12
15. View of Experimental Breeder Reactor I — 115
16. CP-5, Argonne's "work horse" research reactor — 116
17. BORAX I, the "runaway reactor" — 119
18. Explosion of BORAX I — 121
19. ORACLE computer with Margaret Butler and Rudolph Klein — 124
20. The Experimental Boiling-Water Reactor (EBWR) — 145
21. Norman Hilberry — 185
22. ANL Main Administration — 186
23. Albert Crewe and high-contrast electron microscope — 210
24. Xenon tetrafluoride — 226
25. Maria Goeppert-Mayer — 228
26. Robert B. Duffield — 243
27. Argonne's first four directors — 253
28. Robert G. Sachs — 292
29. Hot-Fuel Examination Facility — 296
30. Robert G. Sachs with Dixy Lee Ray — 299

31.	Zero Gradient Synchrotron complex	326
32.	Walter E. Massey	352
33.	The primary system of EBR-II in cutaway perspective	366
34.	Administration Building 201	405
35.	High Voltage Electron Microscope	410
36.	Argonne Tandem-Linac Accelerator System (ATLAS)	417
37.	Argonne's snow-white fallow deer	431
38.	Argonne-East, Illinois	431
39.	Argonne-West, Idaho	432
40.	Alan Schriesheim	435
41.	Integral Fast Reactor	445
42.	Advanced Photon Source (APS)	468
43.	Levitation, by Joseph W. Paulini	476
44.	Ronald Reagan and Alan Schriesheim	477
45.	Sam Bowen and the West Garfield Park Explorers	491
46.	Reactor Tree	515

Foreword

Alan Schriesheim

> Science is intimately integrated with the whole social structure and cultural tradition. They mutually support one another—only in certain types of society can science flourish, and conversely without a continuous and healthy development and application of science such a society cannot function properly.
> —Talcott Parsons (1951)

National laboratories are crucibles of uncertainty. While their scientists eagerly probe the frontiers of knowledge, never knowing where their research will lead, their managers grapple daily with questions of relevance and survival. How to anticipate society's emerging technology needs? How to mount credible research programs to meet those needs? How to secure government funding and public support and an ongoing national commitment to the future? The challenge is daunting, and success is ephemeral.

Throughout its history, the United States national laboratory system has been a microcosm of the nation's evolving science policy and its inconstant political mood. And like its counterparts, Argonne National Laboratory, the venerable elder statesman of the system, has persisted for the past half century by constantly adjusting its course to accommodate vacillating political priorities, changing governmental leadership, and shifting social concerns.

The story of Argonne's first fifty years is replete with scientific accomplishments on behalf of society; they have been well documented elsewhere. The focus of this book is on the political and social *context* of those achievements—on how Argonne and the "whole social structure" have interacted through the years to advance the science and technology agenda, and on how Argonne has struggled to become and to remain a vital national resource.

The establishment of Argonne National Laboratory in 1946 helped lay the foundation for an extensive coast-to-coast infrastructure that would develop new energy sources and pursue scientific knowledge on all fronts. The United

States made a huge investment in its intellectual, scientific, and technology resources. But to be successful, science needs both vision and long-term commitment; progress is measured in years and decades, not months and annual budget cycles. In order to survive, the labs are obliged to take cognizance of and respond to immediate societal demands. They must expend precious energy and capital coping with an environment of constant political and social flux—with programmatic and budgetary fluctuations that may have little or nothing to do with the true merit or purpose of their scientific and technical capabilities. As Jack Holl convincingly documents in this history, the extraordinary accomplishments of Argonne and the other national laboratories are all the more remarkable when viewed in their often chaotic political context.

■

By the time I arrived at Argonne in 1983, the laboratory already had experienced a lifetime of political ups and downs. When Argonne first emerged in the mid-1940s from the University of Chicago's Metallurgical Laboratory and its historic achievement—the world's first controlled, self-sustaining nuclear chain reaction—its work remained under military auspices pending the establishment of the Atomic Energy Commission (AEC). Argonne's initial mission, in fact, was to develop reactors in part for United States military needs. And one goal of that program was to demonstrate to the Soviets—who by then had achieved a nuclear reaction of their own—the superior muscle of American military power.

Yet even in those early years, the laboratory's role and purpose were being questioned: Was it a regional basic research lab, or a government center for engineering development? In 1949, the timid commissioners on the two-year-old AEC were loath to fully define Argonne's responsibilities. Argonne's work on radioisotopes was summarily halted, and its budget cut by 10 percent.

Between 1949 and 1952, a vacillating AEC was hardly the only problem dogging Argonne. The Soviet detonation of an atomic bomb and the Korean incursion drastically altered national priorities and attitudes. An atmosphere poisoned by suspicion and conspiracy bred perceptions of lax security and mismanagement at Argonne, prompting congressional investigations, more budget cuts, and staff shortages.

The election of President Dwight D. Eisenhower in 1952 signaled another shift in the political agendum. His Atoms for Peace program set the stage for commercialization of reactor energy; and it became a central theme in his foreign policy. Nevertheless, Eisenhower did not support large budgets to fund the initiative. Argonne needed a new mission. The Korean War and then the shock of Sputnik restored science and technology problem-solving to its former prominence among the nation's priorities. Argonne received funding to build a new Zero Gradient Synchrotron (ZGS)—the world's leading proton accelerator—and quickly gained a reputation as one of the nation's foremost research institutions for high-energy physics.

John F. Kennedy's New Frontiers program embraced nuclear power for civilian uses as part of his commitment to science and space technology. The AEC expanded its already aggressive breeder-reactor development effort and launched the Fast Reactor Test Facility at Argonne-West in Idaho. But the project was short-lived—doomed by internal AEC politics. Argonne's autonomy in basic reactor research began to crumble under the weight of micro management from Washington; the original postwar concept of government-owned, contractor operated laboratories—designed to operate independently of short-term political pressures—fell by the wayside.

Though embroiled in the Vietnam War and confronting issues of nuclear nonproliferation, Lyndon Johnson and his Great Society also embraced nuclear power. But the public's growing concerns about air and water pollution demanded action on other fronts. The Clean Air Act of 1967 marked the rise of environmentalism to the top of the national agenda along with public health and safety programs. The new "cultural tradition" generated a major new mission for Argonne: the "improvement of man's environment." At the same time, however, the laboratory's basic science programs went into decline. Argonne's high-flux reactor project was terminated in 1968, and its high-energy physics program began to lose its luster with the construction of the Fermi National Accelerator Laboratory in Argonne's backyard. Morale suffered further as a result of ensuing budget cuts and staff reductions in the physical sciences, mathematics, and high-energy physics.

By the end of the decade, increasingly frequent "brownouts" threatened fuel shortages, and rising gasoline prices brought yet another shift in the focus of public debate and a renewed emphasis on Argonne's historic mission: energy research and development. Argonne's breeder funding was again expanded during the Nixon and Ford administrations (although its management came increasingly from Washington); and the ZGS, now fully eclipsed by Fermilab, was replaced by the Intense Pulsed Neutron Source.

When Jimmy Carter took office in 1977, the national focus shifted to conservation and coordinated energy programs; and once again, the breeder program was threatened by partisan attacks under the rubric of a mounting concern about nuclear proliferation. Carter institutionalized the halt to commercial reprocessing and recycling of plutonium. Long-term storage of spent reactor fuel became the only alternative for its disposal. Thus the nuclear waste problem was exacerbated, and public concerns about the environment were rekindled.

The 1980s ushered in another era of close scrutiny and declining funds. In 1981, nearly one out of every eight Argonne staff positions was eliminated. The following year brought government reviews of all the national laboratories; programs and missions were reassessed. The Clinch River Breeder Reactor in Tennessee, conceptually linked to Argonne's breeder program, was finally terminated in 1983 after a long congressional battle. Argonne was criticized for "scientific and institutional introversion." The perception grew that the laboratory had fallen into drift and decline.

Ronald Reagan threatened to dismantle the Department of Energy, the AEC's successor, and move its research activities to the Department of Commerce—foreshadowing years of debate over the size and role of the federal government. Congress resisted, but the argument continued: Who would oversee the nation's reactor program? Reagan thought that the free market should determine energy priorities; that private industry should provide energy technologies. His administration shifted priorities from civilian programs to weapons technology. Argonne's programs fell into ever-deepening jeopardy; with the least political support of all the national laboratories, Argonne was widely rumored to be slated for closure.

■

This, then, was the situation at Argonne when I assumed the post of director in 1984. The laboratory had reached its nadir in terms of both funding and morale. Forty years of buffeting by the capricious political winds had taken their toll on every aspect of the laboratory—its programs, its infrastructure, and especially, its people. The need for dramatic new directions was apparent.

Equally apparent was the necessity of taking a long-term, strategic approach to Argonne's missions and programs. A national laboratory is defined by its initiatives; it must keep moving forward or die. Argonne had to develop fresh initiatives and new objectives to restore its health—more initiatives, even, than could possibly be funded. Its programs had to anticipate, and respond to, the nation's long-term scientific and technological needs. Only by tuning the laboratory's work to meet national needs could we hope to enlist the political support required to ensure continued funding.

With the wise counsel and strong support of my predecessor, Walter Massey—who had moved to Argonne's parent institution, the University of Chicago—of the university's politically astute Board of Governors for Argonne, and of my colleagues at all levels of Argonne's management, we set out to do just that: to make Argonne once again an essential element of the nation's science and technology infrastructure. The advanced nuclear power reactor program was revived with the development of Argonne's Integral Fast Reactor, or IFR. The IFR was designated to be both inherently safe—shutting itself down automatically in case of emergency—and nearly waste-free, with a closed fuel cycle enabling it to recycle its nuclear waste into new fuel. Argonne successfully competed for the Advanced Photon Source (APS), which, with the cancellation of the Superconducting Supercollider, became the crown jewel among Department of Energy research facilities; construction of the APS was completed in 1995, well ahead of schedule and under budget. With users clamoring for access to this state-of-the-art x-ray machine, Argonne's niche in the research firmament was secured for the next two decades at least.

The laboratory also stepped up its collaborations with business and industry, producing not only scientific breakthroughs, but also economic benefits for the Midwest and the nation as a whole. Its environmental and transportation programs grew steadily, and the basic sciences regained strength as Argonne achieved successes and new funding in high-temperature superconductivity, materials science, nuclear and high-energy physics, biology, and chemistry. Long a leader in high-performance computing, the laboratory became a key player in the design of the much-vaunted "information superhighway," devising practical scientific applications for such exotic technologies as immersive virtual reality and telerobotics.

The ups and downs continue, of course. The end of the Cold War occasioned yet another series of reevaluations of the national laboratory system and a demand for clearer, more differentiated, and more cost-effective missions among the individual laboratories. Due to a changing political climate, Argonne's Integral Fast Reactor program was canceled, thereby eliminating the major United States option to destroy completely plutonium and other long-lived fission products and to produce energy at the same time. The political winds have "redirected" Argonne's nuclear expertise to focus on new national priorities: spent-fuel and nuclear waste treatment and disposal; reactor safety, decommissioning, and decontamination.

Thus, as Argonne and other national laboratories enter their second half century, their fate is, as it always has been, in their own hands. To the extent that the labs are able to demonstrate their unique value to society—their relevance to national needs—their future is assured. Knowing the skills, expertise, and experience of the men and women with whom I have had the pleasure of working for the last thirteen years, I have every confidence they will not only survive, but also thrive. The scientific and technological vistas which the national laboratories are uniquely equipped to explore are far too compelling, too promising, and too important to the future of our society for them to be overshadowed by the volatile politics of the moment.

Acknowledgments

This project has enjoyed the support of Argonne National Laboratory since 1991. After reading Richard Rhodes, *The Making of the Atomic Bomb* and inviting Rhodes to offer a seminar on nuclear history at Argonne, laboratory director Alan Schriesheim decided to sponsor a fiftieth-anniversary history of Argonne. Desiring a history that was as objective and scholarly as possible, Schriesheim appointed a laboratory history committee to provide advice and assistance on the project.

On recommendation from the history committee, History Associates Incorporated, a consulting firm from Rockville, Maryland, was commissioned to write the history. In turn, History Associates asked me to serve as the principal investigator and author on the project. I agreed to do so with the understanding that we be given full and complete access to Argonne records and be allowed to interview key laboratory personnel. In addition, I insisted that we be granted unfettered editorial freedom concerning the organization, contents, interpretation, and analysis of Argonne's history.

Our challenge was to write the history of a complex scientific laboratory for a wide and diverse audience. We strove to tell a narrative story that not only was meaningful and useful to the Argonne community, but also would be of interest to scholars, and, hopefully, the general public. The Argonne history committee, which served as our technical advisors, reviewed the manuscript, offering their opinions concerning historical fact, interpretation, and style. We were free, however, to accept or reject the history committee's suggestions. I want to thank the Argonne National Laboratory history committee for their assistance and review of this work. Their historical memories and technical advice have enriched the history. Serving on the committee were: Robert Avery, Ellis Steinberg, Robert Rowland, Joel Snow, Harry Conner, Cindy Wilkinson, Ralph Seidensticker, Paul Fields, James LeFevers, Charlie Osolin, and Russell Heubner.

History Associates Incorporated provided administrative support for the project. I would like to thank Philip Cantelon and Ruth Dudgeon for the hours they spent patiently helping me prepare this history. In addition to extensive research, Ruth Harris drafted some early chapters of the history.

I owe a special dept of gratitude to an old colleague and close friend, Richard G. Hewlett, who provided invaluable assistance, carefully read the manuscript, and virtually redrafted some chapters while recommending important changes in others. Dian Belanger, working closely with Hewlett, helped reorganize and redraft early chapters. Historians at the Department of Energy—Benjamin Franklin Cooling, Skip Gosling, Betsy Scroger, Terry Fehner, and Roger Anders—also offered important suggestions that strengthened the manuscript. Finally, three scholars selected by the University of Illinois Press provided additional review and useful advice that further improved the history.

Special acknowledgments for assistance with the project are due to the following:

Oral history interviews were conducted with Joseph Asbury, Robert Avery, Albert Crewe, Harvey Drucker, Robert Duffield, Paul Fields, Frank Fradin, Walter Massey, David Moncton, Robert Rowland, Robert Sachs, Alan Schriesheim, Charles Till, Albert Wattenberg, and Walter Zinn.

Ruth Harris organized much of the initial research for the Argonne history. Her efficient and exhaustive research was made possible, in part, by the excellent survey of Argonne records that had been compiled previously by Allan Needell. Harris was ably assisted by Ann Deines, Robert Bauman, and Michelle Hanson, all of History Associates Incorporated. Two graduate students, Shawn Woodford and Theodore Tucker, provided me excellent research assistance in Manhattan, Kansas.

The following persons at Argonne National Laboratory provided help and assistance: Technical Information Services: Joyce Kopta and David Hamrin; Records Center: Mary Wilson Haider and Rose Urquiza; Administration: Joseph Asbury, Pat Canaday, Marilyn Wittkovski, and Dianne Hutchison-Wray; Photo Library: Rhoda Shives; Tour hosts: Bruce Brown, Thomas Morgan, C. Arthur Youngdahl, Don Hopkins; Argonne-West: Richard Lindsay, John Sackett, Greg Hackett, and Rodney Harris. I particularly want to thank Cindy Wilkinson for her excellent help and commitment to this project.

I want to acknowledge the help of the following persons from the National Archives and Records Administration: in Chicago—Shirley Burton, Don Jackanicz, Beverly Watkins; in Seattle—Joyce Justice; and in Washington, D.C.—Edward Reese, Terese Hammett, Wilbert Mahoney, John Taylor, Aloha South, Herbert Rawlings-Milton, Jeanne Schauble, Joann Williamson, Ron Servino, Marjorie Ciarlante, and Charles South. In addition, we were assisted by the University of Chicago Special Collections: Richard Popp.

The editorial staff of the University of Illinois Press has greatly improved this history. Richard Martin, executive editor, Theresa L. Sears, managing editor, and Gail K. Schmitt, copy editor, offered patient support for this project.

Finally, I want to thank my wife, Jacqueline, who encouraged me to undertake the Argonne history project. Jacqueline is also the mother of Inga and Mark, professors of chemistry who provided invaluable technical advice, and of Kerstin who gave limitless support.

Introduction

United States national laboratories were Cold War institutions established after World War II to assure continued American preeminence in nuclear science and technology. After the war, the federal government not only found itself deeply involved in fostering science and technology, but, insofar as atomic energy was concerned, the government also claimed an exclusive monopoly. Such extensive government entanglement with science was unprecedented in American history, and the nation's science institutions were unprepared for it. When official sponsorship of science and technology included direct management of basic science research, government involvement transcended encouragement of science and entered the unfamiliar realm of control. The outbreak of the Cold War guaranteed that, for the near future, the government would maintain its control over nuclear research and development for both war and peace.

Following World War II, the civilian Atomic Energy Commission (AEC) inherited a make-shift laboratory system from the Army's Manhattan Engineer District (MED). In the rush to build the atomic bomb the Army Corps of Engineers created the Los Alamos Weapon Laboratory in New Mexico, supported by two major research and production facilities at Oak Ridge, Tennessee, and Hanford, Washington. In addition, the MED adopted two university research laboratories into the Manhattan Project: Ernest Lawrence's physics laboratory at the University of California at Berkeley and the University of Chicago's Metallurgical Laboratory, where Enrico Fermi achieved the first controlled chain reaction in an atomic pile on December 2, 1942. The two university research laboratories provided vital information about the atom (Glenn Seaborg, the discoverer of plutonium worked at Berkeley and at Chicago), but by 1945 both Lawrence's laboratory and the Met Lab were no longer on the front lines of the Manhattan Project research and development. The Met Lab was so decimated after top scientists left for Los Alamos, Oak Ridge, or Hanford that the University of Chicago considered closing the facility. The army, believing that the Met Lab would become a vital nuclear research facility after the war, convinced the University of Chicago to keep the laboratory open.

This book traces the history of Argonne National Laboratory, which was designated the nation's first national laboratory on July 1, 1946. For a half century, Argonne has been an innovative regional, as well as principal national, research and development laboratory. Yet, despite the apparent security of belonging to the national laboratory system, Argonne has continually struggled to define its mission and identity. Heir to the Met Lab, Argonne was organized by the U.S. Army Corps of Engineers six months prior to the creation of the Atomic Energy Commission. The remains of the Metallurgical Laboratory, which had been hastily assembled to meet wartime needs, were housed in temporary quarters scattered over a dozen sites in and around Chicago. Walter Zinn, Argonne's first director, not only had to consolidate laboratory facilities, but he also had to provide strong leadership in developing Argonne's mission and identity.

While Argonne prospered because of the Cold War, as part of the federal science establishment, the laboratory also helped contribute to the growing intensity of the Cold War rivalry with the Soviet Union. The Atomic Energy Act of 1946 directed the civilian AEC to manage and control the development of nuclear science and technology. Supported by Congress, the commission determined that the nuclear arms race with the Soviet Union required a vast scientific establishment for research, development, testing, and production of advanced nuclear weapons. In addition, and almost as important, the commission also promoted the development of nuclear technologies including power and production reactors. Even when national security was not directly at stake, the commission argued successfully that national prestige required the support of advanced research in such esoteric areas of nuclear science as high-energy physics. Thus, although Argonne had a limited and temporary mission of developing military reactors, as a "peaceful" nuclear laboratory it played a major role in building and maintaining American nuclear science as the most advanced in the world.

In the early Cold War years before Sputnik, the Atomic Energy Commission received the majority of the federal government's science research-and-development budget. In support of basic sciences, the AEC initially out spent the National Science Foundation (NSF) and the National Institutes of Health (NIH). Even after the Commission designated Argonne as the nation's principle laboratory for research and development of nuclear reactors, the laboratory continued to devote about 50 percent of its activities to important research programs in basic physics, biology and medicine, mathematics, and computer science.

Although the laboratory prospered during the Truman and Eisenhower administrations, government financial largesse stimulated by the Cold War did not solve all of Argonne's problems. Endless studies and reevaluations self-consciously worried about the laboratory's long-term prosperity; however, it was not Soviet competition that created Argonne's uncertainties about its future. Rather, the domestic politics of big science, not international competition in basic science, principally shaped the laboratory's history. One must be careful, however, not to define science politics simply as an external force promoting or hampering scientific advancement. Laboratory management and scientists frequently complain about the intrusion of politics into scientific affairs as though politics were extrinsic to the scientific enterprise. To the extent that "politics" involves the establishment of priorities or the distribution of scarce resources, then political decisions, perhaps based on sound scientific judgement, are also integral to the structure and execution of scientific research. Thus no "power struggle" or political settlement in this history is described in the simple bipolar language of the Cold War. The laboratory had to cope with national and local politics, rivalry with competing laboratories, jealously of academic institutions, disagreements with AEC managers, and tensions among its own staff.

Nor is the political milieu in which Argonne thrived depicted as necessarily negative or abnormal. This history does not presume that politics corrupts "pure" science. At the end of World War II, Vannevar Bush, a highly regarded engineer and science administrator who had headed the wartime Office of Scientific Research and Development (OSRD), became a leading advocate for government support of science research and development. In his now famous *Science—The Endless Frontier,* Bush tapped into a popular American dream that new frontiers lie just beyond the horizon. The Old West might be long lost in the romantic past, "... but the frontier of science remains," Bush declared. "It is in keeping with the American tradition—one which has made the United States great," he continued, "that new frontiers shall be made accessible for development by all American citizens." Bush advocated government support of the best basic science research in academia and research institutions. Like pioneers of old, the modern science explorer would enjoy the support and protection of the government in research expeditions that would benefit both science and the public at large. But like frontier explorers, Bush and his colleagues expected the government's directions to be general and its interference minimal. That is, normal government involvement would be generous, but otherwise benign.[1]

The scientists at Argonne National Laboratory proved adept explorers on the scientific frontier. In general, they accepted Bush's "frontier ethic" and became impatient when bureaucrats from Washington, D.C., attempted to direct their efforts or demanded close accountability. This was the political interference to which they objected so strongly. They did not see themselves as skilled political operatives within the federal science establishment. Rather, their "frontier spirit" often obscured from the scientists themselves their own deep involvement in the political process.

Argonne National Laboratory, 1946–96

1 | Met Lab Days, 1940–46

Argonne's history began dramatically in the feverish days following America's entrance into World War II. On December 2, 1942, almost a year after Japan's surprise attack on Pearl Harbor, scientists gathered expectantly in the West Stands at the University of Chicago's Stagg Field to witness one of the century's epochal experiments in nuclear physics. Like Columbian explorers sailing uncharted seas, they sensed the historic importance of their next landfall. Word discreetly circulated through the university's Metallurgical Laboratory that the Nobel laureate Enrico Fermi and his associates were about to split uranium atoms, creating the world's first self-sustained, controlled chain reaction. In addition to demonstrating that nuclear fission could be controlled in the laboratory, Fermi's team was also expected to determine the likelihood that similar, larger piles could breed plutonium, which was thought to be a key element in producing powerful atomic bombs. There was hope as well that the newly discovered atomic energy could be employed for peaceful uses, such as the generation of inexpensive electrical power. But as America and its allies emerged from the darkest hours of 1942, poised to roll back the German armies in North Africa and Europe and the Japanese forces in the South Pacific, the scientists gathered in Chicago had but one priority—to beat the Germans in the race for the atomic bomb.

Discovery

Berlin 1938. Following the lead of earlier research by Fermi in Rome, Otto Hahn and Fritz Strassman, scientists at the Kaiser Wilhelm Institute for Chemistry, discovered nuclear fission by bombarding uranium with neutrons. Four years previously, Fermi had conducted similar experiments but missed detecting fission. Word spread quickly through the international science community that Hahn and Strassman had found a barium isotope among their bombardment products. On December 19 Hahn, who immediately understood the significance of his findings, nevertheless wrote his former colleague, Lise Meitner, asking for "some fantastic explanation" about the suspected presence of barium. Meitner, who had fled to Sweden to escape Hitler's persecution, and her nephew, Otto R. Frisch, also quickly recognized the importance of the startling news from Berlin. While on holiday break, Meitner and Frisch hastily worked out the theory that Hahn and Strassman's neutron bombardment had split uranium atoms into two lighter elements. After Christmas, Frisch returned to Copenhagen, where he shared the discovery with Niels Bohr, the Danish physicist.[1]

Shortly thereafter, Bohr left Denmark to attend the fifth Washington Conference on Theoretical Physics and carried the news of the discovery of nuclear fission to America. Scarcely a month had passed since Hahn had written Meitner asking for help to explain his experimental results. American scientists, like those elsewhere, were electrified and rushed to confirm the Hahn-Strassman-Meitner results in their own laboratories. Before the end of the conference, the Carnegie Institution in Washington, Columbia University, Johns Hopkins University, and the University of California confirmed the Germans' success. (Meitner and Frisch, as well as Frederic Joliot-Curie in France, also obtained experimental confirmation.) The discovery of nuclear fission in 1938 was more than an exciting scientific breakthrough; scientists such as Isidor I. Rabi of Columbia University, Eugene Wigner and John A. Wheeler at Princeton, and Fermi, now at Columbia collaborating with the Hungarian refugee Leo Szilard, believed they were on the threshold of achieving a sustained chain reaction. If the process could be controlled, nuclear reactions might provide a new source of virtually unlimited heat and power. More ominously, if uncontrolled, such a reaction might produce an explosion of almost unimaginable magnitude.[2]

Both exhilarating and unsettling, the dawn of the nuclear age was filled with promise and fear. Was a chain reaction, controlled or uncontrolled, even possible? Fermi was cautious and deliberate. It would be best, he thought, to play down the possibility of a chain reaction until further experiments were completed. Even if the reaction were possible, it was not certain that a practical bomb could be built. Szilard, on the other hand, was not as imperturbable. What if the Germans proved the chain reaction first and then pushed ahead successfully to develop an atomic bomb? Szilard's blood ran cold. Even if the prospects were small, Szilard did not want to take the chance that Hitler might obtain an atomic bomb unchallenged by the United States. "Hitler's success could depend on it," he warned Edward Teller, a fellow Hungarian refugee.[3]

As war clouds gathered and broke over Europe in 1939, Szilard's urgency helped mobilize support from the federal government for neutron research. Although the Naval Research Laboratory (NRL) and the U.S. Army Bureau of Ordnance had already expressed interest in the possibilities of atomic energy, procurement and contracting procedures hampered the government's ability to respond quickly with substantial research support. Furthermore, Fermi's careful approach had convinced the navy not to push uranium research until additional experimental results were obtained. For Szilard, and fellow Hungarians Wigner and Teller, such caution in the face of Hitler's menace was exasperating. To secure President Franklin D. Roosevelt's support for their research, they persuaded Albert Einstein, also a refugee from Nazi Germany, to sign a letter to the president summarizing recent research and reviewing the prospects for atomic power and weapons. Although the United States remained neutral, prudently, on October 12, 1939, Roosevelt authorized the government to explore the possibilities outlined in Einstein's letter.[4]

To limit discussions to official circles, Roosevelt initially appointed the Advisory Committee on Uranium, which was chaired by Lyman Briggs, the director of the National Bureau of Standards. Briggs's committee, advised by Szilard, Wigner, and Teller, soon reported back to the president that the chain reaction was a possibility, but still unproven. In February 1940 the uranium committee granted Fermi and Szilard $6,000 to continue their pile experiments at Columbia. Concurrently, they authorized research on isotope separation as well.

Compton's Early Role

At the University of Chicago, Arthur Holly Compton (figure 1), chairman of the physics department and dean of the division of physical sciences, watched these developments with keen interest. The university's new president, Robert M. Hutchins, had recently declared that the university would favor intellectual activities over all others. Supported by his trustees, Hutchins abolished intercollegiate football at Chicago, leaving the famous Stagg Field, scene of football glory from another era, and its West Stands available for different triumphs. By coincidence, during that same period, European scientists had served as midwives for the nuclear age. The idea for Argonne National Laboratory was conceived when Compton resolved to foster atomic research at the University of Chicago.[5]

Regarded as brilliant, compassionate, and diplomatic by his colleagues, Compton received the Nobel Prize at the age of thirty-five in 1927 for discovering that the wavelength of x rays and gamma rays increased when they were scattered by material objects—an observation known as the Compton effect. He maintained an amiable relationship with distinguished scientists in the United States and Europe; his personal contacts ranged from Werner Heisen-

Figure 1. Albert Michelson and Arthur Compton

berg, the German physicist who won the 1932 Nobel Prize for his work in quantum and nuclear physics, to Ernest O. Lawrence, an American who won the 1939 Nobel Prize in physics. At the University of Chicago, Compton, who was also recognized for his cosmic-ray work, could confer with a number of noted scientists, including James Franck, a 1925 Nobel laureate for work in physical chemistry.[6]

On the eve of World War II, the identification of the neutron, the development of particle accelerators, the creation of artificial radioactive elements, and the discovery of nuclear fission made nuclear physics ready for applied research and development.[7] Efforts quickened in June 1940, days before the fall of France, when Roosevelt established the National Defense Research Committee (NDRC) under Vannevar Bush, the president of the Carnegie Foundation. Bush reorganized the uranium committee, which supported the chain-reaction and isotope studies underway at several institutions, including Columbia University, the University of Minnesota, and the University of California Radiation Laboratory at Berkeley. Nonetheless, Ernest O. Lawrence, director of Berkeley's Radiation Laboratory, impatiently complained to Bush that the government was too relaxed about uranium research.[8]

Compton also wanted to bring the University of Chicago into the fission research network. In the summer of 1940 he asked Volney C. Wilson, a young cosmic-ray physicist, to make calculations for a possible chain reaction. Although he completed calculations indicating that it was possible to make a bomb from uranium-235, Wilson, a pacifist and isolationist, declined Compton's invitation to participate further in what could become a bomb-development project.[9]

Compton's connections with prominent scientists already advising the federal government helped him secure participation of the University of Chicago. He expressed interest in uranium research to Harold C. Urey, a Nobel laureate physicist from Columbia University and a member of Briggs's scientific subcommittee. Through Urey's arrangement, Compton discussed potential research with Briggs in October. As a result, Compton obtained an NDRC contract for the University of Chicago to investigate self-sustaining fission in beryllium. To direct the project, Compton recruited Samuel K. Allison, a university physicist then working in Washington, D.C., with Richard Tolman. Allison was expected to collaborate closely with physicists at Columbia, who were investigating regenerative fission in ordinary uranium. Making the agreement retroactive to January 1, 1941, the NDRC signed the contract with the University of Chicago on April 14 and allocated $9,500 for approximately six months of work.[10]

By June 1941 uranium research in the United States was growing in tempo and dollars. Compton expected the University of Chicago to receive approximately one-fourth of the $1 million that the federal government might spend on the project. Allison already wanted $85,000 more for his project but received only another $30,000 to continue his group's uranium-beryllium-carbon experiment.[11]

Meanwhile, Bush tightened control over the research. Using the National Advisory Committee for Aeronautics (NACA) as his model, Bush proposed establishing the Office of Scientific Research and Development (OSRD) with civilians, rather than with military officers, in charge of the effort. On June 28 President Roosevelt approved Bush's plan to mobilize scientific research for national defense. The uranium committee became, first, the OSRD Section on Uranium, then later, the S-1 section. When he authorized cooperation with the British on uranium research on October 9, the president also asked Bush and James B. Conant, the new chief of the NDRC, to estimate the potential and the cost of developing an atomic bomb.[12] Meanwhile, unaware of Roosevelt's decision, Compton, as chair of the National Academy of Science's advisory committee on uranium, urged greater collaboration among scientists in the United States. On November 17 Compton recommended to Bush that some kind of central organization be formed to coordinate uranium research and to meet periodically to "keep the work going at high pitch."[13]

Shortly thereafter, galvanized by the Japanese attack on Pearl Harbor, nuclear scientists intensified their research on the military applications of atomic energy. Scientists such as Volney Wilson cast aside their pacifism and isolationism to join the effort. The coordinated research organization sought by Compton and his colleagues also began to coalesce after the United States entered the war. As Compton told Conant on December 10, "I am trying now to see what can be done to bring definite results out of the work at Columbia, Princeton, Chicago, and California." After consulting his colleagues, Bush presented an OSRD organization plan on December 13 that split the scientific research among Urey, who would explore separating uranium-235 from uranium-238 by diffusion or by centrifuge; Lawrence, who would supervise electromagnetic separation methods; and Compton, who would be in charge of physical research, including plutonium chemistry and the chain reaction tests.[14]

An Early Proposal for a National Laboratory

While uranium research spread among institutions across the country, on December 26, 1941, Henry D. Smyth, chairman of the physics department of Princeton University, proposed concentrating research efforts at a central laboratory for the project. Smyth told Compton that scientists in the S-1 section had already discussed the idea. For himself, Smyth advised that he would feel more comfortable letting faculty such as Eugene Wigner go to a central laboratory instead of granting leave to work at the University of Chicago.[15]

A few days later, Smyth endorsed Milton G. White's proposal to establish a central nuclear energy research laboratory as a permanent government institution. White, a consultant to the National Defense Research Committee, had been an experimental high-energy physicist working with Lawrence's cyclotron group. His suggestion may have been the earliest recognition of the dominant role the United States government would come to play in atomic energy. Captured by White's vision, Smyth forwarded the plan to the highest levels of the atomic bomb research. "It seems to me to be a very clear statement of the situation as it is likely to develop," Smyth predicted. On December 31, 1941, White drew up a detailed outline for a government laboratory "in an industrial region . . . to carry on after the war to develop both power and destructive aspects." To deal with the loose structure of the uranium project and temporary nature of some of the research, White believed that one large laboratory could create the central organization to reassign atomic scientists as they completed their projects. Before preparing his recommendations, White had conferred with Compton's brother, Karl, the president of the Massachusetts Institute of Technology, and Lee DuBridge, director of the MIT Radiation Laboratory.[16]

Consequently, in late 1941, key members of the scientific community already had acknowledged that the federal government should assume responsibility for conducting nuclear energy research in a permanent government laboratory. Moreover, Compton, who was to play a major role in organizing the effort, knew of White's proposal for such a lab before he established the research structure to develop the atomic bomb.

Compton Forms the Metallurgical Laboratory

With the United States now on a war footing, Compton became driven by the atomic bomb project. He was already responsible for coordinating the plutonium studies, but as important as this was, Compton knew that it was more urgent to demonstrate the chain reaction. Experiments and theoretical studies on nuclear reactions were under way at a dozen universities, but the research was badly coordinated. Compton was alarmed that scientists often did not agree on experimental data nor on research priorities. Although he could not centralize all efforts in one laboratory, Compton determined, if possible, to concentrate the chain-reaction studies in one place.

As early as January 1942 Compton referred to his group as the "metallurgical project at the University of Chicago." Adopted as a code name to preserve the secrecy of the uranium research, the project's designation no doubt borrowed from the university's attempts since 1936 to establish a metallurgical laboratory. As Compton explained, "Our Chicago colleagues saw nothing surprising in a wartime metals program." Under the cover of the Metallurgical Laboratory, Compton reported that alterations were already well under way to modify the squash courts in the Stagg Field West Stands to accommodate an atomic pile, so named because the scientists intended to pile materials on top of one another to attempt a chain reaction.[17]

After brief consideration of centralizing the uranium project under Lawrence at Berkeley, Compton decided that scientists working under his supervision at Columbia and Princeton should be transferred to the Metallurgical Laboratory in Chicago. On January 24, 1941, Compton notified leaders of the S-1 project that the work would be "centralized at Chicago as rapidly as possible." Compton later explained to his Princeton colleague, Henry D. Smyth, that "in setting up the Metallurgical Laboratory at Chicago I went against the recommendations of 75 percent of the advisors because I felt a greater confidence that I could see the job through at Chicago better than at any alternative site." In the first of numerous appointments and reorganizations, Compton chose Richard L. Doan, chief physicist of the Phillips Petroleum Company, as director of the laboratory; Fermi as coordinator of research; Wigner as chairman of the theoretical committee; Allison as chairman of the experimental committee; and Gregory Breit, a brilliant theoretical physicist from the University of Wisconsin, as coordinator of fast-neutron reaction studies at laboratories in the eastern and midwestern parts of the country. Doan's duties were primarily administrative and related to handling personnel and financial matters, and some of Wigner's responsibilities included design work. Compton reiterated that he, as project leader, was responsible personally for the project's development.[18]

Emerging as one of the government's largest scientific operations, the Metallurgical Laboratory coordinated ongoing research at the University of Chicago, the University of California at Berkeley, Columbia University, the University of Minnesota, Princeton University, and the National Bureau of Standards in Washington, D.C. Sometimes referred to as the Metallurgical Project, or more simply as the Met Lab, the Metallurgical Laboratory required solid business management that could nurture scientific work. In that respect, the University of Chicago was a logical choice to administer the project for the OSRD. "We will turn the University inside out if necessary to help win this war. Victory is much more important than the survival of the University," E. T. Filbey, the vice president of the University of Chicago, reportedly told Compton in February 1942, when university officials agreed to manage the laboratory; however, the University of Chicago's responsibilities extended only to the fiscal and business needs of the Metallurgical Laboratory. Compton supervised the scientific and technological operation, and although he obviously consulted with Hutchins, he reported to Bush and Conant, who, in turn, worked directly with Roosevelt. The contract was so well written that it proved adequate for the duration of the war.[19]

In creating the Metallurgical Laboratory, Compton adapted practices of both the OSRD and the academic world. Employing a structure similar to that of the OSRD, he organized scientists and engineers in groups by discipline, roughly parallel to those of academic departments. Using his university experience as a guide, Compton established advisory committees of scientists and appointed group leaders with responsibilities similar to those of an academic department chair or laboratory director. Also, rather than starting new laboratories, Compton followed another OSRD practice and left in place some established research groups, such as those at Berkeley, and awarded contracts for work not easily available at the laboratory. Yet Compton's personnel philosophy departed from academic tradition in one major way: assignment to laboratory working groups would be made on a temporary, as-needed basis. Like Bush's treatment of the OSRD, Compton's numerous reorganizations of the Met Lab were governed by work requirements, not an organization chart. Thus, the ever changing nature of the project's needs to solve problems expeditiously required shifting of the physicists, chemists, and engineers.[20]

In recruiting the best people for the project, Compton encountered security restrictions because some of the most eminent scientists were foreign-born refugees from Nazi- and Fascist-controlled Europe. The U.S. Army and Navy initially gave limited clearances to Fermi and Szilard, but by early February 1942 Compton thought these clearances should be broadened.[21]

The structure of the Met Lab became increasingly refined throughout 1942 as Compton and the laboratory research council, led by Fermi, perfected the organization. Compton and the planning board decided on February 24 that the laboratory needed to select a site for the future atomic pile, establish an engineering group, define security procedures, and provide health and safety protection to laboratory workers and the public. To help in this endeavor, Compton hired Norman Hilberry, a physicist from New York University, to serve as his administrative assistant.[22]

Compton wanted to consolidate the work as soon as possible, but without disrupting ongoing research. For much of 1942 he kept informed on the progress of the research through reports from assistants such as Hilberry and through frequent telephone conversations, meetings, and visits with the scientists. Steadily he assembled his team in Chicago.[23] By mid-April 1942 Glenn T. Seaborg, accompanied by Isadore Perlman, arrived by train in Chicago to direct research on separating sufficient plutonium for use in a bomb. Compton also let the pile work at Columbia and Princeton reach an advanced stage before moving those projects to Chicago.

Fulfilling Goals

The establishment of a formal structure did not inhibit the creativity of the Met Lab scientists. Instead, acting as a team, project workers set out to meet the deadlines determined at the beginning of the project. During the winter of 1942, essential work was well under way, with Wigner at Princeton directing theoretical research on production piles, Fermi at Columbia exploring experimental piles, Allison at Chicago also pursuing pile experiments, and Seaborg at Berkeley conducting research on plutonium chemistry. By late spring of 1942, all of the scientists had gathered at the Metallurgical Laboratory in Chicago to concentrate on the ultimate objective of producing nuclear power and plutonium with a controlled chain reaction. Both theoretical and experimental investigations indicated that the prospects were good for maintaining such a reaction.[24]

Fermi and Szilard recommended building a graphite-uranium pile to produce the reaction. Before assembling a full-sized pile, however, the scientists at Chicago and Columbia first obtained exact measurements from a series of intermediate piles. By February 21, 1942, the Metallurgical Laboratory had assembled the first intermediate pile in the West Stands squash court at the University of Chicago (figure 2). Thereafter, the Chicago group completed its first measurements by March 28.[25]

In early April the pace quickened, spurred by the war's urgency. With enthusiastic cooperation from the entire staff, Martin D. Whitaker's group worked day and night at Chicago, first, to build the second intermediate pile and then to complete measurements all within a week. Whitaker, who had left the chairmanship of the physics department at New York University, noted that ordinarily such work would have taken two to three weeks.[26]

In mid-April 1942 Compton, Doan, Allison, and Wigner finally summoned the Columbia group to Chicago. Fermi and Walter H. Zinn, a young Canadian physicist, arrived in late April to build new piles under the West Stands.[27] By May 2 the Metallurgical Laboratory payroll had grown to 153 persons, including about 75 research associates or assistants and 20 high-school boys; about 50 were assigned to the West Stands.[28]

Health Concerns

Concern about harmful exposure to radiation arose immediately after the laboratory started operating. Radiation hazards from x rays and radium were well known when the atomic bomb research started. On the other hand, information was insufficient or lacking about radioactive elements such as uranium and plutonium. Beginning in February, Compton engaged Dr. L. O. Jacobson, a hematologist at the University of Chicago, as a liaison officer between the university and the laboratory to take periodic blood counts and to conduct physical examinations and other tests of personnel.[29]

The most ardent proponents for health protection were the project's scientists. Their lives, futures, and families' welfare were at stake. The project's physicists and chemists engaged in extensive discussions on this matter at a major chemistry conference of the Metallurgical Laboratory on April 22 and 23, 1942. Although the main topic of the conference concerned the chemical

Figure 2. Stagg Field stands (*Frontiers, 1946–1996*, p. 7)

problems associated with successful production of plutonium, the scientists also spent considerable time on the hazards of waste disposal, contamination of air and ground water, radiation exposure, shielding requirements, and protection of workers. For instance, Joseph W. Kennedy, Seaborg's colleague in the heavy-isotope research at Berkeley, suggested placing workers "in shielded cages with tanks to supply their own atmosphere."[30]

After airing concerns about the project's health and safety, the conference participants asked that the laboratory engage several physicians and physiological chemists to look into protection of the workers and study the effects of radiation on personnel. But when Compton tried to expand beyond the health monitoring that was already set up, he encountered difficulties in recruiting additional qualified personnel to staff a new health research division.[31]

Subsequently, several of the chemists and physicists associated with the laboratory conducted their own studies on the possible effects of radiation exposure from the laboratory's research. For example, Wigner concluded that it was "inadvisable to build houses in the immediate vicinity of the pile." The warning was appropriate, but unnecessary, since wartime security precluded locating residences near the atomic works.[32]

Whereas the work of the physicists, chemists, and engineers changed frequently, the mission of the health researchers was of a more constant nature. Supplemental to the Metallurgical Project, the health research was, nevertheless, vital to the success of the project, as the laboratory's scientists had signaled in their push for a health division. By mid-summer of 1942, Compton had found a few physicians and physical biologists with sufficient knowledge and experience to qualify for such work. In July, Dr. Kenneth D. Cole, a biophysicist, left his position at the College of Physicians and Surgeons in New York, to activate the laboratory's health division. Ernest O. Wollan, formerly with the Chicago Tumor Institute, started the health physics section, and Darol K. Froman led the protection unit. After he consulted with Cole and radiologists from the University of Chicago, on August 6 Compton asked Dr. Robert S. Stone to lead the health division and oversee the radiation protection research. Trained as a radiologist at the University of Toronto, Stone had served as a physician at the University of California with Lawrence and the cyclotron. Simultaneously, Compton also invited Dr. Simeon Cantril, a radiologist from Seattle, to head the medical section. During August, Dr. James J. Nickson joined the health division as Cantril's assistant and handled the health problems of employees.

Charged with managing the project's biological research, the division contained two sections: clinical and experimental. The clinical section was responsible for establishing standards and procedures to protect the health of laboratory personnel and the public. Scientists in the experimental section conducted mammalian and cellular investigations of the physical and chemical effects of ionizing radiation to support the clinical efforts and to gain a basic understanding of the biological effects. By July 1942, when the Met Lab's projects were well under way, the work of the health division had barely begun.[33]

Getting Ready for the Army

By the spring of 1942, when development neared the stage requiring production facilities, Bush recommended to President Roosevelt that the U.S. Army Corps of Engineers take charge of plant construction. Consequently, the corps entered the project on June 17, 1942. As Bush and Conant advised, the OSRD continued to supervise research and development while the Corps of Engineers assumed responsibility for construction of the separation plants and the development of the plutonium production project.[34]

To adjust to the needs of the project and a possible take-over by the army, Compton reorganized the Met Lab in June 1942. "In organizing your program, please consider first of all that this is a military job and that we are at war," he reminded Allison on June 5 when the University of Chicago physics professor assumed leadership responsibility of the laboratory's chemistry groups. Compton placed under Allison's direction the research groups led by Seaborg, Frank Spedding at Iowa State College in Ames, and Wendell Latimer at the University of California at Berkeley.[35]

The reorganization reflected more than Compton's desire to meet the expectations of the Army Corps of Engineers; it also followed the OSRD structure by setting up a divisional hierarchy for scientific and engineering fields. Compton gave divisional leaders the discretion of choosing appropriate experiments and appointing divisional councils. During the first year of the laboratory's existence, a principal aim of the chemistry division was to separate plutonium from uranium. The physicists, on the other hand, concentrated on achieving a nuclear chain reaction.[36]

By November 1942 the Met Lab staff had jumped to more than four hundred persons. Fermi headed the twelve physics research groups; Wigner, the fifteen theoreticians; Allison, the five chemistry research groups; and Stone, the health research division. T. V. Moore served as the chief engineer; Edward S. Steinbach was in charge of the draftsmen; Charles Cooper of DuPont was head of the chemical engineers; and Haydn Jones supervised the project's shop. The shop personnel and the scientists designed and built their own instruments, did their own woodworking, and machined the graphite used in experimental piles. In other words, the Metallurgical Laboratory operated in a world of its own by November 1942. The composition and mission of the laboratory, however, were shortly to undergo considerable change.[37]

Selection of the Argonne Site

Meanwhile the project needed a safe, adequate site for a full-sized experimental pile. The secretive work had to be concealed yet located reasonably close to the laboratory. Further, the location could not pose a public health hazard. While on a Saturday afternoon horseback ride, Compton found an attractive site in the Argonne Forest, about twenty-five miles southwest of Chicago. Compton hoped that the land would be acquired by May 1, but the spring passed without action. By June, the pile experiments at the Met Lab had reached the stage at which Moore recommended finding a location for the pilot plant in the Tennessee Valley or the Chicago area.[38]

When the Corps of Engineers entered the project in June, the army assumed responsibility for construction and acquisition of land. Compton now urgently sought approval of the Argonne Forest site for an experimental plant because he wanted the research associated with the plant to be under the close supervision of the University of Chicago. In late June, the army agreed to use Stone & Webster, a prominent Boston engineering firm, for the project's architectural, engineering, managerial, construction, and site development.[39]

On July 6, after Nichols, Compton, Hilberry, and representatives from Stone

& Webster had surveyed what would become Site A in the Argonne Forest location, Nichols confirmed that the Manhattan Engineer District would acquire the tract for the bomb project. Site A was a triangular area of about a thousand acres that included a golf course, a wild-bird reserve, and a Girl Scout camp. By August 13 Cook County's Argonne Forest Commission agreed to lease the forest preserve to the federal government for the duration of the war plus six months. The only other expenditure amounted to about $1,000 to compensate a tenant farmer who had to be evicted from part of the land. Because of its comparatively modest requirements, the experimental pile would be the first plant constructed by the Manhattan Project.[40]

A Breakthrough

On August 20 Seaborg's group succeeded in isolating pure plutonium, the first time a synthetic element had ever been made visible to the human eye. The thirty-year-old chemist found the achievement "the most exciting and thrilling day" he had experienced at the Met Lab. Seaborg reported that his feelings were "akin to those of a father who has been engrossed in the development of his offspring since conception." Now Seaborg and his group needed to explore ways to separate plutonium on a large-scale, industrial basis. Nevertheless, without a large working pile to produce element 94, the Father of Plutonium realized that only uranium would be available for a bomb and that "the dream of atomic power plants" would come to naught.[41]

Seaborg also noted in his diary the bad news from the Soviet Union that Russian troops were retreating before Stalingrad. Anxiously, Compton resolved to move ahead of schedule. En route from California to Chicago on September 12, he wired Hilberry to push ahead with construction of Fermi's pile because he wanted to conduct the "first work" at the Argonne site instead of under the West Stands. Compton planned to "erect pile 1 in the Argonne where the hazards would be minimized."[42] But because of the unpredictability of construction work, possible dangers, and uncertainties about the project's development, indecision continued about the use of the sites and the assignment of research groups. Confusing, too, was the joint supervision of the project by the OSRD and the Corps of Engineers. Under such circumstances and pressures, Fermi and Compton shifted the location for the first self-sustaining pile several times between the West Stands squash court and the Argonne Forest site.

In mid-September the army replaced the project's first director, Brig. Gen. James C. Marshall with Gen. Leslie Groves, who retained Lt. Col. Kenneth D. Nichols as his deputy. Groves and Nichols, who were both West Point graduates, would play major roles in establishing priorities at the Metallurgical Laboratory. Groves had recently supervised the building of the Pentagon, and Nichols, with a Ph.D. in hydraulic engineering from Iowa State College, possessed the empathy helpful in dealing with academics. Together, Groves and Nichols would provide the principal leadership of the Manhattan Engineer District, the corps' code name for the atomic bomb project.[43]

In October and early November, after conferring with Stone & Webster, Groves asked E. I. du Pont de Nemours and Company (DuPont) to assume responsibility for designing, building, and operating the plutonium production plant based on the pile concept that was under development by the Metallurgical Laboratory. Not surprisingly, DuPont decided to review the Met Lab's research before committing the company to such an undertaking. What followed precipitated misunderstanding and strained relationships for some time between the Met Lab scientists and DuPont engineers.[44]

Facing a Crisis

On November 10, 1942, Compton learned that DuPont engineers estimated that there was only a 1 percent chance that the Met Lab's pile could produce pure alloy 49, the code name for element 94 or plutonium. Compounding DuPont's skepticism, the British physicist James Chadwick asserted that it was possible that plutonium would not even work as a weapons component. In a further complication, Stone & Webster reported that the main Argonne building, where Fermi's pile was to be tested and which was scheduled for completion by October 20, could not be ready before December 31 because the construction company had encountered difficulties in acquiring skilled workers and building materials. Finally, labor troubles, caused by the union contention that nonunion men should not be allowed to work on the Argonne building, exacerbated the crisis.[45]

Fortunately, the Met Lab enjoyed the firm support of Groves, who had first visited the laboratory on October 5. Groves learned that, following a successful demonstration of the chain reaction, Fermi, Allison, and Wigner disagreed about the next generation of piles. Fermi wanted to build a small air-cooled pile to produce plutonium samples. Allison and Wigner, who favored larger piles to test cooling systems, argued that bigger steps were required if an atomic bomb were to be developed in time to be used in the war. Affirming the War Department's commitment to the Manhattan Project, Groves shared his own management philosophy. He could not tolerate indecision or delay. A misstep that quickly indicated an error was better than no step at all. If the Met Lab staff were faced with two choices, one good and the other promising, Groves directed them to pursue both. In the race against the Germans, time was more precious than money. Groves left Chicago confident that plutonium provided the surest route to a successful bomb. To be sure, the chain reaction that would produce plutonium still had to be proven, but the chemical separation of plutonium, although expensive and difficult, seemed to Groves to be more in the realm of ordinary industrial processes.[46]

Not wanting to wait longer, Fermi urged Compton to authorize building the self-sustaining pile experiment in the West Stands squash court where Zinn and Herbert L. Anderson had built a series of exponential piles. Fermi calculated that he could build Chicago Pile 1 (CP-1) with low power in order to demonstrate the controlled chain reaction safely within the city limits. Compton agreed but decided not to ask permission from the university's president because Hutchins's only effective answer would have been to say no. Although there was no danger that the pile might explode like a bomb, Compton had to rely on Fermi's engineering talent to assure that there would be no run-away chain reaction with catastrophic radioactive contamination.[47]

Zinn, who was in charge of the day crew, and Anderson, supervising the night crew, were now responsible for the assembly of CP-1. Zinn, an experimental physicist, applied Fermi's techniques and physics calculations to machine 45,000 graphite bricks drilled with 19,000 holes to receive uranium oxide pellets. Anderson, a theoretical physicist, ordered the lumber and designed the construction of CP-1 within an envelope of Goodyear balloon cloth. Albert Wattenberg, Fermi's graduate student, served as a jack-of-all trades for the project and rounded up about thirty high-school dropouts from the back of the Chicago stockyards to help build the pile. "They were trying to earn some money before they got drafted into the Army," Wattenberg recalled later.[48]

Before and during the frantic assembly and measurements of CP-1, the laboratory instruments group, led by Volney Wilson, developed special controls for the first chain-reacting pile, including control circuits and detecting devices. Called "the circuit group," the team of fourteen men built and operated equipment to monitor the pile during construction and operation. In addition, they fabricated cadmium control rods, including the principal safety rod, ZIP, which would operate by gravity. The automatic safety control, which would release ZIP should radiation levels become dangerously high, included relays through a SCRAM line, derived from the slang expression meaning "let's get out of here."[49]

Working day and night from November 16 to the evening of December 1, the physicists and teenagers built an ellipsoidal-like lattice of fifty-seven layers of 16½-inch graphite bricks implanted with uranium metal (or "Spedding's eggs," named after Frank Spedding, who fashioned the cylinders) and uranium oxide. A wooden cradle supported the 20-foot-high graphite structure, which was about 6 feet wide at each pole and 25 feet across at the equator. The estimated cost of the pile—including almost 6 tons of uranium metal, 50 tons of uranium oxide, and 400 tons of graphite—was $2.7 million. To assure correct operation of the pile, beginning with the fifteenth layer, Fermi's team took careful measurements of neutron intensity as each layer was laid down. With a sense of affection for Fermi, who was improving his English by reading A. A. Milne's *Winnie the Pooh,* the scientists, named their instruments after "Roo" and other characters in the Pooh stories.[50]

Meanwhile Groves and Conant were so concerned about DuPont's skepticism that they appointed a committee, headed by Warren K. Lewis of MIT, to report by December 9 on whether the Metallurgical Laboratory should continue its efforts to design a production reactor. Ironically, while Fermi and his colleagues hurried toward their historic destiny, the very future of the Metallurgical Laboratory lay in the balance. Compton marshaled all resources to preserve the laboratory's reputation and existence. To defend the laboratory, the scientists hurriedly prepared a "feasibility report" for the OSRD and MED that not only predicted a 99 percent chance of success but also warned that "we expect the Germans to do it." The most powerful defense of the laboratory, however, would be the successful performance of the pile itself. Just before CP-1 was ready for testing, the Lewis committee arrived in Chicago to review the project.[51]

The Chain Reaction

Glenn Seaborg shivered in the bleak December cold as he walked to his laboratory in Eckhart Hall on the University of Chicago campus. On Wednesday, December 2, 1942, the siege of Stalingrad was in its 100th day—word having just reached Chicago that the Russians had killed 15,000 Germans in a new offensive west of Moscow. Also, the Chicago *Tribune* recently reported that the Germans had pounded Britain with nearly 200,000 bombs, causing 94,000 casualties, including more than 43,000 dead. And on Tuesday, the Office of Price Administration had announced gasoline rationing for Chicago and thirty western states. Seaborg supposed he was lucky that he had no car to add to his worries.

The Lewis committee had just returned from Berkeley, presumably to inform Compton about the fate of the Met Lab. Seaborg did not see anyone from the committee until late Wednesday afternoon, when he ran into Crawford

Greenewalt in the corridor of Eckhart Hall. As Greenewalt approached, Seaborg observed that he was bursting with good news. Greenewalt had just returned from the West Stands to report that Fermi had produced a chain reaction. "The pile is a success!" Seaborg enthusiastically recorded in his diary.[52]

The experiment had begun at 9:45 A.M., according to the notebook entries of Richard Watts of the instrument group. From the balcony, Fermi ordered ZIP and all control rods but one removed from the pile. From the floor, George Weil operated the remaining rod by hand. Fermi asked Weil to pull the rod out one foot—then another—and then another. Throughout the morning as George Weil carefully withdrew the cadmium control rod, Fermi took successive neutron readings and calculated the multiplication factor on his slide rule. Working slowly and deliberately, Fermi checked the mounting neutron intensity at each stage. Impressively, he knew what to expect at each point, conveying the sense that he had everything under control. He even paused at 11:30 for a leisurely lunch after replacing the rods, then returned to the experiment at 2:00 in the afternoon, when Compton and Greenewalt joined forty-two witnesses of the first test of CP-1 (figure 3). In addition to Fermi's crew, they ranged from the senior scientists Szilard and Spedding to students and August Knuth, the young carpenter who assisted in assembling the graphite blocks and cadmium rods for the pile. Leona Woods, a protégé of Fermi, was the sole woman in attendance.[53]

Finally, at 3:00 P.M. Fermi asked Weil to withdraw the control rod one more foot. "This is going to do it," he quietly informed Compton. As the rod was pulled out, Anderson recalled hearing "the counters clicking way—clickety-

Figure 3. 3:22 P.M., December 2, 1942. Gary Sheahan, a Chicago *Tribune* staff artist, painted this reconstruction of the scene when atomic scientists observed the world's first nuclear reactor as it became self-sustaining. Courtesy of the Chicago Historical Society, P&S-1964.0521.

clack, clackety-clack," faster and faster. At 3:42 Watts quickly jotted in his notebook, "We're cookin.'" All eyes were on Fermi. "It was up to him to call a halt," Anderson explained. Tension mounted as the self-sustaining chain reaction intensified. To allay any doubt that the pile had become critical, Fermi waited—confident and calm. "He kept it going until it seemed too much to bear," Anderson recorded, and then Fermi called "Zip in" and the intensity abruptly dropped to noncritical levels. "Everyone sighed with relief. Then there was a small cheer."

CP-1 had completed its historic mission. In celebration, Eugene Wigner broke out a bottle of Chianti that he had purchased months before in anticipation of this moment. A sober crowd silently sipped a dram from paper cups and then signed the straw wrapping on the bottle as it was solemnly passed around (figure 4).

Later that day Compton and Greenewalt confirmed Fermi's success in a meeting with the Lewis committee: the chain reaction worked in a smaller pile than expected, the reaction was "slow enough to be controlled," and "the automatic controls responded perfectly." Fermi had run CP-1 at half watt power for 4.5 minutes. Although significant power could not be achieved until a shield-

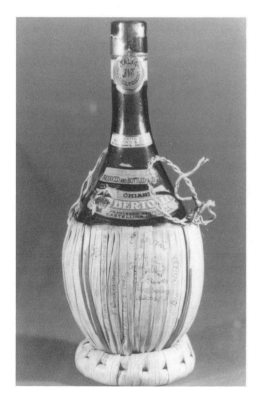

Figure 4. Chianti bottle (*Frontiers, 1946–1996*, p. 12)

ed pile was reassembled at the Argonne site, the Lewis committee departed well pleased with the Met Lab's achievement. After they left, Compton called Conant at Harvard. "Jim, you'll be interested to know that the Italian navigator has just landed in the new world." Conant was obviously relieved.[54]

"Were the natives friendly?" he asked.

"Everyone landed safe and happy," Compton joyfully replied.

Fermi's triumph did not calm Seaborg's apprehension or sense of urgency, however.

> Of course we have no way of knowing if this is the first time a sustained reaction has been achieved [Seaborg confided to his diary]. The Germans may have beaten us to it. I wonder, are they aware that U233 can be made from Th232 and 94 239 from U238 in a chain-reacting pile and that either of these isotopes can be used in a fission bomb? And if they have a pile that chain-reacts, would they use it to generate power or to produce vast amounts of radioactivity as a military weapon? One thing is certain: although Fermi has demonstrated that we now have a means of manufacturing 94 239 in copious amounts, it is the responsibility of chemists to show that the 94 can be extracted and purified to the degree required for a working bomb.[55]

In the Wake of Success

Life at the Metallurgical Laboratory changed markedly after December 2, 1942. Up to that day, the lab had focused primarily on plutonium research and on achieving a self-sustaining chain reaction; afterward, as Seaborg anticipated, the lab's main goal was to produce plutonium for an atomic bomb as soon as possible. But the wartime mission did not obscure the potential for establishing a government-sponsored postwar nuclear research institution somewhat along the lines envisioned by Milton White.

Although the laboratory's technical council had spent the previous months planning for additional facilities, increasingly the lab would have to support projects far away from Chicago and would have to send personnel, and even entire groups, to such sites, often for the duration of the war. Further, Compton and his colleagues had to accept greater direction from and participation by the military and DuPont.

Changes in the use of the unbuilt pilot plant at the Argonne Forest site contributed toward feelings of uncertainty. Originally, the pile and separation plant were to be built at Argonne, and within days of December 1, when DuPont officially assumed broad responsibilities, laboratory personnel started planning for the pilot plant, consisting of three or four piles and two separation facilities, to be established at Argonne.[56] But within a few weeks, doubts arose among all involved about where to build the pilot plant because of concern with its potential hazards. To provide a greater measure of isolation, some of the people considered Hanford, Washington, the remote site recently selected for the full-scale plutonium production piles and separation plants. Ultimately the army, DuPont, and the Met Lab principals compromised. The separation plants would be located at Hanford, but the experimental production pile would be built at Clinton, Tennessee, which was farther from a large population center than Argonne but closer to the Metallurgical Laboratory than Hanford. Much to Compton's dismay, DuPont executives also insisted that the Met Lab operate the semi-works (a small-scale separation plant) about five hundred miles from its base in Chicago. Reluctantly, Compton and officials from the University of Chicago, including Hutchins and the board of trustees, accepted the responsibility as a wartime duty. Groves regarded the arrangement as an example of the cooperative spirit essential to the success of the Manhattan Project.[57]

The Met Lab's status within the Manhattan Project shifted as the emphasis switched from basic research to industrial production. The laboratory's senior scientists previously had led the project while private industry had provided facilities and equipment. The focus on nuclear fuel production, however, placed the scientists in a cooperative position with private industry rather than in a leadership role. By January 1943 Compton understood that the DuPont Company was to design and construct all installations at the Tennessee site in consultation with the Metallurgical Laboratory. As Hilberry saw the laboratory's new role, "Our major responsibility [was] to supply the DuPont engineers and operators with the technical information they needed for their share of our joint endeavor."[58]

The Army Takes Control

Pressured by the army and DuPont for more structure and a clearer line of command, Compton again reorganized the Metallurgical Laboratory with the help of his newly formed laboratory council. By February 17, 1943, Compton was required to report directly to Groves and no longer to the S-1 committee. As Compton told the laboratory council on February 17, the army was charged with ensuring that the "research gets done." Nevertheless, the S-1 committee would continue surveying the technical aspects of the Met Lab's effort. Compton expected that the laboratory would have to furnish more information than in the past to other parts of the project. The Tennessee work now became part of the Met Lab program as the laboratory worked in partnership with DuPont on the construction near the Clinch River. Robert Oppenheimer, who had joined the project in May 1942, would establish his group at a new site in Los Alamos, New Mexico, and operate independently from the Metallurgical Laboratory. Finally, the DuPont company would run the plant in Washington State, where the government was acquiring land.[59]

Army control meant less flexibility than the Metallurgical Laboratory had under the OSRD. Previously, the laboratory had operated rather loosely as a collection of rugged individualists. Now, all general policy questions on the development program had to go through Compton and through him to Maj. A. V. Peterson, Chicago district officer for the Army Corps of Engineers. Current operational matters were cleared through Peterson and sent directly to Groves.[60]

The army's takeover of the project resulted in more rapid progress for the entire Manhattan Project. Groves obtained adequate funding through a high-level military policy committee, a tactic also employed by the OSRD. Groves and his staff persistently insisted that the scientists adhere to the wartime target above all, even though the Metallurgical Laboratory goal had been met.[61]

Argonne Laboratory

In February 1943 the Metallurgical Project scientists prepared to dismantle the CP-1 under the West Stands for reassembling at the new Argonne Laboratory. The origin of the name "Argonne Laboratory" is obscure. At least as early as December 1942, however, work at the "Argonne Laboratory" was discussed among Met Lab personnel as though the subsidiary laboratory was firmly established. At that time, the Argonne laboratory, which was under construction by the army at the Palos Park site in the wooded Argonne Forest Preserve west of Chicago, was already part of the Metallurgical Laboratory. At a laboratory council policy meeting on March 3, 1943, Compton reported Groves's assumption that for military reasons the United States government would maintain a research laboratory for a long time beyond the end of the war. Because the army did not want to construct new buildings at the University of Chicago but was willing to build at Argonne, the ultimate predominance of the Argonne site was evident just three months after Fermi's experiment.[62] Compton also assumed that Argonne Laboratory would continue as a long-term operation although the University of Chicago might not control it after the war. He designated Fermi as the first director of Argonne Laboratory.[63]

Once established, Argonne Laboratory carried on the spirit of teamwork exhibited in previous pile building and served as an experiment station, especially in support of the production efforts elsewhere. While CP-1 was at Chicago, Zinn devised the method of instrumentation used for that and future piles. Fermi and Zinn's crew continued to operate CP-1 under the West Stands until February 28, 1943, when they dismantled and reassembled the pile with a radiation shield at the new Argonne site. By March 20 the pile, now known as CP-2 (Chicago Pile 2) and rebuilt in the shape of a square with the materials from CP-1, started operating so that it could provide information needed for the plutonium production piles at Hanford. Two months later CP-2 reached its critical size at slightly above the fiftieth layer (figure 5).[64]

Changing Attitudes at the Met Lab

Moving the pile out of the West Stands only marginally eased the space problem, which was worsened by the laboratory's increasing responsibilities for training personnel for other Met Lab and Manhattan Project facilities. Through the winter of 1943 Compton and the Corps of Engineers considered additional sites for the Met Lab's research. In March Compton recommended concentrating fission research at the University of Chicago campus, continuing basic

Figure 5. CP-2 assembled in Argonne Forest Preserve

pile research at Argonne, conducting advanced pile research at the Clinton Engineer Works at Clinton, Tennessee, and carrying out secret and hazardous research at Los Alamos.[65]

While Compton and the army worked to solve administrative issues, tensions increased among Met Lab physicists, who were upset by the altered arrangement with DuPont. Eugene Wigner threatened several times to resign. As Wigner later described the differences, laboratory scientists believed that building a chain-reacting pile was "a matter of art," whereas the DuPont engineers viewed pile construction as "a question of mass production." In Wigner's view, the laboratory considered speed more important than did DuPont.[66]

Roger Williams at DuPont held another view. He thought that the Met Lab scientists looked upon the DuPont men as stupid and soft and suspected that the company was trying to dictate the lab's program. Williams also thought that the scientists unfairly claimed that DuPont was seeking a monopoly of nuclear physics developments for postwar profit. Even with Groves's aid, DuPont found it impossible to recruit qualified theoretical physicists to assist the company in Wilmington.[67]

Despite the changing status of the laboratory and the army's strict supervision, there was a positive side to the story. During the winter of 1943 the scientists and Groves recognized that there would be a postwar future for nuclear science. Acknowledging that the immediate task was to win the war, Compton believed the long-term duty of the laboratory staff was to "establish and then to maintain for the United States the scientific and technical leadership in the field." Compton thought that basic research of the kind performed for the Metallurgical project was best done in a university atmosphere. As he told the laboratory council on March 3, Compton believed that such work in the future should be a "government directed program with a main government laboratory plus supplementary research by private individuals . . . through-out the country." According to his vision, the institution should be in a large city, such as Chicago, that had facilities for physical and chemical research.[68]

The Personnel Situation

Despite bright prospects and high salaries, laboratory leaders encountered problems in recruiting personnel. In contrast to other defense efforts, the Manhattan Project had started later, when available scientific and engineering manpower was almost exhausted. Moreover, the cost of living in Chicago was higher than that of most other cities, especially college towns. "I find myself in an almost desperate situation with regard to obtaining personnel," a chief engineer wrote in April 1943.[69] To deal with personnel needs, the University of Chicago adopted a recycling policy similar to that suggested by Milton White. After agreeing to the management contract of April 29, 1943, the university intended, when possible, to reemploy those workers transferred from the Metallurgical Laboratory to other projects once the other work was completed.[70]

Continuing his efforts to strengthen the laboratory, on May 5 Compton appointed two physics professors to lead it—Samuel Allison as the second director and Henry Smyth as associate director. As Hilberry, Compton's deputy, perceived the situation, Doan, the first director, had been an excellent administrator, but most of the staff, who were accustomed to the tradition of a professor as department head, had "never really accepted [him] as laboratory director."[71]

Scientists and the Army

As work became more focused away from Chicago, Groves tightened the security of the Manhattan Project by implementing the army's policy of compartmentalization. Groves interpreted compartmentalization to mean that "each man should know everything he needed to know to do his job and nothing else." Obviously, the security policy hampered the Met Lab's interchange with other project laboratories, including the Chalk River lab in Canada. Far more restrictive than his OSRD predecessors, Groves specified in detail both the means of communication and the exact kind of information that project personnel could exchange with each other. Under his interpretation, for example, a scientist from the Met Lab could only share with a colleague at Los Alamos the particular section of a blueprint that directly fell within the reviewer's expertise or "need-to-know."[72]

After the OSRD S-1 committee dissolved in September 1943, the army assumed complete command of the project. Anxious to retain the participation of scientists, Bush persuaded Groves and Secretary of War Henry L. Stimson to form the Military Policy Committee. Limited by Groves to three members representing the army, navy, and OSRD, the committee included Bush as chairman and Conant as his alternate. The army subsequently received scientific advice directly from members of the research programs, from appointed reviewing committees, and from Conant and Tolman, whom Groves selected as his personal scientific advisers. Another consequence of Groves's security policy was that in December 1943 Allison informed the staff of the Metallurgical Laboratory that Groves wanted to avoid public association of the University of Chicago with the Clinton Engineer Works.[73]

The Fruits of Cooperation

Although security compartmentalization proved cumbersome and sometimes resulted in a duplication of effort, essential information usually found its appropriate way through the security network. In March 1943, for instance, Seaborg discussed with a Princeton physicist that transuranic isotopes, such as plutonium-240, might be produced in a chain-reacting pile. More than a year later, Seaborg's predictions were confirmed by the production of plutonium-240 by the Clinton pile. Subsequent research at Los Alamos on the effects of plutonium-240 determined that the Manhattan Project would have to abandon the gun method for plutonium in favor of an implosion technique for the Hanford-produced plutonium. (The gun method still could be used with uranium, however.)[74]

Again in 1943, despite differing philosophies and some distrust between the laboratory scientists and DuPont engineers, the joint research efforts of Seaborg and DuPont's Charles M. Cooper developed the bismuth phosphate method for chemical separation of plutonium from pile fission products. After the two groups completed successful experiments at the Met Lab, the process was installed in the semi-works at Clinton.[75]

Under pressure to develop practical production and purification processes, the Metallurgical Laboratory chemists also worked successfully with colleagues at the Ames Laboratory of Iowa State College and at the University of California at Berkeley. The Met Lab chemists, in coordination with the technical, physics, and health divisions at Chicago, also studied waste disposal, instrumentation, and health-protection problems confronting the entire production project.[76]

In 1943 most of the Met Lab's research focused on successful completion of the first Hanford and Clinton piles. In October the chemistry division's separations process section moved to Tennessee. The scientists who remained in Chicago investigated questions ranging from the stability of the graphite structure to the decontamination of the bismuth phosphate process and helped the National Carbon Company prepare pure graphite for the project. Under an arrangement with DuPont, the Met Lab scientists reviewed and approved the Hanford blueprints. Wigner and Compton differed on the success of this procedure. Although Wigner was pleased that DuPont's blueprints for the Hanford pile were based on the original design prepared by the Met Lab's theoretical physics section, he was distressed that the company had arbitrarily eliminated the section's design to detect leaks. Compton believed that the process functioned smoothly, resulting in "definite and important improvements in design that might otherwise have been missed."[77]

During this same period, scientists exchanged estimates on Hitler's chances of obtaining an atomic bomb. Fear that Germany might develop a bomb before the Allies constantly troubled scientists, many of whom had been trained in Germany and believed that German science was second to none in the world. Philip Morrison, a research associate in the physics unit, proposed monitoring the scale and location of German work. In yet another patriotic collaboration, Met Lab scientists carefully evaluated what little information they could glean to estimate enemy progress on a fission bomb.[78]

As the Manhattan Project moved into high gear, Compton reevaluated the Met Lab's objectives on November 26, 1943. Citing the technical program as its major responsibility and personnel training for plant operations as a minor function, he mobilized the laboratory to help establish the Hanford operation, finish projects for Clinton, and assist Los Alamos once that laboratory was working fully. In the autumn of 1943, the laboratory's principal technical goals including making the Hanford water-cooled pile and bismuth phosphate separation plant succeed as soon as possible. Although Compton regarded pile development as the laboratory's main responsibility, he also encouraged the Met Lab chemists to obtain a basic understanding of the chemistry involved in the separation processes and to conduct additional studies of the concentration processes. Once work at Los Alamos was under way, Compton expected to add associated research on techniques for purifying plutonium and reducing it to metal. In December Compton assigned the Clinton Laboratories the main responsibility of assisting the project's bismuth phosphate separation plant by studying separation processes.[79]

An Emerging Mission

Because the Metallurgical Laboratory was the home of the first chain reaction, it seemed logical that the laboratory should become a national leader in atomic pile research and development. (After World War II the term *reactor* replaced *pile* in common usage.) Both Groves and the laboratory's scientists anticipated such a role. Acting on the recommendation of a laboratory advisory committee, in September 1943 Groves authorized construction of a low-powered, heavy-water pile at the Argonne facility to test aspects of the Hanford pile at relatively high levels of radiation. Compton regarded this pile project as an integral part of the Metallurgical Laboratory's responsibility. Scientists in Montreal were also developing similar piles, and Groves allowed the Argonne scientists to exchange some technical information with the Canadians. Moreover, Groves later approved Met Lab participation in a cooperative heavy-water pilot plant directed by the British and operated by British, Canadian, and American personnel. Staff members of the Metallurgical Laboratory could collaborate as long as there was no interference with the work supporting Hanford and Los Alamos.[80]

Walter Zinn, Fermi's deputy at Argonne Laboratory, directed construction of Argonne's first heavy-water pile, known as CP-3 (Chicago Pile 3). On New Year's Day 1944, Zinn and his twelve-man crew occupied the pile room. Following the tradition of the Metallurgical Laboratory, Zinn's group worked six and seven days a week until they had installed most of the pile's essential parts. CP-3 reached criticality on May 15, 1944, and on June 23 went into normal power operation. The pile operated satisfactorily throughout the remainder of the Met Lab's existence.[81]

On April 15, 1944, as CP-3 neared completion, planning-group members aired their views on the future of Argonne Laboratory. Smyth, one of those present, had suggested that Argonne Laboratory serve as a regional laboratory for the Midwest and that other regional laboratories be established according to need after the war ended. But unless the hemorrhaging of research talent could be stanched, Smyth's vision would not materialize. The loss of theoretical physicists to Clinton and other sites outside Chicago loomed as a particular threat to the viability of long-term pile research. Building new piles should be a vital part of the Met Lab's future, Compton advocated. But the trouble was that the "top men have been pulled out," Nichols of the Corps of Engineers acknowledged.[82]

Zinn, an experimentalist, urged that Argonne's plight be given special consideration and explained that "at Argonne we are faced with the situation that if theoretical men at Clinton disappear, other men will go. If [our] research program is to be directed toward pile development, we must have a man at Argonne studying piles."

Smyth added that "one cannot do fundamental research without pile development."

"There is no one to direct any research," Zinn continued. Fermi warned that "to build piles . . . on what we know now is not to be recommended."

Zinn then reminded the group, "Argonne Laboratory was built for a job, namely to improve piles."

Especially concerned about the future, Zay Jeffries, a consultant from the General Electric Company, conceded, "Closing down Argonne would be dangerous."

During the next three months Compton worked on measures to uphold the morale and integrity of the Metallurgical Laboratory. He proposed a program for 1944–45 in which the laboratory would give priority to work at Hanford and Los Alamos while enabling researchers to plan postwar studies. Nichols tried to alleviate the situation by authorizing the Metallurgical Laboratory's retention of key personnel for assistance at Hanford and Los Alamos. Further, when priority work allowed time for other pursuits, Nichols authorized the staff to conduct basic research, design, and development of materials and piles appropriate for power and other possibilities. Groves also specified that under these conditions research would be permitted at the Argonne and the Metallurgical Laboratories, but he required the district engineer to approve major construction and engineering designs. Nevertheless, Compton expected the Metallurgical Project's work to diminish as emphasis shifted from development to production.[83]

Change in the Ranks

In the summer of 1944, Groves also directed that research at the Met Lab could continue so that a talent pool would be available for trouble-shooting at Hanford or as needed to protect the war effort. Nonetheless, additional transfers and departures followed. Because the chemistry division's work for Los Alamos was almost finished, T. R. Hogness, head of that division, expected to reduce his academic personnel, which numbered 235 in July 1944. By September, 17 had left the project, and 19 had transferred to other laboratory divisions or to Hanford. Yet the laboratory was able to engage in some hiring, for on September 1, Farrington Daniels, a professor of physical chemistry from the University of Wisconsin at Madison, joined the division as associate director.[84]

The staffing at the Metallurgical Laboratory continued to diminish throughout 1944 as Hanford and Los Alamos drew upon personnel, and Groves forbade hiring most replacements. In September, at Oppenheimer's invitation, Fermi moved to New Mexico to become an associate director at Los Alamos (figure 6). By autumn, 175 technical persons had transferred to installations away from Chicago. The chemistry groups were decimated, and by November, a substantial part of the laboratory, including Allison and most of his instrument section, had moved to Los Alamos. The transfers of Met Lab physicists to Los Alamos in the autumn not only eliminated the general physics section but also limited most nuclear physics research to pile design and instrument development. The new assignments curtailed the theoretical physics group, which then focused on plans for future piles. Ultimately, transfers reduced Argonne Laboratory to a skeleton operation and, in Compton's view, resulted in inefficient use of Argonne's unique equipment. Despite this situation, Compton believed that it was not in the nation's interest to close Argonne because

Figure 6. Ernest O. Lawrence, Enrico Fermi, and Isidor Isaac Rabi at Los Alamos

when personnel returned from other sites, Argonne's research would gradually return to its former strength.[85]

Several executive changes resulted from the 1944 transfers. Joyce C. Stearns replaced Allison as director of what was left of the Metallurgical Laboratory; Zinn became acting director of Argonne Laboratory; and William P. Jesse, a physicist from the University of Chicago and an early member of the Metallurgical Project, took the position of instruments section chief.

Although the future of the Met Lab was not entirely bleak, the summer of 1944 was exceedingly difficult. All of the subsequent work had to be done without purchasing additional equipment or hiring new employees. In addition, Groves's secrecy measures created problems alike for those who transferred or who remained behind. All transfers to Los Alamos from the Metallurgical Laboratory took place under strict security, as specified in 1944 army orders. Laboratory members had to cash checks in Chicago and not in Santa Fe. The Army Intelligence and Security Division ordered that no group movements or automobile caravans should be formed for the trip to Los Alamos. H. G. Hawkins, Jr., chief of the Army Intelligence and Security Division, requested the transferees to avoid discussing the changes with other Metallurgical Laboratory personnel. Compartmentalization sometimes prevented closest friends from discussing their professional plans with one another.[86]

Remaining Work at the Metallurgical Laboratory

With the dispersal of personnel to other sites, the work in the Chicago area decreased considerably during late 1944. Seaborg's golf game was one indicator of the change of pace. In June 1943 Seaborg virtually collapsed from overwork and fatigue, which temporarily knocked him out of the project. Following a hospitalization, an extended recuperation, and a long vacation, Seaborg returned to work in August. After a friend suggested golf as part of his therapy, a grateful Seaborg recorded his "first good night's sleep in months" on August 15. Thereafter, he tried to work in a round of golf weekly. By the summer of 1944 Seaborg, still hard-working and dedicated, often found time for two or three rounds of golf a week.[87]

Argonne Laboratory became a separate unit in 1944, although its employees remained on the Metallurgical Laboratory payroll. In addition to the success achieved with CP-3, Zinn, the acting director during Fermi's absence, and Argonne provided a vital service to Hanford in the fall of 1944. After Hanford's pile failed to operate properly in September, Fermi and Wheeler hypothesized that xenon-135 had poisoned the pile. Zinn ran tests at Argonne on October 2 and 3 that confirmed the xenon poisoning, thus enabling the scientists at Hanford to remedy the problem so that the pile could produce the plutonium needed for the atomic bomb.[88]

Although the work in most Met Lab divisions ebbed in 1944, the health division's load increased. Compton implemented most of the health-protection recommendations that had resulted from a four-month survey of the laboratory by Dr. Cecil J. Watson, associate director of the health division.[89] Additionally, with little known about the toxicity of plutonium, in April 1944, Nickson's Chicago health unit had initiated testing of the Met Lab workers' urine and sputum for plutonium. In collaboration with the army, the division also worked with the laboratory's instrument group to provide radiation detectors for possible use if the Germans unleashed radioactive warfare during the Allied invasion of France.[90]

The demands placed on the health division increased so much that it was the only part of the Metallurgical Laboratory in mid-1944 to be recommended for an increase in staff and space. As a result, the health division increased its personnel from 173 in November 1944 to 196 in March

1945 while all other divisions and Argonne Laboratory lost 20 percent or more of their employees.[91]

Considering the Future

The nuclear community did not wait for the end of the war to plan for the future, and in 1944 two major studies, one originating at the Met Lab and the other with the Manhattan Project's military policy committee, established parameters for postwar planning. With Compton's endorsement, in July Zay Jeffries organized the Met Lab study, which he suggested during a discussion about the decline of the laboratory. Jeffries defined the topic as a prospectus for "nucleonics." Fermi, Franck, Hogness, Robert S. Mulliken, Stone, and Charles A. Thomas served on this committee. Encouraged by Compton, most of the laboratory's scientists contributed ideas to Jeffries's paper, which recommended a wide range of research and development as well as international control of nuclear energy. The Jeffries committee completed "Prospectus on Nucleonics" in November 1944.[92]

A month later the second group, formed in August 1944 and led by Richard Tolman, completed its review for the military policy committee. Henry Smyth, who previously had commended the 1941 White proposal on nuclear energy research, was among the Tolman committee's members. After consulting scientists at the major Manhattan Project research centers, the Tolman committee recommended continuing basic nuclear energy research as well as pursuing technical development. Most importantly, the Tolman committee advocated the establishment of a national nuclear authority that would fund military and civilian research and development in government, academic, and industrial laboratories.[93]

At this time, neither Jeffries's "Prospectus on Nucleonics" nor the Tolman report produced an immediate result.[94] Although the war in Europe had turned in favor of the Allies, primary thinking concentrated on winning the war and developing an atomic bomb to use, if necessary. Nevertheless, the two studies provided a foundation for future action, when postwar planning would require more attention.

The University of Chicago Relationship

The University of Chicago's attempts to woo distinguished scientists to the Metallurgical Laboratory complicated matters. For some time Compton had worked to persuade Fermi to accept a professorship at the University of Chicago, an effort that angered Fermi's colleagues at Columbia University. After hearing a complaint from Isidor Rabi, Vannevar Bush threatened to move Argonne to another site "to take it completely away from Chicago's covetous fingers."[95]

"Our Columbia friends have forced our hand," Compton told Fermi on July 17, 1944. Compton reasoned that the operator of Argonne Laboratory would have to appoint Fermi as research director because it was impractical to move Fermi's experimental piles to another site. It was only logical that Fermi accept a position at the University of Chicago because of the school's proximity to Argonne.[96]

As a result of the uproar over Chicago's efforts to recruit Fermi, Compton initiated discussions over the future of the Metallurgical Project and Argonne Laboratory. At first he proposed that some agency other than the university operate Argonne Laboratory and that Met Lab research projects simply be absorbed into the University of Chicago's departments. As for the Clinton Laboratory, Compton thought that some agency other than the University of Chicago should take over its operation. But Compton's suggestions never reached Groves because the university's advisory board disagreed and asked Hutchins not to approve these recommendations.[97]

Before the war's end, Compton consulted with President Hutchins about the university's peacetime role in nuclear energy research. Hutchins accepted the suggestions that the university establish nuclear science research institutes. Compton also persuaded Hutchins to disengage the university from military research as soon as possible and to retain management of business operations of Argonne Laboratory, where scientific programs were compatible with academic research.[98]

Preparing for a Postwar Laboratory

Early in 1945, Compton and his advisory council were already planning operations for the postwar period. The director of the Metallurgical Project wanted to negotiate a new agreement with Cook County to allow Argonne Laboratory to continue operations at Palos Park. He also sought to increase personnel because dispersals had seriously weakened the project. As of January 15, 1945, the Met Lab numbered 403 persons and Argonne 30, including DuPont trainees. Nichols, however, thought it was unwise to broach the site question not only because of security requirements, but also because of "the present uncertainty concerning our plan for Argonne Laboratory."[99]

Undaunted, Compton asked Fermi and Zinn to propose a program for work after July 1, 1945, the date he thought that the laboratory's wartime mission would end. Fermi and Zinn believed the most enduring problem would be to develop technology capable of transforming uranium-238 and perhaps thorium into valuable fissionable elements. They proposed focusing research efforts on the basics of neutron physics.[100] Compton and the staff also favored strengthening research and development of nuclear processes, especially relating to experimental pile work. Although they did not think nuclear energy research would remain centralized at Argonne, the Met Lab scientists preferred concentrating "the main body of research and development in the Chicago area." They anticipated close cooperation between Argonne and the university once personnel returned from Oak Ridge to continue the work.[101]

As Allied troops moved through Germany and American troops ousted Japan from the Philippines, staffing at the Met Lab remained uncertain. After conferring with Nichols, Compton decided that the project would have to be reduced further after July 1, 1945, the beginning of the new fiscal year. Major tasks had to be accomplished by July 1, but after that the army would not fund as large a research program as that proposed by the Project Council. Still, Compton hoped to save both Argonne Laboratory and, on a scale comparable to that of other centers, research on the Chicago campus.[102] In March 1945 Compton appealed to Groves and Secretary of War Henry L. Stimson to provide federal assistance for research on long-range nuclear problems. "If no funds to support this work are available before the close of this fiscal year, irreparable damage to the atomic research program will result from the dispersal of the scientific staff," Compton warned.[103]

In April, Compton hammered out a provisional research agreement with the army. Biology and health studies would continue with skeleton crews until January 1, 1946. On the other hand, the laboratory had lost so many theoretical physicists and engineers that Compton considered closing down the pile research until Zay Jeffries and an advisory committee persuaded him to keep the research going. After President Roosevelt's death on April 12, 1945, Compton assumed it would be months before the Truman administration could chart the future of nuclear energy research and development. In the meantime, Compton would be leaving the University of Chicago to become chancellor of Washington University in St. Louis. He promised, however, to remain in Chicago until he had established the project on a permanent basis. Thereafter, he agreed to serve in an advisory capacity from St. Louis.[104]

From April through June 1945 Compton prepared the Metallurgical Laboratory for the postwar transition. Stearns resigned as laboratory director to accept a faculty appointment with the University of Chicago. Assuming that he would serve in a caretaker role, Farrington Daniels accepted the position of laboratory director on July 1, 1945. Zinn continued to direct work at Argonne Laboratory. The goals set for the Met Lab centered on research for new piles in addition to service and development work needed by sites outside Chicago. The laboratory kept busy responding to requests for electronic instruments, machine-shop products, and health research and services. Theoretical physics, once conducted by the Met Lab, was assigned to the Clinton and Argonne laboratories. Service to sites outside Chicago was expected to constitute 60 to 65 percent of the laboratory's work; health research and service, 25 to 30 percent, and physical sciences research, 10 to 15 percent. Chemistry was to make up 75 percent of the laboratory's physical sciences research. Overall, the Manhattan District retained responsibility for pile development.[105]

A Proposal for Regional Laboratories

By the summer of 1945 the Truman administration had formed an interim committee to make recommendations to the president on postwar research, development, and military applications of nuclear energy. Compton, Fermi, Lawrence, and Oppenheimer served on the scientific panel advising the committee. The panel, in turn, solicited comments from Met Lab scientists, including Zinn, Seaborg, Wigner, and Stone. Zinn proposed that a national nucleonics authority establish nonprofit corporations to administer regional laboratories. Military applications would be the responsibility of the national agency, which would aid the military services in utilizing discoveries of the laboratories. In each region universities would cooperate in the laboratories' activities. Scientists from these universities would make up most of the board of directors for each corporation. Zinn hoped that the national atomic authority would review results and encourage competition but not attempt to dictate the laboratories' research agenda.[106] Impressed by Zinn's proposal for regional laboratories, Compton suggested placing regional laboratories in Cambridge, Washington, Chicago, and Berkeley. Again, in July 1945, Compton asked Fermi to accept a professorship at Chicago because Fermi would be a key asset if a regional laboratory were established in the Chicago area.[107]

The Atomic Bomb and Its Aftermath

The future of nuclear research, however, depended on the successful completion of the atomic bomb, which had not yet been achieved by July 1, 1945. In addition to mapping out a future technical program, atomic scientists also increasingly discussed obvious postwar ethical questions, such as how they could educate the public about the bomb's destructiveness. Not only was the bomb's destructive power awesome, but its plutonium also proved to be extremely poisonous. In 1945 the laboratory's health division concluded that the toxicity of plutonium was twenty times higher than initially estimated. Stone told his laboratory colleagues that, gram for gram, plutonium was as dangerous as radium when deposited in critical regions of the body.[108]

On the eve of testing the bomb in New Mexico, some of the atomic scientists spoke out about the weapon they had helped to create. Szilard, who had pushed so aggressively to establish the atomic bomb project when he had feared Hitler might obtain the bomb first, now persuaded at least sixty-nine of his colleagues to petition Truman not to use the bomb unless absolutely necessary. Szilard's petitioners urged that the United States not use the atomic bomb in the war unless the Japanese refused to accept public terms of surrender. A subsequent poll by Compton indicated there was no unanimity among Met Lab scientists about the use of atomic weapons.[109]

On July 16, 1945, the first atomic bomb test successfully lit up the desert near Alamogordo. Within a month the United States dropped atomic bombs on Hiroshima and Nagasaki; Emperor Hirohito surrendered on August 14. These events undoubtedly galvanized decisive action by some of the scientists who had helped develop the bomb.

Led by Eugene Rabinowitch, a Met Lab biophysicist, laboratory scientists discussed governance of the atom, including international control of atomic energy. These sessions led to the formation of the Atomic Scientists of Chicago, which later contributed to the successful organization of the Federation of Atomic Scientists, the Federation of American Scientists, and the publication, *Bulletin of the Atomic Scientists*.[110] The establishment of these groups provided nuclear scientists with a collective voice that could speak out forcefully on the structure of postwar nuclear research.

Transition

In August 1945 the War Department published Henry Smyth's brief but startling history of the atomic bomb program. Eight days after the war ended, the Project Council met, surprised at the extent of sensitive information about the atomic bomb the Smyth report revealed to the American public (and not incidently, to foreign governments). The scientists expressed varying opinions on what should have been released, but agreed that, more than ever, international controls were necessary to harness atomic energy.[111]

On August 23 Compton informed Groves that the Met Lab's policy committee unanimously agreed that the government's atomic scientists should be allowed to express their views on the social and political implications of nuclear energy. The scientists believed that their expert opinions could help guide the public debate and correct misconceptions. Nevertheless, under Truman's order, censorship of the scientists continued. Unhappy with continuing restrictions, Stone complained that "many of us believe that the Army is running the show." Peterson, however, assured the laboratory council that the War Department simply was complying with Truman's request.[112] To the distress of the security office, the Metallurgical Laboratory scientists also favored making public all information on the basic chemistry of plutonium. The army feared that such publication might aid unfriendly governments. In October the two sides, with Seaborg assisting, worked on a compromise for the publication of all but essential "bottleneck" information.[113]

As the war ended, Met Lab scientists necessarily became attentive to their to own careers. Uncertainty about the future made the laboratory an ideal hunting ground for institutions eager to recruit former Manhattan Project scientists. For example, after four years leading chemistry research at the Met Lab, Seaborg, who had been offered an attractive position at the University of Chicago, returned to the chemistry department at the University of California, Berkeley, to head the newly organized nuclear chemistry division in the Radiation Laboratory (figure 7). For Seaborg it was a seller's market. Meeting with President Robert G. Sproul of the University of California, Seaborg laid out his terms: "In addition to a full professorship for me, I want Perlman as an Assistant/Associate Professor; English and Orlemann, as Assistant Professors; Cunningham as an Assistant Professor in the Radiation Laboratory; Ghiorso and other research associates, about twelve graduate fellowships."[114]

The future of the laboratory's relationship with the University of Chicago appeared shaky by October 1945. Angered by charges that his university was seeking a monopoly in nucleonics and capitalizing on its connection with Argonne Laboratory, Hutchins threatened to wash his hands of the whole matter. "The University here and now formally repudiates any desire to own, to control, or to operate the Argonne Laboratory," Hutchins wrote Groves. Instead, the University of Chicago already had sponsored a nucleonics conference in the autumn of 1945 with representatives from eight midwestern universities because, as Hutchins explained, the university "regarded itself as a training ground for the country and particularly for the Middle West."[115]

Figure 7. Glenn T. Seaborg in his Met Lab office, January 1946

Groves requested patience and agreed to maintain the laboratory with its existing staff until Congress decided what to do about nuclear energy. The general recommended that the Argonne Laboratory, the information office, and the patent division be consolidated with the Met Lab, and in a move to pacify the laboratory council, the army exempted from censorship speeches and articles dealing with social and political implications of nuclear energy. Finally, Groves agreed to support compilation of the scientific papers produced by the Met Lab staff. Scientists were gratified when Nichols assigned the highest priority to completing this task, known as the Plutonium Project Record (PPR).[116]

Weighing the Fate of Argonne Laboratory

On November 19, 1945, Nichols asked Compton to head a study group to investigate "the continued utilization of the Argonne Laboratory in a manner which would best serve the national interest by furthering research and development in the field of atomic energy." For the rest of the committee Nichols chose R. A. Gustavson of the University of Chicago, Spedding of Iowa State College, Daniels of the University of Wisconsin, F. Wheeler Loomis of the University of Illinois, John T. Tate of the University of Minnesota, and O. W.

Eshbach of Northwestern University as voting members and Major E. J. Bloch, a nonvoting member representing the army. At the time, Argonne Laboratory had two piles operating with associated support facilities. Hilberry suggested to the committee that Argonne Laboratory be converted to a government-owned regional research facility.[117]

Meeting with the advisory committee on Argonne operation on December 2, 1945, Nichols stated that the new laboratory would be a "guinea pig" for postwar nuclear research. Compton's committee told Nichols that the laboratory should continue as a regional research center to serve the north central United States. The laboratory's mission should include basic research that did not duplicate work at midwestern universities, but rather supplemented the programs of these institutions. The committee also favored extending the existing contractual arrangement with the University of Chicago with the ultimate goal of forming a public corporation to operate Argonne in the national interest. Suggesting that representatives of leading area research facilities should belong to the corporation, the committee recommended twenty-four midwestern research institutions for membership on an Argonne Laboratory advisory council. The committee's plan called for election of a seven-member executive board to advise the government on the Argonne Laboratory program and administration.[118]

When Groves visited Chicago in December, plans were under way for establishing a regional laboratory in the Midwest that would include the assets of the Metallurgical Laboratory. Although the laboratory had lost a substantial number of scientists who had returned to universities or industrial companies, Groves assured Met Lab personnel that the laboratories at Chicago and Argonne most likely would continue to operate as permanent facilities.[119] Daniels could only hope that he would be allowed to find suitable replacements because the Argonne staff had fallen to thirty-five scientists and technicians, many of the latter being soldiers who worked for "slave wages." Nichols urged Groves to take immediate steps to "keep our research organizations intact and working on research work rather than playing politics."[120]

On January 21, 1946, Daniels asked his division directors to prepare recommendations to the army for the fiscal year beginning July 1. Despite Groves's assurances, however, the fate of the Metallurgical Laboratory remained uncertain. A poll the laboratory council had taken on February 4 revealed that all but two of twenty members planned to leave by June 30 if the Laboratory did not plan a vigorous research program. Already Stone had left; Hogness was planning to return to his university position in April, and Seaborg would leave May 17, taking many of the most productive scientists in his division with him.[121]

With some urgency to get planning started, on February 11 University of

Chicago chancellor Hutchins and the university's business manager, William Harrell, tentatively agreed with Nichols, Col. Arthur H. Frye, and Daniels that Argonne Regional Laboratory would take over the program, buildings, equipment, and most of the staff of the Metallurgical Laboratory if the latter ended operations on June 30, 1946. They acknowledged that the University of Chicago probably would administer the new laboratory. As a result of these discussions, Nichols called a planning meeting in April of representatives of the twenty-four midwestern universities recommended by the Argonne advisory committee.[122]

Meanwhile, following Nichols's suggestion, Groves again called on Compton to serve on an advisory committee on research and development with six other scientists prominent in the Manhattan Project. Groves asked for advice on both policy and programs and would submit his budget based on the committee's recommendations. In March the group proposed the establishment of a national laboratory at Argonne and another in the northeastern United States—each with a board of directors selected from participating institutions in their respective regions. Groves's advisors also recommended that national laboratories should engage in unclassified basic research requiring the "use of piles and other expensive large-scale equipment" too expensive for university or private laboratories. Further, the committee advised, national laboratories funded by the federal government should not compete with work at universities and private industry.[123]

Nichols expanded the regional concept to include major Manhattan Project facilities across the continental United States. Extending the logic of Groves's advisors, Nichols anticipated authorizing a California regional laboratory to concentrate on high-voltage accelerators; Argonne Regional Laboratory to specialize in piles; a northeastern regional laboratory to work on both piles and high-voltage accelerators; Los Alamos to develop military applications, and the Clinton laboratory to focus on industrial expansion.[124]

With policy for regional laboratories now established, the army approved plans for two reactor projects: Zinn's proposal for a fast-fission pile at Argonne and the construction of the high-temperature pile conceived by Daniels at Clinton. There was a caveat, however. Independent reviewers would have to evaluate whether it was safe to build Zinn's pile at the Argonne site.[125]

In the midst of this planning, Ralph Lapp, then the assistant laboratory director at the Metallurgical Laboratory, dropped a bombshell. Citing legal as well as environmental obstacles, Lapp told Daniels on March 31, that from a long-range viewpoint "Argonne would not serve as a suitable site" for nuclear research and development. After reviewing a topographical map and aerial photographs, Lapp observed that the site had inadequate transportation, lacked gas, had insufficient water, and was too small and too inaccessible for major scientific programs.[126]

Despite Lapp's objections and without apparent discussion with Cook County officials, Nichols tentatively approved establishing Argonne National Laboratory at its existing Palos Park location. On April 5 and 6 the representatives of twenty-four midwestern universities, colleges, and private research institutions met in Chicago to organize the new laboratory. The group established a council composed of one representative from each college and university and a board of seven governors (elected by the council) to serve as an advisory committee to laboratory management. In turn, the newly designated board of governors chose Compton as their chair pro tem. As one of its first acts, the board of governors recommended that the University of Chicago continue as the contracting agent during the first year and that laboratory operations remain at the existing Argonne site; however, the board also agreed that the laboratory should move out of University of Chicago facilities as soon as possible so that the laboratory could occupy its own buildings. Frye, of the Army Corps of Engineers, estimated that new construction for Argonne National Laboratory would cost approximately $2.5 million. Future publications, the board said, should bear the name Argonne National Laboratory (figure 8).[127]

Figure 8. W. H. Zinn and Farrington Daniels examining a blueprint of the proposed Argonne National Laboratory

Generally satisfied with developments, Daniels did not want to exchange one restrictive master for another, however. Candidly, he told board members that he was troubled by the limits the advisors wanted to place on Argonne's basic research programs. Consequently, Daniels persuaded the new board to adopt a liberal interpretation of research policy to enable Argonne scientists to follow research lines of their choosing.[128]

The board's choice of the new laboratory's first director, however, encountered complications. At the April 6 meeting the board decided to establish a rotating directorship appointed for two or three years, but it also determined that the associate directorship should be a permanent one. The board first asked Daniels to continue as director for the next year, but he declined. He wanted to return to the University of Wisconsin on a half-time basis and preferred to spend the remaining time with his research project at Oak Ridge and his responsibilities on the board of governors at Argonne. Next, the directors asked John Tate, but he also refused because of his desire to return to the University of Minnesota. Finally, they asked Frederick Seitz, director of the Monsanto Chemical Company project at the Clinton Laboratories and chairman of the Carnegie Institute of Technology physics department, but he also withdrew from consideration for the position.[129]

By May 6 the board had decided that a rotating directorship would not work. On the basis of a telephone poll of the board and Nichols, the directors offered the director's position to Zinn, Fermi's right-hand man and head of the existing Argonne Laboratory. After reaching an understanding with the board on supporting new pile research and construction and obtaining necessary high-voltage equipment and adequate staff, Zinn accepted the directorship.[130]

On May 27, 1946, with little more than a month left for the operation of the Metallurgical Laboratory, Daniels appointed Norman Hilberry as associate director of the Metallurgical Laboratory. Daniels announced that Hilberry would continue in that position when Argonne National Laboratory opened.[131] Although it seemed assured, the permanence of Argonne National Laboratory as a federally owned but privately operated institution nevertheless depended on the passage of legislation pending in Congress.

The Legislative Foundation

Planning for Argonne National Laboratory took place simultaneously with the government's debate on how best to manage and control nuclear energy research and development following World War II. The future of the Argonne laboratory depended on answers provided by Congress and President Truman. Months before the first atomic bomb was detonated, Vannevar Bush and Harold D. Smith, Truman's budget director, disagreed on how the federal government should support postwar scientific research. Bush sought legislation that would put control in the hands of the scientists, who would be free of political interference. He proposed that a government commission be formed to regulate the uses of atomic energy, but that a foundation be established to support independent research. Smith, on the other hand, believed that the government should manage any large-scale government-funded research. On July 19, 1945, Senator Warren G. Magnuson (Dem., Wash.) introduced a bill supporting Bush's research foundation, and on July 23, Senator Harley M. Kilgore (Dem., West Va.) tossed a bill into the legislative hopper that followed Smith's concept.[132]

In early October, Representative Andrew Jackson May and Senator Edwin C. Johnson offered an alternative proposal. Prepared by the War Department with concurrence of Truman's cabinet, the May-Johnson bill proposed comprehensive government development and control of atomic energy. To promote national defense, public safety, and world peace, the bill required that a part-time atomic energy commission direct "all activities" related to research, production, and release of atomic energy. In accordance with Bush's wish, however, the May-Johnson bill complemented the research-foundation legislation introduced by Magnuson by allowing small-scale nuclear energy research in nonprofit institutions with the least possible interference from the commission and its administrator.[133]

Although Oppenheimer and some of the Manhattan Project senior scientists initially supported the legislation, the May-Johnson bill drew opposition from 95 percent of the scientists who had helped develop the atomic bomb. Manhattan Project scientists, especially the younger ones from the Metallurgical Laboratory, asserted that the bill's restrictions and control would stifle nuclear research by giving too much power to the military. The Atomic Scientists of Chicago joined with groups from other Manhattan Project laboratories to persuade Congress not to pass the bill. The Chicago organization contributed a legal analysis by University of Chicago law professor Edward Levi, who argued that the government should not exercise such rigorous control of nuclear energy that it "would constitute a judgment as to how research can be best carried out."[134] When the White House learned that the May-Johnson bill would create an atomic energy czar not subject to presidential appointment and removal, Truman withdrew his support of the legislation.

On December 20, 1945, Senator Brien McMahon, chairman of a special committee on atomic energy, introduced a new bill that had been written with Levi's help. The drafters of the McMahon bill attempted to satisfy the opponents of the May-Johnson legislation by providing for a civilian atomic energy commission whose chairman and commissioners would be appointed by the president subject to Senate confirmation. The McMahon bill established controls of fissionable material and dissemination of information but permitted the commission to fund private research. Along with their colleagues from Manhattan Project laboratories, the Metallurgical Laboratory Council members unanimously supported the McMahon bill.

While Congress delayed action on the National Science Foundation (NSF) legislation, on July 26, 1946, both houses of Congress passed an extensively amended McMahon bill. On August 1 President Truman signed the Atomic Energy Act of 1946.[135] The passage of this legislation within a year of the bombing of Hiroshima established the federal government's long-term responsibility for the nation's nuclear research laboratories.

The Legacy

The Metallurgical Laboratory and its offspring, Argonne Laboratory, left a variety of challenges for the national laboratory slated to replace them (figure 9).

Among the tasks borne by the old and the new institutions alike was recording the scientific work of the Manhattan Project. The Met Lab scientists spent their last year preparing handbooks and compiling the Plutonium Project Record, a comprehensive library of reports and scientific data from all aspects of the work they had done during World War II. Later, as a part of the National Nuclear Energy Series, the Plutonium Project Record had to be declassified before it could be published. By April 1946 Daniels estimated that the Plutonium Project Record would comprise forty volumes covering the project's biological and biomedical as well as the physical sciences. From the army's perspective, it was of paramount importance to capture a complete record of the Manhattan Project's physics research. The scattering of scientists after the war hampered efforts to produce a complete written record, but the army expected the remaining scientists, including those at Argonne National Laboratory, to finish the project by April 1947.[136]

Argonne National Laboratory inherited another legacy from the Met Lab: the responsibility of providing the weapon and production laboratories with special instruments. Because the work was security classified, the Metallurgical Laboratory's instrument section both designed and manufactured the portable electronic research and survey instruments used throughout the Manhattan Project. The section's backlog of orders was so voluminous that it could not keep up with the demand. Now that the country was at peace, Daniels and his

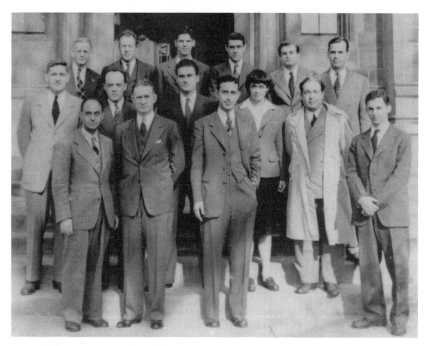

Figure 9. Fourth anniversary reunion of CP-1 scientists

colleagues believed that Argonne should get out of the business of making instruments and let commercial companies take over their manufacturing as soon as the technology could be declassified. Consequently, the laboratory's instrument section was reorganized to operate on a research-and-development basis while still responding to the demand for instruments. Despite these adjustments, the instrument problem remained unresolved at the end of the Met Lab's operation because government customers continued to expect Argonne National Laboratory to service instrument orders.[137]

Shortly before termination of the Met Lab, Daniels and Nickson expressed concern about the environmental hazards left by the Manhattan Project. Both foresaw difficulties. "The disposal of radioactive materials from research laboratories and piles constitutes one of the greatest and most important problems in the development of nuclear research in this country," Daniels cautioned Nickson, who was to continue as senior physician for the new laboratory.[138] The departing director's premonition raised a warning that would echo throughout the history of Argonne National Laboratory.

The principal mission of Argonne National Laboratory, of course, would be to carry on reactor research and development, a responsibility inherited from

Fermi and to be pursued for the next half century. Initially, Daniels expected Zinn's fast-fission pile at Argonne to rank as the laboratory's first priority. Although part of the Met Lab's chemistry research would move to Berkeley with Seaborg, some would continue at Argonne under Winston M. Manning, who would keep in close communication with Seaborg. Argonne would concentrate on reactor chemistry while the Berkeley scientists would emphasize research on particle accelerators and cyclotrons. Daniels considered the health and biology program especially important for the Argonne laboratory because in 1946 almost half of the science budget went to the biomedical projects: "The health and safety limitations seemed to be critical in the development of nuclear reactors." Participating institutions thus would need advice and service on health and protection. Although the University of Chicago expected to take over some of the radiobiology work, Daniels recommended that the Argonne laboratory continue its research at about two-thirds of the 1946 program.[139]

When the Met Lab closed its doors, Argonne National Laboratory was ready to carry on its responsibilities for research in the physical and biomedical sciences as well as on nuclear reactors. Although there were older government laboratories, Argonne would be the first national laboratory established under the Atomic Energy Act of 1946. Following his own project to the Clinton Laboratories, Daniels's farewell informed the staff that after June 30, 1946, "The Metallurgical Laboratory will become the Argonne National Laboratory . . . under a new contract with the Government."[140] With this transfer, Argonne's leadership was turned over to a Met Lab pioneer who had helped construct CP-1, Walter H. Zinn.

2 | A Troublesome Transition, 1946–48

Walter Zinn found the challenges daunting when he took the helm of the new Argonne National Laboratory on July 1, 1946. His attempt to build an institution that would expand knowledge in the nuclear sciences, beyond his own contribution of reactor development, soon collided with Cold War tensions, national and local politics, academic rivalry (and perhaps jealousy), and intergovernmental disagreements, in addition to the anticipated technical problems. These factors would not only intrude but also interact to compound the difficulties of defining Argonne's mission and place in American science and technology. Guiding the direction of and maintaining control over his laboratory in such a setting would take the measure of this strong, demanding leader.

In Walter Zinn the new laboratory had the kind of leader it needed. His education and background provided him with the necessary blend of academic and practical experience that would be essential in overseeing a spectrum of disciplines ranging from theoretical physics to power-plant engineering. Originally attracted by the opportunity to conduct laboratory experiments, the Canadian-born scientist, who had acquired a mathematics degree in 1927 from Queen's University in Kingston, Ontario, left the actuarial department of an insurance company to study physics. His graduate work led to a Ph.D. from Columbia University in 1934, and he then assumed a faculty position at the City College of New York, subsequently collaborating with Leo Szilard and Enrico Fermi to investigate chain reactions. Having taken a major role in pile research and development throughout the lifetime of the Metallurgical Laboratory, Zinn had earned a reputation in reactor engineering matched by few, if anyone, in the Manhattan Project.

Argonne's new director also had the personality that his task demanded. Confident of his abilities as an experimental physicist and comfortable with the role of a group leader, Zinn could be hard-headed and demanding. When he was challenged, his voice took on a raspy, elevated pitch that left no doubt of his determination. He expected the best of his subordinates but was quick to defend them against outside attacks. Walter Zinn was a person to be reckoned with.

The Challenge

Zinn expected to continue to devote much of his time to reactor projects at the laboratory, but his new responsibilities at Argonne extended far beyond reactor development. First, he had to hold together a widely dispersed empire. His inheritance—the remains of the Metallurgical Laboratory and the Argonne Laboratory, both of which had been hastily assembled to meet a wartime need—operated in temporary housing in over a dozen places in and around Chicago, stretching from the Museum of Science and Industry in Jackson Park to the Palos Park site. Although designs for some of the laboratory's promised permanent buildings were almost completed, no one knew then even where they would be situated.[1]

Staffing was another concern. By mid-1946, Argonne's payroll had decreased from a wartime high of some 600 scientific personnel and approximately 1400 other workers to about 250 scientific employees and nearly 950 others. By October the scientific staff had undergone a turnover of almost 40 percent; 98 departures were fortunately offset with 128 new investigators. Most of its prestigious researchers, along with the anti-Nazi passion that had driven the Manhattan Project's successful development of the atomic bomb, were gone. Left were a new generation of youthful physicists, chemists, zoologists, biologists, engineers, and physicians eager to tackle the questions raised by the awesome, emerging field of nuclear energy. Even the remaining Met Lab veterans, many of them in or just out of graduate school on the eve of Pearl Harbor, were mostly young. Zinn fell to the task of organizing this enthusiastic but disparate crew into productive, mature research teams. To bring order to the personnel structure, he asked the staff to set up a wage scale based on surveys of the going rates for similar positions in the Chicago area.[2]

Governance, a key issue, was in limbo. The new Atomic Energy Act provided for the transfer of Argonne's supervision from a military to a civilian entity, but the laboratory was expected to remain under the jurisdiction of the army's Manhattan Engineer District until the new Atomic Energy Commission could be effectively established. The passage of the act just a month after Argonne opened as a national laboratory placed it and Zinn in the awkward situation of having to organize under a departing authority without knowing what to expect from the new one.

Equally obscure were the roles that the University of Chicago and the participating midwestern institutions would play in governing the laboratory. True, Gen. Kenneth Nichols had developed a rough plan of governance that would apply to Argonne and other regional laboratories as long as the army was in control. The laboratory council, representing each of the participating institutions, was already in existence, and the council had elected seven of its members to the board of governors; but just how much of a voice the council and the board would have in setting the laboratory's research agenda was still to be determined. A further complication was that the University of Chicago was expected to manage the government contract only until a permanent arrangement could be made, possibly with a nonprofit corporation created for that purpose.

However these immediate questions might be resolved, Zinn could take some comfort in the fact that General Groves and Nichols had already established the pattern for creating a number of national laboratories, each of which would serve as a nuclear research center in its own region. On July 1, 1946, six months before activation of the Atomic Energy Commission, Argonne was the first of these institutions to emerge officially as a national laboratory. One month after Argonne came into existence, Groves chose Camp Upton, Long Island, as the site for the Brookhaven National Laboratory (BNL), established for nuclear science research in the northeastern United States. Groves was still considering plans for a regional research center in the southeast, which would ultimately emerge as Oak Ridge National Laboratory. There was also talk of a similar laboratory in the West, but for the time being, the University of California would continue to operate Los Alamos and the Radiation Laboratory at Berkeley as government facilities. One thing all the laboratories had in common was that they were "GOCO" facilities—that is, the buildings and equipment were owned by the government but they were operated by private contractors. Zinn could not be certain that the new commission would accept the army structure of national laboratories serving as regional research centers and as GOCO facilities for government research and development, but the concept seemed so deeply imbedded in the laboratories that it would be difficult to change.[3]

Zinn could only wonder what impact the Atomic Energy Act would have on Argonne. Under the new law, Congress gave ultimate control of atomic energy research and development to the Atomic Energy Commission, which was led by a chairman and four other commissioners serving staggered terms. The act included within the commission a general manager, also subject to Senate confirmation; a division of research; and three other divisions. The legislation provided for scientific and technical advice to the commission by establishing the General Advisory Committee (GAC), which was to be composed of prominent scientists and engineers, and the military liaison committee, which was to be staffed by senior military officers and whom the commission was expected to consult regarding military applications of atomic energy.

Through this legislation Congress also conferred extraordinary property-acquisition power on the commission, a responsibility which might speed the acquisition of a permanent site for Argonne. The law allowed the commission the option of exercising eminent domain and of condemning property for government use without following the procedures prescribed under the 1888 law that all other government departments, including the armed services, were required to observe.

The legislation also established the Joint Committee on Atomic Energy (JCAE), composed of eighteen members divided equally between the Senate and House. The Joint Committee was required to make "continuing studies" of the activities of the Atomic Energy Commission and problems related to "the development, use, and control of atomic energy." All legislation and other matters relating to atomic energy in either house were to be referred to the committee, which was authorized to recommend policy to Congress, conduct investigations, and introduce legislation in both houses.[4] The powers of the new committee seemed truly extraordinary, but until it was organized, Zinn could not begin to predict its significance.

Getting Started

Pending the anticipated change from military to civilian control, Zinn had no choice but to continue the projects approved by the army, including work that related to nuclear weapons. Thus, on its very first day as a national laboratory, Argonne stationed some of its scientists at Bikini Atoll in the Pacific to assist the navy in Operation Crossroads, a series of nuclear bomb tests designed to bring the navy into the nuclear era and also to impress the Soviet Union. For Argonne, this participation added to the classified work that could not be shared with regional institutions' researchers lacking security clearances, a continuing source of frustration for the academics.[5]

On August 21, General Groves approved the Argonne National Laboratory program budget proposed by Zinn and his staff for fiscal year 1948. In accord with recommendations made during the first half of 1946, the plan emphasized two fields: the research and development of chain-reacting piles and basic research in nuclear physics, nuclear chemistry, and other sciences related to the release of nuclear energy. Only work that required the unique radiation-protection facilities of the laboratory and that was too expensive for universities and private institutions to pursue on their own would be carried out at Argonne. In keeping with Argonne's prescribed regional role, the plan provided for cooperative work with scientists from the participating midwestern institutions, including supplying them with radioactive materials.[6]

Given these knowns and unknowns, Zinn divided the laboratory into two groups: basic research and pile development. To administer them, Zinn chose two associate directors, Harvard L. Hull, an official with the Tennessee Eastman Corporation at Clinton during the war, and Norman Hilberry, who had been Compton's right-hand man in the Metallurgical Laboratory. Zinn made Hilberry responsible for the following divisions: physics, chemistry, biology, mass spectroscopy and x ray, instrument research, information, patent, and medical and hazard evaluation. Hull was charged with overseeing pile research and construction, metallurgy, instrument fabrication, and the central shops. At the same time, Zinn retained his position as director of pile research and development. He personally hired the pile operators and, according to one source, tolerated their mistakes more readily than those of others in the laboratory.[7] Following a Metallurgical Laboratory practice, Zinn established an executive committee of the division directors, who met weekly on administrative plans and policies, and a laboratory council, which convened to consider research problems on an academic basis.[8]

The Search for a Home

General Groves also recognized the necessity of finding a permanent location for Argonne, a problem of continuing frustration to Zinn and his staff. During the summer of 1946 the general and the Argonne board of governors had proposed establishing the laboratory on the Palos Park site (figure 10), where Argonne was already operating two experimental reactors—the reassembled pile that had produced the first controlled nuclear reaction (CP-1) and the world's first heavy-water reactor (CP-3). Groves and the board clashed, however, over how to respond to the contention by the advisory committee of the Cook County Forest Preserve Commission that the federal government had pledged to restore and return the Palos Park property to the forest preserve after the war. The situation became more confusing after Groves asked the preserve's commissioners for a ten-year extension for 20 to 40 acres. Secretary of War Robert L. Patterson, meanwhile, urged the Forest Preserve Commission to cede permanently 255 acres of the land. The advisory committee countered, offering only to let the laboratory stay on a diminished part of the property until the federal government found the laboratory another location. Senator Scott Lucas of Illinois supported the committee's stand and on July 30 advised the army to drop the quest for the Palos Park location "and seek a site elsewhere."[9]

Within the laboratory, the army's actions also raised concerns. Hilberry believed that the reduced acreage recommended by Groves and Patterson would

Figure 10. Argonne National Laboratory, Palos Park unit

seriously curtail development. He attributed Groves's offer to the general's vision of a future Argonne as "a pure research organization."[10] Arthur Compton, now an Argonne governor, agreed with the superintendent of the forest preserve in opposing any federal seizure of the Palos Park site. Compton urged Under-Secretary of War Kenneth C. Royall to withdraw the request and instead to consider purchasing available land a few miles away from the Palos Park location. The War Department had no intention of condemning the land if local authorities objected, Royall assured Compton.[11] In any case, what would the new Atomic Energy Commission, not yet appointed, have to say of such plans?

Meanwhile Zinn drew up criteria for the new site. He stipulated that the spot should be isolated from population centers yet accessible for commuting and evacuating personnel in an emergency. To accommodate the expected midwestern university researchers, the location should be "at least within an hour's automobile or train distance" from participating institutions.[12] Anxious to get the laboratory settled, Zinn traveled to Washington, D.C., to discuss site negotiations with Groves and Nichols. Both responded favorably to the prospect of a location near Chicago. Consequently, on September 10 Colonel Frye, chief of the Corps of Engineers' Chicago office, asked a local engineering firm to conduct a site survey for the laboratory in Du Page County.[13]

At the same time, Zinn and his colleagues considered five sites in Wisconsin, Indiana, and central Illinois. The laboratory's board of governors, Zinn, and most of the staff rejected them all. As a poll of a chemistry section indicated, many Argonne employees wanted to be near Chicago. Laboratory officials worried about retaining staff; some Argonne scientists already had registered for other employment during the uncertainty.[14]

The army survey found a site in Du Page County. The new choice was about ten minutes away by automobile from Palos Park and about forty-five minutes by car from the Museum of Science and Industry. Concerned about potential radioactive waste in water and air, Zinn considered the site safe because of its physical isolation and favorable wind and other meteorological conditions. On October 7 he advised the laboratory's board of governors to approve the Du Page land, which they did unanimously.[15]

Zinn and Farrington Daniels, who chaired the Argonne board of governors, urged Groves to act soon to acquire the Du Page County site lest more scientists leave the laboratory. On October 24 Frye, too, after investigating all the proposed locations, recommended acquisition of the same land. General Groves personally inspected the site and agreed it was acceptable.[16] Four days after that, however, Truman announced his first AEC appointments, and the army stopped its own condemnation preparations. Instead, it advised the new commissioners to expedite obtaining the Du Page land, which Daniels estimated would cost about $2 million.[17]

The corps continued to play a major role in the laboratory's operations until the end of the year. In an effort to tie up loose ends, Colonel Frye and other officers met with representatives of the University of Chicago on November 22, 1946, to sign the contract authorizing the university to establish, manage, and operate a laboratory in government-owned buildings and facilities at "Argonne National Laboratory near Chicago, Illinois." The estimated cost for the first year was $11.5 million. This first comprehensive contract for Argonne transferred the commitments made during World War II to peacetime arrangements.[18]

Enter the Commission

The president's announcement of the five men he was nominating to be members of the new Atomic Energy Commission removed some of the uncertainty that had dogged Argonne scientists and others in all of the atomic energy laboratories during the autumn of 1946, but the appointments still left many questions. David E. Lilienthal, Truman's designee as chairman, seemed a solid choice. An outspoken liberal but political independent, Lilienthal had served as a member of the Public Service Commission in Wisconsin and as chairman of the Tennessee Valley Authority (TVA). With a national reputation and a flair for publicity, the prospective chairman could be expected to keep the new commission and the national laboratories in the public eye. As a member of the Acheson-Lilienthal committee, which had drafted the administration's plan for international control of atomic energy, Lilienthal knew something about nuclear technology, but he was by no means a scientist. In fact, except for Robert F. Bacher, a young physicist who had worked at Los Alamos during the war, none of the president's appointees could claim any real acquaintance with the atomic energy establishment. Nor could Carroll L. Wilson, the thirty-five-year-old engineer whom Truman named to be general manager. As an aide to Vannevar Bush at the Office of Scientific Research and Development during the war, Wilson had gained some experience in administering research contracts, and he had demonstrated his management ability in heading the staff that was negotiating the transfer of the army's facilities and authorities to the new commission.[19]

Zinn and his colleagues at Argonne could initially take some comfort in learning about the talents and experience of Lilienthal, Bacher, and Wilson, but it soon became apparent that Lilienthal's appointment, and to some extent Wilson's, was far from assured. Under the new act, the new appointees would have to pass muster with the Senate section of the Joint Committee before confirmation by the full Senate. At the hearings, which began in January 1947, Lilienthal immediately came under a vicious attack by Senator Kenneth D. McKellar of Tennessee, whom Lilienthal had offended by thwarting the senator's attempt to exercise his patronage power within the TVA. McKellar considered Lilienthal a radical left-winger who tolerated communist sympathizers on the TVA staff. Conservative members of the Republican majority on the committee broadened McKellar's assault by charging that Lilienthal was a radical New-Dealer who had been advancing the cause of socialism at the TVA and was thus a threat to private enterprise. Now the nominations became mired in a nasty political squabble that seemed interminable to both the nominees and the frustrated scientists in the laboratories. The committee did not send the nominations to the Senate until early in March, and final confirmation did not occur until April 9.

The bitter fight over the nominations all but paralyzed the new commission during the first three months of 1947. Although the agency had full authority over the nation's atomic energy program after January 1, the commissioners were reluctant to make decisions on fundamental issues pending confirmation of their appointments. During the same period, Wilson was swamped with the task of trying to build a full-fledged organization from the few dozen assistants who worked with him during the transition.

The Du Page Site Revisited

For Zinn the weeks of delay meant that it was impossible to get any action from the Atomic Energy Commission on the Du Page site. Although General Groves had urged the commission to act quickly, Lilienthal was not about to be "pushed headlong" into a decision. After visiting the site briefly in November 1946, the commission nominees had agreed with Zinn that the laboratory should be located near Chicago and promised University of Chicago officials that they would act as soon as possible.[20]

By late January 1947 the commission had come around to Zinn's position on the Du Page site after failing to persuade the Forest Preserve Commission to relinquish the Palos Park land for the Argonne laboratory. The Atomic Energy Commission authorized Wilson to obtain the land, amicably if possible for "political reasons," but the impatient Zinn again had to cool his heels.[21]

Still unconfirmed, lacking a Chicago field office, and uncertain about commission-laboratory relationships, the AEC had to rely on the army's district engineer in Chicago to handle the land acquisition. The commissioners settled on 3,667 acres in the southeast corner of Du Page County that were bounded by U.S. Highway 66 on the north, the Des Plaines River and the Chicago Sanitary and Ship Canal on the south, Illinois State Highway 83 on the east, and Lemont Road on the west. The site contained 148 parcels ranging in size from less than an acre to more than 200 (figure 11). The choice omitted several portions within the site, including three forest preserve districts, which were later called the Rocky Glen Forest Preserve, and a small cemetery. Both the commission and laboratory officials indicated that public access would be maintained for these areas.[22]

While the army assumed the mechanics of obtaining the land, the University of Chicago and Argonne took on the local public relations tasks. Prior to making a public announcement, the laboratory and university officials revealed the decision to a large selection of politicians and academics: the governor, local members of Congress, presidents and council members of the participating institutions, the area's state representatives in Springfield, and political leaders in the western suburbs of Chicago. A press release was issued February 20, and area newspapers welcomed the decision. The selection, proclaimed a *Daily News* editor, "holds great significance for the Chicago area."[23]

Unfortunately, the laboratory's public relations planners had ignored the very landowners who would lose their properties. Learning about the commission's decision only through radio broadcasts and local newspaper stories, affected owners were, according to one local official, "too stunned to talk."[24] Those about to be displaced formed protest groups, including Erwin O. Freund, a

Figure 11. Argonne National Laboratory site with farms

wealthy Chicago manufacturer and owner of Tulgey Woods, a large part of the property intended for Argonne. Freund warned, "I propose to use every means at my command for as long as necessary to prevent its being seized from me." Additional opposition came from people who viewed the laboratory as a hazard and from those worried that it would cause an undesirable population increase in certain exclusive residential sections.[25]

Although given extraordinary condemnation powers in the 1946 Atomic Energy Act, the commissioners intended to obtain the land through negotiated purchase if at all possible. After opposition arose, laboratory and representatives of the Atomic Energy Commission met with local people to assure them that landowners would receive a fair price, that Argonne would engage in research and not make atomic bombs, and that employees would spread out in their living arrangements and not cause large-scale increases in nearby villages.[26]

Still, while an advisor to the commission found solid support among the county's leading citizens and businessmen, in a mid-April election in Downers Grove, five miles northwest of the prospective site, voters opposed (1,816 to 803) the laboratory's establishment in Du Page County. The commission rejected requests by Senator Charles W. Brooks and Representative Chauncey W. Reed, both from Illinois, to reconsider the decision and instead on May 6 reaffirmed its intention to acquire the site. Responding to requests by fellow Republicans, Senator Bourke B. Hickenlooper, chairman of the Joint Committee, ordered the staff to reevaluate the selection.[27]

By May, Zinn was desperate. Caught between the imminent return of buildings to the University of Chicago and a laboratory budget based on scientific problems instead of the staff's needs, Argonne was so squeezed for space that Zinn asked division directors not to hire replacements except for outstanding individuals. Even the Joint Committee members who visited Argonne that year wondered how the scientists could get their work done in the limited and scattered facilities they were using.[28]

The final decision came a few weeks later. In a speech in Chicago on May 26, Chairman Lilienthal signaled the commission's eagerness to get the land problem out of the way and push Argonne onto the more defined track of developing reactors. "The national security... and the development of the beneficial possibilities of atomic energy are very considerably dependent upon the Argonne National Laboratory," he asserted as he applauded the laboratory for its relationship with the twenty-nine midwestern universities. By that same day, the Army Corps of Engineers had obtained signed options to purchase slightly over a thousand acres. A week later the university's business manager, William B. Harrell, agreed through a letter contract with the commission to handle the design and construction of the permanent laboratory, a task Harrell rapidly turned over to Ford, Bacon & Davis, Inc.[29]

Environmental Planning

Before the management of Argonne National Laboratory could contemplate a move to Du Page County, it faced sobering environmental problems. The laboratory had to decontaminate the facilities that it was vacating as well as measure background radiation and establish radiation safety measures for the new facility. What radioactive waste was not being shipped to Oak Ridge, Tennessee, was being buried locally, but that practice had to stop before the move.[30]

Disposal of hazardous waste at the Du Page site was the responsibility of Austin M. Brues, director of the Argonne biology division; of John E. Rose, director of the health physics division; and of James Nickson, director of the medical division. Nickson, who had influenced waste-disposal policy during the Met Lab days, noted that most of the Manhattan Project facilities had, "largely by default," buried radioactive waste in the ground, a practice he opposed. By 1947 Argonne was placing all materials contaminated with radioactive isotopes in special containers. Nickson proposed reducing the amount through concentration and storing the remaining material in suitable containers in a guarded, water-free, structurally strong area, such as an abandoned salt mine. Others were suggesting ocean dumping and, first as a joke and later seriously, placing the concentrated unwanted radioactive isotopes in outer space, if appropriate rockets could be built.[31]

Principles of Management

As soon as the Atomic Energy Commission assumed responsibility for the Manhattan Engineer District properties in January 1947, Carroll Wilson began organizing the agency to respond to national research and development needs in nuclear energy. First, he chose James B. Fisk, from MIT and Bell Telephone Laboratories as well as an electronics and radar expert, to head the division of research. Fisk had also been Wilson's roommate at MIT during the early 1930s and would be an important commission contact for the Argonne director.[32]

In February, Wilson appointed a special advisory committee, chaired by John R. Loofbourow, a biophysics professor from MIT, to advise the commission on relationships with its contractors, including those managing the AEC laboratories. Reporting at the end of June, Loofbourow's committee recommended that the government "never provide duplicate management services" in addition to those it purchased from a contractor. Further, the commission should use a field manager as the single official channel of authority with a contractor. By Zinn's interpretation, that meant the commission should "give the money, approve the over-all program and let the laboratories carry on from there."[33]

The Atomic Energy Commission did establish field offices, but they were based neither on Zinn's model nor that of the Manhattan Engineer District, whose physical structure would be generally followed. The new field offices would exercise tighter control than Zinn expected, or probably desired, but would also exert less centralized supervision than the wartime agency had. In fact, the commission delegated broad operational authority to its five major centers: New York, Santa Fe, Oak Ridge, Hanford, and Chicago. The commission inaugurated the Office of Chicago Directed Operations on August 31, 1947, making it responsible for overseeing AEC programs at Argonne National Laboratory, Iowa State College, and the University of California at Berkeley.

The commissioners' choice for manager of the new office, Alfonso Tammaro, was familiar to the staffs of Argonne and the University of Chicago. Previously a lieutenant colonel and then a civilian assistant to Nichols in the MED, Tammaro would lead his staff in negotiating contracts, in assisting prime contractors in carrying out commission-approved programs, and in assuring contractor compliance with commission policies and procedures.[34]

Other significant advice came from the Loomis committee, which had been created by the commission staff to assess personnel practices and policies in large research institutions, including existing commission laboratories. Headed by F. Wheeler Loomis, the chairman of the University of Illinois physics department and a member of the Argonne board of governors, the committee provided guidance on matters ranging from salaries to research planning. At a time when national laboratories were competing with academic institutions for personnel, the Atomic Energy Commission was concerned with its ability to recruit talented researchers. The question of freedom of research bore directly on that issue, and Argonne National Laboratory, even in its first official year of operation, scored rather poorly. The Loomis survey found that whereas Brookhaven allowed maximum freedom and concentrated on basic research, Argonne and Oak Ridge spent proportionally more time on applied work because of assigned projects.[35]

By the summer of 1947 the commissioners were ready to designate five principal objectives for the laboratories: developing more efficient nuclear weapons; improving production of fissionable materials; developing special reactors; effecting large-scale production of power; and advancing basic knowledge. The Atomic Energy Commission intended to support fundamental research for these objectives through its laboratories and through contracts with private institutions. But Argonne's experimental reactors, as the commission noted, were of "particular importance."[36] It appeared that, the Loomis report notwithstanding, the commission was not going to give first priority to basic research at Argonne.

Moving Forward on Reactors

Once Wilson had appointed Fisk as director of research, he assembled all of the AEC's laboratory directors in Washington that January to brief the commissioners and the General Advisory Committee about laboratory operations. Of the two briefings, that with the advisory committee was the more important. Unlike most of the commission nominees, all of the members of the committee had been associated with the Manhattan Project. Robert Oppenheimer, the committee chairman, had been director of the Los Alamos Laboratory; James Conant had helped launch the project in 1942; Enrico Fermi had designed the first atomic pile and helped on bomb design; and the other members, including Glenn Seaborg and Hartley Rowe, had all been major contributors to the project. During most of 1947 the committee, of necessity, took the lead in formulating AEC policy, not only for research and development, but also for production of fissionable materials and weapons.

At the January meeting Zinn learned about the relationships that would govern Argonne and the other laboratories under the new authority. The general manager's thinking on field offices, for example, led Zinn to believe that the Chicago office could approve Argonne work without forwarding the papers to Washington. Further, he expected that laboratories would bear increased responsibility for such matters as accounting and discretionary use of funds and that laboratory directors would have the independence to communicate or collaborate among themselves. Time would prove the Argonne director to have been optimistic, if not naïve. The commission, which would obtain the benefits of the laboratories' research, would itself also be accountable, in particular to the Bureau of the Budget and to Congress through the Joint Committee.[37]

The commission's reliance on Argonne as its center for reactor research may have grown out of this January 1947 meeting. Indeed, Wilson regarded Walter Zinn as the leading expert on reactors, the work at his laboratory being the most promising in this area as well as the first historically. The general manager assigned Zinn the task of preparing a reactor research plan, which would be the largest part of the laboratory directors' report to the General Advisory Committee. Zinn felt that "in general, the Atomic Energy Program deals with reactors." With weapons development to be separated from the rest of the AEC's programs, Zinn pointed out in his first communication to the new committee the importance of developing a variety of reactors. He noted that the commission would need reactors not only to produce plutonium for weapons, but also as radiation sources for testing materials and producing radioisotopes.

In his report Zinn highlighted the critical issues posed by the growing interest in reactors that would generate electric power. Here a fact of supreme importance was the shortage of fissionable material. The stocks of uranium ore at the time seemed scarcely enough to sustain the production of a modest number of nuclear weapons, to say nothing of providing fuel for power plants. Zinn believed that the only hope for power reactors lay in those that would generate more fissionable material than they consumed. This would be possible if the reactor was so designed that if one neutron sustained the chain reaction, the second and the occasional third might be captured by nuclei of fertile material to produce two atoms of fissionable material where one had existed before.[38]

Although Zinn had been studying the idea of a breeder reactor for some time, he warned that the factors to be considered in designing such a device were extremely complicated. For one thing, it seemed unlikely that a breeder reactor could operate using "slow" neutrons, the kind produced in the reactors that had been built so far. As a first step, Zinn was designing at Argonne a "fast" reactor, which would utilize the high-energy fission neutrons without slowing them down with moderators like graphite and heavy water. The new General Electric laboratory at Schenectady, New York, was planning a similar experiment with neutrons of intermediate speed.

Zinn gave equal importance to the high-flux reactor, which would not itself produce useful power but would be essential as a tool in reactor design. The unit would be designed to produce a very large stream or flux of neutrons which could be used to irradiate materials that might be used in future reactors. The very high flux would enable designers to determine the effects of neutron bombardment more quickly than they could with existing reactors. At the time, it was not clear whether the reactor would be built at Clinton or Argonne, but there was agreement that such a reactor should be built.[39]

The General Advisory Committee's priority list for reactor development submitted to the Atomic Energy Commission in March closely followed Zinn's recommendations. Setting aside for the moment the question of additional and replacement production reactors at Hanford, the committee ranked the high-flux reactor first, followed by the Argonne fast reactor, the General Electric intermediate reactor, and the gas-cooled reactor, which Farrington Daniels hoped to build at Clinton with the help of private industry.

The advisory committee's high priority for reactor development was good news for Argonne, but it was a mixed blessing. The more of its resources that the laboratory devoted to reactor development, the less that would be available for basic research in the sciences. The threat to basic research became fully apparent in October 1947, when the committee urged the commission to support a "vigorous program" in basic nuclear science, but through the universities and private institutions rather than through the national laboratories. The scope of fundamental research in regional laboratories, such as Argonne, the committee suggested, should be limited—an emphasis that would, in the minds of some, diminish the importance of the laboratory's basic research and therefore reduce the incentive for regional academic associates to work there.[40]

If anything, the commission was pushing Argonne in the direction that the advisory committee recommended. In October, Fisk and his division of research at AEC headquarters moved to upgrade Argonne's reactor role by identifying reactor research and development as a principal responsibility of the commission. With the approval of the General Advisory Committee, the division established its own reactor development committee, which was headed by Fisk, to plan, coordinate, and promote a national reactor program based on the plan that Zinn had earlier drawn up for the commission.[41]

Zinn was more than prepared to respond. The wartime and postwar pile research of the Metallurgical Laboratory had yielded sufficient information to warrant studies on the feasibility of breeding. Consequently, less than a month after the reactor committee was formed, Argonne requested authorization to design and build a liquid-metal-cooled, fast-neutron reactor. With the recommendation of the division of research, the commissioners approved the project on November 19, 1947.[42]

Tightened Security

A consequence of the growing importance of reactor work at Argonne was greatly increased security, which had become evident by November 1947. The move toward intensive investigations to obtain security clearances for AEC representatives and others in the atomic energy field had been building since the end of the war, especially after the Canadian government revealed that Alan Nunn May, a British nuclear physicist, had passed atomic energy information to a spy ring operating out of the Soviet embassy in Ottawa. Relations between the United States and the Soviet Union further plunged as communist insurgents attempted to take over Greece and as Stalin rejected a plan supported by the United States for international control of atomic energy.[43]

Zinn encountered the policy changes at a 1947 meeting in Washington on security procedures. He learned that it would be "very difficult in the future" to justify a clearance for anyone, employee or consultant, "should there be anything questionable in his background." Feeling he had to remain impartial and could not afford to take time away from his scientific responsibilities, Zinn told his laboratory executive committee that he could not intervene in clearance questions. Moreover, if a scientist's clearance was held up, laboratory officials could not tell the individual the reason.[44] The security restrictions tied to enhanced reactor work were bound to interfere with the freedom of outside scientists to conduct research at Argonne and to complicate the lab's attempts to cultivate the cooperation of regional institutions.

Mission Articulated

On December 11, 1947, the Atomic Energy Commission confirmed that the program at Argonne National Laboratory was to be "focussed chiefly on problems of reactor development, with fundamental supporting research on relevant problems in chemistry, physics, metallurgy, medicine and biology." The commission also noted that Argonne's "special facilities [were] to be made available for the training of students and staff members in the supporting institutions in the Chicago Area."[45] In addition, the laboratory's chemistry division provided most of the experimental work in chemistry and physics needed for the design, construction, and operation of the Clinton Laboratory's high-temperature pile. Argonne's chemical engineering staff spent 80 percent of its time on Redox, a new process for separating plutonium from production-reactor fission products, and 20 percent on Hanford pile problems.[46]

On December 26, 1947, the commissioners formally approved the consolidation of the agency's reactor development program at Argonne, apparently without giving advance notice to Zinn, to the laboratory's board of governors, or to the General Advisory Committee.[47] The decision raised questions about the adequacy of the laboratory's proposed physical structure and Argonne's future course of work and cooperation with the regional research institutions.

Consequences of the Reactor Assignment

Problems of identity, function, and control emerged almost immediately after the Atomic Energy Commission announced on January 1, 1948, that Argonne National Laboratory was to lead the nation in the research and design of nuclear reactors. As originally established, Argonne was to become the principal nuclear science research institution in the midwestern United States. But its designation as both a multiprogram facility and the nation's leading reactor-design center left questions of how the laboratory could live up to the expectations specified at its founding and how it could serve so many masters: several arms of the federal government, the University of Chicago, and the midwestern universities. Zinn and his staff needed a clear definition of Argonne's role to reconcile the intrusion of the reactor work on basic science.[48]

If the commission's announcement disrupted plans at Argonne, it caused consternation at Clinton. For months the laboratory had drifted rudderless without a director or operating contractor. There was even talk of shutting down the laboratory and moving what was left of it to Brookhaven. The one hope of survival lay in the AEC's efforts to negotiate a contract with the University of Chicago to operate Clinton, a hope that would give Clinton the possibility of developing on the Argonne pattern. But at the last minute, in December 1947, the commission had cut off negotiations with Chicago and announced that Union Carbide would operate the Tennessee site as Oak Ridge National Laboratory. With its mission focused on the industrial development of atomic energy, Oak Ridge would no longer be a suitable place for the high-flux reactor. The entire reactor division would now be moved to the Du Page site. The Argonne division directors who visited Clinton found the employees there divided and demoralized over the prospective transfer.[49]

The proposed transfer also posed problems for Zinn. Because facilities at the Du Page site were unfinished, the Clinton power-pile division would not be able to move to Illinois for almost a year. To fund the move, Zinn's staff had to revise Argonne's budget estimate, which had been submitted to the University of Chicago just prior to the commission's decision. The additional request covered what Zinn considered necessary for expansion of the reactor development, metallurgy, chemical engineering, and theoretical physics divisions to meet the new demands.[50]

In 1948 Zinn regarded a strong, integrated reactor research and development program as necessary to the national security and welfare. In his view, the AEC and Argonne had to produce a successful reactor program lest disastrous consequences loom ahead for the country: "At the moment, both the Commission and the laboratory are on trial."[51] Within the reactor assignment, Zinn seems to have interpreted centralization as limited to design. He recognized the necessity for other laboratories to develop reactors to suit their own needs, especially research reactors: It was "not my understanding when Argonne took responsibility for reactor development that this in any way gives it control over the type and quantity of research reactors which might be acquired by any installation other than Argonne."[52]

The surprise designation of Argonne as the center for reactor research caused concern among Argonne's board of governors, who called a special meeting to discuss its implications. They concurred with the commission's decision but worried about the potential impact on the laboratory's basic research program and on cooperation with the participating institutions. Although Zinn personally assured the board that the added work would not affect other responsibilities, he nevertheless asked for and received a written reassurance from Carroll Wilson. The latter pledged support for the fundamental research and for participating institutions' programs and predicted that these activities would be helped rather than hindered by the reactor work.[53] In the meantime Argonne would inherit reactor projects from other laboratories, while Zinn was already making plans appropriate for the new status.

Site Matters: Still Unresolved

As 1948 began, Argonne had its work defined but no place to do it. Several suits challenging the condemnation of parts of the Du Page site were under court consideration, and plans for the laboratory buildings no longer seemed adequate for the added reactor work. The situation eased somewhat on February 24 when a court decision affirmed the commission's powers to appropriate the land. Construction had started, and Zinn expected enough temporary facilities to be ready in Du Page County for laboratory personnel to work there that spring. Anxious to retain support from Congress, Zinn reminded his staff to exercise restraint in travel and long-distance telephone expenses because, as Hilberry explained, "These are items which the Congressmen jump on."[54]

In the meantime, the Du Page landowners fighting the AEC's condemnations found sympathizers among Lilienthal's opponents on the Joint Committee, which established a subcommittee on the Du Page issue. Acceding to the requests of the local protesters, the subcommittee held hearings in Hinsdale, Illinois, on March 18 and 19. The members also inspected the site, only to find construction started and local traffic suspended or redirected by nondeputized commission or contractor personnel.[55]

Upon their return to Washington, the subcommittee members questioned Tammaro about the roadwork and called for site information from the commission. In response to the new responsibility for reactor development and to the subcommittee's inquiry, the Argonne and AEC staffs revised plans for the construction of facilities. To accommodate the reactor program, they changed their estimate for the building area from 840,000 to 1.2 million square feet; the number of personnel from 2,854 to 3,200, and ultimately 4,300; and construction costs from $54 million to 57 million. The Republicans on the subcommittee questioned the choice of Du Page but not the figures submitted. When Zinn, supported by Tammaro, met with the commission on May 12, 1948, he persuaded the commissioners to approve the revised recommendations for increased spending and for laboratory expansion. Argonne was required, however, to obtain the approval of Tammaro and Wilson for the plans of each building before construction started.[56]

The site issue became entangled in political controversy, with the subcommittee members splitting along party lines. In June the Republicans, both from Ohio—Representative Charles H. Elston, the chairman, and Senator John Bricker, prominent in opposing the Lilienthal confirmation—concluded, "Why the best interests of the Nation would not be served by an unfriendly condemnation proceeding in Cook County but would be served by an equally unfriendly condemnation . . . in Du Page County, we are at a loss to understand." Noting that federal law required the armed forces to agree with congressional committees on the acquisition of land, they declared that the appropriation of the Du Page site resulted from the delegation by Congress of too much power to a federal agency. The Democrats, led by Illinois Representative Melvin Price, disagreed in substance. While a majority of the Joint Committee criticized the commission's methods in acquiring the land, they severally approved the site in June 1948, although "begrudgingly," as one journalist put it.[57]

The commissioners ultimately paid a political price for taking the Du Page land, legal though the subcommittee agreed it was. Several years later when a Republican-controlled Congress amended the Atomic Energy Act, the legislators withdrew the AEC's independent power to acquire property, as Elston and Bricker had demanded in 1948, and required the commission to abide by traditional procedures used by other federal agencies to take over privately owned land.[58]

The continuing harassment over the land acquisition was not the only frustration that Zinn faced at the Du Page site in 1948. Much to his dismay, construction lagged far behind schedule. By the summer, delays were so bad that the University of Chicago switched prime contractors. Voorhees, Walker, Foley & Smith was assigned to design all the permanent buildings and the Austin Company to construct them. In July Zinn found the temporary site in sad shape and blamed poor engineering work. Rising costs and limited funds were also to blame for the continuing delays. In August a visiting journalist saw "not specially good looking" Quonset huts among scrubby wild plum and "adolescent oak trees" on the Du Page land that looked "more like an army camp hurrying to get ready to receive men before winter." The following month, these temporary structures housed the first occupants when Argonne's reactor engineering division moved from Palos Park (figure 12). Other structures followed, but as the year drew to a close, the laboratory still struggled to operate as orphaned pieces all over the Chicago metropolitan area.[59]

Argonne: A National Reactor Center?

The AEC's decision in December 1947 to make Argonne its center for reactor development inevitably involved Zinn and his staff in a host of problems that stretched far beyond the confines of the Du Page site. The decision seemed to make Zinn the commission's reactor czar, a position that would thrust him into policy questions concerning not just Argonne, but also the Atomic Energy Commission itself, its reactor development staff, the General Advisory Committee, the reactor safeguards committee, other commission laboratories, the navy, and industrial corporations. If Zinn had any thoughts that he and his staff could settle down in the new year to advance the design of the fast breeder, he was doomed to disappointment. Instead, he would spend much of the year in seemingly endless rounds of meetings at Oak Ridge and at the commission's Washington headquarters.

Figure 12. Quonset huts (*Frontiers, 1946–1996*, p. 13)

The immediate issue was where to build the high-flux reactor. There was no question that the reactor would be an essential tool in testing the effects of intense neutron bombardment on reactor materials and components. Equally important was that Alvin Weinberg, Wigner's successor, and his staff at what was now the Oak Ridge National Laboratory had concluded the reactor should use pressurized water as its coolant and moderator and enriched uranium as its fuel. With no enriched uranium available during the war, physicists at the Metallurgical Laboratory had been forced to use natural uranium as fuel. Because the amount of uranium-235 in the fuel was very small, it was necessary to use a moderator that was the least likely to absorb neutrons and thereby reduce the chances of maintaining a chain reaction. Now, with enriched uranium available in increasing supplies from the gaseous diffusion separation plant at Oak Ridge, it seemed possible to use ordinary water as a moderator (and coolant) rather than highly purified graphite or heavy water, both of which were costly. Ordinary water also had the advantage that it was a substance which physicists, chemists, and engineers had worked with for generations; and its properties, at least in nonradioactive environments, were well known.[60]

The outstanding question in Zinn's mind was whether the high-flux reactor could be operated safely at the Du Page site. Perhaps it would be possible if the power level of the reactor was kept low, but Weinberg was also planning to build a facility for chemically processing the fuel elements as they were removed from the reactor. An accidental release of fission products so close to populated areas posed a real hazard in Zinn's mind. He was more than happy to talk about the safety question with the distinguished members of the AEC's recently appointed Reactor Safeguards Committee at Argonne in February 1948. A few weeks later he discussed the problem with the General Advisory Committee in Washington, telling the members that he favored construction of the high-flux reactor but that he thought it should be built at some remote site.

The lack of consensus within the advisory committee told Zinn that he could not expect a prompt decision on a remote site, but it was still possible that he could work out something with Weinberg at Oak Ridge. Zinn had known Weinberg during the war as a young Ph.D. from the University of Chicago who was also a protégé of Eugene Wigner. Weinberg had risen fast in the reactor group at Clinton and had emerged as a potential leader of the laboratory. Full of energy and enthusiasm, the young physicist was not about to accept the commission's decision to move reactor development to Argonne.[61]

Recognizing Weinberg's capabilities, Zinn was not at all disturbed by his attempt to circumvent the commission's decision. Weinberg's efforts were not a threat to Argonne, and Zinn was no more enthusiastic about building the high-flux reactor at Argonne than was Weinberg, especially given his concern about safety. So in the spring of 1948 Zinn was more than willing to work with Weinberg in finding a solution to the high-flux conundrum. Zinn considered several of Weinberg's suggested alternatives and even went to Washington to discuss with Fisk Weinberg's proposal for a cooperative arrangement whereby both laboratories would build modified high-flux reactors that would meet the criteria of the reactor safeguards committee. In the end, however, it became apparent that building a high-flux reactor at either laboratory would entail too much risk. A remote site seemed the only answer.

Zinn's collaboration did not immediately give Oak Ridge a power-reactor experiment, but Weinberg still had some control of the high-flux project wherever it would be built. More importantly, he had discredited the commission's decision on centralization and thus had gained for Oak Ridge the status of a reactor development laboratory. In his negotiations with Weinberg, Fisk, and the General Advisory Committee, Zinn had made it clear that he considered his authority to be limited to work at Argonne. He did not intend to settle questions that were Washington's responsibility. Furthermore, he had plenty of problems to solve at Argonne without adding more reactor work for his already overly taxed staff.[62]

The Fast-Breeder Reactor

Now that Zinn and Weinberg had thwarted the AEC's efforts to centralize reactor development at Argonne, Zinn hoped to concentrate on the fast breeder, the project he had been nursing along on a small budget with a few physicists and engineers since 1945. Even with limited resources the fast-breeder group, working under the direction of Harold V. Lichtenberger, had been able to devise simple laboratory experiments that gave them a reasonable basis for designing the fast-breeder reactor.[63]

The design Zinn had presented to the commission in the fall of 1947 proposed a reactor core of highly enriched uranium-235 surrounded by a "blanket" of uranium-238 in which high-energy fission neutrons from the core would produce plutonium. To increase the possibility of breeding, the size of the core and the distance that the fission neutrons would have to travel to the blanket were to be kept as short as possible. This feature of the design, however, meant that the heat energy produced by the fission process would be concentrated in a very small volume. The high power density would require the use of a coolant that could efficiently remove very large amounts of heat from the core and would not slow down or absorb the fission neutrons. Theoretically, the best coolant medium would be a liquid metal, but as often happened, theoretical advantages were not easily achieved in practice—very little was known in 1947 about the metallurgy of these substances. After much study, Zinn's group had decided to use a sodium-potassium alloy, which would have thermal properties similar to sodium but would remain liquid at room temperature. Further difficulties lay in the fact that sodium and potassium reacted vigorously with air and violently with water. It was thus necessary to design piping, pumps, and heat exchangers that would be completely leakproof for the long periods of operation.[64]

Although the configuration and general specifications established in 1947 for the fast breeder did not change substantially in the following years, in 1948 Lichtenberger and his staff were just beginning work on the thousands of design refinements that would be required before the reactor could be built. Until a remote site could be found, actual construction would not be possible; but Zinn, impatient to keep ahead of other power reactor experiments at Oak Ridge and Schenectady, kept pushing his reactor staff for results.

The Navy's Intrusion

In December 1947, Zinn had fended off the AEC's idea of making Argonne the center for reactor development, but he was destined to acquire one project that centralization would have brought to the Du Page laboratory. Early in World War II a few officers and scientists in the navy had seen the potential advantages of nuclear propulsion for submarines. Conventional submarines used noisy, smelly diesel engines on the surface, but beneath it, the vessels had to rely on electric propulsion, which restricted them to limited range and low speeds. Nuclear power would give submerged submarines the same capabilities they had on the surface.[65]

Although the advantages of nuclear propulsion were obvious, the navy had played only a minor role in the Manhattan Project and thus had no personnel trained in nuclear technology when the war ended. Not until the spring of 1946 did the navy send a small group of officers and engineers to Oak Ridge to learn the fundamentals of nuclear physics and engineering. Driven by Capt. Hyman G. Rickover, an engineering officer with a reputation for getting results, the group mined every facility at Oak Ridge for technical data, always with a single focus on information needed to design a submarine propulsion plant. Within a year Rickover and his group had a reasonable understanding of the technology and definite ideas about what research and development would be required. In 1947 Rickover took advantage of the doldrums into which the Clinton scientists had drifted and convinced Harold Etherington, the new director of the power-pile division, to apply his plans for a pressurized-water power reactor directly to the design of a submarine propulsion system.[66]

By the spring of 1948 Rickover and the navy were ready to move beyond theoretical studies into actual engineering research and development. Rickover proposed to investigate designs using three different coolants: pressurized water, liquid metal, and gas. Now that Etherington and his division were scheduled for transfer to Argonne, Rickover expected the design of the pressurized-water system to be pursued there. General Electric, already investigating heat transfer with liquid metals, would take the second project. The navy was still seeking a team for the gas-cooled approach.[67]

When Rickover arrived at Argonne in May 1948 to discuss working relationships between the laboratory and the navy, Zinn made it clear that he intended to give the project high priority. Etherington's group was to arrive in August, and Argonne would then push ahead at full speed. Rickover's primary interest, however, was finding a place for Westinghouse in the project. He conceded that Zinn and his scientists knew more about reactors than he did, but they knew nothing about designing shipboard propulsion systems, a technology in which Westinghouse had extensive experience. Just as General Electric was investigating liquid metals under a navy contract, Westinghouse would be doing the same on pressurized water. Rickover was looking for a joint effort in which Argonne would do the basic design while Westinghouse would do the engineering design and then build the reactor.[68]

Hard driving and tenacious, Rickover did not hesitate to tell Zinn what he expected from Argonne; however, Zinn could be just as feisty and determined. He agreed that the Westinghouse study of heat transfer with pressurized water was appropriate for a navy contract, providing the navy understood that Argonne had complete responsibility for the reactor portion of the plant. He was also prepared to accept a contract with Westinghouse under which the company would provide technical personnel and services to the laboratory for the submarine project. But Zinn insisted upon a sharp division of responsibility: Argonne would design the reactor; Westinghouse would develop the heat-transfer system.[69]

Rickover was not about to accept Zinn's definition of responsibility; but, despite all his blustering, Rickover was in no position in the spring of 1948 to make demands of Zinn. The Atomic Energy Commission, still trying to develop a reasoned reactor development program, feared that the navy would divert its power-reactor projects at Argonne and Schenectady to its own purposes. Not until August did the commission, under heavy pressure from the navy and the Department of Defense, agree to accept the two submarine projects. Rickover, now officially designated as the navy's liaison with the commission, was already discussing with Wilson his proposal to establish himself as director of a naval reactors branch in the new division of reactor development, which the commission was in the process of creating.[70]

Zinn hoped that he could work out with the navy an arrangement similar to that which the Metallurgical Laboratory had with DuPont during World War II: Argonne would be responsible for the fundamental design, as well as for "certain design criteria," and for the approval of "certain significant steps" in the detailed design of the reactor; Westinghouse, as a commission contractor, would be responsible for engineering and construction. Just what Zinn meant by "certain design criteria" and "certain significant steps" was not at all clear to Rickover until he met with Zinn at Argonne on October 26. What the words seemed to mean was that Argonne, as the AEC's design contractor for the project, would review all engineering drawings and specifications prepared by Westinghouse. Neither side was completely satisfied with the delegation of responsibilities, but there was enough agreement to proceed. The letter contract that Westinghouse signed with the commission on December 10, 1948, essentially embodied the arrangement Zinn had proposed.[71]

By the end of 1948 Zinn could feel that he had had some success in untangling the knot of confusion that had strangled the commission's and Argonne's efforts in reactor development for two full years. Argonne would have a primary, but not exclusive, role in the commission's program. The fast breeder was now firmly on track, and Argonne would share responsibility with Oak Ridge for the high-flux reactor, both reactors to be built, much to Zinn's relief, at a remote site yet to be determined. Finally, the commission had at last agreed to establish in Washington a division of reactor development, a move that lifted from Zinn's shoulders a responsibility for policy decisions that he had never sought.

University Relations

Perhaps it was inevitable that as Argonne's reactor assignments increased, the laboratory's relationships with the participating universities would suffer. Doubts about the viability of the relationship had arisen almost immediately after the laboratory's opening. As early as November 1946 Nichols, then chief of the Manhattan Engineering District, had told a representative of the University of Chicago that the agreement the eastern universities had made with Brookhaven National Laboratory provided a better structure than the one created for Argonne. Indeed, the midwestern arrangement started with deficiencies on both sides. In the spring of 1947 the participating universities complained that Argonne had done nothing to make necessary materials available to their researchers who desired to work in nuclear science at the laboratory. Yet, fewer than half the institutions sent representatives to a May 1947 meeting to discuss the complaint with Argonne officials.[72]

The designation of Argonne National Laboratory as the center of reactor development led the laboratory's board of governors to take steps in March 1948 to uphold the interests of the participating institutions. The board chose H. Kirk Stephenson to serve as a liaison. Also, in response to a request from the general manager, Carroll Wilson, the board unanimously recommended that Argonne offer a technical educational service to help participating and other educational institutions provide information to the public. Harrell, speaking for the University of Chicago, preferred that this task be turned over to the Museum of Science and Industry, but Zinn defended the decision: "We can't hold the Argonne [Laboratory] over a barrel. The people who work here want their presence known."

Statistics showed the tenuousness of the laboratory-university interaction. In late 1947 cooperation appeared to be improving. Of the 405 scientists participating in research and development at Argonne, 96 were assigned employees or consultants from the midwestern universities or research institutions.[73] But academic participation at the laboratory slipped drastically in 1948 to only 21 university researchers. Although the laboratory lacked sufficient space that year, it also had taken on new reactor projects.[74]

Even in its early years, Argonne had participated in cooperative research with U.S. private industry and with friendly nations, and the wartime industrial cooperation continued. From 1948 through 1951, at least seventy-five private companies worked with Argonne on matters ranging from consulting to joint projects. In accord with the AEC's technical cooperation program, resulting from 1948 agreements to share ores and information with the British and Canadians, Argonne consulted with British and Canadian nuclear experts. In 1948 Zinn participated in the first technical exchange in England.[75]

Research in the Sciences

Although the vagaries of Argonne's reactor assignment commanded much of Zinn's attention during the laboratory's first two years of operation, he could not ignore the issues raised by the rapidly evolving research programs in the physical and biological sciences, nor did he wish to. While the laboratory's dominant role in the commission's reactor development program seemed to assure solid support for the future, Zinn was always aware of the danger of letting Argonne become just a reactor engineering center. He still needed and wanted strong programs in the basic sciences. There were also important contributions that basic research could make to the reactor projects and to the safe operation of the laboratory.

Studies of Hazardous Waste Disposal

At a time when the federal government paid little attention to hazardous waste disposal, the Atomic Energy Commission, while still in its early stages of organization, had trouble arriving at a policy for disposing of radioactive waste and turned to the laboratories for help. With the possible exception of Bacher, the commissioners had little sense of the magnitude of the problem, although they understood that the safety of personnel demanded top priority.

In May 1948 Zinn had disagreed when a group representing all the Commission laboratories voted to let each laboratory solve its own problems of waste disposal. He argued that such an arrangement would dilute efforts to establish effective standards for all of the commission's installations.[76] He warned the Argonne staff in July that "the disposal of radioactive waste materials constitutes one of the greatest and most important problems being faced by this laboratory."[77]

Left to handle the problem locally, Argonne's health physics and chemical engineering divisions started preparations for waste management in 1948 before moving to the Du Page site. Stephen Lawroski, director of Argonne's chemical engineering division, and his group undertook a detailed study based on the quantity of waste generated by production operations. Zinn designated the health physics group to administer the disposal of both chemical and radioactive waste and made it clear that he preferred the chemical waste to be contained in concrete boxes instead of "just putting the stuff in the mud."[78]

Biomedical and Radiation-Protection Research

During the same period, Argonne continued the biomedical research begun by the Metallurgical Laboratory—work that had increased in importance after the war with the growing number of personnel handling radioactive substances. Radiation protection had been a matter of concern since the early part of the century, and in the postwar period the National Committee on Radiation Protection considered revising its tolerance doses. Although independent of the committee, the Atomic Energy Commission maintained a tenuous relationship with it, and commission laboratories like Argonne conducted research covering the same areas.[79]

In 1948, operating out of a former brewery and stable inhabited since the war, the biology division continued work aimed at setting safe radiation exposure levels for universal application. Robert Hasterlik, the laboratory's director of health services from 1948 to 1953, identified the most pressing medical research problems as those of human exposure to beryllium and alpha-particle-emitting radio elements, including radium. Yet the laboratory's biological and medical responsibilities ranged from analysis of U.S. atmospheric nuclear test debris to processing radiation film exposure badges of both the laboratory's employees and those of the midwestern institutions participating in Argonne's activities.[80]

The laboratory's health physics division initiated extensive environmental studies once the new site was in active use. Coordinating work with the U.S. Weather Bureau, the health physicists sampled water, air, rocks, animals, and plants on and off the site to determine existing radiation and meteorological conditions. To study animals on the Du Page property, the laboratory hired a professional trapper, who snared such creatures as muskrats, opossums, herons, hawks, owls, and squirrels for sampling.[81]

At the prodding of Willard Libby, from the University of Chicago's Institute of Nuclear Studies, and of Shields Warren, the AEC's director of biology and medicine, in 1948 Argonne's biology division collaborated with pharmaceutical laboratories to create a radiobiology experiment station in a greenhouse that formerly was part of the Freund estate. The purpose was to find techniques for applying atomic radiation in such fields as medicine, biochemistry, physiology, and agriculture. Within two years the radiobiology experiment station had forty flower and vegetable plants growing in radioactive solutions. Two pharmaceutical firms sent biochemists to the laboratory for a year to determine how substances extracted from these plants could be developed into radioactive medicines.[82]

Cancer Research

In 1948 Argonne National Laboratory renewed a cancer research program that had been initiated by the Metallurgical Laboratory. Several circumstances, including the revelation of new statistics, undoubtedly influenced the Atomic Energy Commission to support expansion of this work. The most significant factor was a rising concern about the disease. In the 1930s cancer had been the seventh leading cause of death in the nation, enough to warrant the establishment of the National Cancer Institute in 1937. In 1947, when cancer was identified as America's number-two killer disease, an AEC medical review board had advocated expanded research and training in the application of atomic energy to medical and biological problems, especially through the cooperation of the national laboratories with universities. Because cancer had caused the deaths of his parents, Commissioner Lewis L. Strauss took a special interest in promoting cancer research. Congress authorized $5 million for commission support of this work in 1948.

A more detailed plan emerged when the commission's Advisory Committee for Biology and Medicine unanimously proposed cancer research units for the Argonne and Oak Ridge hospitals. The Atomic Energy Commission approved the committee's proposal and allocated $1.75 million for both institutions in the 1948 budget. But as Hilberry explained, because Argonne lacked experience in operating a hospital, that role would be undertaken instead by the University of Chicago. Both Zinn and Lowell T. Coggeshall, chairman of the university's Department of Medicine, preferred building the facility as an extension of the university's Billings Hospital instead of placing it on the Du Page site.[83] Just what role the laboratory would have in the future Argonne Cancer Research Hospital was unclear.

Trying to avoid duplication of research conducted by other agencies, Congress financed AEC cancer research in 1948 specifically for the purpose of applying the unique products of atomic energy to cancer treatment. The commission also specified investigations into the uses of radioisotopes for treating cancer and studies concerning the protection of those persons working with nuclear energy. Along with other commission facilities, Argonne participated in all three programs.[84]

Advances in Physics and Chemistry

Once nuclear physicists in the Manhattan Project had helped to fulfill the wartime mission, they looked forward with enthusiasm to investigating the new world of subatomic physics, which they had merely glimpsed during the war. Beyond the minimal studies of nuclear reactions needed to produce fissionable materials and the bomb itself, that new world remained unexplored. Although the commission's laboratories in 1947 had only primitive instruments for studying nuclear reactions, the scientists made the best use they could of existing equipment while better devices were being designed and built.

At Argonne, facilities for nuclear research were relatively limited. Whereas Berkeley and Brookhaven were in the process of building high-energy accelerators that would open new horizons for nuclear studies, Argonne had to rely on the Fermi CP-2 research reactor and on the small CP-3 heavy-water reactor at the Palos Park site. Even these units had limited capabilities when compared to the much larger research reactor at Clinton and the more advanced reactor being built at Brookhaven. Argonne, with its mandate to develop reactors, was not in line for either a state-of-the art accelerator or a research reactor. The best Argonne could expect was a new Van de Graaff machine, which could accelerate a well-defined beam of protons to energies of about one million electron volts (Mev).

Mapping the new world at that time was conceived largely in terms of measuring "cross sections," or the probability that certain reactions would occur when the nuclei of atoms were bombarded with subatomic particles such as neutrons and protons. Measuring cross sections for literally hundreds of nucleons in a variety of reactions would take scores of physicists in several laboratories years to accomplish.

With their limited facilities, the Argonne physicists were able to make solid contributions to the commission's cross-section campaign in 1947 and 1948. Although Argonne did add cross sections to the data bank during those years, its special accomplishments came in developing equipment for controlling and measuring the reactions that took place. Following up on Zinn's wartime work, Argonne physicists developed a mechanical "chopper" that enabled researchers to select neutrons of a specific energy from the spectrum of energies produced in fission reactions. They also fabricated a crystal spectrometer that selected neutrons by reflecting those with different energies at different angles. A third device that they built was the pile "oscillator," an instrument that measured the effects on fission reactions when a selected sample was inserted and then withdrawn from a research reactor.

Argonne scientists, like those of the commission's other laboratories, were eager to move into the new field of electronic computers. Frank C. Hoyt, director of the laboratory's theoretical physics division, foresaw a revolution in science and technology through use of electronic numerical computation. Pressing for acquisition of a large-scale computer in 1948, he expressed his belief that the laboratory was in a "very favorable position for pioneer work in this field." Hoyt purchased an AVIDAC, a digital computer designed to solve complex mathematical problems, and then worked with his staff in building a computer based on a machine developed by the Institute for Advanced Studies at Princeton University.[85]

As part of the reactor mission, Argonne scientists pursued studies of the chemistry and metallurgy of materials likely to be used in reactors. Working with metallurgists in other laboratories, they explored the possibility of removing from metallic zirconium the minute quantities of hafnium that soaked up neutrons and made zirconium, an element with otherwise excellent physical properties, unusable as a structural material in reactor fuel elements. The chemists also continued research on the Redox process, which would make possible the large-scale recovery of uranium from spent fuel slugs in the Hanford reactors. Argonne scientists were also pursuing research in what was then called "hot atom" chemistry, the study of new chemical properties produced in elements under neutron or proton bombardment.

Perhaps the laboratory's most important achievement in basic research in the early years originated in the work of Maria Goeppert-Mayer, a group leader in the theoretical physics division. Building on nuclear theory developed at the Chicago laboratories, on the encouragement of Teller and Fermi, and on intellectual discourse with colleagues David Inglis and Dieter Kurath, Goeppert-Mayer delved into the origin of chemical elements and developed an explanation of an atomic shell structure first observed in 1933. In May 1948, in the first of a number of works on the subject, she gave a seminar, "Closed Shells in Nuclei"; three months later, *Physical Review* published her exposition of shell theory. While scientific colleagues appreciated the significance of her achievement, the Atomic Energy Commission, emphasizing the importance "of providing the men" to do atomic energy research, paid little attention to her achievement. A few years later Goeppert-Mayer, in collaboration with the German physicist J. Hans Jensen, published *Elementary Theory of Nuclear Shell Structure*. For this pioneering work she would share (with Jensen and Met Lab colleague Eugene Wigner) the Nobel Prize for physics in 1963. She would be the first Argonne scientist to win the Nobel Prize and the second woman (after Marie Curie) to receive that honor in physics.[86]

Women and Minorities at the Laboratory

Argonne's hiring of Maria Goeppert-Mayer in 1946 had been fortuitous—given her future renown—but backhanded, not untypical for her time. In 1930 Maria

Goeppert earned a Ph.D. in physics from Göttingen University; married Joseph Mayer, an American chemist studying at Göttingen; and moved to the United States. During the war, Goeppert-Mayer worked on uranium separation research under Harold Urey at Columbia University. In the autumn of 1945, when her husband joined the Metallurgical Laboratory, she was given a part-time appointment and became head of the section of opacity research. With the establishment of Argonne National Laboratory, she became a group leader in its theoretical physics division.[87] Because the University of Chicago prohibited the hiring of both husbands and wives for faculty positions, Goeppert-Mayer held two part-time positions, a paid one at Argonne and a *voluntary* associate professor's post at the university.[88]

At variance with contemporary practices, Argonne employed a modest number of minority and women professionals like Goeppert-Mayer in the postwar years at a time when some laboratories, like most employers, considered minorities only for low-level jobs and replaced women scientists with men. During the same period, Los Alamos Scientific Laboratory, for example, had no black scientists and but two or three women scientists on its roster of more than two thousand. The Bureau of Mines research station in Bartlesville, Oklahoma, fired most of its female scientists once the war ended, and the laboratory chief could not place them in other laboratories because, as one respondent explained, "We would rather have men."[89]

Argonne's slightly more progressive position seems to have reflected the retention of wartime Met Lab attitudes and needs. The 1946 government contract with the University of Chicago carried a clause requiring that contractors for the Argonne laboratory "shall not discriminate against any employee or applicant for employment because of race, creed, color, or national origin." (Gender was not given legal status until decades later.)

Argonne's employment of ten black scientists in the postwar 1940s attracted the attention of *Ebony* magazine, which devoted an article to the three black women and seven black men who held research positions at the laboratory. Five of the group were college graduates and worked as chemists or biologists; the others were technicians. Six of them had worked on the Manhattan Project in Chicago.[90]

Zinn placed one female scientist, Hoylande D. Young, in a senior management position in the very first roster of the laboratory. Young had joined the Metallurgical Laboratory in 1945 after serving as a research chemist, college professor, and wartime researcher at the University of Chicago's Toxicity Laboratory. She edited the plutonium project portion of the National Nuclear Energy Series and then served as director of Argonne's technical information division.[91]

Among other female researchers associated with the laboratory in the early years, Margaret Nickson, a physician, started her study of the effects of radiation on the skin of the hand during her service in the Metallurgical Laboratory. Representing Argonne National Laboratory, she presented some of her findings to the Radiological Society of North America in December 1946.[92] That year, at least seven women worked in the chemistry division, and several more were in the biology division. Two of them had recently received doctorates in zoology from the University of Chicago. S. Phyllis Stearner started as a group leader, and Miriam Finkel, at first an associate biologist, later headed a program testing the effects of internal emitters in mice.[93]

Hope for the Future

In the autumn of 1948 Zinn had some reason to believe that the painful process of transition from army to AEC control was coming to an end. The commissioners had been confirmed and were beginning to focus on policy issues that had been denied much-needed attention. At long last, Wilson had the commission's headquarters staff and field offices in place. The commission was on the verge of establishing a division of reactor development and was in the final stages of selecting a director for the new division. Argonne was now assured a role in at least three new reactor projects: the fast-breeder, the high-flux, and the navy submarine plant. True, the commission had not yet come to terms with the need for a remote test site, but there was little doubt that the growing need would force a positive decision.

Back at Argonne, Zinn was beginning to see how the laboratory would relate to the Atomic Energy Commission and how the contract would be managed. Argonne was now securely ensconced at the Du Page site, and the pace of construction was picking up. Slowly but surely research groups were beginning to move into buildings.

Even in the crazy world of Washington politics, Zinn may have seen signs of better days ahead. One of the battle scars remaining from the prolonged struggle to pass the Atomic Energy Act in the summer of 1947 had been the compromise that limited the first commissioners to terms of two years rather than of five years, which would be in effect thereafter. As the two-year terms approached expiration in 1948, the danger loomed that conservative Republicans in Congress would attempt to prevent Lilienthal's renomination. When President Truman expressed his determination to renominate Lilienthal, Senator Hickenlooper, chairman of the Joint Committee, worked out a compromise that once again limited the commissioners to two-year terms. The president signed the legislation, much to the relief of Lilienthal and Hickenlooper, who expected the Republicans to win both the presidency and control of Congress in the November elections. Perhaps the commission and its laboratories would at last enjoy a period of stability and a chance for solid progress in the years ahead.[94]

3 | Living in a Political World, 1949–52

Walter Zinn's mission as director of Argonne National Laboratory was scientific and technological, but in 1949 he operated in a political milieu— one in which leaders and circumstances were in an unusually high state of flux. The previous November, in the first presidential election since the war, American voters had surprised virtually every forecaster and returned Harry Truman to the White House with a Democratic majority in Congress. Senator Brien McMahon, a sponsor of the Atomic Energy Act and an ardent defender of David Lilienthal's leadership, replaced Hickenlooper as the chairman of the Joint Committee on Atomic Energy.[1] Members from the Republican right, no longer restrained by the responsibilities of leadership, would launch a full-scale attack on Lilienthal and the commission.

There were also changes in leadership at the Atomic Energy Commission and turf battles to mediate within the nuclear energy establishment. Detection of a Soviet atomic explosion and, soon after, a communist incursion in Korea altered national priorities and attitudes. The resulting increased emphasis on military applications and security, in turn, exacerbated existing tensions with Argonne's associated institutions. At the same time, Zinn found his anticipated laboratory space and budgets cut. Individual problems festered and fed on others. It could be thought a wonder that the buffeted and squeezed laboratory director accomplished anything at all. In fact, Argonne achieved notable success, not only in reactor development, but also in basic research during the next few years.

Caught in a Political Web

Although Argonne anticipated a promising future under a settled commission and a friendlier Congress, trouble was brewing early in 1949. Now in the minority party on Capitol Hill and gripped by the anticommunist paranoia of his conservative colleagues, Senator Hickenlooper renewed a campaign against the commission's fellowship program, a crusade that he had begun before the election. Suspicious of pervasive communist spying, the senator charged that the federal government should not fund nuclear research by graduate students who could not obtain a security clearance.[2]

While Lilienthal was defending individual and academic freedom before the Joint Committee, news broke that Argonne National Laboratory could not account for a small amount of fissionable material. Or, as the sensationalist headlines shouted, "Atom Bomb Uranium Vanishes."[3] Hickenlooper now had a weapon—carelessness at Argonne—with which to assault the commission chairman. The laboratory, with a real monitoring problem of its own, was caught in a larger political maelstrom.

Indeed, from its early years, keeping track of nuclear materials had been a constant concern at the laboratory. Because of the complexities in measuring quantities of the unstable substances, the laboratory staff had intermittently encountered difficulties in maintaining accurate records or inventories, and occasionally, minute amounts appeared to be missing. In November 1947 Argonne hired Thomas S. Chapman, a former Manhattan Project and AEC chemist, as an assistant director to supervise the handling of special nuclear materials. A small amount of plutonium had already been "lost" when Chapman arrived; after an extensive investigation, laboratory staff accounted for it by correcting assaying and bookkeeping errors. This resolution, however, could not prevent a second, more serious incident.[4]

The affair that launched Hickenlooper's attack started with the accidental omission of a labeled and analyzed bottle of oxidized uranium chips from a package of enriched material shipped to a storage vault, probably in September 1948. On February 8, 1949, laboratory officials found that thirty-one grams of enriched uranium oxide were missing, about one-sixtieth of the amount then needed for a weapon. Frantically searching for the missing substance, the Argonne staff delayed reporting the loss to the commission for six days—a decision Zinn later conceded was a mistake.

When the story reached the Joint Committee, Hickenlooper publicly blamed Lilienthal's mismanagement for Argonne's slipshod practices and demanded the chairman's resignation. Lilienthal defended his agency, but he admitted that the laboratory had failed to follow explicit commission regulations on handling uranium oxide. Subsequently a comprehensive search accounted for all but one-eighth of an ounce of the missing material. Taking exceptional measures to accommodate Hickenlooper, the commission brought in the Federal Bureau of Investigation (FBI) in March. In May 1949, with a health physicist and an FBI agent watching, laboratory investigators wearing gas masks and disposable clothing dug up waste-containing steel boxes buried on Argonne land. Prying open the contaminated containers, they found the bottle which, through chemical analysis, the FBI confirmed held uranium oxide.[5]

The Joint Committee, however, was not satisfied—especially when twelve grams of plutonium were then reported missing at Argonne. (They apparently turned up in laboratory residues). On May 25, Argonne was obliged to account for the laboratory's complete plutonium inventory dating back to 1944 to the FBI and the congressional overlords. Still upset, the committee summoned Zinn to testify regarding materials security at the laboratory.[6]

"Before the hearings . . . , about two in the morning, Zinn called us out to Argonne," recalled Paul R. Fields, the chemist who had noted the weight of the uranium on the label of the misplaced bottle. The laboratory director produced the bottle retrieved from the waste dump and ordered another scientist who last had handled it to sign his name so that the handwriting could be compared. Fields remembered the fellow was "so frightened that he couldn't hold a pen in his hand [because] Zinn could really intimidate you."[7]

In their June 16 appearance before the Joint Committee, Zinn and seven Argonne staff members reviewed the episode in detail. Zinn admitted that the laboratory accountability procedures were inefficient. The scientists' testimony, however, failed to convince the committee members, who did not understand the intricacies of measuring quantities of radioactive substances, that all of the uranium oxide had been found. Zinn left the hearing certain that some of the congressmen still believed that a spy had entered the laboratory and stolen the material, even though the amount was insignificant. Later he urged the Argonne executive committee to take every opportunity to educate Congress, especially visitors to the laboratory, about the complexities of handling and accounting for nuclear material. But, Zinn knew, Argonne could not avoid censure in such an incident.[8]

The Joint Committee, persisting, engaged Ernest W. Thiele, assistant director of research for the Standard Oil Company of Indiana, to investigate the discrepancies. After reviewing records and interviewing individuals associated with the processing of the uranium, Thiele concluded on July 1 that "the best available data are consistent with the view that no enriched material has disappeared." But, he hedged, because of uncertainties in the analysis, "the data give no absolute assurance that no enriched material was stolen."[9]

The uranium oxide episode brought both the Atomic Energy Commission and Argonne security under review. The result was more stringent controls, including requirements to inventory classified documents and impose stricter regulations on laboratory visitors. It did not help Zinn when the commission ordered the laboratory to take additional security measures without giving it increased funding to pay for such steps. Subsequently, the Chicago field manager made changes in his own staff and requested Zinn to do the same at Argonne.

At their July 11, 1949, meeting, the commissioners finally considered the case closed. For Argonne, however, the incident, blown out of proportion by political hysteria, had more lasting effects. Zinn removed Chapman from any future responsibilities for special materials, assigning him full-time to laboratory monetary matters. Because of the embarrassment, Zinn established a special materials department and an analytical laboratory, units that he had not planned to create because of lack of space and qualified personnel. He placed John T. Bobbitt, formerly employed by the University of Chicago, in charge of the new department. Bobbitt carried out his special materials duties so efficiently that Zinn later promoted him to assistant laboratory director. As an extra precaution the guard force received warnings to be alert for unauthorized persons who might attempt to enter the laboratory.[10]

The furor over the missing uranium oxide also may have chilled relations between Argonne and the Los Alamos Scientific Laboratory. With its concentration on classified atomic weapon research and development, the cautious Los Alamos scientists started dispatching to Argonne only "quite old" physics reports, sanitized with deletions. Stung by such treatment, Zinn told his executive committee to avoid making deals with the Los Alamos people for collaborative experiments or other work.[11]

Scraping for Resources

The continuing political fallout made it more difficult for Zinn to get attention and positive responses for his other pressing problems, many of them of urgent, everyday concerns. Still struggling to build roofs over his programs, Zinn had to face further compromises to win congressional approval for additional construction funds. In February 1949 the commission staff did raise Argonne's building authorization to $63.7 million, an increase of $6 million from the previous estimate. To avoid an even higher completion cost, however, the staff reduced the size of the biology and physics buildings and proposed using temporary facilities for other services for an indefinite period. Agreeing, the AEC forwarded the revision to the Joint Committee.[12]

But even as the structures were being reduced in size, laboratory personnel multiplied. The total staff, working now in ten locations, grew from 1,595 in March 1948 to 2,015 a year later. During that time, the number of employees in reactor development increased from 17 to 90 and from 53 to 92 in associated reactor work. Crowding would remain a problem for years.[13]

In the spring of 1949 Zinn was also frustrated that the commission, after two years of prodding, had yet fully to define the scope of Argonne's responsibilities. At the same time, as Zinn complained to his executive committee, headquarters was limiting Argonne to budget figures that Hilberry had proposed two years earlier without knowing the direction the laboratory would take. In 1949 the commission confronted Argonne with a 10 percent budget cut for fiscal year 1950, a reduction that would impose sacrifices.[14]

Pushing the AEC in late spring to clarify its intentions on the size and character of Argonne operations and the extent of control that commission officials sought to exercise over the laboratory's work, Zinn got less than satisfying responses. The commissioners indicated program support by approving additional construction monies but dismissed the control issue as too complex and interpretive for easy analysis. Instead, they complained that management issues were diverting Zinn's attention from the technical problems that he especially was equipped to solve. Indeed they were.[15]

That summer Zinn received a commission clarification that cost his laboratory. Despite his good fight, Argonne lost most of its radioisotope services assignment when the AEC centralized all radioisotope distribution at the Oak Ridge National Laboratory (perhaps as a compensation for Oak Ridge's loss of reactor work to Argonne). As Zinn feared, the switch "completely upset the Laboratory's service irradiation program for its Participating Institutions." Currently the only source in the Midwest, Argonne in the future would provide just a supplemental radiation service. Midwestern institutions would order materials from Argonne only when their half-lives were too short to survive shipment from Oak Ridge.[16]

The Reactor Program Upgraded

Zinn had some reason to believe that the commission's reactor development program would take a clear direction in 1949 when Lawrence R. Hafstad was appointed director of the commission's new division having that responsibility. Hafstad's record suggested he would take firm control of a program that had drifted without clear-cut leadership since 1947. A physicist and director of research at the Johns Hopkins University Applied Physics Laboratory, Hafstad also served as executive secretary of the Armed Forces Research and Development Board. Both Wilson and Fisk had urged Hafstad to accept the position, but only after encouragement from Admiral Earle W. Mills, chief of the Navy Bureau of Ships, did Hafstad agree. Mills, who had convinced the commission to accept Rickover and the submarine propulsion program, was as delighted with the appointment as were the commissioners. For Zinn's reactor staff, the Hafstad appointment meant closer monitoring of the navy project and a streamlining of communications with headquarters.[17]

The Reactor Proving Ground

Since the summer of 1948, when Zinn realized that the experimental breeder could not be built at the Palos Park site, he had prodded the Atomic Energy Commission to find a proving ground in some remote area of the West. The commission had accepted Zinn's idea in principle, but now the decision would rest with Hafstad. Impressed with Zinn's arguments, Hafstad agreed in February 1949 that the reactor should be constructed at the remote site, yet to be selected.[18]

Already the commission had narrowed the selection to sites in Montana and Idaho. To the consternation of the Montana members of Congress, in March 1949 the commission settled on a navy ordnance proving ground of some 400,000 acres near Arco, Idaho. A scout for Argonne found the terrain satisfactory for reactor construction, the town of Arco "the most desirable one situated close to the site," Idaho Falls and Pocatello acceptable for residence, and the surroundings more scenic than Hanford. Stuart McLain, the Oak Ridge chemical engineer whom Zinn had selected to head the materials-testing reactor steering committee, expected the fully operating reactor to require some 450 workers and thus requested that it be located reasonably close to Idaho Falls.

On April 4 the AEC established the Idaho Operations Office and delegated to it responsibility for the design, construction, and operation of nuclear reactors and supporting services and facilities at Arco. The site quickly became the choice for situating several reactors connected with Argonne, including the navy submarine reactor, the Argonne experimental fast-breeder reactor, the Argonne–Oak Ridge materials-testing reactor, and other experimental reactors unsuitable for operation in Du Page County.[19]

But a turf dispute erupted between Zinn and headquarters over the shift of reactor contractual control from Argonne to the Chicago and Idaho operations offices. An angered Zinn objected to placing decisions regarding contracting for reactor construction with the field-office managers. Argonne, he insisted, had the necessary expertise and should be in charge. "I believe it would be unsatisfactory to ask unqualified people to take responsibility for approvals," he complained upon learning that the field office, not Argonne, would decide who would build the experimental breeder reactor, his personal project. The objections of both Zinn and McLain to a similar arrangement for the materials-testing reactor delayed its construction and led Zinn to urge Hafstad to "stop treating this enterprise as another government post office and to begin to come to the realization that it is a complicated machine which combines unique and novel engineering with the newest advances in atomic energy."[20]

With limitations on both funding and construction, Hafstad, however, considered it necessary on behalf of the commissioners to take more direct control of the reactor program. Although he acknowledged the necessity of depending on Zinn for guidance on technical design details, Hafstad made clear that the Idaho field office and its principal contractors would bear primary responsibility for the engineering design of the reactors going up in the scrub brush near Arco.[21]

Zinn and McLain continued to complain about the incompetence of commission field personnel, and at one point Zinn threatened to withdraw the breeder-reactor project. Word of the dispute eventually reached the Joint Committee, which investigated the situation but found out little. Zinn's fears, however, did not materialize. Hafstad, the compromiser, accepted the Austin Company, the Illinois architect-engineering firm chosen by Zinn, to build the experimental breeder reactor. The commission selected the Blaw-Knox Construction Company of Pittsburgh to carry out architect-engineering services for the materials-testing reactor. Although Hafstad arranged for an information exchange between laboratory scientists and Blaw-Knox representatives through the Idaho field office, the two parties worked together directly on construction plans; they then apprised the Idaho field office of their conclusions.[22]

Without a doubt, Argonne's emphasis on reactors increased tensions within the rest of the laboratory. Some scientists, among them Albert Wattenberg, considered Zinn's support for scientific research too weak. Indeed, Zinn acknowledged that the mission of the laboratory largely revolved around reactor research, which by 1949 consumed almost half of the Argonne budget. In comparison, physical research got about 25 percent of the budget, and biology and medicine somewhat over 17 percent. The studies in the physics, chemistry, metallurgy, reactor engineering, and chemical engineering divisions concentrated on supporting the laboratory's reactor work. All but 16 percent of Argonne's research was classified.[23]

The primacy of reactor projects at Argonne led its board of governors to charge that the laboratory had become a development center with little interest in basic research. In defense, Zinn asserted that approximately one-half of the research program was "fairly decent basic science." He blamed Argonne's changed image on AEC publicity and efforts of its staff to justify a commission reorganization. Caught between the commission and the midwestern institutions' interests, Zinn complained about the lack of a stated commitment from the Atomic Energy Commission that the reactor program was in addition to the laboratory's primary functions. He never had agreed to reduce Argonne's basic work when he took on the reactor mission, Zinn told the governors. "Maybe next time it will be possible to get this in writing," Hilberry reflected.[24]

Repercussions of the Soviet Bomb

Bureaucratic bickering became a secondary problem when the United States detected that the Soviet Union had detonated an atomic bomb in late summer 1949. The Soviet achievement jolted everyone, especially President Truman and the commissioners, who were taken completely by surprise. Combined with other aggressive actions by the Soviet Union in making satellites of the eastern European states it had occupied, the explosion of the Soviet bomb set off a wave of fear in the United States that nuclear war could erupt between the communist countries and the western allies. The Soviet atomic weapon test reinvigorated a debate within the government on whether to proceed expeditiously with the development of a thermonuclear weapon, and the crisis influenced the commission to change both its own and Argonne's reactor schedule.

In view of the Russian accomplishment, Hafstad agreed with Zinn that Argonne would have to finish the submarine nuclear power plant and the materials-testing reactor sooner than originally planned. Already Hafstad and Rickover were displeased with the slow pace of the Argonne work, a circumstance that Hafstad blamed partially on Zinn's giving a low priority to the navy project. The pressure from headquarters forced Zinn to relinquish part of the laboratory's physics and metallurgy hot laboratory building to the navy project. In response to a Hafstad inquiry about laboratory priorities, Zinn was careful to rank development of the submarine reactor at the top in the commission's and Argonne's program. Ordinarily, Zinn explained, he would have given first choice to breeding, but military urgency in the use of reactors influenced the ranking.[25]

Under Rickover's incessant prodding, the naval reactors group was struggling to meet increasing demands for experiments and analyses that would contribute to design of the submarine plant. Fortunately, by September 1949, Etherington's team, which had spent almost a year evaluating reactor types, had confirmed Rickover's assumption that the water-cooled design would be the best choice for the submarine reactor. With that question settled, scientists and engineers at Argonne and the new Westinghouse laboratory at Bettis Field near Pittsburgh pressed ahead on design. Just which laboratory was doing what was always a question. Rickover, always focused more on technical problems than on organization, assigned work where capabilities lay at the moment. As a result, Argonne became involved at times in what seemed like purely engineering activities while Bettis's work often bordered on fundamental research.[26]

Despite the distractions that dogged Zinn in 1949, he had managed to keep in close contact with Harold Lichtenberger, who was leader of the experimental-breeder team. With commission approval of a contract for detailed design of the reactor, the Argonne team had turned to extensive testing of reactor components. Hundreds of hours of testing showed that the fuel elements proposed for the reactor were not distorted when exposed to the hot liquid-metal coolant. Motors and gears in the mechanisms for sharp acceleration and deceleration of the control rods were tested and redesigned. Zinn estimated that by September the Argonne team and the Austin Company had completed 75 percent of the design, and the first excavations for the reactor building at the Idaho site would begin in November.[27]

Preoccupation in late summer with the Soviet nuclear detonation and a possible thermonuclear response meant that Argonne's continuing problems with an inadequate physical plant, of importance to Zinn, hardly commanded the notice of the commissioners or the Joint Committee. The insufficient funding and slow pace of construction of Argonne's permanent buildings hindered laboratory operations and irritated Zinn. By October 1949 construction cost overruns, later blamed on miscalculation of the original estimates, were so severe that William Harrell and Zinn had to take drastic measures with the remaining work to make up for the money already spent. The anticipated administration building, lecture hall, medical building, and guest house were dropped. Temporary structures were reconfigured to serve as permanent housing for the central shop and metallurgy division. Other building sizes were reduced, a step that threatened to reduce the complement of scientists. In December 1949 the laboratory's 450 scientists still worked at three sites: Du Page, Palos Park, and the University of Chicago.

Nearly every program was affected. Construction of the permanent chemistry building was so far behind schedule that Winston Manning, the chief of the chemistry division, divided his chemists, still working in two locations, into two groups, one under his leadership and the other under that of Oliver C. Simpson. The cuts especially inconvenienced the biology division. To understand the effects of ionizing radiation on living matter, the division sought to obtain the biological and medical information necessary to guide the safe design and operation of processes and reactors. But the buildings for this research were reduced so much that the laboratory had to eliminate such important experimental facilities as animal quarters and irradiation space for large-animal studies.[28]

One positive accomplishment in 1949 was that the Atomic Energy Commission approved a revised plan for the Argonne Cancer Research Hospital, a fifty-bed facility to be operated by the University of Chicago. The laboratory would provide the hospital with radioisotopes, and the hospital, as it treated cancer patients with the radioisotopes, was to serve as a clinical proving ground for scientists from the laboratory and the participating institutions.[29]

Zinn and his staff did not see the first permanent buildings completed until 1950, when Argonne personnel occupied Building 316, the Zero Power Reactor Building, and Buildings 301 and 310.[30]

The Question of Identity

For Argonne, the Soviet detonation and the commission's response to it had far-reaching implications, not just in revising reactor priorities and derailing the laboratory's construction plans, but also in threatening to transform the essential character of the laboratory itself. Would Argonne remain a regional laboratory for basic research or would it become a government center for engineering development? Would it lose its original identity?

As government-imposed security precautions at Argonne increasingly limited the access of visitors, the laboratory's relationships with its academic participants had already begun to deteriorate in 1949. Comparing outsiders' involvement at other national laboratories, the Argonne board of governors accused the laboratory of not doing enough to bring in midwestern researchers. Only twenty-one had worked at the installation during the previous year. In addition, with laboratory facilities so inadequate, the board saw little opportunity to expand the cooperation.[31] Later in the year, other federal actions further damaged the relationship, however unintentionally. Changing the date of the federal budget deadline from June to September effectively removed the board of governors from considering proposed budgets before they were submitted to Congress. The new timetable would require the board to meet during the summer, an awkward time for university personnel.[32]

A devastating blow fell at a special board meeting on November 7, 1949, when Hafstad virtually stripped the governors of their perceived effectiveness by telling the body, "It is impossible that any . . . group, elected by outsiders, should be inserted in line of command between myself, Zinn and Hilberry." Agreeing but obviously trying to soften the impact, Zinn explained to the governors that "this is a fast business." Decisions could not wait for occasional academic deliberations, but Zinn assured the board that Argonne appreciated "careful guidance in its relationships to the Government." He considered an advisory board of "elder statesmen" essential to Argonne's functioning, and he claimed that he especially needed the board to encourage participation by the regional institutions.[33]

Hafstad, Wilson, and other senior commission officials tried to improve relations with midwestern universities by proposing commission financing of a small, basic research staff that would undertake projects proposed by the participating institutions. The Atomic Energy Commission also proposed that the board determine, subject to commission approval, the allocation of funding for campus research projects related to Argonne's program. These attempts, however, to make amends with the crippled board of governors did not succeed.[34]

Hilberry thought the commission's offer impractical, and he did not foresee any solution to the worsening state of Argonne-university relations. "Active participation on the part of the Participating Institutions in the *overall* program of the laboratory is utterly impossible," he concluded. The laboratory depended on its own staff and the University of Chicago business manager to determine its research program, draw up the budget, and provide management advice. The commission's proposal would require greatly expanded involvement by the governors, whom Hilberry considered unqualified to judge broad management issues. Yet he wished to retain the advice of the governors and participating institutions for Argonne's educational program.[35]

Hafstad followed his November statement with a governance proposal in early 1950 that omitted the regionally elected board of governors from the direct chain of command that included AEC headquarters divisions, the Chicago operations office, and Argonne. His structure relegated the board to a position of just another laboratory advisory committee to be appointed by Zinn. The beaten board, eliminated from any genuine management function in Argonne's operations on May 1, 1950, dissolved the following day, to be replaced by a council executive board composed of participating institutions. The relationship between Argonne and the midwestern universities would enter a new phase.[36]

Following the board of governors' loss of power, Argonne centered more attention on improving research ties with the regional institutions. Zinn and Hilberry recruited an associate director, Joseph C. Boyce, former chairman of the physics department at the New York University College of Engineering, whose principal duty was to coordinate Argonne-university programs and the laboratory's educational activities. Federal strictures, however, continued to obstruct the cooperative interactions that Boyce tried to encourage.

After six months on the job, Boyce complained of being "considerably hampered" by security regulations. Except for access to the biology division, almost all visitors had to have a security clearance, and uncleared visitors at times were denied access to unclassified equipment and experiments.[37] Others also chafed under the restrictions. In February 1950 the Federation of American Scientists criticized the groups using the regional laboratories for accepting commission contracts to administer AEC fellowships when the research was unclassified but the applicants still had to obtain security clearances. Pervasive fear of communism, including of students who might be communists, had led Congress to pass a law requiring clearances of persons even only proximate to classified work.[38]

At the same time, Argonne's relationship with the University of Chicago, too, was at odds with the Atomic Energy Commission. In November 1949 Harrell recommended to Chancellor Robert Hutchins that Argonne be operated as a division of the University of Chicago. Hutchins regarded the division between the business and scientific management difficult to maintain and observed that the University of California was encountering similar problems with its operation of Los Alamos. Harrell agreed: "The University should be, and I believe is much better qualified to direct the scientific work of the laboratory than are the representatives of AEC (especially local AEC)." Zinn previously had pushed for a distinct identity for the laboratory so that it would not be confused with the university's several institutes; but because he also preferred control of the laboratory by intellectuals (which included the laboratory professionals) rather than by bureaucrats, it is possible that Harrell's proposal either originated with Zinn or at least emerged with his approval.[39] Nothing came of Harrell's suggestion, but the university did gain one small concession. When the commission eliminated its fellowship program for predoctoral students in 1949, the university assumed this responsibility and arranged with the commission to use Argonne to administer these fellowships for the Midwest.[40]

In another government intervention, the Bureau of the Budget launched an inquiry in December 1949 to determine what the Atomic Energy Commission was doing to improve the management of the national laboratories, especially with regard to contractor relations. In response, the commission, itself in a state of leadership turnover and turmoil over the superbomb, identified a need at its headquarters for a central operating unit that would coordinate laboratory programs and thus avoid duplication of effort. In addition, the commission sought mutual agreement among the laboratories in redefining their missions and functions. The general manager called for "better integration" within the laboratories of basic research, which the laboratories favored, and applied research, which the commission promoted.[41]

Finally, the commission assigned Howard C. Brown, Jr., a senior member of the headquarters staff, to direct a study of management practices relating to the national laboratories. When the team members examined the commission's contractor relationships, Brown found "the day-to-day course of business on a muddy track." After spending three weeks at the Chicago office, Brown discerned among the parties, except in Harrell, a lack of mutual confidence in and sometimes little respect for the laboratory's management contract. Although he considered the laboratory's technical program unaffected, Brown identified a lack of cooperation at the working level as a significant problem. Zinn, Brown maintained, "is not confident to let the business affairs of the laboratory rest in other than his own hands . . . [because of the] absence of the backing of and confidence in strong and interested corporate management."[42]

To improve Argonne operations, Brown suggested a new contract and contractor. Short of that, he urged the commission to try to persuade the university's top management and regional advisors to take a greater interest in the managerial and technical operation of Argonne, an option the commission, through Hafstad, had just killed. When the contract for managing the laboratory came up for renewal in 1950, Harrell advocated drafting contract terms to limit the amount of detailed commission supervision.

Brown did see improvement in the laboratory-AEC relationship from 1948, but he considered the headquarters role too vague and the agency's attention to laboratory budget support inadequate. Brown's team recommended that the commission clearly define the channels of responsibility for research direction and a financing system in accord with contracts instead of research areas. The commission adopted several of these suggestions in the summer of 1950.[43]

An Operating Policy for Argonne

After months of attempts to simplify relationships and management at the national laboratories, on June 1, 1950, the commissioners approved a general operating policy that Hafstad had drawn up primarily for Argonne. Hafstad's statement recognized the laboratory as a center for research and development in the atomic energy field and charged it to serve as a stimulus to the regional institutions in nuclear research. The Argonne staff was to "play an important role in the solution of practical problems and in the development of specific military and non-military uses of atomic energy." That is, applied research would be key.

At the same time the policy committed the Atomic Energy Commission to "long-term support of a cadre of carefully selected senior men . . . capable of assuming responsibilities" for leadership or individual research on difficult assignments. The commissioners acknowledged that the basic research projects and methods were the responsibility of Argonne staff but that the commission would determine the developmental objectives and priorities in consultation with the laboratory. The statement required Zinn to appoint a group of recognized experts from universities and—upon the advice of the commission—from private industry to review research and development at Argonne and to advise the director. The commissioners also wanted greater cooperation between Argonne and its neighboring research institutions.[44]

The H-Bomb Decision

The abrupt revision of reactor priorities at Argonne following the first Soviet bomb test was but a small part of the commission's response to the new threat that the USSR posed. Already being planned in the autumn of 1949 was a vast expansion of uranium-235 and plutonium production to support a huge arsenal of diversified fission weapons. And boiling up behind the commission's security barriers was an intense debate over the need to accelerate research on a thermonuclear weapon—the H-Bomb." Opponents, including Lilienthal, several of the commissioners, and J. Robert Oppenheimer and his scientific colleagues on the General Advisory Committee, feared that such an effort would launch an international arms race that would lead to the creation of weapons with a new order of destruction that might well threaten human existence. Proponents, including Commissioner Lewis L. Strauss and a new commissioner, Gordon E. Dean, were convinced that the United States had little choice. Not to proceed, they believed, would give the Soviet Union an unacceptable opportunity to achieve superiority over the United States in nuclear weapons capability. Although Strauss and Dean could file only a minority report with the president on the question, they soon gained overwhelming support for their position from the air force, the secretary of defense, the Joint Committee, and a number of prominent scientists in the AEC's laboratories. Overwhelmed by the opposition, Lilienthal and his supporters were forced to concede defeat. On January 31, 1950, President Truman announced his decision: the United States would proceed to develop a new class of nuclear weapons.

Two weeks later Lilienthal, exhausted and disheartened by more than three years of political conflict, resigned as commission chairman. His replacement was Commissioner Dean, a former law partner of Senator Brien McMahon, chairman of the Joint Committee. Convinced that the appointment represented an attempt by McMahon to gain effective control of the commission and lacking any confidence in the new chairman, Wilson resigned as general manager in August, to be replaced by Marion W. Boyer, a seasoned oil-company executive.[45]

The president's January decision ratcheted up the pressure on the commission to build additional reactors to produce not only plutonium, but also tritium, which would likely be needed in large amounts for thermonuclear weapons. In June, Truman authorized the commission to negotiate a contract with DuPont to design and build two production reactors that would use heavy water as the neutron moderator. The reactors were to be built at a new site, which DuPont selected in August in South Carolina on the banks of the Savannah River.[46]

It was a reasonable assumption that basic design of the Savannah River reactors would fall to Zinn and the Argonne staff. In fact, the idea of working with DuPont on the design of production reactors for military purposes must have seemed for Zinn a rerun of the hectic months the Metallurgical Laboratory had faced in designing the original Hanford reactors during World War II. At least this time Zinn would have some idea of what to expect from DuPont, and he could build on the laboratory's experience with CP-3, the small heavy-water test reactor that had been operating at Palos Park since 1944. The task, however, was still challenging. The laboratory was to design a new type of reactor that would equal those at Hanford in fuel economy but would use cooling towers rather than river water for removing the enormous amounts of heat generated in the fission process. Argonne was to provide the basic scientific design and training of DuPont personnel in design and operation of the reactors to be built at Savannah River.[47]

A New Shock: The Korean War

On June 25, 1950, North Korean forces attacked without warning across the thirty-eighth parallel. President Truman and the National Security Council were as shocked by the incursion as they had been by the Soviet atomic explosion ten months earlier. Two days later the United States entered the conflict. Once more, like all the national laboratories, Argonne had to reassess its program priorities.

The following week, AEC headquarters asked Zinn to identify projects of military significance that could be given increased attention and work that could be set aside to divert staff to the more critical areas. Zinn cited the submarine, materials-testing, and the Savannah River reactors, as well as the laboratory's role in analyzing data from nuclear weapon tests as among its military responsibilities. He was willing to take people away from basic and long-range studies in physics, chemistry, and chemical engineering except, he said, for engineers working on the experimental boiling-water reactor.[48]

Zinn worried about the combined effects on Argonne's program as a result of the changes in commission leadership as well as from the Korean conflict. The first was positive: three days after Gordon Dean became commission chairman, the Chicago operations office authorized the construction and operation at Argonne of the Zero Power Assembly, which would provide critical data for the submarine reactor.[49]

Still concerned about future directions, Zinn questioned Commissioner Henry Smyth on August 15, 1950: would the Joint Committee have a greater voice in commission affairs as a result of Lilienthal's departure? Smyth did not think Dean's friendship with McMahon would have that effect. The commission, Smyth noted, so far had "established a clean pattern of no political interference," and recent sessions with the congressional committee seemed to be going more smoothly. Zinn inquired in particular about how the commission's attitude toward basic research was changed as a result of the Korean conflict. It would be better for Argonne to divert some staff from basic to applied work, Smyth told Zinn. The commissioner personally wanted to keep as much basic research as possible in both the universities and laboratories, he said, and thought the Atomic Energy Commission might accept such a policy. As the nation moved back on a wartime footing, however, escalating demands for reactor development would push the laboratory's research activities into the background.[50]

Spies, Secrets, and Security

An inevitable result of the Korean War was heightened security at all the nuclear laboratories. The predictable anti-communist hysteria was further fueled by revelations in late spring 1950 of Soviet spying on the Manhattan Project by Julius and Ethel Rosenberg and others. Fears of espionage were enough to set off new charges and investigations by Senator Hickenlooper and the Joint Committee. In August an investigation by Senator Hickenlooper's staff took on the appearance of a low-budget movie. The Senate investigators engaged in a series of clandestine meetings with a Chicago-area private detective who had charged the laboratory with partially burning classified waste in a public area. After an extensive investigation, the staff members disproved the detective's allegations but that did nothing to exonerate the laboratory.

Additional accusations burst forth a few months later as a result of a midnight intrusion. Paul Aurandt, who used the name Paul Harvey for his conservatively slanted radio broadcasts, climbed over Argonne's ten-foot barbed wire fence in the dark of February 6, 1951, to demonstrate personally the flaws in the laboratory's security, which he intended to expose and publicize. Guards intercepted the trespasser immediately, fortunately not shooting him. For over a year thereafter, Harvey continued his campaign against Argonne through radio broadcasts, speeches, and contacts with Joint Committee staff. Both the commission's Chicago operations office and Washington security division investigated Argonne's security and found it satisfactory. But the incident did expose a loophole in the Atomic Energy Act: it contained no penalty for trespassing on commission properties. Dissatisfied that Harvey emerged from his escapade "scot free," as the Joint Committee's executive director James T. Ramey put it, Congress later amended the act to incorporate sections providing punishment for such encroachment.[51]

Zinn got better press on a contemporary issue. When the commission initiated the first postwar continental atmospheric nuclear weapons testing program at the Nevada test site in early 1951, it mandated that Argonne and the other national laboratories monitor radioactive fallout in their areas. During the detonations of Operation Ranger, the first Nevada series, Zinn announced that the laboratory was detecting and analyzing local airborne debris. He was quoted as assuring the public that the radioactivity was far below "any level of activity which can cause any concern for the health and safety of any human or animal in this area."[52]

Reactors for Defense

In addition to the three reactor development projects assigned to Argonne in 1948, the commission now looked to the laboratory for design of the Savannah River plutonium production reactors, which Argonne code-named CP-6. In November 1950 the commissioners formally approved DuPont's selection of the South Carolina site. A few weeks later Argonne started training the first group of DuPont engineers, all of whom had advanced degrees and some company experience. The training included studies in physics, physical chemistry, chemical engineering, inorganic chemistry, and metallurgy.[53]

Argonne's excellent relations with DuPont on the CP-6 project were due in part to the leadership of Stuart McLain, who had led the steering committee on the design of the materials-testing reactor. On the strength of McLain's earlier work, Zinn appointed him to coordinate the laboratory's efforts and to serve as the technical liaison with DuPont. Much of the research at Argonne centered on the metallurgy of fuel elements, particularly fabrication techniques and the behavior of alloys under irradiation. Argonne's semi-works research (a small-scale processing plant too small to be considered a pilot plant) would also be significant in DuPont's design of the chemical processing plant at Savannah River. A big headache would be, perhaps, obtaining the twenty-five tons of heavy water that McLain needed for reactor experiments at a time when no significant amounts of the substance were being produced in the United States.[54]

With hostilities continuing in Korea and the United States starting continental nuclear weapons testing in 1951, Zinn in February gave first priority to CP-6 and second to the navy reactor. To keep the navy project on target, however, he found it necessary to add a night shift from Tuesday through Saturday.[55]

The Navy Project

Since the spring of 1950 the navy project had commanded the center of attention at Argonne. By that time, Etherington and his group had completed the general specifications that the laboratory and Westinghouse would use in designing and developing the components for not just one reactor, but two: the land-based prototype, designated Mark I; and the propulsion reactor, Mark II, to be installed in the world's first nuclear submarine. So far, however, Argonne had done almost all of the design while Westinghouse engineers, awaiting completion of permanent buildings, struggled in old aircraft hangars at Bettis Field to develop a process for the mass production of zirconium. Over the next two years, Etherington's group tackled a host of perplexing problems, many of them centering around the complicated drive mechanisms for the reactor's control rods and the composition and design of the fuel rods. Despite Rickover's continuing complaints about slow progress, Etherington kept the project on track and proved a good administrator as well as scientist. At last Zinn and Rickover had reached a workable if not cordial agreement on their mutual responsibilities. Rickover clearly represented the Atomic Energy Commission and the navy, but Zinn insisted on giving orders for all work at Argonne, including that in the navy project. As the burden of design gradually shifted to Westinghouse, however, Rickover's pressures on Argonne gradually decreased. By early 1952 Argonne had few responsibilities left for the submarine project.[56]

The Materials-Testing Reactor

Although the materials-testing reactor would neither produce plutonium for weapons nor propel a submarine, it would be essential in developing reactors for either of these purposes. The cooperative agreement that Zinn had reached with Weinberg continued to work well. By 1950, Stuart McLain had moved to Argonne to monitor research there and would soon move on to the Idaho test site, where the reactor would be built. Zinn had agreed that the Oak Ridge laboratory would design the reactor, whereas John R. Huffman of the Argonne staff would be responsible for all of the plant outside the reactor tank. By March 1950 Huffman and his group had completed 90 percent of the design for the reactor and service buildings, experimental ports in the reactor, coffins for transporting radioactive materials, and the storage basin for irradiated fuel elements. Construction of the reactor facility in Idaho started three months later.[57]

The Experimental Breeder Reactor

Of all of Argonne's reactor projects, the experimental breeder still commanded Zinn's greatest personal interest. Now ranked a poor third behind the submarine and materials-testing reactors on the wartime priority list, the breeder project claimed only eleven members of the Argonne staff. Originally conceived as a small reactor experiment at the Du Page site, the breeder had grown into a full-scale engineering project to be built at the Idaho test station.

There were the usual delays in designing and erecting the reactor building and supporting facilities at Idaho, but the biggest problems were obtaining components from suppliers and fabricating the reactor proper. The last components for the reactor were shipped in January 1951, when Lichtenberger and most of his team moved to Idaho. Installation of the reactor and fabrication of the fuel rods continued at the reactor site during the spring. Attempts to reach criticality failed in June, when Zinn discovered that the reactor core would require an additional 12.5 kilograms of uranium-235. Not until August 24, after the addition of more fuel, did the reactor reach criticality with "zero" power. Lichtenberger ran low-power tests and fast-neutron experiments during the fall until Zinn arrived on the scene a few days before Christmas for what the group hoped would be an historic experiment. On the morning of December 20, Zinn increased the power level in small steps. After checking the heat transfer system, he ordered a resistance load connected to the reactor's electric generator. At 1:23 that afternoon Zinn recorded in his log book: "Electricity flows from atomic energy. Rough estimate indicates 45 kw."[58]

Whether Zinn's accomplishment that winter afternoon was actually the first generation of electricity from nuclear energy is still a matter of dispute, but the event was heralded as such by the laboratory and by the Atomic Energy Commission. Its greatest significance was that it provided a convincing demonstration for American industry and the public that nuclear power could be harnessed for peaceful purposes.

Perhaps Zinn savored the fact that his own project, the experimental breeder reactor with its low priority, was the first of Argonne's reactor assignments to reach criticality. But the laboratory's other projects were not far behind. The materials-testing reactor, the joint Argonne–Oak Ridge project, went critical on May 31, 1952. Within a month it reached full power of 30,000 kilowatts and in August began functioning as a test reactor. The submarine reactor was already taking shape a few miles to the south of the testing reactor. In a large steel building, engineers had assembled a full-scale section of a submarine hull submerged in a tank of water to simulate the radiation configuration of a sea environment. Often working in shifts around the clock, Westinghouse engineers installed the reactor systems and turbine equipment during the spring and summer of 1952. By November the plant was complete except for the nuclear fuel and two heat exchangers. When the reactor reached full power for the first time in June 1953, Rickover ordered it run at full throttle for 100 hours, long enough to take a submarine from the East Coast to Ireland.[59]

Sustaining the Research Mission

Zinn and his frazzled staff at Argonne could be proud of what they had achieved in supporting the nation's defense effort since the summer of 1949. By the summer of 1953, three reactors designed by the laboratory and built with the assistance of its staff were operating in the Idaho desert. At the same time 25,000 construction workers were building the gigantic plant at Savannah River that would include five heavy-water production reactors to be built and operated by DuPont with Argonne's technical support and basic designs.

It was an impressive achievement, but it had come at a price. As the defense projects siphoned away increasing numbers of scientists and technicians, only a few score were left to pursue the laboratory's missions in basic research. Those who remained managed to carry forward research projects with the resources available. Some scientists working on defense projects were given assignments that contributed directly to fundamental knowledge; others proposed research that could be justified as contributing to the defense effort or supporting the operation of the laboratory. Compared to the defense programs, the dollars and staff committed to basic research were small, but in many cases the results of their efforts were significant.

Environmental Research

In one area, environmental research, pioneering studies were performed for the immediate, practical goal of the safety and health of laboratory personnel and the surrounding community. The Atomic Energy Commission itself became more sensitive to environmental hazards associated with atomic energy research as its Reactor Safeguards Committee and health physicists grew more concerned, particularly about radioactivity and explosions. At the commission's request, the U.S. Geological Survey had completed an extensive study of the Du Page site's geology and hydrology in 1949. The commission was especially concerned about possible contamination of area groundwater. Because "little is known about the effect on radioactive liquids of movement through water-bearing materials," the surveyors recommended that for an indefinite period certain wells should be monitored monthly and tests continued to determine if stored and disposed wastes affected earth materials and surface and ground waters.[60]

Even before moving to Du Page County, Zinn had tried to anticipate and resolve the problems of on-site hazardous waste disposal and other environmental effects associated with reactor operations. Environmental health would remain a priority with him, as he pressed his staff and headquarters for advice and safe procedures for handling materials never before encountered. In 1950 the laboratory administration established a committee to delineate procedures for disposing of all solid and liquid laboratory wastes. Later, following the recommendations of the commission's Advisory Committee on Reactor Safeguards, Argonne limited the power levels of its reactors in order to confine radioactive materials within the laboratory boundaries. Similar considerations determined the locations of the chemical and chemical engineering laboratories.[61]

Air monitoring was a significant environmental priority. By spring 1949 a meteorology station and tower were ready for use in the western area of the site. The station enabled the laboratory to monitor meteorological conditions in the immediate vicinity of Argonne and in the surrounding metropolitan area with a special concern for radiation. This capability became especially important when continental testing of nuclear weapons began in 1951.[62] And when the Soviet Union entered the nuclear weapons race, the AEC assigned Argonne and other commission laboratory chemists the task of analyzing debris from Soviet atmospheric tests.[63]

Biology and Medicine

From the earliest days of the Metallurgical Laboratory, the toxicity of radium and alpha emitters in humans was a major subject of health research. Argonne's biology and medicine staff took up the evaluation of radiation toxicity, a project that was to expand nationally. This research was of special interest to the National Committee on Radiation Protection, which was showing increased interest in internal emitters—radioisotopes ingested by humans or animals. Concerned especially with protecting workers exposed to radiation, Argonne continued investigating the effects of radioactivity on mammals. Most of the research was on laboratory animals, but only measurements from humans could provide definitive information.

In the summer of 1950 Argonne researchers extended their radium toxicity studies to humans by taking radium measurements from residents of Ottawa, Illinois, radium-watch-dial painters, patients treated with radium in 1931 at the Elgin State Hospital, and patients given radium by private physicians in the Chicago area.[64] During the next year physicians, biologists, physicists, and engineers sought out hospital records and measured the radium content in some of the same patients. Walking the streets of Ottawa and nearby towns, an engineer in the radiological physics division identified former radium-watch-dial painters through friendly conversation and persuaded many to participate in the Argonne survey. As Robert Hasterlik and Austin Brues, who headed the project, described it, the survey procedures presented no more of a hazard to individuals than did ordinary complete physical examinations. The study added new information to the already large data bank on the effects of exposure to radium.[65]

The possibility of using beryllium as a reactor material in the 1950s required further studies of the toxicity of this substance. Biologists and medical researchers at Argonne took steps to prevent and treat the poisoning that threatened workers machining beryllium in the laboratory's metallurgy shops. To start with, they equipped the shops with completely enclosed, ventilated hoods to inhibit the release of beryllium compounds. In 1951 Argonne scientists successfully used adrenocorticotropic hormone (ACTH) in the first treatment in the United States of a patient with beryllium disease. On the verge of death, the treated patient improved to such an extent that he was alive and active almost thirty years later. In a joint effort in 1952 with the National Cancer Institute, Argonne scientists found that bone marrow transplants might be useful in treating whole-body radiation injury.[66]

Physics and Chemistry

Seeking to bring more distinction to the laboratory, Zinn persuaded Louis A. Turner—an eminent physicist at Iowa State University, a wartime radar developer, and a member of the Argonne board of governors—to join the laboratory. Turner's arrival in 1950 accompanied a research reorganization that consolidated the laboratory's experimental and theoretical physics divisions and the mass spectroscopy and crystal structure divisions into a division of physics.[67]

Most of the research in the physics division continued to support the laboratory's reactor projects, but in return, the physicists did benefit from work on development projects and research in the chemistry division. The heavy-water test reactor (CP-3) at Palos Park was used primarily for reactor design studies, but it did produce samples of tritium (hydrogen-3). The availability of tritium resulted in a significant discovery related to the behavior of materials at very low temperatures. Up to this time, experiments with helium isotopes had been performed only with helium-3 diluted in helium-4. It was now possible to obtain pure helium-3 as a decay product of tritium. Experiments carried out at temperatures close to absolute zero revealed that helium-3, unlike helium-4, showed no evidence of becoming a superfluid capable of flowing through minute cracks in almost any solid. Further experiments revealed that liquid helium-3 could not be frozen by cooling alone.[68]

Improved research facilities became available in 1952, when the new and much larger heavy-water reactor (CP-5) and the 60-inch constant-frequency cyclotron became operational at the Du Page site.

Metallurgy

Along with several other research institutions, Argonne's metallurgy division made strides in producing highly purified uranium fabricated in precise sizes and shapes for the fuel elements in reactors. By 1950 Argonne's research on the effects of small amounts of alloying agents on uranium properties had contributed to improving reactor fuel elements.[69]

An initiative not yet realized was that of W. D. Wilkinson of the metallurgy division, who in June 1951 proposed the establishment of a plutonium metallurgy laboratory. Wilkinson argued that a laboratory such as he envisioned did not exist elsewhere and would help in solving reactor problems. Other Argonne officials supported Wilkinson's recommendations, but almost three years would pass before the commission authorized construction of a facility devoted to plutonium metallurgy.[70]

Regional Relationships

If Zinn had trouble realizing his hopes for a strong basic research program within the laboratory in the early 1950s, his relationships with the participating institutions were even more disappointing. A problem that had haunted him since 1946, it now seemed intractable. Following the demise of the board of governors in 1950, he had agreed to establish a new "council of participating institutions," which would coordinate the use of the laboratory's very limited research facilities that had not been commandeered by defense programs. The new organization, however, seemed to make little difference. By the summer of 1951 only a few dozen area scientists were pursuing research at Argonne, and Zinn began to receive complaints from the AEC that the laboratory was not fulfilling its regional responsibilities. For Zinn it was a complaint he had heard many times before. On the one hand, the commission wanted a vigorous research program at Argonne; on the other hand, engineering and development required top priority in times of international crisis.[71]

Even more disconcerting were several references to Argonne in speeches by Henry Smyth and T. Keith Glennan, both appointed to the commission by President Truman in 1950. Both had described Argonne as primarily a reactor development center whereas they referred to Brookhaven as a basic research laboratory. Because both men were prominent figures in American science and technology, their remarks could saddle Argonne with an image that would be hard to shake. Zinn could only hope that Argonne's record of scientific publication would "eventually determine our strength or weakness rather than a qualifying phrase by any Commissioner."[72]

At the same time, Zinn was facing a renewed challenge to his authority from the University of Chicago. Now that Lawrence Kimpton had replaced Hutchins as chancellor, Harrell had again proposed that the university take over direction of the research program as well as business management of the laboratory. Kimpton, who had worked as a physicist at the Metallurgical Laboratory, took a personal interest in the question and went to Washington in July with Harrell and Zinn to discuss renewal of the Argonne contract. Facing opposition from both the commission and Zinn, Kimpton did not succeed in his quest to take over the scientific management of the laboratory, but Marion W. Boyer, the new general manager, assured him that the commission would give the university as much management freedom as possible within the limits of government requirements. More important for Zinn, the commission agreed to a new five-year contract, which would stabilize relationships with the university at least until 1956.[73]

A New Focus on Education

With the completion in 1951 of more of its permanent facilities, including the chemistry, chemical engineering, and physics buildings, Argonne was able to introduce several important new educational programs. These went virtually unnoticed by the commission, which, focusing on developing a thermonuclear weapon, paid more attention to defense programs.[74]

Zinn had continued the educational projects such as seminars and various courses that had been launched in the Metallurgical Laboratory days. The courses did not lead to academic degrees, but the staff found the training valuable. The laboratory brought in experts such as Glenn T. Seaborg, the University of California chemist, and the New York University mathematician Richard Courant to lecture.[75]

The "special scientific loaned employees" program was an early innovation of 1951. Personnel from private industry, other government agencies, foreign companies and governments, and the United States armed forces participated in on-the-job training, seminars, and special courses. Through 1955 the program trained 209 scientists and "probably contributed more to the development of nuclear energy than any other of Argonne's training programs" up to that time, according to Stuart McLain.[76]

Argonne also offered to the resident research associates benefits that were similar to those available to the special scientific loaned employees. Resident research associates were regional university instructors and postdoctoral researchers, including National Research Council appointees and some foreign scientists and engineers. Under this program seventy-three associates worked at Argonne during the summers of 1951 through 1955 and thirty-two were in residence from six months to two years during that time.[77]

As did other research institutions, Argonne cooperated with universities by taking in degree candidates after completion of postgraduate course work and enabling them to prepare theses and dissertations relevant to Argonne's interests. Some twenty scholars earned their advanced degrees under this program in its first five years. During the 1950s Argonne also employed students from the University of Detroit and Northwestern University in a cooperative program under which the students worked at Argonne part of the year while in school.[78]

Reflections

At the end of 1952 Walter Zinn could look back on the previous three years with a sense of satisfaction as well as frustration. Lack of firm direction from the commission, congressional investigations and second-guessing, charges of mismanagement and lax security practices, endless complaints from the participating institutions, budget cuts, and staff shortages had all made life more difficult than it needed to be. Perhaps, however, the troubles of the past could be considered educational, learning what it took to run a large national laboratory in a political environment. Laboratory management was much more than direction of scientific and technical programs; it involved diplomacy, political tactics, a sense of public relations, skill in bureaucratic infighting, and a tough hide.

Argonne's accomplishments attested to Zinn's skills in all these areas. All the buildings and facilities scheduled for completion at the Du Page site were now in place and operating. The laboratory had succeeded in transforming itself into a reactor development center while still preserving some of its character as a place for basic research. Argonne's part in the four reactor projects assigned to it had been completed, and three of the four reactors were operating.

Whatever the successes of the past, the future presented new challenges. The national political pendulum was swinging again to the right with the election of Dwight D. Eisenhower as president in November 1952. The laboratory, having completed its major assignments for national defense, would need to find new projects to justify its existence, a goal that might prove elusive for an institution focussing on basic research as the nation moved ahead in a determined effort to develop both civilian and military uses of nuclear energy. Zinn, as director of Argonne National Laboratory, would continue to live in a political world.

4 | The Reactor Laboratory, 1953–58

Ten years after Enrico Fermi sustained a controlled nuclear chain reaction with his Chicago pile at the University of Chicago's Stagg Field, a decade after Arthur Holly Compton recruited Manhattan Project scientists to design production reactors for the first atomic bomb, and a decennium since Glenn Seaborg gathered his small group at the Met Lab to study the chemistry and metallurgy of plutonium, Argonne National Laboratory was still on a "war footing." The armed conflict in Korea, the Cold War in Europe, and the espionage war at home focused the AEC's priorities on the military applications of nuclear science and technology. In 1952 nearly all of Argonne's research and development effort was devoted directly or indirectly to building or improving America's nuclear muscle. In one way or another, national security asserted precedence in all areas: reactor development, physical research, biology and medicine, and even in the esoteric field of high-energy physics.[1]

The U.S. government's emphasis on "atoms for war" did not preclude interest in the peaceful uses of atomic energy among the nuclear community. Although they immediately recognized the destructive potential of splitting atoms, nuclear scientists also dreamed of new worlds in which nuclear reactors would produce unbelievably cheap electrical power while nuclear science revolutionized industrial production, medical practice, and agricultural harvest. In 1945, J. Robert Oppenheimer, the "father of the atomic bomb," predicted widespread and profitable application of nuclear power "in the near future." In December 1945, physicist Alvin Weinberg, soon to be appointed the director of Oak Ridge National Laboratory, told the U.S. Senate that atomic energy could "cure as well as kill." While the bomb might drive humans back into the cave, Weinberg also testified that the nuclear age could open broad new horizons for human progress. Almost ten years later, Lewis L. Strauss, Oppenheimer's nemesis on the Atomic Energy Commission, projected that someday nuclear reactors might generate electricity "too cheap to meter."[2]

The President-Elect Attends "Atomic School"

Dwight D. Eisenhower leaned forward to study the stark photograph the Atomic Energy Commissioners brought with them to brief the president-elect on the status of the U.S. nuclear program. Four naked light bulbs strung on a single wire glowed brightly in the picture, casting stark shadows among the metal stairs, bulkheads, piping, and machinery near the turbine-generator of Argonne's Experimental Breeder Reactor Number 1 (EBR-I) system (figure 13). The commission had chosen this simple photograph to illustrate the first known production of electrical energy from nuclear power on December 20, 1951, at Idaho's National Reactor Testing Station (NRTS).

Eisenhower's briefing by the members of the AEC on November 19, 1952, began with introductions by the commission's secretary, Roy Snapp, but quickly moved to a discussion, led by Chairman Gordon Dean, of the recent thermonuclear test in the Pacific and to questions from the president-elect about nuclear security. Dean shared top-secret data about the nuclear weapons stockpile and offered Eisenhower a primer on the principles of implosion-type fission bombs. He also described the AEC's vast complex of plants and labora-

Figure 13. EBR-I lights four bulbs, December 20, 1951

tories that sustained the nuclear weapon effort. Dean made it clear that production of fissionable materials and nuclear weapons was the commission's principal mission.[3]

Dean also reviewed the commission's broad responsibilities for fostering research and development of nuclear science and related industrial, medical, and technological applications. The AEC's reactor programs, Eisenhower learned, contributed directly to military uses as well as to civilian objectives. The glowing 200-watt light bulbs were not part of a long-range plan to develop a power reactor at EBR-I, but rather were ancillary to studies on handling the liquid-sodium coolant at high temperatures and on extraction of heat from the reactor. No doubt Eisenhower also learned that the day after the light-bulb demonstration, ordinary power was cut off from the reactor complex while EBR-I generated 150 kilowatts of electricity to provide the entire power load for the facility. Along with the picture of Argonne's EBR-I, Dean displayed a photograph of the dry-land prototype of a nuclear-powered submarine under construction near EBR-I at the NRTS in Idaho, and proposal summaries for the nuclear-powered aircraft being developed at Oak Ridge National Laboratory. The naval reactor would be installed in submarines where development and operating costs were not the determining issue. Dean was confident, however, that steady improvement in naval reactors would hasten the day when a commercial civilian reactor would prove economically competitive.[4]

Finally, the chairman of the AEC reviewed the commission's programs in basic and applied research in the physical and biomedical sciences. Even in these areas, the commission presumed that progress in the fundamental sciences would prove a good investment for the American economy and national security. The commission's program in biology and medicine, in which Argonne was extensively involved, focused principally on understanding the effects of radiation from day-to-day operations, atmospheric weapon tests, or nuclear war. But the commission also promoted peaceful applications in cancer research and treatment, agricultural and environmental studies, and biological processes and metabolism.

Similar assumptions supported the commission's basic research in the physical sciences. In addition, one could hardly overemphasize the importance of the Cold War as a stimulus to international competition in the natural sciences. Almost no area in the physical sciences escaped comparison with foreign competitors, especially the Soviet Union. Scientists and politicians alike assumed that more than national prestige was at stake in the post–World War II "science race." Eisenhower shared the conviction that discoveries in basic science research laid essential foundations for future technology. Not only was economic progress dependent on fundamental research, but, as the Manhattan Project had demonstrated, the future of democratic government was also shaped in the science laboratory.[5]

Development of peaceful applications of nuclear technology was not only hindered by defense priorities, but also by the fact that the restrictive Atomic Energy Act of 1946 made it almost impossible for the Atomic Energy Commission to share reactor technology with private industry. Eisenhower was aware of the problem. On August 7, 1952, Charles Thomas, president of Monsanto Chemical Company and a former research chemist who had headed the Clinton Laboratories at Oak Ridge during the war, wrote Eisenhower promoting the breeder concept for commercial power reactors. Thomas noted that the commission had made significant progress toward establishing the technical feasibility of commercial nuclear power. "There needs," he reminded Eisenhower, "... to be a modification in the existing Atomic Energy Act before industry can be encouraged to go much further."[6]

Nuclear Reactors, 1952

In 1952, Americans would have been surprised to learn that their president-elect was conversant about problems related to developing nuclear power reactors. Most Americans knew of the horrible power of the atom bomb but understood little else about nuclear science, which remained cloaked in unremitting official secrecy. The Atom evoked such an aura of profound mystery that popular fantasy about the "atomic genie" frequently edged into the realm of science fiction tinged by magic and the occult. "Radioactivity" rolled across the tip of everyone's tongue often with scant understanding of this natural force which could not be perceived by ordinary human senses. Although nuclear science and engineering indeed were as difficult as they were wondrous, ordinary Americans frequently failed to understand that nuclear processes were simply part of the natural order. Commonly, people worried that a runaway nuclear reactor might explode like an atomic bomb, or shared macabre jokes about "glowing in the dark" after radiation exposure; rarely did they understand that nuclear reactors were no more complex, in principle, than their internal-combustion-powered automobiles.[7]

In the fall of 1952, twenty nuclear reactors of various types were operating or under construction in the United States. All were government reactors except for the first private teaching and research reactor, which was being built at North Carolina State College in Raleigh. The plutonium production reactors at Hanford, Washington, and at Savannah River, South Carolina, were the largest of the twenty. Research reactors were located at Brookhaven, Oak Ridge, and Los Alamos, and Argonne Laboratory had four reactors in operation or under construction: CP-2 and CP-3' at Palos Park, EBR-I at the NRTS in Idaho, and CP-5 at the Du Page site (figures 14a–14c).[8]

Argonne's Reactor Programs, 1952

CP-2, which had been reconstructed in 1943 at Palos Park from Fermi's dismantled CP-1 reactor, was a 1,400 ton graphite-moderated reactor that included 10 tons of uranium metal, 42 tons of uranium oxide, and 472 tons of graphite. The reactor core, now shielded inside five-foot-thick concrete walls, was capped by six inches of lead and four feet of wood. A small laboratory atop the graphite pile was the site for limited experiments using neutrons produced by the reactor. The face of the graphite pile also contained ports through which materials could be inserted into the core for irradiation.

Argonne's second reactor, CP-3, became the world's first heavy-water-moderated reactor when it began operation on May 15, 1944. Originally designed as a production reactor when Manhattan Project engineers were uncertain that graphite piles could produce sufficient fissionable materials for atomic weapons, CP-3 was converted to a research reactor when the Hanford production

Figure 14a. CP-2 reactor
(*Frontiers, 1946–1996*, p. 15)

Figure 14b. CP-3 reactor (*Frontiers, 1946–1996*, p. 15)

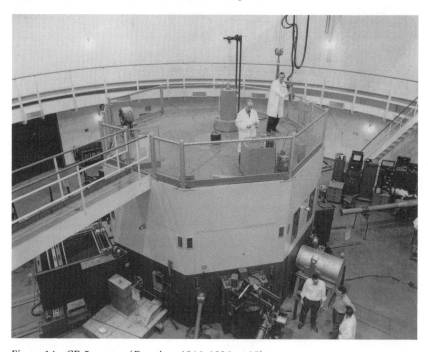

Figure 14c. CP-5 reactor (*Frontiers, 1946–1996*, p. 15)

plant began operating successfully. Much smaller than the graphite-pile reactor, the CP-3 reactor core consisted of an aluminum tank holding 1,500 gallons of heavy water (D_2O) into which six-foot-long natural uranium fuel rods were suspended. The tank was shielded by an eight-foot-thick octagonal concrete wall. Like the graphite pile, CP-3's octagonal face contained numerous ports, protected by removable plugs, which provided access through the shielding to measure neutron intensity, to irradiate materials, or to obtain a neutron beam from the reactor. To improve the reactor's efficiency as a research tool, enriched uranium replaced the natural uranium fuel in 1950. After fuel improvements upgraded the reactor to CP-3', Argonne reactor engineers believed they operated the finest research reactor in the country.[9]

When Argonne finally returned the Palos Park site to the Cook County Forest Preserve, the laboratory was forced to shut down both CP-2 and CP-3'. Commissioner Willard Libby wanted to save Fermi's reactor and asked Argonne to explore the possibility of moving the graphite-pile reactor to a new site, or perhaps even rebuilding it at its original site under the West Stands at the University of Chicago. Norman Hilberry, sympathetic to Libby's nostalgic hope to preserve the historic reactor, nonetheless patiently explained that the great reactor was hopelessly obsolete and could not even meet rudimentary safety requirements. The reactor controls had become dilapidated, Hilberry explained, and persistent problems of venting fission gases would have to be solved. Hilberry could only guess that it might cost $500,000 to rescue Fermi's old reactor. Instead, some of the uranium and graphite from Fermi's reactor became part of an exponential, or subcritical reactor, at Argonne that was used to irradiate indium foils to gather data for designing full-scale reactors. In addition, some of the graphite blocks were reduced to small cubes, encased in plexiglass, and distributed as handsome souvenirs of the first controlled sustained reaction. After salvaging the remaining uranium fuel, graphite, and heavy water from CP-2 and CP-3', the laboratory buried the reactor remains on-site under several feet of earth.[10]

At the National Reactor Testing Station, development work continued with the materials-testing reactor (MTR), and EBR-I. Technically, the MTR was not an Argonne reactor. Originally an Oak Ridge project, in 1948 the MTR was assigned to the reactor testing station when the AEC designated Argonne as the lead laboratory for reactor development. Given high priority because of the Korean War, engineering teams from Argonne and Oak Ridge collaborated on the design of the MTR, whose principal task was to accelerate reactor development by testing components, such as fuel assemblies, and other materials exposed to high-intensity radiation. The reactor was a high-flux heterogeneous reactor whose core contained plates of enriched uranium-235 fuel elements through which light water flowed to provide the coolant and moderator. Surrounded by a beryllium shield, which served as a neutron reflector, the core was submerged in a water tank encased by a secondary graphite reflector, a thermal shield, and a biological shield, all of which formed a cube approximately thirty-four feet per side. To provide access to high-intensity radiation at the reactor core, about a hundred experimental ports penetrated the six faces of the MTR cube.[11]

When the MTR became critical in March 1952, reactor operations were managed by the Phillips Petroleum Company. Although Argonne was not responsible for the operation of the MTR, the laboratory employed the reactor extensively for its materials-testing program. The design of cores and fuel elements of virtually every American nuclear reactor built after 1952 was influenced by studies conducted with the MTR. Rickover's naval reactors program conducted some fifty experiments on the MTR, while the air force and army used the testing reactor to a lesser extent. Argonne used the MTR to test components for both pressurized-water reactors and liquid-metal reactors. Running more than 50,000 hours in its first decade, the MTR performed 20,000 neutron irradiations, at one time loaded with 600 samples for irradiation experiments.[12]

EBR-I Proves the Breeding Principle

At the end of World War II, Zinn and other reactor engineers believed that the scarcity of uranium was a major impediment to the development of commercial power reactors. Of the ninety-two naturally occurring elements, only uranium works as a reactor fuel, and of that, less than 1 percent is fissionable uranium-235. Although uranium is ubiquitous throughout the earth's crust, only a small fraction appears sufficiently concentrated to allow economical mining, processing, and enrichment. Military reactors did not compete economically with the private electrical utilities, but commercial power reactors required fuel supplies that were both economical and reliable. As a government scientist had warned Eisenhower, a nuclear power industry based on the straightforward consumption of uranium-235 would soon exhaust all known or "inexpensive" supplies of this rare uranium isotope. But what of the possibility of creating new fissionable isotopes such as plutonium-239 or uranium-233? Zinn, along with Fermi and others, envisioned using fast reactors to convert unfissionable, but "fertile" uranium-238 or thorium-232 into fissionable plutonium-239 or uranium-233 which could be used for reactor fuel. If this were possible, the available fuel for power reactors would be expanded significantly. The task of the Experimental Breeder Reactor was to determine whether or not breeding could be accomplished.[13]

Dr. B. I. Spinrad, director of the reactor engineering division, described EBR-I as "the first major Argonne pile." Originally designated as CP-4 and nicknamed ZIP (Zinn's Infernal Pile), EBR-I explored the possibility of breeding, gathered data on the performance of liquid metals at high temperatures, and tested the performance of nuclear reactors operating with high neutron flux (figure 15).

Although EBR-I supplied electricity for its reactor complex, power production was only an incidental product of its research mission to provide data on the performance of fast reactors. By early 1953 Argonne chemists were ready to analyze plutonium production in EBR-I's uranium blanket. From blanket and core burn-up samples sent to the chemical engineering division, Meyer (Mike) Novick, director of the Idaho division in 1959, recalled that Argonne technicians "came up with a number showing that we really did have a breeding reactor." On June 4, 1953, just days before he left the commission, AEC chairman Gordon Dean proudly announced, "The reactor is . . . burning up uranium and, in the process, it is changing non-fissionable uranium into

Figure 15. View of Experimental Breeder Reactor I

fissionable plutonium at a rate that is at least equal to the rate at which uranium-235 is being consumed." Dean described the accomplishment as one of the major milestones in nuclear history. EBR-I had proven the breeding principle.[14]

CP-5: Argonne's "Work Horse" Research Reactor

With the loss of both research reactors at Argonne-East, in the fall of 1951 Argonne began construction of CP-5, a new research reactor located at the Du Page site. Nicknamed SIP—Sam's Infernal Pile, after Sam Untermyer, one of the chief designers of the reactor system—CP-5 became operational in February 1954 (figure 16). The new reactor's only purpose was to produce abundant radiation—principally neutrons—for research. Because the reactor would be located in a large metropolitan area, Argonne's engineers needed to build the safest machine possible. Design parameters were clear: the highest neutron flux at the lowest possible power. The success of CP-3' provided a model for CP-5, whose hexagon configuration mirrored that of the smaller reactor.[15]

CP-5 was powered by its enriched uranium core, which was cooled and moderated by heavy water. The core, consisting of seventeen fuel assemblies that resembled 2-foot-long organ pipes, was only 2 feet in diameter. Arranged to allow maximum contact with the cooling water, the entire core assembly sat immersed in 6½ feet of heavy water in a 6-foot-diameter aluminum tank. The

Figure 16. CP-5, Argonne's "work horse" research reactor

tank was surrounded by 2 feet of graphite, which acted as a neutron reflector. The graphite reflector was contained within a steel tank lined with boron carbide, an effective neutron absorber. Gamma rays were significantly reduced by 3½ inches of lead. The entire core structure was encased in concrete walls 4½ feet thick and filled with iron ore to increase density and stop any radiation that escaped the heavy water, graphite, lead, and steel-boron fences. At full power, CP-5 operated at only 5,000 kilowatts, which provided a slow neutron flux of about 8.5×10^{13} neutrons/cm^2/sec in the area of highest flux (compared to CP-3' power rating of 275 kilowatts for a slow neutron flux of 3×10^{12} neutrons/cm^2/sec).[16]

CP-5 was a research reactor, not an experimental reactor for pioneering new concepts in reactor design or materials. The heart of CP-5 was its experimental facilities, in which neutrons and gamma rays were used for research. The reactor was honey-combed with more than sixty ports, beam holes, rabbit holes, columns, thimbles, and trenches all providing access to radiation within the reactor. Horizontal channels, or thimbles, allowed insertion of small samples into the region of highest flux in the center of the fuel elements. Beam holes allowed for the extraction of neutron beams to experimental areas outside the reactor. For intense, but rapid irradiation and analysis of samples, pneumatic tubes called "rabbits" shot samples through the reactor just below the core, while under the reactor two trenches provided space for long-term irradiation. Large "through-holes" with high fluxes provided facilities for engineering tests and some specialized physics. Two 8-by-12-inch biological holes in a low-flux area accommodated experiments to measure the effect of neutron and gamma radiation on small animals. Two 5-foot-square graphite thermal columns provided slow neutrons and, if partially disassembled, space to irradiate large objects. CP-5's great strength was its versatility as a research facility.[17]

Safety was a major consideration in the design of CP-5. Zinn hoped to eliminate almost all hazards associated with reactor operation, especially risks of catastrophic mechanical failure or runaway reaction that released radioactivity into the environment. Four flat cadmium control rods—superb neutron absorbers—swung like railroad semaphore signals between the fuel elements, while a tubular control rod was raised or lowered into the center of the fuel elements to provide "fine" control. The control rods absorbed neutrons in excess of what was required to sustain the chain reaction and to irradiate experiments in the reactor. Once at operating power, CP-5 was placed on automatic control so that the control rods routinely adjusted to minor fluctuations in the chain reaction. In the event of dangerous changes, the reactor automatically shut down and sounded the emergency alarms.

Should the control systems fail, two safety features would help minimize and contain damage. CP-5 operated with an inherent safety factor called negative temperature coefficient. That is, as coolant water heated, it expanded, decreased in density, and retarded the chain reaction. As long as CP-5 did not suffer a loss of coolant accident, there was little risk of a runaway chain reaction. Furthermore, the reactor was designed to withstand thermal and mechanical stress in excess of operating norms. Finally, the entire reactor complex was enclosed within an airtight containment building sealed to prevent radiation escape to the surrounding environment.[18]

BORAX-I: The "Runaway Reactor"

Prior to 1952, reactor engineers believed that bubble formation in the core of a nuclear reactor would cause instabilities. Samuel Untermyer thought otherwise and convinced Zinn that, theoretically, boiling reactors should be safe because formation of steam in the reactor system would result in "very high negative temperature coefficients of reactivity," a feature already incorporated in CP-5. In effect, Untermyer believed boiling-water reactors would tend to shut themselves down should the control mechanisms fail. Zinn, who saw the possibility of generating power from boiling-water reactors, supported a small experiment in Idaho, the *Bo*iling *Re*actor *E*xperiment-I (BORAX-I), to test Untermyer's ideas.[19]

BORAX-I, constructed at the National Reactor Testing Site, was a small, simple, and cheap version of the light-water swimming-pool reactor whose uranium core coolant water was allowed to boil (figure 17). In more than two hundred tests from May 1953 to July 1954, engineers progressively stressed the reactor to test the inherent safety of BORAX-I. For example, when Untermyer suddenly pulled the control rod from the reactor core, BORAX-I immediately boiled and steamed furiously, threw scalding water from the tank, and then shut itself down. Each successive excursion rocked the tough little reactor as Untermyer's team explored the outer limits of reactor safety, fuel-element integrity, and instrument reliability. In every case, the nuclear chain reaction was quenched before the reactor's aluminum fuel plates reached their melting point. To avoid a false sense of security about the safety of boiling-water reactors, Untermyer asked Zinn's permission to perform the ultimate safety test—to push BORAX-I fuel elements to their limit just beyond the melting point.[20]

BORAX-I's final excursion in the Idaho desert had all the drama of a nuclear-weapon test at the Nevada test site. Scheduled for early morning July 21, 1954, meteorologist Paul Humphrey canceled the test at 4:30 A.M. because wind conditions were unfavorable. Not wanting to take any chances, Zinn ordered

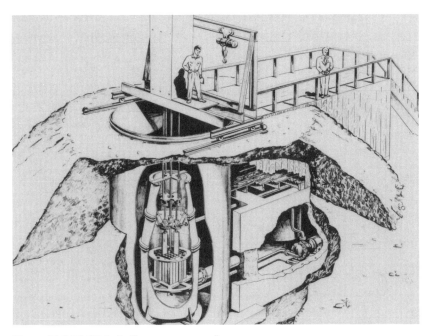

Figure 17. BORAX-I, the "runaway reactor"

caution although he believed there was only a fifty-fifty chance of significant contamination should the reactor core be seriously damaged. Still, he wanted to channel any possible contamination toward the large, unpopulated area south-southwest of the reactor testing station. What if reactor fuel elements melted, or fuel melting was followed by a chemical reaction, or the core ruptured tossing fuel elements into the atmosphere above the reactor? Since most of the radioactivity would be short-lived, Zinn estimated that a maximum of 30 curies of long-lived radioactivity might become airborne. "The worst situation imaginable," he wrote Lawrence Hafstad, "is one in which the immediate Borax site would have to be inactivated for some weeks while decontamination is performed." Zinn assured Hafstad that the experiment would replace speculation with quantitative data on the effects of a reactor catastrophe.[21]

Before dawn flecked the desert sky on July 22, BORAX-I's last operating crew gathered to run a preliminary excursion to ensure that all instruments and controls were in working order. As had happened before, the fuel elements distorted slightly, binding the excursion rod. While engineers replaced the fuel elements and readied the reactor for its final run, smoke pots scattered through the sagebrush tracked the shifting ground winds that blew across the site. At 6:00 A.M. a bus load of dignitaries from Idaho, Chicago, and AEC headquarters, including the commission's Reactor Safeguard Committee, arrived for briefings before taking up positions at the observation point. Only reactor-division personnel and official visitors were present because the entire operation was classified secret. On their way into the site, the observers passed through a ring of radio-equipped safety and security patrols that now surrounded the reactor complex up to three miles from the command post. Health physicists and police from the test site stood by, prepared, if necessary, to close U.S. Highway 20, which passed 1½ miles north of the BORAX reactor. The driver and his bus were placed on emergency stand-by at the observation point, while the division's regular emergency bus was warmed and ready. By 7:50 A.M., with a steady 8-mile-per-hour breeze from the north-northeast, weather conditions were almost ideal for the test.

Zinn ordered the operation to begin immediately. Gently raising the control rods, Harold Lichtenberger held BORAX-I for a moment in a controlled chain reaction for one last time. Then, after removing the safety rods but lowering the excursion rod until it was fully seated in the core, the reactor became subcritical again, ready for its final excursion. A last smoke bomb traced favorable winds. Worried that the excursion rod might stick and so ruin the show for the visitors, the crew decided to "give it everything" they had when ejecting the excursion rod by remote control from the control trailer.[22]

Almost instantly, the reactor blew up with the force of three or four sticks of dynamite, tossing debris and an inky black column of smoke more than a hundred feet above the desert brush (figure 18). "Harold, you'd better stick the rods back in," Zinn shouted. "I don't think it will do it any good," Lichtenberger replied, "There's one flying through the air!" Zinn was surprised. Previous excursion clouds had been silvery white, evidence of discharged steam and water. This dark cloud, with almost vertical sides, indicated a far different, more powerful reaction. "Within the column, and along its edges, and falling away from the edges" Zinn saw "a large quantity of debris," including a large sheet of plywood that sailed off across the desert like a giant playing card.[23]

The reactor was totally destroyed. Within the control trailer Zinn detected a slight tremor, while most observers in the open felt a small shock wave. Outside the trailer, a Geiger counter connected to a loudspeaker startled the visitors with its "frying noise" indicating radiation overload. More sensitive survey meters at the command center showed traces of radioactivity (25 mr/hr). As a precaution, all personnel and visitors at the BORAX site were ordered to take shelter, and workers at a nearby construction site were given an unscheduled coffee break inside the central cafeteria. The cloud drifted slowly southward, reaching "quite high" by 10 A.M., when upper winds pushed it back over the site, where health physicists tracked "non-dangerous but measurable fall-

Figure 18. Explosion of BORAX-I: portions of the reactor tank, control rod mechanism, and reactor core have been hurled 80 feet into the air.

out." Throughout the entire excursion and its aftermath, Zinn, Untermyer, and their crew remained on duty in the control trailer.[24]

Although Argonne heralded the test as a success, Untermyer confessed candidly, "Unfortunately, it is not possible to draw many far reaching conclusions from this experiment." BORAX-I had been sacrificed in a spectacular excursion: the central aluminum fuel plates had been damaged, and the reactor tank had ruptured, contaminating the immediate vicinity of the reactor site, but projecting debris only seventy-three yards from the reactor core. Had Zinn miscalculated? In testifying to the Joint Committee in March 1954, Zinn reported that the 1953 excursion tests "did not produce melting of the fuel and no radioactive contamination of the surroundings whatsoever resulted." On the basis of the 1953 tests, even the *Bulletin of the Atomic Scientists* reported that boiling water reactors were "self-regulating ... shutting themselves down without serious damage in case of trouble." After the destruction of BORAX-I, Zinn was still optimistic about the inherent safety of boiling-water reactors. If Zinn had believed that BORAX-I would not blow-up, he was disappointed. On the other hand, he was pleased that, other than destroying the reactor core, the excursion had caused little permanent environmental damage. Officially, BORAX-I ran away "as planned."[25]

AVIDAC and ORACLE

While dramatic earth-shaking tests with the BORAX-I reactor commanded Zinn's personal attention in Idaho, a small band of applied mathematicians at Argonne-East quietly moved into the vanguard of one of the most profound scientific changes of the twentieth century—the computer revolution. Everywhere after 1945, atomic icons reflected popular American ideas of advanced, modern science. Images of the mushroom cloud, Albert Einstein, and the AEC logo defined the Atomic Age. The future, according to Walt Disney, belonged to *Our Friend the Atom*, a frightening cartoon genie whom wise scientists turned into a friendly, beneficent servant.[26] Only a few scientists anticipated that adding machines were about to transform American life and culture.

Development of the modern computer owed much to the Atomic Energy Commission. Following World War II, the AEC largely, but not exclusively, funded research and development on internally stored program computers pioneered at Princeton by John von Neumann, who served as an AEC commissioner from 1955–57. Von Neumann's Princeton machine, built with AEC assistance at the Institute for Advanced Study, became the prototype for later computer generations that revolutionized all aspects of modern life: science, medicine, politics, industry, entertainment, sports, and education. Rapid expansion of the United States' nuclear arsenal and development of the thermonuclear weapon created unique computing requirements for commission scientists and engineers. In this sense, the nuclear arms race played as great a role in creating America's computer industry as it did in laying the foundations for civilian nuclear power. Unintentionally, establishment of the computer industry became one of the AEC's major contributions to America's postwar economic development. The commission did not yield its leadership in computer research and development to private industry until the late 1950s, when Argonne and the other laboratories generally ceased building their own machines.[27]

In 1950 J. C. Chu dreamed of building his own electronic computer which had first been suggested to him by D. A. Flanders, a senior mathematician in Argonne's physics division. During World War II, Chu had worked as a design engineer for the construction of ENIAC, the world's first all-electronic digital computer, which was built at the University of Pennsylvania. As a designer of the University of Pennsylvania's subsequent machine, EDVAC, Chu arrived at Argonne an advocate of the new computers that could calculate large numbers rapidly and accurately. The "electronic brain" could not think for itself, of course, but its computational advantage in solving large, complex problems in reactor engineering and theoretical physics was obvious. Von Neumann's

Princeton machine, for example, reportedly calculated 100,000 times faster than a trained calculator operating a desktop electric adding machine. With von Neumann's help, Los Alamos built MANIAC I, which was completed in 1952. Because all such machines were still handmade, Chu proposed using the Princeton machine as a model so that he could construct the Argonne computer as quickly as possible.

Chu and his team completed AVIDAC—Argonne's Version of the Institute's Digital Automatic Computer—in January 1953 at a cost of $250,000. Chu was proud of his "compact" machine, which was huge by later standards. AVIDAC was actually four machines that filled a large laboratory room: an arithmetic unit, a memory unit, an input-output station, and a control section that electronically managed the operation. AVIDAC contained 2,500 vacuum tubes, 8,000 resistors, and approximately 3½ miles of electrical wiring. The memory unit could store 1,024 twelve-digit decimal numbers on the inside face of 40 five-inch cathode ray tubes, which were similar to a modern computer screen.

The computing power of AVIDAC was awesome. The machine multiplied two 12-digit numbers (999,999,999,999 × 999,999,999,999) in 0.001 second. Addition of the same two numbers required only 0.000001 second. Problems that might take two mathematicians three years to solve with the aid of an electric adding machine, challenged AVIDAC a mere twenty minutes. Amazing as this was, the most remarkable feature of AVIDAC was that it could be programed to make "decisions" about calculating sequence depending on the results it was generating. In effect, programmers could instruct AVIDAC how to solve complex equations and be assured that the computer would consistently solve the problem the correct way, every time, regardless of the values.[28]

While completing AVIDAC, Chu's engineers designed ORACLE, which in 1953 was the world's fastest computer (figure 19). An improved version of AVIDAC, ORACLE (Oak Ridge Automatic Computer Logical Engine) was constructed with assistance of engineers from Oak Ridge National Laboratory, where the computer was located. Costing $100,000 more than AVIDAC, ORACLE multiplied 12-digit numbers in less than 0.0005 second, and added the same numbers in 0.00000005 second, almost twice as fast as AVIDAC! ORACLE also incorporated a remote-controlled, magnetic-tape memory system to provide Oak Ridge the largest computer memory to date.[29]

At the Atomic Energy Commission, John von Neumann became an aggressive advocate for computer research and development. Not surprisingly, he argued that computers were essential to progress in the weapons program. When he learned that the commission had allocated no funds for this purpose, he demanded and got $1 million for advanced research. AVIDAC and ORACLE were soon surpassed by more powerful and efficient AEC computers (MANIAC

Figure 19. ORACLE computer with Margaret Butler, Argonne mathematician, and Rudolph Klein, Oak Ridge engineer

II, Los Alamos 1956; GEORGE, Argonne 1957; LARC I, Lawrence and Livermore Laboratories 1960; MERLIN, Brookhaven 1961), as well as by developments in the commercial field, but they had established Argonne as an eminent laboratory in the computer field.[30]

In 1956 Argonne formed the applied mathematics division under the leadership of W. F. Miller. One of the largest computer groups in the Midwest, the division provided support and consultation services, which included mathematical analysis and programers, operators, and maintenance technicians for the laboratory's digital and analog computers. Argonne built GEORGE, its last "homemade" digital computer, in 1957. That fall, Argonne also purchased an International Business Machines 704 computer, one of the largest commercial machines, which was soon running twenty-four hours a day, seven days a week. In 1958 the laboratory acquired a large-scale analog computer, PACE (Precision Analog Computing Equipment). Unlike digital computers, PACE did not count digits but, rather, measured phenomena. Very simply, digital computers are like an automobile's odometer, which counts the miles; an analog computer such as PACE, on the other hand, is more like a speedometer, which measures speed. Both are useful, but they perform very different computing functions.[31]

Miller was not modest and believed that his applied mathematics division was engaged in the most important, profound, and far-reaching activity of any group at Argonne. He knew that the development of computer science was one of the most important accomplishments of the twentieth century, and although he could not predict which energy technologies would prevail, he was willing to wager they would be computer-oriented. "We're going to have a lot of fun before its over," he forecast, "and we'll see our lives change a lot before it's over, and if we're mature enough to plot a reasonable course for ourselves it will be all for the good."[32]

Remote-Control Manipulators

From across the Remote Control Engineering laboratory, Bob Olsen leaned forward, lighter in his mechanical hand, and casually lit George Wallace's cigarette. His reach was more than seven feet. This day smoking was part of the job when their boss, division director Ray C. Goertz, supervised a demonstration of Argonne's new electronic remote manipulator. Featured in an Atoms for Peace exposition sponsored by the Nuclear Engineering Congress at Ann Arbor, Michigan, in June 1954, the manipulators developed by Goertz and William M. Thompson were designed to handle highly radioactive materials from a safe and protected distance.

Olsen's reach was actually modest, about the same as he would have had with mechanical manipulators operated by pulleys, cables, and gears. The electronic manipulators could reach several hundred feet and, when coupled with Argonne's three-dimensional television system, performed complex laboratory manipulations with radioactive materials or equipment located in a large "hot cell," "cave," or other contaminated area. Slipping their hands into the handles, operators opened and closed the remote fingers with their forefingers and thumbs. A movement of the operator's hand was duplicated by a reciprocal movement of the mechanical hand. So fine were the electronic controls that operators developed a "sense of feel" for safely performing delicate maneuvers with fragile radioactive or toxic materials and equipment. Combined with television, the electronic manipulators made remote processing and refabrication of radioactive fuel elements practical. They also facilitated performing experiments in sealed laboratories or caves, and moving equipment in "hot" areas unsafe for direct human contact.[33]

Argonne's pioneering work in remote-control manipulators laid the foundation for engineering even more sophisticated machines—mobile robots whose portability freed them to repair and maintain contaminated equipment without the need to unseal containment or decontaminate facilities. The robots could also be dispatched to trouble spots where they might aid investigation and clean-up after a nuclear accident.[34]

Atoms for Peace

In the darkening days of the Cold War, when McCarthyism gripped the land with anticommunist hysteria, Eisenhower hoped for a way out of the nuclear arms race, which he feared inexorably pushed the world toward atomic Armageddon. He was an incurable optimist who believed that "the miraculous inventiveness of man" could be turned from designs of death and consecrated to visions of life. In his Atoms for Peace speech before the United Nations on December 8, 1953, Eisenhower offered a plan to resolve the nuclear-arms dilemma by establishing an international pool of fissionable nuclear materials dedicated to the peaceful uses of atomic energy. Hoping to break the deadlock in stalled arms-control negotiations with the Soviet Union, Eisenhower proposed diverting uranium stockpiled for weapons to an international atomic energy agency sponsored by the United Nations that would foster cooperation in the development of peaceful technologies. Although Eisenhower's proposal for such a pool of fissionable material proved unworkable, his offer stimulated international cooperation through establishment of the International Atomic Energy Agency and the European atomic cooperative EURATOM. At home, the Atoms for Peace initiative prompted a variety of peaceful atomic projects, most notably an accelerated domestic nuclear power program.[35]

Before the United States could launch the civilian nuclear power program, however, Congress had to amend the Atomic Energy Act of 1946, under which the Atomic Energy Commission enjoyed a virtual monopoly on all nuclear sciences, including reactor technology and its applications. While successfully developing plutonium production reactors and military propulsion reactors between 1945 and 1952, the commission had also shown the feasibility of nuclear power in minor, but dramatic demonstrations such as the EBR-I lightbulb display. Despite broad interest in nuclear power, in the fall of 1952 the Joint Committee on Atomic Energy reported in *Atomic Power and Private Enterprise* that American development could not proceed without major changes in existing atomic energy legislation.[36]

Eisenhower encouraged immediate amendment of the Atomic Energy Act as the first step toward establishing a private nuclear power industry in the United States. Even before the President's historic speech to the United Nations, the National Security Council agreed on March 31, 1952, that Eisenhower's desire to foster "economically competitive nuclear power [was] a goal of national importance." But to build a healthy nuclear industry, Eisenhower knew that the federal government would have to promise a substantial subsidy to private investors by assuming obligation for long-term, high-risk research and development. The National Security Council also thought that the government would have to assume responsibility for establishing safety standards for commercial reactor operations.[37]

For more than thirty years, from the 1950s to the 1980s, Atoms for Peace remained the cornerstone of America's foreign and domestic policy promoting the peaceful uses of nuclear energy. Although the United States seemingly held a lead in utilizing atomic energy, Eisenhower believed that the American advantage was only temporary and perceived the AEC's nuclear science and technology establishment to be a "wasting asset"; that is, he knew that, in time, other nations would develop independent nuclear ventures, with or without the help of the United States. While the United States was still the dominant nuclear power, Eisenhower wanted to exploit American expertise, including the national laboratories, to promote American economic and security interests.

Although Eisenhower inaugurated Atoms for Peace, he did not support the large AEC budgets needed to fund his initiative. Already the administration had scuttled the proposed aircraft-carrier propulsion reactor because it was too expensive. Endeavoring to save its project, the commission suggested recouping the government's investment by converting the carrier propulsion reactor into a demonstration civilian power reactor. The commission supported the National Security Council's policy on industrial participation but also doubted that private industry would finance long-term research and development on nuclear power systems even if the government amended the Atomic Energy Act to allow private ownership of plants producing or using fissionable materials.

Agreeing that the government would have to carry the burden of research and development, Eisenhower authorized the commission to explore ideas of salvaging the military reactor for peaceful purposes.[38]

The Five-Year Program, 1954

In March 1954, the same month that the AEC's fifteen-megaton Castle-Bravo thermonuclear test thundered across the Marshall Islands, the Joint Committee on Atomic Energy authorized the commission to launch a five-year civilian reactor research-and-development program. The commission's "new" five-year program incorporated three of the government's existing reactor development projects: Argonne's boiling-water reactors, its fast breeder reactors, and Oak Ridge Laboratory's Homogeneous Reactor Experiment (HRE-1). A small project, the Sodium Reactor Experiment (SRE), conducted by North American Aviation, Inc. in California, represented the only effort by private industry (although the project was largely financed by the federal government).

Uncertain about the future of its experimental reactors, but wanting rapid success, the commission toyed with the idea of making the navy's pressurized light-water reactor, salvaged from the carrier propulsion project, the centerpiece of the five-year program. Zinn agreed that building a full-scale pilot plant was necessary, but he thought it would be a great mistake to modify the aircraft carrier reactor for civilian use. The navy's reactor could not be expected "to come close to being economically satisfactory for central-station power," he argued before the Joint Committee. Zinn noted that the navy and civilian industry required different degrees of reliability and that the costs naval specifications justified could not be tolerated in civilian power stations: "To get nuclear power competitive with coal will take thorough examination of each design feature and a ruthless elimination of any which inflates the cost."[39]

The Joint Committee shared some of Zinn's doubts about the economics of converting the naval propulsion reactor into a civilian power reactor. But all reservations were swept away when Duquesne Light Company of Pittsburgh offered to build the power station at Shippingport, Pennsylvania, a company-owned site on the Ohio River not far from Bettis Laboratory. In addition to providing the site, Duquesne promised to construct the turbine-generator plant, to contribute to the cost of building the reactor, to operate and maintain the facility, and to buy the power from the AEC at rates favorable to the government. For government officials, the deal was almost too good to believe. Best of all, with Rickover in charge, the Shippingport project seemed to offer the quickest, and surest, route to a successful nuclear power demonstration project.[40]

The limited participation of private industry in the five-year reactor development program underscored the need for change in atomic energy legislation. Instead of amending the Atomic Energy Act of 1946, Congress decided to draft an entirely new statute. Especially important for Argonne were provisions which loosened security requirements on restricted data pertaining to peaceful uses of atomic energy. Until 1954 and through most of that year, all personnel working in reactor research and development required security clearances. Outlines of Argonne's participation in the five-year development plan were classified "secret-security information." Routine design and performance data on reactors such as EBR-I were also classified so that meaningful cooperative programs with domestic and foreign scholars were almost impossible. And until the government allowed private ownership of plants producing or using fissionable materials, industrial participation was not practicable. After extensive Congressional debate, on August 30, 1954, Eisenhower signed the Atomic Energy Act of 1954 which enabled the national laboratories to implement the president's Atoms for Peace program domestically and internationally.[41]

Zinn's Approach to Nuclear Power

The key to competitive nuclear power, according to Zinn, was the cost of the nuclear fuel cycle. If the capital costs of building nuclear power stations were less than coal-fired plants, the price of nuclear fuels could be higher than coal. But as long as conventional power plants were cheaper to build, successful nuclear power would have to be based on a nuclear fuel economy less expensive than mining and shipping coal.

One way to lower nuclear fuel costs was through the "long burnup" approach, in which natural uranium or slightly enriched fuels remained in the reactor core until approximately 1 percent of the available uranium was used. In 1948 Argonne had first promoted the idea that acceptable fuel costs could be achieved with sufficiently long burnup of uranium fuels. Three of the reactors featured in the five-year plan, including Argonne's boiling-water reactor, were of this type. As Zinn explained, a number of factors prevented burnups longer than 1 percent: as uranium fuel was consumed the reactor lost criticality; accumulated fission products interfered with efficient operation of the reactor; and the fuel itself became less stable as an increasing fraction of the uranium was transmuted by fission into entirely different elements. Zinn calculated that fuel costs in long-burnup thermal reactors would be 7 mills per kilowatt hour compared to coal costs of 3.5 mills per kilowatt hour for conventional power plants. Although Rickover's pressurized light-water reactor was attractive for military use, Zinn no longer believed it could offer competition to electricity from coal.⁴²

As one of the early proponents of light-water, heavy-water, and boiling-water reactors, Zinn wanted to differentiate his support for the long-burnup idea as a way of getting the nuclear power industry started, as opposed to his opinion about what should or could be done to promote nuclear power "in the long pull." One percent burnup of uranium did not make economic sense to Zinn. According to his calculations, a light-water reactor producing 150,000 kilowatts of electricity per year would also produce 20 tons of uranium-238 contaminated with 200 kilograms of plutonium-239 and fission products as nuclear "waste." He strongly believed that reprocessing "spent" fuel was a better financial risk than mining ever lower grades of uranium ore and processing them for fuel. Of course, the natural uranium yellowcake was only mildly radioactive, whereas the spent fuel slugs were highly radioactive.

By 1954, Zinn believed that the economic future of nuclear power would be determined by regenerative reactors that produced fissionable material in addition to heat.* The great production reactors at Hanford and Savannah River, which produced great amounts of waste heat while creating weapon-grade plutonium, illustrated the principle. Although they were technically feasible, Zinn was skeptical that dual-purpose reactors could be economically competitive. "Inevitably," he told the Joint Committee, "the two products, [weapon grade] plutonium and power, affect the design in different ways and it would be a difficult task indeed to properly distribute the costs to the products."⁴³

Argonne's assignment in the five-year development plan included continued development of boiling-water reactors—the long-burnup configuration—and launching of a major experiment with EBR-II, a breeder reactor. Zinn liked certain advantages of the boiling-water systems: one benefit was that capital investment was moderate compared to that of pressurized light-water reactors. They required no expensive heat exchangers or steam generators, and pumps were smaller. Because the system was not pressurized to prevent boiling, the reactor vessel, piping, and shielding were all designed for relatively lower temperatures and pressures. Lower water temperatures also meant less corrosion of internal parts. Best of all, as tests with BORAX-I showed, reactivity declined in the reactor core as temperatures rose indicating that boiling-water reactors might be engineered to be inherently safe.

---

*Regenerative reactors came in two types: *converter* reactors, in which fissioning uranium-235 partially converted a uranium-238 blanket into plutonium-239; or true *breeder* reactors, in which fissioning plutonium-239 in the core bred plutonium-239 within a uranium-238 blanket, that is burning plutonium to create more plutonium.

But boiling-water reactors had their drawbacks, too, not the least of which was that the direct water cycle put radioactive steam directly into the turbine, complicating routine maintenance and risking radioactive leaks into the environment. Finally, despite the successful operation of BORAX-I without control problems, Zinn conceded that it remained to be determined whether a large boiling-water reactor could operate successfully over an extended time.[44]

Zinn's reactor of choice for achieving acceptably low fuel costs was the liquid-metal cooled, fast-neutron breeder reactor, whose prototype was EBR-I. Although EBR-I was not an engineering test bed, Zinn observed that over two and a half years the reactor had proven "remarkably successful as an engineering device." Except for a serious leak in the heat exchanger, EBR-I operated without serious breakdown. Yet even the leak proved encouraging to Zinn because it was quickly repaired, indicating "that maintenance in a radioactive sodium system [was] not too difficult." Refueling, involving removal of old and insertion of new fuel elements, had gone smoothly. Now Zinn was ready to move ahead with a larger pilot plant to test the feasibility of the breeder reactor concept.[45]

Still working behind a veil of secrecy, Argonne first proposed a development program for a fast-power breeder reactor in February 1953. By May 1954, Zinn envisioned construction of an intermediate-size power breeder, which he called the Experimental Breeder Reactor Number 2 (EBR-II). The reactor would produce about 62 megawatts of heat, with a net electrical generation of at least 15 megawatts. All hardware, pumps, heat exchangers, and valves in EBR-II would be of sufficient size to test full-scale applications. With high capital costs invested in the reactor core and primary systems, Zinn hoped that the liquid sodium coolant would moderate costs by proving highly efficient and relatively noncorrosive.[46]

Although Zinn anticipated that EBR-II would be fueled initially with uranium-235, ultimately the economic feasibility of the breeder reactor would be proven with plutonium. In turn, the plutonium fuel cycle required reprocessing and fabrication of highly radioactive materials as an integral part of the breeder reactor project. To demonstrate the economic competitiveness of the power breeder reactor, Argonne needed to prove that plutonium recycling was practical and did not constitute a significant financial burden. Zinn tossed Argonne's chemists and metallurgical engineers an enormous challenge, but he had supreme confidence that the technical and economic problems would be solved. In fact, he believed that the boiling-water and fast-breeder systems "could very well be complimentary." The long burnup reactor discarded plutonium as a "waste"; the breeder reactor needed plutonium for fuel. Zinn predicted that someday one reactor system would provide fuel for the other.

Reorganizing the Reactor Engineering Division

A month after Zinn appointed Arthur H. Barnes as director of the reactor engineering division on January 1, 1954, Barnes reorganized the division by creating two new sections to support the five-year development plan. Barnes assigned all work on the breeder project to the liquid-metals reactors section with F. W. Thalgott's group at Argonne East heading physical research and Lichtenberger's crew in Idaho responsible for the "fast critical experiments." The water reactors section was responsible for all engineering and development work on water reactors, including boiling-, light-, and heavy-water designs. Assignments included development of the army's portable power reactor and advanced designs for the navy.[47]

Since early 1952, Argonne had had few direct responsibilities for the naval reactors program. Hyman Rickover expected that his two main contractors, Westinghouse and General Electric, would handle all of the navy's requirements for continuing research and development on the submarine reactors. But Rickover and the navy wanted Argonne to sustain a small program ($300,000 annually) exploring long-range applications for naval propulsion, including conceptual studies on flashing reactors, supercritical water reactors, organic-moderated reactors, and boiling-water reactors for larger submarines.[48]

Among the interesting advanced concepts was the "application of thermal reactors to hydrojet propulsion." In laymen's language, the Navy Bureau of Ordnance in 1954 asked Argonne to study the possibility of building an underwater torpedo powered by a compact nuclear reactor. Conventional World War II propeller-driven torpedoes were notoriously slow. Swift hydrojet torpedoes, in contrast, scooped up seawater through their nose, heated it internally, and then shot steam and hot water through a tail nozzle. Chemical fuels, however, severely limited the hydrojet torpedo's range and speed. On the other hand, nuclear-powered torpedoes theoretically might have almost unlimited range. Controlling a nuclear-powered torpedo would be difficult, but probably no more difficult than controlling long-range surface missiles. If it were successful and tipped with a nuclear warhead, the navy could be armed with interoceanic torpedo missiles that rivaled air force dreams for intercontinental ballistic missiles.[49]

Zinn was not enthusiastic about developing small reactors for the armed services, however. With limited budgets and personnel for research on civilian reactors, Zinn did not want to see any of Argonne's precious resources diverted to nonessential military projects. He directed Barnes to mobilize the laboratory's staff and equipment in Illinois and Idaho to support the boiling-water and fast-reactor projects. Argonne's reactor work was the laboratory's most important mission, to which Zinn committed his highest priority and most competent staff. "The Reactor Engineering Division cannot long expect support in any of its work if it cannot succeed in these projects," he warned Barnes. "The laboratory cannot long maintain its leading position in the reactor field if it fails in these projects," he added. "It is clear that all other reactor work must be given secondary position."[50]

The Oppenheimer Case

Simultaneous with launching Eisenhower's Atoms for Peace initiative, the Atomic Energy Commission became embroiled in the Oppenheimer hearings. On the eve of his address to the United Nations in December 1953, Eisenhower ordered the suspension of J. Robert Oppenheimer's security clearance until the commission could review charges from William L. Borden, former executive secretary of the Joint Committee on Atomic Energy, "that more probably than not" Oppenheimer was an "agent of the Soviet Union."

Borden's charges against Oppenheimer and the commission's subsequent hearings and findings, which permanently stripped Oppenheimer of his security clearance, stunned America's atomic community. The Federation of American Scientists and the American Physical Society lodged official protests. Petitions signed by eleven hundred scientists from the national laboratories and universities poured into commission offices. Scientists and their allies were not only upset about the commission's treatment of Oppenheimer, but they were also fearful that the Oppenheimer case would seriously impair their ability to attract and hold outstanding young scientists in government service. *Fortune* magazine reported, for example, that America's outstanding young scientists were "deeply troubled" by the commission's handling of the Oppenheimer case. In their letter to the commission, the Argonne petitioners were not concerned about Oppenheimer personally, although he was highly honored and respected. What would become of scientists holding controversial views? If the United States brands scientists "disloyal" as the penalty for being unpopular or unwise, then who but the most compliant will want to serve the government? Surely employees and consultants can be dismissed if their competence were questioned. But to label nay-sayers a security risk because political leaders disagreed with their views was "to adopt the worst features of totalitarianism," the Argonne scientists charged.[51]

Almost two-thirds of the petitioners were from the Argonne and Los Alamos laboratories, where there was much dark talk about how the best scientists would leave government service for academia. General Manager Kenneth D. Nichols was so worried about morale at Argonne that he offered to explain the government's action in person to the angry scientists. Zinn, enmeshed in getting the boiling-water and fast-reactor programs underway, thought that Nichols's visit was unnecessary and convinced the commission's general manager to stay away from Argonne until the matter had blown over.[52]

While Zinn dismissed out-of-hand concerns that the Oppenheimer case or McCarthyism had any significant impact at Argonne, physical security at Argonne continued to worry AEC officials. Atoms for Peace had promoted extensive declassification of reactor technology which, in turn, created a number of problem areas concerning the protection of secret data. As Argonne became a more open laboratory conducting unclassified research and hosting uncleared students, professors, and foreign nationals, the need for security clearances and fences declined. But, as Argonne became "de-fenceless," Hilberry observed, security problems increased because it was more difficult to avoid security infractions than when everything was classified.

Why should security headaches increase when classification standards are relaxed? In part, the commission had created "gray areas" where the sensitivity of information was ambiguous. One of the most difficult areas was reactor engineering, in which much of the technical data concerning power reactors also included military or production information restricted to employees with proper "Q" clearances. Engineering designs and data concerning plutonium use and production were especially sensitive, but as Argonne became increasingly committed to developing a breeder reactor, the distinction between "military" and "civilian" technology became hopelessly blurred. The problem of protecting vital classified data was compounded by the fact that Argonne's program files bulged with "secret" documents that no longer required classification.[53]

Respect for the security system depended on the belief that the rules were reasonable and enforceable. Perhaps the commission's greatest worry was that Argonne scientists working on unclassified projects in open areas of the laboratory might unwittingly produce classified results. The AEC specifically asked that any experiment that might yield classified data should be conducted in a secure area. Winston Manning failed to follow the AEC's logic. Of course, security-sensitive work would be conducted in secure areas, but how could one anticipate the commission's unlikely scenario of an innocent experiment that produced unexpected classified data? Zinn replied simply that "innocent" experiments could be done as long as laboratory management and the commission's Chicago office were informed in advance. It was difficult to imagine a major activity that did not fall into this category.[54]

The International School of Nuclear Science and Engineering

Eisenhower's Atoms for Peace program rekindled international cooperation in nuclear science and technology. Before World War II, nuclear scientists worked in a strong international community where both theory and experimental data were openly published and widely disseminated. Important discoveries were shared quickly and, in the case of the discovery of uranium fission by Otto Hahn and Fritz Strassman in 1938, almost instantly.

For many scientists, the war abruptly ended a long tradition of scientific freedom that assumed the results of fundamental research would be published and evaluated in the open international literature. The Manhattan Project "compartmentalized" nuclear science so that only security-cleared researchers could obtain vital data on a "need-to-know" basis. After the war, the Atomic Energy Commission allowed few foreigners access to restricted data as defined by the Atomic Energy Act. Until the amendment of the Atomic Energy Act in 1954, the exchange of scientific and technical information with foreign nationals was almost impossible.

Within a month of the amendment of the Atomic Energy Act in July 1954, J. J. Flaherty, manager of the Chicago Operations Office, asked Zinn if Argonne would be able to sponsor a reactor training program for foreign scientists and engineers. President Eisenhower wanted to implement provisions for international cooperation immediately. Was Argonne interested? Zinn did not hesitate and, without approval from the University of Chicago trustees, he replied the same day that it would be "impossible and unthinkable" for Argonne to decline the invitation. The laboratory had already offered training sessions for United States nationals, and with the recent declassification of research facilities, Zinn believed that Argonne provided an ideal location for state-of-the-art reactor training.

The international school presented Zinn serious logistical and security problems. Argonne had no facilities to house students or researchers, and because the laboratory was relatively isolated, there were no suitable nearby, long-term lodgings. At the laboratory itself, physics, biology, and medicine were assigned unclassified buildings, but reactor physics, reactor chemistry, reactor engineering, and metallurgy—precisely the areas of most interest to foreign students—were still located in secure facilities. To solve these problems and establish the training school as quickly as possible, Zinn gave the project his highest priority by asking Norman Hilberry, now Argonne's deputy director, to head the school.[55]

The International School of Nuclear Science and Engineering opened its doors to forty students in March 1955. The first class consisted of thirty-one male scientists from nineteen nations and ten Americans, all from industry.* The seven-month curriculum (four months of classroom study and three months of applied reactor engineering) covered unclassified subjects on the design, construction, and operation of research reactors (especially CP-5), the principles of power reactors, the handling of irradiated materials, and related civilian applications of nuclear energy.

Before arriving in Chicago, the scientists were feted with a State Department reception at Blair House, a luncheon with the Joint Committee on Atomic Energy, and a meeting with the president at the White House. Eisenhower was delighted to celebrate this first harvest of his Atoms for Peace program. Characteristically, he expressed his joy in Cold War rhetoric. "You represent a positive accomplishment in the Free World's efforts to mobilize its atomic resources for peaceful uses and the benefit of mankind," he told the students. Then, in plain language, Eisenhower thanked them for this "heartening sign that we are making progress toward real international cooperation."[56]

Eisenhower affirmed his commitment to international cooperation three months later in his June 11 commencement address at Pennsylvania State University. Nuclear science was not an American science, he told the graduates, but an international cooperative endeavor. For maximum progress in the peaceful applications of atomic energy, the United States must foster "partnership between the world's best minds—in science, engineering, education, business, and the professions." Argonne's international training program was a major step in that direction. But training was not enough. The United States would also have to provide technical assistance to build research reactors abroad so that American-trained scientists and their students had the means to apply and develop their newly acquired knowledge.[57]

Unless the Atomic Energy Commission authorized major expansion of the teaching staff and housing facilities and construction of classroom and demonstration facilities for the school, Argonne could not continue to host the entire training session. Instead of creating a permanent school at the national laboratory, the commission decided to use the reactor school not only to train foreign students, but also to develop strong nuclear engineering faculties at American universities. Commissioner Libby wanted it understood that the AEC intended to close Argonne's school just as soon as university nuclear engineering courses could assume responsibility for international education. Presum-

---

*The nations represented were Argentina, Australia, Belgium, Brazil, Egypt, France, Greece, Guatemala, Indonesia, Israel, Japan, Mexico, Pakistan, the Philippines, Portugal, Spain, Sweden, Switzerland, and Thailand.

ably, this meant that the commission would also assist the universities to obtain research reactors.[58]

Although the school operated on a session-by-session basis, by 1959 Argonne had trained 325 foreign students from more than forty countries and 95 Americans. Beginning with the school's third session in September 1956, students began formal class work at either North Carolina State College or Pennsylvania State University, both of which had research reactors on campus, and then moved on to Argonne for advanced and applied studies on reactor technology and related topics. In addition to classes on reactor design and operations, they studied fuel preparation and reprocessing and could further concentrate on nuclear chemistry, physics, and engineering. They worked in Argonne's "caves," learning how to handle highly radioactive materials. For many of the students, the highlight of the training session was independent research on CP-5, Argonne's principal research reactor. To facilitate nuclear education, the commission authorized development of a training reactor, the Argonaut, designed to be safe enough for students to operate at both home and abroad. While enrolled in the program, students visited Oak Ridge National Laboratory, nuclear laboratories at the University of Michigan, and the Shippingport pressurized power reactor. Their studies concluded with a tour of the National Reactor Testing Station and other western atomic energy facilities.[59]

The international program was the most notable activity of the school but constituted only 60 percent of its effort. Argonne also used the school to improve its troubled relations with the academic community. In addition to providing extended training for graduates of the international program, the school ran summer institutes for American university professors, offered short courses for professors and government officials, and coordinated a two-year postgraduate program for recent university graduates.[60]

The Soviet Challenge

As Eisenhower anticipated, the international school found itself on the front lines of the Cold War. Hilberry, who had championed the establishment of Argonne National Laboratory as a bulwark against "technological emergencies," was an obvious, and appropriate, choice to organize this response to competition from international communism. After the establishment of the International Atomic Energy Agency, the Soviet Union announced its intention to host thousands of foreigners at several training centers where, as Hilberry noted ruefully, the students would "be required to take *all* the courses the Russians give." Following training, the Soviets promised to send out technical teams that would assist recipient countries in establishing reactor programs. The United States would be forced "to offer refuge and training" to the entire free world, Hilberry predicted, and only Argonne was capable of rising to the task. In this sense, the legacy of the Met Lab still burned brightly in Hilberry's vision for Argonne.[61]

The International Conference on the Peaceful Uses of Atomic Energy, 1955

The United Nations' International Conference on the Peaceful Uses of Atomic Energy, which was held in Geneva, Switzerland, offered a showcase for American research and development in peaceful applications of nuclear science and technology. Encouraged by the Eisenhower administration as a means of giving substance to the president's Atoms for Peace proposals, the conference won enthusiastic endorsement from American scientists, who welcomed the opportunity to exchange information on biology, medicine, basic sciences, and engineering. Initiated by the Americans, the conference was endorsed and sponsored by the United Nations General Assembly. All parties agreed that only scientific and technical issues would be discussed—politics would be out-of-order. In August 1955, almost two thousand delegates from seventy-two nations gathered in one of the largest postwar international science conferences. Nuclear experts offered 1,000 papers, 450 of which were read orally, on all aspects of peaceful uses of nuclear science and technology. Cold War competition could not be banned, of course, so the major nuclear powers made every effort to organize impressive exhibits and encourage commercial ones.[62]

Nuclear power and research reactors commanded the greatest interest in Geneva. In planning the United States' participation, the AEC decided to highlight American achievements in developing commercial nuclear power. American papers and exhibits offered impressive breadth and detail about the commission's five-year reactor program. The spectacular centerpiece of the United States' exhibit was a swimming-pool research reactor flown to Geneva from Oak Ridge National Laboratory. The small research reactor, set up on the grounds of the Palais des Nations, was both the first ever built in Western Europe and the first ever exported from the United States. In Geneva, President Eisenhower paid a dramatic visit to the reactor and, with the assistance of Oak Ridge technicians, operated the controls. More important than the president's public relations effort was the profound impression made on thousands of visitors that the age of absolute atomic secrecy was over.

Zinn was sensitive that an Oak Ridge reactor was America's most popular attraction at the Geneva peaceful uses conference. Had Eisenhower stood at the controls of a reactor from Argonne National Laboratory, Zinn surely would have stood beside him. Argonne's director could hardly complain about the American exhibit, but he was disappointed that the "notoriety" of the Oak Ridge research reactor somewhat overshadowed Argonne's more technical presentation. Argonne, naturally, played a major role at the conference, where laboratory scientists presented twenty-four papers (another twenty-six were included in the conference's official proceedings). Zinn wanted Argonne to make a real effort to present new material within the new classification guidelines.[63]

In their presentations to the peaceful-uses conference, Zinn and his colleagues featured Argonne's most recent developments in research, boiling-water, and fast reactors. The papers on CP-5 and the boiling-water reactors offered detailed drawings of fuel-element assemblies, core configuration, and reactor components. Subsequently, visitors to Argonne's reactor exhibit at the Palais des Nations viewed cut-away scale models of four Argonne reactors keyed by color-coded components. The five-foot-square reactor models included two existing reactors, CP-5 and BORAX-III, and the two large reactors developed under the five-year reactor demonstration project, the experimental boiling-water reactor (EBWR) and EBR-II.[64]

Zinn particularly focused on the boiling-water reactor project, describing the recent "run-away" tests and other successes in the BORAX program. Following the destruction of BORAX-I, Argonne built BORAX-II, a reactor similar to its predecessor. To test the reliability of the reactor as a power producer and to investigate the performance of a turbine operating in a radioactive environment, in the spring of 1955 Argonne installed a 3,500 kilowatt turbine generator and renamed the reactor BORAX-III. For more than an hour on July 17, BORAX-III provided the entire electrical load for Arco, Idaho, the first American town to receive all its power from a nuclear reactor. Like the light-bulb demonstration by EBR-I in 1951, nuclear power for Arco was largely symbolic, and temporary, although again as with EBR-I, BORAX-III continued to provide power for its own reactor complex. Zinn proudly showed the conference dramatic movies of the BORAX excursion tests and the lighting of Arco, and he saw no irony in the contrasting desert scenes of the exploding reactor and the bucolic little town. At Argonne, all of these images represented progress toward safe, reliable nuclear power.[65]

Enthusiastically, Zinn, Lichtenberger, and Joseph R. Dietrich described construction of the 5,000-kilowatt EBWR at Argonne-East, the first reactor in the western world designed and built specifically as a pilot plant to advance nuclear power. EBWR was a light-water moderated and cooled, boiling thermal power prototype, employing natural uranium and enriched uranium-235 fuel. Although the EBWR was small and not economically competitive with commercial power production, Argonne's engineers believed that they could readily extrapolate from EBWR to larger, commercial reactors. With ground breaking in May 1955, Zinn and Hilberry planned to have the first of the power demonstration reactors on line by 1957. It would be a source of pride to Argonne for many years that this reactor was built quickly and well within two years.[66]

Design efforts on EBR-II, which Zinn had been promoting since 1953 as the pilot plant of the power breeder reactor development program, were much less advanced than EBWR. Zinn and his colleagues presented four papers on aspects of breeder technology at Geneva but did not offer details on fuel elements, core configuration, or reactor components comparable to the information provided about the boiling-water reactors. Although some old timers wanted EBR-II built at Argonne-East, for safety reasons Zinn strongly urged the commission to place the experimental reactor at the National Reactor Testing Station in Idaho. EBR-II would be higher power than EBWR and would eventually incorporate the entire plutonium fuel cycle.[67]

Most of the American delegates to the Geneva peaceful uses conference did not feel that they learned much from the formal papers and official exhibits. Although official secrecy concerning civilian applications was demonstrably on the wane, a new "industrial secrecy" relating to commercial application was palpable. Nature's laws cannot be stamped "secret," but exhibitors including the British and the Russians were reluctant to give away competitive "trade" secrets, especially those about fuel technology. Little information was available concerning the fabrication of fuel elements or their performance. The Americans surprised the conference by exhibiting for close examination actual "dummy" fuel elements from various reactors. Most startling was the model of Idaho's Chemical Processing Plant depicting recovery of uranium and plutonium from spent fuel. The Russians were completely silent about this issue.[68]

Although the Argonne delegates acquired little new science and technology at Geneva, they learned a great deal about their Cold War competitors, especially the Russians. Everyone agreed that the most fruitful sessions were conducted in the corridors or during informal receptions where scientists met one-on-one to exchange information or get acquainted. Not unexpectedly, the Russian exhibit ranked first in curiosity among Argonne scientists. A little smugly, perhaps, they were reassured that the Soviets trailed the United States in applied technology and nuclear engineering. But the Russians were surprisingly competent. Joseph Dietrich took special notice of their boiling-water reactor, while Ray Goertz was impressed with the sophistication of their master-slave manipulators. Despite the Russians' reticence about fuel technology, Zinn offered the Soviet's a supreme compliment: "The Russian power reactor was interesting to us . . . because of the fuel element. This fuel element is of topnotch performance. It does not have any counterpart as far as I know. . . ."[69] With new respect for the Russians, the Argonne delegation returned to the United States more than ever dedicated to maintaining American dominance in nuclear science and technology.

EBR-I Accident: Fuel Meltdown

As predicted by the AEC's Reactor Safeguards Committee, testing and operating experimental nuclear reactors proved a risky enterprise.[70] Minor "incidents" were common and anticipated at Idaho's National Reactor Testing Station. Between 1955 and 1957, for example, the materials-testing reactor experienced thirty-eight incidents "involving the release of an observable amount of radioactivity" requiring evacuation of the facility for up to eighteen hours (average about five hours) in almost half the incidents. But on November 29, 1955, EBR-I suffered an inadvertent partial fuel meltdown during an experiment to measure transient temperature coefficients. Although Zinn admitted that the partial meltdown was unplanned, he was not alarmed by the incident and assured the public that much had been learned from the experiment, which had threatened neither public health nor safety. Years later, the commission did not even regard the EBR-I incident as an official accident.[71]

Prior to the meltdown, in fact by January 1955, Zinn thought that EBR-I's days were numbered. After three years of operation, he believed the reactor had almost reached the end of its research usefulness because diminishing experimental results could not justify the steady expenditure of money, manpower, and fissionable material. Zinn had planned to place EBR-I on standby in 1956. Although the reactor generally had performed satisfactorily with its second core loading, the engineers had noted an increased positive reactivity temperature coefficient under certain conditions. As a final experiment, Zinn decided to measure the reactor's transient temperature coefficients.[72]

Normally, reactors should have a negative temperature coefficient so that reactivity decreases when temperature of the reactor increases. As demonstrated in the BORAX experiments, this inverse correlation assures that reactors will tend to shut themselves down rather than respond uncontrollably. The opposite relationship, a positive temperature coefficient, can spell trouble because when reactivity increases as the core temperature rises reactors edge toward a *prompt critical* condition, in which fissioning multiplies faster than an operator can control. With both EBR-II and the Power Reactor Development Company's Fermi plant under design, Zinn wanted to measure reactivity in the reactor when fuel element temperatures rose above 500° C. Because he wanted to measure only the temperature coefficient of the fuel, the experiment required shutting off the flow of liquid-metal coolant.[73]

Before leaving for Geneva, Zinn advised the AEC of the risks associated with the experiment, including the possibility of fuel-element melting. Consequently, as in the BORAX-I experiment, Zinn did not anticipate serious trouble but prepared for the worst. He directed his engineers to bring EBR-I gradually to full power while simultaneously reducing coolant flow rates. Operators had orders to "scram" the reactor if the temperature rose dangerously. Even with all caution, Zinn calculated that power would double about every 0.27 seconds. Because of the rapid temperature rise, an emergency shutdown would not only have to be rapid, but also timed at the right moment. As Zinn explained, "an error of one second in initiating the shutdown could be expected to give trouble."[74]

And trouble arrived on a horse called Miscommunication. In the past, reactor operators had successfully ended excursion tests by reversing the motor-driven control rods, which dropped the rods back into the core. In this experiment, Zinn expected the operator to use the fast-acting shut-off control rods to stop reactivity quickly. Instead, when the chief scientist signaled to shut down EBR-I, the operator responded routinely by activating the slower motor-driven rods. Almost immediately the scientist saw the mistake and hit the *scram* button, which instantly dropped the emergency rods. Only two seconds elapsed, but in that time 40–50 percent of the reactor core melted.[75]

Zinn did not know immediately how seriously EBR-I had been damaged. In contrast to BORAX-I, there were no unusual noises, no explosion, no steam, and no smoke. Within fifteen minutes, however, detectors measured abnormally high radioactivity in the reactor building, and all personnel were evacuated. The AEC later reported that no one was injured or overexposed to radiation and that no radioactivity was measured off-site from the reactor testing station.[76]

Although the incident had not been planned, Zinn rationalized that the experiment provided his engineers invaluable lessons. First, he was pleased that EBR-I had survived a dangerous test with "no unforeseen or catastrophic" consequences. The relative ease with which the engineers controlled the radioactivity released into the reactor coolant assured Zinn that melting "highly radioactive fuel elements in sodium [was] not very hazardous to the reactor surroundings." In addition, the incident allowed engineers for the first time to gather data on the melting of fuel elements in liquid metal. This melting and softening of fuel elements, Zinn reflected, could not be observed except under actual operating conditions in the reactor. Finally, Zinn believed the need to replace the damaged core would provide excellent training on handling large, highly radioactive components, perhaps leading to routine and economical techniques for core replacement.[77]

Although Zinn telephoned the commission on November 30 to report the accident, the AEC made no public announcement about the event until forced to do so by the business and industrial press some months later in April. (*Nucleonics,* the industry trade journal, first caught wind of the incident, followed by *Business Week, Time,* and *Science.*) Normally friendly to the fledgling nuclear power industry, the press darkly reported that the commission had tried to hide "the nation's first serious atomic reactor accident." Even after the news became public on April 5, authorities in Idaho allegedly would not allow outsiders to see the damaged reactor. *Nucleonics* affirmed that the incident was "a minor, unfortunate accident with no wide significance"—no worse than normal industrial accidents. So what was the reason for the AEC's secrecy? If the commission officials mistakenly believed the EBR-I fuel meltdown was an internal matter, they not only committed a serious public relations error, but also did not understand that "withholding of news is wrong in principle." Arguing that nuclear accidents were everybody's business, *Nucleonics* predicted that public confidence in the AEC would surely be undermined if the commission continued to operate with so much secrecy.[78]

Business Week also wondered about all the "official hush-hush" and speculated that the EBR-I accident, which ruined the AEC's nearly perfect reactor safety record, raised serious questions about whether the reactor industry was insurable. *Business Week* recalled the December 1952 accident at Canada's Chalk River plant that resulted in melted fuel and featured spectacular pictures of the BORAX-I explosion. Both events had involved small reactors located at remote sites. What worried the young nuclear industry, the magazine reported, was the remote chance that a large power reactor located near an urban center "might scatter deadly radioactive materials over a wide area." If the AEC's reactors were not virtually foolproof, if a technician could cause a serious accident by literally pushing the wrong button, how much liability insurance would be required to cover a single plant? $100 million? $200 million? Even by pooling their resources, *Business Week* doubted that private insurers could provide more than $65 million in coverage per plant. If the Eisenhower administration wanted to promote the nuclear power industry, *Business Week* advised, some form of government insurance would have to take up the slack.[79]

Zinn was thoroughly disgusted with the reporting of the EBR-I incident and the raising of what he considered "red-herrings" concerning nuclear reactor safety and reliability. Notorious for his conservative approach to reactor engineering and his close attention to safety issues, Zinn had created his own Reactor Safety Review Committee in 1952. The proper issue was neither reactor safety nor insurance policies for nuclear power plants, he rebutted. EBR-I was, after all, a test facility, and it was normal for engineers to push the reactor to its limits and beyond. He lectured the readers of *Nucleonics:*

> One cannot expect technologists to undertake difficult tasks if a public debate is to be anticipated whenever everything does not proceed altogether according to plan. Events similar to the EBR-I case have taken place in the past and will again in the future. It would be a disservice to the progress of our atomic energy program if such occasions are not treated as unfortunate penalties exacted by the necessity of getting on with the job.[80]

If it was difficult for Argonne scientists to determine precisely what damage occurred within the EBR-I core, it was almost impossible to detect damage to public confidence in nuclear power resulting from incidents such as the EBR-I fuel meltdown. *Science Digest* headlined the accident, "Reactor Runs Amuck, No One Hurt," and then described the incident in a brief, matter-of-fact report. No one at Argonne National Laboratory saw any connection between the Oppenheimer case and the EBR-I accident, and many scientists would be outraged at the idea that the two events were related. But even before Sputnik, the decline in public confidence in scientific authority was palpable.[81]

Some have linked the increased suspicion of scientists in the 1950s to America's deep-seated anti-intellectualism. Perhaps more to the point was Spencer Weart's observation that Atoms for Peace turned the spotlight of public attention on the AEC, and "people did not like everything they saw." In the most positive light, the commission and its laboratories reflected a single-minded dedication to the promotion of nuclear science in the public interest. In the shadows, however, observers also found arrogance and fallibility not easily forgiven while the scientists seemed to hold the future of the human civilization in their hands. As Weart noted, the AEC was not more or less corrupt than other human institutions, such as academia, but the cloak of secrecy that shielded the commission from public accountability also required higher standards of political acuity than the commission possessed. The Oppenheimer hearing and the EBR-I accident involved two political mistakes in which the commission appeared both self-righteous and patronizing toward those who did not understand the "full facts" of the case. Consequently, on the eve of the Sputnik crisis, the commission suffered important setbacks in trust and credibility.[82]

Lifting the Veil of Secrecy: Dedicating the EBWR

On February 7, 1957, Argonne dedicated the EBWR, the experimental boiling-water reactor (figure 20). Margaret Parker, a reporter for the Du Page County *Daily Journal*, attended the ceremony presided over by Chancellor Kimpton of the University of Chicago, Admiral Lewis Strauss, chairman of the AEC, and Representative Carl T. Durham of North Carolina, Chairman of the Joint Committee on Atomic Energy. "Dedicating the EBWR to the peace and prosperity of all mankind," the dignitaries voiced their usual confidence in America's atomic future exemplified by this first reactor solely devoted to the advancement of civilian nuclear power. At Argonne, the EBWR literally cast light into the darkness on December 23 when the reactor illuminated the laboratory, including a giant Christmas tree, during its first power run.

For Chancellor Kimpton, the EBWR spoke eloquently for itself as a precursor of the social and economic revolution with which nuclear science and technology would transform America. Reminiscing about the Met Lab days, Kimpton contrasted the beautifully and finely engineered reactor with the crude, improvised equipment used to design the first atomic bomb. Although not apologetic about the university's contribution to the Manhattan Project, Kimpton recalled that even while the Met Lab strained every facility in its race for the bomb, visionaries predicted that commercial nuclear power would be-

Figure 20. The Experimental Boiling-Water Reactor (EBWR)

come a reality soon after the war. In a very real sense, the EBWR helped redeem that promise in time to assure American nuclear leadership among the non-Communist world. The reactor not only assisted the welfare and productivity of the free world, but also served as a monument to Argonne's "high dedication" and "creative purpose."[83]

Parker, on the other hand, expressed excited ambivalence about her visit to the laboratory. Exhilarated, yet awed, she observed, "I saw the veil of secrecy that has surrounded 'atomic energy' drawn aside last Saturday at Argonne National Laboratory." It was not simply that Parker gained access to formerly classified information—she may not have recognized former atomic secrets if she had seen them—or that she applauded the commission's new public openness. Rather, at the dedication of the EBWR she had worshiped at the temple of "secret knowledge," an ancient rite often accompanied by feelings of privilege and dread. Although polls consistently indicated that Americans were optimistic about Atoms for Peace, about one-fourth feared the civilian nuclear future. Reflecting her vague unease about encountering the machine in the garden, Parker described her drive to the EBWR through deceptively rural, pleasant countryside ending at a strange, dome-shaped building housing "that mysterious object, a nuclear reactor."

Wide-eyed, Parker entered the reactor building where she became immediately conscious "of the many precautionary measures being taken everywhere." Personnel dosimeters hung on a rack ready to measure individual exposure to radiation. Next to the Coke machine stood a radiation detection device used to check for contamination on hands and feet. On her tour, she counted ten more radiation checking points. The reactor building itself was like a huge, steel-and-concrete-lined, gas-tight shell equipped with an air lock entry more impressive than the vault doors at her local bank. She saw the tanks of pressurized boric acid solution standing by, like huge fire extinguishers, ready to spring into action to quench the chain reaction should the control rods fail. Parker learned that "any abnormal situation" would automatically sound the alarms and, if proper corrective measures were not taken, shut down the remote controlled reactor, whose most important safety feature remained the fact that boiling-water reactors tended to lose reactivity when operating temperatures rose above normal. Still, Parker noted an escape hatch, just in case of an emergency.

In this strange world in which humans rarely came into direct contact with their work but kept watch near the reactor core with the new television medium; in which technicians, like extensions of their robotic arms, deftly handled the nuclear fuel with mechanical hands; in which reactor operators controlled unseen, but powerful, nuclear reactions by monitoring gages and dials on a distant control-room panel, Parker understood the need for fail-safe reactor systems. With the elaborate emphasis on control, "at the same time, one is aware of the calm, confident attitude of the staff," Parker reported. "These men have met and solved problems in which nothing could be left to chance." The public's trust of these experts would be essential for the future of nuclear power.[84]

The Argonaut Reactor

The Argonaut reactor (Argonne Nuclear Assembly for University Training), which first became critical on February 9, the same day as the EBWR dedication, was a low-power, low-cost reactor designed for physics research and instruction. While overshadowed by the power demonstration reactor, Argonaut was as important for nuclear education as the EBWR was for nuclear power. Design criteria dictated not only that Argonaut perform usefully for research and instruction, but also that the reactor prove ultrasafe so that it could be operated by students on university campuses. The cost for all reactor components and the reactor building could not exceed $100,000. Designed by Spinrad and David H. Lennox, who served as project engineer, in April 1957 Argonaut was delivered to R. G. Taecker, director of Argonne's International School of Nuclear Science and Engineering. Borrowing principles from the materials-testing reactor, the basic design originated with the Knolls Atomic Power Laboratory (KAPL) technical training reactor. Spinrad and Lennox built an extremely flexible machine whose power levels did not exceed 10 kilowatts of heat. Because a sustained chain reaction could be achieved by different shapes of fuel loading in Argonaut, students observed the behavior of reactor fuel under numerous conditions, including the introduction of poisons or neutron absorbers. In addition to studying the performance of the reactor core, Argonaut was a flexible research tool and, among other projects, could be employed as a radiation source for calibrating instruments and for neutron activation analysis, a technique by which chemists determine elements within an unknown sample by identifying and assaying radionuclides formed by neutron bombardment.[85]

The Price-Anderson Act, 1957

Ironically, the optimism expressed by reporter Margaret Parker on her visit to Argonne was not universally shared among the nuclear power community. In contrast to the high hopes brought home from the 1955 Geneva conference, two years later industrial leaders and the Democratic congress were discouraged by the lack of dramatic progress in the commission's reactor development program. Democrats on the Joint Committee on Atomic Energy were especially critical of the AEC for not moving more aggressively to implement Eisenhower's Atoms for Peace program. The 1956 presidential elections had provided Democratic Senator Clinton Anderson of New Mexico and his allies an opportunity to attack AEC Chairman Strauss and the commission for favoring private nuclear industry at the expensive of a bold public program. Neither the five-year reactor development program nor the commission's Power Reactor Demonstration Program, in which industry would build, own, and operate the power stations, had yet born significant fruit. The EBWR, although operating at 5,000 electric kilowatts, was but a modest step forward.[86]

To mollify the nuclear power industry, Congressman Melvin Price from Illinois and Senator Clinton Anderson, who was also the Joint Committee chairman, proposed an insurance indemnity bill, which quickly passed both houses. The Price-Anderson Act of 1957 required nuclear power operators to carry the maximum insurance offered by private companies. In addition, the government promised to insure each reactor for $500 million above the limit available from private insurers. The new law, signed by Eisenhower on September 2, 1957, limited public liability for each accident to the total amount of federal and private insured compensation. Political controversy over the insurability of the nuclear power industry was not ended, however, but would continue to be a major issue in the 1960s.[87]

Reactor Roundup, 1958

Between the Geneva Peaceful Uses Conferences of 1955 and 1958, Argonne's major energies were focused on reactor research and development in the design and engineering of EBR-II. In most respects, all other Argonne reactor projects were supplementary or ancillary to this endeavor. When completed at the reactor testing station in Idaho, EBR-II would produce 62,500 kilowatts of heat and 17,500 kilowatts of electrical power. Argonne engineers designed EBR-II to be an "integral nuclear power plant" equipped to handle the nuclear fuel cycle with fuel processing and fabrication facilities as well as the power reactor, heat exchangers, and steam-electric turbine generator. Leonard J. Koch, deputy director of the reactor engineering division and now project manager for EBR-II, explained that EBR-II would test the engineering and economic feasibility of a completely integrated fast-breeder reactor power plant. Preliminary design and engineering drawings nearly complete, in February 1958 the

AEC authorized nearly $30 million to construct the facility, which was to be completed by 1960.[88]

Argonne's scientists built several zero-power reactors, so called because they operated at zero power, so that engineers could assess the performance of various reactor core configurations. Zero Power Reactor Number 1 (ZPR-I) provided basic physics studies for naval reactors, and ZPR-II guided the development of the Savannah River production reactors. In order to test potential core configurations for EBR-II, Argonne researchers built ZPR-III, whose first assembly became critical on October 20, 1955. Through May 1957, ZPR-III analyzed more than thirty different core mock-ups loaded into a dual carriage that split the reactor in half for safety and convenience of loading and unloading. ZPR-III also tested the mock-up of the new Mark III core designed for EBR-I and the core mock-up of the Enrico Fermi reactor being designed by the Power Reactor Development Company (PRDC) for construction near Detroit. To relieve the research overload on ZPR-III, in 1959 two additional zero-power reactors, ZPR-VI and ZPR-IX, were built at Argonne to experimentally examine theoretical estimates on reactor design and performance.[89]

TREAT (the transient reactor test facility), another research reactor to support the fast-breeder program, became critical at the National Reactor Testing Station on February 23, 1958. When Niels Bohr visited Argonne Laboratory that same month, the laboratory presented him a graphite memento sealed in plastic from Fermi's CP-I reactor. At TREAT, some of this same graphite served as a neutron reflector around the reactor's fuel core. Without damaging the test facility, the reactor produced high temperatures with short, high-intensity nuclear energy, which enabled engineers to study the melting characteristics of prototype fuel pins or the stress on components for fast reactors. TREAT also enabled Argonne scientists to conduct safety tests by simulating nuclear excursions, loss of coolant, or loss of coolant flow around fuel elements.[90]

Boiling-Water Reactors

Supported by continuing research with BORAX-IV and BORAX-V, Argonne's boiling-water reactor program approached maturity in 1958. The EBWR performed so well and reliably that on March 20, 1958, the reactor's output was increased to almost 62 megawatts of heat, three times more than the original design. Although its relatively small size made the production of electricity uncompetitive, Spinrad contended that the EBWR provided reliable data for extrapolating construction costs, power ratings, operating costs, fuel cycles, and safety performance. Between August 5 and November 7, 1958, the EBWR provided almost half of Argonne's electricity, and during short periods when commercial service was interrupted for maintenance, the reactor actually serviced the entire power requirement. In over two years there had been no emergency scrams due to a nuclear malfunction, although the plant was deliberately run with failed fuel elements to test core performance. The EBWR suffered only one operating shut-down when a turbine blade failed. Subsequent investigation satisfied Argonne's engineers that the blade failure was caused by stress unrelated to high radiation environments. Spinrad confidently predicted that the General Electric Corporation and other companies which adopted the boiling-water design could accept the EBWR as a reliable working model. Eventually, the EBWR operated at a power level five times its originally designed capacity when it achieved 100 megawatts of heat in November 1962.[91]

Argonne's engineers also developed a portable boiling-water reactor for the United States Army. Constructed at the National Reactor Testing Station in Idaho, the Argonne Low-Power Reactor (ALPR) was not a research reactor, but rather a prototype designed to provide both 260 kilowatts of power and 400 kilowatts of space heating at remote military outposts located in the Antarctic or along the distant early warning (DEW) line in the Canadian arctic. To meet military requirements, Argonne engineered a "full-package" reactor system that could be assembled on almost any terrain, such as tundra permafrost. The system was intended to be simple, reliable, and manageable by a small group. Destined for isolated sites, the reactor was constructed to provide a radar station with power and heat for up to three years without refueling or significant maintenance. The ALPR was dedicated in December 1958 and turned over to the army, which renamed the facility the Stationary Low-Power Reactor (SL-I). Thereafter, the army oversaw SL-I to monitor reactor reliability and to train operating personnel.[92]

In January 1957, the AEC authorized Argonne Laboratory to build a large pressurized boiling-water reactor in Idaho. ARBOR (Argonne boiling reactor) would have been the commission's most ambitious boiling-water undertaking with a thermal capacity of 200 megawatts operating under steam pressures of 2,000 pounds per square inch. But the project never got off the ground because private industry also moved aggressively to develop this promising technology.

Concurrent with the EBWR demonstration, the General Electric Company built its own experimental boiling-water reactor at its Vallecitos (California) Atomic Laboratory. Based on this experience and EBWR's performance, the company built a full-scale plant (110 megawatts) for Commonwealth Edison of Chicago at Dresden, Illinois. In addition, three other commercial boiling-water plants were on the drawing board. Less than a year after the commission's authorization of ARBOR, Spinrad questioned the wisdom of continuing the boiling-water reactor program. Virtually all of the data expected from ARBOR would become available through the normal course of commercial operation. Stated most directly, the ARBOR reactor was simply not needed. Spinrad believed that Argonne had already spread itself too thinly in the reactor field. "It has been clear for some time," he advised Hilberry, "that the laboratory could not . . . accommodate the expanded boiling water program."[93]

Spinrad and Zinn shared the opinion that Argonne ought to focus its top priority on development of the EBR-II or risk losing its leadership in the reactor field. Despite his former enthusiasm for the boiling-water reactor, Zinn privately confided to Strauss that the United States was following the wrong path to nuclear power. Zinn now believed that the AEC had been mistaken to concentrate on water-cooled reactors, whether pressurized or boiling, using enriched uranium fuel. Instead, he favored the breeder concept using liquid-sodium coolant. What was more disturbing to Zinn was the commission's failure to commit the United States to a coherent development strategy. Even if Argonne abandoned its basic research programs, it could not carry the commission's program forward without dramatic laboratory expansion. In 1958 there was no alternative but to concentrate reactor resources on EBR-II.[94]

5 | Accelerators and Basic Research, 1953–57

While Atoms for Peace and the national priority for developing civilian nuclear power assured Argonne a solid position among the commission's national laboratories, there was a price to pay. By concentrating heavily on reactor development, Zinn realized by 1953 that Argonne was in danger of sliding into the backwaters of basic research. True, Argonne scientists were doing impressive work in chemistry, metallurgy, and biology, but the public eye was focused on high-energy physics, which promised to open a virtually unexplored frontier of the physical world: the atomic nucleus. What mysteries might lie in the heart of the atom? Would scientists at last be able to unravel the fundamental secrets of the physical universe?

The Accelerator Principle

High-energy physics, like reactor development, seemed to make a good case that atomic energy could be used for peaceful purposes and human benefit. And like reactor development, high-energy physics required the design and construction of large, complex devices, something physical that could produce tangible results. The size, power, and utility of accelerators and reactors could be readily determined and serve as concrete measures of accomplishment in the Cold War race between laboratories and nations.

Accelerators, however, did present one complexity that reactors did not face. Reactors produced fissionable material and power; accelerators produced only data from reactions that were not in any sense visible to the human eye. Rather, visual images had to be constructed from phenomena that had no direct relationship to vision. Larry Ratner, a physicist at Argonne, used the analogy of trying to "see" a barn, not by looking at it, but by throwing a large number of tennis balls at it. By carefully observing which balls bounced back and which did not, you could, if you threw enough balls, determine the barn's dimensions, and with luck, locate the barn door, hay mow, and silo. By measuring the angle of return, one might calculate the contour and pitch of the roof, and perhaps even learn something about the composition, thickness, and surface characteristics of the barn siding. Every now and again, a tennis ball whizzing through an open window might provide tantalizing data on the mysterious inner structure of the barn itself![1]

Since the discovery of radioactivity late in the nineteenth century, scientists had built a variety of accelerators to study the structure of the atom. In England, Ernest Rutherford, working with Hans Geiger and Ernest Marsden, directed a volley of alpha particles (helium atoms from which the electrons had been stripped) at a thin metal foil. Most of the alpha particles sailed right through the foil, but a few were deflected, and every now and again, one came bouncing right back. Rutherford did not analogize his experiment with tossing tennis balls at a barn but instead described the back-scattering effect much more dramatically: "It was almost as incredible as if you fired a 15-inch shell at a piece of tissue paper and it came back and hit you."[2]

Rutherford's alpha particles obviously collided with something substantial in the heart of the atom. Because natural radioactivity provided limited energy to the speeding alpha particles, their ability to probe the composition of the atom was restricted. But extremely fast moving particles whose speeds had been boosted artificially by accelerating machines might shatter the atomic nucleus, revealing what was inside.

Early Accelerator Development

Scientists were quick to see ways to accelerate atomic particles to ever higher energies. One obvious idea was to use protons, the nucleus of the hydrogen atom, rather than Rutherford's alpha particles as the projectile in the accelerator. Protons (and neutrons) are the constituents of every atomic nucleus in the universe, and because protons carry a positive charge, they can be accelerated by applying magnetic and electrical fields.

The earliest devices, built in the 1930s, stripped electrons from light elements and then accelerated them with a single high voltage through a straight vacuum tube to the target. The energies that could be obtained were limited by the amount of voltage that could be applied to the electrodes. Machines of this type could not accelerate protons above energies of 800,000 electron volts. This limitation could be overcome in theory by placing a series of electrodes along the course of the straight tube, but timing the charge to the electrodes to coincide with the passage of the burst of particles was beyond the technical capabilities of electronic systems of the period.[3]

Ernest Lawrence, an inventive and enterprising young physicist at the University of California at Berkeley, conceived an ingenious but simple solution to the difficulties of using multiple electrodes. By injecting protons at the center of a cylindrical vacuum chamber between the poles of an electromagnet, Lawrence realized that the protons would move in a circular orbit. Only two electrodes, shaped like halves of a round pillbox and discharged alternatively between the magnet poles would be required. Particles introduced near the center of the vacuum chamber would spiral in tight orbits at low energies and in successively larger orbits as they picked up speed. Thus all the protons would resonate with a single accelerating frequency. Lawrence's device, which he called the cyclotron, could attain energies of 10 million electron volts (MeV), which would accelerate the particle to speeds at which they could pierce the electrical barrier surrounding the positively charged protons in the nucleus of the target atoms. Although no one, including Lawrence, had any idea what scientists would find once they got inside the nucleus, by 1939, Lawrence's achievements won the Nobel Prize in physics.[4]

The wartime requirements of the Manhattan Project put accelerator development at Berkeley on hold, but in 1946, General Groves authorized completion of the cyclotron using an electromagnet with pole faces 184 inches in diameter, which Lawrence had procured just before the war. First operated in the fall of 1946, the cyclotron incorporated an idea proposed by Edwin McMillan at Berkeley to synchronize the speed of pulsed particles so as to compensate for their increased mass as they approached the speed of light. The most popular class of cyclotron ever built was the 60-inch cyclotron, which was widely used by hospitals and universities. Within a few years, Carnegie Tech, Chicago, Columbia, Harvard, and Rochester had synchrocyclotrons constructed with funds from the Atomic Energy Commission and the Office of Naval Research.

The enormous size of the 184-inch magnet clearly demonstrated that the scale-up of the cyclotron had reached its practical limits. The next generation of high-energy accelerators would take the form of a ring or "race track" many times the diameter of the single-magnet cyclotrons. Very strong magnetic fields would confine the proton beam to a small diameter and so reduce the dimensions of the magnet and vacuum chamber at any point in the ring. The development of high-powered, high-frequency oscillators during the war made possible the use of multi-magnet systems. Using this principle, William Brobeck began designing a synchrocyclotron at Berkeley capable of accelerating protons to 10 billion electron volts (10 BeV; in modern terminology, 10 GeV). Lawrence would call it the bevatron.[5]

Federal Funding of Accelerators

Aside from their design, a significant new feature of the synchrocyclotrons was that they were funded by the federal government. Before the war, private foundations funded nearly all of the research in high-energy physics, including Lawrence's cyclotron work at Berkeley. But the size and complexity of the synchrocyclotrons that scientists in university and government laboratories were impatiently waiting to build after the war made government financing imperative. No single university or government laboratory could afford to build these accelerators with the modest funds available for physics research. Initially James Fisk, the commission's first director of research, had qualms about funding these projects. He questioned whether the commission was justified in supporting research that was only remotely related to the agency's mission. Furthermore, Fisk understood that government funding of basic research was to be left to the proposed National Science Foundation, but when it appeared that creation of the foundation would be delayed indefinitely, the commission had no choice but to join the Office of Naval Research in funding accelerator projects.[6]

Once the commission resolved the funding issue, the door swung open for the big accelerators. Lawrence was ready with his bevatron proposal and scientists at the new Brookhaven National Laboratory on Long Island were finishing up plans for a 3-billion-electron-volt machine (3 GeV), which they called the cosmotron. The commission authorized construction of Brookhaven's cosmotron in 1947 and Berkeley's bevatron a year later. Immediately, Brookhaven's physicists began to lay plans for a successor to the cosmotron, which would not begin operation until 1952, one year ahead of the bevatron.[7]

A Threat to the Midwest

The federal government's pendulated funding of favored research laboratories in New York and California threatened to relegate midwestern research universities to second-class status in comparison. During the late 1940s and early 1950s, the Midwest had attained considerable strength in physics, especially at the University of Chicago and the University of Illinois. Chicago's Institute for Nuclear Studies had attracted Enrico Fermi and other leaders in the high-energy-physics community. At the University of Illinois, Louis J. Ridenour, dean of the graduate school, had been a gadfly pestering the AEC to loosen its purse strings to fund university research, including help for Professor Donald W. Kerst, Illinois's principal inventor of the electron accelerator, called the betatron.[8]

But by 1952, midwestern university physicists believed they were in serious trouble. According to University of Illinois physicist E. L. (Ned) Goldwasser, the large, midwestern research universities produced about 30 percent of the nation's Ph.D.'s in physics. To maintain the excellence and protect the reputations of their universities, Goldwasser and others believed they would have to stem the exodus to the coasts of the best of the young professors and their talented students. These were impressions rather than hard facts, but they suggested that the future of these graduate programs depended on securing a "strong and large" research facility for the Midwest.

The Atomic Energy Commission was not unsympathetic to the midwesterners' pleas for help. The commission had no intention of favoring researchers on the Atlantic and Pacific coasts, but rather wanted to build strong regional laboratories, including Argonne, in the heart of the Midwest. The trouble was, from the perspective of the midwestern universities, that a dynamic Argonne might prove as much a rival as an ally. Scientists from the Midwest could already conduct research as visiting scholars at Berkeley and Brookhaven. Although Argonne was geographically closer to their campuses, the midwesterners did not want a laboratory where they would be only honored guests. To save themselves and their universities, the midwestern physicists vigorously campaigned for a federally funded research laboratory that they could control after the fashion of Brookhaven.[9]

Zinn Forgoes the Challenge

Zinn had understood the significance of accelerators when the laboratory built its first Van de Graaff generator in 1947, and he explored the possibility of making a strong bid for a large accelerator at Argonne. Both Fisk and Robert Bacher, the scientist member of the AEC, advised him to concentrate for the moment on reactors. Enrico Fermi, his old mentor at Columbia and now the director of the University of Chicago's Institute for Nuclear Studies, cautioned Zinn against duplicating the institute's efforts. Fermi recommended that Argonne confine its facilities to small machines such as the Van de Graaff and a 60-inch cyclotron. As a consolation, he assured Zinn that Argonne scientists could have access to the institute's larger facilities as needed.[10]

By accepting Fermi's advice, Zinn made a decision that would have fateful consequences for the laboratory. Fermi's institute, although a distinguished center for graduate education and research, neither developed into a national axis for high-energy research nor became a major regional facility attracting scientists to Argonne's research program. Administered independently by the

University of Chicago and separated by geography and culture, the university institute and the government laboratory never developed cooperative programs to serve either the midwestern academic or industrial communities. In effect, Zinn had compromised his ability to establish Argonne as a center for academic research comparable to that established at Brookhaven.

Given the commission's focus on weapons and Argonne's designated mission to develop both military and civilian reactors, Zinn probably thought he had no choice but to accept Fermi's advice. Certainly, neither the commission nor the navy would have tolerated academic participation in laboratory management on the Brookhaven model. Still, Zinn knew that eventually Argonne would need a high-energy accelerator if the laboratory was to continue to serve as a regional center for basic research.

A Regional Accelerator: First Overtures

By November 1952 Zinn was convinced that the time had come to push for a large accelerator at Argonne. After talking with Fermi and others at the university institute, Zinn was convinced that such a facility should be built at Argonne so that the laboratory would become the research center for high-energy physics in the Midwest. Such a project, however, would require not only commission approval but also close cooperation with the midwestern universities represented on Argonne's council of participating institutions. To assure good relationships with the council, Zinn had brought Joseph Boyce to Argonne to serve as associate director and secretary of the council's executive board.

Zinn's new resolve had come none too soon. In January 1953 P. Gerald Kruger, a professor of physics at the University of Illinois and a member of the council's executive board, called Boyce to inquire if Argonne had given any thought to acquiring a cosmotron in the Midwest. Zinn's response was positive. On the advice of Fermi and others at the Institute for Nuclear Studies, Zinn called a meeting on January 30 with key physicists from the council. The group promptly agreed to form the Midwest cosmotron committee to promote construction of an ultra-high-energy accelerator in the Midwest. Offering to coordinate the initiative through the laboratory council, Zinn asked Boyce to survey all council members and the chairs of their physics departments to determine their interest in the idea. Eventually seventeen institutions expressed interest, but none made definite commitments.[11]

If the response from the universities was encouraging, even more so was the reaction from the Atomic Energy Commission. Thomas H. Johnson, the new director of the commission's division of research and himself a physicist from the Midwest, was enthusiastic. Not only was the commission committed to "fostering progress in high energy physics," Johnson reported, but Washington also recognized the need to support research facilities whose energy ranges overlapped "to accommodate adequate levels of research." Johnson was excited to learn from the Midwest "that some of our best scientists are ready and eager to press forward into the new field once the facilities are provided." Johnson was perplexed, however, by questions about funding. How often should big, expensive accelerators be constructed? To what extent should their energy ranges overlap? How much priority should high-energy physics have over other research areas? And there was still the ultimate and most difficult question: Was it appropriate for the commission to build a large and expensive facility such as the super cosmotron on a university campus, or should such investments be restricted to the national laboratories? Any effort by the commission to establish a high-energy physics research facility in the Midwest would have to resolve this issue.[12]

Struggling with the Ultimate Question

Within a week, Zinn, Boyce, and Johnson had a sharp reaction to the ultimate question from a prominent university physicist. F. Wheeler Loomis flatly opposed placing the proposed accelerator under Argonne's control. Stating that the lab's management was "wholly inappropriate" for an enterprise in which "pure research has the highest, if not sole, priority," Loomis believed that Argonne's security requirements precluded effective cooperation with the midwestern universities. Instead, he advocated a management scheme based on the Brookhaven model in which a small consortium of universities (including Illinois, of course) would hold the managing contract for the facility. According to Loomis, the accelerator laboratory should have separate grounds, facilities, budget, services, and, most importantly, its own director, who would be independent from Argonne.[13]

Loomis's emphatic statement clearly reflected the difficulty that Zinn had experienced in building a comfortable relationship with the regional universities. Despite the commission's June 1950 definition of the laboratory's role, which encouraged the laboratory to develop partnerships within the council of participating institutions, Argonne's outreach to the academic community had been halting, at best. In part, Argonne lacked space, facilities, and housing to host academic scientists; in part, midwestern scientists often lacked the expertise to take advantage of Argonne's research opportunities. Consequently, on the eve of Eisenhower's Atoms for Peace initiative, Argonne had failed to develop a strong constituency among academic scientists and so lived with a legacy of distrust instead.[14]

At first, Zinn and Boyce tried to minimize the differences between the laboratory and the universities by ignoring the control issue while concentrating on objections that could be more easily answered. One of these was the obstacle posed by security restrictions. The scientists did not openly link McCarthyism with what they considered the government's obsession with secrecy, but they complained that, in contrast to Brookhaven, the commission's strict security rules for Argonne and bureaucratic red tape would make it hard to attract the most creative university scientists to an Argonne-based laboratory.

Security was an issue that the commission could quickly lay to rest. Assured that there would be no classified research conducted on the cosmotron, the commission offered to build the entire facility outside the security fence so that uncleared scientists, and even foreign nationals, could have free, unimpeded access to the facility. Boyce hoped that technical talks could begin before the political issues were settled. Reminding Loomis about new initiatives from the Ivy League, Boyce thought it would be a shame if another machine were built in the East while none was placed in the Midwest.[15]

The issues of location and control were not so easily resolved. Boyce moved into this sensitive area when he met with the council's executive board on March 2, 1953. He suggested that opposition to building the accelerator at Argonne might be met by establishing a cosmotron board as part of the participating-institution program. The board, composed of physicists from the midwestern universities, would have the deciding voice in selecting the director of the accelerator project, which would be located in a separate division of the laboratory. Boyce believed that the arrangement would save the expense of establishing another laboratory and strengthen the participating-institution program. He observed that the arrangement could "guarantee to university scientists the freedom from red tape and security restrictions necessary for the success of the project."

When the full council of participating institutions met the following day, Boyce's suggestion found little support. Frederick Seitz of Illinois reiterated Loomis's declaration "that the cosmotron be operated by a separate inter-university corporation, independent of Argonne National Laboratory." Loomis and Seitz saw the issue as neither scientific nor technical, but rather political: Who would control the policies, priorities, and management of the midwestern accelerator facility—the government or the universities? The two Illinois physicists had drawn a line in the sand from which they would never retreat. Government management of university research was incompatible with collegial traditions of unfettered scholarly research and academic freedom. In the end, the council endorsed the idea of building the accelerator and agreed to sponsor preliminary studies but offered no opinion about the management issue.[16] The ultimate question remained unanswered.

Disintegration Begins

The stalemate on policy issues did not inhibit most university scientists. In March 1953 some of them formed an ad hoc technical working group that met through the spring, some times including Argonne participation, other times not. On April 17 and 18, 1953, more than twenty scientists, including representatives from Argonne, met with high-energy physicists from Brookhaven and Cornell at the Institute for Nuclear Studies for a seminar on accelerator theory and design. The midwesterners agreed to adopt Brookhaven's design for a larger accelerator that featured strong focusing. During the spring meeting of the American Physical Society in Washington, D.C., members of the working group met with AEC commissioner Henry Smyth, Johnson, and Hafstad to request funding for a one-year design study to be made independently from Argonne.

Johnson fueled the hopes of the university professors, not only by acceding to a separate design study, but also by agreeing that selection of the site should be left open until the design studies were more fully advanced. After assuring the working group that there would be no difficulty in funding the design proposal, Johnson could not have been more encouraging to those midwestern scientists determined to set a separate course from Argonne laboratory.[17]

Unfortunately, Johnson may have raised false hopes that only made the university scientists more determined than ever to establish their own accelerator laboratory. In May, the scientists' dreams collided with fiscal reality when Eisenhower's first budget, which cut $4 million from high-energy physics, forced the commission to scale back its projections for building new machines. On May 18, Johnson informed Boyce that the Midwest cosmotron committee would have to combine its funding requests with Brookhaven's. Simultaneously, Brookhaven received the same message. Ironically, the unexpected budget squeeze worked to Argonne's advantage.

On May 20, Lloyd V. Berkner, president of Associated Universities, which ran Brookhaven National Laboratory, proposed to Boyce that the two laboratories jointly design two machines; the first "to be built in the Midwest at the earliest possible date, and the other and larger machine to be built later at Brookhaven." According to Berkner's projection, construction of the midwestern accelerator would begin in 1955 with completion by 1957, while the

Brookhaven machine would follow with a 1960 target date for completion. During the years 1957–60, the Midwest no doubt would produce the highest energy protons in North America. Boyce welcomed Berkner's proposal, in part because it would foster cooperation between the two national laboratories, but mostly because it gave Argonne time to devise a plan for acquiring the machine and an opportunity to correct real deficiencies identified by the laboratory's critics.[18]

Kruger, however, was highly suspicious of a cooperative approach with Brookhaven, which he suspected wanted to keep the midwestern scientists in a secondary position. He was also anxious to neutralize any advantage Argonne might achieve working within the national laboratory system. In response to Boyce's urging that the Midwest cosmotron committee work closely with Brookhaven, Kruger countered that he wanted the Midwest group to work independently as long as possible. Already, the University of Wisconsin had offered to host the design group as well as the accelerator. Kruger endorsed Wisconsin's offer because the Madison campus provided a beautiful site, wonderful summer weather, and housing not available at either Argonne or the University of Chicago.

Frustrated at Kruger's obstinacy, Boyce demanded that Argonne's critics substantiate their charges against the laboratory. Although the machine might operate effectively in Wisconsin, Boyce argued that placing it in Madison would suggest that Argonne had failed in one of its primary missions. Boyce flatly denied that the reactor program had interfered in any way with Argonne's outreach efforts. He wanted specifics on how red tape had impeded university researchers at the laboratory. Not surprisingly, no one offered a long list. Kruger remembered that in 1949 he had waited over four months to obtain a half of a gram of uranium from the laboratory. Someone else recalled a delay in transferring three experimental mice to the custody of visiting scientists. Others complained of being psychologically inhibited by laboratory security that restricted movement and required everyone to wear badges. If the complaints seemed petty, they were also largely beside the point. The key issue remained that a majority of the academic scientists on the cosmotron working group wanted their own facility, a possibility the commission had not as yet vetoed. University of Chicago vice president Harrell replied that Argonne did not want a fight with the midwestern scientists. The University of Chicago was willing to negotiate a separate contract with the cosmotron committee if that would solve the problem.[19]

By the summer of 1953 the midwestern scientists had effectively abandoned Argonne's council of participating institutions by forming the Midwest cosmotron policy group. Ignoring Johnson's desire for collaborative planning, the physicists organized a study group under the leadership of Donald Kerst, which met during the summer either at Brookhaven or Madison.* Johnson was now growing reluctant to support these independent efforts and informed the group that continued funding was contingent on cooperation with Argonne. Responding to Johnson's threat, the policy group, under Kruger's leadership, sought support from the Ford Foundation to fund design studies through the following winter.[20]

Knowing that the Midwest group was having trouble raising money, Willard Libby, a member of both the commission's General Advisory Committee and the council's executive board, urged Zinn to take control of the midwestern initiative. Libby, who would soon assume his new position as a member of the Atomic Energy Commission, saw "a golden opportunity" for Argonne and told Zinn to "grasp this firmly" by appointing John J. Livingood as head of a laboratory task force on high-energy physics. Believing that Zinn had been too passive, Libby wanted him to "stand firm" by dedicating both the staff and facilities to an accelerator project that would consolidate the effort around Argonne.[21]

At the fall meeting of the executive board of the council of participating institutions, Libby continued his campaign to push Zinn out front on the accelerator issue. With Kruger present, the council executive board unanimously resolved that Argonne should "take a vigorous positive position" in support of establishing a high-energy accelerator in the Midwest. The board also encouraged Zinn to seek close cooperation with the Midwest cosmotron group by offering assistance from Argonne's physicists and engineers and by sponsoring "an acceptable" administrative structure to design, construct, and operate the machine. The council said nothing about building the machine at Argonne.[22]

Before acting, Zinn convened his own advisory committee of distinguished scientists and engineers who urged him to work with the Midwest cosmotron group to secure "construction of a cosmotron at the laboratory with AEC funds." Even Kruger appeared to accept the proposition, although if he did, he probably misunderstood its intent. Armed with endorsements from Libby and Argonne's top advisory panel, Zinn asked Johnson for permission to spend $225,000 annually to support an accelerator group led by Livingood and Morton Hamermesh at Argonne.[23]

With the midwestern scientists talking about forming their own corporation and Zinn proposing to establish an accelerator group at Argonne, the

---

*The 1953 summer study group included physicists from the Universities of Chicago, Illinois, Indiana, Iowa, Iowa State, Michigan, Minnesota, and Wisconsin.

Atomic Energy Commission was befuddled. On February 3, 1954, Johnson called Zinn to indicate that the commission was favorable to his request for $225,000 annually on three conditions: (1) that Argonne would be willing to get in the construction line behind Brookhaven's new accelerator; (2) that civilian nuclear reactor development would remain Zinn's personal top priority; and (3) that Argonne would not object if the National Science Foundation granted $25,000 to the Midwest cosmotron group for continued studies. A week later, however, after he had conferred with Commissioner Smyth and others in Washington, Johnson added that the commission would not commit itself to authorizing any funds for accelerator construction in the Midwest, but were it to do so, the machine would be built at Argonne. Johnson also authorized Zinn to establish his accelerator group, but without additional money from the AEC.[24]

Angered, Zinn chose to interpret Johnson's waffling as a refusal of his proposal. Zinn believed that Johnson's ambiguity had undercut the participating-institutions program at a critical time in his attempt to woo midwestern scientists away from Kruger's plan to establish an independent research association in the Midwest. Hurriedly, Johnson asked Zinn not to do anything before he had an opportunity to discuss the matter further with Kenneth Nichols.[25]

Nichols, a champion of the laboratory since the Manhattan Project, wanted Argonne to prosper as a regional laboratory by providing the nation a unique center for nuclear research in the Midwest. Although reactor development was likely to remain Argonne's first priority for the foreseeable future, Nichols did not rule out the possibility that high-energy physics (or some other science, for that matter) might some day become the laboratory's principal program. He acknowledged that the original goal of creating an outstanding regional laboratory had fallen short because of the commission's emphasis on the reactor program, but he did not believe the plan was unreasonable or should be abandoned. As Libby observed, an accelerator project at Argonne would provide a new opportunity to strengthen its mission as a regional laboratory. "Any other course," Johnson advised the general manager, "would further alienate the midwest scientists from Argonne," making it more difficult to return to Nichols's original plan.[26]

After hearing from Nichols and Johnson, the commission had no trouble justifying construction of another particle accelerator at Argonne. Johnson reported that Brookhaven's cosmotron was fully engaged with experiments involving seventy-five scientists, more than half of whom were visitors from the participating northeastern universities. Brookhaven's bevatron would relieve crowding on the cosmotron because the two machines would operate in different energy ranges. Other proposals at Stanford and MIT made Johnson even more uncertain about construction priorities. But this very uncertainty vindicated supporting Zinn's group for two or three years "to evaluate the various possibilities and to take full advantage of the developments at the other laboratories."[27] Zinn now at least had a study project, but the larger question of where the Midwest accelerator would be built and who would control it was still unsettled.

The MURA Proposal

Zinn's failure to get a clear-cut decision from the commission raised the hopes of Kruger and the independent scientists that they could win support for a midwestern accelerator completely separate from Argonne. By the spring of 1954 more scientists had joined Kruger's group, which was beginning to look something like a formal organization, now known as the Midwest Universities Research Association (MURA). On June 10 Kruger and a MURA delegation met with Nichols and Commissioner Smyth in Washington to request more than $250,000 to begin planning and design for an accelerator to be built at Madison, Wisconsin, where MURA intended to establish its headquarters. Nichols was cordial but made it clear that MURA had no choice but to work with Zinn and the Argonne staff. A few weeks later the commission reinforced this decision by diverting $200,000 from the physics research budget to Argonne to begin design studies on a multi-billion volt accelerator.[28]

The Search for a Compromise

Kruger now had no choice but to attempt to come to terms with Zinn. Everyone with a stake in the outcome attended the summit meeting at the Institute of Nuclear Studies on July 16: the entire MURA board; Zinn, Hilberry, and Boyce from Argonne; and representatives of the AEC's Chicago operations office, the University of Chicago, and the executive board of the council of participating institutions. Zinn echoed the commission's decision that any accelerator in the Midwest would have to be built at Argonne, but he promised to consult closely with both the participating institutions and MURA while selecting a director for Argonne's new accelerator division. He would assign the new division an unclassified building outside the security fence and promised support from the computer division and machine shops as necessary. Most importantly, Zinn proposed establishing a steering committee to represent the thirty-two participating institutions, as well as MURA, on the accelerator project. More representative than MURA, the steering committee would play a leading role in advising the laboratory director on program budgets and scientific policy. Zinn envisioned delegating his prerogative of selecting the division director to the steering committee, on which MURA would hold a majority of seats. Beyond this, he would not abdicate his control of the laboratory.

Zinn's compromise was completely unacceptable to Kruger and the other members of the MURA board, who dismissed the steering committee as an advisory committee without real authority. After all, Kruger argued, the MURA scientists represented most of the high-energy physics talent in the Midwest, and they would not participate unless the MURA board had direct control of the "planning, financing, and choice of site for the project." MURA was unwilling to cooperate with the participating institutions unless they individually joined the MURA effort. If the accelerator had to be built at Argonne, Kruger thought that Argonne and the University of Chicago should become members of MURA. In the end, the two sides could agree only to polish their competing proposals, and for Zinn and Kruger to continue discussions.[29]

Zinn found the outcome so frustrating that he offered Harrell his resignation to become effective when the University of Chicago believed Zinn had become an impediment to securing the large accelerator for Argonne. Worse than losing the accelerator, Zinn thought that creating a separate regional laboratory for basic research would squander tax money with obvious consequences for all midwestern scientists.[30]

MURA Resumes the Fight

By the autumn of 1954 it became clear that MURA had just begun to fight. In September the eight MURA institutions (now including Purdue) formally incorporated and promptly rejected Zinn's proposal because the authority and responsibility accorded MURA on the steering committee fell far short of what they could accept. Even though Lewis Strauss, the AEC chairman, firmly supported Argonne, the MURA leaders were determined to take the issue directly to Johnson and the division of research in Washington, D.C.

In describing the "Desirable Characteristics of Such a Laboratory" in May 1955, the MURA group stressed its educational as well as its scientific mission. The best talent could be attracted to the laboratory only if academics were in charge. The laboratory should be committed to basic research in a setting that guaranteed free exchange of ideas and fostered esprit de corps among colleagues and students.[31]

Along with its conception of the new laboratory, the MURA group sent Johnson a technical proposal to build a fixed-field alternating-gradient (FFAG) accelerator in the 10–30 GeV range. Both Livingood at Argonne and Kerst of Illinois, who drafted the MURA proposal, were attracted to the FFAG idea, which the Japanese physicist Okhawa had championed in October 1953 at an international conference in Tokyo. The FFAG would not achieve the high energies of the alternate gradient synchrotron, but it promised high proton intensity with a comparatively simple and reliable design. The MURA five-year price tag to construct the 20 GeV accelerator was $17 million, which included the machine, its building, and staff salaries.[32]

When Johnson abruptly rejected the proposal, the MURA group appealed directly to the commissioners and escalated their own level of representation by bringing in the presidents of the eight MURA universities. In June the presidents tried unsuccessfully to arrange a private meeting with Strauss, who replied that he would see them only if they accepted Argonne as the site for the accelerator.[33]

Before the MURA presidents could gain a formal hearing in Washington, Zinn submitted Argonne's proposal to build a 25 GeV proton accelerator incorporating the FFAG principle. The Argonne team, led by Livingood, tried to compromise the advantage of high-energy with the benefits of beam intensity. At a joint technical meeting of Argonne and MURA scientists, John H. Williams, from the University of Minnesota, encouraged Livingood to hire Lee C. Teng, a bright physicist from Kansas who showed great promise in high-energy research. Teng arrived at Argonne just in time to suggest building a multiunit circular accelerator based on injector principles developed by himself and James Tuck and successfully tested on the synchrocyclotron in Liverpool, England, by Albert V. Crewe. Argonne proposed constructing a two-stage, or tandem, accelerator, in which the smaller 2 GeV accelerator would serve as a particle beam injector into the larger 25 GeV ring accelerator. This approach, using the Teng-Tuck principle, promised exciting results, quickly. The small 2 GeV injector, with an intensity dramatically greater than that of Brookhaven's cosmotron, offered midwestern scientists "unprecedented potency" within six years. Upon completion, planned for about three years later, the 25 GeV accelerator should produce proton beams 100 times more intense than those of the powerful machines under construction at Brookhaven and CERN, the European research center in Switzerland. Zinn estimated that the cost of forging into the lead in this area of high-energy physics would be $32 million, which did not include operating costs.[34]

The Libby Influence

By the time Zinn could get to Washington with his proposal, Strauss, along with many of the AEC's staff and laboratory scientists, had gone off to Geneva,

Switzerland, for the United Nations Conference on Peaceful Uses. Left in Washington as acting chairman was Libby, who just the year before had replaced Smyth as the scientist member of the commission. Libby's appointment provided midwestern science with a powerful voice on the Atomic Energy Commission just as the MURA controversy was coming to a head. No one else at Washington headquarters knew all the personalities better than Libby. No longer a partisan for the University of Chicago or Argonne laboratory, Libby was supremely confident of his ability to broker an agreement among his fractious midwestern colleagues. Just as he had previously coached Zinn on how "to seize the prize," now Libby was in a position to broker his own solution to the problem. Obviously, when he hosted the MURA presidents in Washington, Libby enjoyed Strauss's full confidence as acting chairman.[35]

In Washington, Zinn presented his accelerator governance plan to Libby and his fellow commissioners. They discussed in detail Zinn's proposal to create a separate division for accelerator research, whose director would be recruited and appointed "with the advice and approval" of midwestern scientific leaders. This function, with other advisors providing assistance to the program, would be performed by a nine-member technical steering committee, whose members were recognized for their competence in high-energy physics and related fields. Zinn would go no further in sharing authority for the accelerator program with university scientists.[36]

A few days later the MURA presidents met with Libby and the commission. Although Libby reiterated that the AEC would support only one national laboratory in the Midwest, he appeared willing to entertain ideas for a cooperative structure between Argonne and MURA. Lawrence Kimpton, president of the University of Chicago, assured the assembly that the accelerator steering committee, which Zinn planned as a strong advisory council, would give MURA an effective voice in all aspects of the project. Kimpton believed that all the benefits of Brookhaven management could be achieved for Argonne through the steering committee model, which could be extended to other programs of interest to the MURA scientists. It quickly became apparent, however, that the MURA presidents would accept nothing less than exclusive management of the project. Without asserting any of the conditions upon which the commission had consistently insisted, Libby and his colleagues simply agreed that the presidents should form an ad hoc committee to find a solution to the impasse between MURA and Argonne.[37]

Zinn knew about Kimpton's meeting with the commission before he left for Geneva, but when he returned, he was stunned by Libby's mediation and Kimpton's apparent acquiescence. "Control was taken out of our hands at this point," Zinn told George Norris of the Joint Committee on Atomic Energy. Libby was now convinced that most midwestern physicists would not participate in an accelerator program under Argonne's control. In search of a compromise, Kimpton had agreed to serve on the president's ad hoc committee. Although Kimpton was fully aware of Zinn's unwillingness to abdicate authority to MURA, his temperament and commitment to collegiality compelled him to seek consensus with his fellow university presidents. Only one "compromise" was possible. Working with J. C. Warner of the Carnegie Institute of Technology and Frederick L. Hovde of Purdue, Kimpton agreed to a plan he hoped would be agreeable to Zinn: the midwestern accelerator would be located on the Argonne campus but governed by a second contract independent from Zinn and the University of Chicago. Unfortunately, Kimpton explained to the University of Chicago trustees, he was unable to discuss the settlement with Zinn or other key Argonne scientists, who were attending the Geneva Conference.[38]

Back in his office at Argonne, Zinn promptly tendered his unconditional resignation, throwing the Atomic Energy Commission and the University of Chicago into an uproar. Irritated that Libby had allowed the MURA controversy to spin out of control, Strauss telephoned Zinn not to make his resignation final until they could talk, so Zinn agreed to suspend his resignation until Strauss had a chance to act. Strauss's frustration was evident. After America's triumph in Geneva, where the United States showed off its civilian reactor technology, the commission risked losing the director of its major reactor laboratory over a matter that was important, but secondary, to the AEC's principal mission.

Bypassing Libby for the moment, Strauss asked the AEC's General Advisory Committee for help. Under the chairmanship of Isidor Rabi, the committee met with the full commission on October 31, 1955, to chart a new path for Argonne's basic science program. The members recognized that the commission's first priority was to protect Argonne's reactor programs and to retain Zinn as the head of the laboratory. Beyond that, Rabi saw no reason to modify the committee's earlier recommendation that a large accelerator be constructed at Argonne. He was not highly impressed with the technical merits of MURA's proposal and agreed with Johnson that MURA had underestimated both the time and money required to build its machine. Finally, regardless of MURA's threatened boycott of an Argonne accelerator, neither Rabi nor Fisk thought the laboratory would have trouble attracting regional scientists to design and operate a large accelerator facility.[39]

The Cold War Factor

The General Advisory Committee, however, did offer a bold solution to the impasse. Why not authorize construction of two major accelerators for the Midwest? While in Geneva, American scientists had discovered surprising Russian scientific strength in the field of high-energy physics. The Russians announced that they would soon begin operation of a 10 GeV synchrotron at their high-energy physics research center at Dubna, near Moscow. If the Russian claims were true, the Soviet Union would hold an uncontested lead in high-energy research over Berkeley's 6 GeV accelerator until Brookhaven's 30 GeV alternating gradient synchrotron (AGS) came on line in 1961 or 1962. CERN was also building a large machine that would beat the Russians, but the European accelerator would not be completed before 1960. The committee advocated building a 12 GeV version of Berkeley's old bevatron at Argonne as a stopgap measure to challenge Russia's temporary supremacy in accelerator physics. Even if the smaller machine jeopardized Argonne's chance of obtaining a major accelerator for the Midwest, Rabi recommended a crash program to build a relatively conservative machine whose main task would be to preserve United States leadership over the Russians in the field. Johnson estimated that Rabi's idea to assemble an unscaled bevatron at Argonne would cost only $12–15 million. Libby especially welcomed the Russian challenge as a way of rescuing the midwestern accelerator project from the mud of local science politics.[40] How ironic that the debate now would be settled by international Cold War politics!

Acting on the General Advisory Committee's advice, the Atomic Energy Commission dropped its insistence that an advanced accelerator be constructed at Argonne. The first priority would be to nurture Argonne's reactor program and, if possible, convince Zinn to remain as the laboratory's director. But Commissioners Libby and John von Neumann, supported by Strauss, were also determined to deny the Russians undisputed leadership in the high-energy field. Convinced that the MURA scientists would never cooperate with Argonne, the commissioners on November 8, 1955, informed the surprised but delighted MURA presidents that the commission would ask Argonne to build a relatively straightforward 12 GeV accelerator while MURA designed a more ambitious machine. Libby, who wanted the 12 GeV machine constructed in thirty-nine months, allowed that some innovations might be possible, but the goal was to get the Argonne accelerator running as soon as possible. The MURA project to design and construct a "dream machine" could then proceed at more leisurely pace, with completion scheduled for 1962.[41]

For years Argonne scientists remembered the AEC's decision to build a conventional accelerator at Argonne as a quintessential political decision, made not for scientific reasons but to maintain American "leadership" in high-energy physics in the face of the Russian challenge from Dubna. Science, rather than politics, they believed would have dictated construction of the advanced design, which promised real discovery rather than stopgap research.

The MURA Victory

The MURA presidents, immediately realizing they had won everything they had asked for, were generous in their encouragement of Argonne's Cold War project. Even Kimpton endorsed the idea, believing it would improve relations between the midwestern universities and the commission. The presidents agreed they did not need to identify a site for the accelerator until the design work was completed, but they accepted Libby's assurance that the commission would designate no site without MURA's approval. Strauss also cautioned them to be guarded about their discussions with the commission staff until he and other commissioners had secured concurrence for the plan from the Joint Committee on Atomic Energy. Discreetly, but triumphantly, on November 13, the midwestern university presidents met with the MURA board at the University of Chicago to ratify their agreement with the commission. Libby, Johnson, and Alfonso Tammaro, manager of the Chicago operations office, explained the new priorities. MURA's victory seemed so complete that by the end of the day, Kimpton and Anderson applied for membership on behalf of the University of Chicago.[42]

Both Strauss and Rabi urged a weary Zinn to accept the agreement, but Zinn saw no future in becoming a rabbit for the American high-energy physics chase. Although the commission was anxious that the MURA controversy not distract Argonne from its principal mission to develop civilian reactors, the conflict actually reinforced Zinn's belief that Argonne had no long-range future without a strong basic science program. When Libby and von Neumann tried to console Zinn on the loss of the big machine by reminding him that the participating-institution programs should have second priority to the reactor development projects, Zinn shot back that basic and applied research were never in conflict at Argonne. Ironically, "Mr. Reactor" acknowledged that Argonne's eminence rested disproportionately on its reputation in the reactor field. "The long term well being of the laboratory," Zinn predicted, depended on Argonne's competitiveness in "modern high energy physical research."[43]

Zinn's Last Proposal

On November 15, 1955, Zinn appealed personally to the AEC that Argonne be permitted to proceed with its original plans to build a large tandem accelerator. He was willing to compete with the Russians in designing power reactors, but he believed that the accelerator contest was pointless. Furthermore, he did not want to abandon his own scientists. They wanted to conduct their own investigations, he reminded the commission, and resisted the idea of replicating another's research, especially when they had better ideas of their own. He simply did not want to waste everyone's time building an accelerator that would "only be a clunker." But Libby, now obsessed with the Russian presence, warned that Congress probably would not support any accelerator for the Midwest unless the commission requested "special support" for an urgent project at Argonne.[44]

Livingood and Louis Turner, director of Argonne's physics division, had already informed Zinn that a crash program to build a 12 GeV bevatron was a daydream. Livingood thought that a 12.5 GeV constant-gradient synchrotron might be designed and built for $30 million by 1961 or 1962, but not in time to counter any rivalry from the Russians. One could not simply scale up Berkeley's accelerator in the same manner that one could extrapolate from an experimental nuclear reactor. Regardless of what accelerator Libby wanted for Argonne or elsewhere, the next American machine to compete with the Russians would be Brookhaven's AGS, already under construction. Livingood and Turner resented the fact that politics rather than science had determined the fate of Argonne's proposed tandem accelerator, but they were especially dismayed at the success of MURA scientists in convincing the commission that Argonne was not a "fit place for real scientific work." MURA's apparent victory, Turner feared, had officially relegated Argonne to "the Commission's second string" for physics research in the Midwest. It was galling that a University of Chicago chemist had played such a major role in their defeat.[45]

Argonne had not lost all its friends, however. Most of Libby's old colleagues on the General Advisory Committee did not share Livingood's and Turner's bitterness and actually believed that Libby had Argonne's best interests at heart. Rabi had in fact first suggested the 12 GeV project to Libby. Other members appreciated Argonne's dedication to science rather than to politics, but given Eisenhower's demand for a balanced budget, they also saw the need to obtain construction money from Congress soon.[46]

Determined to make his resignation final by spring 1956, Zinn nevertheless tried to salvage as much as possible for the laboratory's high-energy physics program. On January 24, 1956, Zinn informed Livingood that the commission had asked Argonne's management to submit cost estimates for the new accelerator by February 1. Given twenty-four hours to prepare the budget, Livingood immediately called an emergency meeting of his team to lay out the "ground rules" for Argonne's machine: the machine would be built on a "crash" basis using only well-established principles and technology to achieve energies greater than the 10 GeV accelerator under construction at Dubna. Teng recalled that Livingood's small but enthusiastic group sat stunned when informed that they had been drafted into this "new national and international accelerator ball game." "Do you want to build such a machine or do you want to quit?" Livingood challenged them. Silence—until Martyn Foss (known as Magnet Man to his colleagues) observed that despite wounded pride, the Argonne team could still build one of the most powerful instruments in the world using innovative magnet technology. It was, Teng reflected, "the only sane and positive" encouragement that anyone had offered them in a long time. Livingood's group submitted their budget proposal in "24 hours."[47]

Although Zinn warned the commission that the accelerator facility could not become operational before 1961, Argonne accepted the directive to build an accelerator in the 10 to 15 GeV energy range to compete with the Russian machine at Dubna. Concurrently, the laboratory planned to continue its design studies on an advanced tandem accelerator. Knowing that the commission was unlikely to order Argonne to abandon its theoretical studies, Zinn cared more about keeping up with the MURA group than he did about beating the Russians. Libby, who remained committed to strengthening Argonne's basic research programs once the MURA controversy was settled, saw through Zinn's stratagem by responding that authorization of design studies did not imply that the commission had endorsed the tandem accelerator concept.[48]

The March 1956 issue of *Argonne News* announced both Zinn's resignation and the AEC's authorization of the new accelerator. On April 2, Hilberry named Livingood director of the particle accelerator division "with orders to proceed with all deliberate speed" to develop a "well-considered" plan to outclass the Russian accelerator at Dubna as cheaply as possible.[49]

Zinn Resigns

After ten years as director, Walter Zinn was tired and fed up with the frustrations of running a large and unwieldy federal laboratory. In fact, he had carried such heavy administrative responsibilities since 1942 that "the burden felt more like fourteen years." Zinn complained that there was "complete lack of policy" for Argonne at the top of the government's management echelons. In other words, he felt there were too many bosses—from the president of the University of Chicago and the manager of the Chicago operations office to the commission's division directors, the five commissioners themselves, and the Joint Committee on Atomic Energy. There was no doubt that Libby's concessions to MURA and Kimpton's failure to back Zinn up precipitated his resignation, but it was a step he took with utmost care. In constant touch with Strauss since they returned from the Geneva Conference on Peaceful Uses, Zinn had pledged not to quit without consulting the chairman. At first, he asked for a leave of absence so that he could regain perspective on his job. But when he would not pledge to return within a year, the university and the commission reluctantly accepted Zinn's resignation. Kimpton assured Zinn that he was welcome to return to Chicago as a tenured professor of physics or could continue indefinitely his leave of absence from the university. Rather than seeking refuge in academia, Zinn choose to form his own consulting company.[50]

Everyone believed that Zinn left Argonne because of the MURA controversy. But it was not that simple. Although the loss of the pacesetting accelerator *was* discouraging, other matters disheartened him as well. The commission's failure to fully support the International School of Nuclear Science and Engineering was profoundly disappointing for Zinn. He enthusiastically supported Eisenhower's Atoms for Peace initiative but believed the commission had misused the laboratory by originally setting up the school at Argonne and then transferring much of the teaching program to Pennsylvania State and North Carolina State Universities. When Zinn asked Hilberry to organize the school, he had hopes of establishing Argonne as the leading international center in nuclear education. They instituted the program, defined the curriculum, recruited the teachers, and launched the classes only to learn that the commission decided to turn formal classroom teaching over to the universities, which then called on Argonne for help.

Zinn also heard the alluring call of private enterprise. He was not only exhausted, but he was also tired of training technicians who left for industry to earn much more than his best scientists. Zinn complained he could not hold his best people. All too frequently, the inexperienced nuclear industry sent engineers to Argonne to learn from masters in the field, who made less than their students. While Zinn's physicists increasingly spent their time helping visitors, industry raided his best people. Leaving Argonne enabled Zinn to simplify his life while he established his own company.[51]

The University of Chicago's unsuccessful search for a distinguished outsider to direct Argonne National Laboratory confirmed Strauss's worst fears about losing Zinn. There was no one with Zinn's stature and experience in reactor development who was willing to take the job as laboratory director and who was also acceptable to the laboratory's senior staff. Candidates who turned the position down read like a veritable Who's Who of the American nuclear community. Kimpton wanted to appoint Herbert Anderson of the University of Chicago's Institute of Nuclear Studies, but the commission was reluctant to endorse his appointment when the laboratory's program managers voiced their opposition. Anderson, one of Fermi's Ph.D. students at Columbia University, would have brought dazzling credentials to Argonne. One of the pioneers of the atomic age, Anderson had worked alongside Zinn under Fermi's direction on the construction of CP-1 in 1942. After a stint at Los Alamos in 1944, Anderson returned to Chicago and the Institute of Nuclear Studies where he was promoted to full professor in 1950, when he was thirty-six years old. At the institute, Anderson compiled an outstanding record as an experimental physicist directing the synchrocyclotron program, once the highest energy machine of its type. Seemingly an ideal candidate as Zinn's replacement, Anderson was persona non grata with senior members of the Argonne staff because of his role in the MURA controversy.[52]

From Washington's perspective, Zinn's resignation provided a welcome opportunity to reconcile the differences between Argonne and MURA. Libby suggested the appointment of a joint Argonne-MURA director to promote a spirit of cooperation among scientists in the Midwest. Whereas the commission still believed that the large midwestern accelerator should be located at Argonne, Strauss was willing to build the "dream machine" in another region of the country if MURA did not cooperate. Several attempts to find a mutually acceptable candidate collapsed because neither MURA nor Argonne was enthusiastic about the person. By summer's end in 1956, alarming reports filtered back to the commission that Zinn's new company was luring away some of Argonne's best reactor engineers. It now seemed clear that if a high-energy physicist became director of Argonne to mollify MURA, a top nuclear reactor scientist would have to be hired as Argonne's deputy director. The cost of establishing high-energy physics was becoming unacceptably high.[53]

6 | What Sort of Laboratory? 1957–61

In February 1957 the university and the Atomic Energy Commission agreed to promote Norman Hilberry, who had the unanimous support of laboratory program officers, to be the new director of the laboratory. No new director has ever been more popular among the laboratory community. Argonne's inability to attract top outside leadership following the MURA controversy became evident after the search for a new director dragged on for more than a year. Fred Seitz, a distinguished physicist from the University of Illinois who ultimately became president of the National Academy of Sciences, bowed out of consideration when he failed to secure promises of cooperation from MURA physicists. Although Harrell worried that Hilberry was not tough enough to make hard decisions, the staff was delighted that one of their own had been promoted to the director's chair.[1]

Hilberry, who had served as Argonne's deputy director since the laboratory was established in 1946, was among the group of forty-three present at Stagg Field on December 2, 1942, when CP-1 became critical, and he was frequently teased about his minor role that afternoon. As a safety precaution, Fermi's "suicide squad" stood by ready to douse the reactor with a cadmium sulfate solution should something go wrong. In addition to the main control rod, Zinn had devised ZIP, a weighted safety rod designed to trip automatically if neutron intensity became too high. Finally, Zinn fashioned an emergency ZIP, which he operated by hand and tied to the balcony rail. Hilberry stood by with an ax, ready to cut the rope in the event of a mishap.[2] "Enrico Fermi and others grinned at me when it started," he remembered. "I felt silly as hell. This was a lot of nonsense. We all knew the scientific work would be all right. But I suppose it was good window dressing."[3]

Hilberry willingly played his part and earned his colleagues' gratitude and respect. A 1921 graduate of Oberlin College, he pursued studies in physics at the University of Chicago before accepting an appointment to teach physics at New York University in 1925. Hilberry specialized in research on cosmic rays while he helped develop the applied physics program in New York University's engineering department. His pathbreaking cosmic-ray studies earned him a spot on the

University of Chicago's 1941 expedition to the Peruvian Andes with Arthur Compton. When Compton became director of the Metallurgical Laboratory, Hilberry became his assistant. The University of Chicago awarded him a doctorate in physics in December 1941, the same month he began work at the Met Lab. During World War II, Hilberry was involved in the design and construction of the pilot reactor at Oak Ridge and of the production reactors at Hanford. As deputy director of Argonne, he had a role in all aspects of the government's nuclear programs except weapon research. Among insiders, Hilberry, who had collegially managed the laboratory with humor and tact during Zinn's last year as director, proved a welcome affirmation that Argonne would maintain a steady course in basic and applied nuclear research and development.[4]

Sporting Elements: Argonne and the Discovery of Element 102

Scientists love international competition, especially if they win. From their inception, the nuclear sciences have had an Olympic quality about them. Not only have the nuclear scientists explored the fundamental structure of nature—the realm of the gods—but their quest has had all the elements of a modern international sports competition. With national prestige as much at stake as national security, public treasuries generously contributed to building scientific teams whose quest for discovery, and Nobel Prizes, won prestige for the winners, who customarily earned the prerogative to name nature's elements. From the discovery of radium by Marie and Pierre Curie in 1898 to the detection of element 102 in 1957 and beyond, scientists have raced against themselves and time to seize the laurels of first discovery of radioactive and transuranic elements.

Darleane C. Hoffman of Lawrence Berkeley Laboratory described the years 1940–55 as the golden age for the discovery of new elements heavier than uranium.* Exploration for transuranic elements began in 1934 when Fermi's group, starting with the lightest and working toward the heaviest, systematically bombarded elements in the periodic table with neutrons, inducing, along the way, new "artificial" radioactive isotopes in forty of the sixty elements tested. By the time Fermi and his assistants reached the end of the periodic table, they realized that the heaviest elements tended to capture neutrons, becoming unstable isotopes, which through beta decay transmuted to a heavier element. Thus, when they

---

*The first nine transuranic elements are neptunium (93), plutonium (94), americium (95), curium (96), berkelium (97), californium (98), einsteinium (99), fermium (100), mendelevium (101).

bombarded uranium with neutrons, Fermi initially believed they had created the first transuranic element when in fact they had split the uranium atom into two lighter elements. The discovery of nuclear fission would ultimately lead to the controlled-chain-reaction experiment with CP-1.[5]

The discovery of the first two transuranic elements, neptunium and plutonium, occurred in 1940 at Lawrence's laboratory, at the University of California. That spring, using Lawrence's cyclotron, Edwin M. McMillan and Philip H. Abelson followed Fermi's example and bombarded uranium with neutrons. In addition to the fission products that Otto Hahn and Fritz Strassman identified in late 1938, McMillan and Abelson found another radioactive product, which they determined to be element 93 transmuted by beta decay from uranium-239. They named the newly discovered element neptunium, for the next planet in the solar system after Uranus. After McMillan left the university to join the radar project, Seaborg and his associates also used the cyclotron to bombard uranium oxide with deuterons, producing element 94, which they named plutonium after the planet Pluto. Although both neptunium and plutonium are commonly regarded as "manmade" elements, trace amounts of each actually occur naturally.

Seaborg's interest in discovering additional transuranic elements continued while he studied the chemical and physical properties of plutonium at the Metallurgical Laboratory. As his part in the war effort waned in the summer of 1944, Seaborg, along with Ralph A. James and Albert Ghiorso, arranged to bombard plutonium-239 with helium ions at Berkeley's cyclotron. Conducting the subsequent chemical analysis in Chicago, Seaborg's team found element 96 (curium). Later that fall at the Met Lab, while analyzing plutonium that had been irradiated in a nuclear reactor, Seaborg's team discovered element 95 (americium), created when plutonium isotopes captured successive neutrons. Five years passed until Seaborg's explorers, now headed by S. G. Thompson, produced berkelium (97) and californium (98) by cyclotron bombardment in December 1949 and January 1950. Setting a remarkable record for scientific accomplishment, Seaborg and his collaborators had found five of the six transuranic elements discovered in the past decade. Lawrence's Berkeley laboratory had also been involved in five out of six discoveries, while the Met Lab during Seaborg's residency had played a role in two.[6]

While the Californians pushed ahead with their cyclotron to discover new elements and isotopes, Paul Fields, a chemist at the University of Chicago who had worked in Seaborg's section at the Met Lab, launched his own studies of the transplutonium elements at Argonne in collaboration with Martin Studier. Fields—slight, angular, wiry—personified Argonne's research scientist. With experience at the Metallurgical Laboratory, Los Alamos, Mound Laboratory, and Standard Oil of Indiana, Fields returned to Chicago in 1946 to help Winston Manning train a new generation of nuclear chemists for Argonne. At Argonne, he tried to rekindle the excitement and enthusiasm he had experienced working for Seaborg during the war. In 1952, with a unique high-flux neutron source at his disposal, Fields and Studier, with Manning's encouragement, decided to join the search for transuranic elements.

The Fields-Studier initiative to find new elements became possible following construction in Idaho of the materials-testing reactor. The United States' first high-flux reactor, the MTR provided a neutron source much more intense than that in existing reactors. The scheme was elegantly simple. With help from Los Alamos's metallurgy division, the Argonne group, with Studier taking a leading role, devised a plutonium-aluminum alloy cast as a hollow cylinder that looked very much like a common napkin ring. The "napkin rings" were then strung along aluminum rods inserted in the MTR core in the region of peak thermal neutron flux, where the plutonium-239 was subjected to intense bombardment. In this manner, Fields and Studier planned to irradiate materials up to six years, removing a sample every six months to determine production rates of heavy elements over time. Although no isotope beyond curium-245 had ever been formed in a nuclear reactor, the Argonne scientists set their sights on producing an array of nuclides up to californium-252 and beyond. In the broadly collaborative effort, Los Alamos and Berkeley agreed to assist in analyzing the irradiated rings. Fatefully, the Fields-Studier team inaugurated their research in August 1952.[7]

Before the group could recover the first samples from the MTR, in November 1952 the Atomic Energy Commission tested MIKE, its first thermonuclear device, in the Bikini Atoll at the Pacific Proving Grounds. MIKE not only produced all of the nuclides Fields sought at the MTR but also yielded elements 99 and 100. Fields was part of the group in the chemistry division that analyzed debris from atmospheric nuclear explosions, both American and Soviet. From the debris, or fallout, chemists identified fission products and the distribution of isotopes from the heavy elements, which information enabled analysts to determine the configuration and yield of the bomb.

Samples from the MIKE device, collected on filter paper by airplanes flying through the mushroom cloud, were rushed to Los Alamos and Argonne for evaluation. Led by Sherman Freed, stunned chemists discovered plutonium-244, 245, and 246, which had never before been identified. They informed Seaborg's group at Berkeley, which also became interested in the enormous neutron flux from the MIKE shot. Working twenty hours a day, seven days a week, Fields and his colleagues raced against time to identify the isotopes in the MIKE debris before they disappeared by radioactive decay. Using techniques

Seaborg had developed for separating elements from plutonium, they quickly found californium in the fallout. In less than a microsecond, the MIKE shot generated such an enormous neutron flux that the uranium-238 atoms captured up to seventeen successive neutrons to form unstable uranium isotopes with masses as high as 253 and 255, which decayed by beta-particle emission along mass chains to form elements 99 and 100, respectively. After some disagreement about which group first identified element 99 and element 100, scientists at all three laboratories ultimately shared credit for the discovery of einsteinium (99) and fermium (100).[8]

Because all aspects of America's thermonuclear bomb program were heavily cloaked in secrecy, public announcement of the remarkable discovery of these two new elements did not appear in *Physics Review* until 1955. In April 1954, after the Castle-Bravo test confirmed the shocking power of the hydrogen bomb, *Argonne News* announced that Idaho's materials-testing reactor had produced elements 99 and 100 (not yet named einsteinium and fermium). Details of the original discovery of the elements remained secret, however, so that Argonne could only report that the elements "had been discovered earlier as a result of research work, not yet declassified, at Argonne and other AEC Laboratories."[9] Meanwhile, Seaborg's team, now led by Ghiorso, continued its assault on the periodic table in early 1955 by bombarding less than a picogram of einsteinium 253 with helium ions in Berkeley's 150-centimeter cyclotron. Consequently, the Californians literally produced element 101 (mendelevium) in quantities of one to three atoms at a time. Ghiorso and his colleagues based their claim "on the detection of only 17 atoms."[10]

By 1955, American researchers had identified all nine of the predicted transuranic elements. In one role or another, Seaborg had been involved in every search since the isolation of plutonium. While Seaborg remained the preeminent transuranium explorer, among the laboratories—University of California, Berkeley (UC), Argonne National Laboratory (Met Lab and ANL), and Los Alamos National Laboratory (LANL)—the score stood:

Element		Laboratory		
Neptunium 93	UC			
Plutonium 94	UC			
Americium 95		Met Lab		
Curium 96		Met Lab		
Berkelium 97	UC			
Californium 98	UC			
Einsteinium 99	UC		ANL	LANL
Fermium 100	UC		ANL	LANL
Mendelevium 101	UC			
Totals	7		4	2

For his part, Fields felt no sense of competition. "This was Seaborg's field," he later stated. No one was competing to discover these elements: "Seaborg had the whole deal wrapped up." Nevertheless, that summer, Fields and Arnold M. Friedman, who had participated in the identification of einsteinium and fermium, secured Manning's support to pursue element 102 independently. Their plan was to bombard curium with carbon ions, targeting curium because it was the heaviest element they could obtain in sufficient quantities for testing, and accelerating carbon ions because they were the lightest particles that would yield element 102. They required access to a cyclotron capable of generating intense beams of high-energy carbon ions, however.[11]

A year later in the summer of 1956, chemist John Milsted from Britain's Harwell Atomic Energy Research Establishment visited Argonne for six weeks. Fields and Friedman told Milsted about their quest for element 102 and their need for beam time on a powerful cyclotron. They were in luck, Milsted informed them, because he had recently been offered use of an appropriate cyclotron at Sweden's Nobel Institute of Physics in Stockholm. Negotiations quickly led to the first joint international research team to challenge Seaborg and the Californians in the race for element 102. The Americans supplied the scarce isotopes of curium to be used in the experiments, while the British offered the rare carbon-13 isotope. The Swedes provided the cyclotron, fully equipped laboratories, and a staff of scientists and technicians to run the machine.*

After preparations at Argonne and Harwell, the international team assembled in Stockholm in March 1957. Working night and day for twenty-two days, they bombarded the curium with carbon-13 nine times without producing enough radioactivity to confirm chemically the presence of element 102. Finally, the day before Fields returned to the United States, they tried again. Reminiscent of Fermi's energetic research in the 1930s, the research team needed to work quickly with great precision before decay or transmutation wiped out minute evidence of a very unstable element 102. *Time* magazine described their athletic experimentation:

> Element 102 disintegrated so fast that a major problem was to prove that it had been created at all. The scientists developed a technique that would have done credit to a team of Japanese jugglers. After the curium had been bombarded for about 20 minutes, the Swedes shut down the cyclotron. As the concrete shield opened, a group of scientists, wearing gloves and dust masks against radioactivity, dashed into the cyclotron chamber. One snatched the target from the

---

*The Americans were chemists Paul R. Fields and Arnold M. Friedman from Argonne National Laboratory. John Milsted and Alan Beadle, both chemists, represented Britain's Harwell Laboratory. From Sweden's Nobel Institute for Physics came Hugo Atterling and Bjorn Astrom, physicists, and Wilhelm Forsling and Lennart Holm, chemists.

machine, another took it apart and passed it to a third, who extracted the catcher foil. The fastest runner, generally Swedish Chemist Lennart Holm, then dashed 100 yards to a waiting elevator. Upstairs the foil went into an apparatus that measured alpha particles from the disintegrating atoms. With practice, the scientists cut the total elapsed time to 1½ minutes.[12]

Disappointment

Fields returned triumphantly to the United States believing that his international team had identified element 102. They did not enjoy sweet victory long, however. First, the team in Sweden had difficulty in replicating the experiment, but they eventually produced more element 102 and sent Fields the results. However, the Swedish cyclotron was shortly rebuilt for different types of experiments and was not available to reproduce the results after the team dispersed. Then, the Californians, performing a different experiment, questioned the findings of the American-British-Swedish team. The international team may have produced element 102, but the Californians did not believe that its detection had been demonstrated. Seaborg ultimately dismissed the claims because "the reported discovery ... has never been confirmed and must be considered to be erroneous."[13]

News also arrived from the Soviet Union that Georgii N. Flerov's group at the Kurchatov Institute of Atomic Energy in Moscow found radioactivity that they thought might be related to element 102, but no follow-up chemical analysis was done. Seaborg, Ghiorso, and their colleagues quickly bombarded curium with carbon-12 ions using Berkeley's new heavy-ion linear accelerator (HILAC). They reportedly found fermium-250, indicating that they had produced its parent element, 102, with a mass number of 254, which had transmuted by alpha-particle decay into the fermium daughter isotope. The half-life of element 102 was so short, however, that the Californians had to rely on the indirect evidence of the known alpha-emitting fermium daughter isotope to substantiate their claims. Receiving tit for tat, Berkeley's results were later challenged by the Russians, who entered the contest in earnest. From the Joint Institutes of Nuclear Research at Dubna, a group led by Evgeny C. Donets and another led by B. A. Zager, reporting their own descriptions of element 102 isotopes, claimed that the Californians had incorrectly identified two isotopes. Rival claims between the Russians and Americans over bragging rights for first discovery of manmade elements continued from element 103 to element 106.[14]

The dispute over who really discovered element 102 was not resolved until 1992 when the joint transfermium working group of the International Union of Pure and Applied Physics (IUPAP) and the International Union of Pure and Applied Chemistry (IUPAC) finally recognized the Russian claims. Darleane Hoffman from the Lawrence Berkeley Laboratory would not concede defeat, but Fields, more philosophic about the recommendation, acknowledged the Russians' good work while reaffirming the strong claims of his colleagues in Sweden. According to Fields, "Our results were closer to Flerov's results. It was possible we produced 102^{255} which is a three minute half life as compared to their one minute half life, but not three seconds as Seaborg claimed."[15] As often happens in sport, Fields's front-running team finished out of contention for the honor of discovering element 102. But the race for element 102 had taken an odd twist and an unpredictable turn. According to international custom, the discovering scientists won the right to name the new element. Although they failed to prove conclusive discovery, ironically, the American-British-Swedish team nevertheless achieved international recognition and a measure of immortality when the Berkeley and Russian scientists magnanimously accepted their name for element 102—nobelium, after Sweden's great scientist Alfred Nobel.

Hilberry as Manager

The MURA controversy and the race for element 102 emphasized a new reality for scientific research in post–World War II America: like championship sport, big science required well-heeled sponsors who were not only able to hire the best people but were also willing to make long-term investments in facilities and developmental programs. To prosper, big science required both short-term and long-term game plans. The mythology of the Manhattan Project continually reminded American scientists and engineers of their duty to serve the nation with heroic crash programs in times of national emergency. In the short term, even the most glamorous crash program usually relied more on applied engineering development than it did on basic science research. The Metallurgical Laboratory, for example, was simply an agglomeration of scientists driven by the exigencies of the war. Expediency took precedence over planning, Hilberry recalled, and personalities dominated the organizational structure. In a very real sense, the scientists and their accomplishments defined the Manhattan Project. Although Argonne needed to preserve some measure of the wartime "can-do" attitude, that ethic provided no basis for a permanent, stable program.[16]

In addition to their skill in mounting concentrated attacks on complex problems, the national laboratories also had the capability of dedicating personnel and facilities to expensive, long-term, high-risk research and development that was unsuitable for industry or the universities. The function of this publicly funded research bureaucracy, however, was only vaguely understood by the government, industry, academia, and the national laboratories themselves. Although few persons wanted to dismantle the Manhattan Project laboratories after the war, except for weapons development, no one was certain what the national laboratories should do or how they should be managed.

More than his predecessors, Norman Hilberry reflected philosophically on how to manage this unprecedented scientific enterprise. He firmly rejected management models from industry and business, which produced real goods and services in a competitive market. In almost every way, Hilberry believed, the requirements for scientific research were diametrically opposite to those of industrial production and business profit. He also thought that "the research and development laboratory has but one single product. That product is new ideas."[17]

According to Hilberry, ideas are intangible and can be assigned no meaningful dollar values. In science and engineering, the true value of an idea may not become evident until years after it is conceived. Even mistakes, as exemplified in the search for element 102, can stimulate fruitful research and successful discovery. All that can be said for certain about new ideas, Hilberry believed, is that they are necessary for progress, and without them, stagnation is certain. "This, then, is the single objective of a research and development laboratory—to provide the new ideas upon which progress must depend." Argonne's fundamental mission was the antithesis of a production organization. Like a factory, Argonne Laboratory required investment in machinery, but in contrast to the factory, which mass-produced items for commercial sale, Argonne's machines produced information and data, which were the raw materials for intellectual consumption. The real capital of Argonne consisted of the creative brains of the laboratory staff.

Hilberry did not presume to understand the psychological, physiological, or neurological processes by which scientists grasped new ideas. Education, training, and broad experience probably promoted "efficiency," but Hilberry doubted that external factors significantly modified genius. What, then, was the role of *management* in the scientific enterprise? Hilberry's response, in the face of increasing oversight from Washington, was that "the management of a research and development laboratory can only direct by indirection: it must lead, not order, if it is to be successful."[18] To win the respect and trust of creative scientists, Hilberry believed it was essential for top management to rise through the ranks of the laboratory.

In contrast to Zinn and Lawrence, Hilberry did not regard it his role to provide strong programmatic leadership to the laboratory. The director's principal responsibility was to coordinate laboratory planning, which included assurance that the Atomic Energy Commission received fair return for its research investment. Hilberry believed that the director should assist the staff in establishing research priorities, which in turn defined scientific programs and budgets. Thereafter, the director provided a buffer against outside pressures demanding short-term results at the expense of long-term productivity. Recruitment and retention of top scientific talent was the director's first priority. Regardless of how well the director ran the laboratory, the success or failure of Argonne depended uniquely on the ability of the director to attract outstanding talent. In this regard, Hilberry strove to maintain a working atmosphere that fostered free inquiry and was generously supported with well-equipped laboratories, modern research facilities, and an efficient plant. Finally, Hilberry believed that the director should protect researchers as much as possible from nonscientific activities, such as the commission's reporting requirements. With the assistance of John McKinley, Argonne's business manager through its formative years, Hilberry was responsible for implementing three thousand pages of commission directives while laboratory managers and programs prepared twenty monthly, twelve quarterly, eight semi-annual, twenty annual, and fourteen special reports.[19]

Having rejected an industrial or business model for managing the laboratory, Hilberry adopted an academic paradigm, as if he saw his role as "dean of research" at Argonne. He intended to implement the "university code" as he understood it. "The basic elements of the university code which must be translated into the framework of a successful research laboratory are relatively few and relatively simple," he wrote, but "they are of imperative importance." The first, as noted above, was freedom to think, which included freedom from coercion, distraction, and fear. Although he could not provide his staff with academic tenure, Hilberry believed that researchers without intellectual and financial security lost their most priceless asset—the ability to think, judge, and plan on purely scientific grounds. The second university code essential to a research laboratory was collegiality. Hilberry would nourish solitary genius, but his faith in professional associations reflected the organizational ethic of the 1950s. Modern science required team players, not mavericks, because only through productive associations could creative researchers achieve their individual potential. How then would the organization scientist escape anonymity? The key university code was the guarantee of recognition for the creative researcher. Hilberry did not believe that creative genius required high salaries, fancy offices, or the extensive "perks" common in industry. Rather, "recognition of creative stature" was the principal motivator of the innovative scientist. Because "anonymity and creative productivity are mutually exclusive," he wrote, the large research institution must acknowledge individual achievement in proportion to the significance of each contribution. Finally, the university code provided the essential tools for research, such as libraries, laboratories, and shops. Hilberry knew that excellent facilities were necessary but not sufficient for research success. Spending millions on facilities might not lift a laboratory above mediocrity. "But dollars spent to improve the caliber of staff [were] always much more productive than those spent on the improvement of facilities" (figures 21 and 22).[20]

Hilberry's utopian vision for Argonne was widely shared by the research community, which generally believed that the best science was unfettered by fiscal or political considerations. Of course all scientists, including Hilberry, knew they had to deal with budgetary and policy realities, but these "intrusions" were impediments to be worked around and not an integral or positive aspect of science itself. It was as if the laboratory's chief product, scientific ideas, was apolitical with no intrinsic ethical or social properties. The fact that some of Argonne's research data were classified as secret was a regrettable anomaly that was to be erased as soon as possible. For all his skill as a consensus builder and facilitator of research, Hilberry did not regard his office as a major source of Argonne's research initiatives. Nor did he relish being Argonne's political leader.

Figure 21. Norman Hilberry holding the American Nuclear Society Award to CP-1 participants

Figure 22. ANL Main Administration

While he very much wanted to be named director of Argonne laboratory, Hilberry was most comfortable in the role of a chief operating officer who was well versed in the technical challenges facing the laboratory. As director, he was responsible for coordination of Argonne with the University of Chicago and the Atomic Energy Commission, but he did not envision for himself a larger role as a political leader of American science. He willingly delegated Argonne's liaison with midwestern universities and privately endowed research institutions to the associate director and division directors.[21]

Approving the Rettaliata Plan

In the midst of the MURA crisis, when Zinn threatened to resign, the Council of Participating Institutions asked Kimpton to establish an ad hoc committee to examine the relationship between the laboratory and the university communities. High-energy physics had dominated its attention, but the council reminded Kimpton that the participating-institutions program, representing thirty-two universities, was not only broader than MURA, but that their areas of interest, including chemistry, biology, engineering, and medicine, were also more extensive. Sufficiently discouraged to disband in the face of the seeming hopelessness of the MURA crisis, the Council of Participating Institutions instead asked the University of Chicago to decide whether and how university participation in Argonne's programs would be continued. In response, Kimpton, looking for a respected academic leader who had not become embroiled in the MURA controversy, asked John T. Rettaliata, president of the Illinois Institute of Technology, to head an ad hoc committee (the Rettaliata committee) to study the long-term relationship between the laboratory and the university communities.[22]

Rettaliata surveyed the presidents and representatives of the Council of Participating Institutions to obtain their critique of laboratory-university relations. Although he received perfunctory replies from a few notable universities, for the most part, Rettaliata collected positive feedback. The smaller the institution or the further its distance from Chicago, the more likely that university professors believed their relationships with Argonne were satisfactory. Praising Argonne's scientists for their generosity, the president of Marquette University reported that "all of our men are very pleased with our relationships with Argonne." From the Kansas prairie, the president of the state's land-grant college sent assurance that the laboratory had responded "promptly and effectively" to his faculties' requests for help. Especially important to the agricultural school had been Argonne's conference on the use of isotopes in plant and animal research. Even members of MURA, such as University of Wisconsin president E. B. Fred, remarked that "in general, our departments (other than physics) all speak highly of their associations with the staff at Argonne Laboratory."[23]

Mostly, the Rettaliata committee confirmed what Argonne's managers already knew. Their rapport with professors in chemistry, biology, medicine, and engineering was generally good. On the other hand, their relationship with professors of physics from the large research universities (many of whom had never visited Argonne) was terrible. Suggested improvements reflected old complaints—the need for more housing, less security, more collaboration, and less competition with university scholars. But new issues were also raised. While Argonne's cooperative program was given good marks, its midwestern constituents generally ranked Argonne second to Oak Ridge as a useful partner. In some professors' opinions, Argonne scientists were helpful as long as everyone participated in Argonne's research agenda. Others believed that Argonne's lack of flexibility could be blamed on Zinn's weak commitment to university-related programs. But Boyce, who had moved to Rettaliata's school in discouragement, received high praise for his efforts to promote healthy intramural teamwork. At the University of Illinois's physics department, Fred Seitz was relieved that the criticisms had not been vicious. "It is clear," he commented to Oliver Simpson, a member of the Rettaliata committee from Argonne's chemistry division, "that some of the universities expect too much from Argonne without meeting the laboratory halfway."[24]

The nub of the problem, Rettaliata's committee concluded, was that the relationship between Argonne and the universities had been too passive. J. H. Jensen, provost at Iowa State College, believed that the University of Chicago bore much of the responsibility for the laboratory's mixed reputation. Almost any system could be made to work if invested with trust and priority. If the university encouraged mutual effort, a good cooperative program could be made excellent. Paralleling Hilberry's vision for Argonne, Rettaliata's committee emphasized the desire to enhance Argonne's academic atmosphere with improved education and research programs more oriented toward university needs. But the committee pushed beyond Hilberry's utopian dream to summarize that "most of the work of fundamental significance in the A-bomb program was performed in university laboratories." The implication was obvious. If for no other reason, national security required "an active program of university participation at Argonne. Without such stimulation, research efforts, be they basic or applied, suffer or stagnate."[25] It followed that improving Argonne's academic stature was not only good for science, but also good for the nation.

The Rettaliata committee's recommendations seemed sensible and conservative. The committee encouraged each of Argonne's divisions to promote educational and cooperative research activities with the participating institutions. To ensure that the University of Chicago was kept in the communication loop on important questions, the committee advised the Atomic Energy Commission and Argonne to channel all major policy issues through the university. All of this was a modest corrective to Zinn's exuberant leadership and Strauss's and Libby's penchant for meddling from Washington, but it did not run counter to Hilberry's management style.

More radically, the Rettaliata committee proposed a very unacademic role for the Council of Participating Institutions. First, the committee encouraged the council to create a full-time staff with offices at Argonne to implement a cooperative program with the laboratory. Second, it recommended a policy advisory board, on which the council would have the majority of seats, to advise the University of Chicago on all policy matters pertaining to the laboratory. University officials frequently met with outside constituencies—regents or trustees, accreditation boards, alumni councils—but almost never did they share governance with a self-appointed group that maintained residency at their institutions. To be sure, the policy advisory board would be just that, advisory, but the intent was to establish formal, continuous oversight of Argonne's activities to assure "that the great AEC-sponsored laboratory and the educational and research talent of the participating institutions [were] utilized to the fullest possible extent." Subsequently, at its first meeting on July 17, 1957, the policy advisory board established eight review committees to assess the work of the laboratory.* The review committees were to report in confidence to the president of the University of Chicago and to the policy advisory board. In the mid-1950s, no academic faculty in the Midwest would have accepted similar confidential monitoring from the government.[26]

Establishing the Associated Midwest Universities

Significantly, the Rettaliata committee offered no direct recommendations to the AEC, but as long as the policy advisory board did not interfere with the government's reactor development programs, the commission endorsed the Rettaliata report with the proposed changes. After initial skepticism about the review, Kimpton seemed relieved that the report might serve as a treaty ending the open warfare with the midwestern universities. Unlike Zinn, who did not share a great affinity with university scientists, Hilberry actually welcomed close study of the laboratory's day-in and day-out working relationships. Seeing no contradiction with the need to protect the creative freedom of Argonne scientists, Hilberry confidently predicted that the new arrangement would lead to specific proposals for strengthening Argonne's basic research programs. To show good faith, in 1958 Hilberry established two new associate directorships, first appointing Frank E. Myers, a former graduate dean at Lehigh University and a physicist, as the associate director for education. Next, Hilberry recruited Roger Hildebrand, who represented the University of Chicago on the MURA board, as the associate director for high-energy physics. He also selected Albert Crewe, then at the University of Chicago, as director of the particle-accelerator division to replace John Livingood as supervisor of construction of the ZGS, which was scheduled to begin in 1959.[27]

While Hilberry made adjustments at Argonne, the old Council of Participating Institutions disbanded to re-form as the Associated Midwest Universities (AMU), headquartered at Argonne. Reorganization was necessary so that the association could open a permanent office with paid staff under the aegis of an inter-university corporation. Jensen, from Iowa State, became the first president, and Boyce, Argonne's respected former coordinator of university programs, agreed to serve as secretary-treasurer.[28]

---

*The review committees were biology and radiological physics, physics and applied mathematics, chemistry, chemical engineering, reactor engineering and remote control, the international school, metallurgy, and high-energy physics and electronics.

Geneva, 1958

The second International Conference on the Peaceful Uses of Atomic Energy, which convened in Geneva, Switzerland, September 1–13, 1958, marked the culmination of Eisenhower's Atoms for Peace initiative. The conference was the largest international scientific gathering since World War II. Afterward, Strauss reported to Secretary of State John Foster Dulles: "One cannot examine the statistics of this Conference and the tons of technical papers, reports, transcripts, photographs, newspaper articles, magazine stories which it generated, without becoming aware of the fact that atomic energy has now become part of the fabric of our civilization."[29] Argonne's scholarly contribution to the conference included more than a hundred scientific papers, thirty of which were presented orally.

Although the conference was supposed to feature progress in the peaceful uses of atomic energy, especially nuclear power, not surprisingly, the Soviet Union decided to feature its two *Sputnik* satellites, which had been launched on October 4 and November 7, 1957. The Atomic Energy Commission looked to Argonne to build a spectacular exhibit that would be as important for its propaganda value as for its scientific merit in order to counter Russia's space grandstanding. In a technical tour de force, an Argonne crew headed by David H. Lennox transported to Geneva an Argonaut training and research reactor that was assembled during the first four days of the conference while fascinated delegates monitored the progress like side-walk superintendents. The low-power (10 kilowatts, maximum), thermal, water-moderated reactor had been developed as a training reactor for the international reactor school. Argonne and Swiss technicians assembled the reactor in record time. On the sixth day of the conference, Strauss brought the Argonaut to criticality using a wand containing uranium from Fermi's CP-1. Complementing the Argonaut, the laboratory also presented a nonnuclear working model of the EBWR, which employed computer technology to replicate realistic reactor operation, incorporating "working" control rods and simulated Cerenkov effect. To the delight of the "side-walk superintendents," Argonne also dismantled the Argonaut during the final days of the conference, providing another opportunity for observers to see how the reactor was put together."[30]

Scientifically more important were the exhibits featuring radiation chemistry and biological and medical research. Scientists from the chemistry division at Argonne demonstrated the effects of gamma radiation, the most penetrating radiation, on solids and liquids, specifically water, carbon, organic compounds, and certain crystals. Biologists set up a portable "atomic farm" in which medicinal plants were grown within a sealed greenhouse in a radioactive atmosphere. When the farm's plants breathed carbon dioxide they also took up carbon-14 into their tissues. Because the carbon-14 served as a tracer, scientists were able to follow the path of drugs made from the plants through biological systems. For example, in experimental animals they could watch the ingestion of the drug, follow its path through the organism, determine where and how fast the drug worked, and study the mechanism of its breakdown. In another demonstration with radioactive tracers, rats were fed carbon-14–tagged foods, making it possible to measure the rates at which the food was converted to body protein and nucleic acid. Other exhibits featured the use of tritium in biological studies of the intracellular processes of normal and abnormal cells, providing information not only about cell structure but also about cell kinetics.

The Future of the National Laboratories

Americans returned from the Geneva Conference on Peaceful Uses satisfied that they retained a healthy lead in the nuclear sciences, but they also were aware that foreign colleagues offered strong, vigorous competition in most fields. Leaders in the nuclear community, whether in Congress, at the AEC, or in the national laboratories, resolved to maintain American dominance to avoid the embarrassment suffered by their compatriots in the space sciences. Euphoria over the peaceful-uses conferences, which laid the foundation for lively international cooperation among western and eastern-bloc scientists, did not calm Americans' anxiety about the perils of the nuclear age. On the contrary, Russian accomplishments in nuclear science only confirmed that Americans were in a double race with the communists—the deadly nuclear arms race for strategic supremacy and the friendlier peaceful uses competition for economic advantage. Already jittery about alleged lapses in the United States' nuclear weapons, bomber, and missile programs, as they looked ahead to the 1960s, members of the Joint Committee on Atomic Energy wanted reassurance that the commission's laboratories were prepared to meet what they believed might even be a greater challenge than World War II. In the spring of 1959, the Joint Committee asked the commission for an assessment of the future of the national laboratories.

During the Eisenhower administration, commission expenditures on research and development (exclusive of weapons) ballooned more than 200 percent, to $625 million, with more than half of the increase coming since the Sputnik scare in 1957. The largest increases were allotted to civilian power reactors, but there was also major expansion in high-energy physics and controlled thermonuclear research and significant rises in virtually all other areas. Argonne generally profited from this growth with its own budget growing about 125 percent, to almost $42 million in fiscal year 1960.[31]

In its 1959 long-range plan, Argonne envisioned itself as an American bastion in the Cold War. "To lose this war," the plan warned, "could prove to be fully as disastrous as to lose a shooting war." The war was primarily an economic struggle in which America again was called upon to be an arsenal for the free world. "The battles, to be sure, will rage through every phase of our productive activities, but the atomic energy program is in a particularly acute situation," the Argonne planners declared. For better or for worse, Atoms for Peace had become an integral part of Eisenhower's foreign policy, complementing America's nuclear weapons program. Once again, Argonne found itself on a "war" footing, helping the United States to win a major victory "in the world struggle for economic supremacy": "If we are to win this war, we must make the most efficient use possible of every research and development unit within the country and we must develop each of these units to its maximum potential. In particular we must intensify our basic research, both in science and in engineering." Whereas the Metallurgical Laboratory raced to build an atomic bomb to save the world from Hitler's despotism, Argonne was now enlisted to "achieve some variety of Utopia through atomic energy." Recognizing that utopian dreams were unrealistic, the laboratory nonetheless hoped to contribute to "an observable improvement in the world's standard of living attributable to atomic energy."[32]

At the end of the Eisenhower administration there was no doubt that the national laboratories were vital assets. Neither the Joint Committee nor the commission seriously considered closing them or curtailing their activities. But, in the commission's haste to capitalize on the Manhattan Project, the laboratories had grown large and wealthy without the benefit of a coordinated plan. Zinn had complained bitterly about lack of general policy from Washington. Headquarters program offices usually limited funding and interest to their own laboratory projects. Consequently, all laboratory directors complained that it was difficult to obtain funding for general laboratory needs such as roads, heating plants, service buildings and grounds, and other administrative expenses. For the weapon laboratories, with their relatively narrow mission, the problem was not as great. Both Hilberry and Gerald F. Tape, director of Brookhaven National Laboratory, noted headquarter's tendency to starve the laboratory body as a whole while richly feeding its various members.[33]

Hilberry and Tape were not complaining about minor housekeeping problems but rather were identifying a significant weakness in the national laboratory system (or perhaps one might say, "non-system"). Hilberry had written that "under no circumstances should a laboratory be considered merely as a collection of programs and projects." Yet, with the exception of the weapon laboratories, in 1959, the national laboratories were really regional laboratories with an agglomeration of scientific activities. They were distinguishable because the AEC allocated large, expensive, and unique facilities to each of them. Speaking for his fellow laboratory directors, Hilberry tried to explain why the commission had failed to define their long-range purpose or to provide for the long-term health of the laboratories. The commission may have treasured the laboratories as national assets, but it dealt with them as if they were just another contractor like General Electric or General Dynamics. Naturally, Hilberry and his colleagues thought it regrettable that the commission required the laboratories to bid against one another, universities, and industry for projects.[34]

Indeed, the Atomic Industrial Forum, the nuclear industry's association, complained that its members had not received their fair share of AEC contracts. The forum also worried about the future of the national laboratories. Ironically, forum leaders argued that the laboratories should limit themselves to basic research too risky to warrant industrial interest, a sentiment widely shared among Argonne scientists not involved in reactor research and development. Despite Argonne's triumph at the Geneva conference, the nuclear industry was critical of the commission's development and promotion of the Argonaut research reactor at a time when research reactors could be purchased from private industry. As Francis K. McCune, president of the Atomic Industrial Forum wrote to John A. McCone, the commission's new chairman, "Some of us feel that the nuclear reactor business will not be absorbed into the normal pattern of American private industry as long as part of the research and development work is carried out in government laboratories."[35]

In their own way, the laboratory directors raised a parallel complaint when they alleged that the laboratories received no support from the commission for basic reactor physics. To underwrite the vital basic research that the commission overlooked, the directors buried reactor physics within larger developmental programs such as "water reactors," "gas reactors," or "breeder reactors." Commissioner John F. Floberg, who met with the directors in Chicago about the future of the laboratories, reported unanimous regret that they had to deceive the commission in order to fund some of their best and most necessary programs. With little flexibility in commission budgets, they simply camouflaged basic research within program development. Such subterfuge, whether or not discerned by the AEC, meant that the laboratories followed both manifest and latent research agenda, necessary, perhaps, to attain their goals, but detrimental to their ambition to develop "balanced, integrated, interdependent" national laboratories.[36]

When McCone returned from his October 1959 visit to Soviet nuclear facilities, he noted the Russian ability to focus on and complete highly sophisticated research projects. McCone, who believed that the Americans were generally ahead in peaceful-uses research, advocated greater decentralization of the commission's laboratory management. Hilberry welcomed McCone's assessment, particularly the chairman's emphasis on the need for national laboratories to carry out large, high-risk programs. Optimally, the commission would treat each multipurpose national laboratory as a "complex, interdependent, organic whole." Using a bell curve, Hilberry presented an interesting model that placed the national laboratories at the nexus between university and industrial research:[37]

Although the Atomic Energy Commission did not prepare a comprehensive plan for the future of the national laboratories, its projections for the 1960s would profoundly affect Argonne's management. The commission was proud of its excellent laboratories staffed with highly skilled, well-rounded, and dedicated personnel but regarded them as mature institutions whose growth should be limited to 4,000–5,000 people, a range considered ideal for a research institution (Argonne's complement in 1959 totaled 3,439). The commission anticipated making room for new projects at the national laboratories by phasing out or shifting mature programs to university or industrial facilities. Although they did not formally adopt the "Hilberry model," the commission intended to use the federal laboratories as seed beds for pioneering research which could be transplanted to university laboratories or industrial development as appropriate.[38]

Such transfer was already well under way in the development of civilian nuclear reactors and would be accelerated under the AEC's new ten-year plan for the commercialization of power reactors. To carry out the ten-year plan, announced almost simultaneously with publication of *The Future Role of the Atomic Energy Commission Laboratories,* the commission adopted a "three phase sequence" for promoting commercial nuclear power. First, the commission would not only continue its experimental reactor projects such as the Argonne fast-source reactor (AFSR), designed to support EBR-II already under construction in Idaho in spring 1960, but also phase out Argonne's boiling-water reactor program, which it deemed sufficiently mature for industrial development. Then, under the provisions of the Power Reactor Demonstration Program, the commission planned to cooperate with industry to build prototype reactors of promising designs. Finally, the commission hoped to work with manufacturers and electrical utilities to build full-size power reactors.[39]

The Atomic Energy Commission also anticipated increased support for basic physical and biomedical research in the universities rather than at the national laboratories. Such emphasis, however, was unlikely to seriously curtail programs like Argonne's Janus project, an unusual research reactor with two radiation faces designed exclusively for biological research. Named after the two-faced Roman god, the Janus reactor, costing more than half a million dollars, would enable biological researchers to study the effects of both high and low exposures of laboratory animals to fission neutrons. The JANUS reactor represented the high-cost equipment required for intensive research programs that the commission thought appropriate for Argonne National Laboratory.[40]

The commission planned to keep its laboratories, including Argonne, fully occupied during the 1960s. In their discussion on the future of the national laboratories, the commissioners reaffirmed that the laboratories should support the three major tasks of the AEC: "(1) the development and production of nuclear weapons, (2) development of reactors, and (3) research and development in the field of isotopes"; all three were broadly defined. Nonetheless, the commission recognized that the national laboratory system should also be a national resource not exclusively devoted to nuclear research. The commission did not intend to convert its laboratories into "job shops," but indicated that it was appropriate for the multipurpose laboratories to accept research assignments from federal agencies in response to "national needs that call for out-of-the-ordinary arrangements, effort, and ability." This statement of policy established the basis for much of Argonne's subsequent environmental research in the 1960s.[41]

Undoubtedly most important for Argonne was the commission's reiteration of its commitment to promote research partnerships with American universities. All links between the laboratories and the universities—organizational, personal, project collaboration, and facility sharing—would be encouraged and strengthened. While the commission pledged increased direct funding of university-based research, it also forecast construction of large facilities intended for the cooperative use of government and academic scientists. Recent experience in high-energy physics shaped the AEC's policies.

> Large accelerators and other unusual and expensive research tools, whether provided on university campuses or on commission-owned sites, are intended to serve qualified researchers from other institutions as well as those on the staff

of the operation institution. The commission will insist that these opportunities for broad use of its facilities in the interest of science be maintained.⁴²

The Zero Gradient Synchrotron

After Eisenhower singled out high-energy accelerators as a wise investment for American security, the Atomic Energy Commission announced in December 1957 that Congress had authorized construction of Argonne's 12.5 GeV proton accelerator at a cost of $21 million. With more hoopla than novelty, the commission revealed that Argonne's new machine would produce more accelerated particles than any of those existing or planned, although Brookhaven's and CERN's alternating-gradient synchrotrons would accelerate particles with greater energies. The commission also announced construction of two other accelerators—a joint Harvard-MIT 6 GeV electron synchrotron, and a Princeton-University of Pennsylvania 3 GeV proton synchrotron. As for MURA's hopes, the commission indicated it would continue to fund planning but made no commitments other than to repeat that, if authorized, MURA's machine "should be located at a site where such facilities already are in existence; namely, the Argonne National Laboratory."⁴³

The Milwaukee *Journal* editorially snorted that the whole deal smelled like old fish, noting caustically that Argonne's accelerator was not a new project but a long-delayed one. Defending hometown interests, the *Journal* accused the commission of lying about its criteria for selecting Argonne as the site for the midwestern accelerators (the *Journal* said the commission toyed with "half truths or untruths").⁴⁴

For Roger Hildebrand at the Enrico Fermi Institute in Chicago, however, this unproductive running feud between Argonne and MURA threatened to destroy all hopes to build up the Midwest's physics community. A loyal member of the MURA board, Hildebrand saw no alternative but for the MURA community to embrace the Argonne program. He was worried about Argonne's failure to recruit a distinguished outside director but was more deeply alarmed about the exodus of top physicists from the Midwest. Like it or not, Hildebrand believed that the revival of Argonne, the success of MURA, and the prosperity of physics in the Midwest were linked through commission policy. For this reason, with Albert Crewe's assistance, he agreed to assume responsibility for designing Argonne's Zero Gradient Synchrotron.⁴⁵

One of Hildebrand's first steps was to organize a strong "users group" from among his friends and colleagues, many of them also active in MURA. He hired G. T. Getz to work full time on outreach activities. Ned Goldwasser, another active MURA scientist, shared Hildebrand's view that failure to help Argonne would be counterproductive for midwestern high-energy physics. Goldwasser agreed to organize the ZGS users advisory committee. Lacking a large high-energy physics staff at Argonne, Hildebrand urgently needed help from university scientists regarding the design and utilization of the accelerator. In supplementing their staff, Hildebrand, Goldwasser, and Getz looked for physicists willing to help Argonne design and build the accelerator and supporting research facilities.[46]

The Piore Panel

Meanwhile, the MURA project languished in technical and political uncertainty. In 1958, MURA, already struggling with unresolved design problems while Argonne prepared to move ahead, met another, stronger competitor from the West. The previous year, Stanford University physicists led by Wolfgang K. H. Panofsky proposed construction of a $100 million Stanford Linear Accelerator (SLAC). With mounting costs everywhere in the high-energy physics program, James R. Killian, Jr., Eisenhower's science advisor, asked a joint committee of the President's Science Advisory Committee (PSAC) and the AEC's General Advisory Committee to recommend priorities for construction and operation of expensive accelerators. In what Daniel Greenberg described as "puckering up for what was eventually to be the kiss of death," the Piore panel (named after its chairman, Emanuel R. Piore) recommended that the commission reject MURA's accelerator proposal but continue to encourage theoretical studies.[47]

The midwestern delegation on the Joint Committee of Atomic Energy did not accept the Piore panel's findings, but fading support within the high-energy physics community persuaded midwestern physicists to cooperate more generously with Hildebrand and Goldwasser, who had been elected chair of the users committee. Although the success of the users committee was not exactly built upon scientific altruism, Goldwasser was conscious of the fact that he and Argonne's users committee were breaking new ground. The original accelerators at Berkeley and on various campuses had been built by single individuals or by small teams of in-house experimenters. Brookhaven pioneered with a university consortium that provided scientists from about ten universities with access to its accelerator. Goldwasser's committee, however, became a model for cooperative users groups later established in California, at Brookhaven, and ultimately at Fermilab.

All along, MURA scientists had claimed that Argonne could not independently build its own accelerator and research facilities. The realities of modern big science confirmed that the requirements of the ZGS project stretched the laboratory to its limit. Argonne's users group adopted by-laws and organized committees on bubble chambers, external beams, and electronics and established a general users committee to advise Hildebrand directly on design, construction, and research priorities. Just before completion of the ZGS, Hildebrand created two new committees: the program committee, which he chaired to approve experiments, and the scheduling committee, chaired by Lee Teng to allocate time on the machine. According to Ratner, "As the users group gained in strength, many of the original fears about the feasibility of participation in a research program at Argonne were allayed."[48]

Bubble-Chamber Politics

Hildebrand hoped that the users group could be helpful in the design and construction of experimental facilities. The 12.5 GeV ZGS offered rival laboratories no competition in terms of beam energies, but what the ZGS lacked in power, Hildebrand hoped the accelerator would make up in beam intensity and sophisticated supporting particle detection equipment. Prior to construction of the ZGS, Argonne had no machine capable of producing any particle that had been discovered in the past quarter century. Hildebrand anticipated that the ZGS would produce all thirty of the known or suspected subatomic particles in sufficient numbers to determine their properties. As Hildebrand correctly put it, the ZGS actually advanced Argonne's standing in particle physics more than it enhanced the laboratory's reputation in high-energy physics. Large, state-of-the-art bubble chambers would become key to Argonne's experimental program.[49]

Bubble chambers were one of the most advanced detectors frequently used in conjunction with spark chambers to observe the passage of subatomic particles. In contrast to the cloud chamber, in which water droplets form a visible trail as the particles speed through a supercooled vapor, in the bubble chamber, minute bubbles indicate the path of a particle as it races through superheated liquid. Often employing liquid hydrogen because of the simplicity of the hydrogen nucleus, bubble chambers increased the chances of detecting elusive, short-lived particles.[50]

Hildebrand's search for a bubble-chamber expert ended when R. W. Thompson of Indiana University accepted a joint appointment at Argonne and the University of Chicago to build Argonne's principal bubble chamber. Precise, meticulous, and demanding, Thompson was not a popular choice among midwestern physicists, who objected not only to Hildebrand's selection (in which they had no say), but also to the Argonne–University of Chicago partnership, which might severely curtail access to the research facility.[51]

To assure an opening at the ZGS, MURA scientists proposed constructing in Madison their own bubble chamber, which they could move to Argonne after the ZGS became operational. They intended that MURA "retain prime interest" in the bubble chamber, meaning that it might be moved back to Madison to be used in conjunction with the MURA accelerator. The chamber would be regarded as "midwestern community property." Welcoming even this small interest in the ZGS project, the commission agreed to fund MURA physicists, led by W. D. Walker, to build a comparatively modest, 30-inch bubble chamber, but the commission insisted that the chamber be considered a "national facility," and so available to qualified research groups at the ZGS. Simultaneously, Goldwasser offered to build a small bubble chamber at Illinois for general use at Argonne providing Hildebrand could supply a magnet. Other cooperative proposals were received from Michigan and Northwestern.[52]

These cooperative efforts brought no peace to the laboratory but rather sparked heated debate over Argonne's policies for allocating use of the research facilities. Scientists who devoted their careers to fabricating bubble chambers naturally expected to be assigned priority in their use. Yet if Argonne were to serve as a "national" laboratory, no individual or group should be granted a proprietary interest in any facility. Emotions ran strong, Ratner recalled. Thompson, who had devoted years of his life to the bubble chamber, demanded that anyone wanting to use the large facility join his group. The MURA scientists balked. The commission had already declared their chamber a "national asset." MURA agreed that Thompson and his group should receive a generous share of time, but MURA argued that allocations ultimately should be based on scientific merit. Hildebrand wavered. First he sided with Thompson, then in June 1961 he concurred with the advisory board that all large detectors should be available for general use based on recommendations from the users group. Thompson resigned, and when the AEC failed to fund the large bubble chamber, the whole issue became moot. The scrap ended with the accelerator users committee firmly embedded within Argonne's organizational structure.

Bureaucratically, the accelerator users group belonged neither to the Associated Midwest Universities nor to MURA. John H. Roberson, executive director of the AMU, wanted the users group to work under his jurisdiction, but the physicists, jealous of their autonomy, chose to remain independent of any affiliation. The users group, however, was supported administratively out of the AMU office, which received a contract from the AEC to pay for the committee's expenses. After serving as chair of the users group for two years, Goldwasser stepped down in 1960 in favor of MURA's technical director, Keith Symon of Wisconsin. While chair (1960–62), Symon aggressively represented the interests of the midwestern high-energy physicists but shared no responsibility for strengthening the Argonne-AMU partnership. Symon became effusive in his praise of Hildebrand's cooperation, which he found "almost beyond the call of duty." Hildebrand's generosity only highlighted the weakness of the users group, whose members' role in setting Argonne policy was advisory only. Having no official standing, they relied on the good will of such administrators as Hildebrand for access to accelerator facilities. Although the users were grateful to be welcome guests, they wanted an *official* voice with which to guarantee consideration of their needs and priorities. The accelerator users would not rest until they secured their own contract with the AEC.[53]

Argonne-AMU Partnership

How well did Argonne Laboratory serve the midwestern university community fifteen years after the end of World War II? The answer depended upon perspective. Researchers had long complained about lack of housing and transportation at Argonne. The international school had actually stimulated housing construction, but rooms were few, dining was limited, and public transportation was nonexistent. Government automobiles, available for official business, could not be used for off-site dining, Chicago baseball games, or tours of the Loop. Hildebrand thought these shortcomings provided the AMU with an excellent opportunity to serve its constituency and help the laboratory. Housing, cars, even travel stipends, could be paid for out of AMU funds. "With only AEC funds it is hard to see what AMU can do for us that we can not do for ourselves with university or contract money," Hildebrand wrote Symon. "With private money there is no end to the possibilities we might consider.... But it is futile to think of ... grand schemes if we can not find someone to buy us an automobile."[54] Ultimately, the University of Chicago supplied two cars.

For Hilberry, the laboratory had done a good job considering the fact that none of Argonne's research laboratories or buildings, with the exception of the international reactor school, was built to be a teaching facility. The ZGS, when it came on line, would be available to the universities as a research and teaching instrument, mostly at the postdoctoral and advanced graduate level. But when compared to Brookhaven, Argonne's efforts seemed modest. According to one estimate, with 70 full-time postdoctoral positions and 300 research visitors, academic scientists outnumbered Brookhaven's permanent staff of 300. Argonne employed about 800 professional staff, almost one-third of whom worked in the reactor development programs. Only 98 college and university faculty participated in Argonne's resident research program.[55]

Such statistics for Argonne were probably not comparable with those of

Brookhaven, which undoubtedly had the stronger cooperative program. Still, Argonne's were worth noting if for no other reason than to dispel the myth that the laboratory was indifferent to educational needs. The ZGS users group reported that 162 faculty from 34 universities had participated in various planning sessions. In addition to providing summer employment for more than 50 graduate students, Argonne hosted 13 graduate students working on M.S. and Ph.D. degrees. Argonne also offered summer employment to more than 100 undergraduates from 55 colleges, and part-time employment to 18 students in its co-op program. Thirty-four faculty were enrolled in the long-term residence program (six months or more), and 64 joined the short-term resident research program, spending the summer of 1959 at the laboratory. Specialized summer institutes in nuclear studies enrolled more than 40 faculty (in three years, totaling 146 faculty from 50 institutions). Of the 475 graduates of the International School of Nuclear Science and Engineering, more than half represented foreign universities. Each year, approximately 100 Argonne scientists presented seminars at colleges and universities, while 100 faculty gave seminars at Argonne. Not counting speakers, committee and board members, or casual visitors, more than 300 American students and faculty used the laboratory annually. In addition, since the 1955 Conference on Peaceful Uses, CP-5 provided 1,500 service irradiations for 15 universities and almost 150 irradiated samples, from grasshopper eggs to carbon-14–tagged soybean meal, had been sent to scores of educational and research centers.

Without supplementary funding, Hilberry did not believe Argonne could substantially improve this record, except to expand the summer programs. Already, the laboratory approved applications from 72 percent of the AMU faculty who asked for summer funding (50 percent of non-AMU faculty). Space, not money, was the limiting factor. Summer visitors also required supervision, but they did not place a heavy strain on Argonne's research facilities because students and faculty generally joined existing research groups as assistants. Visiting researchers only became a problem when they requested dedicated time on crowded facilities. In 1961, Argonne's major experimental facilities were almost fully utilized by the laboratory staff. Experiments using CP-5, the workhorse reactor, were scheduled a year in advance. Use of the Van de Graaff generator in the physics division required a four-month reservation. The chemistry division's cyclotron and the laboratory's computer facilities were fully engaged. Group proposals having high technical merit and requiring little machine time were always welcome, but competition was keen. Hilberry especially encouraged long-term university research projects rather than quick summer studies. Once the ZGS was operational, Hilberry anticipated that over 50 percent of the experimental facilities would be available to off-site users.

The Research Group

The lone "walk-on" disappeared from the large scientific laboratory and the major university football team at about the same time. There were exceptions—the solitary scientist working alone at his bench on some exciting problem could still be found in a nether corner of the laboratory—but predominantly physical scientists, who needed continuous access to expensive and labor intensive machines, joined research groups, which provided symbiotic partnership. The prestige of senior scientists was measured by the size, funding, and productivity of their group. After suitable postdoctoral apprenticeship, the future of young scientists depended on their skill in building their own research groups, which, in addition to substantial funding, postdocs, and graduate students, required access to major research facilities. All of these activities exceeded the normal resources of scientists young or old, and increasingly they exceeded the resources of their universities as well. Scientific culture continued to honor the accomplishments of distinguished experimenters, but young researchers discovered that their futures were often determined by anonymous "peer" reviewers at the federal government's large funding agencies.

Argonne's succinct guidelines for proposing an experiment on the ZGS perfectly captured the new realities of modern big science. Research ideas still usually came from "a man working alone," but the creative scientist needed the help of a complex team to be successful. "The task is so great," the laboratory warned, "that a variety of skills must be available in the group." The principal investigator, often a senior scientist with laboratory facilities and graduate students, organized the team, recruiting talent as needed. The research group would need a master theoretician and an expert in electronics or optics. Every group required a computer wizard, and if organizational skills were not the principal investigator's strength, someone had to be found who could manage large, complicated projects. As the users-group policy statement put it, "Together they must seek financial support from a government agency since few experiments are within the scope of university funds. Besides having the necessary skill, the group must be able to work together closely for many months and to withstand the reverses which meet those who follow difficult paths in new areas." To obtain beam time on the ZGS, the applicants submitted a research proposal outlining their objectives, methods, and requirements to the users-group program committee, which evaluated the proposal for scientific merit and feasibility. If the research proposal was approved, the operations committee scheduled the experiment while the research group planned its experiment in detail. Preparation for the experiment, which frequently required researchers to construct their own apparatus, might take six months to a year.

Still, competition for machine time was so intense that only one in three proposals was approved.⁵⁶

When ready, the group was assigned a secondary or "parasite beam" on which to test the apparatus and experimental protocols. After the entire system checked out, the team members were finally assigned a position on the main beam line where the experiment might run twenty-four hours a day for several weeks or might alternate with other experiments simultaneously on line. In any case, running the actual experiment tested the "endurance, ingenuity, and morale of the group," which usually did not know the results for weeks because data analysis itself was often a time-consuming, but cardinal, task. Publication, normally with the principal investigator as first author, staked out the findings, and reputations, of the authors.⁵⁷

In the Midwest, only Argonne National Laboratory had the resources to mount such an effort. Of the wealthiest MURA universities, only the University of Chicago ($2 million) and the University of Illinois ($1.3 million) received more than a million dollars annually for high-energy physics from combined grants from the AEC, the National Science Foundation, and the Office of Naval Research. The rest of the MURA institutions each received less than one million dollars annually from all sources, and over half of these schools actually received less than $100,000 annually in research grants for high-energy physics. The projected annual operating costs for the ZGS exceeded $10 million, more than the combined annual funding for high-energy physics at all MURA institutions.

With so much at stake, it was no wonder that the midwestern high-energy physicists demanded a strong voice in determining research priorities for the ZGS. For those not belonging to MURA's inner circle, membership in the users group was usually required. Long before construction was completed, the ZGS program committee forwarded approved proposals to the operations committee for scheduling, which was predicated on generous AEC funding. By the fall of 1961, sixteen research groups were ready to begin experimentation, and two more were well advanced in their planning.⁵⁸

The Atomic Energy Commission, however, was shocked by the escalating costs of high-energy physics. The National Science Foundation and the Piore panel estimated that by 1963, high-energy physics would expend between $100 million and $150 million annually. SLAC, America's most expensive basic research project to date, alone would cost $100 million. But even during the years of generous science funding following Sputnik, high-energy physics seemed to soak up massive budgets. In 1960, for example, the physicists at Lawrence Berkeley Laboratory asked for an additional $2.8 million in operating and construction funds, an increase that would have had to have been made at the expense of other research programs. Builders of the Princeton-Penn accelerator requested $10 million for modifications and additions, for which the commission had only budgeted $1.3 million. At this same time, Argonne requested an additional $20 million for modifications before the ZGS became operational and another $23.5 million for upgrading the facility after it came on line.[59]

While the commission approved Argonne's construction requests—the ZGS ultimately cost $50 million by the time it began operation in 1963—it slashed prospective operating funds from $16.4 million to $10.4 million. The cut seriously upset the research plans of the users group, which was originally allocated $7 million of the total operating funds. Were it to share the AEC's $6 million cut equally with Argonne, the university portion would have been reduced to $4 million. Accepting that the budget pinch required reassessment of the usefulness of the ZGS for graduate-student training, Hildebrand hoped that the users group would reserve priority for untenured faculty members for whom access was essential. Most distressing, AEC funding would enable the ZGS to run only one shift a day, seriously curtailing feasible research projects. Argonne and the users group did not lose their rosy optimism about the future of the ZGS, but anxiety crept into their councils.

George Beadle, the new chancellor of the University of Chicago, wrote Glenn Seaborg, the new AEC chairman, in October 1961 that his concern was not limited to Argonne National Laboratory. He reminded Seaborg of his colleagues, over a hundred midwestern high-energy physicists, who had been working on the ZGS project for more than two years. "We are fearful of the effect on the morale of these men and on the future of high energy physics in the Midwest if funds are not available to support a reasonably effective use of the accelerator as soon as it is operable." While there would soon be four accelerators on the East Coast, and four on the West Coast should the Stanford accelerator be approved, the midwesterners could not understand why their only machine was not fully funded.[60]

Watchful Waiting, 1961

The year 1961 was one of watchful waiting for Argonne National Laboratory and its constituents. John F. Kennedy's victory in the 1960 presidential election promised important changes at the Atomic Energy Commission. With Democrats in control of both Congress and the White House, Senator Clinton Anderson and Congressman Chet Holifield of the Joint Committee had high hopes for increased emphasis on the development of civilian power reactors and other peaceful applications of atomic energy. What this meant for the national laboratories was unclear.

On January 3, 1961, Argonne received tragic news from the National Reactor Testing Station. At 9:01 P.M. the Stationary Low-Power Reactor Number 1 (SL-1) exploded, resulting in America's first fatalities from a nuclear-reactor accident. Built by Argonne as the Argonne Low-Power Reactor, the SL-1 reactor core had been designed for the army to airlift to remote sites in the Arctic and Antarctic. The prototype portable power reactor had been turned over to the army for testing and training. Following a shutdown for maintenance, during which technicians incorrectly operated the control rods, an explosion destroyed the reactor and killed the three operators. While the accident "sent a bolt of shock throughout the nuclear industry," repercussions for Argonne were rather mild. On the one hand, the laboratory's reactor engineers were surprised that the reactor was not as safe as they believed. The accident evidently was caused by a combination of design flaws, inadequate training, and operator error when the central control rod was extracted too far from the reactor core. In an accident not dissimilar from the BORAX excursion, the SL-1 did not prove to be inherently safe. On the other hand, there was some satisfaction that in the worst-case scenario, there had been no significant release of hazardous radioactivity beyond the accident site, despite the fact that the SL-1 had no containment seal. On the sobering side, the reactor explosion and subsequent core melting destroyed the reactor's safety systems, including the emergency core-cooling system. Evidently, a reassessment of all reactor safety systems was in order.[61]

Two weeks after the SL-1 accident, on January 16, President Kennedy named Glenn Seaborg to be chairman of the Atomic Energy Commission. Alumni of the University of Chicago's Metallurgical Laboratory, such as Paul Fields, had worked with Seaborg on the Manhattan Project. More recently, for two months in 1952 Seaborg had been a summer visitor at Argonne, active in special research and consultation with the laboratory staff. Although rivals in the hunt for new elements and isotopes, the laboratory was delighted that "one of their own" would be the first scientist to chair the commission. With Seaborg at the helm, Kennedy's New Frontier might include pioneering initiatives in reactor development and high-energy physics, Argonne's two major programs. With friends who understood and supported basic science research in power at both the commission and the Joint Committee, the future looked bright for the laboratory despite persistent budget problems that dominated every planning session.[62]

Concurrent with the changes in Washington, Hilberry indicated his desire to step down as director when a suitable replacement could be found. With three years to go before retirement, Hilberry apparently envisioned that the University of Chicago would organize a leisurely search for a new director to ensure a smooth transition and management continuity in the laboratory. To Hilberry's surprise, once the selection mechanism was turned on, events moved much more rapidly than he anticipated. When George Beadle became chancellor of the university in the spring, he moved immediately to establish an ad hoc committee to advise him on the selection of the new director. Beadle showed sensitivity to Argonne's constituents when he named two university vice presidents, two associate laboratory directors, a University of Chicago professor of chemistry, and the president of the AMU to the committee. In addition, the university polled Argonne's staff and the policy advisory board for suggestions. During the search, Beadle kept Seaborg informed of their progress.

As in the search following Zinn's departure in 1956, everyone hoped to lure a world-renowned scientist to lead Argonne. The initial recommendations included over twenty distinguished scientists, none from Argonne. Norman Ramsey of Harvard, Hans Bethe of Cornell, Victor Weisskopf of MIT, and Robert R. Wilson of Cornell headed the list. By summer 1961, when discussions (which Seaborg vigorously assisted) failed to identify a prominent outside candidate, the search committee gave serious consideration to Argonne's senior staff. Whether the candidate came from outside or inside, however, Beadle thought the criteria for the new director were clear. In addition to being an outstanding scientist with broad interests and a positive attitude toward both basic and applied science, Argonne's director would need the skill and experience to work effectively with the commission, the general manager, and the AEC staff. There was also a new element in the job description since Zinn's day—it was also essential that Argonne's director be willing and interested in working with midwestern academic scientists. Beadle realized that such a person might not exist, but he was committed to meet the criteria as nearly as possible.[63]

7 | Targeting Midwestern Science, 1961–67

April 28, 1961. Civil defense officials doubted that Argonne National Laboratory would be targeted at ground zero in a nuclear war. Nevertheless, at Argonne's Emergency Control Center, John Bobbitt, the laboratory's emergency coordinator, waited nervously for news from the Illinois Warning Center. "0400 Zulu . . . from General Homer, State Civil Defense Agency," blared the voice of NAWAS—the national warning system—over the center's loudspeaker.

Flash . . . summary of Nudets [nuclear detonations] to date in Illinois as of 2145 CST: Chicago, 63rd and Pulaski, 20 megaton, surface; . . . Peoria, dud; . . . East St. Louis, 5 miles south of downtown, 10 megaton, surface. . . .

Argonne lay thirteen miles southwest of the 20-megaton blast that obliterated Chicago's Midway Airport at ground zero. Initial damage assessments for the laboratory were sobering: 135 dead; 1,125 injured; 28 percent casualties among laboratory personnel. All wooden structures lay smoldering in ruins, while light steel structures had collapsed, flattened by 3.7 pounds per square inch (psi) of pressure and 125 mile-per-hour winds. Water and transmission towers were toppled by the blast, and persons caught in the open suffered third-degree burns from thermal radiation if they were not mercifully killed outright by flying debris. The woods surrounding the laboratory were set on fire along with any combustible object in the area.

Within fifteen minutes of the hypothetical nuclear attack, Bobbitt's civil defense team assembled a train of 195 trucks and cars to move 900 survivors along Argonne's roads to fallout shelters. Within one hour of the mock nuclear detonation over south Chicago, radioactive fallout drifting over the laboratory presumably measured up to 110 roentgens per hour, sufficient to cause serious injury and possible disability among those unable to find shelter.[1] After a brief "take cover" exercise to familiarize the staff with the location of Argonne's air-raid shelters, the laboratory returned to work while Bobbitt's crew continued to monitor attacking plane tracks, blast reports, national damage, and fallout reports through their link to the Du Page Civil Defense Control Center at Rantoul, Illinois. Bobbitt was satisfied with Argonne's efficient response to the civil defense drill and no doubt pleased that he would have saved many lives in a real nuclear attack. Unfortunately, it was all too evident that Argonne itself, convenient to Midway Airport and downtown Chicago, would not have survived nuclear war as a functioning laboratory.[2]

As the Berlin crisis escalated in the summer of 1961, President Kennedy urged Americans to intensify civil defense preparations. That summer nuclear war loomed more threateningly than at any time since the Korean War. Although the laboratory was as ready as possible, many Argonne personnel realized that they were little prepared at home to survive a nuclear attack. Home shelters, encouraged by the Kennedy administration, might provide significant protection from radioactive fallout. But ordinary citizens were largely unequipped with the radiation monitoring devices necessary to determine when it was safe to evacuate an area following a nuclear blast. In California, the noted physicist Luis W. Alvarez designed a home fallout meter that cost less than 20 cents. Contained in an ordinary water glass lined with aluminum foil, the Alvarez fallout meter employed static electricity generated by a common pocket comb to "power" his instrument constructed of foil discs, nylon thread, Scotch tape, and a tin-can lid. At Argonne, Bill Karraker of Site Administration replicated Alvarez' device to prove that it really worked. *Argonne News* thought the Alvarez fallout meter was so ingenious that it printed detailed plans for laboratory readers.[3]

Albert Crewe Appointed Laboratory Director

Argonne's principal concern during the tense summer of 1961 was not the Cold War, however, but rather the selection of a new director to replace Norman Hilberry, who was retiring after more than four years as Argonne's administrator. New leadership in the White House, at the Atomic Energy Commission, and at the University of Chicago raised hopes of finding a dynamic chief with an international scientific reputation. Glenn Seaborg, chairman of the AEC, and George Beadle, chancellor of the University of Chicago, took personal interest in the recruitment of the new director. Consequently, both laboratory insiders and academic scientists were surprised when the University of Chicago named Albert Crewe, Argonne's new director of the particle accelerator division, to succeed Hilberry.

Only thirty-four years old, Crewe was not only a nonveteran of the Manhattan Project, he was also a British citizen at the time of his appointment. Born in Bradford, Yorkshire, in England, Crewe received his university education and

Ph.D. in physics at the University of Liverpool working with James Chadwick, who won the Noble Prize for discovering the neutron. While teaching at Liverpool, Crewe pursued cosmic-ray research and contributed to the development of England's diffusion-cloud-chamber technology. He won international fame in 1954 when he became the first scientist to extract a continuous proton beam from a high-energy particle accelerator. Using powerful magnets, Crewe deflected protons moving 150,000 miles per second inside the Liverpool accelerator to an external target area without appreciably disturbing the velocity or focus of the proton beam. Before Crewe's achievement, proton bombardment experiments were conducted within the accelerator's vacuum chamber, where space for detecting and recording equipment was severely limited. By extracting the proton beam from the 400 MeV synchrocyclotron, Crewe enabled scientists to expand significantly the range of experiments that explored the fundamental nature of matter and energy.

In 1954, when Roger Hildebrand, associate laboratory director for high-energy physics, visited Liverpool to learn how the English had extracted potent proton beams from a synchrocyclotron, he discovered that "the essential component of their system was a Yorkshireman named Albert Crewe." Back in the United States, Hildebrand persuaded Herbert Anderson to hire Crewe to assist the Americans with a similar project. The following year, Crewe moved to Chicago as an assistant professor of physics and joined the staff of the university's Enrico Fermi Institute of Nuclear Studies. Soon thereafter, he was named technical director of the university's synchrotron program, for which he helped design an underground facility for Chicago's external proton beam. Crewe accepted his first major management position in September 1958, when Hildebrand asked him to direct Argonne's particle accelerator division. This assignment placed Crewe in charge of the 12.5 BeV–ZGS program.[4]

The reasons for Crewe's selection as Argonne's director three years later remain somewhat obscure. Although the University of Chicago sought advice from the AEC, from Argonne senior scientists, and from the Associated Midwest Universities (AMU), the University of Chicago searched for Hilberry's successor without the help of a formal selection committee. Later, Beadle publicly conceded that it had been difficult to find a qualified science administrator to lead Argonne laboratory. Beadle's ultimate choice, Crewe was not only an untenured assistant professor at the University of Chicago and a relatively inexperienced Argonne program manager, but until he could secure American citizenship, Crewe was also a British alien without a "Q" security clearance necessary for access to secret atomic-energy information. On the other hand, at a time when the laboratory was poised to launch new ventures in basic re-

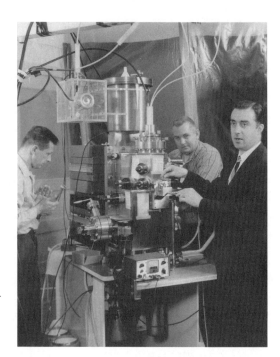

Figure 23. Albert Crewe discusses his design of a high-resolution, high-constrast electron microscope. In the background are J. G. Simmelman and F. W. Reed.

search spearheaded by the ZGS, he was a bright and promising young scientist with a distinguished international reputation in high-energy physics (figure 23). Crewe recalled being recruited personally by William Harrell, the university's vice president, but he himself was rather mystified by Beadle's decision. Years later he speculated that perhaps the university was looking for a new face, someone untainted by the university's long running controversy with MURA.[5]

The ZGS Nears Completion

As the scientist in charge of constructing the ZGS (and recently arrived from England), Crewe had been isolated from the politics of high-energy physics in the Midwest. After the Atomic Energy Commission authorized construction of the ZGS at Argonne, Crewe was left alone politically to get on with the job of building the machine, which John Livingood had planned as senior physicist and first director of the particle accelerator division. Under Crewe's guidance, steady progress was maintained from ground breaking in June 1959 until November 1961, when Lee Teng, who was responsible for the ZGS design concept, became director of the particle accelerator division.

For Teng, construction of the ZGS was a continuous battle. Every day he faced new problems—"some technical, some fiscal, some organizational, some personnel, but all critical." Teng's group completed construction of the machine in July 1963, and on September 18, the $42-million particle accelerator produced a proton beam of 12.7 BeV, surprisingly higher than the nominal 12.5 BeV for which the ZGS had been designed. Occupying forty-seven acres, the ZGS complex included fifteen interconnected buildings grouped around the doughnut shaped synchrotron structure, 210 feet in diameter, which housed the ring of eight magnets weighing a total of 4,000 tons. The $7-million High Energy Physics Building adjacent to the ZGS served as administrative headquarters and offered labs and office space for visiting scientists as well as for Argonne employees.[6]

Since World War II, physicists, especially high-energy physicists, constituted a new elite among American scientists. A powerful "microscope," the ZGS enabled Argonne scientists to join the AEC family of high-energy physicists who explored the realm of atomic particles that would have to be magnified a thousand billion times were they to be seen by the naked eye. For years, the bevatron at the University of California had been America's most powerful accelerator; later, Brookhaven's AGS became the world's most energetic accelerator; MIT's electron accelerator would soon be eclipsed by the Stanford linear accelerator as the world's largest electron accelerator.[7]

From the beginning it was clear to Teng and his colleagues "that it would be difficult to carry out a trail-blazing high-energy physics program on the ZGS" because their machine was about half as energetic as the accelerators at Brookhaven (33 BeV) and CERN (28 BeV), at Switzerland, and came on line more than two years behind. They tried to make up for these deficiencies with "better and fancier" experimental equipment, especially the superb resolution of the 30-inch bubble chamber. Argonne scientists anticipated that the ZGS would produce the most "prolific" (i.e., intense) and "tractable" proton beam, which could be easily extracted to associated experimental apparatus. Already, the ZGS had produced an intensity of thirty billion protons per pulse. With additional adjustment and tuning, the ZGS would accelerate a trillion protons per pulse. Eventually, Argonne scientists believed they could achieve ten trillion protons per pulse. Roger Hildebrand, working with the University of Chicago's cyclotron and bubble chamber, already whetted their appetites for experimental results by investigating whether the mu meson particle was a heavy unstable electron. The laboratory proudly reported that Hildebrand "observed the capture of a mu meson by a proton, creating a neutron and a neutrino. This inverse beta decay is what is to be expected when an electron is captured by a proton." Hildebrand's experiment provided "experimental validation that a mu meson is, indeed, a heavy unstable electron."[8]

Confident that the ZGS could produce all atomic particles known or anticipated, Hildebrand also realized that the ultimate success of the ZGS depended on the involvement of physical scientists from universities throughout the Midwest as well as those from Argonne. Unfortunately, few midwestern high-energy physicists had participated in the early planning and construction of the ZGS, although once the ZGS was operational, Argonne did receive help in fine-tuning the proton beam. Inspired by inter-university cooperation at Brookhaven, and assisted by Ned Goldwasser, Hildebrand set up a ZGS users Committee that became a model for other high-energy physics programs. While Hildebrand could not entirely abrogate his responsibility for running the ZGS program, the users committee was to provide de facto governance for reviewing, planning, and prioritizing research on the ZGS. Crewe assured the policy advisory board in September 1961 that Argonne had followed the AEC's instructions to build a machine as quickly as possible "in order to provide a facility for the Midwest." As the ZGS neared completion, interest grew steadily among the universities, where eighteen groups of physicists (and chemists) planned experiments and constructed equipment for use with the ZGS when it became operational. Despite the promises, some midwestern scientists were skeptical about Argonne's good faith in establishing the ZGS users committee. Crewe would learn later that there was deep-seated fear that, once the machine was completed, the University of Chicago would simply "pull the rug" from under midwestern university access to the ZGS.[9]

Despite the political undercurrent, Albert Crewe was excited, and a little awed, by his selection to head Argonne National Laboratory. Somewhat reluctantly, he relinquished direction of the ZGS project just as it was nearing completion. A devoted research physicist, Crewe would have loved to stay with the ZGS, on which he had spent three and a half years of his young professional life totally dedicated to building Argonne's accelerator. "I wanted to sit in that control room and twiddle all those damn dials and make that machine work," he later reminisced with colleagues. But Crewe also welcomed the opportunity to lead the national laboratory. As Argonne's mission steadily moved away from military-related projects, Crewe opined that "it was a very exciting time" to lead Argonne in research and development of peaceful uses of atomic energy. As a physicist, Crewe knew enough about nuclear reactors to realize that he would be working with some of the world's most outstanding scientists and engineers in the field. With security clearance impediments resolved by October, his appointment as director of Argonne National Laboratory was cleared for November 1, 1961.[10]

Allerton

Crewe's appointment coincided with a reassessment of laboratory-university relationships conducted by the Association of Midwest Universities. Four years after their organization, the AMU board of directors saw only modest progress in establishing university partnerships with the laboratory. Although encouraging, the ZGS users group represented only narrow access to Argonne's varied programs. Through the policy advisory board, the AMU regularly reviewed and advised on Argonne programs, but university participation in research remained limited in other areas. Hildebrand bluntly challenged the AMU to help themselves by constructing housing, providing automobiles, chartering airplanes, and funding research without cash assistance from the Atomic Energy Commission. On January 5, 1962, the AMU board of directors gathered with good intentions at the University of Illinois's Allerton estate to explore how the Argonne-university partnership could be strengthened.[11]

Rather than promoting harmony and goodwill, however, the Allerton meeting unexpectedly exposed the deep suspicion that continued to exist between Argonne and the universities. The AMU acknowledged that the universities needed to do more to make Argonne scientists feel that they were a part of a larger scientific community. But mostly, Crewe thought, the meeting focused on what Argonne could do for university scientists. Believing that as a former academician himself he could speak frankly to the assembled professors, Argonne's new director cautioned that cooperation could not be regarded as a one-way street. While academic scientists seemed to regard Argonne as an "adjunct or extension" of the university, Crewe saw the problem in reverse—he wanted "to see Argonne becoming more involved *in* the universities!" His vision included the possibility of exchanging scientists between the laboratory and the universities.[12]

The Allerton meeting was an eye-opener for Crewe, who was "astonished" at the animosity that existed in the midwestern academic science community against the University of Chicago. Incredibly, a few academic wags teased Crewe that Chicago was not really a university because it lacked an engineering school, a nursing school, or whatever. More seriously, Crewe was dismayed that his frank comments at Allerton were widely distributed in the AMU's published report about the conference. He was further upset that the conference openly disparaged Argonne's biology program, which Crewe privately conceded was weak but as a new administrator felt compelled to defend against outside criticism. In part, Crewe's sensitivity stemmed from the fact that it was unclear just who the AMU board represented. A few of its members were presidents from the participating institutions; others were administrators or professors formally appointed to represent their schools. But some of the AMU board members were simply faculty who spoke only for themselves and not their respective institutions. Because it was unclear just who and what the Association of Midwest Universities actually represented, Crewe was uncertain how and where to answer AMU criticism levied at the laboratory.

Crewe was also deeply troubled by the diffuse academic animus directed toward Argonne National Laboratory. University relations were only a part of Argonne's mission but would, no doubt, become increasingly important to the laboratory. In the immediate aftermath of World War II, Argonne had a clear objective to support the United States defense program in ways not directly related to university research. By 1962, however, Crewe was not at all clear on what Argonne's outlook would be, nor indeed what the future would be for the national laboratories. During the Kennedy administration, Los Alamos and Livermore intensified their commitment to the AEC's weapons program while Berkeley and Brookhaven strengthened basic "academic" research efforts. Oak Ridge also hoped to become more involved in teaching and academically related research. Although Argonne continued to focus much of its research on nuclear-reactor development, Crewe forecast that the laboratory's prosperity was wedded to university participation. But he was frustrated by the tone of the Allerton conference, which seemed to concentrate only on "how the universities could *use* Argonne, its facilities, and its staff." Almost too candidly, Crewe reminded the professors that some Argonne scientists actually distrusted the AMU as well. Tension between the midwestern universities and the laboratory would soon intensify.[13]

The Argonne Graduate Center

The Atomic Energy Commission was also interested in the relationship between its laboratories and higher education. When University of Chicago chancellor George Beadle and the commission opened negotiations to renew the university's operating contract, the AEC asked Beadle about Chicago's real interest in Argonne. Acknowledging the laboratory's importance to midwestern research institutions, Beadle assured the commission that the university wanted to foster intellectual and scholarly work at Argonne to the maximum benefit of regional universities.[14]

On returning from Washington, Beadle asked Crewe how Argonne might best strengthen its relationship with the University of Chicago. Crewe suggested establishing a center to coordinate those graduate studies leading to a Ph.D. in the physical and biological sciences. Crewe's suggestion for creating an educational center at the national laboratory was not original. For years, graduate students at the University of California had received degrees for research conducted at the Lawrence Berkeley Laboratory. When President Eisenhower's science advisory committee suggested that the national laboratories might do more to aid scientific education, Alvin Weinberg, director of Oak Ridge National Laboratory, proposed in January 1962 that the national laboratories might be converted gradually into MIT-type research centers with major responsibilities for educating Ph.D.'s in the sciences. The Argonne graduate center as envisioned by Crewe would not actually grant degrees but, rather, would administer graduate programs as an adjunct of the University of Chicago. Graduate courses would include not only physics, chemistry, biology, and mathematics, but also applied science or engineering, filling what many considered a major gap in the university's professional graduate programs. Crewe projected 500–700 full- and part-time students to be taught by 80 Argonne faculty (600 Argonne scientists held Ph.D.'s). The advantages to the University of Chicago (expansion) and Argonne Laboratory (talented young research assistants) were obvious, but there was no evident advantage for the AMU in Crewe's proposal.[15]

Crewe prepared his plan for the Argonne graduate center in good faith. Because the request originated with George Beadle, Crewe felt constrained to keep his proposal confidential until it was approved by the President and Board of Regents of the University of Chicago. Unfortunately, although Crewe and AMU president W. R. Marshall had exchanged promises to improve communications, rumors of the graduate center reached AMU officials through unofficial channels. Alarmed, AMU executive director John H. Roberson arranged for an October 1962 meeting with the AEC's General Advisory Committee to discuss laboratory-university relations. In anticipation of their discussions with the committee, Crewe sent Marshall a copy of his presentation notes, in which, for the first time, the AMU president saw an outline for the Argonne graduate center. Even this gesture proved in vain when Marshall and his AMU colleagues were unable to study Crewe's notes beforehand. Although conversations with the GAC focused on the graduate center proposal, Marshall professed to have no opinion one way or the other about the plan because he had had no opportunity to study it.[16]

Beadle and Crewe were quick to apologize for not keeping the AMU informed of their plans, but irreparable damage had been done. Crewe tried to explain to the policy advisory board that he had kept plans for the graduate center confidential out of deference to his employers at the University of Chicago. Nevertheless, he believed that the Argonne graduate center would prove beneficial to midwestern universities by strengthening AMU programs and improving Argonne's relations with midwestern research faculty. Potentially, the graduate center could include participation from schools other than the University of Chicago; but now the AMU schools were wary of Crewe's assurances. Why was a proposal intended to benefit all of the schools drafted in secret, the policy advisory board asked. Were the AMU schools not trusted, as Crewe had intimated after Allerton, or was it only a costly lapse in communication, as Beadle had said in order to sooth the board.

Debate over procedures completely swamped substantive discussions about the proposed graduate center. Politely, but coldly, the AMU stopped Crewe's idea. Joseph C. Hirschfelder of the University of Wisconsin and the only member of the board to comment on the merits of the idea, warned that any serious involvement in teaching would necessarily erode Argonne's research output. Questioning the wisdom of national laboratories becoming involved in graduate education, Hirschfelder reminded the board that, after three years, the University of California had abandoned its graduate program at Los Alamos because staff members spent too much time teaching, to the detriment of their research. In the end, John Williams of the University of Minnesota moved that the University of Chicago withdraw its proposal for a graduate center "without prejudice pending further discussion and analysis." In truth, the idea was dead, killed partly by Beadle's and Crewe's inexperience, but mostly by AMU's deep-seated distrust of the University of Chicago. Incredulously, Crewe saw his golden plan to improve Argonne's basic research and graduate studies turned into dross by the AMU and subsequently buried by various midwestern presidents and deans.[17]

High-Energy Priorities

The dust had barely settled over the grave of the Argonne graduate center when controversy renewed over midwestern high-energy physics. While the ZGS moved toward completion in 1962, the General Advisory Committee and the president's science advisory committee jointly agreed to seek expert advice on the United States' priorities in high-energy physics research. Norman Ramsey of Harvard headed the common panel, with Goldwasser, Williams, and Seitz among the ten committee members.* After almost a year of study, in May 1963, less than two months before completion of the ZGS, the Ramsey panel reported its ten-year projection for high-energy physics.

---

*Other committee members included Philip H. Abelson of the Carnegie Institution in Washington; Owen Chamberlain, University of California; Murray Gell-Mann, Cal Tech; T. D. Lee, Columbia; Wolfgang Panofsky, Stanford; and E. M. Purcell, Harvard.

The scientists generally agreed that high-energy physics could move in two directions in the 1960s: toward higher energies or toward greater intensities. The Ramsey panel unambiguously favored higher energies by endorsing construction at the earliest date of the 200 BeV machine designed by the University of California, followed some years later by a 600–1,000 BeV machine proposed by Brookhaven. So that the Midwest would not be left with just the ZGS, the panel also endorsed constructing MURA's high-intensity 12.5 BeV accelerator in Madison, Wisconsin, provided that funding for the MURA machine did not delay construction of the big high-energy accelerators. The Ramsey panel, which had reviewed Argonne's program, also recommended that plans be developed to coordinate use of the ZGS and the MURA facility.[18]

Revival of MURA's 12.5 BeV accelerator, intended to pacify midwestern scientists, whom the Ramsey panel believed would not receive a major machine, instead stirred up old rivalries and quarrels between Argonne and the MURA group. Worried about what would become of the ZGS once the MURA machine became operational, Crewe, Beadle, and Warren Johnson, the University of Chicago's representative on the policy advisory board, challenged the Ramsey panel's assumptions and recommendations. The MURA accelerator not only might jeopardize the future of the ZGS, but even if both machines remained operational, it could also relegate midwestern high-energy physics to second-class status. These arguments, of course, had been heard before. In another reprise of the 1955 debates, Crewe and Beadle argued that if the MURA machine were to be built anywhere, it ought to be constructed at Argonne National Laboratory, where Hildebrand's users group could coordinate experiments on both accelerators.[19]

Seitz desperately appealed for peace. Trying to finesse the question of where to locate the MURA accelerator, Seitz emphasized the need for midwesterners to present a united front in order to secure a second machine. With the East and West Coasts dominating competition for new construction while the Midwest bickered, Seitz predicted that if the MURA machine were not accepted, it would be at least fifteen years before the AEC would build another accelerator in the Midwest. Seitz's prediction turned out to be wrong, but his plea helped unify the midwesterners. The question of location aside, the policy advisory board, including Beadle and Johnson representing Chicago, voted unanimously to endorse the Ramsey panel report by encouraging the AEC to build the MURA accelerator as soon as possible.[20]

Although recommended by the Ramsey panel and endorsed by midwestern scientists, Kermit Gordon, director of the Bureau of the Budget opposed the MURA project. Consequently, the matter landed on President Kennedy's desk. Seitz learned that the president, who favored expanding the government's support for basic research, was likely to approve the MURA proposal. Inopportunely, the president's assassination in November threw the issue into the lap of his successor, Lyndon B. Johnson. Less than a month later, Johnson met with the MURA leaders and their congressmen in Washington and settled MURA's fate with a bombshell. Goldwasser reported this fateful meeting: "President Johnson heard us out, at least in form, but then pulled a piece of paper out of his pocket and read to us a decision which obviously had been made prior to the meeting. The MURA project was terminated."[21]

Applying balm to the sting, Johnson urged MURA's "fine staff" to continue "to serve the Midwest through the universities and at Argonne." He informed Senator Hubert Humphrey of Minnesota that Glenn Seaborg had agreed to use his good offices to promote midwestern cooperation. Johnson hoped that government and university partnership could "build at Argonne the nucleus of one of the finest research centers in the world." With the president's encouragement, Humphrey asked President Elvis J. Stahr of Indiana University to rally the midwestern universities behind Argonne National Laboratory while finding ways to secure "a greater voice" in the laboratory's program management. Understandably, the desire of President Johnson and Senator Humphrey for scientific harmony in the Midwest proved a powerful incentive for all parties to settle their differences. Hildebrand's users group would provide the basis for extended cooperation. As promised, on January 14, 1964, Seaborg mediated a peace conference held in Washington, D.C., to discuss ways by which midwestern scientists might share greater access to the ZGS and perhaps gain some say in establishing research priorities for the laboratory.[22]

The Williams Committee

While Beadle held out the olive branch of reconciliation, Seaborg got to the heart of the problem. The University of Chicago's professed good intentions notwithstanding, the midwestern universities had no official claim on Argonne National Laboratory, as Crewe's ill-fated proposal for the graduate center had clearly demonstrated. More profoundly, many MURA scientists harbored suspicions that Argonne had played a role in the termination of their project. The ZGS users group, although appreciated, held no formal charter and played no firm role in program direction. With the authority of the president behind him, Seaborg directed Beadle to establish an ad hoc committee to resolve these problems.[23]

The following day, Crewe proposed a joint committee representing MURA, the AMU, and the University of Chicago to respond to Seaborg's request. Endorsed by the policy advisory board, the committee, chaired by John Williams of Minnesota, included three representatives from MURA (President Stahr from Indiana, Bernard Waldman who was director of MURA, and Vice President Peterson from Wisconsin); two from AMU (Williams and Goldwasser); Warren Johnson, vice president of the University of Chicago; and Crewe from Argonne. Officially, the Williams committee was directed to study the near-term problem of upgrading the ZGS and enhancing its availability to users and the long-term problem of promoting the growth and development of high-energy physics in the Midwest.[24] Goldwasser, however, believed that there were two additional, and more important, questions for the Williams committee to address: "1) how to increase the confidence of the university physicists in the Argonne program, and 2) how to construct the organization at Argonne so that it would have some chance of attracting MURA personnel to Argonne to participate in this program."[25]

The ensuing discussions of the Williams committee focused on new arrangements for laboratory governance, which would share responsibility for Argonne management. Frequently, the debates became acrimonious, with Crewe and Johnson in the minority pitted against Williams and the rest of the committee, which envisioned creating a governing board modeled after Associated Universities, Inc., the prime contractor for Brookhaven National Laboratory. Crewe and the University of Chicago were willing to share direction of high-energy physics—that is the ZGS—but they opposed community management of the entire laboratory and incorporating reactor programs and other technologies of little academic interest.

Williams and his allies pressed forward, however, advocating a new operating contract for Argonne, which would be managed by a consortium of perhaps fifteen midwestern universities. Just who would be members of the consortium remained vague, but in recognition of Chicago's special interest in Argonne, the majority of the Williams committee was willing to appoint Beadle as the first chairman of the board. The university countered with its own proposal, in which the university remained the prime contractor but was assisted by a new board with expanded responsibilities that would replace the existing policy advisory board. The differences were not subtle. The Williams committee wanted to replace the University of Chicago as the prime contractor with a multiuniversity consortium to manage the laboratory, whereas the University of Chicago offered to work with a new advisory board, which would provide "advice and consent to the appointment of the director" and review program and education policy.[26]

Ultimately, the Atomic Energy Commission stepped in to resolve the impasse when Commissioner John Palfry suggested a compromise: a new tripartite compact among the commission, the University of Chicago, and a new corporation of midwestern universities in which Chicago remained as the laboratory's operating manager while the midwestern universities accepted responsibility for establishing laboratory policy and program direction. Seaborg secured Beadle's agreement "in principal" to the tripartite arrangement, whose details were purposely left vague in order to obtain the consent of all parties. By September 1964 the Williams committee unanimously adopted the tripartite plan, with Crewe and Johnson concurring that "the new tripartite contract shall be such as to assure the University of Chicago that it will be able effectively to operate the Laboratory in a manner responsive to the policies established by the new corporation." This brave promise, of course, would yet require concrete implementation.[27]

The Tripartite Agreement

Almost no one was really happy with the Tripartite Agreement. The University of Chicago administrators, in general, and Crewe, in particular, worried how Argonne's director could manage the laboratory with a board of professors, many of them bitterly critical of Argonne, officially looking over his shoulder and determining laboratory policy. Publicly, Crewe was upbeat, but he and Warren Johnson had just lost a bruising fight and feared that the university physicists would care little about fostering the fast-reactor program, Argonne's first priority in research and development at the time.[28]

Although they had won a major bureaucratic victory, many of the midwestern scientists remained suspicious of Argonne and were skeptical that their role in the Tripartite Agreement would ever be anything more than that of a glorified advisory committee. MURA would lose its identity as a result of the agreement and the AMU, its future uncertain with the expiration of its contract in September 1966, was left in limbo by the Williams committee, which recommended the creation of a new not-for-profit corporation to be organized by a small group of midwestern universities and the University of Chicago. For almost twenty years, since 1946, the University of Chicago had been the prime contractor for Argonne with a variety of arrangements to incorporate the participation of regional universities: the Board of Governors of Argonne National Laboratory (1946–50), the Executive Board of the Council of Participating Institutions (1950–58), and the Board of Directors of Associated Midwest Universities (1958–68). Among the national laboratories, only Oak Ridge was managed by a private corporation, the Union Carbide Corporation. Brookhaven, of course, was run by Associated Universities, Inc., which was organized by nine private northeastern universities. The other large AEC laboratories—Los Alamos, Lawrence Berkeley, Livermore, and Ames—were operated by single universities, as Argonne had been for two decades. Now, stimulated by the politics of high-energy physics, Argonne Laboratory embarked upon an experiment in laboratory management that was unique among the AEC's research facilities. It took two years to organize the Argonne Universities Association (AUA) and to negotiate a new, five-year contract for operation of Argonne National Laboratory.

Despite their agreement in principle, the University of Chicago and the nascent AUA, which was organized in the spring of 1965, remained far apart in their understanding of Chicago's role as "manager-operator" of Argonne National Laboratory. From the university's perspective, the new governing board was to "make or approve" major decisions on *planning* or modification of scientific programs or facilities. The AUA, on the other hand, intended to establish policy for the *operation* of the laboratory *in all its phases*. Key points in conflict included appointment of the laboratory director and employees. The university assumed it would appoint the director, with the concurrence of the board and the Atomic Energy Commission. All laboratory employees would remain employees of the University of Chicago, consistent with AEC requirements and board policy. The Argonne Universities Association, in contrast, was determined to hire the director, deputy director, and associate laboratory directors, and to establish their salaries and duties, making the laboratory's executive staff employees of the board. Under the AUA scheme, the University of Chicago would no longer enjoy direct contact with the AEC but would communicate all management questions through the board.[29]

The AUA's interpretation of the tripartite principle not only upset the university, but it also proved unacceptable to the AEC, which would not accept any management arrangement that precluded direct contact with laboratory personnel on any and all issues deemed appropriate by the government. Still fostering a "spirit of friendly cooperation" to achieve scientific excellence in the Midwest, the commission protected the University of Chicago's operating responsibilities by insisting on a strong program manager directly responsible to the AEC. In the final contract, the University of Chicago retained full fiscal responsibility, and the AUA was expressly relieved of such authority.[30]

After the tripartite contract was signed on October 31, 1966, the old AMU merged its programs into the new organization of the Argonne Universities Association. Proponents of the AUA originally wanted to keep their organization small by limiting membership to the fifteen or so schools that belonged to MURA. But the expected demise of the AMU in June 1968 created pressure to expand the Argonne Universities Association to twenty-six (later thirty) research universities. Although it remained studiously neutral about the matter, the University of Chicago actually welcomed the large organization—the "more the merrier" it felt—which opened the doors for widespread regional participation. Optimistically, Crewe hoped that the Tripartite Agreement would strengthen Argonne's ties with the Midwest's great universities and provide a focal point for unified action in all scientific disciplines represented at Argonne.[31]

Central States Universities

Major research universities were not alone in their desire to benefit from cooperation with Argonne laboratory. In May 1965 twelve small midwestern universities organized as the Central States Universities, Incorporated.* Less concerned than the AUA about securing access to Argonne's major research facilities, the Central States Universities (CSU) sought partnerships that encouraged graduate education in the biological sciences, physical sciences, engineering, mathematics, and especially studies related to the nuclear sciences. Established to promote science education, the CSU also assisted faculty members with their continuing education in contemporary science and with undergraduate honors students preparing for graduate school. George G. Mallinson, dean of graduate studies at Western Michigan and chair of the CSU board, was grateful that Argonne offered faculty and research opportunities not available at the smaller schools. He also noted that "Argonne staff members appear to welcome the flow of ideas and the challenge provided by their association with university professors and students" from the CSU schools.[32]

Responsive to the increasing importance of cooperative education programs, in July 1965 Crewe reorganized Argonne's education division. Frank Myers continued as associate laboratory director for education, while the old Institute of Nuclear Science and Engineering, which was originally created to train foreign nationals under Eisenhower's Atoms for Peace program, became the office of college and university cooperation with the office of educational affairs. Crewe appointed Rollin G. Taecker, formerly director of the institute, to be director of Argonne's cooperative program, which in 1964–65 hosted more than 500 faculty and staff and 1,500 students from more than 200 educational institutions.[33]

---

*They were Western Michigan, Northern Michigan, John Carroll, Bowling Green, Kent State, Miami of Ohio, Toledo, DePauw, Northern Illinois, Southern Illinois, Ohio University, and Iowa State College.

Crewe and Basic Research

Albert Crewe placed basic research foremost among Argonne's priorities. During his tenure as Argonne's director, Crewe pushed basic research budgets ahead of spending for reactor development and other technical programs for the first, and only time, in Argonne's history. He distinguished, however, between "pure" basic research and "objective" basic research. Argonne pursued "pure" basic research, Crewe explained, solely because its intellectual challenge led to a greater understanding of the world in which we live. "Objective" basic research, on the other hand, explored natural phenomena in order to achieve a particular objective. Experiments with the ZGS that explored the fundamental structure of nature illustrated "pure" research. Biological research seeking answers about the effects of ionizing radiation, or Miriam Finkel's study of virus-induced bone cancer in mice, exemplified research directed toward objective medical knowledge. A third category, applied research, included most of the research related to the development of nuclear power reactors for the Atomic Energy Commission. Closely coordinated with the laboratory's engineering and electronics divisions, applied research, for example, employed the results of objective basic research on the behavior of reactor cores and the effect of intense radiation on reactor materials to design working power reactors.

Crewe's lesson was elementary. Unlike the Soviet Union, which he estimated invested little in basic research, pure or objective, Crewe believed that basic research flourished best in a democratic atmosphere. In American universities and national laboratories, Crewe found "some of the best basic and applied research in the world." The only weakness he saw was the need to develop better channels of communication among government laboratories, universities, and industry. He allowed himself to hope that the Tripartite Agreement might serve as a bridge across the "Gulf of Town and Gown" as he knew it.[34]

Sachs Named Associate Laboratory Director

In a major step toward improving relationships between the laboratory and the MURA scientists, on December 4, 1963, during the gala ZGS dedication dinner, Crewe had announced the appointment of Robert G. Sachs as associate laboratory director for high-energy physics.

A student of James Franck and Maria Goeppert-Mayer at Johns Hopkins University, Sachs bypassed his bachelor's and master's degrees to plunge into the study of theoretical nuclear physics. Goeppert-Mayer's husband, Joseph, was on the Johns Hopkins faculty, but because of nepotism rules, she could not serve on the faculty. Nevertheless, she agreed to direct Sachs's dissertation and encouraged her student to seek advice from Edward Teller, then at George Washington University. Consequently, Sachs became Goeppert-Mayers's first student while working on a problem suggested by Teller. He graduated from Johns Hopkins in 1939 with but one degree, a Ph.D. in theoretical physics.[35]

Sachs, a veteran of the Metallurgical Laboratory where he had worked with Farrington Daniels, helped Zinn establish Argonne Laboratory in 1946. Appointed as director of the theoretical physics division, Sachs recruited most of the senior scientists including his former mentor, Maria Goeppert-Mayer, who had moved to Chicago with her husband but still could not obtain a regular position at the university because of the nepotism rules. In 1947 Sachs moved to the University of Wisconsin, where he was appointed a professor of physics in 1948. He worked closely with the MURA group at Wisconsin and looked forward to the research possibilities of the very-high-intensity machine. In August 1963, however, Hildebrand convinced him to return to Argonne to run the ZGS program.

A strong supporter of the MURA accelerator but a stronger advocate for high-energy physics in the Midwest, Sachs assumed in August 1963 that it would be at least six years before the MURA machine became operational. In the meantime, he accepted Hildebrand's challenge to make the most of the research opportunities provided by the ZGS. With his good relations with the MURA group and with other midwestern scientists, Sachs believed he "could help bring things together." Largely successful in this endeavor, he was still surprised to encounter persistent bitterness toward the University of Chicago. "A lot of people felt I was a traitor," he recalled, but eventually "they got over it."[36]

Above all, Sachs was enthusiastic about the research potential of the ZGS and its 7-degree external beam and 30-inch bubble chamber. Thirty years later he still remembered the exciting race for the Ω^- (omega minus) particle:

> Now, there was one experiment that the 7° beam was really designed to do with the 30" [bubble] chamber, and that was to discover the Ω^-, as its called. [Murray Gell-Mann at the High Energy Conference held at CERN in 1962 had predicted the existence of the Ω^-.] It was a very unusual particle, heavier than the others. It was a question of time when one would find it, to some extent, you needed intensity and you needed a very high resolution bubble chamber. The optics in the 30" chamber which were designed by Wisconsin were exceptional.[37]

The ZGS had the intensity and the 30-inch bubble chamber had the optics to do the job. Once the ZGS was up and running, the first order of business was to produce the Ω^-. "And it was a race, a tremendous race, with Brookhaven" to be the first to find the heavy particle, Sachs reminisced. Unhappily, because the Long Island laboratory was also years ahead of Argonne in operating its alternating gradient synchrotron, the Brookhaven bubble-chamber group discovered the Ω^- in 1964, just weeks before the ZGS and its 7-degree beam and 30-inch bubble chamber came on line at operating intensity. Ultimately, the Argonne group produced a large number of Ω^-, but this success would never be as satisfying as the actual discovery of the particle. Despite Argonne's failure to discover the Ω^-, Sachs was proud of the productivity of the group experimenting with the 30-inch bubble chamber. In all, they published 64 *Physical Review Letters* and 114 articles in *Physical Review* and *Nuclear Physics*.[38]

A Not-So-Noble Gas

On the far right-hand side of the periodic table of chemical elements, helium, neon, argon, krypton, xenon, and radon are situated in valence group 8, the rare or noble gases, so called because chemists believed them to be chemically inert. In an age when new discoveries routinely upset familiar scientific truth, most chemists accepted the verisimilitude of the inert gases. Argonne chemist John G. Malm joked that some colleagues actually threatened to quit chemistry the day they found a solid compound of a noble gas sitting on a laboratory shelf.

In July 1962 Malm and his colleagues Henry Selig and Howard Claassen were intrigued to read in the *Chemical Society Proceedings* that a chemist from the University of British Columbia reported a reaction between xenon and platinum hexafluoride. Working with Cedric Chernick, Malm replicated the Canadian's experiment within days but determined that other hexafluorides reacted somewhat differently.

Surmising that xenon just might react with the fluorine directly, Malm's group conducted a very simple experiment using ordinary chemical techniques with equipment already connected to a fluorine tank. After reacting condensed xenon and fluorine at 400° C in a nickel can, they cooled the can with dry ice and pumped off the excess fluorine. To their surprise, the pressure gauge registered zero. Even at room temperature, the new compound registered low vapor pressure. Fearing a mistake or impure xenon, they tried again with similar results. Malm was convinced that the compound was XeF_4, xenon tetrafluoride: "The little square crystals looked like many other crystals except they were very regular and very brilliant. They almost sparkled under the light. The compound was slightly volatile at room temperature. When heated, the crystals grew very rapidly" (figure 24).[39]

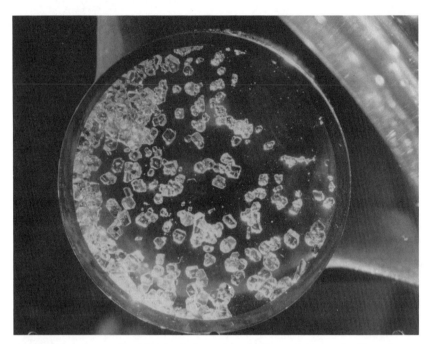

Figure 24. Xenon tetrafluoride (XeF_4), the reaction product of xenon, a "noble gas" heretofore thought to be nonreactive, and fluorine

In 1962 Argonne played a major role in creating a new field of noble-gas chemistry. By the end of the year, laboratory chemists had produced xenon tetrafluoride (XeF_4), xenon difluoride (XeF_2), and xenon hexafluoride (XeF_6), all white solids; and xenon oxide tetrafluoride ($XeOF_4$), a colorless liquid. In addition, they fashioned the fluorides of other noble gases.

The discovery of xenon fluorides typified Crewe's depiction of basic research at its purest—the exploration of nature's secrets for knowledge alone. Along the route, however, they unexpectedly discovered that xenon fluorides react with water to form stable solutions of compounds, such as xenon trioxide (XeO_3), containing xenon and oxygen. Chemists were fascinated that the "inert gas" xenon behaved similarly to iodine. As long as the compound remained in solution, it was very stable but reacted as a powerful oxidizing agent that could release chlorine gas from hydrochloric acid. On the other hand, dried xenon trioxide became a violent and sensitive explosive similar to TNT.

Noble-gas chemistry also quickly opened the door to more "objective" basic research as defined by Crewe. Although industrial applications were uncertain, theoretically, xenon compounds might be used selectively as fluorinating or oxidizing agents. In addition, nuclear reactor scientists had been puzzled for years by the unusual behavior of xenon fission products in certain nuclear reactors. The discovery of xenon fluorides provided important insight into the chemistry of nuclear reactor cores. Within a year of the discovery of xenon tetrafluoride, the laboratory sponsored a major conference on noble-gas chemistry featuring sixty papers by 150 authors; the proceedings were published by the University of Chicago Press as *Noble Gas Compounds*.[40]

Maria Goeppert-Mayer and the 1963 Nobel Prize in Physics

Waiting patiently for the photographer, she stood alone among the men on that warm summer day in 1959. A dozen leading physicists from America and Europe had gathered at Argonne for a summer-long symposium on the problems of nuclear structure. Maria Goeppert-Mayer, a former senior scientist and group leader in the theoretical physics division, quietly bade farewell to her Argonne colleagues. Following the conference she would move with her husband, Joseph, to the University of California at San Diego to accept a regular full-time appointment as a full professor of physics. Four years later, Goeppert-Mayer learned that she had won the Nobel Prize in physics for work she had done at Argonne National Laboratory (figure 25).[41]

Goeppert-Mayer's prize celebrated the laboratory's most distinguished achievement in basic research during Argonne's first two decades. As a theoretical physicist at Argonne, Goeppert-Mayer had contributed to the design of the Experimental Breeder Reactor I (EBR-I). But it was her work "Elementary Theory of Nuclear Shell Structure," completed in collaboration with J. Hans Jensen of the University of Heidelberg, that won her and Jensen the 1963 Nobel Prize in physics, which they shared with Met Lab colleague Eugene P. Wigner.

Inspired by Edward Teller's ideas about the creation of elements, Goeppert-Mayer began her work on nuclear structure at Argonne and at the University of Chicago when she and her husband moved to the Midwest campus in 1946. At this time, nuclear physicists theorized that atoms were composed of a very dense, positively charged nucleus of protons and neutrons surrounded by a distant cloud of negatively charged electrons. The tiny, spinning electrons whirled around the nucleus, often sharing the same orbit. But the laws of atomic structure rigidly determined the number of electrons in each orbit, which the scientists called "shells."

Building on nuclear theory developed at the University of Chicago, on the encouragement of Teller and Enrico Fermi, and on dialogue with colleagues

Figure 25. Maria Goeppert-Mayer (*Frontiers, 1946–1996*, p. 85)

David Inglis and Dieter Kurath, Goeppert-Mayer pondered how best to picture nuclear structure; that is, how were protons and neutrons arranged in the atomic nucleus. Although many physicists, including Goeppert-Mayer, had suggested that the nucleus might be comprised of "shells" of orbiting protons and neutrons also spinning like tops, conventional shell models based on atomic structure did not work out mathematically. A student of Max Born in Germany, Goeppert-Mayer was well grounded in the theory of quantum mechanics and generally far ahead of most physicists in her mathematical sophistication.

One day Fermi dropped by Goeppert-Mayer's office while she was puzzling over how experimental data on nuclear structure could be reconciled with the shell model. They were interrupted when Fermi was called away to the phone.

"What about spin-orbit coupling?" he suggested as he hurried out the door to get his call.

Jolted with sudden insight, Goeppert-Mayer began her calculations anew while her mind pictured how one could postulate spin-orbit coupling of protons and neutrons in the nucleus. By the time Fermi returned to her office, Goeppert-Mayer was euphoric.

"That's it! That's it!" she cheered as her explanation tumbled out faster than Fermi could understand.

"Calm down, Maria, calm down," Fermi gently counseled his friend. "Go home and sleep on it and come back and tell me about it tomorrow." The next day, of course, Goeppert-Mayer's preliminary calculations were just as sound as they had been the day before.[42]

In May 1948, in the first of a number of works on the subject, Goeppert-Mayer presented the seminar "Closed Shells in Nuclei"; three months later *Physical Review* published her exposition of shell theory. Whereas scientific colleagues appreciated the significance of her achievement, the Atomic Energy Commission, engrossed in the nuclear weapon program, paid little attention to her accomplishment. A few years later, in collaboration with Jensen, she published *Elementary Theory of Nuclear Shell Structure*. Goeppert-Mayer's discovery reflected the synergy of modern science. Inspired by ideas from both Teller and Fermi and assisted by Jensen, Goeppert-Mayer's brilliant mathematical solutions established the theory. She was the first Argonne scientist to win the Nobel Prize, and only the second woman (after Marie Curie) to receive the honor in physics.[43]

Argonne National Laboratory Celebrates Twenty Years

In 1966 the Argonne community commemorated two decades as a national laboratory by recalling past triumphs in science and engineering while initiating momentous changes in laboratory management and programs. Although the laboratory's historic leadership in nuclear science and technology was being celebrated, there was little consensus about Argonne's future, its missions, or its programs. Argonne had just passed its twentieth anniversary when President Lyndon Johnson visited the Idaho National Reactor Testing Station in August to dedicate EBR-I as a national historical landmark. The president, accompanied by AEC chairman Glenn Seaborg and congressional and state dignitaries, praised Argonne for its pioneering leadership in developing peaceful nuclear power symbolized by the first generation of electrical energy by EBR-I in 1951. Undaunted when the public address system failed, Johnson predicted that by 1980 more than 20 percent of the nation's electrical power would be supplied by civilian nuclear reactors. Johnson's resolve to support nuclear power, which Richard Nixon reaffirmed in 1971, set the agenda for Argonne Laboratory for more than a decade. Johnson had hoped to fund the Vietnam War without requiring major sacrifices from his Great Society programs. Unfortunately, presidential commitment to power reactor development ultimately eclipsed the federal government's support for Argonne's non-reactor programs.[44]

Argonne and the Breeder-Reactor Project

Since its 1962 report to President Kennedy on the future of civilian nuclear power, the Atomic Energy Commission had supported a growing program for developing breeder reactors. Argonne's EBR-II at the NRTS and the Power Reactor Development Company's Enrico Fermi Fast-Breeder Reactor, near Detroit, were both operating at low power levels by 1964. The commission planned to use these facilities, as well as the proposed Southwest Experimental Oxide Reactor (SEFOR), to test fuels, materials, and reactor safety engineering for the first generation of breeder reactors. To provide an advanced facility for testing reactor components, in 1962 the commission authorized Argonne to build the Fast-Reactor Test Facility (FARET) in Idaho.[45]

Commissioner James T. Ramey, who formerly served as the executive director of the Joint Committee, was impatient with these efforts and argued not only that the commission's program lacked urgency, but also that it was too diffuse to achieve satisfactory results. Consequently, in November 1964 the commission selected Milton Shaw from Admiral Hyman Rickover's nuclear navy program to head the reactor development division. A hard-driving Rickover protégé, Shaw immediately applied his interpretation of the admiral's management philosophy to implement an aggressive breeder-reactor development program. After a year of intense review and debate, he convinced the commission in November 1965 to increase the priority of the fast-breeder program by committing more resources to reactor development and by drastically overhauling both project and management structure. Modeling the Liquid Metal Fast Breeder Reactor (LMFBR) after Rickover's successful Shippingport project, Shaw requested and received from the commission a strong mandate to manage the breeder program from Washington. As the AEC's historians have noted, commission management of civilian reactor projects had come full circle from the government-controlled Shippingport project through the loose government-industry partnership of the power demonstration program and finally back to tight government management under Shaw's determined leadership.[46]

Albert Crewe immediately recognized the threat that Shaw's juggernaut posed to Argonne. At the American Power Conference in April 1965, Argonne's director openly criticized Rickover's methods which were being applied ever more widely to commission programs. Where human health and safety were immediately at risk as in the nuclear submarine and the space programs, high levels of quality assurance through exhaustive design and repeated proof testing were justified. But Crewe did not believe that these same managerial techniques could achieve optimum development in the breeder-reactor program where human health and safety requirements were more long-range concerns. If Rickover instead of Fermi had been in charge of CP-1 (the first graphite reactor), Crewe doubted whether the United States would have achieved the first chain reaction. He reminded the power association that existing nuclear technology rested on the experimental approach; that is, project construction began early and problems were solved as the project progressed. Crewe predicted that the dominant "cult of perfection," obsessed with avoiding occasional or imaginary embarrassment, would so retard the development of the breeder reactor that the United States eventually would end up buying breeder technology from Europe.[47]

For Argonne, the consequences of the commission's centralizing management of the breeder program in Washington were drastic. First, the commission canceled Argonne's Fast Reactor Test Facility in favor of a larger Fast Flux Test Facility (FFTF), to be built by the Pacific Northwest Laboratory at Hanford, Washington. Second, Shaw established an LMFBR program office at Argonne that would report directly to Shaw's staff in the reactor development division. While press releases stated that the changes would strengthen Argonne's role as the primary laboratory for overall research and development of the LMFBR program, in fact, Shaw's reforms shifted responsibility for designing and operating the principal reactor test facility to a rival laboratory and virtually stripped Argonne of all autonomy in planning and managing the breeder project.[48]

The decision to cancel FARET, Argonne's fast-reactor project, in favor of the Fast Flux Test Facility did not necessarily mean that the new test reactor would be built at Hanford rather than at the National Reactor Testing Station in Idaho. Although the commission reported that it would take at least a year to make the final siting decision for the FFTF, the fact that the reactor was promoted by the Pacific Northwest Laboratory left little doubt as to what the final determination would be. The $25-million FARET was designed and ready for construction when Shaw informed Argonne that its project had been canceled. Congressman Melvin Price promised a Joint Committee investigation, but the protest was futile. When the commission made its final decision for siting the FFTF in January 1967, Shaw's staff argued that overlap in the design and construction activities weighed heavily in selecting the Hanford site. Not surprisingly, the commission decided that the Pacific Northwest Laboratory's team, which had designed the FFTF, would build the reactor and then operate it.[49]

Shaw saw little reason for diplomacy when he informed Argonne's director of the changes in management for the breeder program. Henceforth, Argonne would "serve as an extension" of Shaw's office to assist the headquarters staff in the "planning, coordination and evaluation" of the technical activities associated with the LMFBR program. Argonne, once designated as the nation's lead reactor laboratory, became the captive handmaiden of Shaw's Division of Reactor Development and Technology. Shaw allowed no independent initiatives from the field. In November 1965 he authorized creation of an LMFBR program office at Argonne as a "distinct technical organization" staffed with full-time senior scientists and engineers. Shaw directed that the LMFBR program office be separate from other Argonne divisions and that it report to the laboratory director. Taking no chances that he was misunderstood, Shaw ordered the new office "to support and be responsive to" his headquarters staff in upholding the commission's requirement for improved technical management of the breeder program.[50]

The LMFBR program office officially opened on January 24, 1966. Shaw assured the office that it would be assigned as much detailed planning, program assessment, liaison, and technical direction as possible, but he left no doubt that policy decisions and program actions required approval from Washington. Shaw refused to provide the new office with a general charter but, rather, insisted that specific tasks would be assigned when appropriate. Initially, the Division of Reactor Development and Technology in Washington assumed responsibility for preparing plans for fuels and materials, plant design, reactor components, and reactor instrumentation, whereas Argonne was to be responsible for reactor safety, sodium technology, and fuel reprocessing.

Shaw postponed establishing detailed reporting procedures, but borrowing a management tactic from Rickover, he designated two personal representatives from his own staff to set up watch-dog offices at Argonne and in Idaho. He also demanded to be kept "fully and currently informed" by receiving copies of all trip reports, conference reports, and other major program documents, including conclusions and recommendations. He required planning documents to include program objectives, criteria, standards, alternatives, funding, schedules, personnel, facilities, special materials, priorities, responsibilities, functions, liaison, rationale, and industrial participation. Argonne's immediate role was to focus on technical planning and assessments, particularly reactor performance characteristics, materials behavior and requirements, alternative technical approaches, test and experimental requirements, and facilities requirements. Shaw also asked the laboratory to provide special studies and assessments as required.[51]

Argonne's first priority was to produce a general plan by February 1, 1968. To get things started, Crewe named Alfred Amorosi, a twenty-year veteran of reactor development, as director of the LMFBR program office. Amorosi, known to be informal but firm, selected Paul Gast to serve as associate director and Norman Jacobson, an experienced technical writer, to be assistant director. Dan Finucane, an experienced Argonne hand, became administrative officer. Amorosi coordinated directly with Max Jackson, Shaw's site representative at Argonne. The draft plan prepared by Amorosi's office eventually consisted of ten volumes, including an "Overall Plan," which provided a "road map" for breeder reactor development and commercialization, and nine additional volumes corresponding to the nine technical components of the LMFBR.[52]

Although the Pacific Northwest Laboratories continued to lead the FFTF effort, Shaw willingly entertained FFTF assignments for Argonne so as to exploit the laboratory's special resources and experience. Shaw acknowledged Argonne's strength in reactor physics, nuclear safety, fuels and materials testing, and fuel reprocessing, and he anticipated that Argonne's EBR-II would play a major development and testing role in the breeder program. In addition, not all was lost of the FARET efforts. Shaw expected that results from FARET fuel and material design tests, especially on mixed-oxide fuels and advanced cladding, could be applied directly to the FFTF endeavor. Finally, Argonne's plutonium fabrication facility, fuel-cycle facility, and fuels technology center were to be mobilized in support of the breeder-reactor program.[53]

As a postscript, Shaw noted that Argonne's new, or redefined, role in the LMFBR program required consideration of the laboratory's other programs and commitments. Shaw was vague about what he meant and perhaps did not himself sense all the implications of the management revolution he had implemented at Argonne. The Tripartite Agreement was not yet signed when he restructured the commission's reactor development program so, initially, Shaw did not have to contend with participation of the midwestern universities in the management of the laboratory. He offered assurances that programs in physical research, biology and medicine, as well as other programs from the Division of Reactor Development and Technology would be considered along with Argonne's requirements to support the LMFBR. When centralizing reactor management at commission headquarters, however, Shaw did not anticipate that his actions might crash headlong into the regional ambitions of midwestern universities.[54]

Serious resistance to Shaw appeared during a meeting of the reactor engineering division review committee on December 6, 1966. Committee members included Alexander Sesonsk (chair), Manson Benedict, and Gilbert H. Fett. Stephen Lawroski, associate laboratory director for reactor programs, briefed the committee on long-range trends in the relationship between the laboratory and the commission. Argonne had steadily lost its ability to pursue basic nuclear reactor research. Because Shaw required minutely detailed proposals for new undertakings and also required the AEC's approval before any new work could be started, independent exploratory research had become impossible. The laboratory was starved for discretionary research funds and no longer retained flexibility in the assignment of reactor division personnel. The review committee tried to regain a measure of independence in the reactor development field by requesting the commission to allocate discretionary funds for basic research to the laboratory director's office. The committee members, however, were pessimistic that Shaw would support their request for greater budgetary flexibility, and they expressed their determination to take the issue to the general manager or to the commissioners themselves, if necessary. Indeed, they were prepared to rally the other national laboratories behind their cause if they determined that Brookhaven and Oak Ridge also suffered from Shaw's inflexible management.[55]

Shaw brooked no opposition. With strong support from Commissioner Ramey and Senator Henry Jackson on the Joint Committee, he swiftly counterattacked, asserting that Argonne and its management had not given the reactor programs sufficient priority. More seriously, he charged that the laboratory had not been responsive to his direction. Specifically, he believed the "down-time" of the EBR-II reactor was excessive and negligent. With the loss of the Fermi Nuclear Power Plant near Detroit after a partial core meltdown in October 1966, EBR-II became the only operating breeder reactor in the country. Shaw wanted Argonne to centralize all the breeder-reactor activities under the direction of a single project manager. He was frustrated with Argonne's diffuse organization because he could not pinpoint responsibility for specific projects. Rumor swept the laboratory that unless Argonne gave the breeder program appropriate priority, Shaw would find someone who would.[56]

Shaw's frustration bred exasperation at Argonne. With the creation of the LMFBR program office and the new EBR-II project office in March 1967, angry Argonne scientists responded that they had already set up the project-manager system that Shaw requested. But Shaw apparently wanted more than that. Members of the reactor program review committee complained that "he want[ed] to solve today's problems today and to let tomorrow's problems wait." They feared that Shaw's views of Argonne's mission in reactor development threatened to change drastically the complexion of the entire laboratory. Scientists worried that by focusing on immediate results at the expense of long-range planning, Argonne's historic role in developing future reactor concepts or in pursuing exploratory basic research might fall by the wayside. Their questions even went so far as to ask whether Argonne would be involved in reactor development in ten years' time. Although the members answered their self-questioning affirmatively, they insisted that the laboratory must have funding for exploratory research programs. They deplored the fact that the commission had eliminated uncommitted funds, although they knew Shaw acted under pressure from the Joint Committee in this regard.

Transition

While the president and Seaborg fastened the historic landmark plaque to the face of EBR-I and marked the end of one era, the laboratory prepared for transition to new management under the Tripartite Agreement. Albert Crewe predicted that Argonne National Laboratory stood at the beginning of "a new era for science education and scientific excellence in the Middle West." Within the Argonne Universities Association the transition was already well underway. Key AMU people were already serving on the AUA's board of trustees, and the first AUA president, Philip Powers, was Purdue's representative on the AMU board. The merger of the two organizations, with the AUA assuming the AMU's university relations and education advisory roles, only made sense. With the demise of the AMU, Argonne National Laboratory established its center of educational affairs to continue the AMU's educational programs (research, fellowships, and conferences) consolidated with Argonne's other cooperative academic activities.[57]

The Argonne Advanced Research Reactor

Projected at a cost of $25 million, the laboratory's newest reactor, the Argonne Advanced Research Reactor (also known as AARR or A^2R^2) was designed to operate at 100 thermal megawatts, producing a neutron flux eighty times greater than Argonne's "bread and butter" research reactor, CP-5. The high neutron flux—that is, five quadrillion neutrons passing through one square centimeter in one second—would provide abundant neutrons for basic research in nuclear physics, chemistry, materials science, and solid-state physics. Project manager Milton Levenson of the reactor engineering division believed that the reactor would sustain Argonne's research tradition which Enrico Fermi had established with CP-1 and which the laboratory had advanced with the CP-2, CP-3, CP-3', and CP-5 research reactors. Not only would A^2R^2 enable Argonne to pursue experiments impossible with CP-5, but the new reactor would also expedite basic research by allowing scientists to complete their experiments in a much shorter time than ever before. With A^2R^2, Argonne scientists planned to explore the structure and dynamics of matter, including the creation of unique isotopes and new elements in the periodic table. Crewe believed this to be an ideal machine for Argonne's basic science programs.

From its inception in 1961, however, the A^2R^2 was embroiled in controversy. Argonne scientists believed that construction of a high-flux reactor was essential if their laboratory was to remain a major research center in low-energy nuclear physics and nuclear chemistry. Since the establishment of the Metallurgical Laboratory, a research reactor had been the focal point of the physical research program. From CP-1 through CP-5, reactor scientists sought increasing neutron intensity to complete ever more precise and difficult experiments. As early as 1954, with the completion of CP-5, a 4.6-megawatt reactor, Argonne scientists started to work on designs for the next generation of research reactors, which would increase available neutron flux by a factor of ten.[58]

In November 1958, scientists from Argonne, Oak Ridge, Brookhaven, and the University of California met with the commission's research division to discuss the future high-flux research facilities. With a consensus that the United States needed to upgrade high-flux research, the research division authorized Oak Ridge to design a high-flux reactor for isotope production (HFIR), which would support minimum research. At the same time, the division suggested that Argonne's management propose building a high-flux reactor primarily for neutron beam research. Although the Argonne staff initially explored upgrading CP-5, it ultimately proposed constructing a high-flux research reactor, whose costs rose steadily from $13.7 million in fiscal year 1961 to $22.7 million in fiscal year 1964. Each year, either the Atomic Energy Commission or the Bureau of the Budget refused to endorse the project.[59]

No national laboratory can prosper without state-of-the-art facilities. Consequently, when the Joint Committee promised to provide the laboratories with the "essential tools" to keep "pace with new developments and new needs,"[60] Argonne scientists concluded that congressional policy supported construction of the A^2R^2. Yet, they noted with dismay that their laboratory had already fallen behind Oak Ridge, whose facilities included superior research reactors, cyclotrons, and Van de Graaff generators. With the completion of a new high-flux reactor at Upton, New York, Argonne's scientists also predicted that Brookhaven National Laboratory would soon forge ahead of the Illinois laboratory.

The struggle to secure funding for the high-flux reactor at Argonne was unending. In December 1962 Chairman Seaborg reported that the Bureau of the Budget had struck funding for the Argonne reactor from the 1964 budget, despite the fact that the commission had given the project "the highest priority." The situation was especially bleak because President Kennedy himself refused to support the reactor. Nonetheless, Crewe encouraged University of Chicago president George Beadle to apply all pressure possible on Senator Everett Dirksen and Congressman Melvin Price, both members of the Joint Committee, to secure funding for A^2R^2. Although he thought it unwise to badger Seaborg directly, Crewe believed it would be possible "to trigger a local storm" of protest about the lack of federal support for research and development in the Midwest. He was confident that Governor Otto Kerner of Illinois would support the protest vigorously.[61]

Within a month, Crewe and Vice President William Harrell of the University of Chicago secured Melvin Price's assistance. In this and subsequent discussions with the Joint Committee, the Argonne representatives acknowledged their rivalry for scientific preeminence with the other American laboratories, but they emphasized even more the Cold War competition in low-energy physics. Although the Oak Ridge and Brookhaven reactors cost approximately one-half the proposed Argonne research reactor, neither would enable Americans to be competitive with Soviet or European research. Apparently, even the Japanese were not far behind. During the Joint Committee's 1965 authorization hearings, Representative Chet Holifield asked how the proposed Argonne reactor compared to the one visited by Glenn Seaborg and Gerald Tape at New Mellicaz, Russia. Tape replied that the A^2R^2, which pushed the limit of present technology, would exceed the flux design of the Russian reactor. Without their reactor, however, Argonne scientists predicted it would be the United States, not the rest of the world, who would be playing catch-up.[62]

Although Argonne received authorization on June 30, 1964, to construct its high-flux reactor, the project continued under a cloud of financial and technical problems. Despite reassurances from Argonne's management that the reactor would be built within budget, escalating research-and-development costs alarmed managers at the AEC and at the Bureau of the Budget. Originally, the A^2R^2 was designed as a 240-megawatt plant, but Argonne engineers soon determined that their $25-million budget would enable them to build only a 100-megawatt reactor. Yet time and money had already been spent designing components for both reactors.

In March 1966 Shaw agreed to support the research-and-development requirements for the 100-megawatt reactor, but he expressed concern that in three years the costs, projected at $3 million in 1963, had risen to over $14 million. When the Argonne scientists replied that they had encountered unexpectedly high costs in designing the Inconel reactor core, in developing economic fuel-fabrication techniques, and in completing safety studies, Shaw wondered whether Argonne really knew what its research-and-development costs would be. So far the total price tag for the A^2R^2, including research, development, and construction, was running close to $39 million. Apparently, the Bureau of the Budget was even more pessimistic than Shaw that A^2R^2 was worth the expense and suggested to Paul W. McDaniel, the commission's director of research, that the Argonne project be scrapped. Seaborg and McDaniel had been worried about lack of progress on the project and now realized that the A^2R^2 was in serious trouble.[63]

Crewe and his Argonne staff were aghast. Quickly, Crewe reminded McDaniel that A^2R^2 was not just another high-flux reactor but, rather, had capabilities unmatched by the HFBR at Brookhaven or the HFIR at Oak Ridge. But more important than surpassing its rivals, the Argonne high-flux reactor provided the center piece for Argonne's cooperative physical research programs with midwestern universities. Failure to complete A^2R^2, Crewe wrote McDaniel, would strangle Argonne's basic physical research programs except for those in high-energy physics. Loss of the high-flux reactor would not only be a serious blow to the commission's low-energy physics program but would also require a "most searching re-examination" of the purpose of the laboratory's physical research program. From Crewe's perspective, loss of the A^2R^2 would eliminate the physical-science program based on research reactors, thus terminating Argonne National Laboratory's historic mission begun by Enrico Fermi. The impact on laboratory morale, he concluded dolefully, "would be incalculable."[64]

With the Bureau of the Budget sequestering project funds, pending reassurances from the Atomic Energy Commission that budgets and schedules would be maintained, the A^2R^2 project team reluctantly gave in to the inevitable. Having already agreed to reducing design power from 240 to 100 megawatts, Argonne consented in October 1966 to reduce research costs by substituting Oak Ridge's proven HFIR aluminum core and control-rod system for their original stainless-steel system. Crewe thought the substitution could be made easily without changing the primary coolant system, reactor vessel shell, top head, external supports, or other structures. He also predicted that the construction schedule would not be affected and that the reactor vessel design would be completed by January 1967 so that construction and procurement in these areas could proceed.[65]

Unfortunately, the design changes further complicated and extended the reactor's preliminary safety analysis report, which required review by the commission's advisory committee on reactor safeguards and approval from the Division of Reactor Licensing. In addition to a technical review of the reliability of the reactor systems, the review committee explored whether the A^2R^2 containment building and reactor facilities could withstand tornado-force winds, a major earthquake, a terrorist missile attack, and a major core meltdown. Because of greater experience, commission safety experts had more confidence in stainless-steel reactor vessels. Argonne technicians were surprised by the detailed challenges from the commission's safety engineers. They had not anticipated that the Division of Reactor Licensing would apply the same scrutiny to government reactors that they gave to commercial units. In a word, Argonne managers were unprepared to answer questions that the government staff regarded as routine.[66]

As delay and rising costs continued to plague the project, frustration mounted in both Washington and Chicago. Although Argonne managers expressed their willingness to make changes, the project scientists at Argonne were still convinced that their original reactor design was satisfactory. In Washington, an exasperated Milton Shaw complained that Argonne was not responsive to questions from the Division of Reactor Licensing. According to Shaw, Argonne had not only failed to provide necessary information but was also unwilling to cooperate fully with the reviewing organizations. From Shaw's perspective, Argonne's behavior was inexplicably self-destructive.[67]

In May 1967, just prior to the June gala ground breaking, Argonne finally yielded to Shaw's insistence that the laboratory adopt the state-of-the-art stainless-steel reactor vessel. Argonne engineers continued to favor the solid Inconel material but were willing to compromise. Nonetheless, Crewe was outraged by Shaw's accusations that Argonne had been uncooperative and unresponsive and, after attempting to set the record straight, denounced the charges as "not only unfair but completely misleading and inaccurate." Then, conceding that the stainless-steel reactor vessel raised no new safety questions, Crewe warned that the change would have an undetermined effect on schedule and costs.[68]

Whereas the battle for the A^2R^2 galvanized Crewe and the Argonne scientists in a fight to save basic research at Argonne National Laboratory, to Shaw and his reactor development staff, the A^2R^2 controversy was an annoyance that diverted energy and funds from their principal mission to develop breeder-reactor technology for the AEC. His usefulness now totally exhausted in the war against Shaw, Crewe stepped aside in time for Acting Director Winston M. Manning to preside at the A^2R^2 ground breaking on June 12. Shaw did not attend. In his envoi to the A^2R^2, Crewe defiantly proclaimed that "this reactor will be built and will be one of the finest reactors of its kind in the world." Manning, William Harrell, Kenneth Dunbar, Milton Levenson, and Philip Powers ceremoniously turned over the first clump of sod with a five-handled shovel. Unwittingly, the multihandled shovel all too clearly symbolized Argonne's major handicap—divided responsibility with uncertain leverage over key issues.[69]

The A^2R^2 ground breaking did not improve relations with Washington. By midsummer, Shaw was again exasperated because he could not determine whether the reactor design had "been established and documented on a sound technical basis." He was frustrated that the design documentation available to him appeared to be "incomplete, inconsistent, and out-of-date." Consequently, he lacked confidence that Argonne had exercised or would exercise "the necessary disciplined approach" to accomplish the engineering on the high-flux reactor project. Shaw repeatedly asked Argonne to confirm its scheduling of key project commitments, while equally frustrated scientists at Argonne believed that Shaw was continually bombarding them with ever new, changing, vague, or undefined criteria. Although the commission's advisory committee on reactor safeguards finally approved the reactor design on October 12, 1967, the battle for the A^2R^2 was almost over.[70]

Shortly after signing the Tripartite Agreement, Crewe surprised the Argonne community by announcing his retirement as laboratory director. The "new era" which Crewe had heralded just a week before required new leadership with new ideas and programs, he wrote President George Beadle on November 8, 1966. He had served as laboratory director since October 1961, having guided Argonne through some of its most challenging years, and he now wanted to return to full-time research at the Enrico Fermi Institute for Nuclear Studies at the University of Chicago. Under the terms of the new Tripartite Agreement he would have to share policy and program direction with the midwestern university group with whom he had bitterly clashed during the controversies with MURA and the Williams committee. Crewe reminded President Beadle that the AUA had the right to approve his continuing leadership, and insiders believed that Crewe decided to resign rather than face a vote of no confidence from the association. On the other hand, Philip Powers, the president of the Argonne Universities Association, assured the public that the new Tripartite Agreement was not a factor in Crewe's resignation.[71]

Powers's assurances notwithstanding, Crewe was tired of the constant struggles with the AUA and the AEC. Although he would have stayed to battle Shaw had he not been frustrated by the Tripartite Agreement, Crewe was also exhausted by his running conflict with the commission's Division of Reactor Development. He had suffered a major defeat in the spring when the commission transferred the breeder-reactor program to Hanford. Thereafter, problems with the A^2R^2 project also helped to wear him down. Crewe was a fighter, but also a realist about what he could accomplish as Argonne's director. Still a young man, now Crewe genuinely wanted to return to research before he was hopelessly out-of-date.

The search for a new laboratory director ended in August 1967 when President Beadle announced that Robert B. Duffield, assistant director of the John Jay Hopkins Laboratory in San Diego, would assume the directorship on November 1, the first anniversary of the Tripartite Agreement. Duffield, who received a joint appointment as director of Argonne National Laboratory and as professor of chemistry at the university, received the unanimous support of the Atomic Energy Commission, the Argonne Universities Association, and the University of Chicago. Another veteran of the Manhattan Project, Duffield had gone to work at Los Alamos in 1943 after receiving his Ph.D. in chemistry at the University of California. After the war, he served as an associate professor of physics and chemistry at the University of Illinois from 1946 until 1956. He then joined the John Jay Hopkins Laboratory, operated for the General Dynamics Corporation by the General Atomic Division, where he took charge of the research, design, and testing of the TRIGA nuclear research reactor. Thereafter in 1959 he became director of the research and development program for the Peach Bottom (Pennsylvania) nuclear power plant, a high-temperature, gas-cooled reactor (HTGR) constructed by the General Dynamics Corporation as part of the AEC's Power Reactor Demonstration Program. The appointment of a nationally known physicist and chemist with experience in both basic and applied nuclear research appealed to the diverse constituencies supporting Argonne National Laboratory. Perhaps as important as Robert Duffield's outstanding credentials was the fact that he arrived at Argonne with a successful history in dealing with Milton Shaw during the construction of the Peach Bottom plant. This experience seemed invaluable in 1967.[72]

8 | New Research Priorities, 1967–72

Winds of protest swept across the land in 1967. Civil rights protests, urban riots, antiwar demonstrations, student sit-ins, faculty teach-ins, hippie love-ins, among other disturbances, buffeted American institutions as never before in the twentieth century. Distracted by their own internal politics and institutional uncertainty, Argonne scientists scarcely perceived the significance of these events for their own future. They were keenly sensitive to the politics of big science, but they worked relatively insulated from the anti-institutional, antiscience unrest that characterized much of the protest of the 1960s. Astounded by the shocking violence that rocked the Democratic Party's 1968 national convention in Chicago, they did not immediately feel the social tremors that shook the piers supporting the national laboratory establishment—trust in the federal government and faith in scientific progress.

In 1967 Robert Duffield, the new director of Argonne National Laboratory, was most preoccupied with navigating the uncharted waters of the Tripartite Agreement. As he guided the laboratory into the Nixon era, Duffield followed Argonne's previously reliable "Telstars"—basic science and reactor development. Unknown to Duffield at the time, the flood tide of basic research had already begun to ebb at Argonne. Caught between expectations of the Argonne Universities Association and the Atomic Energy Commission, Duffield tried to find a middle course between cooperation with the universities and service to the commission. Instead, he was caught in the strong riptide of declining funds for basic research and also in the rising flood of money for applied reactor research and engineering. After this time of trouble, the staff would long remember being deeply alienated from Duffield and the University of Chicago's vice president William B. Cannon.

Although engulfed in the turbulence that accompanied the storm over the commission's breeder reactor program, like a long forgotten explorer, Duffield would leave his mark on Argonne's landscape. Others warrant credit for establishing Argonne's environmental programs in the 1960s, but Duffield, chosen as director because of his experience with nuclear reactors, possessed an uncommon sensitivity to environmental issues and deserves remembrance for establishing Argonne's center of environmental studies (figure 26).

Figure 26. Robert B. Duffield

"A Challenge to Midwestern Universities"

Following Duffield's selection as the new director, President Philip Powers wasted no time issuing the AUA's policy manifesto, "A Challenge to Midwestern Universities." Powers noted that Argonne was a government-owned national laboratory with established resources and strong traditions. But, in order to exercise its responsibility for policy and program review, he thought that the midwestern universities needed to address five issues: (1) How best could Argonne serve the needs of the Midwest as well as the nation? (2) What should the laboratory accomplish through its basic research program? (3) What role should Argonne play in reactor development? (4) Which programs should be fostered and which should be terminated? and (5) What educational role should Argonne pursue?[1]

Central to Powers's report was the old question of Argonne's purpose and role in servicing its historical constituencies of government, industry, and the universities. Clearly Argonne's primary function was to engage in scientific research and development and, at least to some degree, to support scientific education. In addition, Argonne was expected to provide vital support to AEC's programs, particularly in nuclear reactor development. Finally, the laboratory was regarded as a research resource to be mobilized in a national emergency.

Few would debate such a generalized mission for Argonne, but Powers probed more deeply, ultimately challenging the foundation upon which Argonne had been created.

Powers sent the AUA board nine propositions that could revolutionize the mission of the laboratory. First, he asked, should Argonne limit itself to commission-sponsored programs? Because Argonne was a designated national laboratory, Powers argued rhetorically, the laboratory should have a national purpose which transcended the limited mission of the commission. Just because the Atomic Energy Commission was the principal sponsor for Argonne, Powers asked, was the laboratory required to confine its research program to commission-related activities? Were there not other government agencies to which Argonne could provide program planning, coordination, and evaluation?

When Powers examined the appropriateness of Argonne's relation to a single government agency, he also asked about the laboratory's relationship with the universities and industry. Should Argonne compete directly with midwestern universities for limited research funds, perhaps even diverting resources from the campuses to the laboratory? Or, as Powers evidently preferred, should Argonne and the universities launch cooperative projects that transcended the resources of each party individually? Should Argonne confine itself to areas of acknowledged expertise, or assist industry in commercialization of proven technology? Powers was not opposed to partnerships with industry, but clearly his vision for Argonne was more expansive and dynamic. As a national laboratory serving national needs, Powers believed that Argonne should not be limited a priori in any respect, although as a practical matter the laboratory's programs were determined by the priorities and resources invested by its sponsors. He did not believe, however, that Argonne should remain solely a handmaiden to the Atomic Energy Commission.

What was at stake, Powers implied, was Argonne's future as a basic research laboratory in the physical sciences, mathematics, biology, and medicine. Nuclear reactor research and development, especially the fast-breeder reactor, would continue to receive the highest priority from the Atomic Energy Commission, while high-energy physics, because of political support from the Joint Committee on Atomic Energy, was not threatened by immediate budget cuts. But as recent events indicated, commitment to Argonne's other programs appeared soft. From the commission's perspective, basic research at Argonne was essential to explore areas that promised results pertinent to practical problems confronting the commission, Argonne, or the other national laboratories. Powers challenged the commission's rationale for basic research. It might be even more important, Powers suggested, that Argonne's basic research program strengthen the cooperating midwestern universities by attracting and retaining the most competent scientists and engineers in the region. Perhaps the commission's needs for basic research should have second priority to the goal of building up the scientific and educational reputation of the Midwest: "Should a purpose of basic research be to explore those areas of primary interest to the senior scientists in the laboratory and their peers, and thereby increase the fund of scientific knowledge, regardless of pertinency to AEC 'mission?'"[2]

Powers's challenging questions, of course, had no definitive answers, but they served notice to the commission, to the University of Chicago, and to Argonne's leaders that the AUA under Powers's leadership would not simply accept an advisory role but, rather, fully expected to exercise full partnership in establishing program and management priorities for the laboratory.

Formation of the Argonne Laboratory Senate

In response to Powers's "Challenge," which highlighted the rising influence of midwestern academicians in Argonne affairs, senior scientists at the laboratory organized a laboratory senate in December 1967. Although he had not been present at Argonne during the formative period, Duffield encouraged the establishment of the senate, which modeled its organization on faculty senates common to midwestern research universities. Reflecting their anxieties over Argonne's future, almost all of the senior staff elected to join the senate so they could promote creative research at the laboratory while fostering improved relations with the academic community. Also, when the senators expressed the desire for greater access to Argonne's administrators, especially on matters of performance and operation of the laboratory, Duffield promised he would seek their assistance and advice from time to time as appropriate.

To work with Duffield, the senate established an executive committee composed of 14 elected members and 2 members appointed by the director. Duffield also served ex officio on the executive committee. By the end of their first year, 127 senior scientists (of 128 eligible) were members of the Argonne Laboratory senate. The senate received no formal recognition from Argonne's management, but Manning and Duffield provided assurances that participation would not in any way injure the senior scientists, 26 of whom were also in the laboratory's administration.

Challenged by Powers's manifesto, the Argonne senate explored the purpose, mission, and structure of the laboratory in the context of the Tripartite Agreement. Although the senate affirmed the centrality of Argonne's mission in basic research, the senior government scientists were eager to secure equal recognition with their university counterparts. Prior to the Tripartite Agreement, laboratory management largely had been "homegrown." After the formation of the AUA and during the subsequent negotiations over the Tripartite Agreement, however, the senior scientists at Argonne felt slighted when no one sought their views on how the laboratory should be organized or operated. Under the Tripartite Agreement, midwestern scientists, through their representatives on the AUA board, could profoundly influence the mission and leadership of the laboratory. Argonne scientists did not formally enjoy similar access to laboratory management, and some believed that the staff had become mere pawns in the high-stakes game of big science. Powers's manifesto only served to heighten their suspicions.[3]

Quickly, however, the laboratory senate learned that the Argonne Universities Association was an ally, not an adversary. The AUA warmly welcomed the senate's statement on mission, especially relating to staff-university interactions on cooperative research, joint appointments, and graduate and postgraduate training at Argonne. On the other hand, the AEC responded with alarm when it received a copy of the senate's constitution. On May 13, 1968, Spofford G. English, the assistant general manager for research and development, asked Duffield to explain what the statement meant: What had gone wrong in Argonne management? Had Argonne officials aided and abetted the establishment of the senate? What was the senate really after? What role did the senate expect to play in laboratory policy and program management?

Reminding English that he had not been a member of the laboratory when the senate was established, Duffield sidestepped all questions concerning the laboratory's failure in employee relations or the senior scientists' motives in organizing the senate. He assured the commission, however, that Argonne management was not in any way bound by senate activity. Duffield expected that the senate would aid the "free flow of information and ideas between management and staff," and he did not believe the senate would inhibit management or operation of the laboratory. He thought the senate would help the laboratory, but he promised English that he would remain in charge, subject to the constraints of the prime contract.[4]

The Challenge Met

As the new laboratory director, Duffield could not let Powers's challenge go unanswered, especially when the AUA asked for his response. Duffield conceded that the hardest question to answer should have been the easiest: What was Argonne's unifying mission? As a putative national laboratory, to what extent should Argonne continue its historic focus on nuclear energy research, and to what extent should the laboratory explore broader scientific, technological, and sociological problems confronting the United States?

For his part, Duffield advocated an expanded, more comprehensive, mission. His nuclear physicists and engineers included some of the best systems analysts in the world, and Duffield believed that they could make major contributions toward solving complicated problems in the environmental sciences as well as in reactor technology. He envisioned Argonne leading national research on superconductivity, electrochemical technology, holography, computerized braille readers, and kidney dialysis. He favored strengthening Argonne's commitments to biology and medicine, to applied mathematics and computer sciences, and to alternative energy sources.

But Duffield also reminded the laboratory scientists that they were part of an AEC facility. As long as the commission funded virtually all of Argonne's programs, Argonne's mission would be in concert with the commission's goals. Although he did not criticize Powers directly, Duffield thought it foolish for Argonne to fantasize about establishing bold programs independent from its chief funding source, which defined, guided, and continually reexamined the laboratory's performance. Duffield's civics lesson did not end with the Atomic Energy Commission; he also noted that the commission was subject to the JCAE, which was not "reticent in examining and questioning and criticizing and encouraging" the work of the commission. Support for the laboratory ultimately came from Congress, whose responsibility was to promote programs "to the maximum benefit of United States citizens."[5]

Mindful of the political environment in which he worked, Duffield strongly supported basic research at Argonne. "If Argonne is to prosper as a lively intellectual institution, I believe that it must continue to do basic research," he told the AUA board. Yet Duffield reminded the trustees that support for basic research, always fragile, was under particular pressure as the commission moved its LMFBR program into high gear. Because tight budgets required discriminating priorities, Duffield advised, Argonne had no business expecting support for basic research without rigorous questioning from the universities, the scientific community, and the commission. Indeed, Argonne had to be its own severest critic in assessing the relevance, quality, and efficiency of its basic research.

The basic research areas should be carefully chosen, Duffield believed, using criteria that justified Argonne as a reasonable location for the work. Justification was easy in high-energy physics because Argonne was the promoter, designer, user, and operator of the zero gradient synchrotron. In other fields, such as environmental studies, it was not so easy to define Argonne's special interest or expertise. Argonne had launched a modest environmental program, but the extent to which the laboratory could or should work effectively on environmental problems was a tough question for Duffield. Argonne might make important contributions, but only by starting out slowly and concentrating on areas in which Argonne's traditional scientific and technical skills could be applied. He was wary of plunging too deeply or too fast into sociological research for which Argonne was ill-equipped. With little funding from outside the AEC, Duffield cautiously endorsed exploratory projects in environmental studies, which he acknowledged could immediately mobilize university faculty and students into working groups at Argonne.[6]

Argonne and the Great Society: Environmental Initiatives

On September 19, 1966, Congressman Chet Holifield, chairman of the Joint Committee on Atomic Energy, addressed the Southern Governor's Conference with a call for the Atomic Energy Commission to mobilize its extensive resources to develop a strategy for controlling environmental pollution in American cities. Holifield claimed that he was not merely drumming up work for the commission's well-equipped national laboratories but, rather, was promoting utilization of existing laboratories and personnel for appropriate federal anti-pollution projects. He opposed building costly duplicate laboratories whose resources and expertise were unlikely to match those of the existing labs.[7]

Reacting to Holifield's proposal on November 1, the commission's general manager, Robert E. Hollingsworth, asked how Argonne might engage in environmental pollution research.[8] Responding to Hollingsworth's challenge, E. J. Croke and B. Hoglund of the reactor engineering division offered their vision of Argonne's place in the Great Society. Observing that nuclear science had matured to a point at which the national laboratories no longer had a reason to exist only for nuclear research, Croke and Hoglund followed Holifield and argued that the talent and facilities accumulated at the national laboratories could be redirected to solving national problems. With the same sense of urgency and focus devoted to the development of nuclear energy, the national laboratories were equipped and experienced to study large-system problems such as air and water pollution, waste disposal, crime, transportation, power production and distribution, zoning, alternative energy, and a host of other urban and national issues.[9]

Argonne's study group on environmental pollution, chaired by Harold M. Feder, emphasized that Argonne and its predecessor, the Metallurgical Laboratory, had from their beginnings worked on environmental problems. Before environmentalism reached the national agenda, Argonne's environmental studies program focused on meteorology and the life sciences with work conducted by researchers in biology and medicine, radiological physics, industrial hygiene, safety, chemistry, and chemical engineering. The question for Argonne was not whether to include pollution studies in its mission, but rather whether to deepen and broaden its commitment to environmental research. Not surprisingly, the study committee stressed that the commission's multidisciplinary laboratories offered facilities, talent, and experience ideally organized to accept this national commitment. Quoting from Revelation, "and from the shaft rose smoke, like the smoke of a great furnace, and the sun and the air were darkened with the smoke from the shaft. . . . By these three plagues a third of mankind was killed, by the fire and smoke and sulphur issuing from their mouths . . . ," Feder reported near unanimity of opinion that environmental studies were a "challenging, worthy, and suitable task for Argonne National Laboratory."[10]

The Chicago Air Pollution Systems Analysis Program, a study of sulphur dioxide concentrations in Chicago, was Argonne's first venture into environmental research that was not sponsored by the commission. Aided by the Clean Air Act of 1967, the project was funded for three and a half years by the National Center for Air Pollution Control of the Department of Health, Education, and Welfare and by the Atomic Energy Commission. Chicago's Department of Air Pollution Control assisted the joint venture, but without contribution of city funds.[11]

The project developed the Chicago Air Pollution System Model, which was designed to predict levels and concentrations of sulphur dioxide in the Chicago atmosphere, to establish a pollution alert system, and to develop plans for pollution abatement. Engineers and systems analysts from Argonne's reactor engineering division designed the model, while mathematicians in the applied mathematics division offered data storage and code development. Meteorologists in the radiological physics division provided information concerning local weather conditions to be factored into the pollution model. The project also received vital assistance from scientists at the Chicago Department of Air Pollution Control, Northwestern University, and the University of Chicago.

Developing a reliable pollution forecasting system was a pioneering effort at a time when multivariable pollution-control systems were primitive. Chicago proved an ideal laboratory for the Argonne modeling project because for two years the Department of Air Pollution Control had collected data on temperature, wind speed and direction, humidity, and sulphur dioxide concentration from eight locations around the city. Argonne had access to the necessary raw numbers from which to build an air-pollution database. The project ultimately proved invaluable to the Illinois Environmental Protection Agency, which used the Chicago model to help implement air pollution–control planning guidelines promulgated in 1970 by amendments to the Clean Air Act.

For Argonne, the Chicago environmental project was important for two reasons. First, it helped establish the laboratory's reputation for using systems analysis and mathematical modeling as a management tool for research-and-development projects. More importantly, at a time when the AEC was becoming increasingly impatient with Argonne's management, the Chicago project established good working relationships with a federal agency that "accepted and supported the Argonne approach to problems." Subsequently Argonne signed a second contract with the National Air Pollution Control Administration to conduct research on fluidized-bed combustion using coal and other fossil fuels.[12]

Environmental Studies Authorized

In December 1967, the JCAE authorized the national laboratories to undertake research-and-development programs related to public health and safety. The Joint Committee's authorization enabled the laboratory to pursue projects that were Duffield's pride: the low-cost hemodialyzer (artificial kidney), the high-temperature alkali battery, and the braille-reading machine.[13] The committee's authorization and subsequent amendment of the Atomic Energy Act of 1954 also permitted Argonne to expand its environmental activities, provided this work did not interfere or conflict with the commission's principal mission or require significant funding from the commission.

The AUA enthusiastically endorsed broadening Argonne's mission to include major environmental-studies projects. The group quickly approved of the Chicago air pollution system model project and, during the summer of 1967, organized a committee on environmental pollution chaired by Frederick J. Rossini, a vice president of Notre Dame University. David C. White, an environmental scientist from MIT, was hired as a staff consultant.[14]

Although scant funding was immediately available for environmental research, AUA leaders had high hopes for the initiative. Not only was environmental research new, exciting, and morally imperative, but it also nudged Argonne toward becoming a multipurpose laboratory. As such, Argonne expanded access to the laboratory's staff and facilities for university faculty outside the mainstream of nuclear research or the physical and biological sciences. Murray Joslin, an AUA board member from Consolidated Edison, thought industry would welcome Argonne's venture into environmental studies. Along with its commitment to the fast breeder reactor program, Joslin believed the laboratory ought to be engaged in pollution research. Faculty from AUA universities who were engaged in environmental research observed that pollution was more than a scientific or technical problem, but was also a political, economic, and social issue. Midwestern faculty encouraged the AUA to adopt a broad approach that embraced social, as well as scientific, study of environmental problems.

While coordinating closely with Duffield, the Argonne Universities Association proposed its own agenda for environmental research at Argonne. The committee unanimously recommended creating a center for environmental pollution to foster collaborative research between university and Argonne personnel on environmental problems. The center would not only recruit university faculty to participate in programs at the laboratory but also might support research on the university campuses. The AUA also established a staff office for environmental pollution. Except for the accelerator users group, there was no Argonne-sponsored activity for which universities played a major role in establishing the research priorities. The center for environmental pollution, jointly run by the laboratory and the AUA, would enhance university participation in Argonne research programs.[15]

The prospects of obtaining federal and state funding for exciting environmental research seemed excellent, but questions remained as to how collaborative university-laboratory research could be organized so that all parties equally shared the authority and responsibility. Some members of the AUA even questioned the need to confine their collaborative efforts to Argonne programs. Frederick Rossini favored amending the founders' agreement to permit the AUA to engage in activities independent from Argonne. Convinced that environmental research required social, political, and economic emphasis well beyond Argonne's traditional skills or commitment, Powers asked J. Boyd Page, the dean of the graduate school at Iowa State University, to explore the "desirability and feasibility of establishing a socio-technological research organization" to coordinate environmental studies among the midwestern universities and Argonne National Laboratory. Although Powers was open to suggestions, he believed that Page's study group should focus on two alternatives for the new organization: (1) it could be part of Argonne but under policy guidance from the AUA, or (2) it could operate independently from Argonne and thus be responsible only to the AUA.[16]

Powers's creation of the Page committee on the Socio-Technological Research Organization (STRO) placed the Argonne Universities Association on a collision course with Argonne. From Argonne's perspective, the AUA best served by assisting the laboratory to compete for resources from Washington while concurrently helping to provide a counterweight to excessive management by the commission and the Joint Committee. The midwestern universities, however, regarded Argonne not only as an important regional resource, but also as a potential major competitor with them for limited federal funds. President Johnson's National Environmental Policy Act of 1969 encouraged the AUA to think expansively and not to regard itself as a junior partner with the laboratory. In January 1969 the Page committee recommended the creation of an independent socio-technological research organization sponsored by the Argonne Universities Association and located in the Chicago metropolitan area.[17]

Duffield faced a quandary. He needed support from midwestern universities to fight for Argonne's autonomy in basic science, but he felt his position was undercut by the association's strike for independence, which he knew would never be supported by the commission. To counter the AUA's proposal, Duffield promised to establish a center for environmental studies whose permanent director would report to him. Duffield welcomed broad, interdisciplinary, interorganizational, and intercollegiate projects established by the Argonne Universities Association and allied study groups. But he required that each center project be assigned a project director responsible to laboratory management. He could be creative and flexible in devising working arrangements between the center for environmental studies and the association, but he could not abdicate his responsibilities as laboratory director and did not endorse an independent organization whose rivalry might undermine Argonne's programs.[18]

The O'Hare Conference on the Environment

Recognizing that environmental research required widespread support from midwestern constituencies, the AUA's board decided to organize a regional conference to explore and evaluate the implications of establishing an independent midwestern environmental center. On July 27–29, 1969, the association sponsored the conference "Universities, National Laboratories, and Man's Environment" at Chicago's O'Hare Airport. The conference, which attracted over 250 participants from universities, government, and industry, focused on two themes: "Is Man on His Way to Extinction?" and "Organizing to Shape and Understand Man's Environmental Interactions." The first theme provided an academic attraction for prominent environmentalists and politicians. The second theme enabled the AUA to test sentiment on the establishment of an independent socio-technological research organization. Among the participants were Réne Jules Dubos, a Pulitzer Prize–winning conservationist; Elvis Stahr, president of the National Audubon Society; Henry S. Rowen, president of the Rand Corporation; John L. Buckley, director of ecology, U.S. Office of Science and Technology; Lawrence Hafstad, retired vice president for research, General Motors Corporation; James T. Ramey of the AEC; Chet Holifield, chairman of the JCAE; and Congressman Melvin Price.[19]

The conference concluded almost unanimously that humans were not headed for extinction. Nonetheless, the conferees also agreed that the United States faced an immediate environmental crisis associated with the nation's enormous appetite for energy. The universities were called upon to incorporate environmental studies into their undergraduate and graduate curriculums. Partnerships among government, industry, and academia to address environmental problems were also encouraged. The conference demanded definition of national and regional values and priorities. Although the participants recognized the need to include sociologists, political scientists, economists, architects, lawyers (and even humanists!) in environmental discussions, the conference did not produce a mandate to create an independent environmental center in the Midwest.[20]

Figure 27. Argonne's first four directors. Left to right: Walter Zinn (1946–56), Norman Hilberry (1956–61), Albert Crewe (1961–67), and Robert Duffield (1967–72).

While the AUA continued debate on how best to organize its environmental efforts, the laboratory launched its Great Lakes research programs. Initially funded by the commission to monitor fallout from nuclear weapons tests and deposition of metallic elements in fish, Argonne broadened the Great Lakes study in the fall of 1969 to examine the environmental impact of nuclear power plants on Lake Michigan and the other Great Lakes. Sailing on the *Cisco,* Argonne fishermen netted research specimens while the University of Michigan's research vessel *Inland Seas* collected water samples for analysis of radioactive and stable elements. Research teams from the radiological physics division used neutron activation analysis to evaluate the metallic contamination of the flesh and vital organs of the fish. In addition, Argonne scientists measured the thermal plume from the Big Rock Nuclear Power Plant at Charlevoix, Michigan, and meteorologists from the radiological physics division as well as Canada and Australia calculated the natural cooling process of Lake Michigan. Measurements from the nuclear power plant provided baseline data used in siting large power plants on the Great Lakes. All data on the fish and their location, weather conditions, water temperature and purity, and the presence of metals were entered into Argonne's computer to provide vital information for modeling Great Lakes ecology.[21]

When the environmental conference produced no mandate to establish a socio-technological research center, the Argonne Universities Association reassessed its environmental studies initiatives. Powers shifted from organizational to programmatic concerns by asking John Cantlon, the provost of Michigan State University, to explore what programs the AUA would sponsor. In October the Cantlon committee, adopting an ecological approach to environmental problems, recommended that the AUA and Argonne collaborate on an in-depth study of waste management, which had been identified as the most serious environmental problem. In response, the board affirmed that "the improvement of man's environment should become a major new mission of Argonne National Laboratory," but it did not approve hiring staff to assist Powers or authorize him to seek support for the program from either the midwestern universities or from the federal government.[22]

The Center for Environmental Studies

On December 1, 1969, accepting the AUA's challenge that Argonne make "the improvement of man's environment" a laboratory mission, Duffield officially established the center for environmental studies. To serve as director, he selected

Leonard E. Link, a nuclear engineer who had been providing laboratory leadership in the environmental field for the past two years.[23]

From winter through summer of 1970, Duffield, Link, and University of Chicago vice president William Cannon met with Powers, Rossini, and the AUA environmental studies group to explore how the two groups might best collaborate. At issue was the fundamental relationship between Argonne and the Argonne Universities Association and thus the mission of the association itself. Some members, such as Frederick Rossini, believed that the AUA enjoyed an existence independent from the Tripartite Agreement and that it need not confine itself to matters concerning Argonne. Laboratory environmental projects were not necessarily congenial to university interests. Since the amendment of the Atomic Energy Act in 1967, the laboratory itself was free to participate in programs other than those funded by the commission. Argonne was not only evolving into a multipurpose laboratory, but also into a multisponsor laboratory, a development that transcended the terms of the prime contract. If the laboratory were allowed to establish special funding arrangements outside the terms of the Tripartite Agreement, it only followed that the AUA should be free to pursue an analogous independent agenda. While affirming the centrality of laboratory-university collaboration, in May 1970 the AUA established an office of environmental studies, directed by Lynn Weaver, to assist Powers and the trustees in developing an environmental research program for the midwestern universities.[24]

The AUA Environmental Initiative

By June, Powers had asked Richard Caldecott, dean of biological sciences at the University of Minnesota, to develop a regional environmental research program involving the laboratory and AUA institutions. Caldecott's committee, which included faculty from seventeen universities together with Argonne staff members, proposed the Midwest Regional Environmental Systems Program (MRESP). The project, to be administered by the AUA rather than by the laboratory, proposed to develop models that identified major environmental problems in watersheds of the upper Midwest. The project would study waste management, population structure, hydrology and water quality, land allocation, energy, mineral resources, transportation and communication, employment and regional economics, agriculture, and systems analysis and modeling. The concept was breathtaking and potentially expensive. Although the project as described to the National Science Foundation was open ended, first year start-up money totaled $1.7 million, and long-term costs projected to more than $17 million.[25]

Duffield and Cannon cautiously supported the Caldecott proposal, allowing Argonne staff to contribute secretarial support, photocopying, and telephone access. E. J. Croke, who was appointed director of Argonne's center for environmental studies in August 1971, believed the proposal was too vague. On the other hand, it was well written and visionary and did not conflict with Argonne's mission. In contrast, William Harrell thought that the program, if funded, should not be located at Argonne. Given Argonne's responsibilities for reactor development, Harrell did not want the laboratory to be associated with another large project over which it had no control but for which it might be held responsible.[26] Upon Duffield's recommendation, however, the Atomic Energy Commission ultimately determined that the AUA's proposal, if funded by the National Science Foundation, could be accommodated under the Tripartite Agreement.[27]

Unexpectedly, the association's environmental initiative unraveled from within. On January 10, 1971, Daniel Alpert, the dean of the graduate college at the University of Illinois and a member of the AUA's board of trustees, formally protested the actions of Powers and his special committees in promoting the Midwest regional environmental project. Alpert was perturbed by procedural irregularities and policy mistakes. Procedurally, Alpert complained that Powers had kept neither the board members nor the presidents of the participating universities adequately informed about the executive board's efforts regarding environmental studies. But he was more concerned that the proposal was unmanageable and counter to the interests of the member institutions and of Argonne Laboratory.

Alpert believed the Argonne Universities Association had been established solely to participate in the governance of Argonne National Laboratory through the Tripartite Agreement and that it was not equipped to manage a major research project. Furthermore, an external AUA research program would create a conflict of interest for trustees if it competed directly with programs at Argonne and at member institutions. He noted that for the past two years, the AUA staff, presumably Powers and his committees, had labored under a "conflict of commitment" by excessively promoting AUA activities when they should have been providing support for beleaguered Argonne programs. Quoting Ned Goldwasser, a founder and former trustee of the association who was serving as deputy director of the National Accelerator Laboratory, Alpert warned that "[i]f the Board falls prey to the illusion that AUA may become a research institution with a life of its own, it will only result in a weakening of ANL and an alienation of the member institutions."[28]

Norman Hackerman, chairman of the board of trustees, polled the presidents of the member universities to ask whether the AUA should manage a regional environmental program in which the Argonne and university faculty and staff would participate. At their meeting on November 16, at least thirteen of the trustees, including eight from Illinois, Indiana, Michigan, and Iowa, arrived in Chicago determined to vote in the negative. Although the vote was not recorded, the trustees decided not to authorize submission of the National Science Foundation proposal nor to undertake management of research programs external to Argonne National Laboratory. Individual faculty or research groups were encouraged to submit proposals through regular university channels while the AUA returned to its mission of fostering a strong environmental program at Argonne. In response, the association encouraged the center for environmental studies to increase access to Argonne's facilities and environmental programs, especially for faculty from those midwestern universities lacking environmental centers or programs.[29]

In the end, the center for environmental studies stood alone as the coordinator of environmental science programs throughout Argonne Laboratory. By fiscal year 1973, Argonne's budget for environmental studies exceeded $5 million, with almost half of the funding coming from the Atomic Energy Commission for continued work on the Great Lakes research program and studies on costs, risks, and benefits of electrical-energy generation and power-plant siting. After the Calvert Cliffs decision of July 23, 1971, in which the courts required the commission to assess all environmental hazards of siting nuclear power plants, the commission authorized $1.2 million at Argonne to fund preparation of environmental impact statements. The laboratory also received $658,000 from the U.S. Environmental Protection Agency, principally for air pollution and transportation studies and $580,000 from the Illinois Institute of Environmental Quality for studies relating to air pollution, water quality, land restoration from surface mining, and solid-waste management. Argonne also received several smaller grants from the state of Illinois, the National Science Foundation, and other state and local agencies. Acting on the success of the Chicago air pollution study, the Federal Aviation Administration asked the laboratory to measure air pollution concentrations at O'Hare International Airport and at the Orange County Airport in California to determine the kinds and levels of pollutants to which airport employees and users were exposed.[30]

Crisis in the Physical Sciences: The A²R² Canceled

When Robert Duffield finally became Argonne director on November 1, 1967, he no longer had a chance to save the high-flux reactor project. In mid-November, an AEC freeze on construction brought site preparation to an irreversible halt. Concurrently, Shaw imposed a freeze on procurement. These actions, plus the implementation of Shaw's more stringent reporting and documentation requirements, inexorably drove costs upward. All of these factors proved ominous when the Joint Committee asked the General Accounting Office (GAO) to review the Argonne project as part of an investigation into problems with AEC construction activities.[31]

The GAO noted that the A^2R^2 was one of three high-flux reactors either built or under construction by the Atomic Energy Commission. The reviewers also noted that the CP-5, although not classified as a high-flux reactor, would continue to operate as part of Argonne's physical research program after start-up of the A^2R^2. In its initial review, the GAO noted that the 100-megawatt A^2R^2 was a significantly less capable research tool than originally anticipated. Although provision had been made in the reactor design for future upgrading to 240 megawatts, the congressional auditors noted that "substantial additional funds" would be required to make the Argonne machine "fully operational." Under these circumstances, the GAO suggested that the project be sent back to Congress for reassessment and reauthorization.[32]

Argonne and the AEC managed to convince the General Accounting Office to eliminate its devastating recommendation from the final report published in February 1968. But the problem of cost overruns and missed deadlines, which the GAO highlighted, could not be suppressed. Not only had the high-flux reactor project fallen more than three and a half years behind schedule, but budget estimates also approached $30 million, not including the research-and-development costs.[33]

With the program in dire trouble, Philip Powers tried to rally the AUA behind the project. Although the A^2R^2 was expected to draw international scientists to Argonne, Powers stressed that the new reactor would enhance research partnerships between midwestern universities and the laboratory. At the urging of Commissioner Gerald Tape, Powers appealed directly to the director of the commission's Division of Research. Powers reminded Paul McDaniel that termination of the high-flux reactor at Argonne would have a disastrous impact on physical research at the midwestern universities. Neither the Brookhaven nor Oak Ridge reactors provided satisfactory alternatives for the researchers in the Midwest. The former reactor was not readily available to the midwestern scientists, and the latter did not have the neutron beam ports required by the university users. Most significantly, of course, the loss of the A^2R^2 reactor would severely undermine the premise that the commission, the laboratory, and the midwestern universities working in partnership could advance their mutual interests.[34]

The end came rather abruptly. On March 2, Argonne's top A^2R^2 managers and representatives from the AUA users group lobbied Edward J. Bauser, the executive director of the Joint Committee, to support the project. It was Argonne's last shot, and the members of the delegation burned all their powder. They described their research plans and emphasized how neutron-scattering research would contribute to material sciences, solid-state physics, chemistry, and biology. They pointed to research in fundamental neutron physics, and neutron-diffraction experiments to study actinide and transition elements. They emphasized the importance of transuranic research and noted that in recent years the major achievements in heavy element research had taken place in the Soviet Union. The A^2R^2 would be a major step in reversing this trend. But most eloquently, A. Freeman from Northwestern University voiced the dominant concern of the midwestern scientists. The A^2R^2 was not just another reactor. It symbolized the future of low-energy physics in the American heartland. Cancellation of the project, Freeman predicted, would not only drive the best people from Argonne but, equally important, would also rule out development of strong Midwest university research in this uniquely promising field.[35]

Bauser listened sympathetically, but ultimately the projected cost of the reactor, which had reached $35 million, and the substantial operating and experimental costs when it was completed, proved too much. On April 2 McDaniel and Shaw informed the laboratory that the Joint Committee had rescinded the construction authorization for the A^2R^2 and that all design, construction, and related activities were to be terminated immediately. As a sop, the Joint Committee expressed its hope that the Argonne scientists could work out effective research partnerships at Brookhaven or at Oak Ridge. There would be no high-flux research reactor at Argonne. Profoundly shaken by the finality of the action, Duffield could only express his keen sense of loss for the Argonne community and cooperating midwestern universities. Argonne was left to fill an $80-million hole in the ground.[36]

Crisis in High-Energy Physics: The ZGS

Imagine the irony. While Argonne fought a losing battle from 1966 until 1968 to build the 100-megawatt, high-flux reactor that would enable it to maintain a major role in low-energy physics, Argonne's new rival in high-energy physics, the National Accelerator Laboratory (renamed Fermilab in 1974) in Weston, Illinois, required assistance from Enrico Fermi's old laboratory to build the new 200 BeV particle accelerator. Understandably, although Governor Otto Kerner was distressed about losing the $25- to $35-million A^2R^2, he was elated about winning the $250-million accelerator project for the state of Illinois.[37]

The creation of the new National Accelerator Laboratory in Argonne's backyard raised questions about the future of high-energy physics at the laboratory. Academic scientists from the AUA supported Argonne's high-energy physics program because the laboratory gave them access to large instruments and facilities that were too expensive to be maintained by their universities. Argonne scientists, on the other hand, considered themselves on a par with university investigators. The Technical Advisory Panel, consisting of users of the ZGS, reviewed research proposals, recommended policy and programs, reconciled conflicting schedules, and gave advice on other matters. The user community generally supported basic research that was uniquely suited for the national laboratories. Academic bias assumed that high-energy research that could be conducted at the universities would be both stronger and cheaper. Argonne's high-energy physicists, of course, did not share those assumptions.[38]

The Atomic Energy Commission coordinated Argonne's high-energy physics research through a national program that included management of the major high-energy physics laboratories at Brookhaven, Berkeley, Stanford, and Weston. In turn, the commission's High-Energy Physics Advisory Panel (HEPAP) reviewed the laboratories' research programs, budgets, and long-range plans. The panel of distinguished senior physicists played a major role in shaping the high-energy physics program—the commission relied on the panel in deciding whether to approve improvements in the ZGS or to continue to run the ZGS. Obviously maintaining good relationships with this prestigious panel was vital to Argonne management.[39]

Without hesitation, Argonne offered to provide laboratory space and other facilities for the new accelerator project. After Robert Wilson, a professor of physics at Cornell University, was named the director of the National Accelerator Laboratory, Argonne provided draftsmen, emergency transportation, contract support, and other mundane but essential assistance. The accelerator laboratory was also quickly patched into Argonne's computer.[40]

Initially, space was the new laboratory's most pressing need. Through Shaw's office in Washington, the National Accelerator Laboratory negotiated with Argonne to use the EBWR facility, which had been shut down on July 1, 1967. Even as the commission declared a moratorium on A^2R^2 construction, Argonne was glad to loan space in the EBWR building for Wilson's temporary headquarters. Shaw generously concurred, offering the EBWR facilities to Wilson's project for a term of three to five years.[41]

With the war in Vietnam severely pinching federal budgets, Duffield knew that most of the government's "Midwest" investment in high-energy physics research would go to the National Accelerator Laboratory. The commission had already committed almost 50 percent of its physical research budget to the 200

BeV accelerator. Together, funding for the National Accelerator Laboratory and the Stanford Linear Accelerator left little money for other physical research. Indeed, in the climate of tight federal budgets, Paul McDaniel advocated priority for the national laboratory programs over university projects. Duffield believed that Argonne should not only face the budget reality squarely but also seize whatever advantage it could from its proximity to the accelerator laboratory. Although Argonne would never be a competitor to the National Accelerator Laboratory in high-energy physics, Duffield also knew that there was a great deal of research that could be accomplished with the ZGS while construction and engineering continued at Weston. Scientists at the new laboratory would be itching to conduct experiments while all they had was a doughnut-shaped hole in the ground where the accelerator's racetrack would be built, and Duffield decided to accommodate them in any way possible. He offered Wilson use of the ZGS and its 12-foot bubble chamber, authorized access to Argonne's centralized and expanding computing center, provided logistical help to the accelerator project, and even encouraged accelerator scientists and their families to use Argonne recreational facilities, including the Argonne park and swimming pool. Perhaps teamwork and cooperation were best symbolized when Robert Duffield's wife Priscilla, who had previously worked for E. O. Lawrence and J. Robert Oppenheimer, became Wilson's secretary.[42]

Although cooperation between Argonne and the new laboratory was assured under Duffield's leadership, Victor Weisskopf, chair of HEPAP, worried that the construction of the 200 BeV accelerator so near the ZGS would make it hard to retain a "skilled and enthusiastic" staff to operate the Argonne machine. Weisskopf reported rumors that ZGS scientists feared that they would lose out at the new laboratory if they stayed too long at Argonne. Agreeing with Duffield that Argonne need not leave synchrotron research immediately, Weisskopf emphasized the importance of maintaining a full program on the ZGS until at least 1975 when the 200 BeV accelerator would be fully operational.[43]

Robert Wilson agreed that the continuing operation of the ZGS was important to the success of the 200 BeV program. In fact, Wilson believed an active and vigorous ZGS research group was vital to attracting the highest calibre physicists to the National Accelerator Laboratory. Wilson anticipated that the 12.5 BeV ZGS would prove important to scientists working on the 200 BeV machine in calibrating and debugging their equipment. In addition, high-energy experiments frequently indicated interesting possibilities for research at lower energies. From Wilson's perspective, the advantages to the National Accelerator Laboratory of having a vigorous ZGS program could not be overemphasized.[44]

Both Duffield and Wilson agreed that it would be disastrous to the ZGS program and detrimental to the 200 BeV project to recommend the closing of the ZGS as early as 1975. Wilson thought the ZGS should either be shut down quickly or kept running for the foreseeable future. To close the ZGS in three to five years would result in the impoverishment of both laboratories by the departure of the very best people from Argonne. Duffield bluntly wrote Weisskopf that even a well-intentioned recommendation by his committee to keep the ZGS operating until 1975 "would be regarded as a death certificate for high-energy physics here at the Argonne Laboratory and would lead to the quick collapse of what we think is an excellent experimental program." As recently as 1965, the AEC's "Policy for National Action in the Field of High-Energy Physics" endorsed improvements for the ZGS, including construction of a large bubble chamber, a higher-energy injector, and a new experimental area. Duffield reminded Weisskopf that there was heavy use of the ZGS by university groups whose enthusiasm and spirit could be crushed by precipitous action by his advisory committee.[45]

Duffield and Wilson also thought that the personnel situation had been overemphasized. Some good people had left Argonne for the new laboratory, but Duffield insisted the transfers had not caused problems for Argonne. Wilson explained that the vast majority of engineers at Argonne were "operating," not "design," engineers. The National Accelerator Laboratory had hired some "bored" design engineers from Argonne but would not need to hire operating engineers for some time. In fact, if the ZGS were closed soon, Wilson could not create immediate positions for most of the Argonne staff. Duffield was confident that there were many positive reasons for good people to stay in high-energy physics at Argonne.

Even before he reported to Argonne, however, Duffield learned that Robert Sachs was resigning as associate director for high-energy physics. Losing Sachs was a bitter blow to Duffield, who had to recruit a new director in the shadow of the National Accelerator Laboratory. Sachs had played a major role in revitalizing the high-energy program at Argonne, and he enjoyed the enthusiastic support of his division and the Argonne Universities Association for his wise and energetic administration of the program. Despite intense efforts to convince him to stay, Sachs was determined to return to his research laboratory at the University of Chicago. In January 1968, Duffield asked Philip Powers and the AUA to identify candidates for Sachs's replacement. By July he was delighted to hire Bruce Cork from the Lawrence Radiation Laboratory at Berkeley for the position. In addition to his research on elementary particles and high-energy reactions, Cork had collaborated with colleagues at the Universities of Michigan and Wisconsin on cosmic-ray experiments conducted at high altitudes in Colorado.[46]

Argonne physicists looked to the future with guarded optimism in the summer of 1970. They remained confident that the ZGS still had ten years of life during which several hundred significant experiments should be completed. The energy region of 12.5 BeV remained rich in physics and beckoned to be explored. Since October 1969, for example, the world's largest bubble chamber had enhanced ZGS research by enabling scientists to study particles that were too small and moved too fast to be observed directly. The large bubble chamber measured twelve feet in diameter, stood seven feet tall, and was filled with 6,400 gallons of liquid hydrogen cooled to about -413°F. The bubble chamber was surrounded by the world's largest superconducting magnet, itself a major technological achievement of the high-energy physics program. The superconducting magnet, designed to produce a magnetic field of 18.5 kilogauss using little electrical power, demonstrated the possibility of manufacturing large superconducting magnetic coils that would operate at one-tenth the cost of conventional magnets. High-energy particles, such as cosmic rays, guided through the liquid hydrogen by the large magnets left a "track" of tiny bubbles that scientists photographed with four cameras for identification and analysis. When a particle beam from the ZGS sped through the 12-foot bubble chamber, scientists photographed collisions between the beam particles and the hydrogen atoms in the bubble-chamber fluid.

The elementary particle research program for which Argonne designed the 12-foot bubble chamber was divided into two subprograms: neutrino research, for which the bubble chamber provided a unique and vital tool, and strong-interactions research.[47]

Ironically, Duffield considered moving the 12-foot bubble chamber to the National Accelerator Laboratory. Wilson was open to the idea, but Argonne's bubble chamber was not entirely adequate for his needs. Consequently, Wilson entered into negotiations with Brookhaven, whose design of a new large bubble chamber provided for a 40-kilogauss magnetic field, almost twice that of Argonne's 12-foot bubble chamber. In the absence of a prompt response from Wilson, Cork urged Duffield to keep the 12-foot bubble chamber at Argonne, where the 18.5-kilogauss field was adequate for neutrino physics. The bubble chamber was well engineered and well constructed, Cork reminded Duffield, and was an "optimum" size for neutrino studies at Argonne in view of budget realities.[48]

Although the ZGS remained a major midwestern research facility and ranked fourth in acceleration energy among the world's operating proton accelerators, Duffield and Cork knew that Argonne scientists would be working in the shadow of the 200 BeV accelerator at the National Accelerator Laboratory, a fact that inevitably diverted interest, people, and resources from the

Argonne facility. On the one hand, Argonne enjoyed a temporary advantage because few experimentalists would be able to work at the new laboratory in the near future. Roughly 75 percent of the research projects were conducted by university teams. On the other hand, Argonne scientists, who performed the remaining 25 percent of the research, observed that the user community seemed to be losing interest in the ZGS. They did not need help from users in running the ZGS, but they were dependent on the user groups for planning high-energy research.

The operating budget for fiscal year 1971 grew to slightly more than $17 million, about half of Argonne's physical research budget. With the completion of a new experimental area, the modification of the 30-inch bubble chamber, increased proton-beam intensity, the replacement of the original vacuum chamber, and sustained budget growth, Bruce Cork looked optimistically to the future. By the mid-1970s, perhaps even sooner, Cork expected his program to be strongly influenced by the National Accelerator Laboratory, yet he remained confident that the energy region accessible to the ZGS was "very rich in structure and complex phenomena." Cork planned a vigorous complementary program at Argonne to extend beyond 1976 and even anticipated that the ZGS program would be able to offer the new laboratory precision exploratory experiments at lower energies.[49]

By March 1972 the AUA committee on high-energy physics, chaired by Homer A. Neal of Indiana University, became alarmed that the LMFBR was siphoning off resources and hurting the physical sciences in general and the ZGS in particular. Powers had already complained to McDaniel that inadequate funding was hampering work on the ZGS. The growing effort on the fast breeder was making the situation worse. The high-energy physics committee feared that Argonne's move toward applied reactor development would not only weaken the laboratory's ties to the universities but also would make Argonne less competitive with the other accelerator laboratories.[50]

Neal also worried that the Argonne Universities Association increasingly had only a "ceremonial function" in laboratory planning and policy. Although the physics output of the ZGS was outstanding, the political support that Argonne's high-energy physics program received from the AUA and its member universities was, in Neal's opinion, disappointing. The ZGS users group directly lobbied the Atomic Energy Commission, the Office of Management and Budget, the Office of Science and Technology, and the Joint Committee, but its efforts were rarely coordinated with the AUA or the member universities. Many of the ZGS users did not even know the name of their AUA representative. Neal believed that with improvements the ZGS would remain scientifically competitive, even in the face of the National Accelerator Laboratory. Nonetheless, the morale of the ZGS group hit rock bottom when Argonne's management transferred a substantial portion of its laboratory and office space to the reactor division. As Neal darkly warned, "If this program is not given our full support, our midwestern universities stand to lose their most important collective research facility."[51]

The Experimental Breeder Reactor: Argonne's First Priority

Although reactor development at Argonne received about half of the laboratory's budget, Duffield knew that it required more than half of his attention and time. Since the LMFBR had become the nation's first priority in reactor development and commercialization, the breeder program assumed priority over all other work at Argonne. Almost all the big problems—management indecision, budget shortages, debate over laboratory mission and priorities, tension between basic and applied research, and morale—were rooted in the conflict between the commission's determination to build an LMFBR demonstration reactor and the laboratory's dogged defense of its basic research projects in the physical and biological sciences. When Duffield became director in November 1967, he walked into a laboratory already deeply scarred from its running battles with the commission's Division of Reactor Development and Technology led by Milton Shaw. For example, the AUA resolved to support Shaw's priorities, while at the same time defending Argonne's integrity "as a pioneering, innovating research facility."[52] Argonne's reactor engineering division echoed this strategy by asserting that the laboratory should insist upon maintaining maximum freedom and initiative in planning and executing basic research on new reactor concepts. In December 1967 the AUA's reactor development committee met in Washington with Commissioners Tape and Ramey, General Manager Hollingsworth, and Shaw. Shaw candidly described his plans for the LMFBR project, which he intended to manage from Washington with detailed reports and strict accountability for progress toward stipulated goals.[53]

When Shaw emphasized that inefficient operation of the EBR-II in Idaho lay at the heart of the issue, Powers noted that poor communications between Argonne and Shaw's staff in Washington had made matters worse in Idaho. In fiscal year 1968, EBR-II received about $10 million of Argonne's $41-million reactor development budget. Because EBR-II had been designed as a prototype fast breeder, Argonne's scientists and engineers wanted to proceed experimentally with the fast-breeder reactor, including its fuel cycle, fuel fabrication, reclamation of fissile material, and reinsertion of reprocessed fuel.[54] But Shaw had no interest in supporting EBR-II as a research reactor. Rather, he insisted that the reactor be operated on a project basis as a fast-fuel test facility.[55]

The change of emphasis not only raised questions about Argonne's role in reactor development but also challenged its methods in reactor operation. In its first four operating years, 1964–68, EBR-II had not compiled an outstanding record for efficiency or reliability. The reactor had been hit with "operational incidents" that raised questions about both reactor safety and management. In 1967 EBR-II operated less than three months, or about 20 percent of the time. Results were scarcely better in 1968, when EBR-II increased full-power operation to only 30 percent, or 110 days. To compound routine operating problems, on February 9, 1968, EBR-II suffered a fire in the boiler-plant control room when eighty gallons of sodium burst through a "freeze seal" during the replacement of a valve. Three men received minor burns in the fire, and cleaning, repairing, and replacing the equipment cost $25,000. The entire plant was back in operation thirteen days after the accident. Although relatively minor, the fire was the third recent incident—others included an earlier hydrogen explosion in the equipment airlock and a sodium accident when a water-contaminated plug was reinserted in the primary sodium tank—to undermine confidence in EBR-II operating procedures. From the official report of the sodium fire, it was evident that neither the shift supervisor nor the maintenance foremen were in control or supervising the repair work.[56]

Although there were management problems at EBR-II, Washington, Argonne, and Idaho shared no consensus about what needed correcting. Shaw's staff wanted to require even more detailed instructions and mandatory check-off lists to verify adherence to procedures. EBR-II management in Idaho, on the other hand, used the elaborate written procedures and check-offs for training and reference but did not generally regard them as a step-by-step recipe to accomplish work. From the perspective of Duffield's office, the major culprit in EBR-II inefficiency was none other than Milton Shaw's own staff.

Harry O. Monson, project manager for EBR-II, identified six reasons for the excessive downtime of the EBR-II reactor. Five of the reasons—operator error, equipment failure, faulty procedures, leaking fuel elements, and excessive concern for the power coefficient—were problems that Argonne could correct. But a major cause of reactor downtime, Monson reported, stemmed from the "detailed, day-to-day involvement of the Division of Reactor Development and Technology (RDT) in EBR-II operation and planning." Time was lost in keeping Shaw's office "fully and currently informed," and no action, particularly one relating to reactor start-up or operation, could be taken without approval of Shaw's staff. The on-site representative even required that a gauge be installed in his office so that he could monitor EBR-II's power level without leaving his desk. More time was lost, Monson believed, waiting for communications and directions to wend their way from Idaho to Washington and return via Argonne and the Chicago Operations Office. Yet even if the EBR-II engineers improved both organization and operations, there seemed little hope that Shaw would relax his detailed management of the breeder reactor.[57]

Finally, on June 6, 1968, Shaw leveled a broadside against Duffield and Argonne's management. His indictment was simple: Argonne was not single-mindedly committed to the fast-breeder program. Specifically, Shaw charged that Argonne had not performed effectively on the EBR-II and on other key efforts in the LMFBR program and that consequently the laboratory failed to demonstrate that it could "provide the necessary technical leadership in the overall liquid-metal fast breeder program." Argonne's failure, Shaw argued, shifted inordinate financial and technical burdens to the commission, other laboratories, and industrial contractors. There had been improvements in isolated areas such as the LMFBR program office, Shaw admitted, but he claimed that Duffield's actions had been ad hoc and had fallen short of the drastic and comprehensive measures Shaw thought necessary. Argonne had failed not only to provide competent management, but also to allocate sufficient resources to the breeder project. It was imperative that Duffield take steps to end "the many potentially serious reactor operating incidents."[58]

Targeting the AUA's reactor development committee, which had been critical of his policies, Shaw advised Duffield to end management by committees, which "have caused delays, confusion, and diffusion of technical and administrative effort." How much longer, he asked rhetorically, could the Atomic Energy Commission allow Argonne to spurn the Rickover method, which other organizations had used so successfully in establishing high-priority project management systems? Again Shaw threatened to foreclose on Argonne's participation in the LMFBR project. As clearly and bluntly as he could say it, Shaw informed Duffield that "we continue to be gravely concerned that ANL has abdicated its contemplated leadership role in the breeder program by not making the necessary total commitments to effective prosecution of priority reactor development programs."[59]

Duffield, who had been on the job only slightly more than six months, got the message. He had been chosen director because of his extensive experience with nuclear reactors. Certainly, he was expected to be more effective and sympathetic in promoting the breeder program than Crewe had been. Nor was he naive about how Shaw would achieve his objectives. He had even crossed swords with Shaw earlier while working on the Peach Bottom reactor project for General Atomic. The commission expected Duffield to move quickly and decisively to make the difficult personnel changes that would put the breeder-reactor program on a fast, high-priority track. The recent cancellation of A^2R^2 indicated that Shaw was not bluffing in his determination to focus personnel and resources on the breeder-reactor project.[60]

Reorganization of the Reactor Programs

In September 1968 Duffield announced reorganization of Argonne's reactor development program in three areas: the EBR-II project, reactor physics, and the Idaho organization. Affirming the high priority of the breeder program, Duffield named Milton Levenson, lately director of the ill-fated A^2R^2 project, to head the EBR-II project, reporting to the laboratory director. To provide Levenson more muscle but not entirely emaciate the research divisions, Duffield provided dual appointments for some personnel assigned full-time to the EBR-II project. In addition, he appointed Harry Lawroski from Argonne's Idaho division to serve as superintendent of reactor operations in charge of running EBR-II, with responsibility for the reactor plant, its safety, and its use as an experimental tool. Lawroski reported directly to Levenson. In order to achieve a strong, unified research program in reactor physics to support the breeder program, Duffield consolidated the reactor physics organizations in Illinois and Idaho. The reactor physics division remained under the direction of Robert Avery but included the critical assembly and related analytic work performed by the zero power plutonium reactor (ZPPR) in Idaho as well as reactor physics research in Illinois. Duffield asked Fred Thalgott from the Idaho division to serve as deputy director. To streamline the Idaho office, Duffield selected Meyer Novick, who was director of Argonne's Idaho division, to serve as the assistant laboratory director with general responsibility for activities at the Idaho site. Duffield's goal was to facilitate coordination between Idaho and Illinois and to establish clear reporting lines.[61]

Duffield's changes coincided with marked improvement in EBR-II performance and general progress for the LMFBR project at the National Reactor Testing Station. In fiscal year 1969, EBR-II achieved 199 days of full power operation, 54 percent operation, compared with 30 percent in 1968. By 1970, plant availability at the full-rated power of 62.5 megawatts electric rose to 68.8 percent, whereas only 7.0 percent of the downtime was unscheduled. Plant availability slipped to 53.5 percent in 1972 but rebounded to 65.7 percent in 1973. One reason for improved operations was that EBR-II no longer shut down every time a leaking fuel element was discovered. Rather, EBR-II operators were praised for devising techniques for running the reactor despite leaks from failed experimental fuel elements. Aggressive operation of EBR-II provided the breeder program with a fast-reactor test bed for irradiating fuels and materials, assessing fast-reactor physics, and testing liquid-sodium coolant under operating conditions. As the only operating test reactor, in 1967 EBR-II irradiated more than 380 mixed-oxide fuel elements, providing the commission with not only fast-flux irradiation and fuel-examination service, but also operating and maintenance experience on a liquid-metal fast-breeder reactor.[62]

On April 18, 1969, the ZPPR went into operation. Under construction since August 1966, the ZPPR, which was the forty-sixth reactor built at the NRTS, joined twenty-two other reactors then in operation. Zero-power reactors were designed to achieve a self-sustained chain reaction without producing a significant amount of power. The reactor matrix, which measured 8 feet deep and 10 feet high by 10 feet wide (expandable to 8 by 14 by 14 feet), enabled easy configuration and manipulation of simulated reactor cores. Similar to Argonne's Zero-Power Reactor 3 (ZPR-3) in Idaho and ZPR-6 and ZPR-9 in Illinois, all of which Argonne had modified to use plutonium fuels, ZPPR provided physics data to support the LMFBR program. The reactor's versatility enabled engineers not only to simulate reactor cores of small demonstration plants, but also to test core designs for the 1,000-megawatt-electric commercial breeder reactors. Experiments scheduled for the ZPPR included determination of plutonium fuel requirements to sustain a chain reaction in the LMFBR, assessment of the effects of sodium loss or fuel movement on reactor power, and measurement of the energies of neutrons in the reactor matrix. Finally, mock-ups of LMFBR cores enabled Argonne to verify design engineering before construction of the large reactors. In its first assignment, ZPPR tested the mock-up core of the Fast-Flux Test Facility to be built at Hanford. The FFTF mock-up core contained approximately 1,150 pounds of fissionable plutonium—perhaps the largest amount of plutonium assembled in a reactor core up to that time.[63]

While experiments with the ZPPR proceeded, Duffield announced construction of the Hot-Fuel Examination Facility (HFEF), also built at the NRTS. The HFEF provided a hot-cell complex composed of shielded cells, unshielded laboratories, support areas, and special equipment for handling and examining highly radioactive and toxic materials. Initially the HFEF received materials irradiated in EBR-II, but the facility was also designed to handle materials and test specimens irradiated in TREAT, the ETR, and the Power-Burst Facility (PBF). The large rectangular main cell, which measured 30 by 70 feet by 25 feet high, provided fifteen work stations for handling sodium-wetted materials in an argon atmosphere. Outside the heavily shielded cell, workers used wall-mounted mechanical master-slave manipulators to handle "hot" materials at the work stations within the cell. Electric manipulators, supported by an overhead transporter, moved irradiated fuels and materials from one work station to another. Dedicated by James R. Schlesinger, the AEC's new chairman, on July 5, 1972, the HFEF was completed on schedule and within the cost estimate of $10.2 million. Shaw was pleased and congratulated Argonne's program directors on their completion of this important LMFBR project.[64]

Management Crisis: 1969–72

Improvement with EBR-II and success with the HFEF, however, did not keep the commission's management wolves from Duffield's door. More than anything else, Shaw wanted Duffield to appoint a hard-driving, management "czar" to run the reactor program in the Rickover manner. William Kerr, chair of the AUA's reactor development committee, observed that Shaw believed the national laboratories generally, but Argonne particularly, should adopt a disciplined engineering approach to reactor development. Despite Shaw's continuing complaint that Argonne had failed to develop the necessary urgency about the breeder program, the AUA believed that Argonne had, in fact, implemented the engineering approach. The disagreement was not over goals, but rather over the method to achieve the LMFBR. On the other hand, Shaw asked Argonne to abandon reactor research and instead to engage principally in design and procurement.[65]

From 1969 to 1972, the commission pressured Duffield to revamp Argonne's management by removing the associate director for research, the associate director for reactor programs, and the director of the reactor engineering division—changes he agreed were necessary but was reluctant to make. Reportedly, in 1969, Joint Committee Chairman Holifield, Commissioner Ramey, General Manager Hollingsworth, Shaw, and the federal budget staff warned both Duffield and William Cannon, the University of Chicago's representative, that the commission was close to removing the reactor program from Argonne. Holifield and Craig Hosmer explicitly told Cannon not to expect much help from the Joint Committee for Argonne's other programs until the laboratory moved faster and more successfully on the breeder program. It was no empty threat, Cannon reflected. Oak Ridge's reactor program already had been gutted, and Brookhaven, which Shaw regarded as a "country club," had lost its reactor program entirely.[66]

In June 1969 Duffield met with Cannon, Powers, and several AUA board members to discuss reorganization of the reactor program. According to Cannon, despite the fact that all agreed changes should be made, Duffield did not act. Rather, he stopped communicating with staff and so exacerbated management problems by creating confusion, discontent, and ultimately alienation between himself and the research programs, especially in chemistry, biology, and medicine. Duffield also recognized the need to remove the director of personnel and to restructure the personnel system to incorporate effective performance review, especially of unproductive staff. These were all highly charged issues, and Duffield moved too slowly to satisfy the management above him and too indecisively to mollify those below.

Following the fire at its Rocky Flats, Colorado, plutonium fabrication plant in 1969, safety at the national laboratories became an important agenda for the Atomic Energy Commission. Reacting to the commission's concerns, Duffield established a safety office reporting to him, but he attenuated his action by appointing the former director of the Idaho site, whom he had recently removed, as head of the safety office. Again, despite Cannon's pleas, safety efforts languished.[67]

Only in the area of affirmative action, where Duffield had a strong personal commitment, did his performance satisfy his bosses at the University of Chicago. He accepted Cannon's suggestion to hire Malcolm H. Lee to head the affirmative action office. Lee, a social worker with extensive experience in local government agencies and industrial affirmative action programs, launched a vigorous minority recruitment program focusing on summer-student programs, high-school visitation, neighborhood youth corps, and college recruitment. Argonne also established training and manpower resource programs featuring the laboratory's evening academic program (LEAP), which provided on-site opportunities for night classes; program REACH (refresher education for Argonne clerical help), which enabled staff to qualify for laboratory promotions; and TAT (training and technology), a six-month industrial training course for the unemployed and underemployed conducted at Oak Ridge.[68]

In the face of budget declines and personnel layoffs, Duffield finally acted in January 1970 by implementing his reorganization of Argonne's scientific and engineering divisions. To quiet complaints from commission headquarters about the lack of priority given the breeder program by Argonne, Duffield announced he had taken personal (and temporary) control of the LMFBR program. He also moved Alfred Amorosi from the LMFBR program office to the position of assistant laboratory director for reactor development and technology programs, in which he helped Duffield coordinate the breeder project among the program divisions. Paul Gast remained in the LMFBR program office as the new director. He abolished the old reactor engineering division and created an engineering and technology division under Stanley A. Davis to coordinate all aspects of reactor design, analyses, and engineering. Milton Levenson continued as director of the EBR-II project. Finally, with Cannon's encouragement, Duffield asked Michael V. Nevitt from the metallurgy division to serve as deputy director for the LMFBR program, a move Cannon hoped would bring some administrative "crispness" to Duffield's office.[69]

Duffield cleared his reorganization plan with Milton Shaw, but not until after his administrative changes had been announced by the office of public affairs. He assured Shaw that the changes in organization and personnel were made to accelerate progress across the LMFBR program. His intent was to improve coordination with industrial contractors in order to focus sharply on support of the FFTF. Duffield's problem, however, was program execution, not elocution. Cannon, who believed that Argonne's first mission was to service its patron, the commission, acknowledged that Duffield understood the urgency of the breeder program. But the pattern of administration remained the same: "an acknowledgement of a problem, an acceptance of a solution or at least a visualization of one, delay, strong outside intervention, and, only then, action."[70]

No sooner had Duffield announced his reorganization than the budget ax fell on Argonne. President Nixon's fiscal year 1971 budget, which was announced in February 1970, required Argonne to cut 230 employees. Unfortunately, budget compression continued the following year. Although Argonne's overall budget increased by $2 million in fiscal year 1972, cuts were made again in the physical sciences, particularly high-energy physics, and mathematics. The reduction in force, mostly in the non-reactor programs, further undercut Argonne's sagging morale, and together with reorganization and new personnel evaluations profoundly lowered the staff's already diminished confidence in Duffield's leadership.[71]

Neither Duffield nor anyone else expected him personally to supervise the breeder program for very long. By April, Duffield had selected Robert V. Laney to serve as associate director for engineering research and development, which included responsibility for the LMFBR program. Laney arrived with sterling credentials. He was both a graduate of the U.S. Naval Academy, held a master's degree from MIT, and was a veteran of Rickover's nuclear-navy program. Following discharge from the navy, Laney served as the program manager for the construction of a prototype naval reactor. Thereafter he worked at Bettis Atomic Power Laboratory on naval reactor projects and rose through the ranks of the General Dynamics Corporation to become vice president and general manager of the Quincy Shipyard with responsibility for building submarines for the nuclear navy. In sum, Argonne succeeded in recruiting one of the "stars" of the admiral's nuclear navy program. With Laney's arrival, Argonne finally had reactor-development management that was willing and able to satisfy fully Milt Shaw's demands on resources and personnel.[72]

The appointment of Laney brought Duffield no peace, however. On June 4, 1971, President Nixon sent Congress a special message on energy resources. In the fall of 1969 the United States had experienced the first of recurrent "brownouts" caused by overtaxed electrical generating facilities. Threatened fuel shortages and rising gasoline prices moved the nation closer to energy crisis. In a message highlighted by his proposal to establish an energy administration, the president also affirmed that the fast-breeder reactor was the nation's "best hope" for meeting the growing energy demand. He announced that the breeder reactor was "ready to move out of the Laboratory and into the demonstration phase with a commercial size plant." To that end, he asked Congress for a dramatic funding increase for the LMFBR program.[73]

For Robert Duffield, the president's energy message meant only increased headaches as pressure intensified to raise the priority of the breeder program to meet the president's goal of building an LMFBR demonstration plant by 1980. Laney seized the challenge enthusiastically and promised to develop a comprehensive plan for advancing the LMFBR program into the demonstration-plant stage. To do so, he broadened the authority of the LMFBR program office to mobilize the services of all Argonne divisions as necessary. He also expedited the reorganization begun by Duffield to such a point that the AEC expressed satisfaction by fall 1972. In August, the Joint Committee probed Argonne's readiness to accept a much expanded role in the breeder reactor program. Increasingly the Joint Committee and the commission worried whether they could rely upon industry alone to build the large demonstration plant.[74]

Duffield knew that the president's energy message required him to streamline and tighten Argonne's management even more. Renegotiations of the Tripartite Agreement, which expired on September 30, 1972, had not gone well. The Atomic Energy Commission raised repeated questions about Argonne's management capabilities. First, the commission extended the Tripartite Agreement for three months, and then another nine months, while negotiations continued about the laboratory's future. The commission required subtle, but important, changes in the agreement that gave the government a larger role in reviewing laboratory performance and approving research publications. Consequently, Cannon continued to push for a more rational personnel system dedicated to maintaining and renewing the high quality of Argonne's professional staff. To this end, he brought in Roger W. Jones, former chairman of the U.S. Civil Service Commission, to evaluate Argonne's personnel program, practices, and procedures. Concurrently, the laboratory organized an ad hoc working group to conduct a similar review. The resulting new laboratory-wide personnel evaluation system further discouraged many of Argonne's senior staff.

They suspected, and not without reason, that the new personnel system was designed not only to weed out nonperformers, but also to encourage aging senior scientists to retire.[75]

Between October 1971 and May 1972 Cannon believed the life of the laboratory was at stake. Hollingsworth warned Cannon that Chairman Schlesinger, supported by new commissioners, might even close the laboratory. Although the Atomic Energy Commission was increasingly satisfied with Argonne's progress, there remained intense pressure from the Joint Committee to improve the LMFBR program. The failure of A^2R^2, in which Argonne had sunk over $80 million into a "hole in the ground," still haunted the commission, which worried whether Argonne had the capacity to manage a large reactor project. The AUA's push for an independent environmental laboratory was exhausted, but the vigor of their support for Argonne's mission remained a question. Budget cuts, layoffs, and the loss of independence to do basic research had demoralized much of the laboratory outside the breeder-reactor program. Finally, Duffield, an excellent small-group leader, evidently lost heart in his running battle with Shaw. His absence from the laboratory to attend international meetings, while appropriate, nonetheless literally distanced him from Argonne's continuing management problems.

On April 4, Cannon privately informed Duffield that the university wanted a new director. There would be no rush. Cannon encouraged Duffield to remain indefinitely with the understanding that the university and the Argonne Universities Association would begin the search for a new director. He was assured his tenured position as a professor of chemistry at the University of Chicago for as long as he wanted. In another era Duffield would have made an outstanding laboratory director, Cannon thought, but he was too gentle to cope with the demands of Milton Shaw. Cannon expected to make the change within six months.[76]

Within a week of Cannon's confrontation with Duffield, John Erlewine, the commission's deputy general manager, visited Argonne Laboratory in response to Powers's request for headquarters to brief the AUA on the commission's priorities for Argonne. Erlewine agreed to discuss the AEC's attitude toward the national laboratories in general and their expectations for Argonne in particular. The details of the meeting were not recorded, but Erlewine spoke at length. All present agreed that the commission, or the chairman and general manager, should confer with the AUA to achieve a consensus on program goals and management objectives for the laboratory.[77]

At the National Accelerator Laboratory, Robert Wilson heard rumors that the AEC was pressuring Duffield to resign and he appealed to President Levi at Chicago to support Argonne's director. The amicable relationship that had developed between the two laboratories owed much to Duffield's leadership, Wilson reported. Indeed, Wilson admitted, his laboratory had come to rely upon Argonne for substantial help. If the commissioners always knew best, Wilson reasoned, they would operate their laboratories themselves. But because the commission needed help from an institution of such high standards as the University of Chicago to run Argonne, Wilson hoped the university would stand up to the commission in support of laboratory leadership.[78]

Wilson's single shot had no effect. On July 11, 1972, the Atomic Energy Commission signed a three-year renewal of the Tripartite Agreement while the search for a new director continued. Coordinating with Powers, Cannon made little headway in identifying a candidate acceptable to both the government and the academic community. The commission (or at least Milton Shaw) wanted to focus the laboratory's scientific and engineering divisions on program objectives, which usually did not include basic research, as defined by the United States government. Given Shaw's agenda, Cannon questioned whether the university could effectively manage and operate an AEC laboratory. The answer to this question turned on the selection of the new director.[79]

Behind-the-scenes maneuvering and stress eventually took its toll on Powers, who announced his resignation as president of the AUA on September 19. Officially, Powers stated that after almost six years he was tired of the weekly commute between Chicago and his home in Lafayette, Indiana. But insiders speculated that Powers quit because of criticism from the association's board of trustees. While Powers pursued the mirage of an independent environmental laboratory, he failed to marshall the political potential of the AUA to defend Argonne's basic research mission against the inroads of the breeder program.[80]

Meeting with Cannon on October 16, Duffield finally set his departure date from Argonne for January 1, 1973. Undoubtedly deeply hurt by his firing, Duffield also resigned his academic appointment from the university, severing as quickly as possible all ties to Argonne and the University of Chicago. His pain was evident in his notice to laboratory employees in which he simply said, "I have had an interesting and rewarding five years here, and am looking forward to something else."[81]

It did not take long for the press to catch wind of Duffield's departure. *Science and Government Report* speculated, correctly, that the commission was determined to change the character of Argonne from a relaxed, academic-oriented research laboratory to a hard-driving, disciplined, project-oriented engineering lab. Duffield told reporters that he understood how to manage a research organization, but he admitted he was not very good at detailed business management. And business management was certainly a major factor in Duffield's dismissal, when almost half of Argonne's $88.6-million budget was devoted to reactor development in fiscal year 1972. Notably, the University of Chicago received a management fee of $1.2 million as the prime operating contractor, while the AUA received $450,000 for its services. Perhaps it was coincidental, but two other veteran laboratory directors resigned at this same time: Edwin M. McMillan resigned after fifteen years as director of the Lawrence Berkeley Laboratory, and Maurice Goldhaber, director more than eleven years at Brookhaven, also retired.[82]

After considering more than fifty candidates, the parties to the Tripartite Agreement could not agree on a new laboratory director. Chairman Schlesinger personally favored Allan Bromley from Yale University and General Manager Hollingsworth touted Louis Rosen from Los Alamos. By December the AUA had pared its list down to seven, including Rosen, but there was no likelihood that a successor could be chosen before Duffield left on New Year's Day. Consequently, the university and the AUA squabbled over whom to appoint as acting director. The AUA wanted to name Michael Nevitt. Since his promotion from the metallurgy division, Nevitt had served as deputy director for research. Not surprisingly, Milton Shaw and the commission's reactor division wanted Robert Laney. When Cannon met with Hollingsworth and Erlewine about appointing new leadership for Argonne, the general manager stated that the Atomic Energy Commission expected the university to play an "independent role" in finding and recommending the laboratory director. The commission welcomed cooperation by the AUA. Although Hollingsworth did not want a fight with the midwestern universities, he did not think that the AUA was vital to the laboratory, and he would not let it interfere with or obstruct important laboratory decisions, including the selection of the director. Hollingsworth not only supported appointing Laney as acting director, but he also informed Cannon that he would oppose any other appointment. Hollingsworth also knew that Laney was acceptable to the Joint Committee. If the university were to appoint Laney as permanent director in the face of heavy opposition from the AUA, Hollingsworth promised Cannon he would have the full support of the commission. Nonetheless, the general manager did not want a fight with the AUA. When Cannon mentioned Robert Sachs as a possible candidate for the director's position, Hollingsworth claimed not to know much about him. Sachs's name had not appeared among the more than fifty possibilities for the job, and his background as a high-energy physicist raised concern.[83]

And so it was over. Cannon assumed that the AUA, with support from Argonne's division directors, had backed Nevitt to gain control of the directorship and thus the laboratory. He believed the AUA wanted to seize control from the university and the commission by constructing a political platform that was pro-basic research, pro–academic freedom, antinuclear, and antigovernment. The strategy would not work. Laney's appointment as acting director served notice that the university, and not a coalition of the AUA and division directors, effectively governed Argonne. Nevertheless, even Cannon did not believe that Laney, without a strong commitment to basic research, could lead the laboratory in the long run. As New Year's Day 1973 dawned in Washington, D.C., and across the Midwest there was still no permanent resolution of Argonne's leadership dilemma.[84]

9 | Years of Crisis, Years of Challenge, 1973–77

Argonne National Laboratory approached its third decade with both hope and foreboding. The year 1976 would not only celebrate the nation's two hundredth anniversary but also mark the thirtieth year of Argonne's designation as a national scientific treasure. In the 1960s the laboratory had reached full maturity as a scientific institution with a graying first generation of World War II scientists firmly in command of the laboratory's upper management while gradually making room for the first postwar generation of scientists, including a few women and minorities, at the bottom. The Argonne scientists treasured their Manhattan Project heritage. They were enormously proud of their personal contributions to building one of the world's greatest scientific institutions and were generally unapologetic for their role in fostering big science, which demanded steadily increasing budgets to support the expanding facilities needed to accommodate ever larger machines. They were successful Cold War scientists who skillfully lobbied Washington after the Sputnik crisis for a dramatic increase in federal support for basic and applied science. Yet, while they admired the "can-do" ethic of the Atomic Energy Commission, they were universally contemptuous of any "political" interference in their work. They eagerly trooped to Washington to offer the Congress, the White House, and federal agencies all manner of advice on national science policy and programs but shrank back in horror when science politics inevitably breached the sanctity of their research laboratories. As scientists, they enjoyed a largesse of equipment and resources never dreamed of by their prewar academic mentors, but they were also subject to management and control that many academicians found intolerable. In fact, their riches had exacted a high price of vexing accountability to the government and concomitant loss of prestige among their university colleagues who nonetheless envied their prosperity.

More troubling, the American public, once in awe of all things atomic, had grown increasingly skeptical of the AEC's oft repeated assurances that the country would soon enter nuclear utopia. By 1973, as the nation slid into an ever deepening energy crisis, President Nixon toyed with the unthinkable—the abolition of the Atomic Energy Commission—as part of his larger strategy to organize a Department of Energy and Natural Resources to cope with energy shortages. Under Nixon's plan, all energy would be concentrated in the new department, or perhaps in an energy research and development administration to exist independently of the government's nuclear regulatory function. Separating nuclear regulation from nuclear research and development would escape the AEC's inherent conflict of interest concerning the promotion and regulation of the nuclear power industry. But puzzled scientists at Argonne could hardly believe that the government, in search of a viable national energy policy, might abolish the best, most successful science agency ever established.

If the Atomic Energy Commission was vulnerable, how secure was Argonne National Laboratory? Mature scientists at a mature laboratory had few employment alternatives outside the national laboratory system. No doubt many of the Argonne scientists had originally arrived at the laboratory with the expectation of moving on to a suitable academic or industrial position. Although the revolving door to and from the campus and industry worked to some extent, especially among the upper echelons of the laboratory, Argonne increasingly employed career scientists who did not have the benefit of academic tenure, who lacked professional mobility, and who fully expected to retire from the laboratory after years of faithful service. This "graying" of laboratory personnel was not confined to Argonne, of course, for the commission's General Advisory Committee worried about how to inject "younger blood" into all laboratory programs.[1]

While the war in Vietnam raged on, tight government budgets resulted in accelerated retrenchment of federal support for the laboratory. After a steady and dramatic rise through the Eisenhower administration, in which the commission's budget increased 189 percent, the agency's spending actually declined by 1.6 percent overall in the 1960s. The commission's budget, of course, included military programs as well as reactor development, physical research, and biology and medicine. None of the latter three categories was cut until the end of the decade. During Robert Duffield's years as director, Argonne's budget grew more than 22 percent ($16 million), reaching almost $90 million in 1972. Most of the new money ($13 million), however, was funneled into a 30 percent increase in the commission's LMFBR program. After 1968, when reactor development regained top funding, physical research steadily declined, slipping 5.5 percent in real dollars by 1972. Consequently, for the first time since World War II, when most of the Met Lab employees were reassigned to Los Alamos, Oak Ridge, or Hanford, Argonne was forced to cut its staff. In the three years after 1968, Argonne lost 971 jobs, or 17 percent of its work force. But this time there was no high-priority Manhattan Project to take up the employment slack. Quite the contrary. As the laboratory budgets for research and development decreased, the requirements of the LMFBR program dramatically increased, soaking up whatever dollars it could.[2]

Pulled in several directions by the program's various patrons, friends, and critics, Robert Sachs recalled that in 1973 Argonne National Laboratory was in deep trouble—a virtual "state of shock."[3] The laboratory was fortunate to play a major role in the enormously challenging effort to develop the liquid-metal fast-breeder reactor. But driven by the Joint Committee on Atomic Energy and by the AEC, the LMFBR project provided Argonne its basic mission while at the same time overshadowing the laboratory's other activities. Yet even as the breeder project dominated Argonne's agenda, the laboratory's continued role in the LMFBR project remained uncertain. Although uncertainties about the viability of breeder technology would plague the project by the end of the decade, in 1973 management at both the University of Chicago and at Argonne was principally concerned with satisfying their customers' demands while bucking up sagging morale in the basic sciences and mollifying their partners in the Argonne Universities Association. Ironically, the breeder program drove the laboratory simultaneously toward boom and bust.

At the same time that President Nixon's commitment to the breeder program narrowed Argonne's scientific mission, the nation's burgeoning energy crisis pushed Argonne in the direction of becoming a multipurpose laboratory. In 1973 no one had a precise definition of a "multipurpose" national laboratory, but everyone understood that the advent of the multipurpose lab meant that nuclear programs—whether atoms for peace or war—would no longer monopolize the laboratory's resources or priorities. The General Advisory Committee envisioned that the commission's long-range research goals would focus on the basic structure of matter (high-energy physics), on radiobiology and environmental biology, and on long-range energy resources.[4] Nuclear research and development, of course, remained the largest budget item at most of the national laboratories, but during the 1970s every lab, including Argonne, developed ambitious programs in environmental sciences and alternative energy sources. Typically, annual reports touted these initiatives far out of proportion to their importance, budgets, or accomplishments. Collectively, however, the new programs in the "soft" sciences and energy technology marked an important political shift in government sponsorship of national laboratories.

With the abolition of the AEC in 1974 and the consolidation of energy research first in the Energy Research and Development Administration (ERDA) under President Ford, and subsequently in the Department of Energy (DOE) under President Carter, Argonne could no longer remain exclusively a nuclear laboratory. As frequently happens in political transitions, change was already in the air at Argonne before the passing of the Atomic Energy Commission, and while change was ostensibly accelerated by ERDA and the DOE, Argonne's nuclear identity remained strong through the energy-turbulent 1970s. During the decade between the abolition of the commission and the collapse of the Liquid-Metal Fast-Breeder program, Argonne National Laboratory not only struggled to define its scientific mission but also fought to maintain its very existence.[5]

Robert V. Laney, Acting Director

On January 1, 1973, Robert Laney, Argonne's new acting director, was excited about his assignment to head the nation's oldest national laboratory. In less than three years there, he had moved from associate director in charge of the laboratory's reactor programs to deputy director of operations. Enjoying the confidence of the Joint Committee on Atomic Energy, the Atomic Energy Commission, and the University of Chicago, Laney had become the surrogate director during Robert Duffield's long and painful departure. Although he had support in Washington to be named permanent director when he assumed the position, Laney also knew that he had neither the complete trust of the laboratory's research directors nor the full backing of the Argonne Universities Association. Despite these handicaps, as acting director, Laney did not want to run a caretaker administration but, rather, hoped to reinvigorate the laboratory's basic research program while charting a firm course on breeder-reactor development.

Laney aggressively argued that Argonne must transform itself into a "multiprogram national laboratory" if it hoped to survive the challenges of the 1970s. On the eve of the thirtieth anniversary of the first chain reaction, Laney celebrated Argonne's history with justifiable pride, but he also prophetically warned the laboratory community that there would be lean tomorrows without "complex and painful" rethinking of Argonne's mission. He recalled that the national laboratories had been founded after World War II by a grateful nation flushed with victory and abounding in confidence that wartime achievements in science and technology would be directed to the peaceful service of mankind. Atomic energy, especially, seemed to offer an unlimited horizon for scientific progress, and while the public's faith in science soared to unreasonable heights, the government spent millions of dollars building up the laboratories as "centers of excellence" in creative science. For their investment in scientific research, of course, the public always expected a generous payback in useable nuclear technology.[6]

But times had changed drastically, Laney warned. On the one hand, as the laboratories grew and prospered into large competent permanent institutions, Laney believed that they had lost the "discipline of hard, urgent goals." "Amidst high caliber and imaginative work . . . unhealthy symptoms" became evident. Bureaucratic routine had blunted the creative spirit; job security had replaced task completion; secure funding had dulled research competitiveness. The consequence was not that the laboratories had become complacent, but rather that they had become intimidated by their own history. Unable to recapture the mythic days of the Manhattan Project, they developed a "growing uncertainty of purpose and a beginning of self-doubt."[7]

More seriously, Laney cautioned that the public was becoming increasingly skeptical about the uses and benefits of science and technology. Although Americans had believed almost religiously in progress through science during the 1950s and 1960s, Laney believed that resistance to funding basic research that emerged in the 1970s was tied to a growing public perception that the fruits of science and technology were not invariably good. Between 1965 and 1973, overall federal government spending on science research and development had increased 10 percent, but its share of the federal dollar had dropped from 12.6 cents to 6.4 cents while inflation robbed the dollar of one-third of its value.[8] The AEC had not been the only federal science agency to feel the budget squeeze, of course, but Laney knew that the old days of seemingly unlimited government largess were gone.

Laney's jeremiad identified three interrelated challenges to Argonne's mission. First, the United States confronted "the threat of environmental catastrophe" in which science and technology were not only the villains but also the "only hope for rational solutions." Second, twenty-five years after the end of World War II, the United States suffered growing trade deficits and technological dependency in part because of lagging research and development at home. Finally, the likelihood of severe energy shortages, coupled with environmental problems and trade deficits, would place urgent new demands on America's research-and-development resources—demands, which in Laney's opinion, Argonne could not afford to ignore. It was imperative that Argonne view each of these challenges as opportunities to redefine its research and development goals.

Blackouts, Brownouts, and the Incipient Energy Crisis

The power failure that plunged the Northeast into total darkness sped across New York and New England faster than the shock wave of an atomic bomb. At 5:16 P.M. on November 9, 1965, the malfunction of a relay at a Canadian hydroelectric plant on the Niagara River in Ontario set off a chain reaction throughout the so-called CANUSE (The Canada–United States Eastern Interconnection) power network, eventually knocking out electrical power to 80,000 square miles, including New York City. By 5:30 P.M., 30 million people were caught in the dark at the height of the evening rush hour. Luckily, the late fall evening was rather mild and lighted by a full moon. Although rioting soon broke out in Massachusetts's troubled Walpole prison, the crime rate actually declined throughout the region and the public remained fairly calm. More than 600,000 passengers were stranded in the New York subway system, some for more than six hours, but authorities reported few accidents or injuries, and almost no looting. Airports, hospitals, radio and TV stations, and public utilities were inconvenienced—a few all night because power was not restored to some areas until the following morning. Commuters experienced massive traffic jams when traffic lights blinked out at every intersection across the region. Idling motorists who ran out of gas soon discovered that darkened service stations had no power to pump gas. Those who abandoned their useless cars and tried to find overnight lodging in jam-packed hotels discovered there were no accommodations on the upper floors because those rooms had no elevator or water service.[9]

Although these disruptions resulted in little permanent damage to the Northeast, there were evident worrisome implications of the massive energy failure for both national security and the economy. The Department of Defense assured Americans that communications between the Pentagon and the Strategic Air Command in Omaha and the North American Air Defense headquarters in Colorado Springs had not been interrupted. The Office of Emergency Planning, however, had adopted "a preliminary war stance" just in case the massive power failure had been caused intentionally. The National Warning System alerted civil defense units throughout the region, enabling the Office of Civil Defense to determine rapidly the extent of the power failure and the pace of subsequent restoration.[10]

Failure of the CANUSE electrical grid not only emphasized that Americans were dependent on electrical power even to drive their cars but also revealed how vulnerable they had become to intricately interconnected power systems. As the nation would soon learn, the great northeastern power failure of 1965 was no fluke. Even before the Federal Power Commission could report its findings on the November 9 blackout to the President, a large power failure on December 2 hit the El Paso, Texas, area, knocking out service as far away as Alamogordo and the White Sands missile range in New Mexico. Four days later a small power outage hit ten counties in southeastern Texas. Before the end of summer in 1966, the Federal Power Commission investigated additional failures in Alaska, Nebraska, Missouri, the Pacific Northwest, the Rocky Mountain area, and California. At least twenty-three states had suffered major electrical power losses since the great blackout, and there were more to come.[11]

On the morning of June 7, 1967, a cascading power failure surged through Pennsylvania, New Jersey, Maryland, and Delaware. Within minutes, 13 million persons were without electricity. Philadelphia was hardest hit, where fifteen hundred passengers had to be led to safety from subway tunnels. Uncounted hundreds were trapped in the city's elevators and, as before, traffic snarled when the signal lights failed. Again, people discovered how helpless they were without electricity to pump their water and gasoline.[12]

Less than a month later, on the muggy, foggy afternoon of July 3, 1967, the Kennedy family compound at Hyannis Port, Massachusetts, lost its air conditioning, lights, and electricity when a power failure engulfed Cape Cod. Auxiliary power quickly restored essential service to the Kennedys, but other families vacationing on Cape Cod and Martha's Vineyard had to wait until the early morning hours of July 4th to resume their holiday with lights, water, stoves, refrigerators, air conditioning, and the other "necessities" of the modern age. The Fourth of July outage was the first of two power failures on the Cape that summer. Finally, when Boston and its vicinity were plunged into intermittent darkness for three days during a severe ice storm on December 28–30, Senator Edward Kennedy of Massachusetts lost patience with the electrical utility industry. He acknowledged that accidents and mishaps can befall any highly complicated technology, but he spoke the mind of most Americans by demanding, "It is not unreasonable to expect power companies—private and public—to provide reliable power."[13]

No single cause explained the repeated power failures that had plagued the country since 1965. Reliable power, as well as cheap energy, were virtually an American birthright in the late 1960s, but recurrent power failures or less serious "brownouts" served Americans notice that the era of abundant energy supplies was about to end. Since the end of World War II, American energy consumption steadily outpaced energy production, and now booming consumption strained the capacity of the utilities to deliver peak power loads under stress. When Richard M. Nixon became president in 1969, signs of an impending energy crisis were everywhere. When the winter of 1969–70 turned out to be the coldest in thirty years, the United States experienced shortages in natural gas, heating oil, and liquefied petroleum. During a sweltering heat wave the following summer, "brownouts" rolled up and down the East Coast, pointing toward continued energy shortages for the near future.[14]

Mike McCormack's Task Force on Energy Research and Development

In the fall 1970 congressional elections voters from eastern Washington state sent Democrat Mike McCormack to the House of Representatives. McCormack, regarded as combative, abrasive, but brilliant by his former colleagues in the Washington state legislature, arrived at the United States Congress with a degree in chemistry, an interest in promoting basic scientific research, and a commitment to head off the looming energy crisis. The only bona fide scientist in Congress, McCormack obtained appointments to the Science Committee, the Joint Committee on Atomic Energy, and the Public Works Committee. His assignments positioned him to organize the House Task Force on Energy to identify those areas in which research and development were particularly needed.[15]

While McCormack struggled to organize his task force against entrenched interests in the House, President Nixon, in his first energy message to Congress in 1971, warned Americans that they could no longer take abundant energy supplies for granted. Nixon noted that United States' production of goods and services had been out paced by energy consumption. To develop an adequate supply of clean energy for the future, Nixon asked Congress to establish a new cabinet-level department that would unify all the government's disparate energy research-and-development programs. Nixon's proposals, like McCormack's energy task force, at first made little headway in Congress. Political "gridlock" in Congress was partially the problem, but more importantly, the American public, despite the aggravation of repeated power failures and brownouts, simply did not believe that their cornucopia of cheap, abundant energy was about to run dry.[16]

Although virtually ignored by the press, McCormack's report on energy research and development proved prophetic for federal energy agencies, including Argonne National Laboratory. In February 1973, McCormack's energy task force echoed Nixon's call for energy reorganization by insisting that the focal point for federal energy policy must be in the White House, whereas all government-supported energy research and development should be coordinated by a single agency. In energy policy, the task force stated that environmental protection and energy conservation should be the nation's paramount priorities. In energy research and development, the task force believed the highest priorities should be given to basic science research, materials research, solar energy, geothermal energy, breeder reactors, coal gasification and liquefaction, and nuclear fusion.[17]

Argonne's Initiatives in Energy Research

The recommendations from the McCormack energy task force confirmed Laney's determination to enlarge Argonne's research-and-development mission. Laney, along with most Americans, still clung to the dream that science and technology could develop abundant sources of clean energy, whether nuclear, solar, fusion, or another of the advanced energy systems. But Laney also believed that reordering Argonne's mission would be difficult and painful. On a visit to Washington, Laney learned that the JCAE, worried about McCormack's challenge, contemplated widening its energy responsibilities. It was important, Laney urged, for the laboratory to strengthen its political base by encouraging the Illinois members of the Joint Committee to visit Argonne.[18]

Laney wanted the Joint Committee and others to see that Argonne was already seriously involved in energy research other than the breeder reactor. Argonne's controlled thermonuclear research (CTR) study group in the high-energy physics division coordinated fusion research at the laboratory. While their colleagues in the reactor engineering division perfected technology to split heavy atoms (fission), physicists in CTR research participated in a national effort to create energy through nuclear fusion—joining together two light atoms such as of deuterium, an isotope of hydrogen. When the nuclei of deuterium fuse, the mass of the products is slightly less than the original nuclei. This "lost" mass appears as thermonuclear energy. Scientists estimated that energy available from deuterium in seawater was a thousand million times greater than the heat available from known coal reserves. Although great amounts of energy were also required to fuse the atomic nuclei, scientists were confident that the fusion process would produce a significant net energy gain. According to AEC calculations, the deuterium in a gallon bucket of water could yield the same amount of energy as 300 buckets of gasoline.[19]

Thermonuclear furnaces are commonplace enough, as anyone who has basked under the earth's modest sun might know. The Atomic Energy Commission had already achieved instantaneous thermonuclear reactions in hydrogen bombs, which released vast amounts of destructive energy useless for any peaceful purpose except, as some scientists dreamed, for moving mountains, blasting waterways, or perhaps digging craters. Producing a controlled thermonuclear reaction, on the other hand, presented great challenge because scientists would not only have to heat deuterium to the temperature of the sun, but they would have to confine the superheated plasma long enough for the fusion process to return a net profit of energy. In sum, they sought design configurations through which they could achieve equilibrium and stability of confinement and temperature over time. The challenge was no less than to create a little sun in a bottle from which they could tap excess energy (heat) to fire a boiler to power a turbine to generate electricity. The whole process promised to be wonderfully clean environmentally with few nasty waste by-products.

The AEC's major centers for fusion research were located at Princeton University, Oak Ridge National Laboratory, Los Alamos National Laboratory, and the University of California. Argonne supported research in CTR engineering and design studies at Los Alamos and at Princeton University. The materials science division conducted surface research on the swelling and embrittlement effects of radiation on metals such as niobium and titanium, which might prove suitable for fusion reactor containers. The physics division studied what happened to nuclei when they crashed into the walls of a plasma container and what effects they had on the walls. Chemical engineers at Argonne investigated the properties of tritium, still another heavy hydrogen isotope, while the high-energy physics division, which had built its superconducting magnets for use with the ZGS, studied how superconducting magnets might be used to achieve magnetic confinement of plasma in the laboratory.[20]

Research on fusion energy had renewed the study of magnetohydrodynamics (MHD), the science of electrically active fluids. Magnetohydrodynamics is a long, intimidating name for a very old idea first explored by Faraday and Maxwell in the nineteenth century. When a conductor passes through a magnetic field, it creates an electrical current in the conductor. A fluid conductor flowing through a magnetic field creates a voltage gradient that converts the energy of motion directly into electricity, eliminating both the turbine and the generator normally used to produce electricity. An MHD power plant would require nuclear or fossil fuels to heat a conducting fluid of ionized gas (or perhaps liquid metal) whose expansion would force the conducting fluid through the MHD generator—essentially a tube surrounded by a magnetic field. When the hot ionized gas passed through the generator's magnetic field at extremely high velocity and so created a voltage gradient, electricity would flow between electrodes attached to the generator. Argonne's calculations estimated that MHD power plants could match steam plants in efficiency and, because they lacked moving parts such as turbines, were mechanically more reliable under high-temperature operation and in highly corrosive atmospheres.[21]

Much simpler than the advanced energy systems, but technically more immediately practical, were technologies to exploit America's enormous coal reserves. But the environmental costs of burning coal were high. The General Advisory Committee estimated that proven coal reserves were about thirty-five times greater than proven oil reserves. Albert A. Jonke and his group in the chemical engineering division realized that the fluidized bed technology used to control sulfur dioxide in the production of uranium hexafluoride might also be used to burn coal in a fluidized bed. With support from the National Air Pollution Control Administration (NAPCA), the Environmental Protection Agency (EPA), and Office of Coal Research in the U.S. Department of the Interior, Argonne developed a fluidized-bed combustor designed to burn coal cleanly. In the combustor, a vertical burner was loaded with a finely powered bed of coal and limestone. When air was blown upward through the bed, the coal and limestone behaved like fluids suspended under pressure in the combustor. As the coal burned, the limestone reacted with the released sulphur reducing the emissions of sulphur dioxide, one of coal's most harmful pollutants. Argonne's experiments established that when properly operated the fluidized-bed combustor prevented 95 percent of the sulfur dioxide from reaching the flue gas. Although workable in the laboratory, the fluidized-bed combustor had yet to operate economically as an electricity-generating pilot plant.[22]

Argonne's research on high-performance lithium–metal sulfide batteries was also promising, but shaky. In connection with the reactor program in the early 1960s, the chemical engineering division investigated electrochemical reactions of molten salts and liquid metals. Out of this research evolved the laboratory's battery program. Argonne's experimental battery—which used lithium as the negative pole, molten sulfur as the positive pole, and molten salt as the electrolyte—operated at temperature of 400° C (725° F). Argonne touted the new battery for its environmental advantages. The molten sulfur (iron sulfide) was low in cost, abundant, and nontoxic. Laboratory performance tests indicated that a 600-pound, rechargeable lithium-sulfur battery could satisfactorily power a medium-sized compact car with a cruising range of 150–200 miles. Since automobiles contributed almost half the air pollution in American cities, commercially feasible electric automobiles would significantly reduce urban pollution problems.

Although development of the lithium-sulfur battery looked interesting, competition for funding with the breeder program forced the AEC to drop the program in 1968. As a stopgap, NAPCA, the National Heart and Lung Institute (NHLI), and the army contributed modest funding for the next few years while the research group declined from its peak of forty members to eight. Finally, in 1972 the National Science Foundation agreed to provide $500,000 on the condition that research on lithium-sulfur batteries explore how they could be used to combat brownouts and blackouts caused by excessive power demands at peak hours. The NSF hoped that the lithium-sulfur batteries, arranged in large banks near cities, could store electricity from conventional power plants during periods of slack energy demand and then be brought on line during peak demand hours to contribute electricity to the regional grid. Such load leveling, or "peak shaving," could greatly increase the efficiency of conventional generating facilities. NSF support continued until 1973 when the AEC once again decided to fund the project.[23]

The AEC Slashes Argonne Funding

By mid-January 1973 Laney knew that he would not be appointed as the permanent director of Argonne National Laboratory. Robert Sachs, the director of the Enrico Fermi Institute at the University of Chicago and a former associate director at Argonne, was the odds-on favorite of the University of Chicago. Rumor around the laboratory said that the basic research staff favored Sachs over Laney, although others, especially in high-energy physics, regarded Sachs as too opinionated to guide a national laboratory through the shallows and shoals of big-science politics. At this time, however, Sachs enjoyed the strong backing of William Cannon, who regarded Sachs as the best among the available candidates in the field. Sachs himself was aware that he was not the first choice for laboratory director and that several top candidates had already turned down the position. If Laney harbored any disappointment in not being chosen director, he did not have time to nurse his wounds.[24]

Without warning, the AEC announced drastic, immediate budget cuts for fiscal years 1973 and 1974. The commission's LMFBR program was in deep trouble with serious cost overruns in the development of reliable pumps at Hanford's FFTF. On February 1 Milton Shaw met Laney in Chicago to inform Argonne management that the laboratory would have to cut $3.5 million from its current fiscal budget and anticipate an additional $1.5 million reduction in fiscal year 1974. The cuts required an immediate reduction in force of 375 employees, or about 6 percent of Argonne's work force. Not that Argonne was singled out for budget punishment. Shaw's cuts, which were particularly painful because they occurred in the third quarter of the fiscal year, slashed 10 percent of the employees at Oak Ridge, Lawrence Berkeley, and Sandia; 6 percent at Hanford; and 5 percent at Lawrence Livermore Laboratory. In addition, Brookhaven lost 225 employees, while the Stanford Linear Accelerator released 140. At Argonne, the cuts not only sharply reduced programs in reactor development, high-energy physics, the physical sciences, and nuclear education and training but also devastated morale across the laboratory when the deepest cuts were made in program support. Argonne-West, where work proceeded on the breeder-reactor programs, was spared most of the budget cuts. Consequently, the impact of the recessions was magnified at the Illinois laboratory, where personnel cuts exceeded 10 percent.[25] Laney made every effort to soften the financial blow through retirements, attrition, a hiring freeze, and other savings, but after years of cuts in the basic sciences, he had little room to manage the reduction in force.

As recently as December 1972, Argonne had absorbed an $800,000 reduction, again because problems in the LMFBR program threatened to bleed Argonne's physical research programs. The only good news was in Argonne's small environment, biology, and medicine program, which had received steady increases since 1968 and which suffered no recession in 1973. Argonne also took pains to announce that the reductions were not to be regarded as any lessening in the commission's commitment to develop the breeder reactor as the highest priority in the government's energy program.[26] But the reductions in force of experienced, trained, and skilled employees at Argonne in the face of mounting national energy shortages seemed incongruous to Illinois Senator Charles H. Percy, Congressman John N. Erlenborn, and their constituents. The AEC never hid its reasons. Tight federal budgets and cost overruns in the LMFBR program had left the agency no choice.[27]

In an appointment destined to have profound impact on Argonne National Laboratory, President Nixon named University of Washington zoologist Dixy Lee Ray as chairman of the AEC on February 6, 1973. The second woman appointed to the AEC and the only woman to serve as chairman, Dixy Lee Ray succeeded James Schlesinger, who left the commission to become director of the Central Intelligence Agency. A marine biologist who also directed Seattle's Pacific Science Center until her appointment to the AEC as a commissioner in August 1972, Ray was a close friend of University of Chicago provost John T. Wilson, who had been her mentor when he was deputy director of the National Science Foundation. A bit eccentric, Ray nonetheless possessed a dynamic personality and strong convictions tempered by charm, wit, and informality. In August on her way to Washington, D.C., to assume her new position, she toured the commission's western facilities in her motor home accompanied by her ever present companions, Ghillie, her Scottish deerhound, and Jacques, a miniature poodle. Shortly thereafter in October, Ray toured Argonne, focusing especially on reactor safety, the ZGS, biology and medicine, and environmental research. Although new to the AEC, she shared the agency's pride in its scientific and technical excellence. She also recognized that the commission's mission was increasingly frustrated by public skepticism, and she admirably committed herself to bridge "the gap that so often separates science from society at large." Unfortunately, like so many at the Atomic Energy Commission, Ray believed the commission's political troubles were caused not so much by faulty program and policy, but rather were rooted in weak public relations and public education. As a measure of her commitment to openness, Ray encouraged the commission to declassify over a million documents stashed away in AEC vaults across the country. Although the laboratory had performed little classified work since the end of the naval reactors project in 1955, Argonne declassified 46,000 documents, most of them stamped "Secret—Restricted Data."[28]

Dixy Lee Ray knew that her greatest challenge and first priority was to get the breeder-reactor project back on track. She had to stop the budget hemorrhaging of the Fast Flux Test Facility at Hanford and restore confidence to the field and laboratory management of the Liquid-Metal Fast-Breeder Reactor project. During the search for Argonne's new director, Louis Rosen, Los Alamos's eminent physicist, warned that the LMFBR development was "spinning its wheels" because Shaw tried to run the project as if it were a production program with a few minor glitches rather a research and development program requiring extensive and bold creativity and imagination. Agreeing with Rosen, Hans Bethe, Cornell University's Nobel laureate and one of the Manhattan Project's most distinguished alumni, went further to predict that when (not if) the LMFBR program failed, the University of Chicago would share blame with the Atomic Energy Commission for not insisting on fundamental research in materials, reliability, and safety. Chairman Ray could not ignore that the breeder program was in disarray, requiring strong new leadership both in the field and at headquarters.[29]

Robert Sachs, Director

Robert Sachs seemed a natural choice for director of Argonne Laboratory. He was the very picture of the distinguished, but not controversial, modern science administrator epitomized by former AEC chairman Glenn Seaborg (figure 28). His academic and scientific credentials were impeccable. A professor of physics at the University of Chicago and the head of its Enrico Fermi Institute when he accepted the director's position, Sachs, now fifty-seven years old, returned to Argonne for the third time in his remarkable career. He had received his Ph.D. in theoretical physics at Johns Hopkins University in 1939. After academic appointments at George Washington University, Purdue, and the University of California, Berkeley, he joined the Ballistic Research Laboratory at the Aberdeen (Maryland) Proving Ground in 1943, where he remained until the end of World War II, when he moved to the Metallurgical Laboratory. Although he did not participate in the wartime atomic bomb project, Sachs assisted the Manhattan Engineer District's conversion of the Met Lab into the postwar Argonne National Laboratory. He was director of the theoretical physics division until 1947, when he accepted an appointment to the University of Wisconsin's physics department. After seventeen years at Wisconsin, he returned

Figure 28. Robert G. Sachs

to Argonne in 1964 as the associate director of high-energy physics in charge of the experimental program on Argonne's powerful accelerator, the ZGS.

Sachs's easy manner was deceptive. Tanned and relaxed, an avid sailor who maintained his own boat on Lake Michigan but who also loved to explore unfamiliar waters in the Bahamas and the Caribbean, he preferred to be called Bob, described himself as a "listener," and professed greatest interest in teaching and research, not administration. Sachs and his large family planned to maintain their home in Hyde Park where they could remain close to the collegiality of the University of Chicago. Yet he remained aloof from civic affairs, believing his commitment to science education and research was itself his major exercise of civic responsibility. Nationally, he belonged to the science power elite. Well known for his textbook *Nuclear Theory* published in 1953, he had served on the physics advisory panel of the National Science Foundation (1956–61), the science policy committee of the Stanford Linear Accelerator Center (1966–70), and the high-energy physics advisory panel of the Atomic Energy Commission (1967–69). In 1969, when the National Academy of Sciences organized the Bromley committee to survey the health and needs of physics in the United States, Sachs coordinated the report on elementary-particle physics.[30]

Sachs's name was not among the original fifty candidates considered for laboratory director. While Duffield languished as a lame duck through the long search, Sachs served as a research counselor to the University of Chicago, assisting Argonne management with all of the scientific and engineering programs and personnel. Although an obvious candidate for the position, Sachs was reluctant to leave the Fermi Institute despite his deep concern for the future of Argonne. His apparent modesty and lack of ambition bothered the AEC, which had to be assured that Sachs really wanted the director's job. When Sachs visited Washington for an interview with the commission, he did not "give a damn" whether or not they offered him the position. The general manager, Robert Hollingsworth, who wanted a strong director to manage the breeder-reactor program, worried that Sachs was too nonchalant and perhaps would not disentangle himself from old research interests. But Dixy Lee Ray, with whom Sachs had a very good talk, was enthusiastic about him, and when the University of Chicago promised that Sachs fully understood the urgency of the breeder-reactor program, the AEC consented to his appointment. On March 13, President Levi of the University of Chicago announced that Sachs would assume his duties as laboratory director on April 1, 1973.[31]

Sachs was not naive about the challenge he faced. Morale was low. Constant hectoring from the AEC and perpetual reorganization accompanied by deep cuts in budgets and personnel had almost paralyzed Argonne's self-confidence. In February Sachs had told Cannon that he wanted to play a major role in determining what was cut and who was fired, but he did not want "to take public responsibility for doing so."[32] His first priority when he became director was to lift the morale of the laboratory by personally meeting every employee. For three months, division by division, unit by unit, office by office, Sachs visited laboratory programs. As optimistically as possible, Sachs told Argonne employees he preferred not to look at the budget curve, which showed a negative trend, but rather to focus on the "separate *amounts of money*" in the laboratory budget that represented the "impressive wealth at Argonne's disposal." As if the budget were a tattered dish rag, Sachs expressed confidence the laboratory would "wring the most out of what it has."[33] Whenever possible, Sachs left his office to visit Argonne's labs and shops to reestablish communication between the director and the Argonne staff, which many of them felt had been lost during Duffield's tenure.[34]

Unfortunately, Sachs had scarcely warmed the director's chair when he left for China on May 15 for a month-long tour as a member of the American scientific delegation, whose visit was among the several negotiated in February by Henry A. Kissinger, President Nixon's foreign affairs advisor. Sachs's prestigious assignment could not have come at a more awkward time. Only eighteen months previously, Duffield had been severely criticized for abandoning the laboratory during a crisis in order to tour the Soviet Union. Now Sachs apologized for undertaking foreign adventures while his inspection of Argonne-West waited until his return from China. In the midst of America's energy crisis, workers at the National Reactor Testing Station could not miss the implication that the new director regarded a month in China as more important than a day and a half in Idaho. A friendly sun shone on Sachs during both visits, and he wisely made no comparison between his back-to-back trips. During his whirlwind tour of Idaho, however, Sachs felt like a politician running for office, and he tried to make amends by highly praising the Argonne-West personnel. On his return from Idaho, Sachs paid customary tribute to western hospitality but candidly noted the deep alienation of the Idaho staff. When "they talk about those of us in the 'East,'" he noted soberly, "you would think that we were sitting out on a far promontory of Maine."[35]

The State of the Laboratory, 1973

By 1973 the AEC's network of national laboratories and testing facilities employed 91,000 AEC and contractor personnel at facilities whose capital investment was $9.5 billion. One-third of the commission's annual budget of $1,358 million was spent on the weapons program, whereas 22 percent went to civilian reactor development, 17 percent to physical research, 11 percent to the naval reactor program, and 7 percent to biomedical and environmental research. One of seven multiprogram national laboratories, Argonne's operating budget for fiscal year 1973 was $93 million, of which $3.7 million was for non-AEC work, principally for other federal agencies. In comparison with the other national laboratories, in fiscal year 1973 Argonne received the following: [36]

LABORATORY	AEC ($, MILLIONS)	WORK FOR OTHERS ($, MILLIONS)
Argonne	89	4
Oak Ridge	83	17
Lawrence Livermore	124	12
Lawrence Berkeley	34	2
Brookhaven	50	3
Pacific Northwest	22	16
Los Alamos Scientific	108	12
Totals	510	66

In 1973 Argonne's program funding from the Atomic Energy Commission was divided among reactor development ($42 million), physical research ($34 million), biomedical and environmental research ($9 million), nuclear regulation support ($2 million), national laboratory-university cooperation ($1 million), and applied technology ($600,000). Additional funds in the amount of $3.7 million, primarily for the production and storage of energy and for environmental research, came from the National Science Foundation, the EPA, the Illinois Institute for Environmental Quality, the Office of Naval Research, and others. Funds from non-AEC agencies supported research on batteries, low-polluting processes to burn coal, magnetohydrodynamics, and various environmental programs such as strip-mine reclamation.

The AEC's investment in plant and capital equipment provided still another indicator of Argonne's comparative wealth in 1973. Comparative capital investment—including laboratory space; machine shops; computing services, instrumentation supporting research in biology, physics, chemistry; medical and environmental programs; engineering facilities; and weapon design-and-development facilities—was as follows:[37]

Laboratory	$ (millions)
Argonne	465
Oak Ridge	356
Lawrence Livermore	278
Lawrence Berkeley	156
Brookhaven National	329
Pacific Northwest	58
Los Alamos Scientific	491
Total	2,133

In 1973 Argonne's most prestigious basic research facility, the Zero Gradient Synchrotron, was ten years old. The aging facility, which was capable of accelerating protons to a maximum energy of 12.5 GeV, nonetheless played an important role in the United States' high-energy physics program among the family of proton accelerators, including the 30 GeV Alternating Gradient Synchrotron (AGS) at Brookhaven, and Fermilab's 400 GeV proton accelerator. The ZGS featured the 12-foot deuterium–liquid hydrogen bubble chamber, the largest and most unusual facility of its kind in the world. The ZGS facilities offered marvelous research opportunities to academics, especially young midwestern professors, postdoctoral fellows, and graduate students in the fields of medium- and low-energy physics, solid-state physics, chemistry, and materials science. But the ZGS's days were numbered. In the international competition of high-energy physics, research progress in big science required new and ever more powerful machines. In 1973 the high-energy physics community estimated that the ZGS had only five to eight years remaining of productive research.[38]

Argonne's reactor facilities included the plants and equipment at both the Illinois site and Argonne-West. The LMFBR project was Argonne's largest program. In Idaho, EBR-II continued as the only operating breeder reactor in the United States, gaining operation and maintenance experience while irradiating reactor fuel assemblies, reactor test components, and experimental materials. Also supporting the breeder-reactor program at Argonne-West was the ZPPR, whose mock-ups provided physics data on configuring the reactor core for an LMFBR demonstration plan, and the HFEF (figure 29), a heavily shielded facility for disassembling and analyzing highly radioactive reactor fuels, components, and materials from the EBR-II and other irradiation facilities. As part of the LMFBR safety program at Argonne-West, TREAT simulated quick power surges (transients) to test the design of breeder-reactor fuel elements under abnormal operating conditions.

Argonne also operated four research reactors in Illinois. CP-5 (Chicago Pile No. 5), was a direct descendent of the first reactor, Fermi's CP-1. For twenty years CP-5 had been the workhorse of Argonne's nuclear physics and materials science research program. The zero-power reactors, ZPR-6 and ZPR-9, like the ZPPR at Argonne-West, tested theoretical calculations and operating characteristics of mock-ups of fast reactor core designs. In addition to reactor physics, materials evaluation, and instrumentation, the reactor development program studied sodium behavior and the management of

Figure 29. Hot Fuel Examination Facility and the operation of remote manipulators

gaseous fission products. Associated environmental research evaluated the ecological effects of radioactivity released from nuclear power plants and estimated the environmental impact of breeder reactors. The relatively new JANUS reactor served research in the life sciences.

Comparative employment figures also revealed Argonne's relative strength among the national laboratories. As part of a general reduction in force at the national laboratories, Argonne had reduced its staff below 4,000 by June 1973, down from approximately 5,600 in 1968. Argonne's total employment compared with the other laboratories was as follows:[39]

Laboratory	Scientists Engineers	Technicians	Others	Total
Argonne	1580	765	1637	3982
Oak Ridge	1490	598	1849	3937
Lawrence Livermore	1771	1384	2198	5353
Lawrence Berkeley	645	693	797	2135
Brookhaven	750	466	1221	2437
Pacific NW Lab	516	288	633	1347
Los Alamos	1718	1308	1453	4479
Totals	8470	5502	9788	22,470

It was axiomatic that Argonne's most precious resource was its highly skilled, highly educated professional staff. In 1973 Argonne counted 550 Ph.D., 325 M.S., and 650 B.S. degree holders among its employees. In broad categories, Argonne employed 650 physical scientists, 720 engineers, 765 technicians, 150 life scientists, and 60 mathematicians and computer scientists. An additional average of 550 faculty and students, not paid by Argonne, worked on various research projects throughout the laboratory. Among Argonne's most distinguished scientists, Joseph J. Katz in 1973 became the first staff member elected to the prestigious National Academy of Sciences. Katz had received the Midwest Award of the American Chemical Society in 1969, was a veteran of the Manhattan Project's Met Lab, and had served as secretary of the physical chemistry division of the American Chemical Society and editor of the *Journal of Inorganic and Nuclear Chemistry*. He first established his reputation working on uranium and transuranic elements with Glenn Seaborg and later contributed importantly to the scientific understanding of how chlorophyll uses light in photosynthesis. Argonne's other member of the National Academy of Sciences, Director Robert Sachs, proudly welcomed Katz to the academy.[40]

The high quality of Argonne's staff did not impress Milton Shaw, however. When he was interviewed for the director's position in Washington, Sachs was shocked at Shaw's contempt for Argonne and its personnel. Shaw's bitter attack on Argonne was so appalling to Sachs that he could not believe his ears. Sachs maintained that the only way to approach a job was to be willing to lose it; otherwise one would make compromises when he should be saying "No." Sachs also knew that Shaw was looking for a "yes-man" to fill the director's position at Argonne. With his appointment hanging in the balance, Sachs knew better than to antagonize the AEC's director of reactor development. Yet it was all he could do not to rebut Shaw's tirade against Argonne. The laboratory was for the birds, Shaw charged, not doing the AEC any good service at all. Hesitantly, Sachs countered that EBR-II had been vital to the breeder reactor program. "EBR-II," Shaw reportedly snorted, "Argonne didn't do anything—we did!"[41]

There was nothing for Sachs to do but sit dumfounded as Shaw carried on about Argonne's incompetence. "You know," Shaw lectured Sachs, "right here next to my office are the best engineers in the world." Sachs was flabbergasted by Shaw's assertion that the AEC's top engineering talent sat in the commission's headquarters. If it were true, Sachs reflected, it just proved that Shaw "had absolutely no judgment at all to keep good engineers in Washington." But Sachs believed Shaw was wrong "because those nuclear engineers were not in the same class with the people . . . at Argonne." But he said nothing. What could he say? Sachs was stunned that Shaw could appear so stupid, but he had no information with which to refute Shaw's allegations. Sachs realized that Shaw was trying to mimic Rickover's style, with "all of Rickover's faults and none of his positive qualities." Sachs passed his test by keeping his mouth shut, even to the point of neglecting to tell Shaw that while at Oak Ridge in July 1946 he had taught Rickover what he knew about reactors. If Shaw had known that, Sachs later reminisced, "he really would have had an absolute fit."[42]

But Sachs passed muster with the commissioners and the general manager largely because of Dixy Lee Ray's influence. His discussions about the state of the laboratory with Ray contrasted sharply with his interview with Shaw. Like Shaw, Ray was direct and open, yet Sachs felt good chemistry between them (figure 30). Ray, of course, strongly advocated nuclear power, and Sachs had been associated with reactors since the Manhattan Project. His optimism about the future of the breeder reactor encouraged Ray. Most importantly, Sachs's affirmation of the excellent people at Argonne impressed Ray, who herself had fostered staff and programs at the Pacific Science Center. They both wanted to be tough and independent but open, warm and caring administrators. It was "just one of those things," Sachs remembered, "you look at somebody and they look at you, and you say, this is right—we understand each other."[43]

Sachs returned to Chicago confident of Ray's support, but uncertain about

Figure 30. Robert G. Sachs with Dixy Lee Ray, Atomic Energy Commission chairman (*Frontiers, 1946–1996*, p. 20)

the future of the breeder-reactor program. Although he realized that Argonne's future would be governed to some extent by its personnel and facilities, Sachs told the Argonne Universities Association that he did not want long range planning to be limited to a narrow extension of existing laboratory programs. Rather, he hoped Argonne would have the flexibility to pursue nonfission energy conversion, storage and transmission and the environmental and biological impact of energy production and distribution. Nonetheless, Sachs emphasized that since Shaw had taken over as the director of nuclear reactor research at the AEC (a period that generally coincided with the AUA's participation under the Tripartite Agreement), 50 percent of Argonne's projects had shifted to an industrial-style operation. The commission's drive to achieve a commercial breeder reactor differed greatly from the director's hope to pursue basic science research. Sachs doubted that his academic friends either understood the dynamics of the reactor development program or appreciated how the requirements of the breeder project had nullified priorities in basic research.[44]

Sachs loved the irony that he became director of Argonne National Laboratory on April Fool's Day. Although Sachs claimed to know nothing about the intrigue in Washington, major changes in the breeder program were afoot at the commission. As it turned out, Shaw, Argonne's nemesis, never had a chance to measure Sachs's effectiveness. Within three months of Sachs's appointment Shaw was gone. Rumor spread throughout the laboratory that Dixy Lee Ray had fired Shaw. Officially, Shaw resigned as the commission's director of reactor development on June 8, 1973, after Ray won a major power struggle over Shaw's allies on the commission and on the Joint Committee.

In the spring, under Ray's leadership, the AEC created the new Division of Reactor Safety Research, directed by Herbert J. C. Kouts of Brookhaven National Laboratory, leaving Shaw in charge of a reorganized and diminished Division of Reactor Research and Development. For some time, citizens' lobbies and Argonne's safety engineers had complained that Shaw's drive to develop the breeder reactor sacrificed vital basic research on nuclear reactor safety, particularly that relating to "conventional" light-water reactors. Shaw and his ally on the commission, James Ramey, vainly challenged Ray's reorganization, which ostensibly enhanced the priority of safety research while protecting it from competition with the breeder program for project funds. As Congressman Chet Holifield fumed at the Joint Committee, Ray outmaneuvered Ramey by securing support from Commissioners William O. Doub and Clarence Larsen, both Nixon appointees.[45]

Shaw correctly interpreted the reorganization as a vote of no confidence in his leadership by the commission's Republicans. He complained to the general manager, Hollingsworth, that without support from the commission and without more people and money, he could not do the job. Specifically, Shaw demanded even more control over the program, principally direction of all the reactor people at Argonne. Like his mentor Admiral Rickover, Shaw wanted to command "a strong central mission-oriented organization . . . reporting directly to the General Manager and accountable only through him to the Commissioners." But with the AEC's own future on the line in the looming energy crisis, the Republican commissioners were not tolerant of Shaw's bureaucratic tantrum over the safety issue. Ignoring his demands, they turned a deaf ear to Shaw's warning that the breeder program would lose momentum and perhaps develop severe safety problems. On June 14, two weeks before the end of Ramey's term, the commission simply noted Shaw's decision to retire and moved on to discussions of energy legislation and priorities.[46]

Dixy Lee Ray and the President's National Energy Plan

The energy crisis of 1973 not only heralded the need for a comprehensive national energy plan but also served notice to the national laboratories that they needed to shift research priorities toward developing alternative energy sources. Warning in April that America's energy "challenge" would become an energy crisis if the United States did not change its energy consumption habits,

President Nixon noted that Americans, although only 6 percent of the world's population, consumed one-third of the world's energy. The United States' dependence on foreign oil had increased to the point where 30 percent of American petroleum supplies came from abroad. Unless the United States adopted a comprehensive conservation effort while developing domestic energy resources and technologies, the nation risked the possibility that its vast fleet of cars and trucks might grind to a halt during an international energy crisis. American cars guzzled almost 200 million gallons of gas a day! In addition, Americans sucked dry another 2 million oil barrels per day of diesel fuel, jet fuel, heating fuel, and residual oil to power electricity-generating plants. With demand increasing over 5 percent a year, the government predicted gasoline shortages for 1973 summer vacations.[47]

Among numerous initiatives, Nixon had proposed ERDA to coordinate all federal energy research programs. Under the Nixon plan, the Atomic Energy Commission and its national laboratories would form the core of a new energy research agency responsible for developing energy technologies and applications ranging from fossil fuels and nuclear reactors to solar energy and energy conservation. A separate nuclear regulating and licensing agency would be established in the Nuclear Regulatory Commission (NRC).

When Nixon forwarded his energy legislation to Congress in June 1973, he asked the Atomic Energy Commission for recommendations on how to spend an additional $100 million on energy research in fiscal year 1974, including research and development on coal, advanced energy systems, conservation, environmental control, and gas-cooled nuclear reactors. More importantly, Nixon also asked Ray to develop a five-year national energy plan for the expenditure of $10 billion.[48]

Working under the guidance of Nixon's newly established Energy Policy Office, Ray organized an enormous task force of experts from industry, the foundations, academia, government, and the national laboratories. She set up workshops and panels on all conceivable energy research and development issues. Argonne's former director Walter Zinn participated in the Long Range Nuclear Option workshop. Martin Kyle, Argonne's fossil fuel expert, served on the Coal and Shale Processing and Combustion panel. Under the leadership of John Teem, the AEC's assistant general manager for physical research and laboratory coordination, Sachs and the other multipurpose laboratory directors served as Ray's screening committee to review the energy proposals generated by this hastily assembled and unwieldy planning bureaucracy.[49]

Sachs was simply amazed by the planning process. John Abbadessa, the commission's incomparable controller, orchestrated the all-important budget sessions. Striding to the blackboard, Abbadessa divided the $10 billion into five categories: energy conservation, oil and gas production, synthetic fuels, nuclear, and other advanced energy systems. To Sachs's amusement, after making a rough cut of the $10 billion, Abbadessa went down the list, item by item, asking "does somebody think it should be bigger, the same, or smaller?" "Sixty million dollars disappeared just like that!" Sachs recalled. "There goes my laboratory!" one of the lab directors joked.

But this time they were not slashing budgets; rather they were planning a massive increase in federal expenditures for energy research and development, two-fifths of which would be dedicated to nuclear reactor projects. When Sachs returned to Chicago, he ordered the personnel office to place ads "in all the right places" announcing Argonne's recruitment of engineers with an interest in the environment, coal technology, and solar energy—"all the various energies." When teased by his fellow lab directors that he was crazy to advertise for the positions without money, Sachs assured them, "It'll come—it's coming. You've got to do it."[50]

Dixy Lee Ray's report to the president, "The Nation's Energy Future," arrived on Nixon's desk after the Israeli victory in the October Yom Kippur War. By December 1973 oil supplies fell critically low when the Organization of Arab Petroleum Exporting Countries (OAPEC) placed an embargo on crude oil shipped to the United States. As the United States fell into a genuine energy crisis, Nixon launched Project Independence, which was designed to secure American energy self-sufficiency. Recalling the Manhattan Project, which had built the atomic bomb, and the Project Apollo, which had put a man on the moon, Nixon affirmed his faith that American industry and science could free the United States from dependence on foreign oil. Not since World War II had the federal government established a comprehensive energy policy for the United States. Reluctantly, the Nixon administration established energy price controls and resource allocations to achieve immediate relief. By mobilizing Argonne and the other multipurpose national laboratories, the government hoped to achieve sufficient short-term goals to guarantee American energy independence by 1980. Unlike the race for the atomic bomb during World War II or NASA's Project Apollo, however, the laboratories' objectives and priorities were far from clear.[51]

As Robert Sachs noted, solving the energy crisis was not just a matter of committing resources and having the will to solve the problem. The Manhattan Project and Project Apollo were not constrained by development costs and the goal of commercial feasibility. In both cases, the government was the only customer, the one who willingly paid the entire research and development bill without concern for recovering the taxpayers' investment. Only if the government wanted to operate a for-profit nuclear arms bazaar or to maintain a commercial ferry service to the moon would such analogies be appropriate. To some degree the government could underwrite energy research-and-development costs, but in the long run, Sachs observed, new energy sources developed by the government had to meet not only the tests of "technical feasibility," but also "economic feasibility" and "political feasibility" (public acceptance). Nixon's four energy messages (June 4, 1971, April 18, 1973, June 29, 1973, and November 7, 1973) largely focused on technological fixes to the energy crisis. Neither national politicians nor the American public wanted to believe that the United States faced permanent energy shortages requiring major institutional and policy innovations. Although Nixon's energy policy included the creation of the Federal Energy Agency along with the reorganization of the AEC, the main emphasis of his program was to accelerate development of nuclear power reactors, clean coal technology, fusion energy, and solar technology with an obligatory gesture toward conservation.[52]

For Argonne, the energy crisis meant potential recovery from the budget crunch of 1973. As one of the last to testify before Holifield's House Government Operations Committee in favor of the energy reorganization plans, Sachs invoked Argonne's World War II legacy and the memory of the first nuclear chain reaction at the University of Chicago's Metallurgical Laboratory. Having wrapped himself in the aura of the Manhattan Project, Sachs quickly moved to the main purpose of his testimony: to highlight Argonne's rich potential for solving energy problems based upon the laboratory's extensive background of experience. Holifield's committee did not need a history lesson on Argonne's role in the liquid metal fast breeder reactor project, so after briefly singing the praises of EBR-II ("the only successful operating liquid metal breeder reactor power plant in the United States"), Sachs noted the laboratory's work in "clean coal combustion, high performance batteries for energy storage and automobile propulsion, controlled thermonuclear systems, magnetohydrodynamics, and superconductivity technology." These, along with the laboratory's environmental and ecology programs, defined the assets Argonne would bring to the new Energy Research and Development Administration with its associated laboratories. Recognizing the need for short-term results, Sachs nonetheless reminded the congressmen that the national laboratories were in a position to help solve the energy crisis only because the AEC had made substantial investment in basic research in the physical and biological sciences. "Support of basic science and technology provided in the Atomic Energy Act," Sachs emphasized, "is inherently necessary in the development of all forms of energy, not only nuclear."[53]

While Sachs was defending the high ground of basic science and technology in Washington, D.C., employees back at the laboratory met the energy crisis more prosaically by turning down thermostats, turning off unnecessary lights, and car-pooling to work. "Save A Drop," the *Argonne News* encouraged laboratory employees, who were also assured that, barring unforeseen exigencies, the energy crisis would not cost Argonne more jobs.

The American public not only looked to the national laboratories for technical solutions to the energy shortage but also expected the laboratories to set an example of energy efficiency and conservation. The old Sunday-school song that encouraged Argonne's children to "Let Their Little Light Shine" gave way to cautious advice for trimming holiday trees "to glimmer somehow without lights (encouraged, not mandatory)." Argonne's steam plant switched to low-sulfur coal when available, while the lights in the cafeteria glowed "intimately" for the lunch-time crowd. Throughout the laboratory, hallways, corridors, and common areas dimmed when custodians removed over half the overhead lights. Before the end of the winter of '73 the "layered look" for both men and women became the fashion de rigueur at Argonne National Laboratory.[54]

The AUA Revolt, 1974

In the midst of the 1973–74 energy crisis and government reorganization, the Argonne Universities Association selected a new president, who immediately became embroiled in negotiating a renewal of the Argonne National Laboratory Tripartite Contract, which was due to expire in September 1975. On September 7, 1973, the AUA board of trustees elected Paul McDaniel, another veteran of the Met Lab and the Manhattan Project, as president of the AUA, succeeding Armon Yanders. A physicist with a Ph.D. from Indiana University, McDaniel had served as director of the AEC's division of research from 1966 to 1972. With his permanent residence in Arlington, Virginia, McDaniel informed the board that he intended to establish an office in Washington, D.C., for the AUA President and a staff of four or five. He reasoned that the interests and welfare of the Argonne National Laboratory and the AUA community could best be served by the president's active and close association with the Washington establishment, which set national science priorities. McDaniel hoped the permanent AUA office in Washington would serve as a window through which federal officials and Argonne constituents exchanged information about research-and-development priorities and accomplishments.[55]

The original Tripartite Agreement signed in 1966 had provided for a five-year contract. The renewal in 1972 extended the tripartite contract for only three years, a move which McDaniel interpreted as an AEC test of the AUA–University of Chicago partnership in managing the laboratory. From his perspective as a former AEC employee, McDaniel believed that the commissioners, despite all their other responsibilities, were more familiar with the tripartite contract than were the AUA trustees. Entreating his trustees to devote more time to Argonne's management, McDaniel moved contract negotiations to the top of the association's agenda. He presented five alternatives that he thought the commission faced: continuation of the tripartite contract; a change to a single academic contractor; a change to an industrial management, direct operation, or autonomous laboratory (incorporated).[56]

McDaniel argued that the AUA should decide for itself whether or not it wanted to continue the contract or get out. If the board wanted to stay in, he challenged the association to determine its conditions for continued partnership. At its November 1973 meeting, the AUA executive committee resolved to continue its participation in the Argonne contract and authorized McDaniel to discuss renewal with the AEC.

The University of Chicago's William Cannon immediately sensed the threat from McDaniel. Cannon believed that the university faced no problem as long as Dixy Lee Ray remained as chairman of the Atomic Energy Commission. But should something unforeseen happen, Cannon feared that the AUA under McDaniel would move strongly to exploit "the slightest opening." The University of Chicago affirmed that Sachs was its principal contribution to the contract, and under Ray's encouragement, the university granted Sachs "complete control" over the laboratory, including tripartite negotiations. Cannon, however, was still nervous. Argonne's role in the energy crisis was uncertain, but that uncertainty could turn out to be an advantage. More worrisome to Cannon was the belief that Sachs's management was vulnerable to AEC charges that he had dismantled administrative "improvements" made under Shaw's regime but had not strengthened performance in research or engineering at the laboratory.[57]

On January 11, 1974, Robert Bauer, manager of the AEC's Chicago Operations Office, provided the Argonne Universities Association that "slightest opening" when he asked both the University of Chicago and the Argonne Universities Association for suggestions on continuing or revising the Tripartite Agreement. Believing Bauer's request reflected the commission's deep-seated dissatisfaction over the University of Chicago's contribution to management of the laboratory, McDaniel decided not to equivocate. The Argonne Tripartite Agreement was unique in the AEC. Los Alamos, Lawrence Berkeley, Lawrence Livermore, Oak Ridge, Brookhaven, the National Accelerator Laboratory, and other AEC research-and-development centers were all managed by a single contractor. Only Argonne struggled forward under the yoke of two masters. McDaniel and the AUA executive board seized the opportunity to vent their frustrations about their partnership with the University of Chicago.[58]

Incredibly, while the United States shivered through the winter's energy crisis and Argonne scientists worried about the laboratory's future in the looming energy reorganization, the AUA executive board struck for independence. After private discussions with his trustees (and perhaps off-the-record meetings at the AEC's Chicago operations office), McDaniel denounced the seven-year-old tripartite contract as a defective vehicle for operating the laboratory. Alluding to the laboratory's well-known problems with Shaw, McDaniel described Argonne managers as defensive firefighters rather than as aggressive planners for program stability. He did not fault Albert Crewe or Robert Duffield for their performance under heavy pressures from the AEC, but he observed ruefully that the AUA frequently did not learn of major reprogramming at the laboratory until after the changes had become "bitter history." The association's putative responsibility for formulating, approving, and reviewing Argonne policies and programs was hollow. Rather, the fundamental flaw in the tripartite structure isolated "*policy* from *practice*."[59]

McDaniel believed that the Atomic Energy Commission should choose either the University of Chicago, the Argonne Universities Association, or another organization to be the sole contractor for Argonne National Laboratory. McDaniel convinced the executive board to poll the presidents, AUA trustees, and delegates from the thirty participating midwestern universities for support to negotiate singularly with the AEC. Affirming that Argonne laboratory was "an indispensable national resource" and that Sachs provided "effective and dedicated leadership," the executive board nevertheless advised AUA's constituency, including President Edward Levi of the University of Chicago, that the defective Tripartite Agreement should be terminated in favor of a contract which designed the association as sole manager of Argonne.[60]

In a round-robin to AUA presidents, Levi responded to the association's resolution with measured consternation. Levi, a nonvoting ex officio member of the AUA executive committee, took sharp exception with the view that the tripartite contract should be terminated. Categorically, Levi affirmed the Tripartite Agreement as a practical collaborative arrangement meeting the needs of the midwestern universities, the AEC, and Argonne laboratory. Although polite and restrained, Levi left no doubt the University of Chicago brooked no challenge to the Tripartite Agreement from the AUA.[61]

The AUA was not the only worry to the University of Chicago. Sachs resolutely continued his independent management of the laboratory, now wary of getting caught in the cross fire between the university and the association. Although Sachs never presented the University of Chicago with a serious challenge, William Cannon fussed that the laboratory director was operating out of his depth and far too independently of the university administration. Ironically, it was Cannon who had been most insistent that the university choose a strong leader to provide central direction of the laboratory. Now Cannon worried that Sachs was "running too free," gathering "all power and influence to himself," but not seeking or accepting advice except when he was in trouble. In Cannon's opinion, Sachs, overconfident in his ability to handle the AUA, triggered the 1974 revolt by convincing the AUA executive board that the tripartite contract was unworkable.[62]

Bauer was shocked at the uproar he had touched off. Belatedly, he tried to quell misunderstanding by writing McDaniel that the commission had no preconceived interest in terminating or amending the Tripartite Agreement. Bauer, under heat from AEC headquarters to scotch rumors at Argonne, carried his disclaimer to Sachs in person. Despite Bauer's efforts, however, on February 20 the AUA trustees endorsed the executive board's recommendation to terminate the Tripartite Agreement in favor of a single contractor for Argonne National Laboratory. Their action was the high-water mark of the AUA revolt. According to William Cannon's "score card" kept on the association president's responses to Levi, McDaniel's initiative was in serious trouble both with midwestern universities and the AEC.[63]

Also unknown to McDaniel, Dixy Lee Ray and John Erlewine had already given their unqualified support to the University of Chicago. Ray assured Sachs and Cannon that the commission would neither oust the University of Chicago nor give the Argonne Universities Association any important control over the laboratory. Bauer, now acting in close concert with Erlewine, who had prompted the question, asked McDaniel what alternatives were open to the AUA should the commission not select the association as the sole Argonne laboratory contractor. Caught flat-footed, McDaniel could only reply that the AUA would "help in every possible way" to promote a healthy relationship between Argonne and the midwestern regional universities. To further isolate McDaniel and to ensure that the AUA fully understood the commission's position, in June the acting general manager, Robert Thorne, chastised McDaniel for unnecessarily stirring up conflict between the University of Chicago and the AUA. Thorne unequivocally outlined the commission's position concerning renewal of the Tripartite Agreement: (1) because the University of Chicago was satisfactorily managing Argonne, and (2) since the AUA lacked the capacity to run the laboratory, (3) Thorne concluded that the Argonne Universities Association could withdraw from the contract if it was unhappy.[64] Thorne's ultimatum stopped the AUA revolt in its tracks.

Not only had the University of Chicago turned McDaniel's flank with the AEC, but Levi rallied enough AUA presidents to shatter the association's unanimity. At a special meeting convened on June 25, 1974, some of the AUA trustees were not even certain whether they spoke for their universities or only for themselves. Although the trustees did not accept McDaniel's resignation, a whisker-thin majority refused to endorse the executive board's resolution on the Tripartite Agreement, in part because they believed the board's recommendation was contrary to the founder's agreement of 1965. Rather, the trustees asked the board to offer "viable" alternatives for the management of the laboratory. Instead, McDaniel mounted an unsuccessful campaign to amend the founder's agreement to allow the AUA to pursue any alternative including withdrawal from the Tripartite Agreement. McDaniel's final attack failed when only eight of the thirty participating AUA institutions ratified the proposed amendment.[65]

With national energy reorganization imminent, in July 1974 an impatient Atomic Energy Commission offered the Argonne partners a take-it-or-leave-it two-year renewal of the Tripartite Agreement. Although McDaniel only wanted to extend the contract another year, with insufficient support from the presidents to amend the founder's agreement, the AUA accepted the AEC's offer. The University of Chicago, which asked for a three-to-five-year extension, accepted the commission's "compromise" under protest that an attenuated contact nonetheless encouraged the AUA's continued campaign to capture Argonne management. Prolonged debate over management of the laboratory not only undermined Argonne's stability but also inhibited "progressive, innovative operations" at a time when Argonne should have been devoting its attention to the national energy crisis. Although the University of Chicago easily won the power struggle with the AUA while keeping Sachs in check, the victory was costly both in terms of laboratory morale and AEC confidence in the laboratory. Although Sachs did not lay the blame entirely on the AUA controversy, he believed that he had slipped competitively with other laboratory directors during the contract renewal crisis.[66]

Argonne's managers were pleased with the decision to extend the Tripartite Agreement, but they were hardly overjoyed by the commission's accompanying evaluation of the laboratory's performance. The commission staff had acknowledged significant progress since 1971 in the laboratory's commitment to AEC program objectives, and "marked improvement" in the decisiveness of laboratory management. But overall, the commission evaluated Argonne performance with tepid praise: research-and-development programs were "satisfactory," Argonne scientists and engineers were "technically quite competent," and their work was "above average." The commission rated the performance of the nontechnical staff as "quite good." On the decidedly negative side, the commission described the University of Chicago's involvement in laboratory management as reactive rather than initiative. Although Argonne officials professed a firm commitment to affirmative action programs, the AEC cited unsatisfactory results in hiring minorities and in paying women equitably. Finally, conceding mitigating circumstances that were not entirely the responsibility of the Argonne staff, the commission noted that Argonne's engineering and construction costs ran unnecessarily high, in part because Argonne had recently lost a number of design engineers. Almost parenthetically, the commission admitted that Argonne had waged an uphill fight to maintain plant and facilities because few funds were available for replacing or upgrading aged equipment and buildings.[67]

The Energy Research and Development Administration

The *Argonne News* simultaneously announced the extension of the tripartite contract and the confirmation of Robert Seamans, Jr., as the administrator of the newly created Energy Research and Development Administration. Seamans, a former administrator at NASA, saw little parallel between the Sputnik scare of the 1950s and the energy crisis of the 1970s. Space exploration, including the race to put a man on the moon, remained primarily a government responsibility. Solving the energy crisis, on the other hand, not only was more complex than reaching the moon but also required close partnership between the government and the energy industry. Additionally, Seamans emphasized the international need "to devise new energy resources and power systems." In his confirmation hearing before two Senate committees, Seamans also affirmed the need to strengthen basic scientific research at the national laboratories. Still, Seamans's NASA background worried Sachs, who believed that while the AEC was "science driven," NASA was "politics" driven.[68]

Sachs worked through the summer of 1974 preparing to hitch the Argonne car to the new federal energy train. For more than thirty years Argonne had compiled a proud history answering the nation's call for scientific leadership and innovation. After World War II, Argonne joined the vanguard that developed civilian nuclear power in the early 1950s and then broadened its scientific and technological base when the Sputnik scare revitalized basic science research in the United States. When environmental concerns captured public attention in the late 1960s, Argonne adjusted its mission to embrace environmental research. The energy crisis again challenged the laboratory to accept new assignments to develop and exploit national energy resources with concern for health effects, environmental quality, and conservation. Complementary to the Ray report, Sachs prepared a five-year plan that accentuated Argonne's nonnuclear programs. In addition to Argonne's commitment to the breeder-reactor program, Sachs projected increased emphasis on coal conversion, fluidized bed combustion, magnetohydrodynamics, solar energy, electric vehicles, and biomass utilization as well as strengthened programs in biological and environmental research.[69]

Following the passage of the Energy Reorganization Act of 1974 and the activation of ERDA in January 1975, Argonne became one of ERDA's 55 laboratories, plants, and institutions, which included all of the AEC's laboratories as well as energy research centers acquired from the Department of the Interior. In concert with Edward Teller, Sachs reminded Seamans of the unique contributions of the national laboratories under the Atomic Energy Act. Argonne offered a particularly good example of how the interface between esoteric pure science and practical, hardware-oriented engineering produced important energy innovations even in nonnuclear areas. For example, Argonne's high-energy physics program had produced two potential contributions for advanced energy systems: a parabolic solar energy collector and large superconducting magnets useful for magnetohydrodynamics or fusion technologies.[70]

Seamans personally assured Sachs that ERDA depended on the national laboratories to provide leadership for the nation's energy research-and-development effort. Yet after his visit to Argonne, Seamans chose not to emphasize Argonne's basic research programs or the laboratory's large multimillion dollar development projects in his April address to the American Power Conference at the Palmer House in Chicago. Instead, Seamans featured Argonne's lithium-sulfur battery, which the blasé AEC had supported with uncertain, meager funding. Seamans's political point was obvious—ERDA's priorities would not simply ape the AEC. Seamans literally held up a lithium–metal sulfide battery cell as an example of government-sponsored energy technology ready for industrial-commercial development. "The battery program at Argonne has proceeded through bench testing and appears to have real promise," Seamans reported to the American Power Conference. "It is time to bring industry into the program," he concluded, in order to build full-size prototypes for testing in electric vehicles and energy-storage facilities.[71]

In July 1975, moving to secure command and control over ERDA's far-flung operations, Seamans asked Michael I. Yarymovych, the assistant administrator for laboratory and field coordination, to determine how best to coordinate ERDA's disparate field and laboratory operations with the agency's mission. The national laboratories transferred from the AEC and the energy research centers inherited from the Department of the Interior all brought long histories and proud traditions to the new energy administration. Whereas Seamans had little trouble commanding the immediate loyalty of the headquarters staff, further afield, change often came grudgingly, especially among the old AEC hands.

For years after the establishment of ERDA, along dusty back roads of remote installations, one could still find signs of the Atomic Energy Commission hung like defiant badges on scattered fence posts and forgotten buildings. Even Sachs became peeved by Seamans's order on signage, which directed that "Energy Research and Development Administration" be painted in larger, bolder letters than Argonne National Laboratory at the entrance to the laboratory. Who did this new crowd think they were? Sachs conceded that the government owned the land, but that the land was purchased for the benefit of the laboratory, not the Washington bureaucrats. "We tried to get across to them that they were our guests at Argonne National Laboratory," Sachs recalled. Not surprisingly, ERDA headquarters did not share Sachs's perspective.[72]

Yarymovych's field-and-laboratory utilization review was largely favorable to Argonne, whose combination of programs and funding was among the most balanced of the multiprogram laboratories. In addition to being ERDA's leading laboratory in fast-reactor physics and safety and having programs in fundamental biological, chemical, and physical research, Argonne was noted for its nonnuclear energy research, its strength in computers, energy research and development, environmental studies, and its capacity to design and fabricate total systems. Yet discussions among ERDA planners could not escape the long shadow of Milton Shaw. Laboratory managers, still smarting from Shaw's micromanagement of their programs, successfully recommended that ERDA delegate authority for program execution to the field while retaining policy and funding powers in Washington. To strengthen ERDA's field presence, Yarymovych also recommended creating ERDA centers based upon the AEC's operations offices, which would coordinate ERDA's regional activities with industries, universities, financial institutions, local and state governments, independent inventors, and the public.[73]

Seamans had exempted Rickover's naval reactors program and the nuclear weapons laboratories and production facilities from the field utilization study. For the rest, Seamans wanted a plan of action that not only improved the laboratories' performance and quality but also emphasized ERDA's essential mission to promote commercialization of energy technologies. As he demonstrated when he promoted the lithium-sulfide battery at the American Power Conference, Seamans sought near-term energy applications that could serve as demonstration projects for technology transfer from government to industry. Seamans had been warned that the former AEC laboratories were "far too theoretical" to work successfully with industry in nonnuclear fields such as coal combustion and coal liquefaction. To Robert Gunness, former president of Standard Oil of Indiana, Seamans noted that the "urgency of ERDA's mission" made it imperative to develop effective partnerships with industry to commercialize the laboratories' research-and-development efforts. Seamans hoped the field utilization study would set an effective course for "planning, implementing, executing, and finally commercializing" ERDA's research, development, and demonstration projects.[74]

Sachs assured ERDA managers that Argonne fully understood the need to foster industrial partnerships to assure quick and effective response to the national energy goals. In defining Argonne's role in implementing national energy policy, however, Sachs linked the breadth and scope of Argonne's basic, exploratory research with the emergence of new concepts of energy production and conservation. Fortuitously, in September, Argonne won three awards from among the top 100 technological developments of 1976. The awards, presented at Chicago's Museum of Science and Industry by the publisher of *Industrial Research,* included recognition for the development of liposome encapsulation of anticancer drugs; the fabrication of the in-sodium tritium meter designed to measure tritium concentrations in breeder-reactor sodium coolant; and the invention of the dielectric compound parabolic concentrator (CPC), a solar energy device, shaped to concentrate sunlight onto photovoltaic strips. Sachs reminded Yarymovych that the 1976 awards were the ninth, tenth, and eleventh awards Argonne had received from *Industrial Research* magazine.[75]

The Solar Imperative

ERDA's logo, a blazing yellow sun set among a cluster of fifty stars on a sky blue background, symbolized the new agency's commitment to develop all energy sources, ranging from controlled thermonuclear reactions to the earth's winds and tides. For Sachs, Argonne's principal solar energy project, the award-winning compound parabolic concentrator, also symbolized the ideal synergy between basic science research and applied technology. To ordinary citizens, no area of science seemed more esoteric than high-energy physics. On the other hand, in the popular imagination, no modern technology seemed more promising than solar energy. Sachs took special delight in touting Argonne's compound parabolic concentrator because the new solar collector was based on principles borrowed from high-energy physics experiments conducted by Roland Winston on the ZGS in 1966 when Sachs was associate laboratory director.

Sachs himself had played a major role in seeing the connection between Winston's experiments and solar applications. When he became director of Argonne, Sachs asked for reports on all aspects of the laboratory's research programs. As Sachs read about the problems encountered when conventional mechanical collectors are rotated toward the sun to achieve optimal spot focusing of sunlight, the potential of Winston's research for building efficient solar collectors hit Sachs "like a lightening bolt." Although the solution was obvious, Sachs recalled, sometimes two and two are not put together "until they're right under your nose."[76]

In 1966 Winston built a sensitive counter to collect (and measure) Cerenkov radiation in high-energy physics experiments. Because the blue glow of Cerenkov radiation was so faint, Winston needed to collect all available light. He discovered that the literature on light collectors only discussed forming bright images with either lenses or mirrors. But Winston did not care about getting images as long as he gathered as much light as possible. Consequently, while exploring the field of nonimaging optics, Winston built the CPC, which essentially collected light from all directions. The compound parabolic concentrator, he later learned from Riccardo Levi Setti of the Enrico Fermi Institute, used the same principle for light collection as the compound eye of the horseshoe crab. Until alerted by Sachs, Winston had not connected his research in high-energy physics to solar energy collection. The application, however, was obvious and provided Argonne an important opportunity to sponsor politically attractive solar-energy research and development that promised early commercial use in heating or cooling of homes and businesses.[77]

To work efficiently, conventional parabolic mirror collectors must track the sun across the sky, using expensive and complicated electronic controls to focus sunlight on the collectors. Winston's device collected sunlight broadly along parallel parabolic troughs, whose reflecting sides funneled light onto absorbers at the bottom of the troughs. Although the concentrator with its unique parabolic shape required seasonal adjustment as the sun moved north and south through the year, it did not need to track the sun through the day to achieve higher temperatures than conventional collectors. Initially, Argonne anticipated two applications for the low-cost, lightweight concentrators. First, prototype collectors, when combined with photovoltaic silicon strips, might significantly reduce the cost of obtaining direct electrical energy from photovoltaic cells. Second, the concentrators were also capable of attaining sufficiently high temperatures to make solar-powered heating and cooling practical for homes or small businesses.[78]

Argonne's solar initiative touched four elements of ERDA's mission: basic research in solar energy, applied research in technologies useful for solar applications, engineering development of solar systems nearing commercialization, and demonstration projects of solar devices ready for industrial production. The development of the compound parabolic concentrator as an economical device would require precisely the kind of cooperation with industry that ERDA was anxious to promote. Sachs believed that Argonne was well equipped to evaluate the social and institutional impediments to the deployment of solar energy technology.

In addition to identifying major technical issues, Argonne's systems analysts were trained to assess environmental impacts, economic problems, industry and public acceptance, and government regulation and zoning. The solar program included every element of short-, intermediate-, and long-range planning envisioned in ERDA's efforts to promote the speedy development of energy technologies.[79]

Argonne: An ERDA Laboratory

Although the breeder-reactor program remained Argonne's first priority, Sachs assumed that its importance would decline relative to growth in funding for advanced energy systems, fossil fuel technology, controlled thermonuclear research, and biomedical and environmental programs. By September 1976 Argonne's employment had rebounded from the dark days of 1973 to 4,950 personnel, almost 1,000 additional employees (of whom 840 were scientists and engineers). Even more impressively, Argonne's operating budget accelerated from $93 million in FY 1973 to $168.3 million in fiscal year 1977, whereas support for the breeder program slipped to less than 40 percent of the overall budget. Still, nuclear energy received an impressive $65.5 million, up more than $22 million since Shaw's day.[80]

By 1977 Argonne's budget clearly reflected ERDA's priorities. Research and development in fossil energy, conservation, and solar energy ballooned to 18 percent of Argonne's funding, up from just a few percent in 1973. On the other hand, funding for basic sciences fell to 28 percent of the budget, down from 50 percent in 1967 when Albert Crewe was laboratory director. Although Seamans (and ERDA) repeatedly affirmed the importance of basic research, Sachs watched helplessly as funding for basic science and engineering rose to $47 million, only $10 million more than the best years under Crewe a decade before. Commenting on demands that basic research proposals be subject to competitive bidding on the same basis as commercial contracts, Sachs pointedly warned the Washington bureaucracy that "a major weakness of our industrial society is exactly the short-term emphasis on profit and loss which exerts an overweening influence." Managers not only in industry but also in government stood or fell on "short term performance criteria." Steadfastly true to his faith in basic science and engineering, Sachs argued that only the national laboratories could transcend political impatience for short-term, profit-oriented energy research and development. Establishing the "continuity and stability of selected strong, broadly based scientific and engineering programs with well defined general goals" within the ERDA mission was Argonne's paramount goal for the next five years. In nonnuclear areas—fossil fuels, solar energy, and conservation—the trend was in the right direction. In biology and environmental studies, Argonne expected strong growth under ERDA management. Only in the disciplines of the basic physical sciences was Sachs discouraged.[81]

The Center for Human Radiobiology

According to laboratory tradition, Argonne's Center for Human Radiobiology was one of the few major research programs established at the express wishes of the Joint Committee on Atomic Energy. The laboratory's patron in this regard was Illinois Congressman Melvin Price, who, as vice chairman of the committee, promoted the establishment and expansion of the center at Argonne.[82] The government's interest in nuclear biology and medicine, of course, was long standing. Through the National Academy of Sciences, the AEC funded the work of the Atomic Bomb Casualty Commission, which continuously studied the somatic and genetic effects on human survivors of atomic bombing of Hiroshima and Nagasaki. The commission's division of biology and medicine supported additional research at most of the national laboratories as well as at several medical research centers and universities.[83]

In fact, Robley D. Evans, MIT's distinguished health physicist, had first proposed establishing a national center for the study of human radiobiology at a conference in Sun Valley, Idaho, in 1967. Evans, known to colleagues as "Mr. Radium," firmly believed that only human subjects could provide reliable data on the toxicity of radioactive elements in humans. To provide an "immortal" organization to coordinate research on the effects of radium, mesothorium, plutonium, americium, and inhaled radon decay products in humans, Evans advocated consolidating the commission's scattered projects on radiation effects into a centralized radiobiology laboratory.

In Evans's opinion, research on rats, dogs, and other animals could not be extrapolated reliably to humans. Although it was not morally acceptable to create human experimental populations, there existed in the United States three groups that had received medically significant internal doses of radium and mesothorium between 1918 and 1933. These included luminous-dial painters, whose brush "tipping" caused them to swallow minute amounts of radium; radium chemists, whose use of mouth pipettes also resulted in swallowing radium; and medical patients, whose treatment involved ingestion or injection of radium isotopes. In 1932 Evans began studying these individuals, continuing his research at MIT from 1934 until his retirement in 1970. In 1951 Argonne National Laboratory and the Argonne Cancer Research Hospital initiated an independent, but cooperative research program, which focused especially on radium-contaminated dial painters in Illinois.[84] In addition to the radium toxicity research at MIT and Argonne, the AEC sponsored radium toxicity studies at the New Jersey College of Medicine, New York University, and Georgetown University. Other AEC-funded research in radiobiology included research at Oak Ridge and Brookhaven of whole-body accidental and therapeutic radiation exposures; epidemiological and inhalation studies at the New York Health and Safety Laboratory, Colorado State University, and Los Alamos Laboratory on effects of radon gas on uranium miners; and various human studies at Oak Ridge and Brookhaven National Laboratories, the University of Washington School of Medicine, and the New England Deaconess Hospital Cancer Research Institute. Among the more controversial projects would be Thomas F. Mancuso's study at the University of Pittsburgh School of Public Health of the lifetime health and mortality of AEC and AEC contractor employees subjected to occupational radiation exposure.[85]

With Robley Evans's retirement imminent, the AEC's Advisory Committee for Biology and Medicine agreed that the commission's scattered and uncoordinated radiobiology programs should be consolidated under centralized management. There was little support, however, for the establishment of an independent national research institute such as Evans envisioned. AEC officials recognized the need for continuous, decades-long research on the effects of radiation on human beings, but the advisory committee believed that a new institute created for this purpose alone would slip into scientific isolation and thus would have trouble recruiting top researchers. Nor did the advisory committee believe it was possible to locate all of the research at one location. Research on uranium miners, for example, was best continued in the field at Grand Junction, Colorado, Salt Lake City, and other western sites, while patient facilities in Boston, Brookhaven, and other northeastern sites should be maintained.[86] The commission's advisors supported establishing a "center" at Argonne, where Evans's case records and specimens would both strengthen and broaden Argonne's Illinois studies. Not the least of Argonne's assets in obtaining the center, of course, was the support of Melvin Price. As long-time chairman of the Joint Committee's subcommittee on research, development and radiation, Congressman Price had developed a deep and abiding interest in the human radiobiology program. At Price's urging, and with the endorsement of its advisory committee, in 1969 the commission agreed to transfer Evans's work to Argonne, where it could be carried forward by the young, vigorous, and imaginative Robert E. Rowland, the director of the radiological physics division.[87]

Argonne's advantages promised not only continuity for the program carried forward by Rowland's strong group, but also additional resources available only at the national laboratory. Not only were start-up costs minimized by locating at Argonne, but the new center also, like Fermilab before it, greatly benefitted from access to Argonne's sophisticated computer facilities for data storage, retrieval, and analysis.[88]

With Evans's MIT project safely transferred to competent hands at Argonne, Price continued to press AEC Chairman James Schlesinger to expand the commission's research effort while data could still be obtained from the dial painters, whose average age was sixty-five. "We are in a race against time," he scolded Schlesinger on learning that the Office of Management and Budget had sequestered $770,000 in operating funds earmarked for the new Center for Human Radiobiology. Because radium and plutonium were both bone-seekers, Price understood that the radium studies could significantly contribute to the establishment of industrial radiation safety standards for plutonium in breeder-reactor technology. Price also knew that the health and economic stakes were high for those suffering from inhalation of heavy elements and also for the nuclear reactor industry in general. Although the AEC repeatedly affirmed the importance of studying acute and chronic human radiation syndromes, Price was dismayed to learn that there was not only a one-year backlog of bone samples waiting analysis, but also apparent foot-dragging in Washington on expanding the research program to include more subjects while they were still living or to develop a comprehensive data bank to collate and analyze existing and future data. Price also knew that Argonne's management had not pushed for expansion of the radiobiology program; so he arranged for Rowland to ask the Joint Committee for new facilities. Price was willing to charge the Office of Management and Budget with both bureaucratic myopia and irresponsibility, but he did not believe that the budget office would be insensitive to the urgency of the project if it had they been instructed "by an interested and qualified Commission spokesman."[89]

When Price laid his challenge directly at Schlesinger's door, the commission was uncertain how comprehensive the new "center" should become. As noted by the AUA's review committee, the center could become a simple registry of case reports, or it could develop into a research facility exploring "the biological effects of radiation, the metabolism of nuclides, and their toxicology in man." Further, as originally recommended by Evans, the center could expand beyond radium cases to include research on internal contamination by plutonium, americium, and other radioactive isotopes. Finally, Argonne could establish "a truly *national* Center" by providing advisory service for the treatment and decontamination of persons throughout the country afflicted with radiation illness.[90]

Price's persistence bore fruit in 1974 when Sachs signed a contract for the construction of a $1.3 million facility to house the Center for Human Radiobiology at Argonne. In addition to continuing to monitor former dial-painters, radiation patients, and others cooperating in Argonne's research on bone-seeking radioisotopes, the new laboratory was equipped to provide physical examination of patients, complete with whole-body counting, breath-radon testing, and skeletal radiography. Initially Rowland and his staff monitored about 700 radium cases, but to their surprise, the number grew in time to well over two thousand. As Rowland recalled, when they brought patients in for examination, the examining doctors convinced them that the studies were conducted not only for their health, but also for the health of the nation.[91] In contrast to a research hospital, Argonne was only secondarily concerned about the care and treatment of radiation victims suffering from malignant tumors. Argonne scientists were not indifferent to their subjects' medical problems, but providing therapy for radiation victims was not their responsibility. As Rowland noted, this limitation did not prevent scientists from encouraging patients to seek appropriate therapy or from making recommendations to their families and physicians. Argonne was proud of its new center, which provided the laboratory with high visibility in the sensitive area of nuclear health and safety. The laboratory had picked up a political "hot potato," however, which would soon embroil them in nasty plutonium politics.[92]

Biological and Environmental Sciences

For years, Argonne's research program in biology and medicine had suffered from a lack of clear mission and leadership. Critics charged that the Argonne's biology division was the laboratory's weakest organization, plodding along with unexciting research in the shadow of the more glamorous (and more generously funded) nuclear reactor and high-energy physics projects. Little distinguished Argonne's biological research from the research programs of competing universities. In the physical sciences, Argonne justified its special status because no research university could afford to build, operate, and maintain the large, expensive machines required for modern nuclear research. With few exceptions, however, Argonne's biological projects failed to exceed university research in scale, and Argonne also lagged behind the leading universities in establishing dynamic basic-research groups.

In a sense, there was little tension between applied and basic research in Argonne's biology and medicine division. After World War II, when the AEC named Argonne as the reactor development center for the country, many of the laboratory's more ambitious biologists returned to academia, where the federal government was steadily increasing funding for basic research. While Argonne's research program in biology and medicine grew in support of the laboratory's reactor development program, the division's mission-oriented research significantly weakened its ties to the university research centers. Although numerous university professors spent summers or sabbatical leaves at Argonne, from time to time successfully collaborating with Argonne scientists, by the 1970s, Argonne's programmatic research in radiation biology lay well outside the mainstream of modern biological research. Consequently, Argonne had failed to establish a reputation as an outstanding biological research laboratory among academic scientists. In the opinion of Henry Koffler, the head of Purdue's Department of Biological Science, Argonne had interpreted its AEC mission too literally by narrowly focusing on radiation damage. Not only had Argonne failed to attract the best people, but Koffler did not think its facilities were even needed. Among the national laboratories in biology, Koffler rated Oak Ridge as the "best," Brookhaven as "good," and Argonne "not good" at all. Significantly, the Center for Human Radiobiology was located in Argonne's newly organized radiological and environmental research division, not in the division for biology and medicine.[93]

Why, then, would the Energy Research and Development Administration want to continue a biology program at Argonne? The answer was that the administration's mission required programs in biomedical and environmental sciences whose scope and resources exceeded those of the universities. ERDA continued biological research on the effects of ionizing radiation on biological systems and environmental studies on the effects of radioactivity in the biosphere, but it also moved quickly to establish a "balanced" research program to focus on biological and environmental problems arising from all energy sources. Fossil-energy research received almost two-thirds of ERDA's environmental budget, whereas solar, geothermal, fusion, and conservation vied with nuclear fission for the balance of the budget. Although Argonne maintained important research on bone cancer because most of the radionuclides under study at the laboratory concentrated in bone, Washington was anxious for the laboratory to expand its nonnuclear, environmental activities. Argonne's Great Lakes research program provided the basis for expanded regional studies of the environmental effects not only of radioactive elements, but also of DDT, polychlorinated biphenyls (PCBs), asbestos, lead, mercury, and other pervasive pollutants. In 1975 Argonne assembled a multidisciplinary team of twenty-five geologists, hydrologists, biologists, ecologists, civil engineers, agronomists, and economists to develop a comprehensive national program for reclamation of strip-mined land.[94]

The Plutonium Patients

In 1974 the Center for Human Radiobiology became involved in an embarrassing public relations problem involving eighteen patients injected with plutonium between 1945 and 1947. In 1972 the center's advisory committee had strongly urged the center to expand its studies beyond the radium-dial painters to other groups that had ingested heavy elements. Subsequently in December, at the suggestion of AEC headquarters, a visiting researcher from Lawrence Berkeley Laboratory offered to share her records, which had been originally compiled on plutonium ingestion by the Manhattan Engineer District from 18 "terminally" ill hospital patients in New York, Tennessee, Illinois, and California. To determine how plutonium was deposited in and excreted from the human body, Manhattan Project researchers injected the subjects with small amounts of soluble plutonium. No adverse health effects due to the injections were observed at the time.[95] Almost half of the patients died within a year from illnesses diagnosed at the time of their injection, but at least six patients lived another decade, and four were still living when their records arrived at the Center for Human Radiobiology in December 1972.[96]

Scientists at Argonne, as well as administrators at the University of Chicago, knew immediately that they had a potentially "hot" item on their hands, particularly because three of the patients had been injected at the University of Chicago's Billings Hospital. No one knew for certain how the original studies had been conducted and in five cases, one of those among the living, the center had only estimates on the amount of plutonium injected. Most worrisome from an ethical and public relations perspective, not to mention the potential stress on the living patients and their families, was whether the subjects or their relatives had given informed consent to the experiments. The center could confirm that only the last patient, who was the sole patient injected by the AEC in 1947, gave informed consent to the experiment. Because of the high level of secrecy of the plutonium experiments and the lax guidelines on informed consent involving human experimentation in 1946, the center presumed that none of the other patients or their kin had either been informed about the nature of the experiments or had given their consent to the procedure.

When Rowland traveled to Washington to brief James Liverman, the director of the AEC Division of Biological and Environmental Research (DBER), on the plutonium studies and to ask for additional funding to support the project, he was told there was no AEC money for the studies. As Rowland later recalled, AEC headquarters did not want to become involved with the plutonium patients because of the potential embarrassment to the commission and to the hospitals where the patients were injected. Rowland protested that it did not make any sense to explore plutonium toxicity through radium studies, but to ignore evidence from actual plutonium patients who had carried plutonium in their bodies for more than twenty-five years. Although he never received authorization in writing, Rowland returned to Chicago with what he believed was permission to continue evaluation of the plutonium patients on the condition that he not tell the subjects that they were contaminated with plutonium nor reveal the hospital's original role in the project.

Naively—"very naively," he sadly confessed later—Rowland had not obtained written confirmation of his instructions from headquarters. He directed the staff not to mention the plutonium cases outside the center. In the meantime, the whereabouts of the four living patients were determined—three in Rochester, New York, and the other in Texas. The physician of one of the living patients (and the relatives of one of the deceased) refused to cooperate with further studies.[97]

The center encountered better luck when the physicians of three of the living patients anticipated no difficulty in continuing the research. Arrangements were made with the University of Rochester and Strong Memorial Hospital to examine two of the patients in January and June 1973 with the understanding of their physician "that these patients [would not be] upset by mention of their injection with plutonium."[98] The patients were only told that perhaps they had an unknown radioisotope in their bodies. The Texan, who apparently had been informed of the purpose of the plutonium experiments when injected in 1947, also agreed to follow-up tests in Rochester. Finally, by September 1973 the center arranged to exhume the body of a twenty-year-old woman who had died in 1947 of Cushing's Syndrome (for which she had been diagnosed in 1945) more than two years after she had received her plutonium injection.[99]

While evaluation of three living patients and the single exhumation proceeded in 1973, the Center for Human Radiobiology asked MIT's radioactivity center to arrange for the exhumation of ten additional plutonium patients. Although plutonium would not be mentioned, families of the deceased were to be told that the plutonium patients had received injections of "mixtures of radioisotopes" in experimental treatment and that the purpose of the exhumation was to measure subsequent distribution of radioisotopes. The bones of the woman examined in September definitely contained plutonium-239. As well as continuing efforts to exhume additional cases, Rowland hoped to convince the living plutonium patients to donate their bodies to the center. For the first time, health physicists could measure how fast plutonium deposited on the surface of human bone was buried by subsequent layers of new bone. The data was invaluable to extrapolate plutonium's known toxicity in dogs to humans.[100]

Unfortunately for Rowland and the center's project, when news of the continued plutonium patients study filtered back to the AEC in March 1974, the commission reacted with shock and consternation. Although Dixy Lee Ray and William Doub did not want to beat through ancient history in search of scapegoats, the commissioners realized they had a explosive political problem on their hands. On April 4, James Liverman, the director of the DBER, put Argonne on notice that all further studies of plutonium patients had to be coordinated with the AEC. Liverman also directed Argonne to send copies of all pertinent documents back to Washington, D.C.[101]

The "medical ethics problem," as the commission defined it, required not only follow up by Liverman's division, but also investigation from the inspection division. The issue could no longer be treated as simply a scientific research problem of the Center for Human Radiobiology. The AEC informed both the army's surgeon general and general counsel because investigations would inevitably raise questions about the Corps of Engineers' conduct during the Manhattan Project. Research soon revealed that in December 1946, Col. Kenneth Nichols, the district engineer, had denied authority to inject humans with radioactive substances after all but two of the plutonium patients had been injected. In addition, in March 1947, before the last experiment, the AEC legal division ruled that human clinical testing required informed consent from the patient given in the presence of two witnesses. Until they could sort out procedures and policies, the commissioners suspended exhumations while medical experts evaluated the scientific importance of the plutonium patients project.[102]

The ad hoc scientific committee appointed to advise Argonne and the AEC on the merit of continued studies of the plutonium patients was also keenly aware of the related medical, ethical, and public relations issues. Nonetheless, on the basis of scientific merit, the advisory committee unanimously recommended exhuming "in an orderly manner" as many of the deceased plutonium patients as possible. Although their number was small, and by no means a "normal" population sample, the health scientists believed that analysis of the exhumed remains of the plutonium patients would yield invaluable data on plutonium metabolism and its distribution in human tissues other than bone.[103] As Patricia A. Lindop wrote in *Environmental Pollution,* these data were essential if scientists were not "to over- or underestimate the hazards from environmental plutonium levels" to which the public would become increasingly exposed with the development of the breeder reactor.

Consistent with the original purpose of the plutonium injection experiments, the government's principal interest in continuing the studies was to establish standards for industrial safety. Gathering information for the care and treatment of persons suffering from bone cancer was at best a secondary consideration at this time. Ultimately, the commission decided that all living plutonium patients should be informed of their condition; that attending physicians who objected should refer their cases to their local human-use committee; that the next of kin of exhumed patients should be told of the plutonium injections; but that no purpose was served by notifying relatives of deceased not included in the studies. Finally, the commission ruled that when the surviving plutonium patients died, every effort should be made to discourage cremation so that there would be no risk that their plutonium laden ashes could be scattered to the winds.[104]

Argonne's Plutonium MUF

Ask the director of a nuclear laboratory about his worst nightmare and he will probably tell you that, beside a catastrophic reactor accident or a terrorist attack on his facility, his greatest fear is a plutonium incident resulting in the death of a victim or the loss of special nuclear material. In the arcane nuclear lexicon, missing plutonium is classified as a nuclear MUF—"material unaccounted for."

On May 11, 1975, Argonne public affairs calmly announced that 0.5 grams (or 0.018 ounces) of high-purity plutonium-239 was missing from one of the analytical chemistry labs in the chemical engineering building. The plutonium, no larger than the head of a pin, had been issued to Argonne by the National Bureau of Standards to be used in calibrating highly sensitive instruments. The laboratory did not believe that the plutonium was particularly hazardous, but nevertheless warned finders unequivocally: "Do Not Open the Package."[105]

Although the missing speck of plutonium had no military significance, the local Chicago press had a field day with the plutonium MUF. The Chicago *Tribune* emphasized that the lost plutonium was "one of the deadliest materials on earth." Laney tried to explain to the press that the plutonium was a less dangerous radiation hazard than radium "and wouldn't harm you if it was sitting on your desk." But Laney's reassurances palled with news that the FBI had been brought into the case because the government had not ruled out the possibility of theft.[106]

The Chicago *Sun-Times* described the missing plutonium "as more precious than gold," but "far deadlier than any poison," and "the most carcinogenic ... substance known to man." Noting that plutonium was one of the chief ingredients used to make nuclear weapons, the *Sun-Times* marveled at the lax security which coyly nicknamed lost plutonium as a MUF. The MUFed plutonium weighed only 0.5 grams, nowhere near the fifteen pounds required to make a bomb, but the paper worried that the proliferation of nuclear technology might result in the proliferation of nuclear weapons through similar indifference to security.[107]

Sachs felt compelled to refute the press's innuendo that Argonne had been careless about public safely. First he lectured the paper that plutonium was not the most deadly human poison. Depending upon its chemical form and pathway into the body, strychnine, cyanide, or botulism toxin were "far deadlier poisons by weight than plutonium metal when swallowed." And as Laney had noted, on a weight-by-weight basis, radium salt ingested into the body was "more likely to cause bone cancer than ingested plutonium salt." Sachs's impeccable science lesson was hardly reassuring to the public, however. Regardless of plutonium's relative toxicity, Sachs conceded that plutonium hazards "should not be minimized." If inhaled as a fine power, the MUFed plutonium could produce deadly lung or bone cancer within ten to thirty years.[108] Sachs tried to put the plutonium loss into proper perspective: it was serious, but nothing about which to become unduly alarmed. Teams of scientists and security personnel, equipped with the most sensitive radiation detectors, searched fruitlessly for the lost plutonium. Although the missing plutonium was never found, laboratory management publicly believed that it was accidently thrown into the laboratory's dump.[109]

The press and Argonne management could agree on one fact: there was need for significant improvement in Argonne's safeguards and physical security procedures for handling special nuclear material. Sachs appointed a special investigating committee to recommend appropriate remedial actions. Although the investigators did not preclude the possibility that the missing plutonium had been accidently thrown away, they believed that the most plausible explanation for the loss was that someone had simply walked off with the plutonium standard. The incident was small, and the threat to public health slight, but the implication of the panel's finding was huge for it suggested that Argonne's security problem transcended simple negligence.

The local press, even assuming that the plutonium had carelessly ended up at the dump, drew the largest implications from the affair. With the scientific community on the threshold of providing the nation a breeder-reactor technology designed to produce large quantities of plutonium as a substitute for uranium fuel, the *Sun-Times* editorialized that it was urgent for nuclear scientists to develop a work ethic which gave special nuclear materials "the wary respect and security they require." The plutonium MUF, in the opinion of the paper, was symptomatic of "the failure of the federal government to devise and enforce effective security measures" for safeguarding highly dangerous radioactive materials.[110] On their part, scientists were generally dismayed at what they considered the overreaction and exaggeration of the local papers. The loss of the plutonium was a matter of concern, but the threat to public health was minimal. Research with nuclear materials involved some small risk, but then chemical research in general could be hazardous. More stringent safeguards, while not seriously impairing their work, would smack of unnecessary political management of the laboratory.[111]

Sachs was sympathetic to the perspective of the work-a-day laboratory scientist, but he also understood the imperatives of security and public relations. The plutonium MUF had revealed not only carelessness at the laboratory bench, but also an unacceptable casualness toward safeguards and security up and down the laboratory chain of command. Unfortunately, security questions move into the spotlight only when there is trouble. Not surprisingly, Sachs's investigation determined that veteran scientists and technicians had not adjusted to upgraded standards for storing and handling plutonium. This fact did not excuse the loss, but had the staff been properly trained and supervised, the loss might have been averted. It was a typical bureaucratic dilemma. The government's regulations were clear and sound, but responsibility for their implementation at Argonne was ill-defined and decentralized. Safeguards, safety, and security programs were not integrated, and laboratory audits were unsatisfactory. The laboratory had no coordinated and comprehensive training program for persons handling special nuclear materials. Both Sachs and the Chicago Operations Office accepted ultimate responsibility for the plutonium loss and pledged to take appropriate steps to improve plutonium management.[112]

The Zero Gradient Synchrotron

Nuclear images have dominated much of American thinking, and feelings, about science and technology since the end of World War II. Radioactivity, x rays, the bomb, nuclear reactors, and the mushroom cloud variously conveyed images of hope and fear to scientists and public alike. Plutonium has powerfully integrated public fears of nuclear annihilation from both bombs and reactors.[113] On the other hand, particle accelerators, the principal tool of high-energy or elementary particle physics, have conveyed images of pure science, perhaps esoteric and arcane, but nonetheless the epitome of basic science unadulterated by the priorities of commerce and industry. High-energy physics symbolized the ultimate in scientific research because it sought answers to the most fundamental questions about the structure of matter and energy in the universe and the laws that govern them. In 1963, activation of Argonne's ZGS marked the beginning of large-scale, high-energy research in the Midwest (figure 31), and the formation of a competent theoretical group that enabled the laboratory to claim a legitimate, if modest, place in the prestigious high-energy physics community.[114]

The simple and inexpensive experiments that laid the foundations for modern high-energy physics gave way to experiments requiring ever larger and more expensive machines as scientists probed deeper and deeper into the structure of matter. Until the 1930s, physics explored in the energy range of a few electron volts (eV). (To simplify for the public, Argonne scientists explained that a common flashlight accelerates electrons in the range of 1.5 eV, while a

Figure 31. Zero Gradient Synchrotron complex

thousand electron volts (keV) is the ordinary range of a dentist's x-ray machine.) From the 1930s to the 1950s, accelerators worked in the million electron volt (MeV) range, in which elements such as radium and thorium emit energetic alpha, beta, and gamma rays of several million electron volts. Ultimately, particle accelerators routinely achieved energies of billions of electron volts (GeV). Argonne's ZGS accelerated proton particles to 12.5 GeV.

Richard Feynman supposedly described high-energy physics in terms that any layman could understand: it was like banging two Swiss watches together to find out what comes out; expensive, but gratifying if you want to know how a fine watch works. In a sense, nuclear physicists used high-energy particle beams to hammer out the works of the Great Watch Maker, whose secrets promised to reveal not only the structure of matter, but also the origins of the universe. Yet when scientists bombarded matter with their particle beams, they produced a bewildering array of more than a hundred short-lived particles. It was like the nuclear watch works of protons and neutrons had been smashed to smithereens. But which, if any, of the newly discovered particles were elementary?

In 1964, a year after the ZGS became operational, Murray Gell-Mann and George Zweig theorized that splintered protons and neutrons appeared to consist of quarks, a term borrowed from James Joyce's *Finnegans Wake*. The hundreds of newly discovered particles, according to the new theory, were all species of quarks. When ordinary citizens were still confounded by the atomic and nuclear worlds, high-energy physicists now talked about confusing families of strongly interactive quarks and weakly interactive leptons (including the electron), which occupied a subnuclear realm filled with ephemeral particles. Kameshwar Wali delighted audiences by explaining that quarks and leptons, which might be the ultimate constituents of matter, came in three *colors* and at least five (but perhaps six) *flavors*. Of course when Wali talked about the color and flavor of sub-subnuclear particles, he did not mean color and flavor in the ordinary sense, but rather as a shorthand to the mathematical vocabulary his auditors might not understand.[115]

Although no one could predict what practical applications might result from high-energy physics research, there was no lack of confidence among scientists that basic research provided stimulus to discovery and invention. High-energy physics was not only regarded as the most advanced frontier of physical science, but it also commanded a major portion of the government's total spending on long-range research. In 1968 the Cosmotron at Brookhaven was shut down. During the budget squeeze of 1973–74, the AEC closed two more facilities, the Princeton-Penn Accelerator and the Cambridge Electron Accelerator, leaving five major facilities supported by the commission: the National

Accelerator Laboratory (Fermilab) proton synchrotron, SLAC, Brookhaven National Laboratory's AGS, the Berkeley Bevatron, and Argonne's ZGS. In 1974, the General Advisory Committee reported that the consensus of the scientific community was that Argonne's neighbor, the Fermilab, should be fully supported even at the expense of other high-energy physics programs. Next in priority was SLAC, whose 22 GeV electron accelerator exceeded the capabilities of competing machines in France, Germany, and the Soviet Union. Brookhaven's AGS was strongly supported because it was competitive with programs at CERN and because it was the only regional facility for eastern universities.[116]

The GAC gave its lowest priorities to the ZGS and the Bevatron in Berkeley. Still, the GAC strongly affirmed its support for Argonne's program, helping to preserve the "high caliber of its staff and to stimulate the morale and initiative of the entire laboratory and of the participating universities." Unfortunately the GAC's endorsement of Argonne's staff was problematic. Nowhere did the committee praise the laboratory for the distinctive science of the ZGS facility.[117]

The Atomic Energy Commission had originally supported a "crash program" in 1955 to build a 10 to 15 GeV proton accelerator at Argonne in order to provide the United States "program superiority" over the Soviet Union until the 25 GeV AGS at Brookhaven became operational. Pursuit of basic science was important, but even more important during the height of the Cold War was the commission's determination to maintain superiority over the Russians in all scientific fields. As a Cold War weapon, however, the ZGS was a failure. Instead of coming on line in three years to challenge the rival 10 GeV Russian machine, the ZGS did not become operational until after its Russian competitor had already proved scientifically disappointing. In addition, Brookhaven's 25 GeV accelerator actually beat the ZGS on line by four years, thereby preempting most of the original contributions the ZGS might have made. For more than a decade, on the other hand, the ZGS served as the Midwest's principal high-energy facility, becoming a model for university user groups conducting physics experiments at national laboratories.[118]

The energy crisis staggered Argonne with a heavy blow to its basic research mission in 1974. Static federal funding and sharply rising electric power rates had curtailed research programs at all of the high-energy facilities. At SLAC, Brookhaven, and Argonne, utilization fell from an average of 60 percent in 1973–74 to an average of 45 percent in 1974–75. Yet, while costs for electric power, materials, salaries, and equipment continued to escalate, dramatic discoveries in particle physics encouraged every facility to pursue higher and higher energies with new or upgraded equipment. In the search for greater utilization and efficiency, one possibility was to shut down underutilized accelerators.

Consequently, in addition to rejecting Argonne's proposal to build a heavy-ion accelerator, the Office of Management and Budget asked for a plan to shut down the ZGS "at the earliest possible time." By the spring of 1974 the commission and the National Science Foundation established the Zero Gradient Synchrotron Review Panel to determine the fate of high energy physics at Argonne National Laboratory.[119]

Sachs welcomed the public review, confident that the high-energy physics community would forward a favorable recommendation for continued operation of the ZGS. The physics panel, chaired by Robert L. Walker of the California Institute of Technology, developed the all important assessment of the current and potential physics programs at the ZGS. Walker's committee reminded the commission and the NSF that the ZGS polarized beam capability provided a unique tool for exploring the nature of strong interactions. In addition, the neutrino program utilizing Argonne's twelve-foot bubble chamber importantly complemented the higher-energy neutrino research at Brookhaven and Fermilab. Detectors were as crucial to high-energy physics as accelerators. Argonne's twelve-foot bubble chamber, the first large chamber constructed with superconducting coils, had performed well for four years, and with upgrading, promised to provide unparalleled ability to study neutrino interactions. Finally, the Walker committee described the ZGS hadron (a particle with strong interactions such as the proton) physics program as outstanding. The committee concluded that despite the ZGS's lower energies, it had neither approached "technical obsolescence" nor had it exhausted its "scientific capability." In short, the review committee found no scientific or technical reason to shut down the ZGS. If budgetary constraints forced curtailment of the nation's high-energy physics program, however, the Walker committee recommended that "the ZGS must be considered a leading, but by no means unique, candidate for such elimination." All programs, with the exception perhaps of Fermilab, should be reviewed within two years, and none, including the ZGS, should be eliminated before 1979.[120]

The Walker committee considered transferring ZGS programs to other facilities but concluded that, with the exception of moving the twelve-foot bubble chamber, reassignment was not practical. Relocating the polarized beam research was so questionable that the committee worried about permanently losing its capability. As important as the science that might be lost, the Walker committee inventoried high-energy research at 15 midwestern universities whose programs were dependent on the ZGS. Closing of the ZGS, they predicted, would have serious impact, which could not be offset by Fermilab, on graduate education and basic research among all participating universities. The committee also emphasized the interaction among the ZGS and applied research programs at Argonne, examples of which were noted in nuclear physics research, nuclear chemistry research, medical research, and potentially materials science. The ZGS had even provided useful technology for superconductivity, storage batteries, and solar energy. Not the least of its contributions, the ZGS attracted key personnel who contributed widely to the success of Argonne's mission, the most notable, perhaps, being the former associate laboratory director for high-energy physics, Robert Sachs, who now served as Argonne's director.[121]

No one in the high-energy physics community challenged the scientific contributions or potential of the ZGS. As Daniel Greenberg reported in *Science and Government Report,* the accelerator programs were "like occupants of an overcrowded lifeboat in troubled waters": someone would have to be thrown overboard "to appease the anti-inflationary sharks in government budget offices."[122] Although Sachs tried to read the Walker report as optimistically as possible, it was now clear that Fermilab's soaring budgets had doomed its Illinois neighbor at Argonne. By 1974 Fermilab, the only accelerator laboratory to receive dollar increases since 1969, commanded more than 25 percent of the total $131 million budgeted for high-energy research. Unanimously, the High-Energy Physics Advisory Panel insisted that work at Fermilab should proceed unobstructed. When the panel supported requests from SLAC and Brookhaven for new funding to design colliding beam accelerators, the only place left to save was to cut Argonne's $14 million ZGS budget. The panel was explicit about the political nature of its recommendation, which took into account "the desirability of having the new ventures distributed geographically . . . [assuring] diversity of physics, style, and intellectual input" desirable for excellent science.[123]

While additional reviews and assessments extended operation of the ZGS until 1979, even the most urgent appeals from the Argonne Universities Association could not save Argonne's particle accelerator. Confirming ERDA's decision to close the ZGS facility, John M. Deutch, director of the Department of Energy's Office of Energy Research, acknowledged the "high quality" and "contributions" of the ZGS program, but essential requirements for new high-energy facilities left the government no choice but to terminate the work at Argonne. Until the final shutdown, however, Deutch authorized continued operation of the ZGS as a dedicated polarized-beam facility.

Deutch's small concession recognized that Argonne's unique ability to accelerate a polarized proton beam was not duplicated by any other facility in the world and belatedly may have become the ZGS's most important scientific contribution to high-energy physics. Protons were known to spin like a small top, either right or left about equally. Not surprisingly, physicists surmised that the particle's spin behavior would importantly affect its nuclear interactions. But how? One way to study spin behavior was to polarize the proton beam, that is spin as many particles as possible in the same direction while accelerating them toward a polarized proton target. In collaboration with the University of Michigan, in 1971 Argonne scientists developed techniques for aligning and accelerating the polarized beam. They discovered to their surprise that polarized particles, three-quarters of which were spinning in the same direction, interacted more strongly with one another than nonpolarized particles. And they would continue to explore the cause of this unexpectedly violent behavior until their machine was turned off and the research was assumed by other facilities.[124]

The ZGS finally shut down on October 1, 1979, after completing 276 high-energy physics experiments for eighty-four international research institutions. Without question, the ZGS with its dynamic academic users group had constituted Argonne's most successful outreach program to midwestern universities. Without the ZGS, the lynch pin to the tripartite agreement among the federal government, the University of Chicago, and the AUA was gone.

Sachs's Legacy

Sadly, Sachs regarded the closing of the ZGS as the most important "accomplishment" of his administration. Never before in the history of the national laboratories had the government closed down a major accelerator without replacing it with another facility. Deutch encouraged Sachs to keep the high-energy experiment group together, presumably by securing access to underutilized facilities at Fermilab, SLAC, and Los Alamos. Although discouraged at the loss of the ZGS, Sachs assured the Argonne Universities Association that the high-energy physics group would continue intact, although diminished.

Sachs disdained playing the "political game," and consequently would leave a mixed legacy at Argonne National Laboratory. Because a rising tide floats all ships, Argonne prospered during Sachs's tenure as director when the federal government poured millions of dollars into energy research and development at the national laboratories. While total federal research and development funding in nonweapon areas increased 59 percent, Argonne's funding increased 95 percent between 1973 and 1977. Although support for basic science had declined to 26.3 percent of the fiscal year 1977 budget, Argonne's total budget had climbed to $181.6 million, never to be equaled again in 1977 dollars. During Sachs's last year, employment at Argonne topped 5,100, the historic high for the laboratory. He had successfully guided Argonne through two major agency reorganizations: the change from the Atomic Energy Commission to the Energy Research and Development Administration in 1975 and the transition to the Department of Energy in 1977. In the all-important area of nuclear reactor research and development, Argonne's budget had doubled during Sachs's tenure to $84.4 million in fiscal year 1977, although in a parallel and disquieting development, the nuclear energy budget of Oak Ridge had surged ahead of Argonne by $9 million. Despite the problems and frustrations, however, when measured in terms of dollars and employees, the Sachs era marked the halcyon years of the Argonne National Laboratory.[125]

10 | The Multiprogram Laboratory, 1977–81

Despite lingering shortages, America's energy policy was not a major issue in the 1976 presidential campaign. In numerous ways, energy had been a top domestic priority of President Gerald Ford. In addition to securing establishment of ERDA, under which Argonne National Laboratory prospered, Ford proposed a national energy plan that included dramatic funding increases for research and development of nuclear power as well as permission for private enrichment of uranium. Key to Ford's energy plan was the establishment of an Energy Independence Authority, which was authorized to assist in the private construction of nuclear power plants, coal-fired plants, oil refineries, synthetic fuel plants, and other energy production facilities. Ford was cautious about expanding the federal role in energy management and favored deregulation of natural gas, easing of emission standards established by the Clean Air Act, and opening the United States Naval Petroleum Reserves. His energy program, however, supported long-term research and development of "inexhaustible" sources of energy such as breeder reactors, fusion, and solar power, and promised continued prosperity for Argonne National Laboratory and its varied energy programs.[1]

Ford's 1976 national energy plan emphasized conservation, not nuclear energy, as the best way to buy time to develop alternative energy systems to supplement the declining supplies of fossil fuels. The nuclear option was not ignored, however, when the Ford administration endorsed the Atomic Energy Commission's LMFBR project, including the Commission's decision to build a large demonstration plant on the Clinch River at Oak Ridge, Tennessee. Construction on the Clinch River Breeder Reactor (CRBR) would begin in 1978, with completion scheduled for 1984. Argonne's support of the Clinch River project involved calculating the physics of large uranium and plutonium reactor cores, developing new instrumentation, testing fuels and materials in EBR-II, and assembling full-scale, simulated breeder reactor cores in the ZPPR.[2]

Although Argonne scientists applauded Ford's bicentennial energy initiatives, they became apprehensive in 1976 when the Democratic Party chose one of their own as its presidential nominee, Governor Jimmy Carter of Georgia.

Following graduation from the U.S. Naval Academy, Carter had served in the nuclear navy, where he became chief engineer of the nuclear submarine *Seawolf*. During the construction of the submarine's reactor at General Electric in Schenectady, Carter studied reactor technology and nuclear physics at Union College. Although not a nuclear scientist, Carter made claim to being a nuclear engineer. Carter's nuclear credentials and his expertise on energy issues might have encouraged the Argonne community but for the fact that Carter was also known to be lukewarm about nuclear power. The Democratic presidential candidate was not professedly antinuclear, but Carter assigned his lowest energy priority to the development of the nuclear option.

During the presidential campaign, Carter sharply criticized Ford's energy record. The United States was the only developed nation that did not have a "comprehensive energy program or policy," Carter argued. Predicting that there was only "about thirty-five years worth of oil left in the whole world," Carter bluntly warned: "We're gonna run out of oil." In addition to his proposal to consolidate all government energy programs in a cabinet-level energy department, his short-term solution to the energy crisis was to increase coal production while intensifying conservation efforts. Long term, he did not rule out nuclear power altogether, but as a nuclear engineer, Carter noted the "limitations of atomic power." He advocated focusing America's research efforts on developing clean-burning coal, solar energy, and strict conservation measures. "And then as a last resort only, continue to use atomic power."[3] Worries over nuclear reactor safety and nuclear weapons proliferation underlay Carter's cool support for nuclear power.

While Argonne National Laboratory closely monitored the presidential candidate's energy statements, the electorate as a whole seemed uninterested in the energy issue. To the extent that Americans focused on the energy problem at all, they believed that energy shortages were temporary and largely the responsibility of Arab oil producers, American oil companies, the federal government, or all three. Integrity in government, inflation and unemployment, and United States foreign policy were the major political issues of the 1976 campaign. Jimmy Carter's *Playboy* interview in which he confessed that he had lusted in his heart and Gerald Ford's stumble over his estimate of Soviet influence in eastern Europe excited voters more than abstract debate over America's energy future.[4]

Following the November election, in which Jimmy Carter triumphed in one of the closest presidential elections in American history, the winter of 1976–77 turned bitterly cold. As temperatures and supplies of natural gas fell to record lows, energy shortages forced the closing of schools, plants, and businesses, especially in the northern industrial states. Yet another energy crisis greeted

Carter on his frigid inaugural day. Immediately, the new president appointed James Schlesinger, the former chairman of the Atomic Energy Commission, as his "energy czar" to hone the President's energy policy and reorganization plans, which included establishment of a new cabinet-level agency, the Department of Energy. Evoking the memory of Franklin Delano Roosevelt's calm reassurance of the American people during the Great Depression, Carter in a televised "fireside chat" on February 2, 1977, proclaimed a national energy emergency that called for sacrifice, conservation, and patience. Dressed in a comfortable cardigan sweater, which conveyed more the image of Mr. Rogers than of FDR, Carter promised the nation a comprehensive energy plan by early spring.[5]

Within one hundred days, Carter presented his national energy plan to Congress. In doing so, he warned the nation that the energy challenge not only tested the mettle of American character, but also the ability of the president and the Congress to govern. With the exception of winning the Cold War, Carter described the energy crisis as America's greatest challenge, and borrowing from the philosopher William James, he declared that America's struggle for energy self-sufficiency was the moral equivalent of war. The president's dramatic rhetoric was important because only during actual war had the government imposed the energy controls now being advocated by the White House.[6]

War-weary Americans, disillusioned by failure in Vietnam and jaded by the Great Society's war on poverty, were skeptical about Carter's call to arms. Just what kind of energy battle, figurative or otherwise, were Americans girding to fight? Like Richard Nixon before him and Ronald Reagan afterward, Carter had won the presidency by campaigning as much against Washington's entrenched bureaucracy as against the Republican Party. As Burton Kaufman has observed, "A revolt against Washington was what the [1976] election was all about."[7] Americans wanted less intrusion into their lives from Washington, D.C., not more. When Congress deadlocked over passage of Carter's energy plan, public opinion failed to rally behind the president. With almost no public demand for passage of the administration's program, the well-financed energy lobby successfully sidetracked tough legislation designed to raise the energy taxes that would have lowered oil imports and stimulated conservation. With no constituency demanding higher energy taxes as the cornerstone of Carter's national energy plan, Schlesinger ruefully noted that "the basic problem is that . . . there are many constituencies opposed."[8]

Few of the tactical elements in Carter's energy plan were original; many were similar to proposals made by Ford, others were counterproposals advanced by the Democrats. The difference between Carter and his Republican predecessors was that Carter developed a strategic plan for a comprehensive energy policy encompassing all energy sources: fossil, hydroelectric, nuclear, advanced, and alternative energy systems. Energy policy was not new for the federal government. Nixon and Ford, for instance, worked hard to increase domestic energy supplies. What was new in Carter's plan, which placed great emphasis on conservation, was an energy policy that self-consciously coordinated federal programs involved in production, consumption, pricing, research, development, and commercialization of all energy resources.[9]

Carter's energy battle plan included more than a hundred actions, ranging from enacting new taxes, legislation, and regulations to establishing new administrative structures, especially the Department of Energy, which consolidated most of the government's energy programs in one agency. The DOE acquired about forty regional and field offices, laboratories, research centers, and university programs from its predecessor agencies. In the new department, Argonne would have to compete for funds not only with the former AEC laboratories, but also with the Department of the Interior's energy technology laboratories at Bartlesville, Oklahoma; Morgantown, West Virginia; Laramie, Wyoming; and Pittsburgh, Pennsylvania, as well as with ERDA's new Solar Energy Research Institute (SERI) in Golden, Colorado. The big question in Chicago was how would Argonne and its mission be mobilized to support Carter's energy program.

Initially, veterans at Argonne were hopeful that laboratory programs would prosper in Carter's administration. With the Atomic Energy Commission beyond resurrection, Argonne scientists welcomed Carter's proposal to place the national laboratories in a cabinet-level Department of Energy under Schlesinger's leadership. Robert Sachs believed that the national laboratories themselves had already "provided eloquent answers" to their future. Although Argonne continued to devote 40 percent of its research-and-development effort to the liquid-metal fast-breeder reactor program, since the end of the AEC, Argonne's commitment to research and development in fossil energy, conservation, and solar energy had grown to 15 percent, whereas environmental and system studies and policy evaluations had become 10 percent of the total effort. Under Sachs's leadership, Argonne maintained its strong role in nuclear technology and the basic sciences, but diversified into both analytical and technical programs of immediate concern, such as waste management and nuclear-proliferation studies.[10]

Today at Argonne

On December 20, 1976, Argonnians celebrated the twenty-fifth anniversary of the day EBR-I lighted a string of four light bulbs at Argonne-West, marking the first time in history that a nuclear reactor produced useable electricity. Nuclear power advocates believed that EBR-I, which was dedicated as a national historical landmark by President Lyndon Johnson in 1963, symbolized the promise of nuclear technology, which by 1976 included 454 power reactors—capable of more than 340,000 megawatts of electrical power—operating, under construction, or planned worldwide.[11] After President Carter's inauguration in 1977, however, nuclear power critics presumed that EBR-I embodied the hubris of the old energy creed, which looked to big science and risky technology to solve America's energy problems.

The April gloom could not have been more disappointing to NBC anchorman Tom Brokaw and the *Today* show staff. Their prerecorded taping of behind-the-scenes testing of the lithium–metal sulfide battery and the bench-scale fluidized-bed combustor would not be affected by the threat of an early-morning rain. But Brokaw's plans to broadcast a live demonstration of Argonne's solar collector were frustrated by dark clouds over the laboratory. He assured the national television audience that Argonne's solar energy research was "a little more sophisticated than rubbing two sticks together" in the laboratory's woods! Patiently, Sachs described the compound parabolic concentrator, which in principle could be used to heat or cool homes, and the "dielectric" CPC, which converted sunlight into electricity. Improvising, Brokaw turned on a small lamp which substituted for the sun. Instantly, a radio receiving electricity from a solar panel blared forth wake-up music. Then, playing a high-tech Zeus, Brokaw extended his hand, cloudlike, between the "sun" and the solar panel to silence the anonymous radio station.[12]

The eight-minute visit of the *Today* show was great fun for Argonne's staff and dramatically showed off the laboratory's innovative work in alternative energy systems. But the implications of the show were not reassuring to Argonne's top management. Public interest in solar energy and the other technologies was all well and good, but they remained a small part of the laboratory's overall effort and were not likely to command large expenditures from the federal government. Unfortunately, the television crew had shown no interest in featuring any of Argonne's nuclear research, except for a small segment on the ZGS. As Sachs and others painfully realized, the major cloud hanging over Argonne laboratory was political, and it threatened to dampen seriously the breeder-reactor program.

Together, Sachs and Robert Laney, the deputy director of the laboratory, launched a veiled attack on Carter's nuclear energy policy. From the "Director's Corner" of the *Argonne News,* Laney reviewed the troubling potential of worldwide nuclear weapons proliferation coupled to the international growth of nuclear power reactors. Simply stated, the nuclear fuel that powered the reactors also produced plutonium, a key element in nuclear bombs. The spread of nuclear power reactors would be followed by the multiplication of nuclear-fuel reprocessing facilities, which, in turn, increased opportunities for diverting nuclear materials to weapon production. During the presidential campaign, Carter called for new international efforts to halt the spread of nuclear weapons. The greatest danger to nuclear weapons proliferation, Carter argued, was the sale or transfer of uranium enrichment and spent-fuel reprocessing technology to nonnuclear powers, and he called for a voluntary moratorium on international traffic in such technology. On October 23, 1976, President Ford also had warned the nation that the reprocessing of plutonium should not proceed in the United States until reliable technical safeguards eliminated the risks of proliferation. Although Laney did not challenge Ford's and Carter's calls for a moratorium on the export of nuclear reprocessing technology, he did question "to what extent should our domestic policy, say on plutonium recycle, be influenced by an international objective to inhibit weapons proliferation?"[13]

Nuclear-fuel reprocessing to extract plutonium from spent fuel was essential for the success of breeder-reactor technology. In this sense, to attack reprocessing technology was to strike at the Achilles heal of the breeder economy. Without plutonium reprocessing, the breeder program would be crippled. Alvin Weinberg, the former head of the Oak Ridge National Laboratory and a vigorous advocate for nuclear power, warned Argonne scientists that as a consequence of the fall elections "nuclear energy is in trouble." Although voters in seven states had rejected a nuclear moratorium by a two-to-one majority, Weinberg bleakly predicted, "I don't believe that a primary energy system that is feared or rejected by 33 percent of the public can survive in the long run." On the hopeful side, Weinberg believed that with rigorous safeguards and security, including isolating reactors in remote energy "parks," perhaps five hundred to one thousand breeder reactors would eventually replace all light-water reactors.[14]

Laney launched his direct assault on Carter's nuclear policy under the guise of a debate with environmentalist David D. Comey, the executive director of Citizens for a Better Environment. Meeting at a conference of the American Association of Physics Teachers at Chicago's Palmer House hotel, Laney criticized opponents of nuclear power for being disingenuous about the energy crisis. According to Laney, there were five irreducible facts that dictated United States' energy policy: (1) America was running low on oil and gas while (2) becoming increasingly dependent on foreign imports; (3) America's "yawning" energy gap (4) would have to be filled in a future dominated by increasing energy uncertainties; consequently, (5) the United States had no choice but to develop a sensible national energy plan "based on multiple, diverse, and reliable options." Obviously, Laney concluded, one of those options would have to be "safe, economical, and reliable" nuclear energy. Laney especially emphasized the safety record of the nuclear industry. Citing the Rasmussen reactor safety study, he argued that there was little risk to the public from "potential accidents in nuclear power plants."[15]

Comey countered by challenging Laney's assumptions about nuclear reactor safety and economy: neither had been proven through research or operating experience, Comey argued. His principal objection, however, echoed President Carter's fears about nuclear proliferation. Whereas Laney doubted that American abandonment of nuclear fuel reprocessing would have much influence over the nuclear policies of foreign countries, Comey countered that the United States' reprocessing policy would have a tremendous impact abroad. It was necessary for the United States to lead the world away from the inherent dangers of breeder-reactor technology. "What other countries are planning to do with the plutonium produced by the civilian power reactors is very frightening," Comey admonished. Yet, he was not nearly as worried about what sovereign states might do as he was about what terrorist organizations would do if they got their hands on weapon grade nuclear material. "What happens when Idi Amin gets his hands on a nuclear weapon?" Comey asked rhetorically.[16]

When Senator Charles Percy visited Argonne two months later to discuss world energy supplies, Laney tried his best to assure the senator that, while there was no "proliferation-proof" nuclear option, Argonne was working on technology that would reduce the risk of proliferation. Laney then turned Percy's attention to Argonne's research related to advanced energy systems and other programs: the CPC, fluidized-bed coal combustion, magnetohydrodynamics, lithium–metal sulfide battery, and conservation and environmental studies. Carefully, Laney couched the nuclear option within the prevailing rhetoric of the Carter administration. With help from the national laboratories, the United States would make the transition from energy dependency on oil and gas to energy abundance with "inexhaustibles"—solar, fusion, and geothermal energy. But transition to the inexhaustibles might take as long as one hundred years. In the meantime, Laney proposed, the United States needed to build a long, stable energy bridge supported by triple piers of conservation, coal, and uranium. As insurance against the limitations of these three piers, Laney reasoned, the United States urgently needed to develop the breeder-reactor option.[17]

While Laney took the point in defending Argonne's breeder-reactor program, Sachs gently assessed Carter's nuclear power policy announcement of April 7, 1977, in which the president indefinitely deferred commercial reprocessing and recycling of plutonium produced in American civilian power reactors. President Carter based his decision on a recent Ford Foundation study, "Nuclear Power: Issues and Choices," which warned that plutonium used as reactor fuel would needlessly increase the risk of nuclear weapons proliferation.[18] Sachs, however, put the best possible face on the president's announcement, which also included Carter's short-term commitment to nuclear energy. Sachs applauded Carter's emphasis on energy conservation and on energy conversion from oil and natural gas to coal. He was especially enthusiastic about the president's dedication to a vigorous research program to develop "renewable" and "inexhaustible" energy technologies. Depending upon how budget priorities were adjusted, Sachs anticipated an increasingly important role for Argonne's coal and conservation programs, and although he acknowledged that Argonne's nuclear programs were in trouble, he urged the Argonne community not to panic. Even if the president succeeded in canceling the Clinch River demonstration breeder reactor, Sachs anticipated redirection of Argonne's nuclear research programs, not their death. With the plutonium-breeder program gone, Sachs predicted renewed interest in alternative nuclear systems, such as the thorium–uranium-233 fuel cycle. Accentuating the positive, Sachs envisioned that the search for different breeder fuels and materials might actually increase the demand for irradiations in EBR-II. Depending upon Carter's energy priorities and budgets, it was possible that EBR-II could be even busier than otherwise planned.[19]

Robert D. Thorne, ERDA's administrator for nuclear energy programs, was less optimistic than Sachs about the future of nuclear research at the national laboratories. Whatever the outcome of Carter's breeder policy, as nuclear technology became "commercially mature," Thorne anticipated a decreasing role for the laboratories, especially Argonne and Oak Ridge. In the long run, Thorne predicted a significant decline in the government's support for nuclear energy programs. His estimate was based not only on the Carter administration's emphasis on conservation, fossil fuel, and solar energy programs, but also upon the administration's policy to emphasize short-range applied research and development rather than long-range basic research.[20]

New Directions with the Department of Energy

The big question, both in Washington and Chicago, was how could Argonne, along with the other national laboratories, adjust its historic programs to coincide with the DOE's new priorities. When activated on October 1, 1977, the Department of Energy inherited eight multiprogram laboratories, thirty-two specialized laboratories, and sixteen nuclear materials and weapon production facilities capitalized at nearly $15 billion and employing more than 105,000 people.[21] As the Congressional Research Service asked, "How much science and technology" should the government support? "What is the precise role of the laboratories" in shaping energy policy? "How many laboratories," in fact, does the government need?[22] One academician, favored by the Congressional Research Service, believed that the former AEC laboratories had prospered by pursuing their own interests within the framework of government policy. Sensitive to policy shifts that affected budget levels, each laboratory sought autonomy to set its own agenda and to manage its own affairs. Under the AEC and ERDA, the national laboratories constructed a "substantial bureaucratic empire," striving "to shape policy according to their own interests."[23] Carter had already served notice that he intended to "bring immediate order" out of the fragmented system of federal energy programs. Creation of the Department of Energy, the president predicted, "will allow us, for the first time, to match our research and development to our overall energy policies and needs." And Carter left no doubt that United States' energy policy and needs would drive future research-and-development activities. If nothing else, under the Carter administration, the national laboratories would have to justify their research programs in terms of the government's overall energy policy.[24]

Among scientists at Argonne, the notion that the government laboratories had been unresponsive and unaccountable to national goals and priorities was ludicrous. From the days of the Manhattan Project through the establishment of the Atomic Energy Commission, from the period of Milton Shaw's micromanagement from Washington until the current emphasis in ERDA on both nuclear and alternative-energy research, Argonne continuously adjusted its program to fulfill national energy goals. Unnoticed from Washington, perhaps, was the extent to which individual scientists and engineers had redirected their work to serve the national agenda. Argonne claimed that its effectiveness depended upon the ability of high-quality staff, well schooled in the fundamentals of science and engineering, to solve new problems. Between 1973 and 1977, for example, research scientists at Argonne regularly shifted from basic research projects to specific energy-related programs. This move from basic science and engineering to applied programs involved almost one-quarter of the researchers in materials science; one-third of those in molecular, mathematical, and geosciences; and one-half of those involved in nuclear sciences. "The capability to bring fundamental knowledge and innovation to bear on applied programs is one of the major reasons to maintain the national laboratories' strong programs in basic science and engineering," Argonne rationalized. But the government could not draw indefinitely upon Argonne's store of basic research without depleting this fragile resource. Renewal in fundamental research was urgent to restore the thinned ranks of Argonne's basic scientists. Not all of tomorrow's problems can be identified today, the laboratory's management asserted. The Department of Energy had a responsibility to invest in new knowledge and new understanding so that Argonne could attack unresolved and new problems in the future.[25]

Cheerfully, Sachs assured both DOE managers and the Congress that Argonne had already made major shifts in the direction of nonnuclear research and development. Fossil fuel, conservation, and solar technologies had grown to more than 20 percent of Argonne's efforts, whereas nuclear research had dropped to 40 percent. Although Argonne received more funding from ERDA's nuclear energy program than any other multiprogram laboratory, it also received more money from the fossil and conservation programs than its national laboratory competitors. Including the small, but steady, growth in biomedical and environmental studies, since 1972 Argonne's nonnuclear research programs had enjoyed impressive prosperity; but prosperity at the expense of basic research in the physical sciences, which had slipped to 26 percent of Argonne's employee efforts. Sachs's optimism faded when he reported to Congress that the decline of Argonne's physical research program was worrisome because basic research had provided the foundation of the laboratory's "entire program." Strength in physical research, Sachs emphasized, was the reason Argonne had been able to respond so quickly to changing national needs. While George H. Vineyard, director of Brookhaven National Laboratory, complained bitterly of Washington's increasingly "fine-grained and detailed" management of his basic research programs, by 1977 Sachs was worrying less about Argonne's research autonomy than about how he could stanch the hemorrhaging of his basic research budget.[26]

The DOE's Organization and Budget

Initially, the organization of the Department of Energy left Argonne National Laboratory in bureaucratic limbo. Under ERDA, Argonne, along with all of the other national laboratories, had reported to Administrator Seamans through the assistant administrator for field operations (AFO) and the Chicago Operations Office. The field administrator, in turn, coordinated all programs between the laboratories and ERDA's major research and development offices, including nuclear energy, fossil energy, solar energy, fusion energy, geothermal energy, and conservation. The Department of Energy abandoned ERDA's traditional fuel-oriented approach in favor of an organization that not only emphasized the evolution of energy technologies from research and development through engineering and demonstration to commercialization, but also established economic regulation to encourage the use of certain energy sources while curtailing that of others. DOE's bureaucracy reflected Carter's promise to develop a "comprehensive" energy policy rather than simply to organize the department to promote favored fuel technologies.

For institutional needs under the DOE, the laboratories reported directly to a program office rather than to field and operations offices. Replacing the assistant administrator for field operations, a field and laboratory coordination council, composed of all line administrators and chaired by the under secretary of energy, strengthened headquarters's oversight of the laboratories and field operations. Not surprisingly, the weapon laboratories (Los Alamos, Livermore, and Sandia) reported to the assistant secretary for defense programs. Because of their emphasis on basic science, Brookhaven, Lawrence Berkeley, Fermi, Stanford, and Ames Laboratories were assigned to the director of the office of energy research, who funded physical research programs. The Hanford Engineering Development Laboratory, the Idaho National Engineering Laboratory, the Savannah River Laboratory, Princeton's fusion laboratory, and the naval reactor laboratories were placed under the assistant secretary for energy technology, whose responsibilities included implementation of nuclear technologies. Argonne, Oak Ridge, and Pacific Northwest Laboratory, viewed principally as nuclear energy and life sciences laboratories, reported directly to the under secretary, ostensibly because these multiprogram laboratories did not fit neatly into any of DOE's program categories. Ominously, the assignment left Argonne without a "godfather" among the program officers but instead placed it under the care of the undersecretary of energy. The arrangement intensified uncertainties for the laboratory's future should the Clinch River Breeder Reactor project be canceled.[27]

With all the talk about Carter's support for conservation and "inexhaustible" energy sources, the stark reality of DOE's fiscal year 1977 budget for the multiprogram laboratories was hardly impressive. Since 1975, budget outlays for nonnuclear research had increased 262 percent but remained only 8.9 percent of the laboratories' total budget of nearly $1.5 billion. The multiprogram laboratory budget by effort in fiscal year 1977 included:

Effort	Outlay ($, millions)	Percent
National security	632.4	43.5
Nuclear energy	323.0	22.2
Nonnuclear energy	129.2	8.9
Environment and safety	126.3	8.7
Basic science	226.6	15.6

Program support	5.3	0.4
Work for others	10.3	0.7
Totals	1,453.1	100.0

Even in the area of environment and safety, 31 percent of the effort was devoted to nuclear energy. According to an audit from the General Accounting Office, nonnuclear research and development was fragmented within the national laboratory structure with over 90 percent of the growth in nonnuclear programs having been initiated by the national laboratories themselves.[28]

What was the likelihood that the multiprogram national laboratories could broaden their research and development base in nonnuclear energy research? On the positive side, Sachs convincingly demonstrated that national laboratories like Argonne possessed exceptional technical competence readily transferable to the development of such technologies. America's investment in nuclear technology, like the investment in space technology, would have payoffs reaching into numerous fields of energy technology. On the down side, the nonnuclear programs were fragmented, ad hoc, and in competition with other DOE energy research centers for funds. The Carter administration certainly appreciated the fact that the national laboratories offered exceptional talent in long-term, high-risk, big-science research. The administration's emphasis, however, was on the development and demonstration of near- and mid-term energy technologies. There had been strong advocacy under the Energy Research and Development Administration to reform the national laboratories into NASA-style job shops in which the laboratories would perform very little in-house research and development but, rather, would act as contract managers of program funds distributed to industry and universities. Although the Department of Energy did not convert to the NASA model, which used the management and technical abilities at its research facilities to monitor the work of its contractors, Argonne's contract management responsibilities, especially in nonnuclear research, grew steadily under ERDA. At the same time, this development was partly offset by the fact that Argonne itself picked up "work for others," especially from the Nuclear Regulatory Commission, but also from the Departments of Defense, Interior, and Health, Education and Welfare; NASA; Oak Ridge, Pacific Northwest, and Sandia laboratories; the Electric Power Research Institute; Exxon Nuclear; and the Japanese government.[29]

The Deutch Appointment

James Schlesinger selected John Deutch, the distinguished chemist from MIT, to direct DOE's Office of Energy Research. Although he did not oversee all laboratory activities, Deutch played a major role in setting research-and-development priorities for the new department. In concert with the administration, he emphasized partnerships with industry to accelerate short-term commercial development of energy technologies and encouraged modest grants to universities to support small-scale basic and applied research projects. For the immediate future, Deutch intended to maintain the national laboratory structure, but he did not propose to expand the mission or funding of the multiprogram laboratories.[30]

Perhaps the national laboratories had become muscle-bound by their science and engineering strength. Not all of the important problems of energy research and development were subject to an engineering solution. Carter's promise to develop a comprehensive energy policy for the United States included his commitment to consider the environmental, economic, political, social, and legal costs of continuing old or implementing new energy technologies. For years, the Argonne Universities Association had unsuccessfully promoted the establishment of a social science group at Argonne to evaluate the socioeconomic consequences of energy use in the Midwest. But traditionally, the government's engineering laboratories avoided such research, which not only seemed "soft," but also appeared politically risky. As of March 1977 the eight multiprogram laboratories employed almost 13,000 professionals, of whom only 120 (less than 1 percent) were trained social scientists. The comptroller general of the United States observed that the national laboratories lacked "the educational background to assess and provide solutions to the socio-economic issues that may impede with the widespread use of nonnuclear energy technologies assigned to them."[31]

After being goaded by the General Accounting Office to place the multiprogram laboratories under unified DOE management, Deutch added Argonne and the other "orphan" laboratories to his responsibilities in the Office of Energy Research. But this new "safe haven" did not offer immediate reassurance to the Chicago laboratory—while he evaluated Argonne's institutional plan, Deutch coolly appraised the laboratory's mission and programs. Aware of Argonne's weakness in the social sciences, Deutch discouraged the laboratory from becoming heavily engaged in socioeconomic research. To the extent that the DOE required analysis of the economics of energy production, the environmental impact of energy distribution, the sociology of energy consumption, or the political acceptability of energy policy, Deutch anticipated that the government would seek help directly from university scholars. Furthermore, Deutch made it clear to Henry V. Bohm, the recently elected president of the AUA, that he expected to shift funding for basic energy research toward the universities and away from the national laboratories. He did not plan to reallocate basic energy-research funding by slashing laboratory budgets, but rather by providing incremental increases to the universities' funding. Bohm came away from his conference with Deutch convinced that DOE was not going to make a capital investment in Argonne in the foreseeable future. Bohm's "gut impression" was that neither Argonne's leadership nor the AUA had been effective on the political level in lining up support for the laboratory's programs.[32]

In July 1978 Deutch conducted a comprehensive review of Argonne's programs based on a five-year (FY 79–FY 84) institutional plan prepared by the laboratory. His critique was not gentle. Deutch noted that the Argonne plan contained "fine descriptive write-ups on individual ongoing activities," but that overall Argonne did not present a "clear view" of its base programs or propose realistic planning budgets. Deutch was dismayed that Argonne lacked "a firm grasp of what the future role of the laboratory should be in the DOE structure." He flatly told Sachs that the department would not even consider construction of a new physical research facility before fiscal year 1983. Yet in the next breath, he questioned why the ZGS was not being phased out faster and wanted specific projections on the employment for former ZGS personnel. He also asked Sachs for concrete estimates of the effects of substantial cuts in the breeder program at Argonne. For nonnuclear research, Deutch required a "detailed presentation" justifying Argonne's "assumed" expectations in environmental studies, solar and geothermal research, and conservation activities. Deutch concluded that in energy research Argonne should proceed more vigorously to phase out older, less productive programs in order to make room for new initiatives such as materials research related to solar energy and nuclear waste disposal.[33]

The Search for a New Director

As he approached his fifth anniversary as director of Argonne National Laboratory, Sachs decided to return to full-time research and teaching at the University of Chicago. He had been heard to quip that he first decided to resign on April Fool's Day 1973, his first day on the job. More seriously, with the establishment of the Department of Energy in October 1977 and the election of Hanna Holborn Gray in December 1977 as the new president of the University of Chicago, Sachs felt that the time had come for an administrative change at Argonne as well. Although he assured the AUA board of trustees that his decision to resign, effective September 1, 1978, had not been hasty or prompted by any one problem or issue, Sachs had become increasingly frustrated in the job. In ERDA, he had direct access, albeit not always satisfactory, to Administrator Robert Seamans. During his few months in the Department of Energy, Sachs saw Secretary James Schlesinger only once, at a cocktail party. In the Department of Energy, Sachs complained, "everything was so political; there wasn't anything for me to do." He had had enough, and so he decided to return to his laboratory.[34]

In March 1978, when he announced his intention to resign as director, Sachs gave the laboratory six months to find his replacement. With Sachs on board until the end of September, John Wilson, who was about to retire as the university's president, was hopeful that the search for the new director could be concluded by June. Wilson asked D. Gale Johnson, the provost; William Cannon, the vice president for administration; and Nathan Sugarman, the distinguished University of Chicago chemist who had played a key role in recruiting Sachs, to serve as the search committee on behalf of the University of Chicago. The AUA asked its planning committee, chaired by Nunzio J. Palladino to review candidate credentials on behalf of the AUA.[35]

The appointment of the new laboratory director marked a major watershed in the history of Argonne National Laboratory. For the first time since its establishment as the AEC's principal reactor laboratory in 1946, Argonne would have to find a new leader and a new mission within the broad mandate of the Department of Energy. With the future of the laboratory uncertain and far from secure, there was no heir apparent, no obvious choice, or even compelling candidate to whom the laboratory could turn. The search was wide open, and as it turned out, quite unusual.

In recruiting a new director, the University of Chicago sought the advice of all laboratory constituents. Seldom had the process for selecting new laboratory leadership been so open, so democratic, and so confused as the search for Sachs's replacement. The search that Wilson hoped would end in three months took over a year to complete. Almost literally, the search committee met with everyone interested in the appointment of the next director.

History rarely captures the inner dynamics of institutional bull sessions, but when Argonne's directors, managers, scientists, and engineers met to discuss their criteria for a new director, they revealed the laboratory's inner soul, including its hopes, fears, confidence, insecurity, generosity, and bias. Individually, they were mostly realistic in their expectations for a new leader; collectively, the laboratory managers reflected Argonne's pride and frustration in adjusting to the DOE's new demands. In the process, they arrived at a remarkable consensus about who they were, where they were going, and who they wanted to lead them.[36]

Not surprisingly, Argonne's managers wanted a "tough" leader, someone who could "stand up to DOE"; who would take risks; who was willing to be unpopular and sometimes to say "no." A weak director would be "shredded" by conflicting interests or might withdraw from conflict, thus setting the laboratory adrift. Only easy questions can be delegated; the director must handle all difficult issues personally. The managers' desire for toughness, however, was ambivalent because it was not clear against whom, among their constituents, they expected the director to stand firm. The DOE? The AUA or the University of Chicago? Competing laboratories? Themselves? Would they chose a "strong manager" over a "charismatic manager"? In truth, Argonne's managers, who sought strength, stability, and clear direction for their own programs, needed a leader who could guide the laboratory through an unfamiliar and dangerous bureaucratic landscape.[37]

Argonne was in trouble, the managers agreed. With some justification, the Department of Energy had criticized Argonne "for being an aging institution that has passed the peak of its vitality and has developed to an undesirable degree a sense of scientific and institutional introversion."[38] In addition, basic science programs were not only under attack from Washington, but also from the universities and industry. "Argonne National Laboratory is [still] basically an AEC lab," another scientist noted, and consequently suffered from "an identity crisis." The DOE was steadily pushing Argonne into project management and paper studies and regarded Argonne's scientists as "managers" rather than "doers." The new director would have to lead in reconstruction of the laboratory by securing large research facilities compatible with the DOE's mission. Argonne should have "multiple" thrusts for major facilities, rather than seeking new machines one at a time. The DOE fiscal year 1979 budget contained "significant funding for basic research facilities," but somewhere along the road, Argonne lost its machine-building experts. Not only strength, but also aggressiveness was required in the director's office.[39]

Was experience as a skilled, tough manager more important for the director than a reputation as an outstanding scientist? Not surprisingly, Argonne scientists placed a high priority on hiring a renowned scientist to be their leader; they literally could not think in terms of a different leadership paradigm. They were uncomfortable with the notion of hiring a director who was better known for management than for science. Furthermore, it was inconceivable that they would be unable to find a prominent scientist with extensive management experience. Given Argonne's mix of basic research and project engineering, however, should the laboratory limit its candidates to persons with a Ph.D.? Not necessarily, but regardless of the new director's educational background, in order to set laboratory priorities, the director had "to understand science."

The Argonne managers did not define what they meant by *science*—no doubt that refinement seemed unnecessary. Yet in their conversations, *science* conveyed a variety of meanings. It most often referred to a discipline, such as high-energy physics or solid-state physics. The word also meant a methodology, often rooted in higher mathematics. But frequently when Argonne managers talked about *science,* they meant their culture, especially its values and priorities. A director who "understood science" was not only a leader who possessed special knowledge but also one who belonged to their science culture. For example, the new director would have to be sensitive to several scientific disciplines for which he had no interest, expertise, or involvement. As Argonne moved away from "basic physical, biological, and nuclear engineering fields," the laboratory had not always acquired the in-house expertise to evaluate the soundness of new programs. The Argonne managers even realized that the laboratory would have to embrace the "soft" social sciences, although they continued to distinguish between the scientific and technical aspects of their research and its social and political consequences.

The next director would also have to be a salesman, diplomat, and promoter on behalf of the laboratory. "We need a director who can be a spokesman and 'sell' us to places other than DOE," one scientist advocated. Although Argonne managers indicated that the laboratory needed to improve its public relations efforts with industry and academia, most believed that Argonne had a good image within the industrial community. Under DOE, however, much of Argonne's work fell between the abilities of industry and the universities. One manager wanted to find a director with connections to industry so that Argonne could strengthen its technology-transfer programs. Another looked for a director capable of building the laboratory's "unique capabilities" that distinguished it from industry and academia. More commonly, the managers wanted the new director to build a solid political base for the laboratory in Washington, D.C. However they characterized their leadership needs, Argonne staff clearly wanted a director who could reverse their sense of drift and decline.

Finally, it was imperative that the new director be sensitive to affirmative action requirements at Argonne. The search committee interviewed Argonne's Assembly on Equal Rights (ANLAER), a civil rights group outspoken in its advocacy of a candidate who would be committed to affirmative action so that the assembly would "not have to start from scratch" sensitizing a new director. Regardless of Argonne's immediate postwar employment record, civil rights leaders stated that there was growing unrest among the laboratory's minority workers. They reported some harassment of assembly members, alleged lack of equal pay for equal qualifications, and questioning of the fairness of the laboratory's promotion practices. They praised Argonne's start at making the laboratory and its buildings handicapped accessible while urging the laboratory actively to recruit qualified handicapped workers. Good intentions, however, did not always filter down to the level where hiring decisions were made. Above all, the civil rights advocates argued, the laboratory needed more minorities "at decision-making levels."[40]

Revealingly, of all the criteria outlined by the Argonne community, no one suggested that the new director should be politically acceptable to the Carter administration. Of course, Argonne leaders wanted a director who was savvy in the ways of Washington's bureaucracy because they knew that energy research and development had to be compatible with national energy policy. But their self-image of being part of one of the nation's premier scientific and engineering laboratories would not allow them to legitimate political criteria for the selection of Argonne's next director. Consequently, although their search for a director did not run counter to the requirements of the Carter administration, at first, the laboratory scientists were none too sensitive to the new political realities at the DOE.

In June 1978 President John Wilson and William Cannon from the university met with John Deutch and Robert Thorne, assistant secretary for energy technology, who supervised the breeder-reactor program. Point blank, Deutch told Wilson that the Department of Energy envisioned a future for Argonne different from that pursued by the laboratory's current management. Deutch explained that the department was reaching for innovation and change in the laboratory because it wanted dramatic advancement of basic and applied energy technologies. But he was skeptical that Argonne's leadership could move the laboratory out of the rut it had been in for so many years. Afterward, Cannon concluded that the DOE would not invest heavily in Argonne unless it recruited the "new" type of leadership sought by the department.[41] Deutch cautioned Cannon that Schlesinger himself was worried about the recruitment of Argonne's director and, therefore, would personally review the university's selection.

By October 1978 all of the outside candidates had either withdrawn or had been eliminated, often in consultation with Deutch. The DOE was particularly insistent that Argonne target young scientists in the thirty-five to forty-five age group. When the search for Sachs's successor turned inward, however, Robert Laney emerged as the overwhelming favorite of Argonne's scientific and engineering community. He possessed poise, intellect, perspective, outside stature, and the respect of both the scientific and engineering divisions. He knew the laboratory well and had served faithfully as acting director before Sachs. Laney had everything going for him, except his age and the confidence of Washington.[42] Deutch became livid when he learned that the search committee considered Laney the front runner of the internal search. He vowed that Laney would become Argonne's director "over his dead body" and threatened to transfer Argonne-West to a private contractor and close the laboratory within three years if the university appointed Laney to the position. To make sure that Cannon understood that Deutch was not making an idle threat, he called in Thorne to confirm their "alliance" to raise up Argonne's performance or disband it. Like Milton Shaw before them, Deutch and Thorne could not wait for Argonne's uncertain transition to a better laboratory. "Things are so bad," Deutch reportedly complained, "the lab is so poor that it has to be handled right away"; which meant in Deutch's view, a thorough housecleaning at the top levels of Argonne's management. It was not Laney's thinly veiled pronuclear attacks on Carter's energy policy that had made him politically persona non grata in Washington. Deutch and Thorne regarded themselves as strongly "pro-nukes." Rather, Laney's handicap was that he epitomized the old spirit of Argonne, which Deutch was determined to sweep away.[43]

With Laney out of the picture, and no other viable internal candidate available, in late fall 1978 the search committee accepted a list of twenty nominees (only three of them were under the age of forty-five) from the Argonne Universities Association. Among the candidates was Walter E. Massey, described by Henry Bohm as "a well qualified university dean of great possible interest."[44] Massey did not move immediately to the top of Argonne's "short list," however. Donald Hornig, the presidential science advisor under Lyndon Johnson, praised him as a well organized, solid physicist. Although he had no real technical experience, Massey was regarded as a good theorist who dealt well with people. Was he tough enough? No one was sure. But Massey's biggest handicap was that, at first, Argonne did not believe it stood a chance of enticing him away from Brown University.[45] Consequently, while informal telephone conversations continued between Chicago and Providence and Argonne pared down its list, which included three women, Peter A. Carruthers, leader of the theoretical division at Los Alamos and a top contender, officially withdrew his name on January 30, 1979.

Carruthers's declination was a hard blow because it struck at some of the sorest issues troubling the laboratory. From his perspective, the Department of Energy managed Argonne too closely and yet failed to "assume proper responsibility for the health of the laboratory through consistent funding of those large facilities necessary for focus of research work." DOE budgets allowed the director less discretionary money than Carruthers controlled at Los Alamos. Sadly, the director seemed to have "little real opportunity for creative restructuring of programs in the present situation." Finally, Carruthers found the history of AUA "replete with frustration on all sides." Prophetically, he observed that "the relation with the AUA needs radical revision."[46]

Walter Massey

Argonne's long search for a new leader ended on May 22, 1979, when President Hanna Gray, after appropriate clearances with John Deutch in Washington, announced that Walter Massey, dean of the college at Brown University, had been selected to be Argonne's sixth director (figure 32).[47] Massey's appointment was a personal triumph for Gray, who worked hard to bring him to Argonne. Only forty-one years old when Gray appointed him to head the laboratory, Massey had been a mere child in Hattiesburg, Mississippi, in 1942 when Fermi achieved the first controlled chain reaction with CP-1.

Although not a veteran of the Manhattan Project, Massey's credentials neatly fit into the profile for Argonne directors. He was well known to the Argonne community as a physicist and academician. After his sophomore year in high school, Massey entered Morehouse College to study physics, graduating with honors in 1958. Thereafter, he attended Columbia University and Howard University before matriculating at Washington University in St. Louis to begin graduate studies, which he completed in 1966 with a Ph.D. in physics. Fresh out of graduate school, Massey received a postdoctoral fellowship from Argonne's chemistry and solid-state science divisions in 1966–68 to work on the low-temperature properties of quantum liquids and solids. After his postdoc-

Figure 32. Walter E. Massey

toral fellowship, Massey continued to serve as a consultant to the laboratory until 1975. He served on the faculty of the University of Illinois until 1970 when he moved to Brown University to become an associate professor of physics. He was thirty-seven when he was promoted to full professor and named dean at Brown in 1975.[48]

Recently named by President Carter to the National Science Board, which shaped policy for the National Science Foundation, Massey was especially renowned for his contributions to science education. As the chief academic officer for undergraduate education at Brown, he was proud that Brown ranked second only to Harvard as the most sought-after college in the Ivy League. In 1974, he received the Outstanding Educator in America award. The following year, the American Association of Physics Teachers presented him their distinguished service citation for his exceptional contributions to physics teaching. By 1978 Massey was recognized as one of the hundred most respected leaders in higher education by the American Council on Education and *Change* magazine. A distinguished educator, Massey also managed to maintain his research in the low-temperature properties of liquid helium. Optimistically, he hoped to continue his research at Argonne.

Massey's research interests, of course, were welcomed at Argonne, but of more immediate concern to Argonne's scientists was Massey's commitment to the laboratory's existing research programs, especially those in nuclear energy. He acknowledged the acute frustrations of scientists and engineers who had devoted their entire "lives and careers" to developing a nuclear technology whose future enjoyed uncertain support from the Carter administration. The nuclear question, Massey told readers of *Ebony* magazine, was more political and sociological than scientific. "People are simply afraid of things they don't understand," he mused, "and people do not understand the process of nuclear energy." He believed that scientists and engineers needed to be more forthcoming about the dangers of nuclear energy so that the public could realistically weigh the benefits against the risks. Personally, Massey believed that nuclear energy would have to play a major role in America's future. Regardless, Argonne's broad mission in energy research and development was absolutely clear to the new director: "It's to make sure that the technology and research base is there whatever social and political decisions are made."[49]

If Massey brought any weaknesses to Argonne, it was his lack of experience managing technical and engineering programs. Although Deutch supported Massey, he insisted that the University of Chicago and the AUA not rely entirely on Massey for technical leadership of the laboratory. As a condition to his appointment, Massey and Hanna Gray agreed to recruit a "top person" to serve as Massey's principal deputy for applied research and reactor engineering programs. Although the university could not promise to hire Deutch's own candidate for the principal deputy position, Gray promised that "we will continue our search for the kind of person described until we have found a person who can do the job to your satisfaction, AUA's and ours."[50]

Redirecting the Breeder-Reactor Program

Throughout Jimmy Carter's presidency, the Congress and the White House stood at loggerheads over the future of the nuclear breeder-reactor program. In the long run, the debate concerned the role of nuclear energy in the twenty-first century; in the short run, it focused on the fate of the Clinch River Breeder Reactor project. The LMFBR program had been the nation's top priority in energy research and development for more than a decade when, on April 7, 1977, President Carter announced his restructuring of the breeder program to provide higher priority to alternative breeder designs and to postpone the target date for commercialization from 1986 to about 2020. The long extension of the breeder program reflected several reservations of the Carter administration. In addition to the administration's predominate concern about nuclear weapons proliferation, Carter also worried about the safety of the LMFBR. In the long run, DOE planners forecasted reduced growth in projected electricity demands. With development of the breeder reactor less urgent, they doubted whether the LMFBR would be economically competitive until the twenty-first century. In the short run, they argued that the Clinch River demonstration breeder reactor was "too small, too costly, and technically obsolete."[51]

On April 5, 1977, over fifty congressman, led by Marilyn Lloyd of Tennessee, wrote President Carter insisting that "the prompt development of the U.S. breeder program is essential to our national energy policy." For three years, Carter had proposed energy budgets that eliminated the Clinch River project, and just as stubbornly Congress renewed funding for the CRBR. No one, of course, wanted to promote nuclear weapons proliferation, but studies since 1977 by the Congressional Research Service, the Office of Technology Assessment, the General Accounting Office, and the Ford Foundation all concluded that the development of commercial nuclear power plants were unlikely to contribute significantly to the proliferation problem.[52] For those who dismissed the specter of nuclear weapons proliferation stimulated by the commercial breeder reactor, the administration's safety and economic questions about the breeder program were arguable.

Regardless of the merits of the disagreements between the Carter administration and Congress over the future of the LMFBR program and the fate of the Clinch River Breeder Reactor, the protracted debate profoundly unsettled Argonne's nuclear scientists and engineers. No one could predict with any confidence what their fate might be. The General Accounting Office reported in 1980 that the breeder program, which lacked a "clear mission and focus," was in disarray throughout the ten government laboratories and among the dozens of contractors who were working on pieces of the project. Caught between the president and Congress, the Department of Energy continued to support the breeder program under congressional mandate but could not establish a comprehensive technology-development plan endorsed by the White House. According to the General Accounting Office, LMFBR technical planning—including fuels, materials and chemistry, physics, safety, and reactor components—was virtually nonexistent. Argonne, for example, was responsible for engineering reactor components. But without development deadlines or reactor plant construction schedules, it was impossible for Argonne to ensure that reactor hardware would be ready when needed.[53]

Argonne was proud of its "quiet" story associated with more than a decade of successful operation of EBR-II. EBR-II's work as the major LMFBR irradiation facility for testing materials and fuels was important. Since 1964, EBR-II had run more than nine thousand irradiation experiments on fuels and fuel assemblies and on control and structural materials. Tests involved irradiating fuel assemblies to determine when and under what conditions holes or cracks appeared in their cladding—the stainless-steel jackets that contained the fuel elements. Less dramatic, but increasingly vital, was the development of reactor instrumentation. As reactor systems became larger, more complex, more powerful, and potentially more dangerous, the development of reliable and user-friendly instrumentation was essential to the safety of all nuclear reactors, including breeders. Argonne pioneered experiments with quick-responding instruments to measure reactor heat, power and pressure levels, and coolant flow; and tests with slower-responding instruments (in seconds or minutes rather than thousandths of a second) that monitored fission-product gases and chemical processes in the reactor system. Some instruments depended on periodic sampling or laboratory analysis, but the most successful were *on-line* devices that reported the reactor's operation minute by minute, especially fuel and coolant performance. In the LMFBR, special care was taken to measure radioactivity in the sodium coolant loop and to monitor for serious sodium-water chemical reactions due to leaks in the steam generators. EBR-II provided space within the reactor core to insert instrument bundles for in-core testing (INCOT). INCOT enabled experimenters to test the performance of neutron detectors, thermocouplers, pressure sensors, and flowmeters operating within the breeder environment. Argonne even developed an acoustic monitoring system for EBR-II, which listened for mechanical problems in the reactor's pumps and heat exchanger.[54]

Most importantly, while EBR-II served as an irradiation facility for the breeder-reactor project, Argonne engineers continued research and development on EBR-II itself. Although the DOE did not believe that EBR-II, with its metal-fuel driver, had a long-term future in the development of commercial power reactors, headquarters tolerated a certain amount of experimental work on the reactor to improve EBR-II's efficiency, reliability, and safety as an irradiation facility. Into this opportunity gap, engineers at Argonne-West steered their own developmental research on EBR-II, focusing especially on designing more efficient metal-fuel elements. EBR-II's metal fuels had been notoriously inefficient, initially achieving only slightly more than 1 percent "burnup," because swelling and overheating, which threatened to damage fuel cladding, truncated the time a metal-fuel pin could remain in the reactor core. For more than twenty years, ceramic (oxide) fuels with their greater "burnup" efficiency, had been the fuels of choice for the breeder program, in part because of the extensive technological base already established for oxide fuels used in the light-water reactor. While EBR-II irradiated hundreds of oxide-fuel pins for the LMFBR program, Argonne-West engineers quietly worked to improve the performance of their own metal driver fuel. Steadily, they had increased metal-fuel efficiency by fashioning a uranium-plutonium-zirconium alloy fuel which achieved more than 10 percent "burnup," basically solving the metal fuel problem of the 1960s and 1970s.[55]

Ironically, Carter's assault on the Clinch River project changed the government's plans to shut down EBR-II in 1980. Originally, the ERDA believed that the useful life of EBR-II would end as new generations of test facilities came on line. But with the construction of new experimental reactors in doubt, the Department of Energy believed it would be prudent to maintain EBR-II, the nation's only complete, operating LMFBR power station. The Fast-Flux Test Facility was still under construction at Hanford, but that facility had not yet proven a reliable, successful experimental reactor. On the other hand, the EBR-II, located at the National Reactor Testing Station, operated with a closed fuel cycle, including reprocessing, refabrication, and waste control, in a setting not unlike the nuclear-park concept advocated as an anti-proliferation design. Because of general uncertainty in the Carter administration's breeder-reactor program, it was clearly worth hanging onto the bird in the hand (EBR-II) until the bird in the far-western bush (FFTF) was actually on-line and productive. Argonne's EBR-II review committee believed that if the reactor were closed down, however, the government would send unmistakable signals that it was abandoning the breeder-reactor concept, and the Carter administration was not yet ready to fight that politically explosive issue.[56]

DOE's decision to operate EBR-II indefinitely raised morale among Argonne's staff, but it did little to instill confidence about the nuclear future. From Argonne's perspective, Carter's national reactor program was no program at all, but "only a list of things allowed and prohibited."[57] Argonne leaders believed they had a solid breeder-reactor development program, which the administration rejected, forcing them to look at alternatives such as the thorium-uranium fuel cycle. Previous development policy had encouraged reactor economy and breeding efficiency, but under President Carter, design priorities emphasized proliferation-resistant fuel cycles. Fully cognizant of the fact that the Clinch River Breeder Reactor might never be built, but in the absence of clear direction from Washington, Argonne continued its nuclear research programs confident that "at least half" of its effort would prove useful for future nuclear power development. Anticipating that the FFTF would assume the experimental irradiation program, EBR-II operators planned to increase their focus on operational safety, as well as on reactor component and systems research.[58]

Of profound disappointment to the Argonne community was the failure of Senator Charles H. Percy to support the Clinch River Breeder Reactor project. While congressional delegations from Tennessee, Washington, and Idaho fought hard to save the program, Percy favored terminating the LMFBR project, which he believed committed the United States to an inappropriate breeder-reactor design. Although he had been well briefed about the LMFBR program by Argonne officials, Percy did not want to pour funds into a single energy option but, instead, favored research on other breeder options, fusion power, and solar energy. Believing that much could be accomplished through conservation and that uranium supplies were adequate to fuel light-water reactors well into the twenty-first century, Percy saw "no reason to rush commercialization of the breeder." In addition to economic concerns, Percy also shared Carter's worry about nuclear proliferation. He assured Argonne that he continued to favor funding for breeder research, the FFTF facility, and related activities, but he would not vote to continue the Clinch River project.[59]

Frustrated by Percy's lack of political support, Argonne nevertheless enjoyed a temporary reprieve for the breeder program in the so-called McClure compromise. After months of deadlock between the president and Congress, Senator James McClure of Idaho, who single-handedly had blocked the administration's natural-gas deregulation initiative, met at the White House with Carter, Vice President Walter Mondale, and Secretary Schlesinger to strike a compromise on the LMFBR program. In return for a de facto moratorium on the

Clinch River project, Carter agreed to fund the annual LMFBR base budget at about $500 million for three years to include conceptual design studies for a demonstration LMFBR reactor, for fabrication and testing of reactor components, and for enhancement of safety research and engineering facilities (SAREF) at the Idaho National Engineering Laboratory. Although the McClure compromise was never formalized, it reflected the consensus to maintain a viable breeder option. Given the administration's implacable opposition to the CRBR, the McClure compromise, proved a windfall for the EBR-II. Although debate over the Clinch River project was not resolved, EBR-II received a new lease on life when it was included in the SAREF program.[60]

Inevitably, protracted uncertainty in the breeder program not only affected morale but also created problems in recruiting and retaining highly qualified staff, especially experienced reactor operators. Because of the length of time required to train certified operators of EBR-II, their loss threatened Argonne's ability to maintain a full experimental program. Many of the most experienced people at Argonne-West had been in the LMFBR program for almost thirty years. While age and discouragement accelerated personnel attrition, managers reported they were having a difficult time attracting "new, young talented people" to the breeder program. The complaint was universal among the national laboratories, and predictable. Officials at Argonne projected dire personnel shortages if the Carter administration forced the LMFBR program into "hibernation." Alarmed, the GAO estimated that it might be necessary to close many test facilities, resulting in the permanent loss of still more skilled nuclear workers. Defenders of the LMFBR program predicted that Carter's policy could devastate America's nuclear talent, setting the nation back five to ten years in its ability to assemble trained and talented technical personnel.[61]

Three Mile Island

Sometimes small events have very big consequences. By all accounts, the nuclear reactor accident at Three Mile Island (TMI) on the Susquehanna River in south-central Pennsylvania began with what seemed a minor problem. At four o'clock in the morning of Wednesday, March 28, 1979, Metropolitan Edison's Unit 2 nuclear reactor scrammed when a malfunction of the reactor's cooling system automatically "tripped," or shut down, the power plant's huge steam turbine. At first, the reactor's built-in safety procedures worked as planned when the pipes that carried the river's cooling water became unacceptably laden with undissolved minerals and silt. As designed, the main feedwater pumps shut down which, in turn, tripped the turbine. As the turbine slowed, steam pressure rose in the reactor's core cooling system. A pressure-relief valve then opened automatically to allow the steam to escape while, simultaneously, control rods slammed into the reactor's core halting the nuclear reaction.

In the control room, warning lights and alarms alerted the reactor operators of trouble with Unit 2. Although not blasé about the flashing lights and ringing alarms indicating that both the turbine and the reactor had tripped, the operators responded with professional coolness to ensure a safe shutdown of the reactor. The procedure was routine and largely automatic. They had been trained to cope with all anticipated problems and had handled similar situations many times before. Unfortunately, in the early morning of March 28, unexpected small troubles were about to cascade into America's most serious commercial nuclear accident. First, the pressure relief valve did not close as expected, but instead remained open, and undetected, for two hours, allowing radioactive coolant to spill out of the reactor containment building into the plant. Second, control room instruments initially indicated that the emergency core-cooling system was working normally pouring water into the reactor core, when it was not. In addition to the instruments that gave false readings, others were blocked by large yellow caution tags hanging from the control panel or were obscured by the operators themselves.

But the problems were not limited to mechanical and instrument failure. Confused by signals from the instrument panel, which seemed to indicate that too much water had flooded the reactor core, one of the operators took manual control of the reactor. Mistakenly, he slowed the pumps to prevent flooding, when in fact the reactor was running dry because two valves controlling emergency feedwater lines were closed. Eight minutes into the accident the operator realized his error, opened the two emergency values, restarted the pumps, and then watched in dismay when the control panel lit up like a Christmas tree as multiplying problems set off one alarm after another. "I would have liked to throw away the alarm panel," the operator later remarked. The hundreds of flashing red and green lights and insistent alarms provided the beleaguered operator no useful information on the condition of the reactor. Once emergency cooling water was pumped into the reactor, thousands of gallons of water rushed into the reactor core, through the system, and out the stuck pressure release value, soon filling the dump tank and spilling onto the floor of the auxiliary building. By 4:15 A.M., when the Nuclear Regulatory Commission later believed Metropolitan Edison should have declared a site emergency, the reactor operators at Three Mile Island were virtually "flying blind."[62]

As men of action, the control-room operators were not only drilled in procedures to maintain control of a malfunctioning reactor but also schooled not to panic or overreact during an emergency. They were all familiar with the remote possibility of a China Syndrome, a complete meltdown of the reactor core. They had trained to combat a major loss-of-coolant accident and other accident scenarios envisioned by safety engineers. But no amount of training could prepare operators to control a reactor when they did not know what was happening. Unless intentionally built into the response system, it is extremely difficult for trained professionals in any field to declare a general emergency when they do not know the nature and magnitude of the problem with which they are dealing. Premature or unnecessary alarm not only can make an operator look silly, but also can have severe career-limiting consequences. Because they were uncertain about the seriousness of the accident in the predawn hours and because they could not determine when they crossed the threshold from the routine to the extraordinary, the TMI operators were slow to declare a site emergency.

At 5:18 A.M. the first radiation alarms began to sound in the containment building. Momentarily distracted by a false fire alarm, technicians began to monitor radiation readings in the reactor containment building. By 6:30, rapidly increasing radiation levels were detected in both the containment and auxiliary buildings, indicating that the reactor core had suffered serious damage. Radiation monitoring alarms sounded throughout the containment building and, most ominously, in the containment dome, signaling almost certain rupture of the reactor fuel elements. Almost three hours after the start of the accident, Metropolitan Edison finally declared a site emergency at 6:55 A.M.[63]

The call notifying Pennsylvania's civil defense authorities that there was trouble at Unit 2 activated the federal government's interagency radiological assistance plan designed to coordinate state and federal emergency response to radiological accidents. By prearrangement, the DOE's radiological assistance teams across the country were placed on stand-by alert. Argonne's health physics office received its first call Wednesday morning asking that the laboratory be prepared "to respond to a serious incident at the Three Mile Island nuclear power plant in Harrisburg, Pennsylvania." Argonne's radiological assistance team comprised health physicists, specialists in occupation safety, and a member of the electronics division—all trained to respond to nuclear emergencies. Typically, Argonne's team responded to six "incidents" a year over a ten-state area, where they evaluated the seriousness of the radiological hazard, took steps to alleviate immediate danger, and advised local authorities on proper cleanup procedures. Normally, they dealt with slightly damaged shipments or radioactive spills in hospitals and industrial plants.[64]

While Argonne stood ready with reinforcements, radiological assistance teams from Brookhaven and Bettis Atomic Power Laboratory, located closer to Three Mile Island, had already been mobilized and were on their way to Harrisburg. Quickly, aerial and ground measurements detected that small amounts of radioactivity had escaped from the plant, creating minor fallout in the surrounding countryside. On Friday, however, an unplanned release of significant radioactivity into the atmosphere prompted officials to advise local citizens to take shelter. A wailing air-raid siren caused brief panic in Harrisburg. Ultimately, Pennsylvania governor Richard Thornburgh closed the local schools and advised pregnant women and small children to evacuate a five-mile radius around Three Mile Island. With offsite radiation problems apparently becoming more serious and the future uncertain, the DOE requested immediate help from Argonne's radiological assistance team, which was mobilized along with a similar team from Oak Ridge.[65]

A chartered plane rushed Argonne and Chicago Operations Office personnel to Pennsylvania while the main team followed in a specially equipped Winnebago recreational vehicle outfitted with sensitive instruments to detect, measure, and analyze radiological hazards. Arriving at Three Mile Island Saturday afternoon, Argonne's mobile response laboratory was dispatched to the west bank of the Susquehanna River, where the Winnebago set up near the town of Goldsboro, about three-quarters of a mile from Unit 2. Through cold rain and warm fog, Argonne technicians worked around the clock in eight-hour shifts monitoring the weak radioactive plume streaming from the stricken plant. Six times a day they moved through the largely abandoned community to sample air, soil, vegetation, and rain puddles at twenty-two points along the river. Across the river, terrestrial and aerial monitoring continued while scores of engineers and technicians worked to bring Unit 2 safely to cold shutdown. Anxious citizens who remained behind under evening curfew welcomed the Argonnians roaming freely through the town as a reassuring indication that radiation levels could not be too high. Dairy farmers voluntarily offered milk samples to be tested for radioactive iodine. After a week of monitoring in the Goldsboro vicinity, the Argonne team found no iodine, nor any other radioactive residue, at ground level other than the radioactive "shine" of the overhead plume clouds. The public had received a very low radiation exposure, and health physicists anticipated no significant health effects. The main health effects had been stress and anxiety, and even these were expected to be temporary. On April 9, when Governor Thornburgh advised that it was safe for pregnant women and children to return to Goldsboro, Argonne's radiological assistance team also packed for home, satisfied that they had performed their limited mission well.[66]

Argonne laboratory had played a minor, but important, role in containing the most serious commercial nuclear power accident in American history. In the aftermath of Three Mile Island, Argonne assisted the Kemeny commission with assessment of the accident and the Nuclear Regulatory Commission with the evaluation and assessment of plant recovery, including preparation of an environmental impact statement on decontaminating and disposing of radioactive wastes from Unit 2. The study included environmental assessment of the cleanup of the auxiliary building where radioactive water had spilled on the floor, decontamination of the containment structure, opening of the reactor vessel, recovery of damaged fuel elements, decontamination of the primary cooling system, and removal of radioactive wastes. Initially, Argonne estimated that it would take two to four years to cleanup the site.[67]

The far-reaching political and economic consequences of the Three Mile Island accident for the nuclear power industry could not be so easily estimated, however. The Kemeny commission, which was appointed by President Carter to investigate the causes of the Three Mile Island accident, concluded that the accident resulted from a combination of mechanical, institutional, and human failures with operator error being the decisive factor. The Kemeny commission cited deficiencies in operator training and emergency procedures, but its most serious finding from the perspective of Argonne safety engineers was the allegation that reactor-operator interaction during the accident was demonstrably faulty. The reactor had no will of its own, unlike HAL in *2001: A Space Odyssey*, but the interface between man and machine had proven needlessly complex and unreliable. Communication between the reactor and operators ultimately failed when key instruments gave false or ambiguous readings, and numerous warning lights, alarms, and plant signals distracted and confused technicians with information overload.[68]

Robert Avery, Argonne's leading authority on reactor safety, and his committee of reactor experts at Argonne assessed the Kemeny commission's findings. The Avery committee acknowledged that a major lesson of TMI was that "operator intervention is not necessarily always a force to the good, but [could] make the situation worse." The solution, however, was not to improve the machine-operator interface but, rather, to make the reactor as "people-proof" as possible. Some of the most serious problems at TMI might have been averted if it had been impossible for the operator to interfere with the safety systems. The good news from Three Mile Island was that no one, neither TMI personnel, emergency workers, nor the public, was seriously injured in the reactor accident. The best news, according to Argonne specialists, was that the reactor design—the containment—had actually "saved the day at Three Mile Island."[69]

For years, the government's nuclear-safety research and estimates had been dismissed by critics as too optimistic and self-serving.[70] The debate over nuclear reactor safety initially focused on the light-water reactor and, prior to

Three Mile Island, had culminated in the publication of the Rasmussen report in August 1974. After a comprehensive three-year study on risks to the public from large nuclear power plants such as Unit 2, sixty scientists and engineers led by MIT professor Norman Rasmussen, concluded that risks from nuclear reactor accidents were small in comparison to other risks to the public in America's technological society. Using a "fault-tree analysis" to predict the effects of mechanical failures in large, complex systems, the Rasmussen group evaluated thousands of computer-simulated reactor failures and concluded that the "maximum credible" accident would occur only once in every ten million operating years. With fewer than two hundred nuclear reactors operating worldwide, a catastrophic reactor accident seemed unlikely. Lesser accidents, of course, could be expected with greater frequency. Before the Three Mile Island incident, critics claimed that the Rasmussen report minimized both the likelihood and the dangers of a nuclear reactor accident. Ironically, the Kemeny commission found the Rasmussen report prophetic for projecting the probability of the "minor" type of accident, such as occurred at Three Mile Island.[71]

For Avery and his committee, Three Mile Island underscored Argonne's need to advocate expanded safety programs for both light-water and breeder reactors. In addition to enhanced studies of reactor core meltdown (Argonne was already helping to assess TMI core damage), the committee proposed risk-analysis studies of the light-water system through the entire fuel cycle. The Kemeny commission criticized the government for excessive concentration on emergency procedures to combat a sudden, massive loss-of-coolant accident. The Argonne committee agreed that more study was required on the dynamics of slowly evolving incidents, such as that at Three Mile Island, to analyze the kinds and quality of signals that operators received. In this regard, the committee suggested that Argonne's computer scientists develop two related reactor-safety programs: the first would train operators to cope with TMI-like accidents in a simulated control room; the second would provide operators with control-room computer programs that determined the appropriate actions to take during abnormal situations and perhaps even interdicted the operator's actions when necessary.

Nuclear Reactor Safety Research

Because the light-water reactor did not have a long-term future, in Argonne's view, the most pertinent TMI lessons should be applied to the breeder program, for which the laboratory was already engaged in extensive safety studies at Argonne-West. Reactor-safety research traditionally pursued three tasks: the design of safe reactors, the development of safety systems to minimize damage during reactor malfunctions, and the analysis of accident data indicating the success or failure of the first two goals. These objectives included studying reactor physics, modeling the causation and consequences of reactor accidents, and testing the accuracy of the mathematical models and analytical predictions. Based on their experience, Avery's group believed that the breeder might actually be less susceptible to operator error in a slowly developing incident than was the light-water reactor under similar circumstances.[72]

The fast-breeder-reactor safety program, a lead mission for Argonne after 1980, emphasized four objectives—called "lines of assurance" (LOA)—for safety research and development. The lines of assurance organized research into logical program categories: accident prevention, limitation of core damage, maintenance of containment integrity, and attenuation of radiological hazards. The first line of assurance objective, for example, would establish that breeder reactors could be "designed, constructed, and operated" with an extremely low possibility of accident. Hopefully, the other objectives would demonstrate that with breeder technology, even when stressed by accidents, damage would be minimal and danger to the public nil.[73]

Argonne's main experimental reactor devoted to LMFBR safety research was the Transient Reactor Test Facility, the twenty-year-old veteran located at Argonne-West. TREAT simulated potential breeder accidents under actual operating conditions, some of which were first modeled by Argonne's computers. Safety engineers believed that pumping sodium through stainless-steel loops containing fuel capsules provided a realistic, but controlled, environment to test probable breeder accidents. In more than six hundred experiments simulating two thousand transients (or power surges), TREAT demonstrated the effects of loss of coolant flow in sodium loops, the dynamics of fuel failure in a reactor transient, and the consequences of fuel failure in a loss-of-flow accident. Using a device called a neutron hodoscope, which used neutrons much like an x-ray machine used x rays, Argonne engineers were able to "see" into the reactor's core to watch the behavior of fuel elements during a simulated accident. In a particularly important test, TREAT simulated an operator or mechanical error in which the controls rods were accidentally pulled from the reactor core. Encouragingly, coolant flow was not interrupted while the fuel dispersed from the center of the core, suggesting that even under the most severe conditions, the breeder reactor would probably shut itself down. Under the SAREF program, TREAT would be significantly enlarged to test fuel assemblies for the FFTF, the CRBR, and the British prototype fast reactor (PFR).[74]

EBR-II also offered a superb facility for expanded safety research under the SAREF program. The accident at Three Mile Island not only raised general questions about the future of nuclear power but also rekindled interest in the inherent safety of breeder reactors such as EBR-II. The Kemeny commission criticized the Department of Energy for its overemphasis on studying "low probability-high consequence" accident scenarios and encouraged the government to increase its research on minor, low-consequence incidents that nonetheless had a high probability of occurrence. Although continued analysis of catastrophic accidents remained important, EBR-II proved an ideal reactor with which to study mechanical problems that were most likely to happen during routine operations. In addition to being a research reactor, EBR-II normally produced 70 percent of the electricity used at Argonne-West, or about 40 percent of the power consumed by the entire Idaho National Engineering Laboratory (INEL). In 1980 DOE designated EBR-II a "cogeneration facility" because the reactor also produced 60 percent of the steam that heated Argonne-West's buildings. Although EBR-II was frequently shut down to change experiments and load fuel, for five years the reactor had operated at 70 percent capacity, a record that Argonne believed compared favorably to commercial power plants.[75]

EBR-II: The "Self-Protected Plant"

By 1980 the United States' only operating fast-breeder reactor became the prototype of a "self-protected," inherently safe power reactor. In the aftermath of Three Mile Island, engineers renewed interest in the pool-type, sodium-cooled design. EBR-II's reactor core and the major system components, including the pumps that maintained coolant flow, sat submerged in a large pool of molten sodium operating at about 700°F (370°C), well below sodium's boiling point (figure 33). Because of the thermal hydraulics of its basic design, in which the sodium provided extensive cooling should the core overheat, Argonne's engineers believed it was not possible for EBR-II to suffer an accident similar to Three Mile Island. What would happen if EBR-II totally lost its primary and emergency cooling pumps? Project engineers believed that if the coolant flow were interrupted and the reactor shut down as designed, no damage could occur in the reactor core. But even if all emergency systems failed and the reactor continued to operate, two inherent thermal characteristics of the reactor would protect the core from serious damage. First, even with coolant flow interrupted, natural convection currents in the liquid sodium would maintain sufficient cooling to protect the reactor. Second, as the sodium coolant heated, it would also expand and allow neutrons to escape from the reactor core. With neutrons fleeing the core, power levels would automatically drop enabling the reactor to cool somewhat. In a reactor accident of this sort, everything would be exceed-

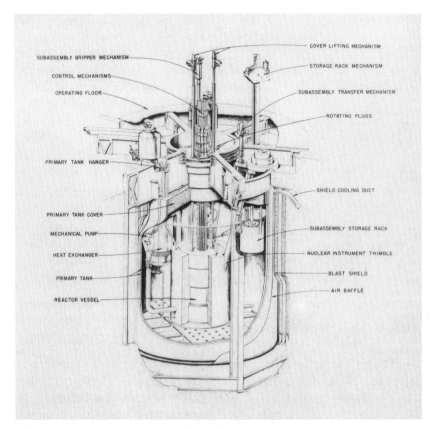

Figure 33. The primary system of EBR-II in cutaway perspective

ingly hot, but well below the levels for serious core damage or threat to public health and safety.

The differences between this breeder-reactor scenario and the Three Mile Island light-water reactor accident were dramatic. The Unit 2 core was cooled by ordinary water under very high pressure, while the EBR-II core sat in a large pool of liquid sodium, where it was cooled under comparatively low pressure. At Three Mile Island, the stuck pressure release valve allowed water to flash to steam and create a sudden loss of coolant. The resulting vapor lock, which blocked the coolant flow, exposed the reactor core and caused the reactor to overheat. With EBR-II, in contrast, the sodium coolant was under such low pressure that loss of coolant through a stuck valve would produce little, if any, sodium vapor. During the accident at Three Mile Island, Argonne's emergency response team at Goldsboro carefully monitored for radioactive iodine gas released into the atmosphere from ruptured fuel pins. If breeder fuel assemblies were to rupture, Argonne engineers predicted that any iodine released from the damaged fuel pins would react chemically with the sodium coolant and not escape as gas into the atmosphere. In order to avoid severe core damage should the primary cooling system fail, light-water reactors require both rapid reactor shutdown and an emergency core-cooling system. The EBR-II, on the other hand, would suffer only minimal core damage, with no risk beyond containment should the plant lose coolant flow without being able to shut down the reactor. If the contrast between a light-water reactor loss-of-coolant accident and a breeder reactor loss-of-coolant-flow accident seemed like comparing apples and oranges, Argonne's safety engineers believed the comparison was fair because of the virtual impossibility that a pool-type reactor like EBR-II could lose its sodium coolant. Skeptics might wonder about the government's assurances, but the Argonne staff believed that EBR-II provided an excellent model for an inherently safe, "people-proof" reactor.[76]

Continued FFTF construction at Hanford, Washington, required keeping EBR-II on-line to prolong irradiation tests of materials and fuels and so create a data bridge between EBR-II and the FFTF. Meanwhile, Argonne's Operational Reliability Testing program (ORTP) and related safety research steadily moved center stage as EBR-II's principal research-and-development mission. The ORTP involved a variety of safety experiments with transient power bursts in the reactor core, loss of coolant flow and related heat decay in the reactor vessel, and the so-called run-beyond-cladding-breach tests, which evaluated the effects of fuel-cladding failure and chemical reactions between the damaged fuel elements and the liquid-sodium coolant. The breach tests demonstrated that cladding breaches and fuel failures could be detected, located, and accommodated. The safety tests usually posed minor problems in each area that gradually were amplified in intensity and duration as EBR-II simulated increasingly serious accidents. These mechanical experiments incrementally gathered data on breeder-plant dynamics.[77]

The Three Mile Island accident stimulated renewed "philosophic" debate on the principles of reactor safety. Argonne designers favored passive, self-protective safety systems built into the reactor design: systems that relied upon the ordinary laws of nature and not on extraordinary actions of operators to ensure reactor safety. But even the most inherently safe reactor required interactions between humans and the machine at some operational level. Much had been learned about the science of ergonomics—the study of human interface with machines—with the advent of the space age. Space and aeronautical engineers had worked especially on designing ergonomically friendly cockpits in space ships, military airplanes, and commercial airliners. In modern cockpits, jammed with flight controls and instrumentation, the ergonomic puzzle was to fashion avionics that operated within the tolerances of human attention, comprehension, reaction, coordination, strength, comfort, and endurance. After the control-room nightmare at Three Mile Island, when the instruments, warning lights, and alarms confused and stressed operators rather than helped them during a slowly evolving event, the nuclear industry realized it faced a similar ergonomic challenge. In this regard, Argonne's specialty was the design and testing of reactor instrumentation.[78]

In 1981, after extensive study, EBR-II began transient tests as the first step of a comprehensive program that combined loss of coolant flow, breached cladding, and transient testing in the ORTP. Nearly thirty operators and observers crowded into the control room on February 9 for the first test of a major power burst on EBR-II. Setting the reactor's power level at 24 megawatts throughout the tests to ensure uninterrupted supplies of power and steam for Argonne-West, in less than a minute EBR-II's operator rapidly ran the reactor up to more than 62 megawatts before powering back. In February and March, EBR-II performed fifty-six transient tests in which "the reactor's power was run up and down like a yo-yo." Sometimes the transient tests were performed three times a day during which EBR-II was run up to 62 megawatts and held there for twelve minutes before running back down to 24 megawatts. If this stress test does not seem terribly serious, something like repeatedly red-lining the rpm tachometer on an Austin Healy Sprite, the idea was to learn about the performance of reactor fuels and components under accident conditions most likely to occur in the real world.[79] Ultimately, EBR-II tested metallic uranium and experimental uranium-plutonium oxide fuels in transient bursts that ran the reactor up to 62 megawatts in just six seconds.[80]

Massey Takes Command

Walter Massey officially became director of Argonne National Laboratory on July 9, 1979, during a period of profound change at the laboratory. In May, just after Three Mile Island but just before Massey arrived at Argonne, Deutch conducted a general review of Argonne. Ostensibly the DOE carried out the review as a prelude to renewing the Tripartite Agreement in October. But no review of this sort, which undertook a comprehensive assessment of Argonne programs and activities, was ever routine. Headed by James Ling from the Office of Field Management, the DOE team included Deutch's assistant, Douglas Pewitt, among the reviewing officials from Washington.[81]

In general, the review team rated Argonne's performance "above average when compared to the broad range of federally supported laboratories" with which the reviewers were familiar. What the "above average" rating really meant was impossible to determine, however, because there was no objective frame of reference or baseline data against which to compare Argonne's performance. The review team acknowledged that Argonne had strong applied and technical programs but believed that Argonne's management wanted basic research to be the laboratory's principal mission. "The prevailing environment is more academic than industrial," they observed with thinly veiled criticism.[82]

In the all important nuclear energy program, the DOE stated that Argonne's technical performance was "above average," although the government believed that Argonne's staff was too academic and the AUA reviewers were not fully qualified to assess Argonne's technical and engineering projects. In addition, the DOE believed that the nuclear energy program lacked new talent and fresh ideas, while the laboratory only reluctantly cooperated with headquarters in defining program objectives. They noted that Argonne management was not committed to quality assurance, nor was the laboratory effective in securing international exchange of technical information. Finally, the DOE charged that the management of the nuclear energy program was both top-heavy and excessive. If the Department of Energy believed that this represented "above average" performance, one can only wonder what "average" or "below average" programs must have looked like. Overall, the department concluded that Argonne needed more vitality and urged the University of Chicago and the AUA to accept responsibility for invigorating the laboratory.[83]

Massey masterfully used the review to consolidate his new position as director of the laboratory while securing support for his administration from DOE headquarters. Skillfully, he distanced himself and the laboratory from the DOE's harshest criticisms by pointing out that Ling's committee had evaluated the old laboratory management, not the new. On the other hand, he embraced the committee's positive assessments, promising to improve Argonne's already strong programs. While affirming many of the Ling committee's evaluations and recommendations, Massey conceded nothing substantial to the Department of Energy. Most importantly, Massey carefully chose his ground for discussions with Deutch, knowing what he had to concede, what he had to promise, and what he could reasonably expect to achieve.

Diplomatically, Massey praised Ling's committee "for carrying out a difficult task in a professional and competent manner." He refused to denigrate Deutch's reviewers but, rather, announced he would use the Ling report as a "source document" to identify areas at Argonne requiring his attention. "My overall impression," he wrote Deutch's deputy, "is that the report correctly identifies many areas in which the laboratory's performance could be improved." He continued firmly, however, that the report also emphasized the negatives at Argonne, which might incorrectly leave the impression "that our problems outweigh our strengths."[84]

Massey bridled at the insinuation that Argonne's "academic atmosphere" hindered effective research and development. The former dean from Brown defended the laboratory's "academic atmosphere" if that meant that "creativity, innovation, and serendipity" were the norm at Argonne. If the government's innuendo implied that Argonne lacked disciplined, hard-hitting staff committed to meeting deadlines, Massey asked the DOE to look forward with Argonne's new management rather than backward on the old. "New ANL management does not perceive the fundamental mission of the Laboratory to be basic research," he categorically announced. Carefully, Massey crafted a succinct mission statement:

> Argonne's role is to apply its broad-based scientific and technological resources to the timely solution of problems of energy supply, energy utilization, and environmental protection. We fulfill this role through a variety of programs including those both in basic physical sciences and engineering and in energy technology and development programs. Basic research is an indispensable, but not predominant, part of our mission.[85]

Writing personally to Deutch, Massey outlined his plan for the laboratory. Recalling their earlier conversations, he agreed with Deutch that Argonne's basic research programs were not as good as they should be, or as they could be. Ironically, given the DOE's earlier review and Massey's response, both men actually wanted improvement in basic research to be a top priority, but for external and internal political reasons, they knew they would have to move quietly in this area. From his academic perspective, Massey advised Deutch that the quality of basic research at Argonne was dependent on the quality of scientists the laboratory could recruit and retain. Massey's goal was to attract "first-rate" people in fields where Argonne was weak, and then only to promote truly outstanding scientists.

Argonne's applied programs provided a different kind of challenge. On the whole, Massey speculated that the personnel in the applied programs were better than those in the basic programs. Attracting excellent individuals was a lesser problem than securing first-rate "leadership, teamwork, and coherency of programs." "What we need most on the applied side," Massey conceded to Deutch, "is a sense of direction and a feeling of vitality." Here Massey's top priority was to fulfill his promise to Deutch to hire a deputy director who could provide leadership in the technical fields from nuclear to fossil energy. On October 31, less than four months after Massey became laboratory director,

Argonne's deputy director of operations, Robert Laney, retired after nine years of service.[86] Thereafter, Massey moved quickly to hire Eric S. Beckjord for the newly created position of deputy director for science and technology. As the director of the headquarters division of reactor development and demonstration, Beckjord had managed the LMFBR program for the Energy Research and Development Administration until 1977. He had also served as the DOE's director of the Division of Nuclear Power Development, where he managed near-term development programs. Beckjord, who came to Argonne on March 30, 1980, after serving as coordinator of the U.S. International Nuclear Fuel Cycle, an international study to reduce risk of nuclear proliferation from commercial nuclear power, brought an impressive balance of government experience and technical expertise.[87]

Perhaps Massey's toughest job was to polish Argonne's image at DOE headquarters while laying to rest old myths about the laboratory's lackadaisical management. Institutionally, ingrained perceptions are hard to change even when there is no objective evidence to support them. Exasperated by Argonne's reputation among Washington bureaucrats for a lack of responsiveness to program direction, Massey tried to ferret out the sources and origins of persistent criticism of the laboratory. Not surprisingly, the trail led to the nuclear reactor programs, where Massey discovered long-held grudges, some more than fifteen years old. John W. (Jack) Crawford, an old Rickover hand working in the headquarters nuclear energy division, helpfully opened his files and laid out before Massey the sorry story detailing years of infighting between headquarters and Argonne over direction of the breeder program. Massey was shocked. Everything was documented, nothing was forgotten. Yet even in Crawford's office, Massey shared the perspective of his immediate predecessors. The conflict had less to do with Argonne's lack of responsiveness than it did with the fact that Argonne's engineers honestly differed with headquarters bureaucrats about the appropriate technical direction of the reactor program.[88]

Determined that he would not get caught up in cyclical rounds of recriminations, Massey confronted Deutch directly with the rumors. Boldly, Massey demanded documentation that Argonne lacked responsiveness, that the laboratory would not or could not manage programs, and that it was indifferent to quality assurance. Massey himself tried to track down the complaints and document them. "I must say," he reported to Deutch, "I have been less than successful in tracing these complaints." The quality assurance issue, a long holdover from the Milton Shaw years, was a case in point. Evaluated independently, but at about the same time, by a quality assurance task force from the DOE, nuclear energy programs at Argonne-West received good marks including praise for "a high degree of competency. . . ." On the other hand, the Ling committee's comments that Argonne was not "committed to quality assurance" were based on unsubstantiated conversations with headquarters staff. Massey assured Deutch that he would continue to look into the matter, but he wanted facts, not innuendo.[89]

Astutely, Massey affirmed Deutch's basic criticism that the laboratory needed direction and vitality. But even on this point, he did not let Deutch completely off the hook. A major source of Argonne's problem was in Washington, not Chicago or Idaho. How could Argonne strike boldly ahead on the breeder-reactor project when nuclear power policy was in disarray in Washington? "You realize, of course, Jack," he jabbed at Deutch, "that much of what may appear to be lack of direction or lack of vitality in our nuclear programs stems from the national policies on nuclear energy." Argonne's major problems in this regard were not technical, but psychological. Massey accepted the challenge "to maintain morale, commitment and creativity in a national atmosphere of uncertainty and lack of direction." But the challenge was no less his than the DOE's. "In this respect, I need your help," he confronted Deutch.[90]

Reportedly, Massey had turned Deutch around "90 percent." Deutch was encouraged by Massey's candor about Argonne's problems and he believed that Massey's "drive for excellence at Argonne" was exactly what the laboratory needed to restore its sense of purpose and vitality. "I hope you have unpacked your winter clothes," Deutch teased Massey, "because you will need them before this job is done."[91]

The Future of the Laboratory

Massey also asked Hanna Gray for more direct support and involvement from the University of Chicago. Massey was blunt with Gray, who had played the major role in recruiting him to Argonne. In contrast to the close relationship between the University of California and the Lawrence Berkeley Laboratory, Massey did not believe the University of Chicago really cared whether Argonne was an excellent laboratory or not. He did not accuse the university of neglect but observed that there was not the same kind of pride and identification with Argonne Laboratory as there was with the university's medical and law schools. Consequently, the university did not make equivalent demands of excellence on Argonne as it did on its medical and law schools. Of course, Argonne did not enjoy the same relationship with Chicago as the Lawrence Laboratory did with the University of California. Argonne was not an intimate part of the university like its California counterpart but, rather, was both physically separate from the campus and managed in concert with another corporation, the

AUA. The geographical distance was a minor problem in promoting closer ties, but the Tripartite Agreement, which was extended for another three years in September 1980, importantly stood in the way of Massey's vision to make Argonne Laboratory an integral part of the University of Chicago.[92]

William Cannon sympathized with Massey's frustration. He admitted that throughout the years (and especially since the tripartite arrangement) the university had almost despaired over the question of how to build up "a first class laboratory." Cannon also speculated that perhaps the university, influenced by the initial success of Walter Zinn, had placed too much reliance on the laboratory director. That approach had not worked well and was analogous to placing all of the responsibility for the success of the college in the hands of the dean. Massey could not miss Cannon's point. Chicago supported Argonne over the years not because the laboratory provided the university unique opportunities for basic research, but because the laboratory enabled the university to participate in solving a major social problem, that is, developing major energy resources for the United States and the world. Cannon, who had always believed that Argonne's first priority should be to serve its patrons—the Department of Energy and its predecessors—endorsed Deutch's goal to upgrade Argonne's scientific and technical personnel. Hiring Beckjord was a good start, and Cannon looked forward to additional appointments and promotions to key technical positions in the breeder-reactor program.[93]

The budget news from Washington, however, was far from encouraging. Carter's budget cuts again targeted the breeder program. Although the White House planned a modest increase in Argonne's basic science and alternative energy programs, the president's budget would cut nuclear programs by 22 percent and slash breeder development by 50 percent, zeroing out the program by 1982. Massey expected Congress to restore at least partial funding for the LMFBR, but no one could predict how Carter's budget would ultimately affect the laboratory. Fearful that the breeder program might be saved but crippled, Massey was also worried that the ensuing horse trading in Congress might also cut back the modest gains in basic research, resulting in losses for everyone.[94]

Massey's smile seemed genuine and his handshake warm, but his photo opportunity with President Carter enabled him to lobby the beleaguered president on behalf of Argonne's nuclear reactor programs. Other guests at the President's White House luncheon had raised tough questions about the economy and the American hostages in Iran. Massey thanked Carter for his increase in funding for basic research but expressed concern for America's energy future. "I told the President," Massey reported later, "I believe our energy policies will require a strong research and development base in nuclear fission." Carter's wan smile offered no concessions, but Massey, honing his political skills, knew his encounter with the president would be well received in the *Argonne News*.⁹⁵

In 1980 the Department of Energy once again pushed hard for each of the national laboratories to define their roles and missions. But with the national energy policy and related budgets so volatile, Massey questioned publicly whether the nation as a whole, let alone the national laboratories, could plan intelligently. The continual tug-of-war between the Carter White House and Congress over national science and energy policy and related funding kept everyone at the laboratories off-balance and uncertain about the future. The Carter administration especially emphasized the importance of planning, and each year conducted an "on-site program review" not only to assess the past year's accomplishments, but also to discuss Argonne's future plans. Despite fluctuating budgets and the uncertainty of the approaching presidential election, the DOE informed Massey that a major objective for 1980 was to develop a *definitive* role statement and five-year plan for Argonne.⁹⁶

Massey, who was quickly learning the ropes, accepted the challenge good naturedly and then settled down to wait and see. In mid-summer, he told the Argonne community that "we will not change drastically the mix of technologies in which we are now engaged, nor the distribution between applied and basic research." Massey believed that any realistic plan for Argonne must build on a stable nuclear program that focused on reactor analysis and safety, reactor physics, and components development. EBR-II would most likely be the centerpiece in reactor research. Basic research should emphasize existing strengths in nuclear physics and materials science that were shaped to support applied efforts. Growth areas for Argonne no doubt would include fossil fuels, solar energy, conservation, and environmental research. He was willing to develop a regional mission for Argonne, building on the research-and-development strengths of the Midwest universities and industries, but assured everyone that these plans were tentative. In planning, Massey emphasized the importance of "mutual obligation" between the department and the laboratory. The laboratory must be responsive to the department's priorities, but at the same time the department should have the responsibility to continue programs at Argonne once the government had made a commitment. Who could miss the message? Argonne would wait until after the 1980 presidential elections before undertaking any major reprogramming.⁹⁷

Three Mile Island had changed everyone's perspective on the future of nuclear power. At Argonne, other changes were more prosaic but no less significant. Aging facilities did not run forever. On September 28, 1979, CP-5, Argonne's old heavy-water, enriched-uranium research workhorse, finally retired after twenty-five years and 536 million kilowatt hours of operation, during which it irradiated almost 27,000 samples. CP-5, a direct lineal descendent of the Manhattan Project's first reactor, CP-1, had provided abundant neutrons for studying the structure and behavior of a great variety of materials, both solid and liquid.[98] Three days later on October 1, 1979, the Zero Gradient Synchrotron permanently ceased operation. By Thanksgiving morning, the ZGS's 107-ton superconducting electromagnet completed its seventeen-day trip to the Stanford Linear Accelerator Center. These large-scale, sophisticated facilities, used by university as well as laboratory staff, had played a key role in fostering basic research while developing the laboratory's technical base. The government terminated both facilities without provision for replacing them with new large research projects. Alarmingly, under Carter's projected budgets for fiscal year 1981, the laboratory faced static or reduced funding across the board.[99]

Massey not only reflected on what these changes meant for Argonne, but also on what they meant about the state of American science and technology. For more than a decade, scientists, engineers, and industrialists had worried about the relative decline of federal funding for scientific research and development and the consequent slippage of America's international leadership in scientific and technical know-how. It seemed that the best young minds no longer flocked to nuclear sciences or engineering. Argonne bore the brunt of some of this decline and watched angrily as the French appeared to forge ahead in breeder technology, with the Japanese and the British not far behind. Was America losing its capacity for innovation? Was the federal government adequately supporting research and development? Had Americans become disillusioned about science and technology?

Massey was uncertain about the answers to these large questions, but he believed that if discouraged scientists unilaterally withdrew from the public arena, the questions themselves would become self-fulfilling prophecies. Scientists, engineers, and technicians had a public responsibility to define the role of educating the public about "both the realistic benefits *and dangers* of technological development." The laboratory's Center for Educational Affairs established in 1968 provided broad-based educational and research programs for students and faculty from high school through graduate school. But Massey's vision was much larger. Without becoming "glib-promoters of a scientific-technical nirvana," he wanted Argonne to educate the public, in language the public could understand, about the pros and cons of major scientific and technical issues confronting the United States. For too long, Massey warned the laboratory, scientists had taken public support and esteem for granted. "We must now begin to rely less on the *public's faith* in science and technology," he concluded, "and more on their *understanding* of science and technology" if Argonne were to continue to prosper as a public institution.[100]

In fall 1980 Massey named Charles E. Till as associate laboratory director for engineering research and development, placing Till in charge of Argonne's fast-breeder-reactor program. Till's promotion not only helped Massey fulfill his promise to Deutch and Cannon to advance the best people to technical leadership, but it also reflected Massey's determination to find effective public educators for top management positions. A Canadian, Till had been named director of Argonne's applied physics division in 1973 after only ten years at the laboratory. Holding a Ph.D. in nuclear engineering from the University of London and an M.B.A. degree from the University of Chicago, Till was among Argonne's most dynamic reactor engineers. Till thought he was Massey's first choice for deputy director, Beckjord's job, but believed he was vetoed in Washington because he was too pronuclear for the Carter administration. Indeed, not since Walter Zinn had Argonne found a more effective manager and advocate for nuclear power. As will be seen, under Till's leadership, Argonne actually strengthened its reactor research-and-development programs during a period of major budget cuts and retrenchment.[101]

The election of Ronald Reagan as president of the United States in 1980 further confused prospects for breeder-reactor research at Argonne. On the one hand, the Reagan administration was known to be "pronuclear," committed to vigorous promotion of the nuclear energy option. On the other hand, the administration also wanted to slash federal spending by abolishing the DOE and many of its research-and-development programs. Massey, who had no crystal ball before Reagan's inauguration, advised the Argonne community that it was not clear what the new administration really intended when Reagan pledged to "abolish" or "dismantle" the Department of Energy. Political rhetoric aside, Massey did not believe the Reagan administration would be "anti-science and technology." The Reagan Republicans wanted to curb the DOE's regulatory activities and cut back large demonstration projects, but there seemed to be consensus that the national laboratories should pursue long-range, high-risk research and development. Massey anticipated a close scrutiny of Argonne's budget and mission by the new administration, but such reviews were now commonplace. As Argonne prepared to celebrate its thirty-fifth birthday, Massey anticipated significant reorganization under Reagan, but stopping well short of dismantling the department.[102]

11 | On the Brink, 1981–84

Hattiesburg mayor Bobby Chain, the key to the city in hand, smiled broadly when he congratulated Rowan High School's newest graduate. Twenty-five years earlier, following his sophomore year, Walter Massey left high school to pursue an accelerated science program at Morehouse College. On April 9, 1981, Mississippi's gentle spring welcomed Argonne's director home to receive his high school diploma before a proud congregation including the mayor, the president of the local university, civic leaders, and ordinary students and folks gathered to honor one of Hattiesburg's most illustrious sons. Although Massey's diploma, unanimously voted by the school board upon the recommendation of the former high school principal, represented a personal tribute to the genial science administrator, the award also celebrated America's continued faith that education remained the surest path for minority advancement.[1]

American science leaders, however, were increasingly worried about the state of science education in the United States. For more than a decade Scholastic Aptitude Test scores in math and science had declined alarmingly. Frank Press, the president of the National Academy of Sciences, convened an unprecedented conference to discuss precollege education in science and mathematics. While educators assumed that science education in American high schools lagged woefully that in behind the Soviet Union, George Keyworth, President Reagan's science advisor, thought the nation's declining industrial competitiveness with Japan and West Germany was a more serious problem. Observing that the National Science Foundation budget for science education had been sharply cut from $100 million to less than $20 million, Massey asked how Argonne could assist precollege science education. In the summer of 1980, Argonne launched a six-week summer course to encourage minority high-school students to study mathematics, science, and engineering. Through the "adopt a school" program, Argonne cooperated with local industry to assist with upgrading science education in the Chicago public schools. Massey wrote about the crisis in American science education, saying "I feel very strongly that it is a legitimate part of Argonne's mission to play a central role in the effort to improve the quality of pre-college science education."[2]

Massey's belief that Argonne had a special responsibility to foster science education reflected more than an educator's commitment to academic excellence. Long gone were the days when the laboratory toiled in secrecy on the Manhattan Project or focused narrowly on nuclear reactor development and basic research in the physical sciences. Since the environmental movement of the 1960s and the energy crisis of the 1970s, the national laboratories increasingly served as vehicles for social change as well as centers for scientific research and development. As recipients of hundreds of millions of federal dollars, these federal institutions mirrored the social, economic, and political agendas of their public sponsors and their employees. With high hopes and occasional shortcomings, Argonne became a showcase for federal affirmative action programs hiring women, minorities, and the handicapped.[3] Additional initiatives hosted blind and deaf students in Argonne's 1981 Physically Handicapped in Science Program. Environmental research encouraged ecological awareness on the Argonne campus and the surrounding community. Foreign-exchange students sponsored by the American Nuclear Society fostered international cooperation. Not surprisingly, Argonne employees were also leaders in more than eighty communities in the Chicago area. They were most active in church, youth, and service organizations, but Argonne employees also provided leadership in educational, political, professional, and social-reform groups.[4]

Directly and indirectly, the laboratory became a major local educational, social, and economic institution, a fact that the federal government occasionally recognized explicitly. For example, Representative Cardiss Collins (D-Ill), a member of the House Committee on Energy and Commerce and the chair of the Congressional Black Caucus in 1979–80, warmly congratulated Massey for receiving the DOE's minority business award. By contracting with minority businesses, the laboratory strengthened all aspects of the minority community. The award praised Argonne for exceeding its goal by signing more than $2.5 million in contracts with minority-owned business—over 3 percent of all contracts awarded in 1979.[5]

Ironically, while the laboratory's community-based educational programs expanded at all levels, Argonne's mission came under close scrutiny from both the federal government and the Argonne Universities Association. On October 1, 1980, the Department of Energy extended the Tripartite Agreement for only three years, an indication that no one was happy with the arrangement. Although the closing of the ZGS in 1979 did not foreclose continued participation in Argonne management, once the synchrotron was shut down, the continued role of the AUA became increasingly problematical.

The AUA Quits

Henry Bohm, the AUA president selected in April 1978, did not believe that the Department of Energy, the University of Chicago, or Argonne's laboratory directors wanted the AUA to be an effective partner. Bohm knew that John Deutch, DOE's director of research, intended to set national laboratory priorities, but he was discouraged that the AUA played almost no role in helping the energy department establish policies or review programs at Argonne. By the same token, Bohm doubted that the University of Chicago wanted a strong AUA to share management responsibilities under the tripartite contract.[6]

Bohm's frustration came to a head during Argonne's search for a new director in 1979. Bohm believed that the laboratory's directors had followed the lead of the university and the government by minimizing the management role of the AUA. According to Bohm, Argonne's directors were historically content with an inactive AUA. He was grateful that the AUA had been asked to approve the acting director, but Bohm felt that the association had been ineffective during the search for Robert Sachs's replacement. Although the AUA first identified Walter Massey as a candidate for director, Bohm believed it had failed to influence the process, complaining that the association simply did not have the "clout" to influence the recruitment process.[7]

Shortly after his appointment as Argonne's director, Massey met with AUA trustees to discuss the future of the laboratory. When he asked what could be done to improve relationships between the laboratory and midwestern universities, the meeting degenerated into a general gripe session. The AUA trustees complained that they were ignored and isolated and had no role to play in planning and spending at Argonne. With the closing of the ZGS and the decline of the reactor program, Argonne did not appear to have "a *major* lead role in a *major* DOE program." Argonne seemed to have evolved into a diffuse, amorphous, multiprogram laboratory, "which would do anything as long as there are dollars." Board chairman George A. Russell, the chancellor of the University of Missouri–Kansas City, summarized the AUA's assessment of Argonne management by noting that "the laboratory has not been a place which has generated big, new scientific ideas." Russell's harsh judgement reflected the AUA's discontent with Massey's predecessors, but the trustees offered few constructive ideas on which Massey could act.[8]

By 1980 the Argonne Universities Association saw that Argonne was losing ground to rival national laboratories but believed that there was little the midwestern universities could do to reverse the decline of "their" lab under the Tripartite Agreement. The pact was irreparably flawed because Argonne's management could not be divided between "two separate, independent and dissimilar institutions"—the AUA and the University of Chicago. No other government-owned, contractor-operated laboratory was encumbered with bifurcated management. From Bohm's perspective, neither the association nor the University of Chicago had sufficient leverage to check the DOE's bureaucracy, which tried to run the laboratory from Washington. Ominously, Bohm reported to the board that the congressional General Accounting Office would investigate Argonne's Tripartite Agreement as a "typical case-study" of national laboratory management. No one was fooled by the GAO's euphemism. Argonne was to be singled out for congressional scrutiny.[9]

On July 15, 1980, in the midst of a presidential election year and in the face of the Accounting Office's investigation, the AUA board of trustees asked Bohm to assess the future of the Tripartite Agreement. The board's request was not naive because it expected Bohm to oppose continuation of the contract beyond 1983. Just two days before Christmas, Bohm recommended scrapping the AUA's partnership with the University of Chicago. Discouraged that he had helped neither the laboratory nor the midwestern universities, Bohm felt a deep sense of failure. He had been hired in 1978 to strengthen the AUA so that the contract functioned as intended. He now concluded that the interests of Argonne, science, and the universities would be best served if the Tripartite Agreement were terminated and the stewardship of the laboratory were consolidated under a single contractor.[10]

The University of Chicago was aghast at the association's move to terminate the Tripartite Agreement three months after its renewal in October 1980. From the university's perspective, the AUA's timing could not have been worse, or perhaps more calculated. Following Ronald Reagan's election as president in November, the Argonne community waited anxiously to learn how the new administration's energy policy would affect laboratory funding. With Democratic majorities in both houses of Congress, it appeared unlikely that the president would be able to fulfill his campaign promise to abolish the Department of Energy. Severe cutbacks in domestic spending, including funding for energy research and development, were in the offing, however.

At the University of Chicago, it was a rerun of 1974, when the AUA had attempted to scuttle the Tripartite Agreement during the transition from the Atomic Energy Commission to the Energy Research and Development Administration. Although Bohm publicly claimed that the AUA had not initiated another attempt to become the sole contractor of Argonne, William Cannon, the vice president for business and finance for the University of Chicago, informed President Hanna Gray that the players were different but the issues were the same: the contract was not working because the AUA could not fulfill its responsibilities. The AUA's attack on the Tripartite Agreement exasperated the university, which feared that disruption of the status quo could jeopardize its affiliation with Argonne. Worst of all, as Cannon feared, the GAO seized on Bohm's report as evidence of the laboratory's mismanagement.[11]

Planning from the Bottom Up: The GAO Reviews Argonne Management

Fundamentally, the General Accounting Office wanted to know who determined Argonne's mission as a national laboratory and who established the laboratory's research priorities as a DOE facility. GAO investigators believed that the Department of Energy had failed to define Argonne's program objectives within the broad context of the department's energy research-and-development mission. Rather than determining what the government wanted Argonne to do, the department reacted to proposals from the laboratory. Of course, Argonne was not alone in this regard, according to the GAO.

Since 1978 the congressional investigators had been complaining that the Department of Energy had failed to define "the roles of its multiprogram laboratories in nonnuclear energy research, development, and demonstration."[12] Instead, research priorities were largely established during the annual budget process, when the laboratories initiated proposals that program managers at headquarters endorsed for funding. Subsequently, the Department of Energy, the Office of Management and Budget, and finally Congress, after review and adjustment, approved funding for selected projects. The GAO believed this "bottoms-up" approach had created a hodgepodge of nonnuclear research and development among the national laboratories. Although the department had initiated a five-year institutional planning process, in practice, the department permitted, even encouraged, each laboratory to develop its own plan, which course of action inevitably emphasized maintenance and/or expansion of current projects.[13]

Argonne had not challenged the need for a mission statement that defined the laboratory's unique role among the DOE's multiprogram national laboratories. But Argonne scientists were apprehensive that the department might define the laboratory's role too restrictively. Although no laboratory had a monopoly on good ideas, they warned, a narrow mission could make Argonne less responsive to changing national needs. The trouble was, when allowed to set their own research agenda, the scientists were naturally inclined to perpetuate the laboratory.

The General Accounting Office believed that the tripartite contract had outlived its usefulness, and investigators criticized the Department of Energy for renewing the three-year agreement. On what basis had the department made such a decision? they asked rhetorically. Henry Bohm's report obviously indicated that the contract was flawed, but the department had not conducted "consistent and timely" evaluations at Argonne based upon established criteria. The GAO noted that Deutch had made belated efforts to review Argonne's performance but also observed that the Deutch's review generated "vague and undocumented criticisms" of laboratory management, resulting in unwarranted tension between the Office of Energy Research and the laboratory and a decline in laboratory morale. Unless the Department of Energy defined Argonne's mission as consistent with national energy priorities, the General Accounting Office did not believe that the government could determine whether noncompetitive renewal of the Argonne contract with the University of Chicago and the AUA was warranted. The General Accounting Office recommended that the government suspend more than $400 million in planned renovations and expansion of facilities and equipment until the laboratory's role was better defined. Finally, the GAO accepted Bohm's view of the AUA's ineffectiveness and recommended ending Argonne's tripartite contract.[14]

Massey and the laboratory felt under siege during spring 1981. With the Reagan administration threatening to slash research-and-development budgets, with the General Accounting Office questioning the procedures by which the Department of Energy awarded the Argonne contract, and with the Argonne Universities Association calling for an end of the Tripartite Agreement, Massey and the University of Chicago did their best to control the damage through the summer and beyond.

Teetering on the Brink

Spring 1981 also seemed especially dreary to the Chicago *Tribune*'s economic writer, Richard C. Longworth. Reviewing Chicago's shrinking tax base, the flight of citizens from the city to the sun belt, the city's deteriorating services, and a troubled school system, Longworth called for action to save the City on the Brink. Responding in the Chicago *Sun-Times,* Massey suggested exploiting the region's research-and-development establishment to help solve Chicago's economic problems. Research was a proven asset that benefitted local industry and economic activity, Massey argued, while promoting a new alliance among Chicago's academic, scientific, and industrial communities.[15]

Excited by Massey's vision, Chicago mayor Jane Byrne established a task force on high technology development and named Massey as chair. The mayor's task force recruited twenty-five members from city government, labor, large and small business, the financial community, local universities, and Argonne. Byrne asked the task force to identify ways in which existing and prospective industry could take greater advantage of Chicago's research and academic community. How, she pointedly asked, had cities such as Boston successfully lured high-technology industries to their areas? Chicago, with its superb transportation network, its financial base, its skilled labor force, its abundant energy, and its outstanding educational and cultural assets, should certainly become a leader in the high-tech industry. In addition to strengthening cooperation among the research laboratories, the universities, and industry, the mayor's task force discovered a need for general education about the links between science, technology, and economic development.

At the national laboratories, efforts had been underway since 1979 to improve relations between the laboratory and industry, but progress had been slow because industry was unaware of opportunities for collaborative research and because government laboratories had little experience with market-driven industrial research. During the Carter years, the DOE had been anxious to push government and industry partnerships in commercializing energy technologies such as Argonne's solar and battery projects. To promote partnerships, the department encouraged establishment of the National Laboratory–Industrial Research Institute Task Force. The Industrial Research Institute represented 277 of the Fortune 500 companies having major research-and-development activities. Argonne hosted the first institute workshop held at a nonweapons laboratory. Nearly sixty executives from companies such as Gould, Inc., Amoco Oil Company, Borg-Warner Corporation, the Dow Chemical Company, Deere and Company, and U.S. Steel Corporation gathered at Argonne to assess research, development, and commercialization opportunities. Despite the ballyhoo about laboratory and industry cooperation, however, the Reagan administration seemed reluctant to encourage anything that strengthened the Department of Energy or its nonweapons laboratories.[16]

On July 29, 1981, the House Committee on Science and Technology, which had oversight over most of the DOE research-and-development programs, reviewed the role of the national laboratories and their relationships with university research and private industry. The Department of Energy was responsible for fifty-seven laboratories, twelve of which, including Argonne, were designated multiprogram laboratories. In 1981, with a total operating budget of about $2 billion, the multiprogram laboratories employed more than 50,000 workers, about half of whom were scientists, engineers, and technicians.

Throughout the Cold War, universities and industry had recognized the government's legitimate interest in maintaining strong national laboratories for defense research and development. In addition, the national laboratories were regarded as the proper locations for large research facilities that were too expensive for individual campuses or companies to build and operate. During the national energy crisis, however, when the government appropriated millions of dollars for unclassified energy research and development, the universities and industry competed for grants and contracts to fund energy projects that were as appropriate for campuses and industrial parks as for national laboratories. Not surprisingly, these competitors for federal energy money frequently complained that the national laboratories received a disproportionate share of the government's largesse.

The Committee on Science and Technology asked two questions: Have the national laboratories enjoyed an unfair advantage in competition for research funds with universities and industry? Have they been effective in transferring government-developed technology into the private sector? Arthur M. Bueche, the vice president for corporate technology at General Electric and a science and technology consultant on the Reagan transition team, sharply questioned the government's role in energy research and development. Noting that American industry had a strong tradition of research and commercialization, Bueche favored a substantial shift of federal funding from the national laboratories to private enterprise. From 1974 to 1979, roughly the Carter years, Bueche testified that industry's share of federal research dollars had fallen from 64 percent to 38 percent. "I think it's time we take a hard look at these laboratories," Bueche concluded, "and begin to think about transferring the work and people to the university campuses and industry."[17]

Did Bueche's views indicate that the Reagan administration had plans for massive cutbacks or even elimination of the laboratories? Perhaps not, but George Keyworth, the president's science advisor, and D. Allan Bromley, a distinguished physicist at Yale and president of the American Association for the Advancement of Science, favored reallocation of federal science funding. Keyworth thought that the national laboratories had performed well, but he still believed that they received too large a share of energy research funds. With the expansion and diversification of the national laboratories during the Carter administration, there had been a concomitant blurring of role and purpose at the labs, Keyworth asserted. Although he did not indicate specific targets for reallocation, Keyworth wanted to seize this "critical time to redefine the national laboratories' mission."[18]

Bromley, who had served on program review committees at Argonne, was not a spokesman for the Reagan administration, but he also wanted a reevaluation of laboratory funding. He noted that in 1981 the Department of Energy had spent $736 million in energy research funds at the national laboratories but had allocated only $201 million to universities. Bromley conceded that the national laboratories provided facilities for university users, but research cutbacks had fallen disproportionately on the academic community. Formerly, research at the laboratories had complemented work at the universities, he observed, but increasingly they engaged in "precisely the same research," with the scientists at the national laboratories holding a competitive advantage because they had "more time, better facilities, better technical support, fewer distractions, and essentially no competing obligations." Despite such advantages, Bromley criticized the huge spending at the laboratories because they lacked satisfactory institutional mechanisms for self-renewal that universities offered through their graduate and research programs.[19]

Massey was the only witness to defend the national laboratory system. During his two years as director of Argonne, Massey had discovered that the "GOCO" (government-owned, contractor-operated) laboratories and their missions were not well understood by the university community and were even less appreciated by private industry—and this despite the fact that one out of twelve American Ph.D. scientists and engineers worked at one of the government's laboratories. Massey reported that the twelve multiprogram laboratories employed more than 8,600 scientists and engineers. Yet in some measure, he believed that the legacy of the Manhattan Project and continued national security requirements had obscured the importance of this vast reservoir of scientific and engineering talent available to the nation.

Massey acknowledged that the federal government could spend its energy research-and-development money at universities or in industry instead of supporting research projects at the national laboratories. But in the long run, he believed that such a policy would be unhealthy for research institutions and the nation. As a professor and former dean, Massey opposed creating "pseudo-national laboratories" on university campuses, where government research dollars would entice schools to engage in activities inimical to their educational mission. He also believed that it was inappropriate for government planning to override market forces by subsidizing industry to develop new products of dubious profitability.[20]

Massey's plan for allocating federal energy research funds not only followed familiar guidelines well tested in Congress during the Cold War but also was congruent with the Reagan energy policy. First, Massey reiterated the government's undisputed need to conduct basic and applied research for national defense, especially related to the development, testing, production, and proliferation of nuclear weapons. During the Cold War the government science establishment never failed to invoke this mantra. Second, the national laboratories should assume responsibility for research on large energy systems, such as the LMFBR, that were likely to have major consequences for public health or the environment. Third, the government's laboratories should engage in research and development of long-range, high-risk energy systems with high potential but low immediate payoff, such as magnetic fusion. Finally, and related, the national laboratories should operate large, complex research facilities, such as high-energy accelerators, that were too expensive or unique to be run by a single university or industry. Within these guidelines, Massey believed there was extensive latitude for cooperation among the laboratories, universities, and industry.

In response to charges from Bromley, Bohm, and the AUA that Argonne neglected its relationship with universities, Massey documented extensive partnerships between laboratory and academic scientists. Contemporary, multidisciplinary research on problems that transcended the boundaries of classical science required the assemblage of research teams with wide-ranging talents. By 1980 more that 60 percent of the publications from the national laboratories included collaboration with scientists from universities and industry. In addition, Argonne and Oak Ridge each appointed about 1,200 visiting faculty and students, while Brookhaven hosted 1,000 academic scientists from 200 colleges and universities. Lawrence Berkeley Laboratory, whose staff included 160 scientists with joint appointments at the University of California, provided support for more than 500 resident Ph.D. students. Even a weapons laboratory such as Lawrence Livermore appointed 350 visiting academics. At Argonne alone, over 80 outside user groups performed experiments at a dozen research facilities. Conferences, short courses, workshops, and seminars brought another 1,200 faculty to Argonne in 1980. University scientists and engineers were also directly involved with Argonne's research programs through formal contractual agreements ranging from $4 million to $7 million a year. Finally, in cooperation with the AUA, Argonne had initiated a new program for faculty research leave at Argonne to enable faculty on sabbatical to pursue research at one of Argonne's major research facilities.[21]

Regardless of efforts to incorporate faculty participation, Massey knew that affluent national laboratories could not afford to appear smug concerning their assistance to relatively poorer, and frequently envious, academic scientists. Publicly, Massey offered the traditional apology that the laboratories had not done enough to advertize their availability to the academic community, and he pledged a new campaign through the Argonne Universities Association to promote Access to Argonne.

Massey understood that neither laboratory scientists nor academics fully appreciated the competitive process through which the other acquired and maintained support. For Massey, this insight explained much of the jealousy and misunderstanding between the two communities. He had discovered that many laboratory scientists were not aware of the intense, detailed, and uncertain process of proposal writing and peer review required of faculty to justify support (and for many, to gain academic tenure). On the other hand, faculty were frequently indifferent to the extensive, systematic, public reviews of programs conducted by the Department of Energy, Congress, and the AUA—reviews that not only decided the fate of research programs, but also determined the careers of project scientists. Massey believed that laboratory and university cooperation would become more productive if scientists acknowledged that research at all levels faced stiff competition and accountability for decreasing federal dollars. During the Reagan administration, Massey cautioned his colleagues to stop fighting among themselves for the same government money and to form imaginative and creative partnerships with industry to encourage new funding from the federal government.[22]

The Reagan Energy Policies and Budget, 1981

Scientists throughout the national laboratory system were worried about President Reagan's science policy and its effects on laboratory programs. Dismay rippled through the labs when Reagan appointed James B. Edwards as the third secretary of energy. Edwards, a dentist and a former governor of South Carolina, was an outspoken advocate of the free-market economy who pledged to dismantle the Department of Energy. On the positive side, Edwards was also an avid proponent of nuclear power and supported the government's breeder reactor research-and-development programs. While governor of South Carolina, Edwards had established the South Carolina Energy Research Institute, chaired the nuclear energy subcommittee for the National Governor's Association, and presided over the energy committee for the Republican Governors Association.[23]

Edwards and the Reagan administration moved swiftly to restructure the DOE, mostly through budget reduction and the elimination of government regulations and price controls. On the whole, Reagan wanted to end government activity in areas where private industry and the free marketplace could better establish energy priorities. The new administration wanted private capital, not the federal government, to prove the commercial viability of new energy technologies. Edwards believed that the federal government's proper role was to invest in long-term, high-risk energy research and development in which private industry would not invest. "Only in areas where these market forces are not likely to bring about desirable new technologies and practices within a reasonable amount of time," he testified to Congress, "is there a potential need for federal involvement."[24]

Although Edwards made clear that nuclear research and development was not threatened by the Reagan administration, many of Argonne's newest programs in fossil and renewable energies, in conservation, and in solar energy were in jeopardy. In March, Edwards toured the laboratory, but he offered scant reassurance. That same month, the president's first science budget proposed increasing defense research and development by $4.7 billion while reducing spending for the DOE's nondefense research programs by $344 million. By September the president modified his proposed budget to increase defense allocation by only $3.8 billion, but at the same time to cut deeper into energy programs by an additional $663 million. If the budget were approved by Congress, total cuts to energy research-and-development programs would exceed $1 billion. In March, Reagan's proposals precipitated layoffs and other spending reductions at the national laboratories, which prepared for additional layoffs, perhaps as high as 25 percent. Anticipating major retrenchment, by December, Argonne had cut 600 positions.[25]

Concerned about maintaining morale at the laboratory, Massey tried to reassure the Argonne community that the president's budget cuts were not antiscience. He conceded that the budget, if sustained by Congress, would have a severe impact on Argonne's programs along with those in other national laboratories. But Reagan's cuts were selective. While reducing dollars for conservation, solar, environmental, and fossil energy programs, the Reagan administration generally sustained funding for nuclear and physical research. The pattern was the same in other agencies. At the National Science Foundation, the National Institutes of Health, and the Department of Defense, the physical sciences received support, whereas other areas, including social and behavioral sciences and science education, were cut drastically. Massey warned that it would not be easy to adjust to Reagan's economic priorities, but he believed that Argonne scientists could match the laboratory's strength with the president's energy agenda. Massey acknowledged that there was "a lack of focus or sense of mission" within the national laboratory structure and that the laboratories frequently engaged in demonstration projects more properly left to private industry. Under scrutiny from the Reagan administration, Massey anticipated that Argonne, through internal planning, would have to set new research priorities and improve the quality of its programs.[26]

Massey was not so diplomatic in responding to the General Accounting Office, whose critical report was finally released in June. Simply put, he argued that the GAO staff understood neither the nature and purpose of the national laboratories nor the organization of the Department of Energy. Furthermore, he was haunted by the persistent inaccuracies and unsubstantiated rumors about Argonne's management that the GAO had gleaned from old reports and Bohm's critique. With Argonne under Reagan's budget gun, Massey was furious about the GAO's recommendation that the DOE suspend $400 million in renovations and improvements in laboratory facilities until the government defined Argonne's mission more clearly. He curtly dismissed the recommendation as "unwise and self-defeating" unless the government planned to abandon the laboratory.[27]

On the Brink: Rumors of Closing Argonne National Laboratory

Normally upbeat, Massey's optimism steadily wore down through the summer of 1981. While chairing Mayor Jane Byrne's Task Force on High Technology Development, Massey parried complaints from prominent Chicago industrialists that local political leaders were indifferent to the relationships among science, technology, and economic productivity. Massey acknowledged that scientists could always improve their public relations with politicians, but he was surprised by the industrialists' unanimity that area scientists, including those in government and industry, had failed in educating Illinois's politicians about the vast economic potential of the region's science and engineering resources. By September's end, Argonne's director felt that support for his laboratory was melting away. In Washington, he heard rumors from Alvin Trivelpiece, the DOE's new director for research, that Reagan planned to cut a billion dollars from the department's budget, half from defense and half from energy programs. Trivelpiece did not vouch for the accuracy of the rumor, but urging the laboratory directors to prepare for the worst, he warned that 1983's budget would be more draconian than 1982's. On October 1, Massey wrote Hanna Gray that he believed Argonne lacked essential political support from the Illinois congressional delegation to survive the funding collapse.[28]

Massey was also upset by rumors from Washington concerning continuing investigations of the laboratory. Charles Till, Argonne's director of nuclear reactor research, learned from a friend at Los Alamos who was reputedly close to the Reagan administration that Argonne would be shut down when the DOE was eliminated or reorganized.[29] Subsequently, while attending a laboratory director's meeting in Washington, D.C., Massey learned that the department's Energy Research Advisory Board (ERAB) planned new studies of the role and missions of the national laboratories. Concurrently, George Keyworth's Office of Science and Technology Policy (OSTP) scheduled a similar review.[30] Brave-ly, Massey professed no concern about "objective" reviews from the ERAB or the White House, but he was deeply worried by persistent rumors that the Reagan administration intended to close one of the national laboratories.

Massey feared that Argonne was the "targeted laboratory" because it had the least political support in the national laboratory system. The weapon laboratories were untouchable in the Reagan administration. Oak Ridge enjoyed the support of the Senate's most powerful Republican, Howard Baker from Tennessee, and counted two area representatives on the House Committee on Science and Technology. Brookhaven depended on solid support from the New York congressional delegation, which also included representatives on the science and technology committee. With 12 percent across-the-board budget cuts on the horizon, the Department of Energy unofficially notified Argonne scientists that they would bear a disproportionate share of the cuts because the laboratory was an easy target. "Lack of interest" from the Illinois delegation and on the part of other influential people from Chicago became a major political liability for the laboratory. "A cold-hearted process of elimination makes Argonne a very vulnerable institution," Massey warned Gray. He had two messages for Gray: Argonne not only needed the university's help in dealing with the DOE, but also required Gray's assistance in generating public and political support for the laboratory. Otherwise, Argonne was in serious danger of being closed.[31]

Somehow, a copy of Massey's letter to Hanna Gray leaked to the Chicago press with predictable consequences. While the Chicago *Tribune* and the *Sun-Times* played up Massey's worst fears that Reagan's budget cuts could close Argonne National Laboratory, Gray dismissed the rumors as "speculation." Nonetheless, both the *Tribune* and the *Sun-Times* reported Massey's worry that the Illinois congressional delegation would not fight effectively to protect Argonne's programs. Perhaps Massey was surprised by the public furor he had created, but wittingly or not, he was playing a dangerous game. Quickly, Massey notified his congressional delegation that he had been misquoted in the press concerning their alleged lack of support for Argonne. He soothed the politicians by confessing that the scientific and technical communities had not effectively provided Congress the basic information needed to understand and defend the importance of government funded research in the Chicago area. While Massey tried to ease the politicians off the hook, the *Tribune* snorted that the Illinois delegation should not have to be "spoon fed" about the importance of Argonne, its 4,600 jobs, and its $250 million budget. Sometimes, the *Tribune* observed, the Illinois delegation stood around "with thumbs in the ears" instead of using their considerable political clout to secure funding for government research facilities in the state.

Hanna Gray Leads the Charge

The Argonne Universities Association and the General Accounting Office forced the University of Chicago and the Department of Energy into a reexamination of Argonne's management. With criticism from the AUA that the Tripartite Agreement did not work and with charges from the GAO that the contracting process was flawed, the Department of Energy was under irresistible pressure to correct the situation. Fearful that the AUA might try to become the sole contractor or that the Department of Energy might decide to select a new contractor, President Gray decided to seize the initiative.[32]

Have you ever met Hanna Gray? If so, Massey recalled, perhaps you discovered that she was formidable once she decided to take action. After discussions with Massey, Gray personally assured the DOE that the University of Chicago would assume a stronger role in managing the laboratory. She also quietly talked with some of her colleagues at other AUA universities and discovered there was no strong support among the presidents for an enlarged AUA role in running Argonne. As in 1974, enthusiasm for AUA leadership was largely confined to AUA board members and staff who were faculty members at their respective institutions. The presidents did not need more headaches, but they pressed Gray to search out new ways to expand midwestern university participation in Argonne research.[33]

After reviewing the options, Gray decided that the University of Chicago would offer to manage Argonne alone. As much as possible, the university's plan accounted for AUA sensibilities, but distrust ran so deep that it was not possible for the university to satisfy all objections of the AUA board. On the one hand, the Argonne Universities Association criticized the University of Chicago for its indifference toward Argonne, especially at the highest levels of the university. On the other hand, some members of the AUA did not want the University of Chicago to exert stronger leadership. If the AUA could not become Argonne's sole contractor, they favored establishing a not-for-profit corporation modeled after Associated Universities, Inc., which managed Brookhaven National Laboratory.[34]

Argonne's Board of Governors

Ingeniously, Gray proposed a compromise that did not mollify all members of the Argonne Universities Association but nonetheless proved politically irresistible. At Massey's recommendation, the University of Chicago invited Gerald Tape, a former AEC commissioner, to help draft an acceptable plan.

First, Gray accepted the criticism that the University of Chicago had not provided Argonne with strong leadership under the Tripartite Agreement. She pledged to devote more of her own "energy, time, and expertise" to promoting Argonne before Congress and the Department of Energy. In addition, the university would play a direct role in defining Argonne's scientific and programmatic goals and would become systematically involved in the management of the laboratory.[35]

Based on the AUA objections, Gray also saw the wisdom of distancing the university somewhat from daily operation of the laboratory. But how could she fulfill her promise to increase university involvement and yet promote autonomy? Gerald Tape's clever suggestion was to modify the AUA's proposal of a not-for-profit corporation. Gray offered to establish a board of governors for Argonne, modeled after Brookhaven's governing board, that would report to the university's board of trustees, who would "delegate full responsibility and authority . . . to act for the University of Chicago on all laboratory matters, consistent with general university and Department of Energy policies." To the extent possible within the law, the governors would be "independent, autonomous, and self-sustaining." Argonne's board of governors would replicate the strengths of Brookhaven's management without establishing an independent corporation. Gray herself would preside as chair, and the university would establish a new position, vice president for research, with specific oversight of Argonne. While the new position would have responsibilities broader than Argonne's, the vice president for research was to represent the university on issues pertaining to the laboratory. "In that sense," Gray explained, "the new vice president would be my alter ego with respect to Argonne matters."[36]

Argonne's board of governors, who were recruited for their ability to provide strength and leadership for the laboratory as well as for their outstanding scientific and technical expertise, would represent three constituencies: the University of Chicago community of trustees, officers, and faculty; academic scientists, engineers, and administrators mainly from schools represented in the old AUA; and scientists, engineers, and corporate leaders from high-technology industries. Although not self-perpetuating, the governors established their own nominating committee, whose recommendations were to be approved by the university trustees. Gray envisioned that Argonne's board of governors would operate with the "fullest possible independence" with authority and responsibility for the laboratory's long-range planning, programs, budget, facilities, and staff within guidelines established by the university and the government.[37]

Perhaps the board's most important responsibility was to be the selection and review of Argonne's director and other top management officials at the laboratory. Performance review and evaluation, salary, and other matters pertaining to employment were all to be determined through the governors' executive committee. Although the chain of command and management responsibilities were well defined, encouraged by the AUA to replicate Brookhaven's management structure, the university actually hoped to strengthen the role and visibility of the laboratory director as manager of the scientific, technical, and administrative functions of the laboratory. But the university also intended to appoint a principal deputy director with responsibility for the day-to-day operation of the laboratory, thereby liberating the director to concentrate on Argonne's major problems—development of long-range planning, definition of role and mission, and effectiveness in national policy councils. With the director also serving ex officio on the board of governors, however, the lines of authority and responsibility between the laboratory director and the new vice president for research were unfortunately blurred.[38]

Finally, Gray reminded the AUA that the University of Chicago, as both a national and a regional university, felt a major responsibility for the vitality of scientific research and development in the Midwest. The University of Chicago had been associated with Argonne since the laboratory was founded on the campus in 1946, and Gray reaffirmed that the university's historic ties with the laboratory were an important asset for Argonne and the Midwest. The university had a significant institutional stake in the success of Argonne; a commitment, Gray presumed, that could not be matched by a new prime contractor.[39]

After vacillating about whether to support the University of Chicago's plan, the Argonne Universities Association finally gave its consent on January 18, 1982. Some members of the AUA, who remained suspicious of Gray's motives, believed that Argonne could never prosper as a lesser satellite of the University of Chicago. Others, noting that Chicago did not have an engineering school, doubted that the university could muster the necessary expertise to manage research and development in applied technology. The most cynical might have believed that Gray was fighting to save the university's $1.5 million annual management fee, but John Walsh, writing for *Science,* doubted that money was a major issue. At the heart of the controversy, he observed, was the long-standing reluctance of scientists from rival research universities to confirm Chicago's dominance in setting their research priorities. Only after being convinced that the University of Chicago would seek a "capable, distinguished," and nationally respected scientist or engineer to serve as the new vice president for research did the AUA board of trustees finally endorse the University of Chicago's plan to become Argonne's sole contractor. Under the circumstances, Walsh described the AUA's decision as a "dignified," and considering recent history, "a statesman-like exit."[40]

A New Contract

The DOE welcomed the end of the Tripartite Agreement and moved quickly to make the contract change. Robert Bauer, manager of the Chicago Operations Office, enthusiastically endorsed the University of Chicago's plan to establish a board of governors to oversee Argonne. In line with Reagan administration policy, Bauer stressed that the new arrangement enhanced access by industry, as well as by universities, to Argonne's management and programs. Bauer effusively praised Massey, noting that "overall performance of the contractor has improved significantly during the past 2 years."[41]

Ironically, after more than three years of constant criticism, the Department of Energy concluded that Argonne's institutional plan was one of the best among the multiprogram laboratories. In other areas, the laboratory was rated from satisfactory to excellent. Louis Ianniello, director of the Division of Materials Sciences in the DOE's Office of Basic Energy Sciences, rated Argonne's technical performance with the Intense Pulsed-Neutron Source (IPNS) as "excellent" but noted prophetically that Argonne's main problem was "the *image* they have of being a laboratory without a mission." To be sure, there were still some major deficiencies, some of them in Argonne's plant management, where cutbacks had created serious problems in deferred maintenance. But Massey was given high marks for sustaining productivity and morale despite budget revisions, personnel reductions, program cuts, and threats of reorganization. Because there were no rivals to the University of Chicago, the Department of Energy decided not to compete renewal of the Argonne contract. On October 1, 1982, the department designated the University of Chicago as the sole contractor for the operation and management of Argonne National Laboratory.[42]

Concurrently with discussions ending the Tripartite Agreement, the Reagan administration recommended shifting all the DOE's nonweapon research-and-development activities to the Department of Commerce. Although most of the Washington establishment knew that Reagan's proposal was "dead on arrival," Department of Energy officials projected major adjustments in national laboratory programs and funding. Deputy Secretary of Energy W. Kenneth Davis and Alvin Trivelpiece assured worried scientists at Argonne and Brookhaven that their laboratories would not be closed. But they also served notice that the administration intended to "revitalize" the laboratories along with the nation's economy. Regulatory and commercial demonstration programs were to be gutted from the DOE while the nonweapon laboratories were to focus on "long term, high-risk, high payoff research and development," just as Massey had projected.[43] Although the Reagan administration disdained the federal government's management and regulation of the national economy, the Republicans had a strong penchant for blue-ribbon review panels and administrative micromanagement. The former were intended to show the public that the administration had taken bold steps to control the costs of government, whereas the latter no doubt drove up bureaucratic costs without significantly increasing productivity. The president's private-sector survey on cost control in the federal government, headed by J. Peter Grace, chief executive officer of W. R. Grace & Company, was privately funded through a nonprofit, tax-exempt foundation but cost the government hundreds of hours to service thirty-five task forces looking for government overlap and inefficiency. Understandably, the Department of Energy and its contractors, already targets of Reagan's political rhetoric, feared that the Grace commission might be used as a hammer to dismantle the department and its programs. DOE officials had grown so paranoid since Reagan's inauguration that Secretary Edwards tried to calm the staff by relaying White House assurances that "there will be no attempts to surprise anyone. . . . It is not an attempt to criticize individual performance but rather to improve overall efficiency and cost control."[44]

Senator Percy Rallies behind Argonne

Reagan's assault on the Department of Energy helped Massey mobilize unprecedented political support behind Argonne National Laboratory. On March 22, 1982, Senator Charles Percy's government operations subcommittee met at Argonne ostensibly to explore requirements for long-range energy research in the United States. But everyone knew that this key Republican senator brought his committee to Illinois to serve notice at the White House that Argonne would not be carved apart on the administration's sacrificial platter. Straightforwardly, Percy told Secretary of Commerce Malcolm Baldridge, the hearing's first witness, that he had not made up his mind about Reagan's proposed restructuring of the Department of Energy and would reserve his judgement until he was assured that "critical energy R&D programs" were secure at Argonne and other midwestern laboratories. Percy noted that budgets for the three midwestern energy laboratories—Ames, Argonne, and Fermilab—were down by less than 1 percent from fiscal year 1982, but that this mild overall cut did not reveal how seriously Argonne had been savaged. Whereas Ames had suffered only a 2 percent cut and Fermilab had actually increased funding by 21 percent, Argonne had been hit by a crippling 17.7 percent reduction of $36.8 million and dropped to $154 million for fiscal year 1983. Although energy secretary James Edwards had promised Percy that the midwestern laboratories would receive "their fair share" of federal dollars, the Illinois senator worried that Argonne might cease to be a vital research institution.[45]

Percy, with Governor James R. Thompson of Illinois, had just officiated at the dedication of Argonne's new high-sulfur dry scrubber demonstration project installed at the laboratory's main boiler plant, which could now burn high-sulfur Illinois coal with atmospheric emissions lower than Environmental Protection Agency standards. With hyperbole, Percy asserted that the high-sulfur dry scrubber demonstration was to coal what Fermi's chain reaction under the Stagg Field stands was to nuclear energy. Illinois was the "Saudi Arabia" of coal, but the high sulfur content of local coal rendered it costly and environmentally unacceptable. Argonne's dry scrubber promised to solve that problem. What frustrated Percy, of course, was that Argonne's environmentally inspired, cost-effective demonstration project was just the sort of program that the Reagan administration intended to zero out. It did not make any economic sense to Percy to "cut the heart out" of a program that could create employment, aid the economically distressed Midwest, and earn valuable foreign exchange while removing sulfur from Illinois coal at an affordable cost. With no direct answer to Percy's challenge, Trivelpiece, Secretary Edwards's spokesman at the hearings, lamely replied that the high-sulfur coal programs were the responsibility of another assistant secretary.[46]

Although Percy could not forestall substantial cuts in Argonne's fossil-energy programs, he wrung from Secretary Edwards repeated pledges not to close the laboratory. Just in case Edwards failed to get the message, however, Percy also organized the entire Illinois congressional delegation to protest the fragrant discrimination against the Illinois research facility. At the very time California's and New Mexico's weapon laboratories were prospering under Reagan's defense buildup, Percy was incredulous to learn that the Department of Energy contemplated transferring Argonne's pulsed-neutron scattering research to Los Alamos, already the beneficiary of substantial federal spending. Fifteen Democrats and eleven Republicans from the Illinois congressional delegation petitioned Edwards "to block any further disintegration" of the Argonne complex. Edwards, of course, could not promise anything in the face of congressional demands, but without pledging his support to a specific program, he reiterated his promise to maintain Argonne's viability as a research institution.[47]

Public congressional hearings, in themselves, rarely change public policy, yet sometimes they yield unintended surprises. Time, place, witnesses, and issues were carefully chosen and orchestrated by the committee staff. As the participants joked, everyone knew what everyone else was going to say. Many of the key witnesses who testified before Percy's subcommittee, including Trivelpiece, Massey, and Hanna Gray, submitted extensive prepared statements which, as commonly practiced, were simply printed as part of the record. The purpose of the hearings at Argonne was not to debate the role of the midwestern laboratories in energy research and development, but rather to send a clear signal to the Reagan White House that Massey and Gray had successfully mobilized political support behind the laboratory.

Now and again, the right questions were raised. At the conclusion of Trivelpiece's testimony, Percy asked the DOE's spokesman to remind Secretary Edwards that his support for energy reorganization was conditioned on the ability of the Reagan administration to establish forward-looking priorities. "Where are we going?" Percy asked. "What kind of program are we going to be carrying out?" Amazingly, these were the same questions that the Department of Energy, the Office of Science and Technology, and the National Academy of Sciences were asking about Argonne and the national laboratory system.[48]

Where Are We Going?

In October 1981, more than a hundred scientists meeting with the National Academy of Sciences joined with the president's science advisor, George Keyworth, and Secretary Edwards in calling for a review of the federal government's system of research institutions. Testifying before the House Committee on Science and Technology, Keyworth questioned whether the nation was receiving adequate return on its research investment in the national laboratories. Many of the laboratories, such as Argonne, were more than thirty years old; some of them had fulfilled their original missions "long ago," others were simply "outdated." Keyworth advocated a reassessment of the entire national-laboratory structure and missions. Similarly, Edwards asked the ERAB, chaired by Louis Roddis, to assess the future of the department's multiprogram laboratories, including

> review, appraisal and recommendations regarding the scientific and technological capabilities of the laboratories, the appropriate roles and missions of the laboratories vis-à-vis industry and the universities, the Department's policies and procedures with respect to the laboratories, and the effectiveness of the Department's use of the laboratories in meeting our energy objectives.[49]

Massey asked Robert Rowland, the interim associate laboratory director for biomedical and environmental research, to head Argonne's task force that defined the laboratory's research-and-development priorities for ERAB's multiprogram laboratory panel. The Rowland report, "Argonne Laboratory: A Resource for the Nation," wonderfully captured Argonne's vitality and diversity and remains a landmark study of the laboratory's historic contributions to the national laboratory system. But while Rowland's colleagues celebrated past and current research accomplishments, they found it difficult to chart Argonne's future on any basis other than modifying the laboratory's existing programs.

Materials science at Argonne provided an excellent example of how the laboratory's programs evolved from their original mission. Initially, nuclear scientists and engineers in the reactor development programs needed information on radiation effects in metals and graphite exposed to intense radiation environments in the reactor core. In addition, Argonne's scientists wanted to study the basic properties of the newly discovered actinide elements and their compounds, research that they knew could have far-ranging consequences. Interesting results were not limited to nuclear engineering, however, and as the government's research activities broadened during the energy crisis, Argonne's materials science program expanded correspondingly to include research related to fossil, fusion, solar, conservation, and energy-storage systems.[50]

Rowland's task force highlighted Argonne's pioneering materials research and state-of-the-art user facilities, the Intense Pulsed Neutron Source and the High Voltage Electron Microscope (HVEM)–Tandem Accelerator facility. These major machines, and the Argonne Tandem-Linear Accelerator System (ATLAS), which was under construction, secured Argonne's immediate future as a national user center for materials research. Materials science had become the largest program in basic research and base technology. In recognition of Argonne's deepening commitment to this area, in October 1982 Massey created a new materials science and technology division to provide a coherent, rationalized focus for materials research scattered throughout the laboratory. Projecting current programs and responsibilities, Argonne's future seemed clear. Yet Rowland's task force was ambivalent when it concluded that "the future is uncertain," except that in the foreseeable years, Argonne would surely become a smaller, but more sharply focused, laboratory working on the development of nuclear power and on basic research in the physical and biological sciences.[51]

The foundations, however, were shaky. The breeder-reactor program, the mainstay of Argonne's research structure, was in deep trouble with Congress and under increasing scrutiny from a cost-conscious White House. Competition with universities, industry, and other laboratories for steady-state funding in the basic sciences had intensified. And the Reagan administration had already served notice that draconian cuts should be expected in the applied and alternative energy systems formerly championed by President Carter.

The major discordant note in an otherwise upbeat review was the laboratory's complaint that "rapid and unpredictable oscillations" in federal funding made long-range planning virtually impossible. Because Argonne had no budget "fly-wheel" to smooth out drastic changes in federal funding, the laboratory was subject to regular financial shocks that not only wasted resources and sapped morale, but also produced poor work which undercut the laboratory's reputation. Highest on Argonne's wish list, besides a major new research machine, was a new funding strategy that would avoid disruptive fluctuations.[52]

The funding problem, however, was related to another issue that Rowland's task force did not address—how did the laboratories, with little or no independent, discretionary money, establish new research priorities? If the laboratories' critics complained of "on-going-sameness" in research programs, how were the laboratories to achieve self-renewal when the federal government held all of the purse strings? John Walsh of *Science* observed that the laboratories were well regarded, but that their constituencies held widely varying expectations concerning their missions. The White House and the DOE viewed the laboratories as handmaidens of the administration's prevailing energy policy. Congress, on the other hand, looked to the laboratories to provide a technological "fix" to the nation's energy problems. State congressional delegations and local politicians, of course, knew the laboratories were a major source of employment and economic stimulation, as well as magnets for high-tech industry. Industry remained worried that the laboratories would capture a disproportionate share of federal commercialization dollars, and the universities fretted over achieving equitable access to the federal research facilities. For their part, Argonne's scientists and engineers only wanted to pursue their research projects undisturbed by political or academic intrusions. Under these circumstances, innovation and internal renewal were further discouraged by the fact that expert advisory panels, such as the Energy Research Advisory Board and the High-Energy Physics Advisory Committee (HEPAC), often established national research priorities among the competing laboratories. In this milieu, while rhetorically pursuing "high-risk" research without discretionary funds, Argonne simply could not afford to become too daring in setting an independent research-and-development agenda.[53]

Argonne's problems were not unique. The federal laboratory system had grown without plan or direction since the end of World War II. By 1982 the federal government spent $15–20 billion, about one-third of its research-and-development budget, at 755 laboratories. Government employees ran most federal laboratories, but contractors operated many of the largest, such as Argonne, which was a perfect example of a laboratory originally established to pursue research that could not be performed by universities or the private sector, but whose mission had changed significantly over time. Although the Reagan administration acknowledged that the accomplishments of the laboratories had been impressive, it also believed that the research capabilities of the universities and industry had also grown impressively. It was time to reassess federal research-and-development expenditures.[54]

It would be difficult to convince the scientists that it did no good to overreact to rumors that the Department of Energy planned to close one of its multiprogram laboratories. Trivelpiece, who tired of reassuring nervous scientists that no laboratory would be abandoned, admitted that the Department of Energy "would have to invent the national laboratories if we didn't already have them." Reagan's administration soon learned that the federal laboratory network could not be attacked on a broad front without unacceptable expenditure of political capital. From the White House, however, Keyworth was determined to press forward with his reassessment of federal research-and-development expenditures by examining laboratory missions, evaluating historic performance, and identifying unique strengths.[55]

Ironically, despite Republican determination to cut wasteful duplication in the national laboratories, the Reagan administration sponsored three management studies of government-sponsored research and development. Pyramiding one upon the other, every review reached similar conclusions about the vital importance of the national laboratories. Following visits to the laboratories, the ERAB panel found "the laboratories' scientific and technological capabilities impressive and laboratory leadership to be of high quality."[56] But the White House studies also advanced the president's political agenda by focusing criticism on the beleaguered DOE instead of on the labs. For example, David Packard, chairman of the White House Science Council's federal laboratory review panel, found that competent laboratory scientists were poorly supervised by the Department of Energy. Although it cited a few problems in the sixteen labs surveyed for Keyworth, Packard's panel did not identify a facility that should be shut down. The laboratories did not need more money, Packard reported. What they needed was "better management."

"Reagan Orders Overhaul of 'Poorly Run' Labs"

Packard's report provided President Reagan the ammunition he wanted to blast the federal bureaucracy once again. According to the Chicago *Sun-Times,* in another move to discredit the energy department, the president "ordered an overhaul" of the federal laboratories because tax dollars were being wasted through poor management and program duplication. Packard's findings and the president's rhetoric were short on specifics, however. Conforming to the rule, the more aggregate the review, the less detailed was the attendant analysis. Principally, the Packard panel discovered that the multipurpose laboratories did not have clearly defined missions to guide the DOE and the labs in setting goals and assessing performance. "Preservation of the laboratory is *not* a mission," Packard proclaimed, trying to talk tough about the issue. Of course, he had not discovered a new problem, and unfortunately he had no ready prescription, no mechanism, with which to propose mission renewal.[57]

The White House Science Council offered a number of recommendations, which even if adopted, hardly added up to an "overhaul" of the national laboratory system. Reflecting influence from the ERAB, the Packard panel listened carefully to laboratory unhappiness about the Department of Energy. For example, the laboratories complained bitterly that they were micromanaged from Washington, D.C. The ERAB reported to the White House that

> Argonne National Laboratory... is required to be responsive to 137 different DOE orders and policy issuances, entailing dedication of substantial manpower and funding for compliance, reports, and audits. Equally disturbing is the fact that Argonne receives its funding through 129 separate accounts which are noninterchangeable and require controls. Micromanagement to this degree deprives the laboratory of virtually all flexibility for better management of its resources.[58]

Funding uncertainties were aggravated by the existence of 410 budget-control levels at Argonne. Massey had no authority to reprogram funds from one category to another without permission from program officers at headquarters and, occasionally, without approval from Congress. Unpredictability and inflexibility in funding, the Packard committee added, impeded both rational planning for and effective conduct of research-and-development activities.[59]

Micromanagement from headquarters was more than just a pain in the neck according to a University of Chicago chemist, R. Stephen Berry. Despite claims from Milton Shaw, Berry believed that DOE project managers were rarely technically competent in the research-and-development areas they supervised. More importantly, regardless of scientific and technical uncertainties, headquarters' managers too frequently assumed that high-risk research-and-development programs would succeed if the laboratory staff were competent. Consequently, not only did managers sometimes meddle in programs far beyond their competence, but they were also biased in supporting projects they understood and whose success was nearly assured. Understandably, because ambitious bureaucrats wanted to manage successful programs, Berry believed they pushed the laboratories into programs more appropriate for private industry, which had been a major criticism of the Reagan administration. The laboratories had received poor marks from the ERAB, the Packard panel, and the Grace commission because the reviewers questioned the appropriateness of some laboratory activities rather than their usefulness.[60]

The laboratories needed new vision—and new blood. Although no one documented the declining quality of new recruits, science administrators widely believed the myth (which may have been true) that laboratories such as Argonne no longer attracted the cream of young scientists as they had in the 1950s and 1960s. While the laboratories "aged," few new jobs were created, and those that became available seemed less attractive to young talent. There were no hard statistics, and it was difficult to pinpoint where there was a talent shortage in the national laboratory system. But deeply ingrained perceptions were hard to contradict when scientists believed they enjoyed decreasing mobility between the laboratory, academia, and industry. Although the national laboratories contributed importantly to the education and training of scientists and engineers, the comparatively low salaries tied to civil service scales reportedly made it tough for laboratories to compete with private industry for the best and brightest college graduates or for qualified scientists and engineers.[61]

Ultimately, President Reagan's "overhaul of poorly run" national laboratories did not require wholesale replacement of leadership, restructuring of management, redefinition of mission, or reprogramming of budgets. The White House Science Council recommended that national laboratory directors be given greater autonomy to initiate innovative research and increased flexibility to allocate 10–15 percent of the budget in discretionary exploratory spending. Most importantly, the council supported proposals for multiyear congressional funding and carryover from year to year to provide fiscal stability for long-term research and development projects. While praising the laboratories' excellence and lamenting micromanagement from Washington, the Packard committee could not resist suggesting that each national laboratory establish an external oversight committee comprised of academics and industrialists. Through regular reviews, the oversight committees would assure high quality and relevance of research, while somehow reducing micromanagement.[62]

Roger Batzel, Massey's counterpart at Lawrence Livermore National Laboratory, viewed the whole process with a skeptical eye. Batzel agreed that it was important to clarify missions, improve planning, and increase interactions with universities and industry, but he believed the recommendations were all about money as well. He suspected that there was a hidden agenda behind endorsements for greater cooperation between the laboratories and universities and industry. Implicit in the suggestion was the belief that more federal dollars should be spent at the universities and industries with the laboratories increasingly becoming "job shops" for the pass-through of the government's money.[63]

President Reagan gave up on the idea of abolishing the Department of Energy in November 1982 when he named Donald P. Hodel as James Edwards's successor at the department. The new secretary of energy, who had served as the director of the Bonneville Power Administration, was the first energy expert to lead the department since its inception. He shared the administration's doctrine that commercialization of energy resources was not the federal government's business. On the other hand, he believed that one of the DOE's essential roles was to explore uncertain and expensive frontiers of energy research. He compared the government's responsibility in exploring the energy frontier with its role in opening the American West. Explorers such as Lewis and Clark, Fremont, and Bonneville had been supported by the United States government. In turn, settlers were offered free land under the Homestead Act (1862), and transcontinental railroads were built in the West with government incentives. Some prospered; others failed. But the West was settled without creating government farms or towns or government railroad corporations. Similarly, Hodel believed the government should sponsor energy exploration on the frontiers of basic research and high technology related to America's military and industrial future but should stop short of financing commercial ventures. Private citizens should be encouraged to "homestead" the energy frontier. Just as land transfer was a major task of the government in the nineteenth century, so technology transfer from government laboratories to the private sector would be an important agenda for the Department of Energy in the twentieth century.[64]

With the easing of the energy crisis, Hodel tried to tamp the heated reorganization rhetoric. He realized that abolition of the DOE and reassignment of the national laboratories to the Departments of Commerce, Interior, or Defense was politically impossible. When Hodel met with laboratory directors to discuss the Packard report, he stated candidly that if the national laboratories did a better job in defining their missions, they would be less troubled with micromanagement from Washington. Hodel favored multiyear funding for the laboratories, which almost everyone endorsed in principle, but doubted whether the Office of Management and Budget and Congress could ever work out the details.[65]

Argonne Renewal

Short of abolishing the laboratory outright, Massey anticipated correctly that Argonne was too large and complex to be reformed dramatically by the Reagan administration. He predicted that the various government reviews would identify some "critical changes" required at Argonne, which he advised Hanna Gray would "actually be implemented gradually over the next two to three years." But if Reagan's overhaul of the national laboratories was flabby, the end of the Tripartite Agreement with the AUA and the reorganization of laboratory oversight and management under the University of Chicago's laboratory board of governors pushed Argonne in much the same direction as recommended by the ERAB and the White House Council on Science and Technology. Fortuitously, the organization of the laboratory's new board of governors in 1982–83 dovetailed with the changing of the guard in the secretary of energy's office in Washington and the appointment of Hilary Rauch, replacing Robert Bauer as manager of the Chicago Operations Office.[66]

Argonne became the first national laboratory to operate under a board of governors that offered laboratory leaders experience and support from industry as well as from academia. In a friendly setting, the board asked the same questions about planning, budgeting, programs, and personnel that had been raised repeatedly by review panels for the past several years. How was "success" defined at Argonne? How were goals set? How free was Argonne to establish its own research agenda or to change direction? Now, however, these basic questions about laboratory mission and management were asked in the spirit of sympathetic renewal.

Top laboratory leadership was the first issue addressed by the board of governors. In a move that caused considerable distress among the AUA, Hanna Gray had appointed Massey the university's first vice president for research, which meant that he wore two hats; one Argonne green, the other Chicago maroon. The intent of this structure borrowed from an industrial model was to enable Massey, the laboratory's chief executive officer, to represent the laboratory externally in his capacity as a university officer. The university would soon begin to search for a senior deputy director to serve as the laboratory's chief operating officer. The governors saw no inherent conflict of interest in the arrangement, but they wanted assurance that Massey, for all his talent, could handle both jobs and that the university would move quickly to hire a first-rate senior deputy whose authority and status would not be compromised by Massey's dual appointment.[67]

Building 201: A Space Symphony

Argonne's building 201 symbolized the beginning of "a new era for Argonne," Massey proclaimed at the gala dedication of the new administration building just two weeks after the University of Chicago became the sole contractor for Argonne in October 1982. Autumn was well advanced in Chicago, but inside the stunning award-winning building, Argonne employees and guests were giddy with spring-like fever celebrating the laboratory's renewal. In general, government buildings are distinguished neither for their esthetics nor for their efficiency. Except for Freund Lodge, Argonne's architecture was stark, functional, and unimaginative—more suggestive of an industrial park or a military base than an academic or research campus.

Building 201, set in the middle of the laboratory, provided architectural distinction, focus, and definition for Argonne (figure 34). Constructed of lustrous polished aluminum and glass, the north side of the three-story building curved gracefully east and west through a young pine wood, while the front facade described a hard-edged line connecting the ends of the gentle crescent. To soften the entrance, a small reflecting pond offered a muted counterpoint to the straight lines and dramatic arcs of the adjacent building. Here, in time, a small flock of red-winged blackbirds sang their konk-lore-rhea chorus among the cattails while green-backed herons jigged across lily pads to the amusement of fellow diners on their noon break. Together, building and pond formed a circle that mirrored the laboratory's circular road system, and according to the architect, evoked a symbolic association with the sun's nuclear and solar energy.[68]

Building 201 was more than an aesthetic triumph, however. The architects, Murphy/Jahn Associates of Chicago, also won major awards for their energy-efficient solar design. The sun provided most of the interior illumination. Skylights, glass panels around the perimeter, and a three-story "light well" in the center of the building "opened up" the entire building to daylight. Along the line of the southern exposure, awnings deflected the direct summer sun, while light from the semicircular reflecting pool danced through floor-to-ceiling

Figure 34. Administration Building 201

windows. The building's solar design also incorporated an advanced heating and cooling system. Ironically, the prize-winning solar heating and cooling system proved too complex and expensive and was never completed.

Of central importance to Argonne's new administration building, of course, was the work space. For thirty-five years Argonne's administrative offices had been scattered among World War II Quonset huts and other temporary structures in the laboratory's east area. Before they moved into building 201, the DOE's representative and the laboratory director had to commute more than a mile to discuss laboratory matters in person. Building 201 gathered almost 750 Argonne, Department of Energy, and University of Chicago employees into an "open" semicircular administrative office that enclosed maximum space with a minimum of walls. Gone were the stacked and boxed cubbyhole offices jammed with government-issue gray steel desks and filing cabinets. The atrium foyer and irregular central "light well" created open space among the three-tier office galleries, where the staff worked in partitioned "offices" trimmed in bright Argonne green, gray, and off-white and appointed with compact, coordinated furnishings.[69]

Lewis Mumford once noted that each civilization proclaims its basic values with monumental buildings as well as through commemorative statuary and memorials.[70] The same is true for society's institutions. When the University of Chicago built the magnificent Regenstein graduate library on the site of the historic Stagg Field stands, the university trustees unambiguously announced that the university's commitment to academic excellence far surpassed its interest in athletic achievement or historic preservation. One might expect to find a similar expression of values via structures at the national laboratories, where major research facilities dominate the central landscape like Gothic cathedrals. Indeed, such was the case at Fermilab and at the Stanford Linear Accelerator, where the particle accelerators both clearly defined the physical dimensions of the facility and boldly proclaimed the laboratories' basic mission. At Argonne-East, however, the shell of the old ZGS facility stood empty, waiting expectantly for a new assignment.

Pavilion/Sculpture for Argonne

Multiprogram laboratories, like sprawling multi-universities, face difficult challenges in defining their boundaries. When space becomes diffused, function can also lose focus. In this regard, Massey recognized the importance of using public space to define institutional purpose. To commemorate Argonne's thirty-fifth anniversary, the laboratory commissioned an outdoor sculpture that would assist employees and the public in understanding Argonne's mission. Massey was excited that for the first time, the National Endowment for the Arts, through its Art in Public Places program, granted matching funds to a national laboratory for placement of an outdoor sculpture planned in conjunction with construction of building 201. Additional funds were raised from private sources, including laboratory employees, industrial contributors, and patrons of the arts.[71]

On October 3, 1982, following chamber music performed by the Arriaga String Quartet of the Chicago Symphony Orchestra, the laboratory dedicated *Pavilion/Sculpture for Argonne,* by New York artist Dan Graham. Constructed from glass and mirrored panels supported by steel frame, Graham's work conceptually complemented building 201. The sculpture, which was divided diagonally with clear glass, featured open spaces where mirrors reflected the morning sun, creating a prism-like image of changing sky and clouds as the laboratory woke to the day's work. According to the artist, both structures captured the sun; and both "worked" best when people entered interior spaces to observe and be observed. Participant-observers separated by the clear diagonal wall were "two different audiences that can't touch each other but can see each other," artist Graham explained.[72]

Massey may not have cared for Graham's veiled reference to C. P. Snow's two cultures (one scientific and technical—the other humanist and artistic) divided so that they could observe but never touch, or presumably, understand one another. Massey knew that Argonne was a cultural island as well as a scientific laboratory. Although the laboratory shared similar values, language, and worldview with the larger scientific community, it also possessed a unique history that gave Argonne its singular identity. But Massey resisted any drift toward tribalism when he affirmed that art and science were simply different ways of structuring reality to enhance understanding of the universe, "and hopefully by that to better understand ourselves." For Massey, *Pavilion/Sculpture* helped resolve the tension of being both a part of and apart from the larger society that universally shapes the identity of all institutions. He was delighted that the sculpture project had enlisted support from the arts community at a crucial time in Argonne's history. Massey believed that Graham's work helped interpret Argonne's mission to people who normally had little interest in science and technology. *Pavilion/Sculpture for Argonne* symbolized the unity of the creative spirit, because art, music, and scientific research reflect "different facets of the same universal creative impulse," affirmed the former dean of arts and sciences.[73]

Massey's Triumph

Walter Massey, like other directors before him, worked indefatigably to secure for Argonne a new multimillion dollar research machine, the instrument that kept big-science laboratories alive. Why, then, was Argonne's only major construction in fifteen years an administrative center and not a large research facility? The Energy Research Advisory Board had already reported that the multiprogram laboratories, which increasingly served as "job-shops," were overburdened with administrative red tape from Washington. Was it symbolic that building 201 was Argonne's first facility dedication since the closing of the zero gradient synchrotron?

If nothing else, building 201 indicated that science administration would play an increasingly important role in Argonne's mission. Massey observed that the building not only symbolized Argonne's proud past, but also promised a more distinguished future. Joining the collective euphoria, Jan Mares, the DOE's acting under secretary, announced that the department had entered discussions with the University of Chicago for a new five-year contract—evidence of DOE's faith that Argonne would "retain its technological vitality." The standing-room-only crowd warmly applauded Mares's announcement that Argonne's immediate future was secure. But the victory really belonged to Massey, who had finally put to rest speculation that the laboratory would be closed. At the conclusion of the building 201 dedication ceremonies, a hundred green and white balloons drifted upward through the three-story atrium. No one misunderstood this symbolism of rising hopes and spirit at Argonne National Laboratory.[74]

The Argonne laboratory board of governors lost no time organizing for the future. Argonne's budget for fiscal year 1984 was $226 million, an 11 percent decline from the peak 1981 Carter energy budget. Employment was down 20 percent (1,000 positions) from its peak in 1978. Refusing to be discouraged by these setbacks, the governors urged Massey to press ahead with new initiatives in materials science, supercomputing, and neutron scattering. Learning that the Department of Energy's HEPAC had recommended construction of a multi-tevatron particle accelerator (the Superconducting Super Collider—SSC), the governors agreed that Argonne might be considered as a lead design center for the accelerator.[75]

To oversee all scientific and technical work at Argonne, the governors established the Scientific and Technical Advisory Committee (STAC) to coordinate all peer-review committees and to recommend changes in laboratory procedures and policies. The review committees supervised by STAC initially included reactor development, high-energy physics, the Intense Pulsed Neutron Source, the medium-energy electron accelerator (MEEA), the fusion program, biological and medical research, and nuclear engineering education. Frank Putnam, a distinguished biologist from Indiana University, a member of the National Academy of Sciences and a former member of the AUA board of trustees, agreed to serve as chair.[76]

In addition to the future of the breeder and fusion programs, the Science and Technical Advisory Committee was especially interested in reviewing programs involving user facilities and industrial interactions, the most notable of which were in the growing area of materials science research. Argonne hoped to be designated a Department of Energy lead laboratory for materials science. Rejuvenated after World War II by the advent by new analytical and characterization techniques that enabled scientists to probe the molecular and atomic structure of matter, materials science had enjoyed a boom in the development and modification of materials with enhanced mechanical, electrical, corrosion resistance and wearing properties. Argonne excelled in the operation of major national facilities for materials research: the High-Voltage Electron Microscope–Tandem and the IPNS Facilities.[77]

The High-Voltage Electron Microscope

On September 21, 1981, Argonne dedicated the High-Voltage Electron Microscope–Tandem facility, the only electron microscope in the world that enabled research scientists to record radiation damage at the molecular level while it actually occurred (figure 35). Electron microscopes employ a beam of electrons, instead of a beam of light, to magnify objects. High-energy electrons have very short wavelengths in comparison to ordinary light. Consequently, electrons scanning across a surface can detect variations thousands of times smaller than a conventional high-powered microscope. The greater the accelerated energy, the deeper electrons penetrated the target, the smaller the object they detected, thus the higher the magnification. The HVEM was able to magnify objects from sixty-three to one million times, enabling surface scientists to analyze sample areas as small as ten atoms in diameter.

Argonne modified a commercially manufactured English electron microscope to create the world's most powerful HVEM-Tandem. The uniqueness of this facility was its ability to bombard target samples in situ with beams of heavy ions injected into the viewing area by one of two accelerators (300,000 volts and 2,000,000 volts). In 1983, improvements to the HVEM-Tandem facility enabled researchers to study radiation effects on the molecular level using the

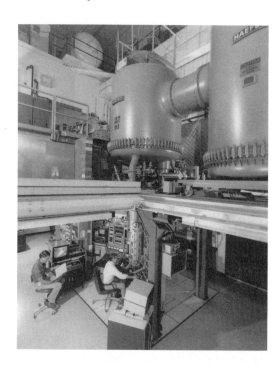

Figure 35. High Voltage Electron Microscope

two accelerators simultaneously. Among the observed phenomena, researchers could "watch" radiation damage in "real-time," detect the onset of crack formation, and measure corrosion as it occurred.[78]

The HVEM-Tandem attracted research scientists from across the nation to study the fundamental chemical microstructure of materials. One of the most important discoveries at the facility clarified understanding of radiation-created defects in metals. In high-radiation environments such as are found in a nuclear reactor, radiation can knock atoms away from their normal place in metal crystals, creating voids where atoms should normally reside. These vacancies often caused distortions, embrittlement, or weakening of the structure. Conventional theory held that the voids would be "frozen" in place at temperatures close to absolute zero, but under higher temperatures would probably collapse, weakening the metal. Argonne studies of radiation effects on gold-copper alloy and iron, however, determined that temperature made almost no difference. In other words, calculations predicting radiation damage made under the old theory actually underestimated the number of voids created.[79]

In practical terms, the research led to the development of new or improved materials for use in nuclear reactors, fusion devices, and fossil-energy systems. Research with the HVEM-Tandem provided scientists with a better understanding of sintering, surface change, corrosion, and mechanical failure, all physical processes important to improving materials used in industry.[80]

The high-voltage electron microscope provided the cornerstone for Argonne's electron microscopy center, which established a unique, world-class, user facility for exploration of materials in a variety of gaseous, electron-, and ion-irradiation environments. In addition to the HVEM, the center maintained conventional and advanced electron microscopes especially well suited for analytical projects requiring highest resolution under ultra-high vacuum.

The Intense Pulsed Neutron Source

The major facility in Argonne's materials science program was the Intense Pulsed Neutron Source, another important national user facility. As Massey explained to the board of governors, the IPNS generated an intense beam of neutrons, constituents of the atomic nucleus without electrical charge and thus able to penetrate matter more easily and deeply than charged particles, such as electrons. Experiments with pulsed neutrons provided data on the positions of atoms within molecular structures and assisted in understanding the behavior of atoms under extremes of pressure, temperature, and magnetic fields. Thus, neutrons, along with x rays, which provided some of the same information, had become important tools used by physicists, chemists, material scientists, and biologists for studying condensed matter, i.e., liquids, solids, polymers, and organic materials, but not gases.[81]

"Weak" interacting neutrons had an advantage over "strong" interacting x rays, however, because the lower-energy neutrons did not significantly disrupt the material being studied. As Massey explained, "by projecting a beam of neutrons into a specimen and looking at modification of the momentum, energy, polarization and intensity of the beam as it emerges from the material, one can infer the most intimate structural detail of the specimen." One of the greatest challenges in developing advanced energy systems from solar to fusion power, was to understand how materials behaved under high stress. In addition, neutron experiments enabled researchers to study radiation damage to materials similar to that produced in neutron-rich environments found in fission and fusion reactors.[82]

Since the end of World War II, fission reactors had provided the principal source of high-flux neutrons for research. Until it was shut down almost simultaneously with the ZGS, Argonne's CP-5 reactor (1951–79) was one of three high-flux research reactors in the United States: the others being Brookhaven's

High-Flux Beam Reactor (HFBR) and Oak Ridge's High-Flux Isotope Reactor. These research reactors helped pioneer neutron scattering research in the United States using high-flux reactors to provide intense "steady-state" neutron beams. Technically, the Brookhaven and Oak Ridge reactors had virtually reached the limit of their flux (10^{15} neutrons/cm^2/sec, depending on how they were operated), which restricted the range of problems researchers could pursue. Smaller research reactors at the University of Missouri and the National Bureau of Standards had neutron fluxes of about 10^{14}.[83]

As early as 1975, with the closing of the ZGS in sight, Jack Carpenter and David Price of Argonne's solid-state physics division proposed constructing a $62 million Intense Pulsed Neutron Source that would have unique capabilities for exploration of the structure and dynamics of condensed-matter and radiation-damage research. The pulsed-neutron beam would be produced with an accelerator from the old ZGS rather than from a reactor. Accelerated protons bombarding a heavy-metal target, such as uranium, dislodge a stream of neutrons with energies close to that of the accelerated protons in a process called *neutron spallation*. Because the accelerator pulsed protons to the target in bursts, the spallation of neutrons also produced high-intensity neutron pulses for scattering experiments. The IPNS generated fewer neutrons than the competing reactors, but because it delivered its neutrons in rapid pulses Argonne scientists expected that they could obtain a concentrated flux 10 to 100 times greater than the "steady-state" flux from a reactor. The accelerator-assisted neutron scattering technique promised to provide important insights into the "microscopic properties of liquids, crystal dynamics, diffusion in solids, molecular vibrations, structure of biological molecules, structure and dynamics of glassy solids, behavior of magnetic materials, and in such possible applications as new studies of chemical reaction rates, protein solutions, surfaces, time-dependent phenomena, and liquid crystals and membranes."[84]

In 1977 the National Research Council of the National Academy of Sciences, acknowledging the limitations of "steady-state" neutron sources from reactors, recommended the development of a pulsed-neutron source with a high-flux (10^{16} neutrons/cm^2/sec) pulsed-spallation facility. In April, Argonne proposed a pulsed-neutron source in two stages, building on the success of a prototype pulsed-spallation facility called ZING-P. Although the total estimated cost of the IPNS had now crept up to $70 million, Argonne believed it could build the first stage, IPNS-I, for only $9 million because the buildings and supporting equipment, including the all-important accelerator worth more than $20 million, were available as surplus from the defunct ZGS. The Department of Energy did not approve total funding for the project but in 1977 authorized construction of the laboratory's first stage, IPNS-I. Simultaneously, the department also gave Los Alamos National Laboratory the go-ahead for development of a pulsed neutron source, the Weapons Neutron Research facility (WNR-PSR). Since the department could not decide which program to back, it encouraged friendly competition between the laboratories in developing the intermediate machines, leaving to a later decision when, where, or whether the "second generation" or 10^{16} machine would be built. Meanwhile, Massey reported, Argonne was prohibited from spending any money, either programmatic or discretionary, on IPNS-II design.[85]

The Department of Energy inherited neutron-scattering research from the Atomic Energy Commission, which promoted the field in the 1950s as part of Eisenhower's Atoms for Peace program. At virtually no expense to the weapons program, the AEC fostered neutron-scattering research at the national laboratories and elsewhere as part of its charter to explore the peaceful uses of atomic energy. The Department of Energy carried forward most of the AEC research agenda, but as budgets tightened, DOE program managers increasingly were forced to question the relevance of these old commitments. Neutron-scattering research had some energy applications, but much of the research was not pertinent to the department's principal mission. With promises to build new machines at Argonne and Los Alamos and the need to refurbish research reactors, Department of Energy managers fretted over their growing commitments to neutron-scattering research, which threatened to jeopardize other programs sponsored by the Office of Energy Research.[86]

Choices: The IPNS Challenged

As it tightened research budgets, the DOE welcomed the help of science experts to sort out the Reagan administration's priorities with minimal political repercussions. Peer review in allocation of government grants had become universally accepted in the science community and was now used to establish broad directions in research and development as well as to review individual research proposals. But when the Department of Energy asked William F. Brinkman of Bell Laboratories to head a neutron-scattering review panel to evaluate the department's programs and facilities, Massey feared that the deck might be stacked against Argonne. The community of researchers working on neutron scattering was small, and given expectations that funding would remain level, he anticipated that recommendations would be based on perpetuating existing reactor-based programs.[87]

The Brinkman panel considered two alternatives: (1) growth in funding for neutron-scattering research, and (2) no growth in available dollars. Under the growth scenario, the Brinkman panel obviously saw no problems in funding everyone and recommended upgrading of the Brookhaven and Oak Ridge reactors and support for development of IPNS-I and the Los Alamos facility. Under the no-growth option, which projected constant funding at about $15 million a year (1981 dollars) from DOE's Division of Materials Science, the Brinkman panel endorsed the projects at the other three laboratories but recommended the termination of Argonne's IPNS-I so that remaining laboratories would not have to sacrifice their programs to restricted budgets. Under limited funding, the Brinkman panel was unwilling to endorse more than one experimental pulsed-neutron source. Because Argonne's project heavily taxed the neutron-scattering budget, whereas the Los Alamos program was largely supported by defense programs, it was easy to sacrifice the Argonne program, which would save money for upgrading the research reactors at Brookhaven and Oak Ridge.[88]

Massey was frustrated by the Brinkman panel report, which, if adopted, would zero out Argonne's most important basic research program. He was also tightly boxed in. Unable to complain about DOE management directly or to attack publicly his rivals at Brookhaven, Oak Ridge, and Los Alamos, Massey had no choice except to critique the Brinkman panel. To do so, however, meant he had to challenge the peer-review system. Whatever the merits of his case, Massey also knew he could not win at DOE headquarters with only a scientific or technical rebuttal. Saving the IPNS project fully taxed Massey's political skills.[89]

While Massey cautiously argued that the Brinkman panel "missed important differences" between IPNS-I and its rival at Los Alamos, he built his case more on institutional grounds than on scientific justification. Implementation of the Brinkman panel recommendations would seriously damage Argonne's "health and vitality," he reminded the Department of Energy. For three years, the Intense Pulsed Neutron Source had been Argonne's highest priority project; for three years the department had approved and endorsed IPNS planning and construction efforts. Together, the laboratory and the DOE had invested thousands of man-hours and millions of dollars on the neutron-scattering project. Massey recalled that "pulsed neutron scattering in the United States was essentially invented at Argonne," and that IPNS-I was the culmination of development that began with the ZING and ZING-P facilities. To help counter accusations that he sidestepped peer review, Massey reminded the department that public scientific conferences and frequent project reviews had repeatedly demonstrated that Argonne had the necessary "scientific, technical, and managerial" talent to build IPNS-I and beyond.

What would be the consequences of closing IPNS-I? In addition to crippling Argonne's materials science research program, Massey predicted that "Argonne, and the basic energy sciences effort in the Midwest, would be left without a major research facility, thereby adversely affecting a large segment of the university community, and the relationship between the laboratory and the universities."[90]

Midwestern science would not be the only loser, according to Massey, who again played on Cold War insecurities to defend local science programs. Americans had fallen behind the Europeans in neutron-scattering research, he estimated, with Great Britain, West Germany, and France each spending more than the United States. The French facility, Institut Laue-Langevin (ILL) at Grenoble, received more funding than the entire United States program. Both Great Britain and Germany planned to build pulsed-neutron facilities. Yet until the British machine came online at the Rutherford laboratory in the mid-1980s, the United States, through Argonne, would reign supreme by operating IPNS-I, the finest pulsed-neutron source in the world.[91] The Brinkman panel's recommendation to abandon Argonne's Intense Pulsed Neutron Source, Massey concluded, was scientifically and politically untenable.

IPNS-I Gets the Green Light

Massey deftly bought time for IPNS-I by not directly challenging the Department of Energy or the peer-review system, but rather by presenting a strong scientific and institutional case that did not require a repudiation of the Brinkman panel or an "all-or-nothing" commitment on the part of the department's Office of Energy Research. Because the department had terminated both the ZGS and CP-5 in 1979 and had not given Argonne anything in return, the Department of Energy agreed that it was politically impossible to scrap Argonne's Intense Pulsed Neutron Source just six months before it was scheduled to begin operation. Meeting with the laboratory directors, Douglas Pewitt, the acting director of the Office of Energy Research, struck a compromise—Argonne's IPNS would be funded through 1985, after which Argonne would have to find sources of money other than the Department of Energy. Because funding levels only allowed Argonne to operate IPNS-I on a part-time basis, however, recruitment of outside users began immediately. Although Massey had no idea where he would get the money to run IPNS-I in the long run after the facility became operational on May 5, 1981, Argonne's best alternative was simply to continue and worry about crossing the next funding bridge when the time came.[92]

Department of Energy
Materials Sciences Neutron-Scattering Funding Plan
($, millions)

LABORATORY	1981	1982	1983	1984	1985
Ames	0.33	0.36	0.36	0.36	0.36
MIT	0.20	0.24	0.25	0.25	0.25
ANL	6.50	6.50	6.00	4.00	2.20
ORNL	2.60	3.13	3.50	3.89	4.40
BNL	3.00	3.91	4.44	5.00	5.50
LANL	0.85	1.00	1.10	1.40	1.70
New source	0	0	0.25	1.00	1.49
Totals	13.48	15.14	15.90	15.90	15.90

The DOE's 1981 five-year neutron-scattering funding plan conveyed both hope and disappointment to Argonne's materials sciences division. The division was happy to have won the political battle for survival, but it was distressed by budget projections indicating that Argonne could no longer hope to promote IPNS-II as the laboratory's first priority for research and development. With constant funding projected until 1985, the plan clearly indicated that after the Los Alamos facility became operational in the mid-1980s, Argonne would cease to be a center of pulsed neutron research. John Walsh of *Science* was surprised by the ingenuity with which Argonne constructed IPNS-I at a bargain cost out of "hand-me-down equipment" rescued from the old ZGS. Yet for all their resourcefulness, Walsh also reported that Argonne's neutron scientists were worried that they did not have enough money to keep their machine running until the Los Alamos facility was ready.[93]

ATLAS

"From the very beginning," Lowell Bollinger recalled, "our new machine [the Argonne Tandem-Linear Accelerator System (ATLAS)] had everything against it." Bollinger, the director of the accelerator project, explained that ATLAS employed a difficult and unreliable technology, struggled along on limited funds and a tight schedule, and relied on a young and thinly staffed development team. But the world's first superconducting, heavy-ion accelerator, although not as large a facility as IPNS-I, became a distinguished world-class research facility. ATLAS was designed to probe nuclear structures with heavy-ion projectiles, particles heavier than a helium nucleus. During the 1970s, scientists using tandem Van de Graaffs and cyclotrons reached the limits of size and energy at which they could accelerate heavy particles (figure 36).[94]

Rather than scrap their research accelerators, scientists at Stanford and other

Figure 36. Argonne Tandem-Linac Accelerator System (ATLAS)

universities in the 1960s and early 1970s explored ways to increase the power and efficiency of the tandem Van de Graaff by boosting energies with innovative linear accelerators (linac). One group, led by Kenneth Shepard at California Institute of Technology, invented a device called a split-ring resonator, which offered interesting possibilities for accelerating positively charged heavy-ions to sufficient energies to penetrate positively charged target nuclei. An obvious, but bold, approach would be to link a superconducting version of Shepard's linac to a tandem Van de Graaff that injected heavy-ion beams into the system. Superconductivity is achieved when metals, at temperatures close to absolute zero, lose their resistance to the flow of electrical current.

When Bollinger first presented his ideas for a superconducting linac at a 1973 meeting of the American Physical Society, reportedly his vision was rejected with a "scathing attack." Nevertheless, work on a heart-shaped niobium resonator began at Argonne in the spring of 1975 and by April 1976, when Shepard's team successfully tested its first prototype, converts had rallied behind Argonne's innovative program. Recognizing that this promising technology might be used to upgrade tandem accelerators across the country, the Department of Energy invested a modest $2 million in 1975 to keep development on track.[95]

After many failures due to faulty welds, the project managers asked for help from Argonne's materials experts in the reactor program. Demonstrating the strength of a multiprogram laboratory, within weeks the nuclear engineers solved the welding problem. By November 1977 the accelerator group successfully tested the niobium split-ring resonator, in which a powerful charge of electrons oscillated 100 million times a second between the resonator's two doughnut-shaped drift tubes. When the pure niobium split-ring resonators were cooled by liquid helium to near absolute zero (4.6° Kelvin or -450° F), they lost all electrical resistance. The superconducting resonators produced rapidly alternating electrical fields that accelerated the heavy-ion beam using one-tenth the energy of a conventional accelerator. By September 1978 Shepard's team successfully accelerated a beam of oxygen ions through two modules of the prototype superconducting linear accelerator. That success justified pushing ahead to complete the linac by 1982 with a series of twenty-four split-ring resonators. The aggressive program won support from Congress in 1980 when the laboratory designated its program the Argonne Tandem-Linear Accelerator System. Even before completion of the facility, whose individual modules could be used independently for research, ATLAS attracted scientists from universities and research institutes around the world. Almost 50 percent of the available beam time was dedicated to visiting researchers, making ATLAS one of Argonne's major user facilities.[96] Following Argonne's disappointment in securing endorsement from the Department of Energy for IPNS-II, the laboratory had remained without a big-science research program other than support of the Clinch River breeder-reactor project. The development of the tandem-linear accelerator was promising with the successful test of the split-ring resonator on June 26, 1981. But full-scale operation of ATLAS, with three additional resonator modules, was at least three years away, provided Argonne received the necessary funding from Congress.[97]

Paul Fields, at the time the director of chemistry, remembered the agony that went into the IPNS and ATLAS projects. Without start-up money for new projects, Bollinger and Fields had to tax their respective divisions (to the tune of a half a million dollars in the case of ATLAS) to obtain funds just to get the projects underway. Although the laboratory director did not have discretionary authority to move money around, the division directors did. With Fields's agreement, the ATLAS project started in the cyclotron building of the chemistry division, where all the accelerator people were located. "In order to maintain the accelerator people, I had to let my chemists go," Fields recalled years later, the painful choice still fresh in his memory. The price of staying at the cutting edge of science, Fields knew, too often required sacrificing valued programs and people. It was never easy, he lamented.[98]

The GEM-SURA Controversy

Of great interest to the high-energy physics community in 1981 was exploration of the inner structure of the atomic nucleus. The atomic nucleus consisted of protons and neutrons, called nucleons. Scientists believed that nucleons, in turn, were made up of quarks, incredibly small motes of energy with varying characteristics that combined to form nuclear particles. In some fashion, the quarks, in turn, were held together by "gluons." Although researchers could not determine their size, shape, or behavior, it remained possible that quarks themselves were not even the fundamental "building blocks" of nature. Electron accelerators provided one possible means to study quarks and their nuclear environment.[99]

In April 1982 the Nuclear Science Advisory Committee (NSAC) to the Department of Energy and the National Science Foundation proposed building a powerful electron accelerator with a peak energy of four billion electron-volts (4 GeV) to explore the deep inner structure of matter. When the department approved the committee's recommendation, Argonne, which desperately wanted a large physics project, proposed to build the world's most advanced electron accelerator to serve as an international users facility. Potential competitors were the MIT, the Southeast Universities Research Association (SURA), the University of Illinois, and the National Bureau of Standards.[100]

Argonne's entry into the multimillion dollar competition was an innovative design called GEM (GeV Electron Microtron). The GEM machine was not a conventional linear accelerator, which is constructed to accelerate electrons in a straight path. Rather, GEM consisted of three identical short linear accelerators connected by three identical pairs of magnets that bent the electron beam along a six-sided "hexatron" path. Accelerated electrons would race around the hexatron thirty-seven times to reach an energy of 4 GeV in 22 millionths of a second. Small versions of the microtron had already been tested successfully at Argonne, which had pioneered magnet technologies for both basic and applied programs. The laboratory estimated a construction cost of $107 million, which included saving $35 million by rehabilitating old ZGS buildings.[101]

The manager of the hexatron project, Harold E. Jackson, poetically described his proposed accelerator as a "gentle" explorer of nuclear structure. Proton accelerators broke nuclei apart with great force so that scientists could study the pieces. In contrast, GEM was like a fine jeweler's tool used to probe the inner structure of the nucleus without smashing it. Electrons, two thousand times less massive than protons, were a "gentle force," Jackson explained, because they bounced off atomic structures, creating detailed "shadows." As science writer Dennis Byrne illustrated, GEM was like a powerful flashlight illuminating a target: "The shadow cast by the object reveals some information about the object. But instead of a weak, visible beam of light [only 1.5 eV], the accelerator shoots a very powerful beam of electrons [4 GeV] at an object, also producing an informational 'shadow.'"[102]

Of the five entries, Argonne was not worried about the challenge from the University of Illinois and the National Bureau of Standards. More serious because of its potential political clout was the SURA proposal to build a National Electron Accelerator Laboratory (NEAL) at Newport News, Virginia. Sponsored by a consortium of twenty-two southeastern universities, SURA offered a safe design, but with the existing national laboratories under close scrutiny from the White House, Massey could not believe that the Reagan administration would authorize still another independent laboratory. The real competition seemed to come from MIT, which proposed to upscale its existing Bates Linear Accelerator. Massey acknowledged MIT's strength in intermediate-energy physics, which might have given it the edge except that the Bates facility had a reputation for being badly managed. Massey noted privately, "We have the advantage . . . by having a really innovative, revolutionary machine."[103]

Argonne's GEM team and Massey felt good about their proposal, which had been assembled in almost record time for consideration by the Panel on Electron Accelerator Facilities (of the DOE-NSF Nuclear Science Advisory Committee), chaired by D. Allan Bromley. Not since the MURA controversy in the 1950s had Argonne gone head-to-head in competition with a university consortium to obtain funding for a major research facility. There were a few rough edges in the proposal, especially regarding coordination with area universities, whose representatives believed that Argonne should hire a permanent senior electron physicist to work at GEM instead of relying on prominent visiting scientists. They also knew that their proposal included some "unknowns and risks" but were confident that Bromley's peer-review panel would welcome their project which pushed technological limits.[104]

SURA: The Chosen

Argonne and its friends could hardly believe the findings of Bromley's Panel on Electron Facilities. Neither its chief rival, MIT, nor Argonne were chosen for the site of the DOE's new electron accelerator. Instead, the new consortium of southeastern universities with its conservative design won the committee's recommendation. Their endorsement was not unanimous, however, and Bromley's committee stated that Argonne's concept was feasible, high praise for the Illinois laboratory's proposal. With an ironic cut at the midwesterners, however, the panel predicted that SURA's laboratory "will be constructed and managed as national rather than a regional facility." The panel carefully established SURA's technical "superiority" not only in its more conventional design, but also in its potential to be upgraded to 6 GeV. Bromley's panel was not enthusiastic about SURA's site at Newport News and asked the consortium to explore another site. On the other hand, the panel was ecstatic that SURA promised to establish five Commonwealth professorships, twenty-five tenured professorships in experimental nuclear physics, and at least five tenured professorships in theoretical physics.[105]

Having succeeded in neutralizing the Brinkman panel, Massey politely but vigorously sought to overturn Bromley's recommendation. Confident of his political skills, diplomatically he asked for a change of venue on the decision concerning the electron accelerator. SURA's pledge to create thirty-five new professorships was impressive, but on purely technical grounds, Massey continued to believe that his laboratory excelled in cost, support, and experience. Massey wanted Bromley to forward his report without recommendation to the Department of Energy, where "the institutional and policy questions can be discussed within the proper framework." It was the only chance he had to save Argonne's bid for the electron accelerator.[106]

Meanwhile, Argonne's allies, headed by Senator Charles Percy, appealed directly to Secretary Hodel, who was reminded that the ERAB had opposed creating another new laboratory. Argonne and Chicago had every advantage in laboratory facilities, skilled employment pool, and community infrastructure over SURA and Newport News, which even lacked a major international airport. If the Department of Energy really believed the SURA proposal was scientifically superior, Percy offered to construct SURA's machine at Argonne where it could be built at least cost.[107]

Argonne's battle to save the GEM project was short-lived. At the White House, presidential science advisor George Keyworth was furious with Massey for undermining the integrity of the peer-review process by playing pork-barrel politics in what looked like an attempt to steal the SURA machine for Argonne. Keyworth, who had studied undergraduate physics with Bromley, dismissed Argonne's appeal as narrowly "institutional." "If we are going to improve excellence and expand opportunities in American science, the single greatest threat to progress is parochial institutional concerns," he chided. Keyworth respected effective, aggressive competition, but he demanded to keep science decisions within the science community and out of the political arena.[108]

The GEM Postmortem

In history, sometimes events that do not happen are as instructive as those that do. Argonne's loss of the GEM project not only reflected how the laboratory was perceived within the science community, but also revealed how Argonne staff viewed themselves.

Keyworth's rhetoric notwithstanding, Massey was sensitive to the nuances of science and politics, and by late spring, he knew he had lost the fight. Meeting privately with Bromley on May 17, Massey assured him that Argonne would not press the issue to the point that no one had the accelerator. Bromley assured Massey that politics had not played a direct role in what he described as a "strange" decision. Bromley confided that he had favored Argonne's proposal and was surprised that the Illinois laboratory did not win the contest. One problem was that Argonne's top scientists in the physics division did not seem fully committed to GEM, but instead were focused on the successful ATLAS project. SURA's promise to create thirty-five professorships signified that the organization was "totally behind the project," in contrast to the "half-hearted commitment" of Argonne's best scientists.[109]

Bromley's assessment of Argonne's failure to win the electron accelerator was no more frank than the internal evaluation of J. P. Schiffer, the associate director of the physics division. Under the best of circumstances, the GEM design would be difficult to sell in competition with SURA's conservative approach. Although imaginative, innovative, and sound, the microtron was untried at higher energies. Schiffer looked for lessons learned rather than recriminations, and he concluded that Argonne's failure was institutional rather than scientific.

In general, Argonne's GEM team should have been more focused in its presentation and response to Bromley's panel. Acknowledging Argonne's poor record of university relations, Schiffer thought Argonne should have lined up "some support" from regional universities and cultivated the accelerator community "so that they understood [GEM's] problems and solutions in detail." As it was, he reflected that his own people were good workers, but an "absolute disaster" in relations outside the laboratory; or were very good technically, but gave the "impression of slick arrogance"; or had excellent judgement, but did not follow through on assignments; or worked very hard, but did not seek advice; or possessed good instincts, but had limited experience and few contacts in the field. Schiffer tagged himself with naivete for not anticipating traps in the review process with Bromley's panel. Lastly, he held Massey responsible for failing to assert his authority on priorities and management and to mobilize outside support, especially with the University of Illinois, soon enough to be helpful.[110]

Massey threw in the towel on the GEM project on July 19, 1983. Although the Illinois congressional delegation was willing to block funding for the electron accelerator if Argonne did not get the machine, he knew such a "victory" in the long run might hurt the laboratory more than losing GEM. Under pressure from Keyworth in the White House and Trivelpiece at the DOE, Massey withdrew from the stalemate, which could hurt American science if the project were delayed or lost. Although rumors circulated that the department had promised Argonne a consolation prize, no official announcement ever confirmed any such horse trading.[111]

Alan Schriesheim Appointed Senior Deputy Director

In the midst of the GEM-SURA controversy, Argonne's management hired Alan Schriesheim to the newly created position of senior deputy director. Schriesheim's recruitment fulfilled the University of Chicago's two-year-old promise to find a technically savvy administrator to serve as chief operating officer of the laboratory. Officially, Schriesheim was responsible for Argonne's day-to-day operations, which included chairing the management council, developing plans and budgets, and coordinating the laboratory's relations with industry and government. At least initially, he was to concentrate on internal affairs while Massey promoted Argonne's interests in Washington, D.C., and beyond.

When Schriesheim left the Exxon Research and Engineering Company, where he served as general manager of the Engineering Technology Department, to become second in command at Argonne, he became the first director of a major corporate research program to move to a top management position at a national laboratory. Born in 1930, Schriesheim grew up in New York City where he developed a precocious interest in chemistry. Not satisfied with his A. C. Gilbert chemistry set, young Schriesheim haunted local chemical supply houses where he purchased "all kinds of chemicals" to bring home to his family's "very small apartment" in Rockaway Park, Queens. Schriesheim graduated with a bachelor of science degree in chemistry from the Polytechnic Institute of Brooklyn and a Ph.D. in organic chemistry from Pennsylvania State University in 1954.

Schriesheim had no interest in teaching, and after a two-year stint at the National Bureau of Standards, at age twenty-six he took a job as a research chemist with Esso Research and Engineering Company, a recently established research division in New Jersey. Within five years he had published more than fifty papers and established a reputation in the field of homogeneous and heterogeneous catalysis. As a leader of one of Esso's research groups, he soon discovered he had considerable organizational talent. In his early thirties, Schriesheim chose to pursue the company's management track instead of remaining in a technical position.

By 1980 Schriesheim was restless. After a distinguished career at Exxon, where he reached the top of the senior executive ladder, at fifty-three he wanted a new challenge. He knew Washington, D.C., intimately, having served on the Department of Energy's Energy Research Advisory Board and as chair of the petroleum division of the American Chemical Society, among several boards and committees with which he was affiliated. He was well acquainted with Trivelpiece, and as a member of MIT's chemistry visiting committee, he became good friends with John Deutch, the chair of the MIT chemistry department. In addition, when Paul Fields was director of the division, Schriesheim served on the chemistry division review committee. From these contacts and others in industry, Schriesheim knew that Argonne was looking for a chief operating officer. Indeed, Fields and others in the chemistry division had recommended him to be Argonne's chief operating officer.[112]

Well acquainted with the situation at Argonne, Schriesheim was encouraged by the appointment of a strong board of governors well represented by industry. Although he was enticed by a joint appointment in the University of Chicago's chemistry department, the chief attraction was the opportunity to run a major government laboratory. The only position that really interested Schriesheim was Massey's job. No commitments were put in writing, but Schriesheim's "gentlemen's agreement" with Hanna Gray, Massey, and Trivelpiece assured him of promotion to the directorship if they all liked one another. When he joined Argonne on September 15, 1983, Schriesheim was uncertain how long he would serve his apprenticeship, but he anticipated becoming director of the laboratory within a year.[113]

The transition from Massey to Schriesheim was not easy. When Schriesheim arrived at Argonne he discovered "a much bigger challenge" than he expected. The laboratory had just lost the GEM project, and the Clinch River Breeder Reactor program was in serious trouble. Argonne had no other major research-and-development initiative. Although building 201 was architecturally stunning, "the place itself [was] depressing," Schriesheim remembered, run down and shabby because little money had been allocated to maintaining facilities. Reagan's budget cuts, which reduced Argonne personnel to fewer than four thousand for the first time in a decade, had also battered employee morale. In Washington, Schriesheim learned, Argonne was called "the sick man" of the national laboratory system.

Schriesheim understood Massey's outreach responsibilities and appreciated his political skills but chaffed under Massey's continued involvement in Argonne's internal operations. In retrospect, Massey himself confessed that he probably withdrew too slowly from day-to-day management of the laboratory. But Massey enjoyed working in his new office in building 201, where Argonne employees and the Chicago community continued to regard him as the laboratory's director. Schriesheim perceived himself as a builder, not a hatchet man, but it was impossible to implement his own agenda to pare away the weak and build up the strong as long as he was playing second fiddle to Massey. A dynamic, risk-taking personality, Schriesheim remained professionally friendly with Massey while he waited impatiently for the transition he was promised.[114]

Cancellation of the Clinch River Breeder Reactor

Suspense over the fate of the Clinch River Breeder Reactor ended suddenly on October 26, 1983, when the Senate stopped funding. The House had previously refused to continue authorization for the breeder, which had been the first priority for the nation's nuclear energy policy since the Nixon administration. But research and development had been troubled by technical delays, cost overruns, and an easing of the energy crisis. The Reagan administration gave the breeder reactor its highest energy research-and-development priority, but major questions about breeder economics among Congressional fiscal conservatives doomed the Clinch River project as wasteful.[115]

What were the effects of the collapse of the Clinch River Breeder Reactor project on Argonne? In 1983, the breeder program accounted for 40 percent of Argonne's total budget, and virtually all of Idaho's. As Charles Till, the associate laboratory director for reactor research and development, explained, for several years the CRBR program received about half of the DOE's $600 million reactor development budget. Argonne's funding totaled $80 million of the $300 million base program, which was approved by Congress. Thus the immediate impact of the cancellation of the Clinch River project was small since Argonne's research-and-development work in Idaho remained, for the time being, almost fully funded.

Till and his associates, however, were deeply worried about the future of their work because so many nuclear critics in and out of Congress believed that the Clinch River project *was* the breeder program. Even the friends of nuclear power in industry were likely to lose interest in base breeder technology after the CRBR was gone. Till conceded that continued support for breeder research and development would be difficult without a prototype reactor project such as the CRBR. What was the point of spending millions of taxpayer dollars on advanced research that did not support construction of a demonstration reactor?[116]

Till's frustration was palpable. The United States' nuclear power program was in deep trouble, he believed, in part because it had embraced two reactor designs that had no long-term future, the light-water reactor and the CRBR loop-design breeder. There had been no new orders for light-water reactors since 1979, and several projects were in financial difficulty. Since the days of EBR-I, nuclear power proponents at Argonne believed that the breeder reactor would be the ultimate power-generating system of choice. Nuclear critics, however, linked development of the breeder with the future of the light-water reactor. Until light-water reactors exhausted uranium supplies, according to the antinuclear argument, there would be no need for breeders. With the light-water industry in the doldrums, uranium supplies were going to last a very long time.[117]

Released by the Senate's action from his obligation to support the administration's CRBR policy, Till expressed relief to be freed from the handicap of the Clinch River project. Even with the CRBR dead, there remained in Congress substantial support for continued research on breeder technology as a long-term source of very large amounts of environmentally clean electrical power generation. Till was happy to sweep away the rancorous debate over the CRBR; the conflict had obscured America's very real need to develop large-scale energy alternatives. The Clinch River project was not only too costly, but Till and his Argonne colleagues also believed it promoted the wrong version of the breeder technology. He had never been enthusiastic about the CRBR, and now Till saw an opening for Argonne to reclaim its leadership in reactor design. Argonne's patient and continuous work on EBR-II, once thought out of the running as America's preferred reactor, now positioned the pool design, featuring metal fuel and metallic fabrication and reprocessing, to compete for research-and-development funds. Optimistically, Till announced, "With the CRBR gone, a new era begins."[118]

The Board of Governors and the Future of Argonne

Argonne's "new era," of course, would be charted by the board of governors with the advice of the scientific and technical advisory committee. First on the agenda for the governors was to answer the findings of the Energy Research and

Advisory Board (September 1982), of the Grace commission (April 1983), and of the Packard committee (May 1983).[119] All three reviews had emphasized the need of the national laboratories to establish clearly defined and appropriate missions. Argonne's governors agreed that this was a high priority, but they also responded that laboratories could not adopt a mission that supported the Department of Energy unless the department itself clearly knew its role and mission. Vagueness of mission was as much of a problem in Washington, D.C., as it was in Chicago. Further chiding the government, the governors predicted that writing mission statements was useless without "commitment for continued and predictable support." If program funding were to be determined by laboratory mission, then funding decisions could not be made by the program managers at headquarters, who were ignorant of, or hostile to, mission-defining programs.[120]

Although the governors sent a testy response to Washington, they confessed that the laboratory reviewers had a point: Argonne was not guided by a comprehensive role or mission. Argonne's centers of excellence and of research strengths were well documented, especially in nuclear reactor technology, biology and medicine, and materials science. Programs in physics, high-energy physics, and nuclear physics were world renowned. But despite the multiplicity of distinguished programs and personal excellence, the governors observed, "an overall absence of *program focus* [was] a major deterrent in presenting Argonne as DOE's technical leader." In other words, Argonne was not a "lead laboratory" for any of the DOE's major program initiatives.

Laboratory managers were handicapped in establishing a defining mission because they could not unilaterally combine projects into thematic programs, assign old programs to new areas, or assume leadership for research priorities in areas where they had unique talents. Yet, the governors believed it was essential for Argonne to seize the initiative somewhere if the laboratory were to establish a recognized national mission.[121]

Working closely with Massey and Schriesheim, the board of governors developed a strategic plan that sharpened Argonne's identity through program and facility enhancements. Drawing from the laboratory's strengths, the governors recommended expansion in the breeder program (the integral fast reactor), fusion technology, and advanced materials-science research. The governors also suggested that serious consideration be given to creating major new programs in biotechnology and advanced computing research and to establishing a Midwest coal technology development institute.

Reviewing existing or developing facilities, the governors acknowledged the excellence of the IPNS, the electron microscopy center, ATLAS, and the reactor program facilities at Argonne-West. These facilities constituted a unique national resource offering research opportunities not available elsewhere, fulfilling a major role as defined by Secretary Hodel for multiprogram national laboratories. But the governors also observed that much could be improved in this area. Although Argonne operated outstanding facilities, they were largely stand-alone programs not integrated into a larger laboratory plan and, because of limited funding, IPNS only operated at 60 percent capacity.[122]

In defending Argonne against the Packard committee's criticism that the national laboratories did not sufficiently encourage university and industry scientists to use their facilities, the governors recorded that in 1983 Argonne hosted more than 225 users from academia, industry, and foreign countries. Others benefited from Argonne facilities through collaboration with laboratory staff. In all, more than two thousand faculty and students participated in laboratory programs annually.

Nevertheless, in their self-assessment, the governors echoed previous critics from the Argonne Universities Association and the Energy Research and Advisory Board, who claimed that Argonne had not aggressively fostered an academic user community of scientists who felt "ownership" of Argonne's facilities. Below the level of top management, relationships between the laboratory and the University of Chicago were almost nonexistent. Collaboration was unusual, as faculty and their students did not often visit the laboratory. Except for Massey and Schriesheim, academic appointments for Argonne staff was the exception. Relationships with other universities were no better. The governors shared the opinion of ERAB's university-programs panel that Argonne should broaden user participation where appropriate, for example, by helping scientists discover state-of-the-art electron microscopy through workshops and symposia. Fellowships, joint appointments, exchanges, and enhanced collaboration were imperative. As the governors looked to the future, they confirmed that the relationship between Argonne and Midwest industry had not been close. Schriesheim was expected to change this. The governors encouraged industrial exchanges, collaboration, advisory groups, and joint programs, and they requested greater industrial representation on the board of governors. They also knew that the Reagan administration wanted to curtail the laboratories' commercial ventures while enhancing industrial access to developed technologies. The board endorsed the laboratory's proposal to establish an Argonne development center, funded by venture capital, to facilitate transfer of mature technology from Argonne to the private sector. Technology transfer for the center would include engineering development, economic analysis, market research, and product research and development. The governors looked forward to receiving a specific proposal at an early date.[123]

Schriesheim Named Argonne Laboratory Director

On May 10, 1984, less than nine months after moving to Chicago, Alan Schriesheim was named director of Argonne Laboratory. Superficially, there was a great contrast between the urbane, Ivy League former director and the new, tough industrial executive whose résumé included more than seventy scholarly papers, twenty-two United States patents, and the chemistry award of the American Chemical Society. Both men were politically astute, however. Schriesheim had already learned that there was not much difference between Argonne and a major industrial research laboratory. The mix of scientists and engineers was about the same in either the public or the private sector. Whether at an industrial, academic, or national laboratory, Schriesheim believed that the formula for success was the same: top management must foster and preserve a "spirit of excellence" as part of the laboratory culture. He reflected that one could manage either growth, decline, or the status quo—all could be painful. "Consistently I have found that life in an organization is difficult when you're at a status quo or when you're declining," he reflected in his "State of the Laboratory Address," given jointly with Walter Massey. "I'd much rather suffer the agonies of growing," he continued, "than I would the agonies of shrinking. Growth is what makes an organization vital." Obviously, renewed growth was Schriesheim's principal agenda for Argonne.[124]

12 | Challenging the Nuclear Option, 1984–94

White *Dama dama* deer roam wild at Argonne-East. Glimpsed along the tree line through a morning fog, found on a knoll during an evening rain, or spotted in headlights near a road at night, the white deer have become ubiquitous in the research park. As many as two hundred white deer, a rare variety of fallow deer native to North Africa and southern Europe, range freely over the research reservation. In contrast to the indigenous white-tailed deer with which they do not interbreed, the white deer are grazers thriving on Argonne's vast expanse of lawns and open fields. A gift from Chicago clothier Maurice L. Rothchild to Erwin O. Freund in 1936, the original herd of thirty-eight animals could not be contained by six-foot fences, and so were allowed to graze about Freund's estate. When the federal government acquired the Freund property in 1947, the Atomic Energy Commission was unaware that it was about to create a refuge for a small herd of the unique fallow deer (figure 37).[1]

The white deer were only the most visible evidence of the government's unintended success in establishing important wildlife refuges congruent with its national laboratories and production sites. Security and safety requirements dictated that the national science laboratories occupy large tracts protected from populated areas by extensive buffer zones. Generally, within the laboratory boundaries itself, human activity was confined to a relatively small sector, leaving an undisturbed environment for regional flora and fauna. At Hanford, for example, towns were actually abandoned to the desert along the only wild, free-flowing stretch of the Columbia River, where eagles and migrating curlew returned to find refuge.

Argonne National Laboratory, East and West, proved no exception in creating invaluable natural habitat (figure 38). The laboratory first realized the ecological importance of the buffer zones in 1967, when biologists at Argonne-East conducted a field survey that identified nearly two hundred species of plants and more than eight hundred species of animals, including insects and twenty-five species of mammals. The 1967 preliminary survey, which found eighty-six species of birds, was revised upward in 1990 to include forty resident summer species, plus another one hundred migrating or wintering bird

Figure 37. *Dama dama:* Argonne's snow-white fallow deer (*Frontiers, 1946–1996,* p. 4)

Figure 38. Argonne-East, Illinois

species. Among threatened or endangered species, Argonne-East potentially offered refuge to the bald eagle, peregrine falcon, piping plover, interior least tern, and Kirtland's warbler during migration or winter. The black-crowned night heron, listed by the state of Illinois as an endangered species, was a documented resident of Argonne's wetlands. The only large predators to threaten the laboratory's larger mammals were coyotes and dogs.[2]

Wildlife was abundant at the Idaho National Engineering Laboratory, which surrounded Argonne-West, because hunting was prohibited there (figure 39). In 1975 the reservation was designated as a National Environmental Research Park. That designation was not surprising because this desert habitat hosted forty-four species of mammals and more than two hundred species of birds, including the bald eagle and peregrine falcon.[3]

While the white deer thrived at Argonne-East, Americans became increasingly wary of allowing nuclear power plants to be built in their communities. Although the dangers of living near a nuclear plant were much less than those from air pollution, storms, crime, driving, flying, swimming, falling, having sex, or even drinking coffee, many Americans rejected nuclear power because they believed it was too risky. This perception puzzled and frustrated nuclear power advocates at Argonne because Americans willingly tolerated electrical production from fossil fuels which were far more damaging to human safety and health. While the Nuclear Regulatory Commission estimated that a nuclear

Figure 39. Argonne-West, Idaho

power plant might cause one human death over its lifetime, a coal-burning plant might cause more than a thousand deaths, primarily from air pollution. Initially, nuclear power debates thrashed over statistical estimates of risk. But after more than a quarter century of experience (including Three Mile Island and Chernobyl), with 112 nuclear power plants operating in the United States and another 200 similar plants operating around the world, nuclear power advocates had solid, historical data on which to base their argument in favor of the safety and reliability of western nuclear technology. Hans Bethe, Cornell University's Nobel laureate, noted the remarkable safety and environmental record of Western (non-Chernobyl) power reactors: "No death or serious injury has been caused by radiation from any American-style light-water reactor anywhere in the world in thirty years of commercial nuclear power."[4] The white deer, which had shared the research reservation with Argonne's staff for a half century, offered no direct testimonial in favor of nuclear power. Biologists reported, however, that their reproductive habits and success differed little from their old-world cousins.[5]

Reagan's Energy and Science Policies

By 1984 the Reagan administration had implemented most of its energy reorganization goals, making it unnecessary to deliver a political coup de grace to Carter's energy department. Deregulation of gasoline and oil prices, abolition of heating restrictions, elimination of most commercialization programs, reduction of personnel, and restructuring of the energy budget accomplished most of the administration's energy-management priorities.

The Reagan administration's research-and-development policies in energy and science were similar. In both areas, the administration sought to reduce federal spending where it believed the private sector could conduct research and development as well or better than a government agency or laboratory. At the same time, the administration maintained or increased funding for activities it believed vital for national security or too risky or expensive for private enterprise. As the President's science advisor, George Keyworth, explained, the Reagan administration wanted to promote growth for basic research, especially in the universities; increase and improve America's trained technical talent; and encourage cooperation between industry and academia while drawing a clear line between the research-and-development responsibilities of the federal government and the private sector.

Keyworth believed that the energy crisis of the 1970s had taught the American taxpayer an expensive lesson. Billions of dollars had been spent by the government on the development of new energy technologies, he reflected, but these technologies had failed to relieve energy shortages in any significant way. Instead, conservation stimulated by higher prices, new energy supplies, and the partial collapse of the oil cartel brought an end to the crisis. The nation would have been much better served, he argued, had the money wasted on "foolish" federal development projects been invested in basic research that strengthened the American economy and society over the long run.[6]

Although overall spending on nondefense research and development remained flat during Reagan's first term, Keyworth emphasized that the relative funding of basic research had actually increased because of the drop in government development programs. Much of the difference, according to Keyworth, went to universities and colleges, which gained a larger share of the federal nondefense research-and-development budget after consistent declines in basic research funding since 1968. Keyworth's message to the academic science community was crystal clear—stick with the Reagan administration, which was committed to real and sustained support of basic research.

Where did this leave the national laboratories, which employed about one-sixth of America's scientists and engineers with a combined annual budget of about $18 billion? Although the Reagan administration was determined to eliminate wasteful development programs, it had no misgivings about encouraging promising technologies that not only strengthened America's military muscle, but also enhanced the nation's economic competitiveness. Keyworth believed the national laboratories were "superb" but "under-utilized" resources. "We darn well better be finding ways to get substantial industrial benefit out of a federal investment of that magnitude," he lectured science writers in San Francisco.[7]

The new watchwords became "technology transfer." Casual observers might be excused if they could detect little difference between Carter's "commercial development" and Reagan's "technology transfer" projects.

Indeed, at Argonne, favored programs in nuclear technology, materials sciences, chemical sciences, and computer sciences enjoyed an easy transition from the one rationale to the other. Keyworth wanted "to stimulate the flow of ideas, expertise, and people between the federal laboratories, universities, and industry." While the Reagan administration promised to reduce micromanagement of the laboratories from Washington, D.C., in return it expected laboratory directors to revise their missions to emphasize more cooperation with universities and industry. In spite of billions of government dollars spent for research in space, biotechnology, health, and agricultural sciences, Keyworth thought America often floundered in its attempts to adapt laboratory research to industrial markets: "We respond so poorly, that we're in real danger of letting other countries assume the industrial lead in profitable new fields of technology that American scientists have done most of the research to establish—and that American taxpayers have underwritten."[8]

Ever alert to shifting political fashions, Argonne Laboratory soon established the Technology Transfer Center to upgrade its old Office of Industry Interaction and Technology Transfer. Argonne's administrators, of course, had long been sensitive to the need for close ties to the midwestern industrial and business community. (Acting Director E. Gayle Pewitt had established two task forces in 1979 to improve industrial interactions, and Walter Massey had served as chair of Mayor Jane Byrne's high-technology task force.) Rhetoric aside, what was different about Alan Schriesheim (figure 40), newly arrived from industry, was that he regarded Argonne as a virtually untapped storehouse of science and technology. Schriesheim did not believe that the midwestern economy could wait for the slow evolution of high technology to work its way from the laboratory to industry. In its drive for greater prosperity, Schriesheim believed Midwest industry needed a shot of scientific adrenaline, which Argonne could provide. Task-force studies had their place, but Schriesheim wanted to see hard, concrete initiatives.[9]

Figure 40. Alan Schriesheim (*Frontiers, 1946–1996*, frontispiece)

Schriesheim Takes Command

Alan Schriesheim visited every program at Argonne National Laboratory—division by division—while he was senior deputy director. The staff learned that he was not a glad-hander, but had come to learn what they did, what their problems were, and what their outlook was. And as he poked into every corner of the laboratory, he put it all into his "computer" so he could figure it out. Although anxious to make needed changes, Schriesheim was methodical and thorough. When reflecting on his own management strengths, Schriesheim thought he was good at ordering institutional priorities. "A person has only so much nervous energy," he liked to say, "and they need to figure out what's important, what's not important, and spend . . . time working the important issues." On his walkabouts, he noted shabby buildings and threadbare carpets. He found worn lawns and broken walkways. He saw discouraged weariness in the eyes of Argonne personnel. Mostly, he noted the lack of new initiatives, the ravages of budget cuts, the uncertainty of laboratory management, the dependency on nuclear programs, and the absence of strategic planning.[10]

After becoming director on May 10, 1984, Schriesheim's first step was to spruce up the laboratory. As often happens at institutions stressed by financial crisis, maintenance and repair had been deferred at Argonne while deep cuts were made in nonessential services such as groundskeeping. "Save people and programs now—worry about paint and plaster later" became the guiding assumption of distressed managers. But Schriesheim believed that it was hard to get first-rate work out of people unless they felt they were working in first-rate facilities. Lower morale always accompanied deferred maintenance. Although most of his managers complained that they did not want to take money out of programs and spend it on facilities, Schriesheim saw no other choice. High among his priorities was enhancement of the laboratory site, an effort he asked his wife Bea to head, without pay, of course. Because first impressions are so important, an attractive entrance to the laboratory was constructed, a modern visitor's center built, and attractive, useful signage placed around the laboratory. Also, the historic Freund Lodge was refurbished, and the auditorium, offices, and meeting rooms were upgraded. In time, laboratory employees were proud of Argonne's beauty, and even the Washington bureaucrats had to admit that the laboratory "looked like a first-rate institution."[11]

Like his predecessor Walter Massey, Schriesheim also used the arts to foster a sense of community at Argonne. Again, Bea Schriesheim assisted by helping to create an Arts at Argonne program with Hans Kaper, director of the mathematics and computer sciences division. Kaper assumed the major role in organizing and maintaining high-quality music and art programs that were enthusiastically received by the Argonne community.

The improvement of the work environment was a necessary morale builder, but Schriesheim's main thrust was to develop new program initiatives for the laboratory. Schriesheim was very clear in his own mind about how to recover lost ground in laboratory funding and personnel. While managers complained about the DOE's failure to support Argonne's programs, Schriesheim's response was counterintuitive. "Look," he told his staff, "we've got to stop complaining. . . . We've got to put more initiatives on the plate than there's funding for. Don't worry about the funding, we'll work the funding issue, but first things first."

The only way that Schriesheim knew how to recover lost funding at Argonne was to generate new initiatives. His first step was not to launch another round of budget appeals in Washington, but rather to create an initiative-generating apparatus at home. Argonne was not going to survive with its collection of "little projects thrown up there on the table," Schriesheim warned the scientists he selected for the laboratory's new strategic planning group. The laboratory needed major initiatives on the table as well. Schriesheim did not agonize over identifying Argonne's role and mission. That was what strategic planning would accomplish: "Ultimately, the institution is defined by its initiatives."[12]

Implementing the "Future of Argonne"

To help the laboratory follow up on the board of governors' report, the "Future of Argonne," the board appointed an implementation group co-chaired by Massey and W. Dale Compton, the vice president for research at Ford Motor Company. The implementation group, accompanied by Hilary Rauch, the director of the DOE Chicago Operations Office, met with Al Trivelpiece on November 19, 1984, to report their findings and to recommend initiatives to the Department of Energy. Trivelpiece, whose Office of Energy Research had just completed its appraisal of Argonne's management by the University of Chicago, welcomed this opportunity to review the past performance and future direction of the laboratory. Although his office had discovered some softness in Argonne's staff support, Trivelpiece reported happily that he found the laboratory's program performance generally excellent with superior ratings in high-energy physics. When Trivelpiece promised to forward his recommendations to Secretary Hodel, the implementation group concluded that Trivelpiece had endorsed the board of governors' priorities designating the integral fast reactor as Argonne's principal development program. (A week later, Massey, Schriesheim, and Wallace B. Behnke, Jr., the vice chairman of Commonwealth Edison Company, also briefed John McTague of the Office of Science and Technology Policy concerning Argonne's emphasis on the integral fast reactor.)[13]

One of Schriesheim's first organizational moves was to strengthen Argonne's central planning. In October 1985 he appointed Joseph Asbury to head the laboratory's strategic planning office. Before this assignment Asbury had led a multidisciplinary group in the energy and environmental systems division. The group, which specialized in technology assessment and energy research-and-development policy analysis, had achieved national visibility. It was best known—sometimes to the chagrin of laboratory management—for having eviscerated several sacred cows of United States' energy policy in the late 1970s and early 1980s.[14]

Schriesheim's strategic planning efforts quickly bore fruit. With the assistance of Asbury's group, by summer 1986 Schriesheim reported a number of strategic initiatives. These included a new organizational structure for the laboratory, revamped productivity and cost-control systems, and, most importantly, a centralized strategic planning process to identify and support major program initiatives. Schriesheim's administrative changes laid the foundation for the programmatic thrusts identified by Asbury's central planning process. The program initiatives would include

- development of the Advanced Photon Source;
- commitment to the Integral Fast Reactor;
- development of the Uranium Beam Facility (ATLAS upgrade), Advanced Computing Research Facility, Structural Biology Center, nonphase materials, and HERA-ZEUS detector, and;
- strengthening of Argonne's outreach to industry and the academic community.

Although Trivelpiece seemed to support Argonne's priorities, the mounting deficit made him pessimistic about funding Argonne's reactor programs in fiscal year 1986. While program funding in the physical sciences, biomedical research, and energy and environmental technology would remain constant (although eroded by inflation), Trivelpiece predicted that reactor development would be hit hard by a reduction in Argonne's program from $31.9 million to $9.9 million. The projected cut would virtually eliminate reactor programs at Argonne-East and enable the laboratory to maintain only a minimal nuclear-safety program. Facilities could be maintained at Argonne-West, but there would be no development work on the integral fast reactor.[15]

Despite the gloomy forecast, Schriesheim and Massey (who now served as the vice president for research at the University of Chicago) were confident the budget picture would improve substantially. Not only was the Illinois congressional delegation informed of the impending financial crisis, but Schriesheim also took special care to alert Representative Daniel Rostenkowski about the impact of Reagan's budget on his district. Meanwhile with Till's encouragement, Schriesheim ordered the staff to gear up a broad, integrated fast reactor program with all of the key developmental areas covered. Characteristically, Schriesheim and Till refused to panic over where the money would come from but, rather, concentrated on creating a strong IFR initiative. They were encouraged that fabrication of metal fuel looked very promising and would soon be available for EBR-II. Had the Office of Science and Technology Policy been notified? Was the Illinois congressional delegation being mobilized? Were others informed? Anxious to move ahead quickly, Schriesheim did not wait to resolve funding problems before appointing Yoon Chang, one of Till's key nuclear scientists, as general manager for the reactor fuel program.[16]

Unflinching in his aggressive defense of the breeder reactor, Schriesheim also resolved to lessen Argonne's dependency on the nuclear programs. At the White House, George Keyworth agreed that Argonne needed fresh initiatives, and he urged Schriesheim and the implementation group to chart a new course that would tap funds available at the Department of Defense or secure funding from private industry. In line with the Packard panel and the Grace commission, Keyworth would have preferred that the DOE determine Argonne's research agenda. Because bureaucratic realities dictated otherwise, he accepted Argonne's participation in the administration's steel initiative, which Keyworth thought might serve as a model for similar activities elsewhere.[17]

The steel initiative originated with the president's Commission on Industrial Competitiveness, which suggested that federal laboratories could help the United States' ailing steel industry develop "leap-frog" technologies to enhance the international competitiveness of American steel. The American Iron and Steel Institute took the lead in forming a joint industry-federal laboratory steering committee with representation from Argonne, Oak Ridge, and the National Bureau of Standards. To match industry contributions of $2.85 million, the group asked the federal government for $10 million in FY 1986 to explore processes to make liquid steel directly from iron ore, replacing expensive coke ovens and blast furnaces, and to use electromagnetic technology to cast molten steel.[18]

The steel initiative fit neatly into Schriesheim's search for new initiatives and the Reagan administration's agenda for the national laboratories. This government-industry partnership promised not only to strengthen the United States economy and international trade, but also to augment the administration's defense build-up. In addition, Keyworth liked the biotechnology and advanced computing initiatives outlined by Schriesheim and the implementation group. Although continued funding for the breeder-reactor program was in serious jeopardy, Keyworth assured Schriesheim that the Department of Energy would support some of Argonne's new initiatives. In the end, the steel initiative did not work out as intended, but it served usefully to educate Argonne personnel how to work with industry.[19]

The Department of Energy had its own concerns about Argonne's future. In April 1985 Trivelpiece approved the laboratory's five-year plan for fiscal years 1985–90 but warned that "Argonne must make a fundamental shift in its missions." Offering Schriesheim whatever help he could in re-ordering Argonne's priorities, Trivelpiece forwarded his own suggestion that Argonne secure a lead role in chemical research by establishing a new center for the fundamental chemistry of energy processes.[20]

Schriesheim's strategy of developing new initiatives paid off well in terms of soliciting renewed interest from Washington, but it was severely tested by Reagan's first budget of his second term.

Reagan Budget Redux

President Reagan's appointment of John S. Herrington in February 1985 to replace Donald Hodel as secretary of energy renewed uncertainties at the laboratories. Herrington, a graduate of Stanford University and the University of California's Hastings College of Law, was unknown to the energy community. Prior to joining the president's cabinet, he had served as assistant secretary of the navy for manpower and reserve affairs, as special assistant to the White House chief of staff, and as assistant to the president for personnel. Herrington's close ties to the White House were evident, and it was rumored among nervous DOE employees that he was one of Reagan's "hatchet men." But he brought expertise in organization, administration, and personnel to the department, and as the White House announced, "a combination of the knowledge of defense and civilian management and organization."[21]

Herrington arrived at the DOE with energy and science priorities similar to those of his predecessors, but with a thin wallet. Although Reagan's and Carter's energy budgets were roughly the same ($12.6 billion for Carter's projected budget in FY 1982 compared to $12.8 for Reagan in FY 1985), they contrasted sharply in that Reagan halved Carter's budget for energy research and development while doubling expenditures for the nuclear weapons program. The Office of Management and Budget had repeatedly slashed the department's research-and-development funding, especially for conservation, fossil energy, and solar and other renewable energies. Overall, the administration anticipated saving $858 million in three years with 10 percent cuts in fossil energy, solar, conservation, nuclear fission, and fusion research. Appeals restored some funding, but, following the collapse of the Clinch River Breeder Reactor project, by fiscal year 1985 Argonne had suffered critical losses at the hands of Reagan's budget cutters.[22]

During Reagan's first term, the defense laboratories almost doubled their operating funds with Lawrence Livermore Laboratory, the headquarters for Reagan's Star Wars initiative, gaining 112 percent. Alarmingly, while other energy laboratories grew slowly during the same period, Argonne was the only national laboratory to lose in terms of real dollars during the Reagan years. (Brookhaven, in contrast, was the only energy laboratory to gain funding, when corrected for inflation). To add insult to injury, during this same period when Argonne lost ground, the defense laboratories, especially Livermore, actually gained in nondefense operating funds. From its peak in fiscal year 1978, when Argonne received $414.3 million, the laboratory operating budget had fallen to $226.9 million in fiscal year 1985. Although Argonne requested $265 million for fiscal year 1986, the Reagan administration proposed cutting another $50 million. Following cuts of nearly $13 million the year before, the laboratory faced a reduction in force of 1,335 employees, or 34 percent of Argonne's personnel.[23]

The secretary of energy soon discovered that Schriesheim had successfully rallied the Illinois congressional delegation. Presiding over his appropriations subcommittee, Congressman Sidney Yates asked Herrington who was out to kill the laboratory. Indicating Trivelpiece, Yates bluntly asked, "Is he the one who is trying to kill Argonne?"

"Negative!" Trivelpiece replied nervously while laughter rippled through the hearing room.

Yates bore in. "I just happen to have a letter . . . from Dr. Schriesheim" outlining Argonne's accomplishments, he baited Herrington. "Aren't you impressed, Mr. Secretary?" Yates teased Herrington.

"I have been impressed with Argonne from as far back as I can remember," Herrington responded dryly.[24]

The secretary of energy was furious that Schriesheim had not sent him a courtesy copy of the Yates letter. Instead, Herrington sat before the congressman unprepared while Yates recited a litany of Argonne's achievements. Remarkably, Schriesheim escaped this fiasco without serious injury. Herrington, who actually admired Schriesheim because of his experience in heavy industry, was willing to overlook the director's faux pas as unintentional. In Washington, however, the incident enhanced Argonne's reputation for political maneuvering.[25]

Herrington tried to mollify the Argonne community by assuring Illinois Republican Congressman Harris W. Fawell that the laboratory "had not been singled out for disproportionate cuts." Unfortunately, Argonne had depended too heavily on funding from advanced nuclear reactor programs. This dependence seriously hurt the laboratory when Congress reduced the nuclear program after Reagan cut Argonne's demonstration projects. Herrington repeated Trivelpiece's encouragement for Argonne management to redirect the laboratory's long-term mission towards "broadened technologies." Yet he also strongly encouraged Argonne to press on with reactor research and development so that America could preserve the breeder option for nuclear power development.[26]

State of the Laboratory, 1985

Although the budget picture was not rosy, Schriesheim remained optimistic about Argonne's future. He was confident that Argonne would recoup much of the president's budget cut, and the administration's draconian figures provided him an opportunity to galvanize fourteen members of the Illinois congressional delegation into a block supporting the laboratory. Assurances from Herrington and Trivelpiece lessened his anxiety, but not his resolve to use the fiscal crisis to secure support for Argonne's new strategic initiatives.[27]

By December 1985 Schriesheim reported a change in mind-set—a change in attitude at Argonne. The laboratory was no longer "instinctively defending old programs and old turf." Rather, Argonne now instinctively sought "offensive advantage in leading the way into new areas of national need." His bravado was part cheerleading, of course, but was welcomed by a laboratory eager for good news. What was most important for Schriesheim was that the laboratory had begun to think and feel like a winner.[28]

His imagination captured by Chicago's stunning skyline, Schriesheim found an analogy for Argonne. For the past five years, the hard work of Walter Massey and the Argonne staff had laid the foundations for a new skyscraper—an intellectual one—to grace the Illinois prairie. In its first two years of operation, the IPNS hosted two hundred experiments, about three-quarters of which involved visiting scientists, including researchers from Exxon, Du Pont, and Amoco. Demand for access to the IPNS, which provided the world's most powerful beam of fast neutrons for materials research, continued to grow and created such competition for beam time that Argonne rejected more than half of the research proposals. Unfortunately, Department of Energy funding allowed IPNS to operate only four thousand hours, or six months each year. Scientists planned to enhance the IPNS neutron flux 300 percent by replacing the depleted uranium target made from uranium-238 with an enriched uranium target made of 83 percent uranium-235, the rare and readily fissionable uranium isotope. The new target would produce 80 to 85 percent of its neutrons through fission and the remainder through spallation. In addition, Jack Carpenter, technical director of the IPNS, and his group developed a new device—the solid methane moderator, which produced low energy neutrons at about 20 degrees above absolute zero. These "cold" neutrons, whose low energy implied long neutron wavelength, enabled scientists in solid-state physics, materials science, chemistry, and biology to analyze neutron scattering at small angles while studying voids, clusters of atoms, and viruses in solution.[29]

In another technical triumph, Argonne's Tandem-Linear Accelerator System was completed on schedule and within budget in June 1985. ATLAS was the world's first superconducting accelerator for heavy ions. It could accelerate beams of particles from across the periodic table. Scarcely had the machine been dedicated when ATLAS manager Lowell Bollinger announced plans to upgrade the facility. Like the IPNS, ATLAS became a major user-facility with more than a hundred scientists from thirty-five research institutions worldwide using the accelerator to study collisions of atomic nuclei. Although half of those working at ATLAS were non-Argonne scientists, requests for beam time, like those for the IPNS, were twice what the accelerator could accommodate. The machine proved so successful that at least two universities borrowed ATLAS technology to upgrade campus research facilities.[30]

Schriesheim was especially proud of the peer recognition received by the laboratory. For example, in 1984 five Argonne scientists—chemist Darrell C. Fee, metallurgist Nestor J. Zaluzec, high-energy physicist Ray Hagstrom, biophysicist N. Leigh Anderson, and chemical physicist Patricia Dehmer—were named among the one hundred outstanding young scientists in America by *Science Digest*. In addition, *Science Digest* selected Lowell Bollinger, head of the ATLAS project, and Philip Horwitz, project leader of the TRUEX radioactive waste reduction process, among the top one hundred science innovators in 1985. Finally in 1984–85, Argonne won six prestigious I-R 100 awards given annually by *Research and Development* magazine to the 100 most significant technical products. In the twenty years since the magazine started the annual award, Argonne had collected thirty-three I-R 100 awards.[31]

Argonne's First Priority: The Integral Fast Reactor

Argonne's award winning programs were gratifying and its numerous small initiatives were encouraging, but Schriesheim knew that Argonne's future depended on funding the laboratory's major, long-term big-science initiatives. Historically, nuclear reactor research and development had provided 40–50 percent of Argonne's budget, but Congressional cancellation of the Clinch River breeder-reactor project in 1983 raised fundamental questions about the future of the United States' reactor development program. Optimistically, Charles Till believed that in the long run, the loss of the Clinch River project would prove fortunate for the nuclear industry.

Till was a man with a mission, a "true believer" who prophesied that nuclear energy would be the salvation for humanity's energy needs. No dour Jerimiah, Till nevertheless was driven by the crusader's fervor to serve the world as well as his laboratory. Affable, handsome, gregarious, and politically skilled, Till had steadily worked himself up Argonne's management ladder since joining the laboratory as an assistant physicist in 1963. He watched Argonne's leadership in reactor engineering gradually slip away during Milton Shaw's regime in the 1960s, and he was determined to move Argonne back "to the absolute center of international reactor development."[32]

From Till's perspective, the demise of the Clinch River project offered Argonne an opportunity to recapture leadership of America's nuclear program. Congress had killed the CRBR, but not research on breeder-reactor technology. When the Carter administration first raised concerns that plutonium produced in a Clinch River–type breeder might be diverted or stolen for use in nuclear weapons, Till and his colleagues Yoon Chang and Robert Avery explored ideas on how to make the breeder fuel cycle proliferation-proof but still economical. Knowing that waste disposal was second only to fears over nuclear proliferation, they also discussed means of closing the fuel cycle. Following the Three Mile Island accident in 1979, they focused on ways to capitalize on efforts to demonstrate that EBR-II was an inherently (or passively) safe reactor. By the early 1980s, Till recalled, "most of the pieces for the IFR [integral fast reactor] were floating around in our heads." The key to making the breeder reactor economical, proliferation-proof, environmentally acceptable, and inherently safe lay in designing a new reactor fuel and its related reprocessing cycle.[33]

Till and his colleagues found their answer in metal fuels, which had been abandoned years before because metal fuel rods had burned poorly and tended to swell, disrupting the geometry of the core and frequently breaking the cladding. Before ending the metal-fuels work, however, Argonne's metallurgists had fashioned fuel rods of uranium, plutonium, and zirconium alloy and had tested them in EBR-II. These pins offered some hope: they seemed to perform well when containing useful concentrations of plutonium (a long-standing problem with metal fuels); they conducted heat excellently; and they promised high burnup. Basing a fuel program on such limited data was a gamble, but with the encouragement of the chief metallurgist, Leon Walters, Till decided to move ahead. The gamble paid off handsomely when the first fuel assemblies fabricated by Walters's group achieved a 19 percent burnup, which was as good as or better than that of ceramic assemblies and twice what Argonne thought necessary for feasibility. Commercial reactors averaged only 3 or 4 percent burnup, and utilized only 0.5 percent of the uranium mined. Advantages in longevity, breeding, and recycling of the metallic fuels potentially increased fuel efficiency a hundred times over that of reactors loaded with ceramic fuels. Till argued that a dramatic increase in fuel efficiency was necessary because the world's supply of uranium was not sufficient to support nuclear power in the long run. According to his calculations, if current nuclear reactors replaced only 40 percent of the world's fossil-fuel electricity-generating capacity, uranium supplies would last only about thirty years.[34]

The IFR concept married the venerable EBR-II (with its metal fuels) to the Fuel-Cycle Facility (FCF), which had been modified to reprocess spent fuel with a new technology called electrochemical pyroprocessing (figure 41). Joined together physically, the reactor and the reprocessing facility integrated power production, fuel reprocessing, and waste treatment at one site, thereby elimi-

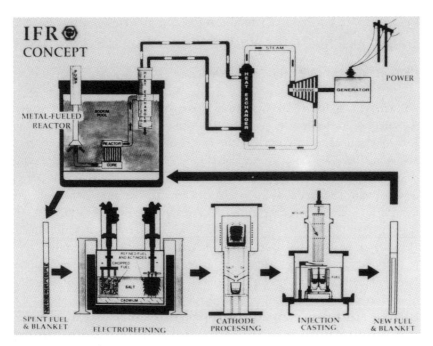

Figure 41. Integral Fast Reactor

nating the need to transport spent fuel and wastes from one location to another. In addition to EBR-II and the Fuel-Cycle Facility, Argonne's test facilities in Idaho offered major support for proving the technology. Engineers used the HFEF to examine IFR test fuel pins, operated TREAT to subject fuel elements to severe power transients, and employed the ZPPR to test the physics and safety of various large IFR core designs.

Strongly committed to developing the nuclear option, Schriesheim readily embraced the IFR as Argonne's number-one research-and-development priority. Till's vision fit neatly into Schriesheim's strategy to support initiatives in areas of proven strength with aggressive leadership. As one of the principal "thrust areas" in the 1985 strategic plan, the IFR program capitalized on Argonne's "tradition, history, skills, and facilities." In short, it was Schriesheim's only viable option for preserving Argonne's nuclear reactor program. At EBR-II's twentieth anniversary celebration on August 28, 1984, Schriesheim and Till introduced the IFR as the "next logical step in breeder reactor research." Almost defiantly, Schriesheim told the more than three hundred guests that "contrary to the hopes of our critics, the demise of Clinch River did not sink the breeder program." Idaho Senator James McClure enthusiastically endorsed the IFR as a shield against a future energy crisis. Scornfully, McClure dismissed the antinuclear lobby as "fat, dumb, and happy. They are living in a fool's paradise," he scoffed. "We will eventually need that [breeder produced] energy."[35]

EBR-II: An Inherently Safe Reactor

The accidents at Three Mile Island and Chernobyl jolted public confidence in the safety of nuclear power reactors. The two events, so contrasting in their consequences, came to symbolize the failure of the nuclear power industry. Following the 1979 accident near Harrisburg, Pennsylvania, only a few politicians like McClure gave their unconditional endorsement to nuclear power development. The partial meltdown of the Fermi reactor near Detroit in 1966; the fire at the Brown's Ferry plant in Alabama in 1975; and the short-circuiting of part of the electrical system at the Rancho Seco reactor in California in 1978 helped erode public trust in the nuclear industry well before the Three Mile Island accident. By 1980, according to a survey commissioned by the Department of Energy, 60 percent of Americans opposed further construction of nuclear power reactors. But public distrust of nuclear technology did not go so far that Americans wanted to abandon nuclear power. In contrast to Sweden, referenda to ban nuclear power failed in Maine, Massachusetts, and Oregon.[36]

Till acknowledged that Americans instinctively knew that all mechanical systems ultimately fail, either from physical wear or through human error. To recapture public confidence in nuclear power, Till and his associates proposed to build an inherently safe reactor system: "To the maximum extent, [reactors] should be foolproof." No design could be absolutely risk free, but Till advocated developing a liquid-sodium-cooled reactor that would passively shut down should the cooling systems fail. "Sodium is a marvelous coolant," Till informed Richard Rhodes. "It has the disadvantages that it reacts with water and burns in air." But those were the only disadvantages. Liquid-sodium coolant conducts heat wonderfully at atmospheric pressure. Without question, observed Ronald Teunis, Argonne-West's site manager, EBR-II's most significant contribution in twenty years was the plant's safe, reliable operation: "The right kind of fuel plus sodium coolant would ride right through an accident like TMI or Chernobyl."[37]

The right fuel, of course, was the new uranium-plutonium-zirconium alloy. Argonne engineers intended to exploit the fact that metals expand when hot. If designed correctly, overheated fuel elements would expand and lose reactivity as the uranium and plutonium atoms moved farther apart. Under extreme conditions, in which the core lost all coolant from the pumps (as at TMI and Chernobyl), the reactor would simply shut itself down without intervention from either an operator or mechanical safety devices. Passive, benign reactor shutdown during a loss of coolant accident was what Till meant by "inherently safe."

On April 3, 1986, three weeks before the Chernobyl accident, Argonne engineers put EBR-II to the ultimate safety test by simulating a blackout at the nuclear power plant while operating at full power. Two tests would replicate accidents involving a loss of cooling capacity in the reactor core: loss of coolant flow when coolant pumps fail, and loss of the heat sink when heat exchange is blocked between the reactor core and the electrical generator. Almost sixty representatives from the electrical and nuclear industries and United States and foreign governments gathered to witness the historic tests.[38] Till stood in the back listening to the countdown—waiting for the pumps to "fail." When the coolant flow stopped, Till remembered, "the first thing we saw was the temperature going straight up! Not a pretty sight to anyone who's had anything to do with a reactor." Then *bang!* A pressure valve popped open in a nearby heat sink line causing the assembled dignitaries to jump from their seats to see if the Argonne staff had already run from the building. In retrospect, it was a good, if unintended, joke. Shortly thereafter, the core temperature spiked, and within ten minutes the reactor's power dropped to zero while temperatures returned to normal. In both tests, reported Pete Planchon, the manager for reactor plant analysis, EBR-II regulated its own power and temperature without intervention by human operators or emergency safety systems.[39]

Nature was impressed with the audacity of the safety demonstration. "It worked on the blackboard, it worked in computer simulations, and the engineers present were willing to bet their lives that it would work in practice." Similarly, after the death of the CRBR, the engineers had boldly bet their careers that the IFR would prove an important contribution to America's energy future.[40]

Ironically, the Chernobyl accident in Ukraine on April 26, 1986, actually drew public attention to Argonne's dramatic reactor demonstration in Idaho. Till noted that Chernobyl, not Three Mile Island, focused favorable national coverage on the IFR program. Not surprisingly, the nuclear industry expressed keen interest in IFR research. After *National Geographic* featured the IFR in its essay on the future of nuclear power, California utility executives as well as public utility commissioners wanted to know what they could do to help. Importantly, a few environmentalists, more concerned about problems of acid rain than with potential risks from nuclear power, also warily expressed interest in Argonne's experimental reactor. According to *National Geographic,* Jan Beyea of the Audubon Society emphasized the need to phase out fossil fuels "if we are going to save the planet." While continuing to support solar energy, Beyea agreed that it made sense to invest research dollars in "testing these so-called idiot-proof reactors."[41]

Despite dramatic success of the EBR-II safety tests, the IFR could not shake persistent opposition, especially from congressional staff and others in Washington who had fought against the plutonium breeder reactor since the Carter administration. While Argonne's other initiatives generally found support in Washington, the IFR encountered deep skepticism among staff on the House Appropriations Committee and in the White House Office of Science and Technology Policy. In private meetings with Nancy M. O'Fallon, who was acting as a liaison to Washington, congressional staff expressed their frustration with Argonne's lobbying, which they characterized as the "worst" among national laboratories. From the staff's point of view, the IFR was a "political reactor" without any merit, and they were "outraged" because pressure from Yates and Rostenkowski forced them to provide Argonne funding for the breeder reactor. Under pressure of the Gramm-Rudman budget limitations, staff at both the Office of Management and Budget and the Department of Energy reportedly proposed contingencies for terminating the breeder program and closing EBR-II and Argonne-West. Trivelpiece, who had already warned Schriesheim that the DOE would cut "every pork barrel program from the FY 1987 budget," suggested that Argonne find refuge for the IFR program in the Department of Defense budget.[42]

IFR, Nuclear Wastes, and Weapon Proliferation

Instead, Till and Schriesheim worked tirelessly to build support for the IFR in Congress and the DOE. In addition to inherent safety, Till emphasized the IFR's fuel-recycling features which, if less dramatic, were important to a new reactor design. When he won Herrington's support for the IFR during a visit to Argonne-West, Till made a special effort to explain the IFR's solution to the nuclear waste problem.[43]

After President Carter discontinued American commercial reprocessing and plutonium recycling in 1977, the United States had no alternatives for disposing of spent reactor fuel other than long-term storage in a nuclear waste facility. Although spent uranium-ceramic fuels could be reprocessed, the technology was expensive, requiring a facility as large as an oil refinery, and hazardous, because transport of the highly radioactive wastes from reactor to reprocessing facility risked accident, diversion, or theft. To make matters worse, selecting long-term, nuclear-waste storage sites had proven environmentally controversial and politically daunting. Of necessity, the IFR initiative had to ameliorate, not exacerbate, the nuclear waste problem.

Fuel irradiated in a nuclear reactor created two kinds of "waste": fission products and transuranic elements, or actinides. Of the two, the transuranic elements (plutonium, americium, neptunium, curium, and others) were heavy "man-made" radioactive materials whose long half-lives required safekeeping for ten thousand years or more. On the other hand, the fission products consisting of hundreds of isotopes decayed at a much faster rate, becoming less radioactive than natural ore in a few hundred years. While it was mind-boggling to envision how one might safeguard transuranic wastes longer than all recorded history, it was certainly possible to conceive of ways to isolate fission products for two or three times "four score and seven years."

A big advantage of the IFR's metal fuel was the ease with which spent fuel could be recycled in a facility attached to the reactor plant. At EBR-II, spent fuel would be removed from the reactor directly to the contiguous Fuel-Cycle Facility where the spent metal fuel was to be reprocessed and refabricated on site. The electrochemical recycling process (or pyroprocessing) was simple and relatively inexpensive. Spent metal fuel was chopped into small pieces and then dissolved in a layered cadmium and molten-salt solution through which flowed an electrical current. As the uranium, plutonium, and actinides accumulated on an electrode, the fission products collected in the cadmium and molten salt. Most of the uranium, plutonium, and actinides could then be refashioned into fuel elements, while the "shorter-lived" fission products were removed to nuclear waste storage sites where they required protection for only a few hundred years.[44]

In addition to recycling its own spent fuel, the IFR could also burn plutonium from dismantled nuclear weapons and actinides from the spent fuel of commercial light-water reactors. Although the IFR would not solve the nuclear waste problem, Argonne's promoters believed their reprocessing technology could contribute significantly to lessening the burden of long-term waste management.

According to Till, the IFR electrorefining pyroprocess also discouraged nuclear proliferation. Critics had raised major objections to earlier metal-oxide breeder designs because the reactors could produce plutonium that was usable in nuclear weapons. Countries adopting such breeder technology for power production would also have ready access to weapon-grade plutonium. But not so with the plutonium produced in the IFR, which Till argued was actually proliferation resistant. A pure plutonium product would not be possible. IFR plutonium would be mixed in with uranium and residual actinides, making it highly radioactive, and unsuitable for weapon use without additional, offsite reprocessing. Because the spent fuel was extremely radioactive, the entire fuel cycle was performed in a large "hot fuel cell" with remotely controlled manipulators. High radiation levels and physical inaccessibility at the reactor site, Till believed, reduced opportunities for theft or diversion of the plutonium to almost nil. He estimated that fifteen hundred IFR plants, accepting "waste" from existing light-water reactors and old weapons, would extend existing uranium reserves two thousand years, while producing a large fraction of the world's electricity.[45]

Evaluating the IFR

At Schriesheim's request, the University of Chicago's board of governors established a special scientific and technical advisory committee to evaluate the IFR program, focusing special attention on the contributions of Robert Avery's reactor analysis and safety division. The advisory committee was favorably impressed with the IFR technical program but warned the laboratory against betting its future on the unproven system. The committee expressed "agnosticism" about the advantages of metal fuel and "skepticism" about the laboratory's claims that IFR technology was "proliferation resistant." Propinquity of the reactor and fuel reprocessing in itself did not enhance security. Both of these issues were important for public advocacy, the advisory committee recognized, but they were unrelated to the technical adequacy of the IFR program.

Furthermore, should the laboratory's reactor analysis and safety division promote IFR technology? How could the division offer the Department of

Energy and the nuclear industry unbiased evaluation of reactor safety systems if it were also involved in boosting the IFR? Although Avery saw no conflict of interest between advocacy of the IFR and his professional ability to evaluate base safety technology, the committee suggested that Schriesheim assess the independent judgment of the reactor analysis and safety division.[46]

Following the committee's review, the DOE asked the National Academy of Sciences to assess reactor safety at the Idaho National Engineering Laboratory. The academy's panel evaluated safety at the navy's advanced test reactor and at Argonne's EBR-II. Conceding that the risk of a serious accident at the two test reactors "may be appreciably less than at commercial reactors or at the DOE-owned defense production reactors," the panel nevertheless recommended improvements in safety procedures. The panel acknowledged that EBR-II possessed a number of "attractive passive characteristics" that made a loss of coolant accident "extremely remote." But it criticized the laboratory for discontinuing its acoustic monitoring system, which was designed to detect cracks and other problems. Then, echoing the concern of the special advisory committee, the panel urged the laboratory to remove a potential conflict of interest by ending the management of quality assurance and safety teams by reactor operations. Till promised to implement the academy's recommendations.[47]

Helping the Russians with Reactor Safety

While the National Academy of Sciences reviewed Argonne's safety programs, the laboratory was engaged in another project to assist reactor operators in Russia and Eastern Europe to avoid another Chernobyl-type reactor accident. Bruce Spencer, from reactor engineering, led the DOE's efforts to write operating procedures that would reduce the risk of accidents in Soviet-designed power reactors. "Argonne's facilities are unique," he noted, the best in the world to simulate various accident scenarios. With assistance from Brookhaven and Oak Ridge National Laboratories, Spencer and David Wade, director of the reactor analysis division, reconstructed from Soviet reports the sequence of events that occurred during the Chernobyl accident. Spencer believed that their study was not only the first, but also the best, analysis of the accident. Spencer's group determined that the Soviet reactors were vulnerable to serious accidents because reactor pressure vessels had become brittle after years of irradiation, the emergency core-cooling systems were ineffective, and the porous confinement buildings would not prevent the escape of radioactivity during a severe accident.[48]

Following the break-up of the Soviet Union, eastern European countries such as Bulgaria, the Czech Republic, Slovakia, Hungary, Lithuania, and Ukraine lost assistance from former Soviet technicians in maintaining their power reactors. The situation so alarmed the international nuclear community that it organized the Lisbon Nuclear Safety Initiative to help these countries and the Russians reduce their dependence on old Soviet reactors and reduce the risk of operating their flawed reactors until they could be replaced or shut down. When Ivan Kuznetsov, director of the Russian Institute of Physics and Power Engineering, visited Argonne-West in 1990, he admitted that Russian scientists had lost the confidence of the people following the Chernobyl accident. Inviting Americans to help improve reactor safety was part of the effort to regain the confidence of the Russian people.[49]

Argonne assisted with risk reduction. Working with teams from the Electric Power Research Institute (EPRI), Brookhaven, and Pacific Northwest Laboratories, Spencer's group focused on short-term "hardware fixes" that would immediately up-grade reactor safety at twenty-seven reactors. Picking three reactors as demonstration sites, major improvements were made in instrumentation, electrical power backup, piping integrity, emergency core cooling, and containment. Cables entering into the reactor area were sealed, and vents were redesigned. In all, the United States spent almost $20 million on Russian reactors, two-thirds of it in the United States on engineering services, equipment, and hardware. Spencer's goal was to complete the work in two years. "There's a lot of motivation for speed," he reported, because the reactors required remediation before they got any older.[50]

The Battle for the "Green Machine"

Among antinuclear groups in America, making reactors safer in Russia and eastern Europe did not make them safe enough for the United States. Public critics of the IFR had long disparaged the government's nuclear research program. In 1989 consumer advocate Ralph Nader called for redirecting funding for advanced nuclear reactors toward safe, clean, and more socially acceptable renewable energy technologies. More pointedly, Robert Pollard, a nuclear safety engineer for the Union of Concerned Scientists, turned Argonne's safety claims topsy-turvy and charged that any nuclear reactor was "inherently dangerous." The IFR might survive a loss of coolant flow, he conceded, but it could not withstand a rupture of the sodium lines. In contrast to Nader, Pollard would not go so far as to say the IFR should not be explored, but he cautioned, "Argonne ought to be very candid about pointing out the drawbacks."[51]

Charles Till first realized that the IFR program would encounter trouble in a Clinton administration during the New Hampshire primaries. Listening to one of the debates, Till recalled the transparent joy with which Bill Clinton tagged Michael Dukakis for being pronuclear. "Oh my," Till groaned, remembering the same elation rising in Jimmy Carter's voice when he canceled the Clinch River Breeder Reactor project. During the presidential campaign, Till tried unsuccessfully to obtain assurances from the Clinton organization that their candidate was not antinuclear. After Clinton's inauguration, Till watched in dismay as former antinuclear activists were appointed to key positions in the White House, the vice president's office, the Office of Management and Budget, and the Department of Energy.[52]

Still, everyone at Argonne was stunned when President Clinton announced in his first State of the Union address that "we're eliminating programs that are no longer needed, such as nuclear power research and development." As noted by Thomas Lippman in the *Washington Post*, among opponents of nuclear power within the Clinton Administration, the IFR was

> Son of Clinch River—a mutant offspring of the government's long abandoned plan to build a reactor at Clinch River, Tennessee, that would produce more radioactive material than it consumed. To these critics, it is more than just nuclear pork: It is an unnecessary and potentially dangerous program that could lead to the use of plutonium as a commercial fuel.[53]

The battle for the IFR was now joined, with the survival of Argonne's principal program at stake. In pursuit of spending cuts and deficit reduction, Clinton was the first president of the nuclear age not to support nuclear reactor research and development. Since the end of World War II, the United States had spent an estimated $24.2 billion on reactor research and development, although funding levels had dropped precipitously since 1981. Clinton inherited three power-reactor development programs from the Bush administration: the IFR, the power reactor innovative small module (PRISM, which was based on IFR technology), and the modular high-temperature gas-cooled reactor (MHTGR). In his initial economic prospective, "A Vision for Change in America," Clinton decided to discontinue all of them. At stake for Argonne were about $117 million and five hundred jobs at Argonne-East and one thousand jobs at Argonne-West.[54]

Without scolding the president by name, Schriesheim accused the federal government of throwing away its $700 million investment in IFR technology, which "could generate electricity safely into the foreseeable future at the same time that it gets rid of long-lived nuclear waste and avoids producing greenhouse gases." Schriesheim nicknamed the IFR "the green machine" because of its ability to generate electricity from dismantled nuclear weapons and commercial spent fuel and destroy them in the process. Although the National Academy of Sciences cautioned that proliferation and physical security required that "special attention" be paid to breeder programs, Schriesheim emphasized that the academy also recommended that the IFR "should have the highest priority for long-term nuclear technology development." Utilities agreed, Schriesheim noted. Till had secured $2 million from a California utility to support research, while Japanese utilities promised $46 million to help IFR development through 1996.[55]

Intending no sarcasm, Schriesheim thought "the age of miracles" may have come to pass. Israelis and Palestinians signed a peace treaty; Democrats pledged to cut the size of government; and the environmentalists' best friend turned out to be a nuclear reactor. He was frustrated, however, by critics with a 1960s mindset who persisted in equating the IFR with the Clinch River Breeder Reactor, which foundered on the problems Schriesheim believed the IFR had solved. "Some Jews and Arabs fight the peace effort," he wrote. "Some Democrats fight government downsizing. And some environmentalists fight funding for the Integral Fast Reactor."[56]

Schriesheim and the IFR were not simply butting heads with aging environmentalists, however. As Schriesheim himself had acknowledged, Clinton's new democratic coalition had brought to Washington freshman congressmen determined to cut wasteful government spending. Led by Democratic congressman Sam Coppersmith of Arizona, who teamed up with Republican congressman Martin Hoke of Ohio, a bipartisan company of twenty-two members of the House Science, Space, and Technology Committee petitioned their chair, George E. Brown of California, to eliminate the IFR and other reactor programs.[57]

Schriesheim had minimized the fact that escalating costs had been one of the major reasons Democrats and Republicans agreed to terminate the Clinch River project in 1983. A decade later, in 1993, a second coalition of taxpayer, environmental, and nuclear-weapons-control groups ganged up to kill another major power reactor project. Jill Lancelot of the National Taxpayer's Union declared that elimination of the IFR would test the president's resolve to reduce the deficit: "We just look at this as a waste from an economic point of view. How many billions are we [ultimately] talking about?" More bluntly, Gene Karpinski, director of the U.S. Public Interest Research Group, contemptuously campaigned to dump Argonne's "boondoggle reactor."[58]

Catching the budget-cutting wave, Scott Denman, the executive director of the Safe Energy Communication Council, challenged the nuclear industry to finance its own technology development: "Since when is that the taxpayer's responsibility?" Denman, who along with the Sierra Club, the League of Conservation Voters, and Public Citizen, had lobbied fiercely against the IFR, welcomed alliance with the taxpayers groups. A champion of developing renewable resources such as solar energy, Denman believed that the IFR with its emphasis on fuel recycling was part of a "Cold War culture" at the national laboratories that needed to change. Believing that America was on the cutting edge of a "renewable culture," Denman thought it made neither economic nor environmental sense for the government to subsidize the development of 1,000-megawatt power plants. According to Denman, Argonne should be directed to develop low-cost solar and other renewables "to reflect the new culture." But if Argonne and the other national laboratories could not adapt to a new post–Cold War mission they were of little value. Echoing Trivelpiece, who for different reasons had sent Schriesheim a similar message, Denman saw no future for the laboratories if they could not change.[59]

Schriesheim, in contrast, was worried about the long-term economic health of the United States. He did not doubt that solar and other advanced energy systems could play a future role, but unless a reactor like the IFR were ready in the next century, Schriesheim forecast that Americans would have to trade environmental caution for electricity. There was no other foreseeable option for generating huge amounts of electricity except fossil-fuel power plants, which meant increased pollution from greenhouse gases. During the Cold War, even a hint that the Soviet Union might challenge American nuclear superiority in either military or peaceful applications usually opened generous Congressional pocketbooks. But now, with one of America's chief economic rivals, Japan, actually investing in IFR technology, Americans seemed blasé about maintaining their historic lead in nuclear science and technology. Echoing concerns of the entire reactor-program staff, Argonne's director wondered how their grandchildren would be able to forgive the generation that threw away America's most promising energy technology. Schriesheim and proponents of the IFR were especially frustrated because they believed they needed only $300 million to bring the program to the decision point for technology deployment.[60]

The debate over the IFR roller-coastered through Congress after the president's budget message. With more support in the Senate than in the House of Representatives, funding for the IFR was voted up and down several times while the Clinton administration's energy budget limped its way along the appropriations track. Although Illinois congressman Sidney Yates was no friend of nuclear power, and Republican Harris Fawell was an outspoken critic of Democratic "pork barrel" projects, both stood firm with the Illinois and Idaho congressional delegations defending the IFR program. And Congressman George E. Brown of California, chairman of the House Science, Space, and Technology committee and who had been one of the chief opponents of the Clinch River project, did not want to cancel the government's modest research on advanced reactors. One of the IFR's principal friends was J. Bennett Johnston, chair of both the Senate Energy Committee and the Senate Appropriations Subcommittee on Energy and Water. But Johnston, an author of the 1992 energy bill who led the floor fight in the Senate to save the IFR, was also realistic that the federal budget had its limits. Compromise was in order.[61]

When Clinton's secretary of energy, Hazel O'Leary, first arrived at the Department of Energy, she did not understand the IFR program and remained neutral through most of 1993 while she sized up the situation. She discovered the scientific, technical, and management dimensions of the IFR project to be "first rate," but she also encountered strong opposition to the program at the White House. O'Leary related to Schriesheim and Hugo Sonnenschein, the new president of the University of Chicago, that the White House staff literally danced in the halls whenever a decision went against nuclear programs. Despite pleas from the department's Office of Nuclear Energy to champion the IFR, O'Leary switched from neutrality to opposition when President Clinton decided to cancel the program. (It was not inevitable that Secretary O'Leary would follow the president's lead in this matter. Back in the Reagan administration, when David Stockman and others at the White House laid out plans to abolish the DOE as the president had promised, first, the deputy secretary of energy, W. Kenneth Davis, and then the secretary of energy, James Edwards, fought successfully to save the department). Upset by the failure of laboratory management to accept President Clinton's decision to cancel the IFR, O'Leary coldly informed Schriesheim and Sonnenschein that she expected them "to get on with it." O'Leary pledged to arrange a "full-employment" budget for Argonne in return for Schriesheim's commitment to define alternative missions for the laboratory.[62]

The Clinton Compromise: Save the Actinide Recycling Program

Regardless of the technical merit of Argonne's reactor or of the political clout of its promoters, it is almost impossible to save a major development program opposed by a determined president. On the other hand, with razor-thin majorities in the Senate and House, Clinton could not afford to alienate completely his pronuclear Democratic constituents. In August 1994 Congress terminated the reactor program while providing $84 million to shut down the IFR operations. But also, the Clinton administration agreed to fund the fuel reprocessing program—renamed the "actinide recycling project"—while closing down the venerable EBR-II and other supporting facilities at Argonne-West. For the moment, at least, the White House was willing to explore the benefits of actinide recycling. O'Leary promised that no jobs would be lost at either laboratory site in FY 1995.

As reported in the press, the Clinton administration authorized reprogramming of $33.2 million for research on nonproliferation technologies, spent-fuel studies, fuel-cycle and reactor safety, and decommissioning and decontaminating techniques, which included extending Spencer's cooperative programs with the Russians. Added to the IFR close-out funding, few jobs would be lost immediately in either Illinois or Idaho.[63]

Almost no one was happy with the Clinton compromise. Antinuclear activists charged that funding any portion of the IFR was a step backward. Citing studies from the National Academy of Sciences and Lawrence Livermore National Laboratory, they claimed that the reactor was not needed for disposing of weapon-grade plutonium. Till agreed that the IFR offered no short-range solution for disposing of nuclear-weapon waste, but he believed that the IFR was the only reactor suitable for stabilizing plutonium stocks in the long run. Exasperated, Till thought it pointless to fund the fuel recycling program while closing down EBR-II, which was designed to burn the fuel. Yoon Chang, Till's close colleague, simply did not understand how the nation that led the world in nuclear science could walk away from developing socially responsible reactor technology.[64]

From London, the editors of *Nature* shared Yoon Chang's confusion. What was Clinton trying to accomplish, they wondered? Why was the IFR budget abruptly trimmed, but not cut to zero? Did it mean that the Clinton administration, while hostile to nuclear power, was also worried about jobs? That it wanted to keep open America's nuclear options, but on a scale that would not give offense to environmental activists and proliferation worriers? It was incongruous to *Nature* that an environmentally sensitive American president would turn his back on "the most important source of energy that does not emit carbon dioxide." Thanks to Clinton's "boundless capacity for compromise," *Nature* speculated, the United States might not realize its mistake until it was too late.[65]

Cancellation of the IFR program and the shutdown of EBR-II signaled the end of the reactor era. Not since the reconstruction and start-up of Fermi's reactor in 1943 had Argonne been without a major reactor project. After fifty years of pioneering in the nuclear field, the president and Congress decided it was time to stop Argonne's exploration of advanced reactor technology. Consequently, hundreds of employees at both Argonne-East and Argonne-West faced layoffs and an uncertain future. Till grimly held to his belief in

IFR technology and counselled that the setbacks must be borne stoically. The Chicago *Tribune* characterized the end of reactor research as "an unscientific defeat for Argonne." Always supportive of regional science, the *Tribune* believed the decision was neither about money nor the merits of technology. Rather, termination of IFR development reflected post–Cold War misgivings about big science coupled with the Clinton administration's antinuclear sentiment. Vulnerable in this climate were big projects (with the exception of the space station) that advanced knowledge but only vaguely promised commercial gains.[66]

13 | New Paths to the Future, 1984–94

"Without question," Alan Schriesheim soberly reflected, "the biggest single impetus to change in American research came when the Cold War collapsed in the rubble of the Berlin Wall." In the half century since the end of World War II, the Department of Energy's research establishment had grown from the scientific remnants of the Manhattan Project to a national system of thirty laboratories employing close to 30,000 scientists and technicians. Nine of the facilities, including Argonne, were multiprogram laboratories that received funding from several sources in addition to that from the Department of Energy. During the Cold War years, the interests of the scientists, the government, and the public were largely congruent.

"We wanted national security, and we needed energy," Schriesheim recalled. "Our highly visible foes of World War II had been replaced by an equally visible—and just as determined—Cold War foe, the Soviet Union." National security requirements had always been double-edged—military and economic. Despite the unprecedented prosperity of the booming postwar economy, American's high standard of living also required vast and reliable sources of inexpensive energy. For fifty years, the twin demands for national security and abundant energy guaranteed generous congressional support for the government's weapons and energy research laboratories. With high regard for science reflected in solid public support, Schriesheim inferred, laboratory scientists regarded their research "almost as an 'entitlement,' a self-evident public good which surely would produce results of value to the nation."[1]

In the aftermath of World War II, America's faith in progress through science seemed boundless. The irony of war, whether hot or cold, was that it stimulated scientific discovery. President Eisenhower had caught this same spirit in his 1953 Atoms for Peace speech when he pledged that the United States would devote "its entire heart and mind to find the way by which the miraculous inventiveness of man shall not be dedicated to his death, but consecrated to his life."[2] What did it matter if every laboratory project did not yield immediate economic benefit—surely even basic research would ultimately produce some public windfall.

Although the laboratories were never indifferent to public accountability, as they matured through the 1960s to the 1980s each developed a distinct "personality" based on its differing objectives, talents, structures, and constituencies. In time, they became so different that "no one even superficially familiar with the national labs could mistake Argonne for Oak Ridge or Berkeley." No longer bonded to a common mission such as the Manhattan Project, the national laboratories eventually carved out separate identities, but as they did so, public support for their endeavors gradually fragmented. "Most of us did not notice it initially," Schriesheim admitted, "the divergence of political, social, and scientific agendas happened slowly."[3]

It was, he observed, perhaps the chemists who came up against it first—in the environmental firestorm erupting from the pages of *Silent Spring* and the anger of Love Canal. The energy labs got a wake-up call from the OPEC oil embargo, smog warnings, Three Mile Island, Chernobyl, oil-soaked shore birds, and the so-called ozone hole. Researchers in high-energy physics encountered it somewhat more recently, in the scuttling of the Superconducting Super-Collider, the SSC.

Bill Clinton's election as president in 1992 punctuated the transition already underway at Argonne. Noting that most of the DOE's budget was dedicated to nuclear work, the President-elect demanded "a different direction and a different policy" in the department's research and development programs. Secretary O'Leary explained that the administration's priorities required renewed emphasis on energy conservation, efficiency, and alternative energy sources, all of which should promote environmental quality and economic health. But changes were needed in the national laboratory establishment. In addition to promoting reform of the laboratories' nuclear culture, O'Leary pushed for greater emphasis on applied research, which might create more jobs. Quite literally, the administration's commitment to "reinvent government" included a resolve to "reinvent laboratories" so that they no longer pursued outdated Cold War objectives. The DOE's new mission would emphasize equally the department's responsibility for fostering industrial competitiveness, developing energy resources, improving environmental quality, and maintaining national security. According to O'Leary, science and technology provided the "lynch pin" that united the department's various responsibilities into a common theme. The core of that scientific and technical talent was found in the national laboratories, which O'Leary encouraged to explore new, customer-oriented paths to the future.[4]

New Directions in High-Energy Physics

After the shutdown of the ZGS and the loss of the GEM project, Argonne's high-energy physics program languished without a machine of its own. As it turned out, closing the ZGS and losing the GEM project were essential to Argonne's securing the Advanced Photon Source (APS), the machine that became the laboratory's principal research facility on the threshold of the twenty-first century. In 1983, with the breeder-reactor program in deep trouble and Argonne faced with redefining its research mission, no one anticipated that the laboratory would have an illustrious future with one of the brightest x-ray synchrotrons in the world.

In 1983, acknowledging Argonne's past contributions, the Department of Energy's High-Energy Physics Advisory Panel recommended modest support of the laboratory's high-energy physics group but presumed that the future of Argonne's program would be centered at nearby Fermi National Accelerator Laboratory (Fermilab). HEPAP believed that closer cooperation between Argonne's and Fermilab's high-energy physicists would greatly benefit both laboratories and strengthen American high-energy physics generally.

The panel suggested that the Department of Energy might designate Argonne as a high-energy physics support center. Although Argonne did not have a major accelerator, HEPAP reminded the energy department that it had made a considerable investment in high-energy physics at the laboratory. Argonne maintained tall bay areas suitable for the assembly of large equipment, as well as numerous specialized shops that supported accelerator programs. In addition, as Massey had argued when Argonne sought the GEM machine, the laboratory possessed the whole range of technology associated with accelerator research, including expertise in conventional and superconducting magnets, pulsed-power sources, vacuum technology, and detectors. Although some large universities also maintained extensive support facilities, the HEPAP knew it was impractical to duplicate Argonne's excellent mechanical shops and large and proficient engineering staff at every university that wanted a high-energy physics program.[5]

As a stop-gap activity following the closure of the ZGS, Argonne established a high-energy physics user support center, which provided essential help to university researchers while enabling the laboratory to hold on to some of its talented staff. Frank Fradin, who would become Schriesheim's associate laboratory director for physical research, recalled those "lean" years when the high-energy physics group accepted major projects to build detectors at other accelerator facilities. "Now you say, well, gee, that sounds like pretty small change," Fradin smiled knowingly. But some of the detectors, such as the instrument at

Fermilab and the ZEUS detector at the HERA facility in Germany, were multi-million dollar projects that not only sustained Argonne's group, but also kept the scientists active within the international high-energy physics community.[6]

Argonne and the Superconducting Super Collider

It was clear that Argonne scientists would not be indefinitely content to play a supporting role in this prestigious field. Argonne's new initiative, however, evolved in a most unusual way.

In July 1983 a HEPAP subpanel headed by Stanley Wojcicki of Stanford University met at Woods Hole, Massachusetts, to discuss how the United States might reclaim the lead in high-energy physics. With the recent discovery of the so-called W and Z particles at CERN in Geneva, European researchers threatened to outpace the Americans. According to *Science,* high-energy physicists believed the American program had been "lean but healthy" through the 1980s. But unless Americans tried something dramatic, *Science* predicted "a future of progressive mediocrity and a not-so-gradual brain drain to Europe."[7]

The Woods Hole committee proposed a daring, but risky, plan to build a huge accelerator forty times more energetic than Fermilab's new tevatron, the largest and most energetic machine at that time. In comparison to the tevatron, which was designed to accelerate protons to 1 trillion electron volts (1 TeV) within a ring of superconducting magnets four miles in circumference, the Superconducting Super Collider (SSC) would accelerate two beams in opposite directions at about 20 TeV, smashing protons together in a ring that would be about sixty miles in circumference (a ring that could encircle Washington, D.C.). Scientists hoped that an accelerator of this magnitude would experimentally yield insights into a unified theory of fundamental physical forces. The plan called for abandoning Brookhaven's partially finished colliding beam accelerator (CBA or ISABELLE) and launching the largest, most costly science project in history. The recommendation to discontinue the Brookhaven project, dubbed WASABELLE by wags, was astounding in itself. But physicists now considered Brookhaven's colliding beam accelerator, whose design was similar to the SSC, unnecessarily conservative. Without doubt, European discoveries would render the CBA obsolete by the time Brookhaven brought the machine on line. Still, *Science* considered the HEPAP proposal a breathtaking venture: "The [high-energy physics] community is gambling its future on a program for which there is no explicit proposal, no design, no site, no research and development plan, no management plan, no management team, no director, no budget, and no guarantee of long-term federal support."[8]

Argonne eagerly tried to exploit this leadership vacuum. Within a week of the Woods Hole meeting, President Gray of the University of Chicago offered Argonne, with its experienced staff and superb facilities, as a planning and coordination center for the SSC. Nearby Fermilab would also prove an invaluable asset. Although no one at the laboratory thought Argonne could compete for the SSC, Argonne could provide a centrally located, neutral headquarters for the SSC project until a permanent site was selected. Argonne's dream of hosting the planning and design of the SSC project was, no doubt, unrealistic, but it reflected the renewed aggressive confidence that Schriesheim had instilled in the laboratory. Within a year, however, it was evident that Argonne would play only a small, supporting role for the Superconducting Super Collider. Overall, Argonne's budget slipped to less than 2 percent of the Department of Energy's total high-energy physics funding. When Trivelpiece designated Lawrence Berkeley Laboratory as the project center with regional offices at Brookhaven and Fermilab, Kenneth Kliewer, the associate laboratory director for physical research, advised Schriesheim of the "unfortunate truth" that Argonne's small cadre of high-energy experts could not be expected to compete successfully with those at the large accelerator centers. Kliewer suggested that Argonne offer five-year employment contracts as an enticement for talented accelerator specialists.[9]

Argonne's own high-energy physics review committee, headed by Steven C. Frautschi of Cal Tech, bluntly advised Schriesheim that it was crucial for Argonne to be involved in the SSC project if the laboratory wanted to remain in the forefront of the field. But Frautschi's committee urged the laboratory to find a significant, yet realistic, role in the SSC program appropriate to Argonne's expertise and organization. Developing computer codes, examining injector design, and evaluating site and environmental criteria were all suitable activities for the laboratory. The review committee, finding morale higher than at any time since the shutdown of the ZGS in 1979, politely complimented the laboratory on its "strong," "healthy" high-energy program. Nevertheless, if the laboratory hoped to remain a major competitor in high-energy physics, it would be necessary to expand the staff and reassemble the Argonne accelerator experts who had dispersed after the shutdown of the ZGS.[10]

The Synchrotron Radiation Source

Argonne National Laboratory would never be more than a bit player in the scientific drama of the ill-fated SSC project. Fortuitously, having missed a major role in the super collider program, Argonne was available to seize a more viable opportunity that materialized at this same time. At the Department of Energy, Trivelpiece's advisors on synchrotron radiation sources recommended constructing a 6 GeV ring at a national facility. In November 1983, while Argonne was still seeking to affiliate with the SSC project, the president's science advisor, Keyworth, asked the National Research Council to recommend top priorities for new facilities in materials science research. As part of its recommendations, which totaled $1 billion, the committee, headed by Fred Seitz and Dean Eastman of IBM's Yorktown Heights Laboratory, placed top priority on building a 6 GeV synchrotron x-ray source estimated to cost $160 million. In addition to evaluating scientific significance, the Seitz-Eastman committee had assessed the potential for improving America's technical and commercial competitiveness, supporting the national defense effort, and enhancing education for scientists and engineers.[11]

In July 1984, exactly one year after Massey withdrew the GEM proposal, Schriesheim informed the board of governors that Argonne had decided to pursue the 6 GeV synchrotron as one of the laboratory's major initiatives. Schriesheim set aside $1 million of discretionary funds to develop Argonne's proposal and assigned Kenneth Kliewer to work full-time on the project with a handpicked staff of ten dedicated researchers. By August 9, he flew to Washington, D.C., to inform Trivelpiece personally that Argonne would be a major contender for the proposed synchrotron facility.[12]

Learning from the GEM debacle and the failure of their IPNS-II proposal, Schriesheim encouraged the University of Chicago to recruit a special 6 GeV advisory committee (including top industry executives as well as distinguished academics*) headed by Jerry B. Cohen, a materials scientist and engineer at Northwestern University's technological institute. Cohen identified committee members with experience in synchrotron radiation work. Then, to avoid scattershot dialogue, he divided the advisory committee into four discussion groups on detectors, insertion devices, beam-line design, and accelerator issues. By no means overbearing, Cohen nevertheless organized his committee carefully, precisely—like a coach preparing for the big game.

Accepting a broad mandate, Cohen's committee did not limit itself to technical matters. Cohen asked for volunteers from each state (Illinois, Indiana, Iowa,

---

*Academic representation included: University of Iowa, University of Illinois-Chicago, University of Illinois-Urbana, University of Michigan, Northwestern University, University of Wisconsin-Madison, Michigan State University, Case Western Reserve University, Northern Illinois University, Iowa State University, Ohio State University, University of Missouri-Columbia, University of Wisconsin-Milwaukee, University of Minnesota, Purdue University, University of Chicago, and IIT. Industrial representatives came from Signal UOP Research Center, General Motors Research Laboratories, Ford Motor Company, Amoco Research Center, McDonald Douglas, and Gould Incorporated.

Michigan, Minnesota, Missouri, Ohio, and Wisconsin) to alert their congressional delegations about the importance of the project not only for Argonne, but also for midwestern science and industry. While the volunteers beat the political bushes, Cohen wanted each university to identify potential users of the facility. The strongest proposal would include agreements for joint appointments and promises from the universities to hire new faculty. Because the new synchrotron would likely become a Ph.D. factory, Cohen thought Argonne should reserve time and space on the synchrotron for university research and set aside dollars, office space, laboratories, and housing to help support graduate students working at the facility. Although no institution was willing to make firm promises unless Argonne got the machine, a survey indicated that at least thirty positions, most of them new, would be dedicated to the synchrotron facility. Schriesheim asked the University of Chicago to establish fourteen University–Argonne Synchrotron Radiation Professorships (UASR, pronounced "user") for temporary, but long-term renewable appointments from participating universities.[13]

Cohen next considered how the industrial members intended to support the project. Although midwestern corporations could favor the project with goodwill and political counsel, the most effective endorsement would come through substantial contracts to conduct industrial research at the facility. To obtain demonstrated support from industry, the corporations proposed a workshop at which industrialists could define their interests and requirements for the synchrotron. The target companies were Honeywell, Control Data, 3M, Motorola, Abbot, Searle, and Lilly. When Trivelpiece recommended that the laboratory line up solid backing from industry, Argonne formed an industrial subcommittee of the special 6 GeV committee for this purpose. Most of all, whereas the special committee promised to do all it could to support the 6 GeV synchrotron project, Cohen advised Schriesheim "that the (truly) full time staff on this project needed considerable expansion."[14]

Schriesheim and Kliewer stepped up their efforts to recruit accelerator specialists and went hunting in the most obvious place—Fermilab. Quickly, they lured back to Argonne their former director of the particle accelerator division Lee Teng, who accepted a joint appointment as an Argonne fellow to be codirector of the design group with Yanglai Cho, who had major responsibility for designing a German synchrotron. Frank Cole and Alessandro Ruggiero signed on as well, while resident scientists Gopal Shenoy, Gordon Knapp, and Frank Fradin agreed to serve in key design positions. With dramatic recruiting successes, Schriesheim also assured the Argonne community that planning for the synchrotron would be protected from Reagan's budget cuts.[15]

When ERAB endorsed construction of the 6 GeV synchrotron, Trivelpiece established an Advanced Photon Source steering committee chaired by Peter

Eisenberger of Exxon Research and Engineering Company. In March 1985 Eisenberger invited American and European accelerator experts to a workshop at the National Bureau of Standards. Cohen and David Lynch of Iowa State University, both members of Argonne's special 6 GeV committee, participated in discussions on machine design, location, and costs. Win or lose, this time around Argonne was not going to falter because it had failed to coordinate planning efforts with its constituents, colleagues, and rivals.

Initially, Argonne's competitors for the 6 GeV accelerator, which would produce high-energy "hard" x rays, included Brookhaven and Stanford, with Harvard-MIT and the University of Michigan as less likely contenders. In addition, Lawrence Berkeley Laboratory sought authorization for a 2 GeV synchrotron designed to create radiation in the ultraviolet and "soft" x-ray regimes. Both machines would complement and extend the capabilities of Brookhaven's national synchrotron light source (NSLS) and Stanford's synchrotron radiation laboratory (SSRL).* When the National Science Foundation decided not to upgrade Wisconsin's Aladdin machine, competition for the DOE's 6 GeV synchrotron was effectively narrowed to Argonne and Brookhaven, the laboratories most capable of satisfying Department of Defense research requirements as well. Speculating with Leon Lederman, the director of Fermilab, Schriesheim doubted that Illinois would get both the SSC and the 6 GeV synchrotron. If the choice came down to one or the other, Schriesheim thought it more likely that Illinois would acquire the 6 GeV synchrotron than the SSC—"for environmental reasons if nothing else."[16]

While Eisenberger's committee completed its studies, Argonne pushed ahead assembling its own proposal in collaboration with eight midwestern universities.† In November 1985 Schriesheim traveled to Madison to tour the university's facilities and to solicit Wisconsin's support for the 6 GeV project. The bitter rivalries now forgotten, University of Wisconsin chancellor Irving Shain thanked Schriesheim for Argonne's recent generous collaboration with University of Wisconsin's synchrotron program. Ebullient and expansive, Schriesheim outlined Argonne's plans for the new synchrotron project. Ultimately, the laboratory would commit more than fifty accelerator scientists,

---

*Existing synchrotron radiation facilities in the United States were located at National Bureau of Standards, SURF II (0.28 GeV); University of Wisconsin, TANTALUS (0.25 GeV); University of Wisconsin, ALADDIN (0.75 to 1 GeV); Stanford University, SPEAR (3 to 3.5 GeV); Cornell University, CESR (5 to 5.5 GeV); Brookhaven NL, VUV (0.75 GeV); Brookhaven NL, XRAY (aka NSLS) (2 to 3 GeV).

†Iowa State University, Northern Illinois University, Purdue University, University of Chicago, University of Illinois-Urbana, University of Minnesota, University of Missouri-Columbia, University of Wisconsin-Madison.

materials scientists, physicists, chemists, biologists, and engineers to the design effort. In dramatic contrast to Argonne's battle with MURA in the 1950s, Shain wrote Trivelpiece on behalf of twenty-six midwestern universities and industries requesting the Department of Energy to build the synchrotron facility at Argonne. In addition to the University of Chicago's UASR professorships, Shain reported that twenty-one participating universities identified fifty-two prospective visiting appointments.[17]

Schriesheim's strategy to develop an effective scientific and political constituency while fostering improved relations with the Department of Energy paid off. In January 1986, Trivelpiece called to alert Schriesheim about the department's funding plans: Argonne would get the 6 GeV synchrotron; Berkeley, the 2 GeV synchrotron; Brookhaven, a heavy-ion collider; and Oak Ridge, a reactor. With concurrence of the laboratory directors, Berkeley would be the first to receive construction funds in fiscal year 1987. The rationale for placing Berkeley's small synchrotron first was the belief that if Argonne were given priority, the Berkeley machine would never be built. Assuring Argonne that he would ask for construction funds in the fiscal year 1988 budget, Trivelpiece warned Schriesheim to be content with design and planning money for the present and not to lobby for construction funds in fiscal year 1987.[18]

In February 1986, long before Eisenberger's group finished its work, Louis Ianniello, the director of materials sciences for the Department of Energy, announced that Argonne would manage the 6 GeV synchrotron project, which would probably be constructed at Argonne as well. Delighted that the DOE budgeted $3 million for the initial design and development of the synchrotron x-ray source, Schriesheim reveled that winning the synchrotron project was "recognition of the world class work our people have already put into this project in cooperation with national leaders in the materials science community and DOE." Schriesheim's self-assessment was right on the mark.[19]

The Advanced Photon Source: A Most Bright Light

As planning progressed, Argonne's design group recommended increasing the energy of the synchrotron facility, which became known as the 7 GeV Advanced Photon Source, or more simply, the APS. The Advanced Photon Source was designed to accelerate positrons (particles similar to electrons but with positive rather than negative charge) to energies in excess of 7 GeV. The accelerated positrons would then be stored in a half-mile circumference ring in which they would circulate at nearly the speed of light to produce synchrotron radiation, or photons (figure 42). As the positrons sped around the ring, their paths

Figure 42. Advanced Photon Source (APS)

would be bent by powerful magnets called "wigglers" and "undulators." When the positrons "wiggled" or "undulated" in relatively shorter or longer wavelengths they would emit tightly focused x-ray beams available for experiments. Theoretically, the synchrotron radiation source could produce energies ranging from the infrared, through the visible and ultraviolet, and into the spectra of x rays. Argonne's Advanced Photon Source was expected to produce x-ray beams ten thousand times brighter than Brookhaven's powerful synchrotron radiation source.[20]

Most Americans were familiar with medical and dental x rays which produced pictures of bones and teeth. Modern x rays employed the same principles pioneered by Roentgen and Coolidge at the turn of the twentieth century. Schriesheim explained to Congress,

> "Recent scientific applications have put a premium on a property called 'brilliance.' High brilliance requires that many x-rays emerge from a point-like source, flow in a pencil-like beam, and have nearly the same wavelength. Remarkably, several recent advances in accelerator technology have made possible x-ray sources with just these properties—sources that are more like lasers than like flood lamps."[21]

The advantage of using bright light to explore dark spaces is obvious. The pinpoint brilliance of the APS would enable scientists to illuminate extremely small objects with x rays ten billion times brighter than medical x rays. X rays, whose wavelengths are one angstrom (about the diameter of an atom), make excellent tools for studying the structure of atoms in molecules. (In contrast, the wavelength of visible light is about 5,000 angstroms—far too large to detect individual atoms). Researchers actually expected to observe molecules undergoing chemical reactions. The molecular world, of course, is measured in microns (one millionth of a meter, 0.00004 inches, or $\frac{1}{175}$ the size of a period). For example, Brookhaven's synchrotron could obtain information about objects 6 microns square. Using the APS, scientists anticipated obtaining data about crystals as small as 1 cubic micron. Beaming bright light onto structures so small is almost impossible to comprehend except by analogy. One scientist compared 1950s synchrotron technology to studying Manhattan Island from ten miles up. By the 1970s, they could study individual skyscrapers. With the APS in the 1990s, they could not only discover match boxes lying on the street, but also describe their color and determine which way they are turned.[22]

The APS promised to open broad research opportunities in atomic sciences for materials scientists, chemists, geologists, biologists, engineers, physicians, and physicists. The research emphasis would be on the study of molecular behavior and the development of new materials. In addition to being able to observe smaller structures and perform experiments in much less time, the APS would enable scientists to record fast chemical and biological reactions as they happen. For example, chemists anticipated studying fast reactions in forty billionths of a second, effectively enabling them to obtain "stop action" data about ongoing reactions. To achieve more efficient production, petroleum engineers wanted to explore the process by which little-understood catalysts promoted chemical reactions. Similarly, using instruments developed at Argonne, biochemists hoped to take three-dimensional photographs of enzyme reactions. By understanding the reaction process, scientists might develop "designer drugs" to influence digestion, blood clotting, blood pressure, or other metabolic functions. Cardiologists saw potential for developing diagnostic technologies for fast, safe, and inexpensive screening of large populations for heart disease. Medical researchers hoped to observe the behavior of viruses. For the first time, scientists might visualize the chemical reactions in coal combustion which could yield clues for the development of cheaper, cleaner-burning coal. Geologists expected to detect ultralow concentrations of elements in meteorites and geological samples. Perhaps most exciting, the APS promised to provide insights into the structure and behavior of new and old materials, ranging from semiconductors and polymers to steel and nylon.[23]

The chemical industry developed a special interest in the Advanced Photon Source. There was much to learn about the basic chemical reactions involved in the manufacture of such common materials as nylon, plastics, and polyurethane foam. In an important cooperative agreement among DuPont, Dow Chemical Company, Northwestern University, and Argonne Laboratory, the chemical companies agreed to pay $10 million in equipment costs and $1 million annually in operating fees to reserve two x-ray beam lines and six experimental stations for research dedicated to the Du Pont–Dow–Northwestern project. Du Pont proposed constructing a spinning machine at Argonne so that its engineers could study the effects of spinning nylon fibers at rates of thousands of feet per second. Using APS bright x rays, Du Pont also wanted to study cracking and bubbling inside plastics undergoing mechanical stress, whereas Northwestern's researchers wanted to watch how plastic melted. Dow Chemical hoped to improve the quality of its foam by examining the "moment of creation," when molecules chemically formed polyurethane. Although applied research dominated proposals for the APS, Kenneth Kliewer noted, "when you explore new territory in science, progress occurs in unanticipated areas and in unanticipated ways."[24]

APS: "A Tub on Its Own Bottom"

While Schriesheim worked indefatigably to secure the Advanced Photon Source, he harbored doubts that Argonne was capable of constructing a billion-dollar project, which was what he projected the APS (including beam lines) would eventually cost. The laboratory had neither the organization nor the leadership to pull it off. Rather than trying to patch together a new division under uncertain management, Schriesheim decided to establish the synchrotron program as a "tub on its own bottom"—that is, to create a laboratory within a laboratory whose head reported to the laboratory director. Using a business model, Schriesheim intended to set up a "bottom-line operation" with sufficient responsibility, authority, and resources to accomplish its mission. By his own assessment, Schriesheim's most critical challenge was to identify and recruit a dynamic science administrator to run the project.[25]

Schriesheim "found" David E. Moncton at the Exxon Research and Engineering Company in May 1987 when Exxon, also vitally interested in the success of the APS project, agreed to loan their top solid-state physicist to the laboratory. A graduate in engineering physics from Cornell University, Moncton earned a master's degree and a Ph.D. in physics from MIT. Both men were intense, driven, and totally committed to excellence, but in common with the avuncular Schriesheim, Moncton was also a builder who laughed easily, often at himself.

Moncton's scientific background fit perfectly with Argonne's needs. In 1985 he won the DOE's materials research competition in recognition of his outstanding contributions to the study of magnetic structures using synchrotron radiation techniques. Before his assignment as the interim associate laboratory director for administration of the APS, Moncton managed Exxon's synchrotron facilities at Brookhaven National Laboratory and at Stanford Synchrotron Radiation Laboratory. Recognized nationally as a leader in the field, he had succeeded Eisenberger as chair of the Department of Energy's synchrotron users group. Shortly after accepting Argonne's interim appointment, Moncton won the department's prestigious E. O. Lawrence Memorial Award for his outstanding achievements in atomic energy. Ultimately, Schriesheim successfully prevailed on Moncton to accept a permanent appointment as director of the Advanced Photon Source. The APS "tub," sitting on its own bottom, finally had its own captain, so to speak. "We now have a world leader in X-ray studies to head a permanent team experienced in managing construction, accelerator development and service to scientific users of the facility," Schriesheim reported, well satisfied and relieved that he had concluded this critical search so successfully.[26]

In contrast to the SSC, which served a relatively small scientific community of high-energy physicists, Argonne's Advanced Photon Source, which was endorsed by a large and varied constituency of scientists, engineers, and industrialists, enjoyed reliable support from the White House and consistent backing in Congress. Moncton believed that synchrotron politics involved a simple, straightforward strategy. After identifying about ten key committee members in Congress, Moncton's synchrotron group recruited potential users in or adjacent to those congressional districts, such as the pharmaceutical company Eli Lilly in Indiana, a constituent of Representative John Meyers on the House Appropriations Committee.[27]

Moncton's chief political problem was not lining up support for the APS in Congress but, rather, worrying about the preoccupation of the Illinois delegation with the Superconducting Super Collider. Although he did not lobby against the SSC, privately Moncton hoped that the huge project would not come to Illinois. Even should Congress authorize both the Superconducting Super Collider and the Advanced Photon Source for Illinois, the larger project was certain to dominate Illinois politics. Moncton quipped that when he worked at Stanford, high-energy physicists owned the ring; when he worked at

Brookhaven, high-energy physicists owned the laboratory. "If the SSC came to Illinois, they'd own the state," he laughed.[28]

By the summer of 1988 the Advanced Photon Source had replaced the IFR as Argonne's highest priority. Funding for the APS hit a temporary snag, however, when Congressman Tom Bevill of Alabama and Senator Bennett Johnston of Louisiana agreed to eliminate APS construction money from the fiscal year 1989 Energy and Water Appropriations Bill. To help reduce the federal deficit, they agreed there would be "no new construction starts." At first, Moncton was in a state of shock, but he quickly recovered when Congressman Harris Fawell succeeded in restoring $6 million for detail "design." The Congress need not worry about unauthorized construction, Moncton thought, since there was little one could build with only $6 million.[29]

The following year, when Congress considered Argonne's fiscal year 1990 budget, $40 million was included for construction of the Advanced Photon Source under the guise that this was not a new start-up, but the continuation of an existing program. Although Congressional rationale was transparent, broad support for the APS increased funding to $90 million in fiscal year 1991 and $167 million by fiscal year 1994. In addition, the state of Illinois contributed $5 million in development funds, including $1.5 million to construct housing for visiting scientists and $2 million to support beam-line instrumentation at the APS.[30]

Although confident that they would eventually receive construction funds, Schriesheim and Moncton were anxious to get the APS venture underway. Foreign science competition was not as intense as during the height of the Cold War, but rival high-brilliance x-ray synchrotrons were being built in Grenoble, France, and Tsukuba, Japan. In the United States, other major projects beside the giant SSC also competed for attention: the $100 million Advanced Light Source at Lawrence Berkeley Laboratory, the $400 million Relativistic Heavy Ion Collider at Brookhaven National Laboratory, and the $430 million Continuous Electron Beam Accelerator Facility at Norfolk, Virginia (Argonne's former rival in the SURA controversy). But Moncton, who described himself as "one of the first traveling condensed matter physicists," believed that the APS reflected an important change in post–Cold War science. From his perspective, the scientific goals of the Advanced Photon Source project took precedence over institutional loyalty. Argonne had learned well the bitter lessons of trying to "go it alone" with the GEM machine. Without question, the APS was Argonne Laboratory's most successful collaborative project with outside users from both academia and industry.[31]

In June 1990, after Congress finally authorized construction funds for the $456 million Advanced Photon Source, Argonne lost no time in officially breaking ground for the Department of Energy's second largest basic science project (when Congress canceled the Superconducting Super Collider, the APS became DOE's largest project). During a celebration that resembled a local fair with big tents, barbecue, and red balloons outlining the location of the synchrotron ring, Moncton reflected that when Argonne completed the project in 1995 on the centennial of Wilhelm Roentgen's discovery of x rays, the APS would increase the brilliance of Roentgen's x-ray tube by a trillion fold. With the world's most intense x-ray beams for research, Moncton anticipated historic and rapid breakthroughs in materials science studies. The *Argonne News* reported: "A single machine has never been built with as broad a mission as the APS, Moncton told the assembled dignitaries. But that mission is broad in another dimension which is less often recognized, yet equally important. That other dimension is the time scale for the return on the taxpayers' investment in the APS."[32]

User groups for the Advanced Photon Source allocated beam time even before construction of the facility was completed. In addition to the Du Pont–Dow–Northwestern University partnership, a consortium of pharmaceutical firms invested $6 million in equipment to study the crystal structures of proteins that should help in designing effective drugs. Within a year of ground breaking, 19 research groups had proposed studying the structure of materials ranging from metals to human tissue. When construction was nearly half complete, the APS users group had granted access to the beam line to 15 research groups, called collaborative access teams (CATS). Two-thirds of the collaborative teams involved industrial participation. The more than 500 prospective investigators came from 28 industrial firms, 18 national or private research laboratories, 8 medical schools, and 72 colleges and universities.[33]

Argonne's Structural Biology Center

Argonne's structural biology center, which opened in October 1992, was established as a prototype of a larger research center to be established in conjunction with start-up of the Advanced Photon Source. Directed by Edwin Westbrook, the "mini" structural biology center, which operated two beam lines at the national synchrotron light source at Brookhaven National Laboratory, was still the nation's most advanced facility of its kind. Ultimately, the center would run two x-ray beam lines at the APS while continuing to operate two beam lines at Brookhaven. Funded by the DOE's Office of Health and Environmental Research, the $21 million center would become a magnet for biotechnology research in the Midwest by enabling scientists to collect data on disease processes at a pace hundreds of times faster than ever before.[34]

The center's research efforts focused on protein crystallography, which explored the structure of viral or bacterial proteins by bouncing x rays off their crystals—complicated structures containing tens of thousands of atoms, but smaller than a grain of salt. After growing a crystal of the protein to be studied, crystallographers showered the crystal with intense x rays, producing unique scattering or diffraction patterns characteristic of the molecular arrangement within the crystal. Using diffraction techniques, researchers could "see" the dense electron clouds that surrounded each atom. With the assistance of computers, they analyzed each pattern mathematically to produce an electron density map which revealed the protein's molecular structure in three dimensions.

Crystallographers used the technique to produce a detailed electron relief map of the structure of cholera toxin, a protein produced by the bacterium *Vibrio cholerae*. Cholera, caused by consuming food or water contaminated by *V. cholerae*, can kill swiftly and terribly by producing devastating diarrhea that can drain 60 percent of a body's weight in a day. The cholera toxin, which binds to cells in the wall of the small bowel, forces host cells to pump massive amounts of water continuously into the bowel. In the 1850s, unlucky travelers who drank polluted water on the Oregon Trail frequently became ill with cholera in the morning and were dead by sundown. In 1991–92, a cholera epidemic caused by contaminated shellfish ravaged South America's Pacific coast, killing 3,600 of 350,000 victims.[35]

At the structural biology center, computer analysis of the cholera toxin map produced a three-dimensional, doughnut-shaped model of the toxin's atom clusters. From this modeling, Argonne researchers, working in collaboration with scientists from Yale and Brown Universities, examined and mathematically confirmed how the cholera toxin protein recognizes and binds to receptors in the infected cells. Knowing that injected vaccines are generally ineffective against cholera, understanding the disease process could enable scientists to design more promising oral vaccines.[36]

Harvey Drucker, the associate laboratory director for energy, environment, and biological research, could hardly wait for completion of the APS. With the ability to determine the three-dimensional structure of proteins, Drucker was anxious for the structural biology center to visualize the immutable protein structures of the AIDS virus. It was currently impossible to make an AIDS vaccine, Drucker explained, because the virus continually mutates. But, if Argonne could identify the immutable parts of the protein structure of the virus from a three-dimensional model, Drucker believed it would be possible to design an effective AIDS vaccine, which in historical significance would rival Fermi's controlled chain reaction in 1942.[37]

Eliezer Huberman, director of the Center for Mechanistic Biology and Biotechnology, launched the initiative to develop a genome program at Argonne. In 1989 Huberman had recruited two Yugoslav scientists, Radomir Crkvenjakov and Radoje Drmanac, who were developing a new method of sequencing DNA by hybridizing target DNA with thousands of oligomers and using computer algorithms to determine the target's sequence from the individual bits of data. With the blessing of Schriesheim and Drucker, Huberman recruited Crkvenjakov and Drmanac, who created the foundation for Argonne's work in the Human Genome Project. As noted by computational biologist Ivan Labat, the outcome of the genome project "could have far-reaching effects on human and veterinary medicine by allowing the diagnosis and treatment of genetic disorders."[38]

Levitation

Joseph W. Paulini only received honorable mention for one of the most extraordinary photographs of the twentieth century. *Industrial Photography* magazine no doubt had good reasons for bypassing Paulini's startling picture for its highest awards. But seldom has such a significant advancement in science been so dramatically illustrated. Paulini's stark photograph showed a small black magnet hovering in a nitrogen cloud above a ceramic superconducting disc. Telltale reflections on the magnet indicated that it had been spinning while suspended, unattached to anything in the air. Much more than a magician's trick, Paulini's picture captured a buoyant magnet levitated by the effects of electrical superconductivity (figure 43).[39]

About the same time, a young graduate instructor performed the same demonstration live for his blasé freshman chemistry class at Cornell University. Theatrically, he placed the small magnet on a superconducting ceramic disc; with a grand flourish, he doused the magnet with liquid nitrogen; and then with all the excitement of a novice scientist, he waited for the magnet to levitate above his laboratory bench. "I couldn't believe it," he later reported, "no one could have performed this demonstration of the Meissner effect a year ago, but most of my students just sat there like we had been performing the experiment for the past 100 years." Unfortunately, the Cornell students had no frame of reference from which to understand the significance of their laboratory demonstration.[40]

In Washington, D.C., Schriesheim replicated a similar demonstration for a more attentive audience. President Reagan, accompanied by his science advisor William Graham, Secretary of Defense Caspar Weinberger, Secretary of State

Figure 43. Levitation (Joseph W. Paulini photo)

George Schultz, Secretary of Energy Herrington, and National Science Foundation Director Erich Bloch, gathered around Argonne's director to see for themselves this phenomenon, which had excited the imaginations of scientists, engineers, and politicians (figure 44). The occasion was part of a ceremony during which the president named Argonne the nation's Superconductivity Research Center for Applications.[41]

Breakthrough in Superconductivity

A major breakthrough in superconducting technology in 1986–87 had created this excited attention. Ordinary conductors, such as copper wire, can perform two tasks: carry electrical current and create controlled magnetic fields. But conductors also resist the flow of electricity, creating heat and wasting energy. Energy experts estimated that as much as 5 percent of all electricity generated by American power plants was lost in the transmission lines before it ever reached customers. All along the line, resistance and heat dissipated the efficiency of electrical motors, appliances, and lights. The equation was simple enough: reduced conductor resistance equaled increased electrical efficiency.

In 1911–12, the Dutch scientist Heike Kamerlingh Onnes discovered that

Figure 44. President Reagan watches Alan Schriesheim demonstrate superconductivity.

certain metals, such as mercury, tin, and lead, lost all electrical resistance when cooled to near absolute zero degrees Kelvin (-459° F). At 4.2° K, electrical current flowed almost unimpeded. This virtually frictionless flow of electrons through supercold conductors was called superconductivity. For five decades, research gradually pushed the critical temperature upward until 1973, when the critical temperature (T_c) reached 23° K (-419° F). By 1986 the use of a ceramic raised the T_c to about 90° K (-297° F), enabling researchers to use liquid nitrogen to cool the ceramic.[42]

The advantages of using liquid nitrogen to achieve critical temperature seemed almost limitless. In contrast to liquid helium, nitrogen when liquefied was easily handled and cheap, thus significantly reducing the cost of refrigeration. Imaginations soared over speculations that high-T_c superconductors would revolutionize electrical production and transmission, transportation and communications, electronics and computers, and medical research. "Genuine scientific breakthroughs occur only rarely," Schriesheim observed, comparing progress in superconductivity to the discovery of the laser and transistor. Whole new industries waited just over the horizon. "By the year 2001," the editor of Argonne's *logos* predicted, "large-scale uses could appear in the form of power generators, large magnets and motors, energy storage devices, and levitated trains." Once again, Americans were tantalized by the quest for the holy grail of cheap and unlimited energy.[43]

Argonne built the first motor using the new superconducting technology "fueled" by liquid nitrogen. The motor, employing the Meissner effect, was crude and too small and weak for any practical use, but Roger Poeppel, one of Argonne's top ceramicists, explained that it effectively demonstrated superconductivity at work.[44]

Argonne had been a leading laboratory in superconductivity research since the 1960s, a quarter of a century before development of the amazing new ceramics. Argonne scientists had developed superconducting cables and pioneered in the construction of large superconducting magnets essential for the 12-foot bubble chamber used in connection with the ZGS to track particles in high-energy physics research. In a cooperative research program with the Soviet Union, Argonne built a 40-ton superconducting dipole magnet to conduct magnetohydrodynamic experiments in which electricity was generated by passing charged gases through a magnetic field. The Argonne Tandem-Linear Accelerator System was the world's first superconducting accelerator of heavy ions, using forty-two split-ring superconducting resonators.[45]

Argonne aggressively pursued superconducting research after the announcement of the high T_c. Remarkably, within months of the discovery of the superconducting ceramics, Argonne scientists achieved an historic first on April 1, 1987, when they successfully conducted electricity along a ceramic wire made from the new superconducting material. The flexible ceramic wire, however, which was seven thousandths of an inch in diameter, was fragile and carried only a small current. Other high-T_c superconductors made from the new ceramic, such as films, tapes, crystals, and cylinders, were also very brittle and carried little current. In order to develop functional superconductors, both strength and current capacity would have to be significantly improved.[46]

In 1987 the Department of Energy designated Argonne, Brookhaven, and Ames Laboratory at the University of Iowa as centers in the government's national superconductivity initiative. Argonne's national center for superconductivity applications was given the task of developing practical superconductors that might be available for commercial applications within five years. To lead Argonne's efforts, Schriesheim named Frank Y. Fradin as associate laboratory director for physical research with special assignment to manage the laboratory's programs in high-temperature superconductivity. Fradin, a senior scientist and fellow of the American Physical Society who joined the laboratory in 1967, had formerly directed the materials science division. As a member of the international advisory board to the 1988 conference on high-temperature superconductors, Fradin also brought additional international distinction to Argonne's quartet of associate laboratory directors.[47]

In the euphoria over recent breakthroughs in superconductivity, Fradin's colleague, David Moncton, warned that too much ballyhoo might create unrealistic public expectations of achieving quick commercial benefits. The challenges were formidable. While the press focused on critical temperature, improving current density was more important to Argonne's engineers. Higher current density meant that superconducting magnets could be reduced in size and weight, and/or superconducting wire could be made thinner and thus more flexible, improving reliability and durability. In general, the smaller the superconductor, the more economical its manufacture.[48]

Argonne's high-temperature superconducting research focused on two areas: developing superconducting materials having commercial application and understanding the mechanics that enable superconducting materials to carry electrical current with no resistance. Researchers especially examined ways to increase the critical temperature of materials operating in magnetic fields (vital for use in motors) and to improve the chemical stability and flexibility of superconducting materials.

While researchers continued to improve the strength and purity of the best-known superconducting material—called YBC 1-2-3 because it contained yttrium, barium, and copper in a ratio of 1:2:3—Argonne scientists also experimented with a new compound of mercury, barium, copper, and oxygen whose superconducting properties could be investigated by the IPNS. Unfortunately, as magnetic fields increased, the critical temperature at which YBC 1-2-3 became superconducting decreased. Findings determined that the mercury material conducted electricity in magnetic fields at higher temperatures and with less loss than other materials, offering a promising application for electrical motors.[49]

Argonne pressed ahead on the vexing problem of manufacturing practical superconducting wires. Individual crystals of YBC 1-2-3 were superb electrical conductors, but when they were fired together, electrical current had difficulty passing from one grain to another. A solution to this grain-boundary problem involved loading superconducting powder into a silver tube that was rolled or drawn into a thin or flat wire. Developed in partnership with Intermagnetics General Corporation and Superconducting Products, the "powder-in-tube" process eliminated the weak connections that inhibited current in ceramic wires. Producing uniform wire proved difficult, however, because, when drawn, the powder-in-tube tended to bunch into "sausages" within the silver tube rather than to distribute evenly along the length of the wire. These density variations, which researchers detected by x-ray imaging, formed "sausage links" that interfered with electrical current as it traveled along the wire.

Although the technique was imperfect, the "powder-in-tube" superconducting wire still set records for carrying high current. According to the president of Intermagnetics General Corporation, Carl H. Rosner, the goal was to manufacture practical superconducting wire at least one hundred yards long to be wound into coils for magnets or other uses.[50]

The superconducting magnetic bearing was another remarkable device, which was developed jointly by Argonne and Commonwealth Edison. The magnetic bearing, which employed the Meissner effect, levitated above a superconducting plate, spinning with a friction coefficient of 0.0000009, according to Commonwealth engineer Robert Abboud. "That's three times lower than that of any other [magnetic] bearing we know of," confirmed Argonne's John Hull. Or a thousand times less friction than produced by roller bearings in American cars and trucks. When connected to a flywheel, the "super bearing" might solve an electrical energy storage puzzle that had long challenged DOE scientists. Batteries were incapable of handling large, continuous demands, but no other technology existed for off-peak storage and later distribution of vast amounts of electrical energy to communities and industry. Consequently, power companies presumed that generating capacity throughout the electrical supergrid must meet peak demands.[51]

Abboud and Hull believed that utilities using a system of super bearings and flywheels would overturn the assumption that peak demand required equivalent generating capacity. Edison's twelve nuclear power plants reduced their load at night when customer demand was low. But if the utility were to transfer power to the flywheels at night, consumers could draw on that power the following day without Edison having to increase power plant capacity. The super bearing–flywheel system would not only lessen Edison's need to build new power stations but would also reduce the utility's dependence on fossil fuels. After fabricating a small but operating flywheel, the Edison-Argonne team estimated that commercial flywheel devices storing 5 megawatt hours could be built the size of a large semitrailer. Such a device could provide a small town with electricity for an entire day or could supply emergency electricity to hospitals and other vital installations during power blackouts. Argonne–Commonwealth Edison researchers scheduled a field test of a 50-kilowatt-hour energy storage unit for 1995.[52]

ARCH

Even when the national laboratories generated commercial ideas, it was difficult to move the government's inventions into the marketplace. Unless companies could obtain an exclusive license to the concept, they were understandably reluctant to invest capital to develop a consumer or industrial product. Recent federal legislation, the 1980 Stevenson-Wydler Technology Innovation Act (which mandated that 0.5 percent of national laboratory funding be dedicated to technology transfer) and the 1984 Bayh-Dole Act, encouraged federal laboratories to share proven technology with private industry. As a result of a proposal by Massey and Schriesheim, on October 20, 1986, the University of Chicago created the Argonne–University of Chicago Development Corporation (ARCH) as a not-for-profit entity to hold selected patents, inventions, and copyrights from Argonne and the University of Chicago.

Industry-laboratory partnerships turned to ARCH for help in finding commercial markets for promising superconducting devices and other advanced technology. Although Argonne was a DOE laboratory, the federal legislation stipulated that the intellectual property created by the laboratory, with certain exceptions, belonged to the University of Chicago or to individual inventors who could form their own companies to market technology with commercial potential. In general, inventors received at least 25 percent of gross royalties. The balance of the profits was paid into ARCH's development fund. Should royalties exceed 10 percent of the inventor's Argonne salary, the Department of Energy was to be assured that the inventor was not engaged in any conflict of interest or in private consulting that interfered with work at Argonne.[53]

ARCH became the principal agent through which the University of Chicago reviewed ongoing research and development for the purpose of promoting the commercialization of Argonne technologies likely to find reliable markets within five years. ARCH also provided legal and business support for developing technologies, such as the steel initiative and metal reactor fuels, whose commercial potential was more long-range. Finally, through direct negotiations with industry, ARCH provided a conduit to the laboratory for "work-for-others" funding.[54]

Not all of Argonne's interactions with industry directly involved ARCH. Argonne's Technology Transfer Center, created in 1985 under the leadership of Brian Frost, was jointly funded by the Department of Energy and the state of Illinois. The unit, designated by the state of Illinois as one of fourteen technology commercialization centers (I-TEC), provided a link between the state's research and business communities. Argonne was still not allowed to compete directly with private industry or institutions or to bid directly for work. Under the aegis of the center, however, Argonne could perform work-for-others if its unique facilities or expertise were employed. If expenses were fully reimbursed, the sponsor maintained all rights to intellectual property. If costs were shared, industrial partners could still obtain nonexclusive use of the research.

The best known of the center's activities was a superconductivity commercialization pilot project that coordinated development agreements with industry. Another of the center's major tasks was to alert ARCH about promising inventions and patents.[55]

The first laboratory-university partnership under the Bayh-Dole Act, ARCH was in many respects similar to the university-industry technology transfer arrangements at MIT and Stanford. Joint funding proposals between Argonne and industry and cooperative industrial research programs that utilized Argonne's facilities usually did not concern ARCH until the joint ventures required commercial licensing. Steven Lazarus, president and CEO of ARCH, also held a joint appointment as the associate dean of the Graduate School of Business at the University of Chicago. Acutely sensitive that it did not have an engineering school to help with commercializing technology, the University of Chicago offered the expertise of its Graduate School of Business to aid ARCH with management, financing, planning, and marketing advice.[56]

Before ARCH, scientists who received a patent dispensation from the government and paid legal costs, sometimes as high as $10,000, could leave Argonne's employment to establish their own companies. In 1978 Stanley Cohen established his own business to market SPEAKEASY, a public-domain computer language he had helped to develop. Another scientist, Charles F. Ehret of Argonne's Division of Biology and Medicine, designed Argonne's Anti-Jet-Lag Diet while studying the biological rhythms of animals. Subsequently, he formed a consulting firm to advise industry on the effects of biological rhythms on shift workers.[57]

Joseph Stetter, winner of an I-R 100 award for his portable "sniffer," which located and identified toxic gases, also left Argonne to form his own company but licensed the rights to the hazardous-gas monitor through ARCH. Developed with the help of a million-dollar grant from the United States Coast Guard, Stetter and his associates in Argonne's environmental divisions constructed a light-weight, portable, battery-powered instrument designed to assess hazardous chemical spills or to survey suspicious cargo holds. Stetter called his device a "poor man's spectrometer." Rather than waiting hours for the results of laboratory analysis, with the hazardous-gas monitor, fieldworkers could quickly identify the atmospheric concentration of toxic gases and take immediate action.

Stetter's sensor had obvious potential for a broad range of industrial applications. Robot sniffers could investigate and report on suspected toxic-waste sites and spills. Space sniffers, similar to smoke detectors, could monitor the interior of space shuttles, alerting the crew to the build up of inappropriate or dangerous gases. With improvements to reduce size, extend battery life, and expand to over 100 detectable industrial chemicals, potential applications of the Stetter sniffers as investigators, detectors, monitors, and warning devices seemed unusually promising. By 1991 Stetter's young company, Transducer Research Inc., grossed more than $1 million in annual sales.[58]

By 1994, under the leadership of Steven Lazarus, ARCH had assisted in founding thirteen companies, which then created some 720 jobs. In return, licensing royalties earned Argonne and the University of Chicago $2 million. Although three companies were dormant, enterprises such as Illinois Superconducting Corporation, which developed superconductor filter products to enhance cellular communications, proved successful.

Argonne's Environmental Programs

Argonne's environmental activities included programs in both research and management. In research, Argonne focused on atmospheric studies, ecosystem dynamics, subsurface geochemical processes, global climate change, and remote sensing. For example, Argonne scientists assisted the United States Army by evaluating the environmental impact on Ft. Riley, Kansas, of training exercises involving heavy tanks and other motorized units. Based on computerized analysis of Landsat images following training maneuvers, Argonne developed dynamic landscape models that projected changes in forest, shrub, and grassland vegetation at Ft. Riley. Argonne's models helped the army to predict long-range environmental damage at the fort and to project the effect of remedial actions such as plantings, controlled burns, and soil stabilization.[59]

Directed by Senior Scientist Ruth Reck, Argonne's global climatic change project studied an old, and tricky, question. For millennia, people had believed that human actions could profoundly affect the weather, but it had been very difficult to establish a scientific relationship. In the nineteenth century, for example, frontier farmers on the Great Plains actually believed that "rain followed the plow." While such fanciful views were later discredited by widespread drought and dust storms in the Midwest, beliefs that industrial activities affected global weather seemed more plausible to urban residents all around the world, many of whom were choked by noxious killer smog. In 1896, the Swedish chemist Svante August Arrhenius predicted that gradual global warming might result from increased carbon dioxide spewed into the atmosphere by modern industry (the "greenhouse effect").[60]

By the 1990s some climatologists believed they could link global warming, as well as depletion of atmospheric ozone at the poles, with the combustion of fossil fuels, among other industrial processes. Whether these phenomena reflected fundamental worldwide climate change or could be traced to human activities such as increased production of greenhouse gases remained scientifically controversial. From the 1940s to the 1970s, when atmospheric pollution rapidly increased, climatologists noted little change in global temperatures. To explore this puzzle, Argonne's search for data on the causes of global climate change included studies of deep ice cores from the polar caps and analysis of growth rings from the earth's oldest trees.

Argonne's research indicated that several variables might accentuate or mask the "greenhouse gas effect." Natural climatic variability caused by El Niño, fluctuations in solar radiation reaching earth, short-term cooling from volcanic fallout, and cooling resulting from sunlight reflecting off atmospheric pollutants complicated the analysis of climate change. Indeed, Argonne scientists suspected that since the end of World War II, the "pollution effect," which reflected sunlight back into space, may well have counterbalanced the "greenhouse effect," which captured the sun's radiated heat within the earth's atmosphere. As reported by environmentalists Mark Fernau and David South in 1992, "scientists have not been able to sort out all these uncertainties in the global temperature record and say with certainty that the enhanced greenhouse effect has been observed." It was so difficult to evaluate or project climate change because important global physical processes that produce climate were not well understood, while other processes that were well known were difficult to model on existing computers.[61]

Closer to home, the environmental impact of the Cold War on American production and scientific facilities was much easier to measure but expensive to ameliorate. While numerous programs, such as construction of the Advanced Photon Source, proceeded with enhanced environmental awareness, Argonne's office of environmental restoration and waste management developed strategies for assessment, cleanup, and monitoring at both defense and nondefense work sites. The two largest environmental problems associated with the production of nuclear weapons were disposal of high-level and low-level wastes and the cleanup and decontamination of fabrication sites. Argonne's wide-ranging programs included developing waste disposal, site cleanup, and environmental restoration technologies. In addition, the laboratory acquired skills in evaluating health risks and preparing environmental impact assessments that were utilized by other agencies besides the Department of Energy.[62]

The cleanup of Weldon Spring, Missouri, presented Argonne's environmental assessment division a difficult challenge in proposing a safe and effective plan. During World War II, the United States Army had manufactured explosives at Weldon Spring. Following the war, the Atomic Energy Commission used the site to process uranium and thorium. Then, from 1966 until 1969, the army reacquired Weldon Spring to produce herbicides. Anthony Dvorak, the director of the environmental assessment division, stated simply that "these operations each had an environmental impact on the site." In 1985 the DOE asked Argonne to formulate a cleanup plan acceptable to environmental agencies and the citizens of Weldon Spring. The plan, approved in 1993 by the department and the Environmental Protection Agency, included cleanup of waste pits, buildings, and soil. According to industry sources, Meredith Hunter of St. Charles Countians Against Hazardous Waste praised Argonne's plan for its "completeness and candor." Dvorak anticipated that the Weldon Spring plan would serve as a model in developing cleanup proposals for other sites.[63]

Although Argonne helped others solve their environmental problems, the laboratory faced its own cleanup problems. In the half century after Enrico Fermi built CP-1 under the West Stands at the University of Chicago, environmental standards and regulations had changed profoundly. Just as the Manhattan Project and the AEC initially followed prevailing industrial safety standards, so Argonne followed normal industrial procedures in disposing of laboratory wastes. Although the practice was discontinued in 1978, for more than twenty years Argonne had dumped thousands of gallons of waste chemicals into twenty-foot-deep holes called French drains. In 1981 the laboratory opened a new landfill to collect Argonne's trash, but by 1990 the facility no longer met the state of Illinois's upgraded environmental standards. At one time, radioactive wastewater was confined in an unlined lagoon; later radioactive wastes were temporarily stored in concrete vaults at an on-site storage yard while waiting transfer to a licensed nuclear-waste dump. In 1986 monitors discovered that a drainage pipe was leaking radioactive water. The pipe was capped, but environmental watchdogs such as the Natural Resources Defense Council worried that radioactive materials and other pollutants had escaped from the laboratory into Sawmill Creek, which flowed through the Waterfall Glen Forest Preserve and into the Des Plaines River.[64]

Estimated costs for cleaning up the DOE's national laboratories varied, but in 1989 the Environmental Protection Agency projected the figure to be a staggering $14 to $18.6 billion. Estimated costs for cleaning up Argonne ranged up to $668 million, which included $236 million to monitor the chemical waste pits, prevent seepage from closed landfills, clean up contaminated soils, and decontaminate CP-5, which leaked traces of radioactive gas into the atmosphere.[65]

The Palos Park Forest Preserve Revisited

The laboratory's most serious environmental problems, at least politically, were not located on the main campus at Argonne-East, however. After revelations of environmental contamination at Brookhaven, Los Alamos, Oak Ridge, Hanford, Lawrence Berkeley, and Lawrence Livermore laboratories, as well as at Argonne, nervous local officials wondered what problems lurked below the surface at former AEC sites. In the spring of 1990 inspectors from the Illinois Department of Nuclear Safety rummaged through the weeds and brush around Site A in the former Palos Park Forest Preserve, where the Manhattan Project had reconstructed Fermi's reactor as CP-2 in 1943. Returned to Cook County in 1956 after Argonne decommissioned and buried two research reactors on the site in fifty-foot pits, the area, now called Red Gate Woods Preserve, had been monitored periodically by laboratory specialists. Nevertheless, the Illinois inspectors found a uranium pellet similar to those used as fuel in Fermi's first reactor. An intensified search uncovered blocks of graphite, probably used to moderate a reactor pile, metal scraps, and bits of broken glass, perhaps from the old laboratory. None of the found debris was radioactive except the uranium, which produced faint readings.[66]

Local environmentalists were alarmed by the discovery of nuclear-age trash in a forest preserve visited by children, joggers, picnickers, cyclists, bird-watchers, church groups, and others seeking respite in the suburban woods. Unfortunately, traces of radioactive strontium 90 were found in the sediment near Site A-Plot M, a separate dump for high-level wastes that was used until 1949. Trace amounts of tritium, well within federal drinking-water standards, were also found in two wells, which were subsequently closed. In addition, plutonium, technetium, cesium, and uranium were detected here and there in the woods. The overall radioactivity was so faint that it was impossible to determine whether the elements had escaped from Plot M or whether they had been deposited as fallout during the atmospheric nuclear tests in the 1950s. Despite government assurances that the forest preserve posed no threat to public health or safety, activists backed by the Illinois Sierra Club and Broken Arrow, a local antinuclear group, demanded a thorough survey and decontamination of the area.[67]

Almost everyone, friends and critics of the government alike, understood the dynamics associated with the heightened environmental concerns at the Palos Park site. Manhattan Project and AEC scientists had not been unconcerned about nuclear safety, but they had worked under circumstances far different from those surrounding DOE researchers in the 1990s. Although the atomic bomb project created its own insistent priorities, the Metallurgical Laboratory did move its reactor to the Palos Park site in 1943 partly because of concern for public safety. Although the old site was unfenced, before closing the Palos Park operations, the government had placed four large granite "tombstones" to mark the reactors' grave. Thirty-five years later, advanced technology enabled investigators to detect minute traces of radiation in the forest preserve.

Broken Arrow, disdainful of the Department of Energy's reassurances, encouraged Greenpeace to conduct an independent survey. The Greenpeace-sponsored study found additional graphite blocks, detectable radiation in soil, apples, and leaves, and more than a dozen "hot spots" that Greenpeace believed warranted cleanup. But, even the antinuclear community knew there was no environmental emergency; Linda Josephson of Greenpeace admitted that persons could sit on contaminated sites in the forest preserve without danger. Still, Broken Arrow demanded that the Department of Energy fence the entire nineteen acres and lobbied Cook County board president Richard Phelan to close the forest preserve until the government removed all traces of radioactive contamination. In a related action, Broken Arrow also advocated that Willow Spring declare itself a "nuclear free zone." When the Department of Energy recovered the nuclear wastes, Broken Arrow did not want them transported through the village of Willow Spring.[68]

Politically, Phelan had little choice but to order the forest preserve closed "out of an abundance of caution" until the Department of Energy conducted another survey and removed all harmful contaminants. Reporter Tom Houlihan doubted that Site A posed much danger. After interviewing a naturalist at the Little Red School House, he wrote that "there is no evidence that anyone or any animals have ever gotten sick because of what's buried there." Like everyone else, however, Houlihan, who would rather be ultra-safe than sorry later, facetiously asked for "assurances that 15-foot-long salamanders and garage-sized deer aren't hiding somewhere in Red Gate Woods."[69]

With distrust of the government pandemic, Argonne's long monitoring of Site A made no difference to environmentalists who demanded remedial action, however costly. Frustrated, Argonne agreed to spend $3.4 million to secure and survey the site, clean up debris, monitor abandoned wells, and remove underground storage tanks. After fencing twenty-six acres, crews cleared brush and scrub trees so that Argonne could map the area after locating building foundations, former roads, and the like. One can only guess how many nesting sites were lost for migratory songbirds as thirty years of environmental healing was stripped away. Ultimately, the DOE committed $24.7 million to rehabilitate the former Manhattan Project site.[70]

In 1994 the Department of Energy unfortunately listed Red Gate Woods as one of twenty-six possible national sites under consideration for disposing of low-level mixed radioactive wastes. The fact that the federal government did not even own the land seemed to have escaped notice of the bureaucrats in Washington. The announcement triggered another flurry of protests from local environmentalists and the revitalized Broken Arrow, whose founder pledged to remain vigilant until the government made good on its pledge to clean up the site and abandoned all thought of turning Red Gate Woods into a nuclear dump. Unhappily, the announcement further undercut Argonne's credibility among local environmentalists. It made no sense to spend millions to cleanup Red Gate Woods only to turn it into a nuclear wasteland. Besides, if the site were as environmentally benign as Argonne had always maintained, why would the federal government list Red Gate Woods as a possible nuclear dump along with the Nevada Test Site, Hanford, and Rocky Flats—facilities known to have serious contamination problems? At best, the government was fumbling; at worst, environmentalists feared they were not being told the truth. Ultimately by the fall of 1994, Argonne reported that tritium levels around the dump had dropped below the detection levels of the instruments used to measure them.[71]

Argonne and the Inner City: Bethel New Life, Inc.

While Argonne received a pounding from community activists in the suburbs for alleged environmental carelessness and indifference, in West Garfield Park on Chicago's west side, laboratory engineers worked with Bethel New Life, a $7.6 million economic development corporation operated by Bethel Lutheran Church, to rehabilitate the urban neighborhood. A United Way charity, Bethel New Life invited Harvey Drucker's environmental and energy experts to see for themselves how their contributions were spent.[72]

West Garfield Park suffered numerous symptoms of inner-city rot. Since World War II, most whites had fled to the suburbs. During the civil riots following the 1968 assassination of Martin Luther King, West Garfield Park lost almost 20 percent of its housing. Afterward, the neighborhood declined precipitously. Unemployment rose, crime increased, and poverty spread. Almost two-thirds of 16- to 19-year olds were high-school dropouts. Between 1970 and 1980, 200 houses were abandoned or demolished. By 1990 the neighborhood had lost almost 30 percent of its 43,000 inhabitants. Where industry once thrived, almost forty factories closed, their sites frequently contaminated with dangerous and toxic wastes. Vacant lots dotted the landscape with little more prospect for productivity than scores of West Garfield Park residents. Fast-food restaurants and liquor stores dominated retail trade.

Bethel New Life introduced the suburban high-tech laboratory to inner-city poverty. David Poyer and his group in Argonne's socioeconomic research and analysis program had already determined that poor black and Hispanic households were hit harder than white households by the rising costs of non-electric energy (natural gas, fuel oil, and liquid petroleum gas). But other than this macro-computer modeling, Argonne thought it had little to offer inner-city communities—until Harvey Drucker's engineers encountered Bethel New Life. "National labs don't really think about city neighborhoods in old, rust-belt cities like Chicago," Drucker commented. "But on this occasion, we realized that a lot of the problems in West Garfield are not just sociological, they are technical." In what Drucker called a "personal labor of love," the visiting engineers brainstormed on how Argonne could directly help the West Garfield Park community.[73]

Argonne's subsequent partnership with Bethel New Life established the first agreement between a national laboratory and a community development corporation to employ technical expertise to revitalize Chicago's inner city. The innovative program involved five joint initiatives. The most important of these was the industrial-site reclamation program in which the laboratory developed plans for cleaning up abandoned industrial sites. With assistance from Lawrence Graham, his Bethel New Life partner, Argonne's Jack Ditmars checked closed factories for PCBs, dangerous chemicals that would have to be removed if sites were to be rehabilitated. The object of the Argonne survey was to determine the costs and methods required to bring decrepit properties into compliance with federal, state, and city building, safety, and environmental codes.[74]

In a city whose housing problems seemed epitomized by the Robert Taylor high-rise housing project, Drucker also wanted to salvage abandoned apartments and single-family houses in West Garfield Park at a cost of no more than $45 a square foot. Drucker noted that the Department of Energy had spent large sums of money developing conservation techniques, windows, doors, heating systems, and insulation for medium- to high-priced housing and commercial buildings but had invested little in heating, cooling, and lighting systems for low-cost housing. Argonne's team proposed to design and test efficient, small apartments whose energy costs would not exceed $1,000 a year.[75]

Argonne supported two programs to promote jobs and economic growth. The first evaluated Bethel New Life's waste recycling plant for paper, aluminum, and plastics. Drawing on Argonne's own research in recycling plastic and non-metallic auto parts, experts such as Sam Jordy worked with Bethel's Ben Robinson to increase output from the recycling plant and to find new markets for its products. A busier and more efficient recycling plant, of course, meant more jobs in West Garfield Park. The second program assisted Bethel New Life in the development of small businesses. The small-business initiative sought to improve the use of steam produced at the City of Chicago's nearby waste-to-energy facility.[76]

Finally, Argonne helped West Garfield Park with job training and adult literacy programs, which augmented the lab's ongoing assistance to local schools with math and science education. Through the urban engineering program, Argonne provided summer employment for five Bethel youths. Working with Chicago Science Explorers, the target was to recruit twenty-five students a year to attend area engineering schools.[77]

The Science Explorers

Since 1989 the laboratory had sponsored the Chicago Science Explorers, a remarkable educational program inspired by newscaster Bill Kurtis of WBBM-TV. Schriesheim had been captivated by one of Kurtis's productions, *The New Explorers*, a public television series featuring the work of exciting scientists. Obtaining the support of Secretary of Energy James Watkins, Chicago's public television station WTTW, the Amoco Corporation, and Waste Management Corporation, Schriesheim encouraged Argonne's division of educational programs to incorporate *The New Explorers* series into an innovative outreach program designed to attract Chicago students to careers in science and engineering. The Chicago Science Explorers program was built around videotapes of scientists at work, supplemented by imaginative laboratory demonstrations and visits to Argonne and other Chicago institutions such as the Shedd Aquarium, the Adler Planetarium, the Field Museum of Natural History, the Lincoln Park Zoo, the Brookfield Zoo, the Chicago Museum of Science and Industry, the Chicago Academy of Sciences, Chicago State University, the Fermi National Laboratory, Commonwealth Edison, the Chicago Police Department Fifteenth District, and the University of Chicago Lab School. In addition to its programs for college and graduate students, in the first two years the New Explorers program reached an estimated 36,000 Chicago students in grades 4–12 (figure 45).[78]

As Argonne expanded its educational outreach, the laboratory also intensified its science education effort among those underrepresented on the laboratory staff—minorities and women. The work in West Garfield Park represented a major effort to encourage minority students to pursue science-related careers. Argonne's Women in Science program not only provided information about careers in science, engineering, and related fields, but more importantly provided role models for Chicago schoolgirls uncertain about their future in a male-dominated profession.[79]

Figure 45. Sam Bowen (right) leads a lively question-and-answer session at a monthly West Garfield Park Explorers meeting

Women in Science

Schriesheim was delighted by programs initiated by Women in Science and proud of the accomplishments of women at Argonne. The only problem was, there were so very few of them. If one were to recall the role of Argonne's most distinguished scientists, Maria Goeppert-Mayer would share top honors with Argonne's other Nobel laureates, Enrico Fermi and Glenn Seaborg, both of whom won their prize for research conducted elsewhere. But as Argonne neared its fiftieth anniversary, the laboratory employed only one female scientist division director and six senior women scientists, the highest numbers in history.

During the Schriesheim decade, the status of women had gradually improved at Argonne. Responding to an initiative from Argonne's women scientists, Schriesheim agreed in 1990 to establish a new position called the Women in Science program initiator (WPI). The initiator's position, first held by biochemist Maryka Bhattacharyya, coordinated Argonne's burgeoning Women in Science program of outreach, recruitment, education, communication, and awareness. Women in Science provided a forum through which women voiced their opinions, identified female candidates for laboratory employment, reviewed programs affecting women, and advocated changes in Argonne's family-leave policy. Their best-known program was the annual conference "Science Careers for Women."[80]

Argonne received well-deserved accolades from the government and from the national laboratories for its sparkling Women in Science initiative. The raw statistics, however, were still troubling. Despite Schriesheim's generous support, from 1988 to 1993, the overall employment of women at Argonne had risen only slightly, from 27 percent to 29 percent of the workforce. Women were strongly represented among Argonne's administrative personnel, where 41.5 percent of the staff were women. Schriesheim increased the role for women on committees that influenced hiring and promotions and assured that women were represented on most of Argonne's committees (93 percent). This step was positive, but whereas women composed about 20 percent of the science faculty in academia, at Argonne only 11.4 percent of the research staff were women. Almost 10 percent of the male scientists and engineers held the top grade of senior scientist, whereas the six women senior scientists represented 4 percent of female technical employees. Of the technical staff, one-half of the men held the rank of scientist, while only one-quarter of the women did. On the other hand, one-third of the women were assistant and support scientists as compared to one-fifth of the men.[81]

What did these disparate numbers in grade distribution add up to? For some women the answer was glaringly obvious—Argonne laboratory was run like an "old boy's club" whose corporate culture was chilly, even at times hostile, to women scientists. They complained that women scientists were deliberately excluded from some activities and professionally were often shoved into corners. They believed that salaries were low, promotions were slow, and during the recent layoffs, some divisions dismissed women in disproportionately large numbers. In this alleged "old boy system," critics acknowledged that hiring followed the letter of the law, but women frequently did not even know a position was open until after it was filled. There was no disputing the fact that Argonne women were largely absent from top management, although two each sat on the scientific and technical advisory committee and the board of governors.[82]

Not surprisingly, male scientists generally did not share the perception that Argonne was dominated by an "old boy" culture that disadvantaged women. Indeed, one woman, Betsy Anker-Johnson, had served as associate laboratory director in the 1970s. From the perspective of Argonne's merit system, most evident differences between men and women scientists could be explained by differences in their education, experience, and accomplishments. Among professionals (salaried and exempt), women enjoyed less experience in every grade.

While not denying that women were sometimes paid less than men for substantially the same work (in 1989, sixteen female scientists received merit adjustments for this reason), Argonne's internal reviews indicated that women were promoted at about the same rate as men, and during reductions in force were not laid off in disproportionately greater numbers.[83]

During his ten years as director of Argonne National Laboratory, Schriesheim did not compile an outstanding record of appointing women to management positions. In supporting the Women in Science program, Schriesheim recognized that the laboratory should not only strengthen its affirmative action office, but also, in the words of senior scientist Marion Thurnauer, should foster a commitment to developing a diverse workforce throughout the laboratory: "As positions open up at all levels, including higher management positions, "there must be open and aggressive searches for qualified candidates." But women scientists were as scarce outside the laboratory as they were at Argonne. Whereas 98 percent of all secretaries and 64 percent of all salesclerks were women, the Women's Bureau of the U.S. Department of Labor reported that women did not exceed 10 percent of all physicists and engineers.[84] Long-term, the key to improving the lot of women in science was to increase their numbers dramatically. Prophetically, Thurnauer became director of the chemistry division in October 1995.

Women in Science sponsored an annual conference on science careers for women. As many as four hundred students from Chicago-area high schools gathered at Argonne to learn about technical careers in science and engineering. Workshops included advice on selecting colleges and graduate schools, descriptions of pathways toward a technical career, discussions of problems related to working in male-dominated fields, and the projection of new fields attractive to women scientists. The conference provided students an opportunity to visit with representatives from business and government as well as to tour the laboratory. Most important, however, was the girls' chance to interact with more than seventy-five successful women scientists, who shared their personal stories and encouraged the girls to think seriously of careers in science.[85]

Providing role models for young scientists did not end with high school programs. Women in Science also encouraged graduate students who were already on the road to technical careers. At Argonne's technical women's conference, women scientists and engineers from Argonne hosted graduate students from area universities to offer guidance and motivation by featuring the accomplishments of Argonne women in basic research, energy technologies, and environmental sciences.[86]

Technology Transfer in the Clinton Administration

In addition to its emphasis on environmental cleanup and affirmative action, the Department of Energy under the leadership of Secretary O'Leary also attempted to move the national laboratories toward greater commitment to applied research through cooperation with private industry. In concert with President Clinton's pledge to increase jobs and strengthen the economy, O'Leary asked the department's laboratories to devote at least 10 percent of their budgets to technology transfer.

Technology transfer encompassed three activities. Best known to the public were commercial spin-offs of work originally performed for the government. The space program was legendary for its spin-off of useful commercial products ranging from computer technology to Teflon. Historically, Argonne's principal contributions in this area included the broad range of nuclear reactor systems, but since 1985 much of the spin-off technology had been marketed through the technology transfer center and ARCH. Technology transfer also involved private use of the laboratories' facilities and employment of laboratory scientists as consultants. Ford, Motorola, Amoco, Caterpillar, Boeing, and Packer Engineering had sent scientists to Argonne to conduct research in fields concerning supercomputers, superconductivity, and transportation. The Advanced Photon Source would be a large user facility that encouraged technology transfer through industrial as well as academic participation. Whether working in West Garfield Park or in the former Soviet Union, Argonne consultants promoted technology transfer by exporting their expertise at home and abroad. Finally, in a new departure for Argonne, technology transfer incorporated partnerships between the government and private corporations to develop market-driven technology.[87]

Cooperative Research and Development Agreements

Cooperative Research and Development Agreements (CRADAs), authorized by the Federal Technology Transfer Act of 1986 and augmented by the National Competitiveness Technology Transfer Act of 1989, established new ground rules for technology transfer from the federal government to private industry. Whereas the laboratories previously had not been allowed to play favorites in sharing government-funded research, under CRADAs they were encouraged to form exclusive partnerships in which the laboratory pledged to protect corporate proprietary interests. In the "Decade of CRADA," as Harvey Drucker called it, "invention [was] not a spinoff, but rather the intended consequence of [research] efforts. Technology transfer [became] *the* purpose of the specified work, and the prime contracts of all national laboratories [were] altered so that technology transfer [became] an official mission." During the Bush administration, the Department of Energy signed more than three hundred cooperative research and development agreements. In O'Leary's first year as secretary of energy, the number doubled.[88]

Argonne's first CRADA, signed with Baxter Healthcare Corporation in 1991, enlisted the laboratory in the fight against AIDS. Under the agreement, Baxter put up $45,000 while the Department of Energy provided $35,000 to develop a new technology using light-sensitive chemicals to sterilize donated blood against AIDS, hepatitis, and other viral diseases. The research project tested the use of light-sensitive, virus-killing drugs. Argonne's research task was to study the virus-killing effectiveness of different wavelengths of light and to assess any by-products produced by the chemical reaction.[89]

Also in 1991, America's Big Three automakers (Chrysler, Ford, and General Motors), in collaboration with the Electric Power Research Institute and the DOE, organized the U.S. Advanced Battery Consortium (USABC) to develop light, powerful, and long-lived batteries that would make electric vehicles competitive with gasoline-powered vehicles in some markets. The battery consortium, a four-year, $260 million joint government-industry venture, arranged research-and-development partnerships between national laboratories and private companies. Environmentally friendly electric vehicles were clean and quiet, but because they were equipped with conventional lead-acid batteries, they could not match the range and performance of vehicles driven by an internal combustion engine.[90]

Argonne had been a leader in battery research since 1976, when the Energy Research and Development Administration authorized its battery analysis and diagnostic laboratory. Under the CRADA, Argonne collaborated with Saft America Inc., Cockeysville, Maryland, to commercialize an Argonne invention, the lithium–metal sulfide battery. Saft America, a leading supplier of lithium batteries to the Department of Defense, believed a battery that used a molten lithium salt had potential to power airport vehicles, lift trucks, and hybrid vehicles, as well as to provide standby energy storage and backup power for computers and telephone switching equipment. Tests of lithium–iron disulfide battery cells indicated that they could be recharged 1,000 times, and in proper configuration energize a vehicle for 100,000 miles before replacement. Saft America agreed to perfect the battery hardware and manufacturing methods while Argonne worked to refine stack design and cell performance. Argonne would also test Saft-built lithium–iron disulfide batteries to establish simulated power profiles for urban and suburban driving. In a separate agreement with the USABC, Argonne agreed to test other experimental batteries sponsored by the consortium.[91]

Inspired by the success of the battery consortium, in January 1992 General Motors organized a "garage show" at which all the national laboratories were invited to exhibit their most innovative technology.* Argonne featured its work on batteries, fuel cells, flywheels, recycling, and high-temperature lubrication. The fair was not exactly a high-tech Wal-Mart, but the American automobile manufacturers asked how they could gain access to the rich store of national laboratory technology. Eagerly, the Clinton administration endorsed an ambitious collaboration among the DOE's weapons laboratories and the automobile industry to improve United States' international competitiveness. The administration-backed program involved three objectives—advanced manufacturing technology; improved automobile safety and emissions; and a full size, economical family car. "Our goal," President Clinton stated, "is to develop affordable, attractive cars that are up to three times more fuel-efficient than today's cars . . . and meet strict standards for urban air pollution, safety, performance and comfort." Dubbed the "Supercar," the New Generation Vehicle Initiative easily excited public imagination.[92]

In the autumn of 1993, the Clinton administration proposed its Partnership for a New Generation Vehicle Initiative, a program neatly attuned to the president's desire to reinvent government and to the vice president's worry about global warming. Seeking a model of government-industry collaboration that stimulated employment and enhanced American competitiveness while improving the environment, the New Generation Vehicle Partnership seemed a perfect marriage of Detroit's know-how with the laboratories' technical expertise.[93]

USCAR

In December 1993 the Department of Energy allocated almost $20 million to the U.S. Council for Automotive Research (USCAR), a "master" CRADA organized to design and develop a light, nonpolluting, usable supercar by 2005. Argonne received generous funding from USCAR, which established the laboratory as one of the centers of its activity. By 1994 the laboratory had executed ten CRADAs with the automotive industry, five under the aegis of USABC.

---

*Laboratories represented were Argonne, Brookhaven, Idaho National Engineering Laboratory, Lawrence Berkeley, Lawrence Livermore, Los Alamos, National Renewable Energy Laboratory, Oak Ridge, Pacific Northwest Laboratories, and Sandia.

Harvey Drucker explained that USCAR was not so much a program as it was a collection of projects, including USABC, now under the USCAR umbrella. Argonne, for example, became the lead laboratory developing techniques to recover plastics and metals from junked automobiles. The laboratory also worked on instrumentation to monitor supercar performance and emissions and became involved in the "intelligent vehicle" initiative. Drucker noted that Argonne even operated one of the largest fleets of alternative fuel vehicles in the country, testing natural gas, ethanol, and methanol-powered vehicles, a fuel cell bus, and a hybrid delivery van in which compressed natural gas turned a generator that charged advanced batteries.[94]

Under sponsorship of USCAR and the Partnership for a New Generation Vehicle, Argonne hosted a forum to alert small automotive suppliers in the Chicago area to the research-and-development opportunities at Argonne. Three thousand small manufacturers and suppliers learned how to set up joint research projects, apply for government funding, and protect their patents and proprietary interests. Don Walkowicz, executive director of USCAR, explained that Detroit could not mass-produce the supercar unless small suppliers were prepared to manufacture the components of the future.[95]

Argonne's supercomputers proved among the laboratory's most important contributions to the USCAR initiative. As part of the $2.7 million CRADA, the Argonne-USCAR supercomputing project developed complex programs to design and evaluate supercar performance in three key areas: aerodynamics, safety, and power. Argonne was the first laboratory to obtain the 128-processor IBM POWER parallel computer system. Named SP1 by Argonne, the machine was IBM's largest, most powerful parallel supercomputer.[96]

The SP1 Parallel Supercomputer

How fast was the new SP1 parallel supercomputer? Rick Stevens, director of the mathematics and computer science division, remembered that in the seventeenth century, astronomer Johannes Kepler needed four years to calculate Mar's orbit, a problem that contemporary microprocessors can solve in four seconds. SP1, each of whose processors could store 128 million characters (or 128 megabytes), could make 16 billion calculations per second. In contrast to conventional computers that solved problems a step at a time, parallel computers divide the problem into segments and solve each simultaneously, something like producing a complete fast-food meal of burger, fries, and beverage every few seconds. The SP1 operated at a level near one tera-flop, that is, one trillion floating point operations a second (about a million times faster than this historian's personal computer). The speed gained by solving problems in tandem, or parallel, enabled researchers to solve problems in a day that once took weeks to solve on a supercomputer and to obtain answers in a week on problems that ordinary fast computers would require more than a year to solve.[97]

Collaborating with AlliedSignal, Argonne adapted a computer program that modeled a molecular system, that is, a program that simulated the behavior of thousands of atoms under all conditions. On SP1, Argonne could run the AlliedSignal program a hundred times faster than on a single-processor computer. Success with this adaptation enabled Argonne and AlliedSignal to develop a computer model that simulated wear and heating of automobile disc brakes under various braking conditions and by applying computerized "friction," researchers estimated mechanical and thermal effects to the brake pads and rotor. Ultimately, SP1 enabled the Argonne-AlliedSignal collaboration to develop virtual reality software creating realistic, three-dimensional modeling of working disc brakes. None of this would have been possible without the IBM parallel processing facility.[98]

The SP1 parallel supercomputer worked on numerous problems in addition to that of supercar design. In cooperation with Pacific Northwest Laboratory, Argonne modeled molecular processes that predicted the environmental impacts of pollutants. NASA's Lewis Research Center obtained access to SP1 to study propulsion systems. High-energy physicists from Lawrence Berkeley Laboratory, the University of Illinois at Chicago, and the University of Maryland employed SP1 to evaluate huge volumes of data generated from their accelerator experiments. Engineers at the Illinois Institute of Technology studied turbulence, and researchers from the University of Chicago used SP1 to produce advanced scientific imaging in medical studies, geophysical modeling, and astrophysics. Molecular modeling with SP1, when combined with empirical data such as x-ray crystallography obtained from the Advanced Photon Source, promised to accelerate dramatically the commercialization of new drugs, polymers, advanced materials, and other biotechnology products. Pleased with the robust success of the SP1 in both commercial applications and scientific research, Schriesheim celebrated: "There has never been a better example of joint research that advances scientific knowledge at the same time it helps improve the competition of American industry."[99]

By May 1994 Argonne had executed, with large and small companies, fifty-four CRADAs, whose total value approached $35 million, including contributions from both government and industry. With the demise of the Integral Fast Reactor, Harvey Drucker's program of energy and environmental science and technology, which negotiated most of the CRADAs, became the second largest part of the laboratory with an annual budget of $104 million and a staff of 950. With about 28 percent of the laboratory's budget and 18 percent of its employees, the energy and environment area supported the largest number of programs (about seventy-five). Unlike the Advanced Photon Source, however, Drucker's area, which turned projects over, did not enjoy a focused mission. In that regard, energy and environmental programs lived in a very different kind of world from the rest of the laboratory. Hazel O'Leary's strategic plan for the Department of Energy placed the laboratories' scientific and technical expertise at the center of the department's revised mission to promote industrial competitiveness, develop energy resources, and enhance environmental quality. Argonne's broad array of energy and environmental programs were most in tune with the Department of Energy's post–Cold War mission, but at the same time, most vulnerable to political pressures on the Clinton administration to cut taxes and the federal deficit.[100]

Although Schriesheim envisioned a bright future for industry-laboratory partnerships, he also offered a word of caution about foreseeable pitfalls. First, he warned that the national laboratories and the federal government were "rotten" at picking commercial winners. Consequently, he believed that all commercial initiatives should originate with industry. Second, Schriesheim advised that companies should expect additional paperwork and delay when spending tax dollars in partnership with the laboratories. There was no avoiding the accountability demanded by both Congress and the White House when taxpayers' money was on the table. Finally, he reflected his staff's concern that overemphasis on applied development could hurt basic research. At many corporate laboratories, demands for short-term payoffs had ultimately squeezed out long-term research.[101]

Laboratories in Crisis

For different reasons, America's great research laboratories, both private and public, were in serious trouble at the end of the Cold War. Along with their parent companies, AT&T Bell laboratories, Xerox PARC, and IBM's T. J. Watson Research Center, which had pioneered in communications and computing, struggled in the 1990s to adjust to rapid changes in their industry and international markets. Facing increased economic pressures, companies expected immediate help from their research programs, which were pushed toward more applied work. Furthermore, these corporations discovered that their competitors, without investing in high-priced, long-term research and development, easily obtained innovative technology from other sources. Consequently, the private laboratories curtailed fundamental studies at the same time that the national laboratories were under pressure to cut back on basic research.[102]

The Cold War had assured all of the national laboratories relative prosperity. If not directly involved in building America's nuclear arsenal, national laboratories such as Argonne helped maintain America's scientific and technical "fitness." For more than four decades the laboratories essentially had one customer—the federal government and its agencies. The Sputnik scare of the 1950s and energy crisis of the 1970s validated the belief that the laboratories were irreplaceable assets essential to the federal government during national emergency. While the threat of the Soviet Union loomed over the United States, scientists at the national laboratories who credibly invoked the specter of Soviet competition generally won Congressional support for their research.

As the grip of nuclear terror eased, federal bureaucrats became increasingly disillusioned with the GOCO laboratory system. Established during World War II under emergency conditions when accountability was subordinated to outcome, the **go**vernment-owned, **co**ntractor-operated laboratories were well suited to the "crash" mentality of the early Cold War, when the Atomic Energy Commission delegated great latitude to the laboratories to develop nuclear technology. Most scientists looked back nostalgically on this "damn the torpedoes, full speed ahead" era when political interference from Washington, D.C., seemed minimal. But the sad story of radiation patients and a habit of environmental carelessness at the production facilities were also an unfortunate legacy of the "good ol' days."

In contrast, by the 1990s, an ever-tightening web of laws, regulations, and orders from Washington had undermined laboratory autonomy to the extent that scientists believed excessive bureaucratization had stifled discovery and innovation. With Congress requiring ever more accountability for the expenditure of tax dollars, DOE managers demanded increasing oversight of laboratory programs. Commitment to the GOCO principle declined as administrators with no institutional memory managed the laboratories at arms' length. In 1982 the department's Energy Research Advisory Board reported a growing adversarial relationship between the department and its laboratories. In 1992 the Secretary of Energy Advisory Board also noted an erosion of the GOCO partnership that undercut trust in the labs, increased bureaucratic management and overhead costs, and decreased mission efficiency. When the end of the Cold War forced the department to revaluate its mission, not surprisingly, the secretary of energy also reexamined the GOCO concept.[103]

Following the collapse of the Soviet Union, Americans not only anticipated a "peace dividend" from deep cuts in defense spending, but they also looked for major tax savings by decreasing the size and spending of the federal government and its institutions. "Porkbusters," such as Congressman Harris W. Fawell of Illinois helped lead an assault by conservative Republicans and new Democrats against federal spending that supported "special interests" rather than the national interest. "Big science" was subjected to the same criticisms as big government—it was controlled by elites, out of touch with the needs of people, and wasteful. The difficulty was in determining whether large research-and-development projects, such as the Superconducting Super Collider and the IFR, actually served the national interest more than special interests. Without question, if private industry had been willing to invest heavily in either of these projects, they would not have been killed by a cost-conscious Congress. By 1994, however, even the existence of Japanese financing did not rally Congress to protect "made-in-America" nuclear technology.[104]

"Refitting Cold War Science"

Ronald Yates, the financial reporter for the Chicago *Tribune* summarized Argonne's challenge: as the DOE reoriented its programs, the laboratory would have to "refit Cold War science" for competition in the civilian economic sector. No Argonne director, of course, had been insensible to the recurrent need to define the laboratory's mission. The federal government had been the laboratory's exclusive customer until the mid-1950s, when President Eisenhower's Atoms for Peace initiative first encouraged the laboratories to share unclassified technology with American industry. Thereafter, no administration overlooked the commercial potential available at the national laboratories. Over the years, seemingly countless blue-ribbon panels (actually about fifteen) returned recommendations on the mission of the national laboratories. The panels inevitably encouraged increased collaboration between the laboratories and the private sector.*[105]

In 1992, however, the Council on Competitiveness headed by Erich Bloch, the former director of the National Science Foundation, recommended a fundamental reorientation of the laboratories away from developing products for old government customers toward promoting technology-transfer programs directed to United States industry. Bloch believed that the laboratories' post–Cold War challenge was similar to that of the United States as a whole—to become more broadly competitive. Unfortunately, according to the Council on

---

*The latest being the Secretary of Energy Advisory Board (SEAB) Task Force on the Future Roles and Missions of Departmental Laboratories established by Secretary of Energy James D. Watkins (1992) and the Galvin Task Force on Alternative Futures for the DOE National Labs organized by Secretary of Energy Hazel O'Leary (1994).

Economic Competitiveness, the laboratories were more geared to the Cold War era than "to the era of intense international economic competition in which American business and industry find themselves."[106]

Schriesheim touted Argonne as a "reservoir of scientific and technical talent" that could help American industry to compete in international markets. His assertion, however, was a variation of an old argument used repeatedly to defend the laboratory since the abolition of the AEC. Almost no one disputed that the national laboratories were a valuable public asset of comparable importance to the major research universities and great private laboratories. But even if one accepted Schriesheim's claim, how could the significance of Argonne's research be evaluated independently?

Schriesheim worried that in a politically charged environment, the balance between "political science" and substantive science could be distorted. Because of his industrial background, he was cautious about the argument that the national laboratories could play a pivotal role in helping American industry compete in the international marketplace. He believed that Argonne's industrial linkage should be strong but not the main rationale for the lab's existence. Indeed as the nation's industrial long-range research base contracted, the Argonne's role in high-risk, long-term research had increased.[107]

But excellence in research did not necessarily translate into smooth technology transfer, as Argonne discovered in its superconductivity programs. The laboratories were not technology supermarkets where industry could shop for useful gadgets or manufacturing ideas. Argonne, which had emulated the university research model years before under Norman Hilberry, had recently developed a portfolio of technology transfer mechanisms: ARCH, the technology transfer center, and the applied superconductivity center among others. Argonne, however, was not generally geared to respond to the market-driven demands of private industry. It was not that the laboratories were "ivory towers," but rather that only a small fraction of their resources were dedicated to technology transfer. Traditionally, the national laboratories' mission was to develop long-term, high-risk technology such as the IFR, or to maintain expensive, one-of-a-kind facilities like the APS, neither of which private industries or the universities could afford.

The New Uncertainty Principle

Uncertainty wracks institutions that lack fixed points of reference. Schriesheim mused that during the "Roaring Twenties" German physicist Werner Heisenberg unsettled physical scientists by declaring that one cannot calculate accurately both the position and the momentum of subatomic particles. Heisenberg won the Nobel Prize for his famous Uncertainty Principle, which taught Schriesheim and his colleagues to think of the physical world in relative rather than absolute terms.

Challenged by contrasting feast and famine in Argonne's scientific programs, Schriesheim recalled Heisenberg's uncertainly paradigm. He discovered that Argonne faced a New Uncertainty Principle—that is, he could not "forecast with much accuracy the position, the momentum, or the probable intersection of society's goals, the nation's political agenda, and the scientific merits" of America's research enterprise. Mastering the New Uncertainty Principle would not win Schriesheim a Nobel Prize, however. His reward would be survival of Argonne National Laboratory in the twenty-first century "as a vital and essential American institution."[108]

■

Alan Schriesheim retired as director of Argonne National Laboratory in June 1996. His emphasis on strategy and focus had served Argonne well in preparation for the changes brought on by the end of the Cold War and the shrinking of the nation's science and technology budgets. Argonne's future, like its past, was linked to the fate of the national laboratory system, in which Argonne ranked sixth in budget and fifth in staff among the eight multipurpose laboratories.* Caught in the uncertainties of national politics and budget cutting, the entire national laboratory system faced profound changes.

For fifty years, Argonne's nuclear reactor programs provided the central, fixed reference point that had defined the laboratory's identity and mission. But after 1994, the nuclear era was as passé as the Cold War. For the first time since the Eisenhower administration, no nuclear power reactors were being researched, developed, or constructed. Stripped of the integral fast reactor project, Schriesheim knew that Argonne was fortunate to have the Advanced Photon Source, a major facility that provided a solid anchor for the laboratory's research programs. From Walter Zinn to Walter Massey, each laboratory director in turn struggled to clarify Argonne's mission and champion its accomplishments, but none faced Schriesheim's daunting challenge. In a political climate that often seemed indifferent, or even hostile, to America's scientific preeminence, Schriesheim and the laboratory searched for a new mission—not just an odd assortment of excellent projects—to define Argonne's identity. It had been his most difficult initiative—one that was unexpected and uncertain, but one that was vital to the future of the laboratory.

---

*See Appendix 1 for modern Argonne budgeting and staffing.

Were scientific frontiers—once thought endless—gone? Hardly. But the frontier rhetoric and ethic was lost, replaced by a new metaphor of uncertainty borrowed from science itself. The outlook was healthier and more realistic for the laboratory. A century before, a morose Henry Adams had found historical understanding of American decline in the second law of thermodynamics. In contrast, as Argonne approached a new fin de siècle, the laboratory faced a reordering of its traditions and culture with considerable optimism. The future, although uncertain, offered scientific discovery as exciting as that found on the old nuclear frontier.

Appendix 1: Argonne National Laboratory Funding and Employment, 1979–94

Year	Number of Employees	Operating Costs ($)	Construction Costs ($)	Equipment Costs ($)	Work for Others ($)
1979	5300.0	403.2	89.7	28.6	26.4
1980	5155.0	410.8	36.5	25.1	28.5
1981	4996.0	331.7	26.8	16.2	34.8
1982	4339.0	298.6	33.5	15.4	30.8
1983	4188.0	269.6	33.4	10.7	40.3
1984	4117.0	269.5	22.5	16.1	41.6
1985	3893.9	261.5	27.6	22.3	30.1
1986	3689.1	242.0	11.2	15.9	57.0
1987	3692.7	241.3	15.5	28.7	63.9
1988	3758.1	257.7	25.9	23.6	71.6
1989	3812.0	273.4	32.3	25.1	72.8
1990	4015.0	303.1	69.1	29.3	59.5
1991	4352.0	308.8	104.3	29.9	60.0
1992	4609.7	346.4	119.9	22.1	63.7
1993	4839.8	373.0	137.9	35.1	65.6
1994	5271.4	367.3	119.1	47.5	68.8

Source: Joseph G. Asbury, Deputy to the Director, to Jack M. Holl, November 10, 1994.
Note: Funding in millions of "then-year" dollars.

Appendix 2: Budgets and Staffing of DOE Multipurpose Laboratories, 1995

Laboratory (Operating Contractor)	Fiscal 1995 Budget ($ millions)	Staff
Argonne (University of Chicago)	$500	4,300
Brookhaven (Associated Universities Inc.)	451	3,471
National Engineering (Multiple [Idaho])	746	6,178
Lawrence Berkeley (University of California)	287	2,598
Lawrence Livermore (University of California)	983	7,310
Los Alamos (University of California)	1,050	6,865
Oak Ridge (Martin Marietta Energy Systems)	534	4,399
Pacific Northwest (Battelle Memorial Institute)	536	3,590
Sandia (Martin Marietta Corp.)	1,495	8,500
Totals	$6,582	47,211

Source: Will Lepkowski, "National Laboratories Enter New Era of Hope Mixed with Uncertainty," *Chemical and Engineering News,* Aug. 14, 1995, p. 24.

Appendix 3: Argonne National Laboratory Key Personnel

The University of Chicago (Chief Executive Officers, 1946–96)
  Robert Maynard Hutchins, 1929–51
  Lawrence Alpheus Kimpton, 1951–60
  George Wells Beadle, 1961–68
  Edward Hirsch Levi, 1968–75
  John Todd Wilson, 1975–78
  Hanna Holburn Gray, 1978–93
  Hugo Sonnenschein, 1993–

Argonne National Laboratory (Directors, 1946–96)
  Walter H. Zinn, 1946–56
  Norman Hilberry, 1957–61
  Albert V. Crewe, 1961–67
  Robert B. Duffield, 1967–73
  Robert G. Sachs, 1973–78
  Walter E. Massey, 1979–84
  Alan Schriesheim, 1984–96

Atomic Energy Commission, Energy Research Development Administration, and the Department of Energy (Chief Executives, 1946–96)
AEC (Chairmen)
  David E. Lilienthal, 1946–50
  Gordon E. Dean, 1950–53
  Lewis L. Strauss, 1953–58
  John A. McCone, 1958–61
  Glenn T. Seaborg, 1961–71
  James R. Schlesinger, 1971–73
  Dixy Lee Ray, 1973–75
ERDA (Administrators)
  Robert C. Seamans, 1975–77
  Robert W. Fry, 1977–78 (acting)

DOE (Secretaries)
    James R. Schlesinger, 1977–79
    Charles W. Duncan, Jr., 1979–81
    James B. Edwards, 1981–82
    Donald P. Hodel, 1982–85
    John S. Herrington, 1985–89
    James D. Watkins, 1989–93
    Hazel O'Leary, 1993–97

Argonne Universities Association (Presidents, 1966–82)
    Philip N. Powers, 1966–73
    Paul McDaniel, 1973–78
    Armon F. Yanders, 1976–77
    Henry V. Bohm 1978–82

Appendix 4: Selected Argonne Projects and Technology Highlights, 1946–96

1946 Radiation studies conducted to examine effects of radiation on all living organisms.
1947 Argonne designated as principal laboratory for U.S. reactor research.
1948 Pressurized-water thermal reactor developed for Westinghouse.
1948 Nuclear structure studies led to invention of the nuclear shell model.
1949 Master-slave manipulator developed for handling "hot" isotopes by remote control; provided basic research into robotics.
1950 First submarine reactor prototype, the Zero Power Reactor I (ZPR-I), built and operated.
1951 Experimental Breeder Reactor (EBR-I)—the world's first heavy-water-moderated reactor; produced first nuclear electricity.
1951 Materials Testing Reactor, the first high-flux test reactor in the U.S., designed.
1952 ZPR-I critical experiments provided physics data for first submarine (*Nautilus*) reactor.
1952 Einsteinium and fermium, elements 99 and 100, codiscovered.
1953 Engineering design of Materials Testing Reactor completed.
1953 EBR-I demonstrated the possibility of generating more fuel than is consumed.
1953 Series of BORAX experiments undertaken to evaluate and study nuclear heat and super-heat concepts.
1953 Zero Power Reactor II (ZPR-II) demonstrated design feasibility of the Savannah River Production reactor.
1953 AVIDAC digital computer, an electronic brain, designed and constructed.
1955 International School of Nuclear Studies and Engineering established to fulfill Atoms for Peace program.
1956 Argonaut, a low-cost training and research reactor based on a series of multiplication experiments, completed.
1956 Experimental Boiling Water Reactor (EBWR), the country's first reactor built solely for research on electricity generation, proved the feasibility of a direct-cycle boiling-water reactor system.

1957   GEORGE, a large-scale computer that operated around the clock, designed and built.
1957   Element 102 codiscovered.
1957   Codemonstration that parity conservation does not apply in radioactive decay of neutrons.
1958   Transient Reactor Test Facility (TREAT), a versatile irradiation tool for studying fuel behavior during sudden power surges, achieved criticality.
1959   Fuel Fabrication Facility, the first large-scale plant for making nuclear reactor fuel elements from plutonium, completed.
1959   Mossbauer line in iron-57 discovered.
1960   Juggernaut, a versatile tool for scientists and students, completed.
1960   DNA synthesis studies led to first demonstration of early (aminothiol-mediated) gene expression and its effects on DNA synthesis.
1960   Liquid Metal Reactor (LMR) studies undertaken to demonstrate safety of liquid-metal-fueled reactors.
1962   Studies of chemical bonding led to discovery of xenon tetrafluoride, a combination of the "noble" gas xenon (previously thought to be inert) and fluoride, and to the discovery that radon (another noble gas) is capable of making chemical compounds.
1962   Leukemia studies led to discovery that white corpuscles in blood stream represent only a minute portion of those manufactured in the bone marrow.
1962   EBR-I achieved chain reaction with plutonium.
1963   Radiation chemistry studies led to codiscovery of a "new" hydrated electron.
1963   Zero Gradient Synchroton, one of the world's major particle accelerators, led to greater understanding of the spin dependence of particle interaction through its program of polarized target development and exploitation.
1963   The 12.5 GeV Zero Gradient Synchrotron, the world's leading proton accelerator, designed and built.
1964   EBR-II, a fast integrated, closed-cycle breeder reactor system for power production, designed to test fuels and evaluate their performance after long exposure produced electricity; its fuel reprocessing plant demonstrated potential advantage of using fast reactors for central station power plants.
1964   Janus, world's first reactor designed exclusively for biological research, built.
1964   First use of a superconducting magnet in a full-scale, high-energy physics experiment.
1965   Search for bone cancer virus led to discovery of a filterable agent that produced cancer tumors in mice.
1965   Research at temperatures near absolute zero solidified the isotope helium-4 for the first time.
1966   EBWR demonstrated that plutonium recycle operation of water reactors could generate usable electricity.
1966   Radioactive element astatine measured for the first time.
1967   Radiation-induced voids in metals discovered.

1968	Braille reading machine and inexpensive hemodialyzer developed.
1969	World's largest (12-foot) hydrogen-deuterium bubble chamber with superconducting magnet for high-energy physics research became operational.
1969	Zero Power Plutonium Reactor placed in operation for physics studies.
1969	Center for Human Radiobiology established to study effects of radiation on humans.
1969	Studies of human immune response at the molecular level led to isolation and crystallization of two related abnormal blood and urinary proteins from a single bone cancer patient.
1970	World's first observation of neutrino particle tracks in a hydrogen bubble chamber at ZGS.
1973	Prototype testing and development of the superconducting linac concept for accelerating heavy ions.
1973	Method for removing radioactive xenon and radon gases from contaminated atmosphere discovered.
1973	ZGS produced first useful polarized proton beam.
1973	Circadian rhythm studies demonstrated, for the first time, the existence of a biological clock in single protozoan cells; led to 1982 development of Argonne's Anti–Jet Lag Diet and regimen.
1974	Synthesis of a new chlorophyll molecule that mimicked natural chlorophyll in green plants.
1975	Hot-Fuel Examination Facility (HEFF) placed in operation to examine highly radioactive experimental reactor fuel elements.
1975	Two viruses isolated in search for their role in the formation of cancerous bone tumors.
1977	First automated system for continuous growth of human cells in culture developed.
1977	World's first large superconducting magnet designed and built.
1979	First successful experiment to use high-energy neutron beams.
1980	Automated reasoning studies included OTTER, the world's most powerful general-purpose automated reasoning system.
1981	High Voltage Electron Microscope–Tandem Accelerator, the only research facility in the world that permits scientists to "watch" effect of damage to materials as it occurs.
1981	Intense Pulsed Neutron Source (IPNS), the world's most advanced source of spallation neutrons, provided information about atomic and molecular structure of matter to help develop improved composites, ceramics, and metals.
1983	Lemur, an innovative parallel processor system, developed.
1983	First measurement of the reaction rate between fluorine and water made.
1984	Integral Fast Reactor, an advanced concept, designed to reprocess its own fuel and to burn up its own long-lived atomic wastes.
1984	Advanced Computing Research Center established.

1985 Argonne Liquid Metal Engineering Experiment (ALEX) began operation. The largest facility in the world for studying the effects of magnetic fields on liquid metal.
1985 Electron Microscopy Center, primary research facility for the use of high-voltage electron microscopes in the study of atomic structure and materials.
1985 Argonne Tandem Linear Accelerator System (ATLAS), the world's first superconducting linear accelerator for nuclear physics research, dedicated.
1986 Liquid Metal Fast Breeder Reactor (LMFBR) confirmed passive safety characteristics of liquid metal reactors.
1987 World's first flexible ceramic superconducting wire extruded.
1987 First observation of Wakefield acceleration structures and plasmas achieved.
1987 World's first motor based on properties of high-temperature superconductors built.
1987 First successful attempt to put electrical current through newly discovered yttrium-barium-copper oxide wire.
1987 Most powerful superconducting dipole magnet built.
1989 RESRAD, a computer model for developing residual radioactive materials guidelines, developed.
1989 New class of algebras with applications to both quantum and fluid mechanics discovered.
1990 Chemical process developed for removing strontium-90 from liquid nuclear waste.
1990 Transuranic Extraction (TRUEX) process developed for extracting and recovering transuranic elements from radioactive wastes.
1992 Structural Biology Center established to provide advanced instrumentation for crystallographic research.
1992 Tiny, rugged ceramic-metallic gas sensor that identifies gases in harsh industrial environments invented.
1992 "Biochip" to sequence genomes developed.
1992 RISKIND, computer program for calculating the radiobiological consequences and health risks from transportation of spent nuclear fuels, developed.
1993 World record for "giant magnetoresistance" achieved; could lead to faster computers and more precise robots.
1993 Method invented to detect trace amounts of curium generated during clandestine weapons production.
1993 Meson containing both bottom and strange quarks discovered.
1993 New ceramics invented that could be stronger welds for bonding some hard-to-join materials.
1994 Technology invented for harvesting diamonds six times faster than the current method.
1995 CAVE, a computerized virtual reality environment in which information is seen, heard, and manipulated, became operational.

1996 Advanced Photon Source, a powerful synchrotron research facility, became the source of the nation's most brilliant x-ray beams in the "hard x-ray" portion of the spectrum.

Appendix 5:
Argonne National Laboratory Nuclear Reactor Program

Argonne's heritage in the development of nuclear reactors began with CP-1, the world's first nuclear reactor, which was built by Enrico Fermi and his team and was tested on December 2, 1942, under the West Stands at Stagg Field on the campus of the University of Chicago.

Dr. Walter H. Zinn, one of Fermi's close colleagues working on CP-1, became Argonne National Laboratory's first director in July 1946. In 1948 the United States Atomic Energy Commission (AEC) transferred the major portion of the nation's nuclear reactor development to Argonne. Under Dr. Zinn's vision and leadership, Argonne extended the program to include all types of reactors. They were designed to produce electric power, nuclear materials for military purposes, and isotopes and for many kinds of research. No other reactor development program has been as productive and diverse as the Argonne program.

The many "firsts" and accomplishments of the Argonne nuclear reactor program are shown in figure 46, which graphically depicts the scope and diversity of the Argonne program. The "apple tree" shows all of the reactor systems worked on by Argonne, most of which were built and operated by the laboratory. Some of the reactors were constructed and run by outside researchers, and some were canceled for a variety of reasons. As seen in the figure, there are several main branches from the "tree" that characterize the depth and scope of the Argonne reactor program:

- Pressurized water reactors
- Production reactors
- Research reactors
- Boiling-water reactors
- Liquid-metal reactors

Many of the major reactor systems developed by Argonne are described below.

- *Submarine Thermal Reactor (STR):* a pressurized-water reactor used aboard the nation's first nuclear-powered submarine, the *Nautilus.* Argonne was re-

Figure 46. Reactor Tree

sponsible for the complete concept and basic design of the reactor for the *Nautilus*.

- *Zero Power Experiments (ZPR):* a series of essentially zero-power, full-scale reactor-core mockup assemblies used to gain understanding of a variety of reactor concepts and to assist in the engineering design of these reactor systems (e.g., STR, Savannah River, EBR-I, EBR-II, and others).
- *Experimental Reactor I (EBR-I):* a sodium-potassium-cooled fast reactor that proved the principle of breeding nuclear fuel. The energy from EBR-I was used to produce electricity from nuclear power for the first time.
- *Argonne Research Reactor (CP-5):* a heavy-water cooled and moderated reactor that was the workhorse for over twenty years in providing a source of high neutron flux for a wide spectrum of basic and applied research programs.
- *The Savannah River System (CP-6):* the five reactors built at the Savannah River Plant for production of materials for military use. Argonne designed this larger and more advanced system using the lab's backup design for the Hanford production reactors (CP-3).
- *Boiling Water Experiments (BORAX-I):* the first of a series of boiling-water experiments. BORAX-I proved the inherent safety of this type of reactor, which is manifested by the forceful ejection of boiling and steaming water from the reactor core, resulting in termination of the nuclear chain reaction.

- *BORAX-III:* an advanced boiling-water concept, the first nuclear reactor in the world to produce large quantities of power. Its electrical output lighted up the entire town of Arco, Idaho in 1955.
- *Experimental Boiling Water Reactor (EBWR):* a prototype of future boiling-water reactors, which demonstrated how to design, construct, and operate a complete nuclear reactor power plant. Ground breaking for the construction of EBWR at Argonne's Illinois site was in May 1955. Electrical power was produced just nineteen months later, on December 29, 1956.
- *ARGONAUT:* a low-power (10 kw) research reactor, moderated by light water, designed for training and research purposes. ARGONAUT served as a model that was copied widely here and in other countries. In 1958 it was disassembled, shipped to Geneva, Switzerland, and reassembled (as an operating reactor) for hands-on display at the historic Geneva conference on President Eisenhower's Atoms for Peace program.
- *Nuclear Rocket Reactor:* used to investigate feasibility of a compact, high-power space propulsion reactor. Tungsten-urania fuel elements for a 200-megawatt fast reactor successfully tested in flowing hydrogen at temperatures up to 2700°C. Critical experiments were conducted to support the design.
- *Experimental Breeder Reactor II (EBR-II):* a sodium-cooled fast reactor in which all of the primary coolant system components—pumps, heat exchanger, fuel handling equipment—were contained in a single large vessel containing sodium at atmospheric pressure. For three decades, EBR-II operated safely and reliably, producing and selling electric power to Utah Power and Light, as well as serving as an irradiation facility for hundreds of advanced fuel and materials development tests carried out in its core. Using the EBR-II system, Argonne initiated the Integral Fast Reactor (IFR) program in 1984, developing reactor, fuel-cycle, and waste processes as a single entity. The IFR program continued until 1994. It provided engineers with information essential to the design and operation of advanced liquid-metal reactor plants. The research and development on spent fuels and waste processes have continued. As part of the current program, spent fuel from the EBR-II reactor (now shut down), is being processed in the Fuel Cycle Facility collocated with the EBR-II reactor plant at the Argonne-West site, at INEL.

The inherent safety of the sodium-cooled reactor, operating at atmospheric pressure, was demonstrated with the EBR-II by deliberately cutting off the primary sodium pumps and preventing the insertion of control rods into the reactor (which action normally would terminate the excursion). The increasing temperature of the fuel resulted in its expansion and subsequent termination of the chain reaction—a completely passive phenomenon. The importance of this, and related IFR development, will become apparent if nuclear power plays a substantial role as a future energy supply.

Appendix 6

Notes

Abbreviations

AHC	Arthur Holly Compton Papers, Washington University, St. Louis, Missouri
ANL	Argonne National Laboratory Records
ANL CES	Argonne National Laboratory Records, Center for Environmental Studies
DDE	Dwight D. Eisenhower Presidential Library, Abilene, Kansas
DOE	Department of Energy Archives
FPC	U.S. Federal Power Commission
FRC	Federal Records Center, Seattle, Washington
GLFRC	Great Lakes Federal Records Center, NARA, Chicago, Illinois
GPO	Government Printing Office
HSTL	Harry S. Truman Library, Independence, Missouri
JRO	J. Robert Oppenheimer Papers, LC MSS
LC MSS	Library of Congress, Manuscript Division, Washington, D.C.
MHD	Manhattan District History
NAPNW	National Archives, Pacific Northwest, Seattle, Washington
NARA	National Archives and Records Administration, Washington, D.C.
NARA-GL	Great Lakes Archives, NARA, Chicago, Illinois
OPA	Office of Public Affairs, Argonne National Laboratory
OSRD B-C	Records of the Office of Scientific Research and Development, Bush-Conant Files
PD	Physics Division, Argonne National Laboratory
PIO	Public Information Office
RG	Record group
TIS	Technical Information Services, Argonne National Laboratory
UCVP ANL	Office of the Vice President for Argonne National Laboratory, University of Chicago
UCRL SC	University of Chicago Regenstein Library, Special Collections
VPSP	Office of the Vice President for Special Projects, University of Chicago
WNRC	Washington National Records Center, Washington, D.C.

Introduction

1. Vannevar Bush, as quoted in George Mazuzan, *The National Science Foundation: A Brief History,* NSF 88-16 (Washington, D.C.: NSF, 1988), p. 2.

Chapter 1

1. Richard G. Hewlett and Oscar E. Anderson, Jr., *The New World, 1939–1946,* Vol. 1 of *History of the United States Atomic Energy Commission* (University Park: Pennsylvania State University Press, 1962), pp. 10–11; Richard Rhodes, *The Making of the Atomic Bomb* (New York: Simon and Schuster, 1986), pp. 251–67.

2. Hewlett and Anderson, *The New World,* pp. 10–11; Rhodes, *The Making of the Atomic Bomb,* pp. 251–67.

3. Szilard quoted in Rhodes, *The Making of the Atomic Bomb,* p. 281. See also Hewlett and Anderson, *The New World,* pp. 14–15.

4. Hewlett and Anderson, *The New World,* pp. 15–17.

5. The course of the work that led to the development of the atomic bomb has been described by many insiders and observers, including Henry D. Smyth, Compton, General Leslie Groves, Richard G. Hewlett, Oscar Anderson, Vincent Jones, and Richard Rhodes.

6. Rhodes, *The Making of the Atomic Bomb,* pp. 363–64; Arthur Compton, *Atomic Quest* (London: Oxford University Press, 1956), pp. 5, 11, 37–38.

7. F. G. Gosling, *The Manhattan Project: Science in the Second World War* (Washington, D.C.: DOE, 1990), pp. 1–2.

8. Hewlett and Anderson, *The New World,* pp. 19, 24–25, 36; Gosling, *The Manhattan Project,* pp. 5–8.

9. Volney C. Wilson, "Instrumentation and Control of the First Nuclear Reactor," (paper presented at the American Physical Society Symposium of the Division of History of Physics, "The Birth of the Nuclear Age," Washington, D.C., April 23, 1992); John A. Wheeler, "Notes on Conferences," Jan. 3, 1942, p. 3, Folder "Notes on Conferences J. A. Wheeler (Secretary)," Box 1 of 4, F. Daniels Office File, Box 7, RG 326, ANL, NARA.

10. Hewlett and Anderson, *The New World,* p. 24; Lyman J. Briggs to Arthur H. Compton, Sept. 27, 1940, Met Lab Compton Chron. File 11/40–11/43, Box 34X, RG 326, ANL, NARA; Compton to Briggs, Nov. 26, 1940, and attachment "Regenerative Fission with Uranium and Beryllium" and Briggs to Compton, Nov. 28, 1940, both in Folder "1940," Chron. File "Misc. Docs. 1940–May 1944," Box 1X, RG 326, ANL, NARA; "Proposed Work on Regenerative Fission at University of Chicago," Jan. 20, 1941, W. B. Harrell to S. K. Allison, April 28, 1941, and Contract no. NDCrc-101, all three in Folder "1941," Chron. File "Misc. Docs. 1940–May 1944," RG 326, ANL, NARA; "Meeting of the Advisory Committee of the National Academy on Uranium Disintegration," File "HILSPEC 55462, Minutes of meetings of National Academy Committee," ANL, TIC.

11. Compton to E. T. Filbey, June 24, 1941, and W. B. Harrell to Lincicome, Nov. 19, 1941, Folder "1941," Chron. File "Misc. Docs. 1940–May 1944," Box 1X, RG 326, ANL, NARA.

12. Jeffrey K. Stine, *A History of Science Policy in the United States, 1940–85*, Science Policy Study Background Report No. 1, report prepared for the House Task Force on Science Policy Committee on Science and Technology, 99th Cong., 2nd sess., 1986, p. 17; Hewlett and Anderson, *The New World*, p. 41; Vannevar Bush to Conant, October 9, 1941, AEC Historic Document 17, Folder 2, OSRD B-C, RG 227, NARA.

13. Compton to Bush, Nov. 17, 1941, Met Lab Compton Chron. File 11/40–11/43, Box 34X, RG 326, ANL, NARA.

14. Wilson, "Instrumentation and Control"; Compton to Conant, Dec. 10, 1941, and Bush to Compton, Dec. 13, 1941, Folder "1941," Chron. File "Misc. Docs. 1940–May 1944," Box 1X, RG 326, ANL, NARA.

15. H. D. Smyth to Compton, Dec. 26, 1941, Folder "1941," Chron. File "Misc. Docs. 1940–May 1944," Box 1X, RG 326, ANL, NARA.

16. *American Men and Women of Science*, Vol. 6 (New York: R. R. Bowker Company, 1976), p. 4822; Milton G. White, "Outline for Organization of Nuclear Energy Research Laboratory," Dec. 31, 1941; Smyth to Compton, Jan. 9, 1942; K.T.C., memorandum, Dec. 29, 1941, "Concentration vs. Dispersion of NDRC Research Activities," Folder "1941," Chron. File "Misc. Docs. 1940–May 1944," Box 1X, RG 326, ANL, NARA.

17. Compton quotation from Compton, "Progress Report. Metallurgical Project at University of Chicago," for week of January 4–10, 1942, File "HILSPEC 56119, A. H. Compton—Progress Reports," boxed material, ANL, TIC; "Tentative Group Organization of Metallurgical Project at University of Chicago," Jan. 6, 1942, Folder "Jan.–Feb. 1942," Box 4 of 4, Chron. File "Misc. Doc. 1940–May 1944," Box 1X, RG 326, ANL, NARA; Compton, *Atomic Quest*, 82.

18. Compton, "Special Report to Members of the S-1 Physics Project," Jan. 26, 1942, Folder "Notes on Conferences J. A. Wheeler (Secretary)," Box 1 of 4, F. Daniels Office File, Box 7, RG 326, ANL, NARA; and Compton to Smyth, Sept. 1, 1943, Folder "September 1943," Box 34X, RG 326, ANL, NARA; Compton to Gregory Breit, Jan. 28, 1942, Folder "Jan.–Feb. 1942," Box 4, Chron. File "Misc. Doc. 1940–May 1944," Box 1X, RG 326, ANL, NARA.

19. Irvin Stewart, *Organizing Scientific Research for War. The Administrative History of the OSRD* (Boston: Little Brown and Company, 1948), pp. 192–93; Robert P. Crease and Nicholas P. Samios, "Managing the Unmanageable," *The Atlantic Monthly*, January 1991, p. 84; Compton, "Operation of the Metallurgical Project by the University of Chicago," July 28, 1944, 080 (Argonne–University of Chicago), MED Decimal Files, RG 77, NARA. Stewart pointed out that the navy and army only had production and procurement contracts at the time; consequently, the OSRD had to draw up its own form of contract. Originally the OSRD contracts dealt only with research, but by the time the Metallurgical Project appeared, experience showed that development had to be included in the contract.

20. "Organization of Metallurgical Laboratory," Folder "Jan.–Feb. 1942," Box 4 of 4; and Lt. Col. Thomas T. Crenshaw to Compton, Feb. 24, 1943, Folder "Feb. 1943," Box 2 of 4, both in Chron. File "Misc. Docs. 1940–May 1944," Box 1X, RG 326, ANL, NARA; Stewart, *Organizing Scientific Research for War*, p. 120.

21. Compton to Briggs and Conant, February 11, 1942, Folder "Jan.–Feb. 1942," and Conant to Compton, July 21, 1942, Folder "July 1942," both in Box 4, Chron. File "Misc. Docs. 1940–May 1944," Box 1X, RG 326, ANL, NARA; Compton, *Atomic Quest*, p. 83. The February 11 memorandum was not sent to Briggs and Conant, but in this document Compton expressed his concerns that such compartmentalization would hinder the work in the long run.

22. J. A. Wheeler, Notes on Conferences, Feb. 24, 1942, Folder "Notes on Conferences J. A. Wheeler (Secretary)," Box 1 of 4, F. Daniels Office File, Box 7, RG 326, ANL, NARA; Compton, "Report of Physics Tube Alloy Project," March 18, 1942, Folder "Jan.–Aug. 1942," Box 1 of 4, Met Lab Compton Chron. File 1/42–1/43, Box 34X, RG 326, ANL, NARA; "Organization of Metallurgical Laboratory," Feb. 25, 1942, Folder "Jan.–Feb. 1942," Box 4 of 4, Chron. File "Misc. Docs. 1940–May 1944," Box 1X, RG 326, ANL, NARA.

23. See, for example, Compton, "Progress Report. Metallurgical Project at the University of Chicago," for week ending March 7, 1942, File "HILSPEC 56119, A. H. Compton—Progress Reports," boxed material, ANL, TIC.

24. Henry D. Smyth, *A General Account of the Development of Methods of Using Atomic Energy for Military Purposes* (GPO: Washington, D.C., 1945), p. 64. The story of the first successful sustained nuclear reaction has been told many times by participants, historians, and journalists, including Enrico Fermi, Richard G. Hewlett and Oscar E. Anderson, and Richard Rhodes. Consequently, I shall avoid repeating a detailed account but shall focus on the institutional aspects of the work leading up to and including the successful test of Chicago Pile 1, known also as CP-1.

25. A. H. Compton, "Progress Report, Metallurgical Project at University of Chicago," for weeks of Feb. 16–21, March 23–28, and March 30–April 4, 1942, Folder "CP-1, Report Data," Box 3 of 4, Reactor Hist. Files, Box 20X, RG 326, ANL, NARA; Smyth, *Atomic Energy for Military Purposes*, p. 43.

26. A. H. Compton, "Progress Report, Metallurgical Project at University of Chicago," for weeks of March 30–April 4 and April 6–11, 1942, Folder "CP-1, Report Data," Box 3 of 4, Reactor Hist. Files, Box 20X, RG 326, ANL, NARA; Compton, *Atomic Quest*, p. 83.

27. Compton to Fermi, April 16, 1942, Folder "April 1942," Box 3 of 4, Chron. File "Misc. Docs. 1940–May 1944," Box 1X, RG 326, ANL, NARA; Glenn T. Seaborg, *History of Met Lab Section C-1: April 1942 to April 1943*, Pub. 112 (Berkeley: Lawrence Berkeley Laboratory, 1977), p. 30.

28. Seaborg, *History of Met Lab Section C-1: April 1942 to April 1943*, p. 1; A. H. Compton, "Progress Report. Metallurgical Project," for week ending May 2, 1942, Folder "C-55 Declassified Progress Report for week ending May 2, 1942—A. H. Compton Leader," Box 2 of 4, Met Lab C-Series Reports, Box 35X, RG 326, ANL, NARA.

29. Barton C. Hacker, *The Dragon's Tail: Radiation Safety in the Manhattan Project, 1942–1946* (Berkeley: University of California Press, 1987), pp. 10–29; Seaborg, *History of Met Lab Section C-1: April 1942 to April 1943*, pp. 9–11; April 23 and 24, 1942, Report no. CN-111, Box 3 of 4, Box 35X, RG 326, Report nos. CH-137 and CH-187, Box

4 of 4, Box 35X, RG 326, and Report no. CH-237, Box 1 of 4, Box 36X, RG 326, all in ANL, NARA; R. L. Doan to Irvin Stewart, Feb. 11, 1942 (dictated Feb. 10), Folder "Jan.–Feb. 1942," Box 4 of 4, Chron. File "Misc. Docs. 1940–May 1944," Box 1X, RG 326, ANL, NARA; Louis Schwartz, F. C. Makepeace, and H. T. Dean, "Health Aspects of Radium Dial Painting. IV. Medical and Dental Phases," *Journal of Industrial Hygiene* 15 (Nov. 1933): 447–55; J. A. Wheeler, "Notes on Conferences," Feb. 24, 1942, Folder "Notes on Conferences J. A. Wheeler (Secretary)," Box 1 of 4, F. Daniels Office File, Box 7, RG 326, ANL, NARA; Meeting of Engineering Council, May 28, 1942, Report no. CE 106, "Declassified Meeting of Engineering Council 5–28–42," Box 1 of 4, Met Lab C-Series Reports 75–132, Box 35X, RG 326, ANL, NARA; Conference on Chemistry, Chicago, April 22–23, 1942, p. 16, Report no. CN-111, Box 3 of 4, Box 35X, RG 326, ANL, NARA; Robert S. Stone, ed., *Industrial Medicine on the Plutonium Project: Survey and Collected Papers*, Vol. 20 of *National Nuclear Energy Series* (New York: McGraw-Hill, 1951), p. 15.

30. Conference on Chemistry, Chicago, April 22–23, 1942, Report no. CN-111, Box 3 of 4, Box 35X, RG 326, ANL, NARA.

31. Ibid.; Stone, *Industrial Medicine*, p. 15.

32. Obituary of Mary Wigner, *Physics Today*, July 1978, p. 58; E. P. Wigner, "Protection against Radiations. Protection against X-Rays," undated but probably written c. mid-June 1942, according to the report number, Report no. CH-137, Box 4 of 4, Box 35X, RG 326, ANL, NARA; E. Teller and F. L. Friedman, "The Intensity of Radiation Near a 100-Watt Pile after Cessation of Operation," July 14, 1942, Report no. C-187, Box 4 of 4, Box 35X, RG 326, ANL, NARA; Metallurgical Project Report for month ending August 15, 1942, Report no. CH-237, Box 1 of 4, Box 36X, RG 326, ANL, NARA. Teller and Friedman used the calculations of Lyle B. Borst, who in 1943 became a research associate in physics at the Clinton Laboratories. Borst prepared "Preliminary Estimates of the Radiations from Fission Products" in CC-40 of April 12, 1942, and later revised his calculations. See L. B. Borst, "Preliminary Estimates of the Radiations from Fission Products," Report no. C-11-74, declassified, Box 2 of 4, Met Lab C-Series Reports, Box 35X, RG 326, ANL, NARA.

33. Meeting of Planning Board, June 6, 1942, Report no. CN-112, "Declassified Meeting of Planning Board 6/6/42," Box 3 of 4, Met Lab C-Series Reports 75–132, Box 35X, RG 326, ANL, NARA; Compton, Report for month ending September 15, 1942, Report no. CH-259, "Report for Month Ending 9–15–42," Box 2 of 4, Box 36X, RG 326, ANL, NARA; Hacker, *The Dragon's Tail*, pp. 29–30; Compton, *Atomic Quest*, pp. 176–77; Stone, *Industrial Medicine*, p. 15. Seaborg, *History of Met Lab Section C-1*, p. 198; Metallurgical Project, Report for month ending Aug. 15, 1942, Report no. CH-237, Box 1 of 4, Box 36X, RG 326, ANL, NARA; E. O. Wollan, "Diffusion of Fission Products," in Metallurgical Project Bulletin for week ending Aug. 29, 1942, CA-247, Box 2 of 4, Met Lab C-Series Reports, Box 36X, RG 326, ANL, NARA.

34. Vincent Jones, *Manhattan: The Army and the Atomic Bomb* (Washington, D.C.: Center for Military History, United States Army, 1985), pp. 37–39.

35. Compton to Allison, June 5, 1942, Folder "June 1942," Box 4 of 4, Chron. File "Misc. Docs. 1940–May 1944," Box 1X, RG 326, ANL, NARA; Meeting of Planning

Board, June 6, 1942, Report no. CS-112, "Declassified Meeting of Planning Board, 6/6/42," Box 3 of 4, Met Lab C-Series Reports 75–132, Box 35X, RG 326, ANL, NARA.

36. Compton to division and group leaders, July 15, 1942, Folder "July 1942," Box 4 of 4, Chron. File "Misc. Docs. 1940–May 1944," Box 1X, RG 326, ANL, NARA; Conference on Chemistry, Chicago, April 22–23, 1942, Report no. CN-111, Box 3 of 4, Box 35X, RG 326, ANL, NARA.

37. "Metallurgical Project Personnel," November 23, 1942, Folder "November 1942," Box 3 of 4, Chron. File "Misc. Docs. 1940–May 1944," Box 1X, RG 326, ANL, NARA.

38. Compton, *Atomic Quest*, pp. 110–11; Meeting of Planning Board, June 6, 1942, Report no. CS-112, "Declassified Meeting of Planning Board 6/6/42," Box 3 of 4, Met Lab C-Series Reports 75–132, Box 35X, RG 326, ANL, NARA; Meeting of Planning Board, August 28, 1942, Report no. CS-248, Box 2 of 4, Met Lab C-Series Reports, Box 36X, RG 326, ANL, NARA.

39. Jones, *Manhattan*, p. 49.

40. Chronology of District "X," entries of June 25 and 26, July 6 and 17, and Aug. 13, 1942, Colonel Marshall's diary, pp. 11, 12, 15, 33, 34, 60, 98, MED Top Secret Documents of Special Interest to General Groves, 1942 to 1946, Box 15, Misc. Records, RG 77, NARA; Jones, *Manhattan*, pp. 19, 43, 49; Rhodes, *The Making of the Atomic Bomb*, pp. 425–27.

41. Seaborg, *History of Met Lab Section C-1*, pp. 192–93.

42. Compton to Hilberry, Sept. 12, 1942, Folder "September 1942," Box 3 of 4, Chron. File "Misc. Docs. 1940–May 1944," Box 1X, RG 326, ANL, NARA; Chronology of District "X," entry of Aug. 15, 1942, Colonel Marshall's diary, p. 102; Marshall, visit to site, Dec. 15, 1991; Cyril A. Tregillus, "Historical Notes and Reminiscences of Area 6405 . . . Alias Palos Park Site," *Argonne News*, May 1954, p. 4; Chronology of District "X," entry of Sept. 18, 1942, Colonel Marshall's diary, p. 153; Jones, *Manhattan*, p. 78; Compton to Fermi, Sept. 14, 1942, File V-23, Hist. File–1942, ANL, TIC.

43. Jones, *Manhattan*, pp. 43–44.

44. Hewlett and Anderson, *The New World*, pp. 105–6; E. P. Wigner to Compton, "Brief History of Planning for W and Status of Blueprint Reviewing," Jan. 7, 1944, Folder "January 1944," Box 2 of 4, Met Lab Compton Chron. File 12/43–8/11, Box 34X, RG 326, ANL, NARA.

45. Hewlett and Anderson, *The New World*, pp. 106–10; Metallurgical Project Report for month ending Nov. 15, 1942, Report no. CE-344, Box 4 of 4, Met Lab C-Series Reports, Box 36X, RG 326, ANL, NARA; statement by Fermi in Technical Council meeting report, Nov. 5, 1942, Report no. CS-335, Box 3 of 4, Met Lab C-Series Reports, Box 36X, RG 326, ANL, NARA. On Nov. 5, Fermi thought it "best to do experiment in West Stands as originally planned. No one knows whether or not Argonne buildings may have eventually to be abandoned anyway."

46. Hewlett and Anderson, *The New World*, pp. 180–81; Rhodes, *The Making of the Atomic Bomb*, p. 431.

47. Rhodes, *The Making of the Atomic Bomb*, p. 432.

48. E. Fermi, "The Experimental Chain Reacting Pile and Reproduction Factor in

Some Exponential Piles," in Metallurgical Project Report for month ending November 15, 1942, Report no. CP-341, Box 3 of 4, Met Lab C-Series Reports, Box 36X, RG 326, ANL, NARA; H. L. Anderson, "The First Chain Reaction," in Robert G. Sachs, ed., *The Nuclear Chain Reaction—Forty Years Later* (Chicago: University of Chicago, 1984), pp. 33, 51; Albert Wattenberg, interview by author, April 24, 1992, Washington, D.C.

49. Seaborg, *History of Met Lab Section C-1*, p. 182; H. Barton, T. Brill, S. Fox, R. Fox, D. Froman, W. Hinch, W. R. Kanne, W. Overbeck, H. Parsons, G. Pawlicki, L. Slotin, R. Watts, M. Wilkening, and V. C. Wilson, "Appendix 2. Monitoring and Controlling the First Pile" in Enrico Fermi, "Experimental Production of a Divergent Chain Reaction," *American Journal of Physics* 20 (December 1952): 550–56.

50. Fermi, "Experimental Production," pp. 356–57; Data Sheet CP-1 (West Stands Reactor), Norbert Golchert Files, ANL; Corbin Allardice and Edward R. Trapnell, "The First Pile," in Department of Energy, *The First Reactor* (Washington, D.C.: DOE, 1982), pp. 10–13; Technical notebook 86, p. 2, Technical Notebooks of the Metallurgical Laboratory, RG 326, ANL, NARA-GL. The Pooh names for instruments appear in the National Archives documents. A memoir on tape cassette of a retired Metallurgical Laboratory employee discusses Fermi's use of the Pooh stories to sharpen his command of English. Dr. Shirley Burton of the National Archives, Great Lakes Region has custody of the cassette.

51. Quotation from Seaborg, *History of Met Lab Section C-1*, p. 374; Compton to Bush and Conant, Nov. 23, 1942, B-C 16, 11/23/42, Roll 3, RG 227, NARA; Compton, *Atomic Quest*, pp. 134–37; Sachs, *The Nuclear Chain Reaction*, p. viii.

52. Sachs, *The Nuclear Chain Reaction*, p. viii. "Gus Knuth, Man Who Saw the Future Born," *The Carpenter*, February 1963, pp. 14–16.

53. Information in this and the following two paragraphs is from Richard Watts, laboratory notebook, p. 29, Technical Notebook 150, Technical Notebooks of the Metallurgical Laboratory, 1943–1964, RG 326, ANL, NARA-GL; Argonne National Laboratory, *Controlled Nuclear Chain Reaction: The First Fifty Years* (La Grange Park, Ill.: American Nuclear Society, 1992), pp. 3–8.

54. A. V. Peterson to Gen. Groves, Dec. 7, 1942, Entry C&TR, Box 4 of 13, Classified Records, Box 43, Met Lab–Univ. Chicago, MED, OSRD, RG 227, NARA; The Compton-Conant conversation is taken from Rhodes, *Making of the Atomic Bomb*, p. 442.

55. Seaborg, *History of the Met Lab Section C-1: April 1942 to April 1943*, p. 391.

56. C. M. Cooper to J. C. Stearns, Dec. 14, 1942, Folder "CP-2," Box 3 of 4, Reactor Hist. Files, Box 20X, RG 326, ANL, NARA.

57. Roger Williams to E. B. Yancey, Jan. 12, 1943, Folder "Jan. 1943," Box 4 of 4, Chron. File "Misc. Docs. 1940–May 1944," Box 1X, RG 326, ANL, NARA; Compton to J. C. Marshall, March 6, 1943, Entry C&TR, Box 4 of 13, Classified Records, Box 43, Met Lab–Univ. Chicago, MED, RG 227, OSRD, NARA; Seaborg, "Plans for Chemical Extraction of 49 from Argonne Pile Material," Jan. 16, 1943, Folder "8. Piles, General," Box 1 of 4, Met Lab Subject Files 4–11, Box 32X, RG 326, ANL, NARA; Hewlett and Anderson, *The New World*, pp. 191–93; Leslie R. Groves, *Now It Can Be Told* (New York: Harper & Row, 1962), p. 69.

58. Compton to Groves, Jan. 19, 1943, Box 1 of 5, MUC-ABG-AJD, Box 66X, RG 326, ANL, NARA; Norman Hilberry, "Organization of Metallurgical Unit, Development of Substitute Materials Division OSRD," MUC-NH-3, Box 3 of 4, MUC-LWA-OCS, Box 68X, RG 326, ANL, NARA.

59. Laboratory Council policy meeting, Feb. 17, 1943, File "CS-476, Declassified Laboratory Council Meeting 2/17/43," Box 3 of 4, Met Lab C-Series Reports 458–498, Box 37X, RG 326, ANL, NARA; J. C. Stearns to Allison et al., Feb. 23, 1943, MUC-JCS-8, Box 1 of 4, MUC Files JC-RSM, Box 68X, RG 326, ANL, NARA.

60. "Organization of Metallurgical Unit," Jan. 25, 1943, MUC-NH-43, MUC Files, Box 68X, RG 326, ANL, NARA.

61. Jones, *Manhattan,* pp. 115–16.

62. Seaborg, *History of the Met Lab Section C-1: March 3, 1943,* p. 565; J. C. Stearns to Wilson, Dec. 8, 1942, Folder "December 1942," Box 2 of 4, Chron. File "Misc. Docs. 1940–May 1944," Box 1X, RG 326, ANL, NARA.

63. Laboratory Council policy meeting, February 19, 1943, Folder "CS-479 Declassified Laboratory Council Policy Meeting—2/19/43," Box 3 of 4, Met Lab C-Series Reports 458–498, Box 37X, RG 326, ANL, NARA; ANL press release, Nov. 24, 1950, and "A Brief General Description of the Argonne Uranium-Graphite Pile (CP-2)," Report no. CP-2459, p. 3, Norbert Golchert Files, ANL; Hewlett and Anderson, *The New World,* p. 207.

64. ANL press release, "Background Information on the Argonne National Laboratory's Uranium-Graphite Nuclear Reactor," Nov. 24, 1950, Norbert Golchert Files, ANL, NARA; "The Argonne Uranium-Graphite Nuclear Reactors," *Argonne Low-Power Research Reactors* (Chicago: Argonne National Laboratory, 1950), p. 3, Folder "CP-3," Box 2 of 4, Reactor Hist. Files, Box 20X, RG 326, ANL, NARA; "Exhibit 'C' Biographical Data of Key Administrative Staff Members of the Argonne National Laboratory," ANL History 1946–59, UCVP ANL, ANL.

65. Compton to J. C. Marshall, March 6, 1943, Entry C&TR, Box 4 of 13, Classified Records, Box 43, Met Lab–Univ. Chicago, MED, OSRD, RG 227, NARA.

66. Hewlett and Anderson, *The New World,* pp. 200–201; Quotation from Wigner to Compton, "Brief History of Planning for W and Status of Blueprint Reviewing," Jan. 7, 1944, pp. 3, 8, 12–16; Compton to Wigner, July 23, 1943, and Wigner to Compton, Aug. 5, 1943, Folder "7/23/43 and 8/5/43," Box 2 of 4, Met Lab Reading File, Box 5X, RG 326, ANL, NARA; A. V. Peterson to Groves, Aug. 13, 1943, and enclosures, Series 5, 080 (Laboratories), MED Decimal Files, RG 77, NARA.

67. Roger Williams to file, Aug. 21, 1943, Folder "August 1943," Box 1 of 4, Met Lab Compton Chron. File 11/40–11/43, Box 34X, RG 326, ANL, NARA; C. H. Greenewalt to file, Aug. 19, 1943, Folder "August 1943," Box 1 of 4, Met Lab Compton Chron. File 11/40–11/43, Box 34X, RG 77, ANL, NARA.

68. Laboratory Council policy meeting, Report no. CS-497, Box 3 of 4, Box 37X, RG 326 ANL, NARA; Compton to J. C. Marshall, March 6, 1943, Entry C&TR, Box 4 of 13, Classified Records, Box 43, Met Lab–Univ. Chicago, MED, OSRD, RG 227, NARA.

69. Compton to H. T. Wensel, Jan. 12, 1944, Folder "Jan. 1944," Box 1 of 4, Chron.

File "Misc. Docs. 1940–5/44," Box 1X, RG 326, ANL, NARA; James Franck and T. R. Hogness to J. C. Stearns, Oct. 22, 1943, Folder "10/22/43," Box 2 of 4, Met. Lab. Reading File 10/6–11/8/43, Box 7X, RG 326, ANL, NARA; M. C. Leverett to John R. Suman, April 10, 1943, Folder "4/10/43," Box 1 of 4, Met Lab Reading Files, Box 5X, RG 326, ANL, NARA; and Charles D. Coryell to J. C. Stearns, Sept. 9, 1943, Folder "9/9/43," Box 2 of 4, Met Lab Reading File, Box 5X, RG 326, ANL, NARA.

70. W. B. Harrell to district engineer, Feb. 3, 1945, Folder "Jan.–April 1945," Box 3 of 4, Chron. File "Misc. Docs. June 1944–1949," Box 2X, RG 326, ANL, NARA; Contract no. W-31-109-ENG-38, UCVP ANL. Later the Army Corps of Engineers wanted the university to agree to a clause guaranteeing that the employee returning to his original place of employment would receive a salary equal to the one for the position just vacated. William B. Harrell, the university business manager, objected to this provision.

71. Compton to Allison, May 5, 1943, MUC-AC-194, Folder "April–May 1943," Box 1 of 4, Met Lab Compton Chron. File 11/40–11/43, Box 34X, RG 326, ANL, NARA; Hilberry to L. R. Hafstad, Nov. 2, 1949, p. 5, Folder "In Date Order," Box 3 of 4, Chron. File Misc. Doc. 6/44–1949, Box 2X, RG 326, ANL, NARA.

72. Jones, *Manhattan,* pp. 269–71.

73. Compton, "State of the Nation," October 20, 1943, MUC-AC-419, Box 1 of 5, MUC-ABG-AJD, Box 66X, RG 326, ANL, NARA; Hewlett and Anderson, *The New World,* pp. 81–83; Allison to all division chiefs, et al., Dec. 20, 1943, Folder "December 1943," Box 2 of 4, Met Lab Compton Chron. File 12/43–8/44, Box 34X, RG 326, ANL, NARA.

74. Report for month ending March 15, 1943, "Special Chemistry of 94," CK-514, Report no. CK-514, Declassified Special Chemistry of 94, Report for month ending March 15, 1943, Box 4 of 4, Met Lab C-Series Reports 499–537, Box 37X, RG 326, ANL, NARA; MDH, VIII-2, IV-16-18, VI-8, VII-11, VIII-5; Hewlett and Anderson, *The New World,* p. 251; G. T. S. to Franck, Aug. 24, 1943, File V-74, Drawer 2, loose material, ANL, TIC.

75. Hewlett and Anderson, *The New World,* pp. 182–85, 204–5.

76. Hilberry to C. A. Thomas, July 27, 1943, Folder "7/27/43," Box 2 of 4, Met Lab Reading File, Box 5X, RG 326, ANL, NARA.

77. H. W. Bellas, Annual Progress Report Engineering Development Section, MUC-HWB-30, Box 4 of 4, MUC Files, Box 67X, RG 326, ANL, NARA; Allison to A. V. Peterson, Sept. 17, 1943, Folder "9/17/43," Box 2 of 4, Met Lab Reading File, Box 5X, RG 326, ANL, NARA; Compton, "State of the Nation," Oct. 20, 1943, MUC-AC-419, Box 1 of 5, MUC-ABG-AJD, Box 66X, RG 326, ANL, NARA.

78. P. Morrison to Allison, Sept. 23, 1943, MUC-PM-7, Box 4 of 4, MUC Files, Box 68X, RG 326, ANL, NARA. Allison passed on this recommendation to Groves. See Allison to Groves, Oct. 11, 1943, Folder "October 1943," Box 1 of 4, Met Lab Compton Chron. File 11/40–11/43, Box 34X, RG 326, ANL, NARA.

79. Compton to Allison, Nov. 26, 1943, MUC-AC-494, Box 1 of 5, MUC-ABG-AJD, Box 66X, RG 326, ANL, NARA; Compton to Allison, Nov. 26, 1943, Folder "November 1943," Box 1 of 4, Met Lab Compton Chron. File 11/40–11/43, Box 34X, RG 326, ANL, NARA; Compton to M. D. Whitaker, Dec. 7, 1943, Folder "December 1943," Box 2 of 4, Met Lab Compton Chron. File 12/43–8/44, Box 34X, RG 326, ANL, NARA.

80. Smyth, *Atomic Energy for Military Purposes*, pp. 107–8, 111; Compton to Allison and Smyth, Sept. 22, 1943, MUC-AC-352, Box 1 of 5, MUC-ABG-AJD, Box 66X, RG 326, ANL, NARA; Compton, "State of the Nation," Oct. 20, 1943, MUC-AC-419, Box 1 of 5, MUC-ABG-AJD, Box 66X, RG 326, ANL, NARA; Project Council policy meeting, April 19, 1944, Report no. CS-1597, "Project Council Policy Meeting 4/19/44," Box 4 of 4, Met Lab C-Series Reports, Box 45X, RG 326, ANL, NARA; Compton to Groves, Jan. 12, 1944 (draft), Folder "January 1944," Box 2 of 4, Met Lab Compton Chron. Files 12/43–8/44, Box 34X, RG 326, ANL, NARA. The general, however, prohibited the Metallurgical Project scientists from disclosing to the Canadians any information on the graphite plants for producing plutonium or the heavy-water plants except the technical aspects concerned with the piles that both groups were discussing.

81. W. H. McCorkle, J. D. Richards, and A. Wattenberg, "The CP-3, CP-3' Argonne Heavy Water Reactor," p. 7, ANL-WHM-26, Folder "CP-3," Box 2 of 4, Reactor Hist. Files, Box 20X, RG 326, ANL, NARA; Zinn to Wigner, April 25, 1952, Folder "CP-3," Box 2 of 4, Reactor Hist. Files, Box 20X, RG 326, ANL, NARA.

82. Conversation in this and following paragraphs taken from Minutes of the Advisory Board meeting April 15 at 11:00 A.M., Folder "April 1944," Box 2 of 4, Met Lab Compton Chron. File 12/43–8/44, Box 34X, RG 326, ANL, NARA.

83. Compton, Allison, and Whitaker to Groves, April 10, 1944, MUC-NH-728, Box 3 of 4, MUC Files, Box 68X, RG 326, ANL, NARA; Project Council policy meeting, June 7, 1944, Report no. CS-1779, Box 1 of 4, Met Lab C-Series Reports, Box 47X, RG 326, ANL, NARA; Project Council policy meeting, Sept. 20, 1944, Report no. CS-2155, Box 3 of 4, Met Iab C-Series Reports, Box 50X, RG 326, ANL, NARA; Compton to Allison, June 6, 1944, MUC-AC-991, Box 1 of 5, MUC Files, Box 66X, RG 326, ANL, NARA; J. C. Stearns et al. to K. D. Nichols, June 30, 1944, MUC-JCS-112, MUC Files, Box 68X, RG 326, ANL, NARA; Compton to Allison, July 10, 1944, MUC-AC-1068, Box 1 of 5, MUC Files, Box 66X, RG 326, ANL, NARA; "Proposed Program for Met," Folder "March 1944," Box 2 of 4, Met Lab Compton Chron. File, Box 34X, RG 326, ANL, NARA.

84. T. R. Hogness to Allison, July 25, 1944, File V-23, Drawer 1, loose material, ANL, TIC; Farrington Daniels to T. R. Hogness, Sept. 18, 1944, File V-23, Hist. File–1944, Drawer 1, ANL, TIC.

85. Oppenheimer to Fermi, August 24, 1944, Box 33, Fermi, Enrico, JRO, LC MSS; Hewlett and Anderson, *The New World*, p. 312; Charles M. Cooper, Project Council policy meeting notes, Oct. 4, 1944, Report no. CS-2239, Box 2 of 4, Met Lab C-Series Reports, RG 326, ANL, NARA; Groves to Compton, Oct. 31, 1944, 201 General, MED Decimal Files, RG 77, NARA; Metallurgical Laboratory Report for November 1944, MUC-JCS-9, Met Lab Progress Reports 5-43-6.46, Drawer 2, Cabinet 24, ANL, TIC; Compton to Groves, Dec. 7, 1944, MUC-AC-2458, Series 5, 201 General, MED Decimal Files, RG 77, NARA; Compton to Groves, Dec. 7, 1944, MUC-AC-2458, Box 1 of 5, MUC Files, Box 66X, RG 326, ANL, NARA.

86. H. G. Hawkins, Jr., to Wayne W. Johnson, Oct. 24, 1944, Folder "October 1944," Box 1 of 4, Chron. File Misc. Doc. June 1944–1949, Box 2X, RG 326, ANL, NARA.

87. Seaborg, *History of Met Lab Section C-1*, passim.

88. "Proposed Program for Metallurgical Laboratory for 1944–1945," July 28, 1944, Folder "HILSPEC 5511," TIC boxed material, ANL, TIC; J. C. Stearns to Groves, Oct. 9, 1944, MUC-JCS-AD-29, MUC Files, Box 68X, RG 326, ANL, NARA; MUC-LAK-254, September 4, 1944, Box 2 of 4, MUC-JC-RSM, Box 68X, RG 326, ANL, NARA. The independent status apparently came earlier than September 1944. See also J. C. Stearns, Met Lab Report for April 1945, MUC-JCS-305, Met Lab Progress Reports 5.43-6.46, Drawer 2, Cabinet 29, ANL, TIC; Seaborg, *History of Met Lab Section C-1: May 1944–April 1945*, Vol. 3, p. 260; Compton to Groves, Oct. 3, 1944 and Zinn to Compton, Oct. 3, 1944, 400.12 (Experiments), Entry 5, Box 66, MED Decimal Files, RG 77, NARA; Lt. Col. F. T. Matthias to Maj. Gen. L. R. Groves, Oct. 3, 1944, 319.1 (Misc.), Entry 5, Box 56, MED Decimal Files, RG 77, NARA; A. H. Comas to Groves, Oct. 4, 1944, Folder "October 1944," Box 3 of 4, Compton Chron. File, Box 34X, RG 326, ANL, NARA; Conference on Chemistry, Chicago, April 22–23, 1942, Report no. CN-111, Box 3 of 4, Box 35X, RG 326, ANL, NARA.

89. C. J. Watson to Compton, Jan. 4, 1944, MUC-HG-337, Box 3 of 4, MUC Files, Box 67X, RG 326, ANL, NARA; F. W. Albaugh to Seaborg, Jan. 31, 1933, MUC-GTS-437, Box 1 of 4, MUC Files, Box 67X, RG 326, ANL, NARA; French Hagemann to Seaborg, Feb. 18, 1944, MUC-GTS-467, Box 1 of 4, MUC Files, Box 67X, RG 326, ANL, NARA. Watson cited radiation exposure and metal or halogen poisoning as among the chief hazards. He recommended a number of measures that Compton implemented, including keeping Met Lab personnel under closer than normal observation and establishing a laboratory health service at Billings Hospital.

90. Daniels to area engineer, April 29, 1946, Folder "OK for Declass.—E. N. Pettit," Box 3 of 5, Hilberry Special and Important Historical E. N. Pettit, MUC Memos, Box 37, RG 326, ANL, NARA; Compton, Project Council information meeting (Health), June 6, 1944, p. 2, Report no. CS-1773, Box 1 of 4, Met Lab C-Series Reports, Box 47X, RG 326, ANL, NARA; Hacker, *The Dragon's Tail*, p. 48.

91. Proposed Program for Metallurgical Laboratory for 1944–1945, July 28, 1944, p. 4, Folder "HILSPEC 55111," TIC boxed material, ANL, TIC; University of Chicago organizational charts for Nov. 5, 1944, and March 12, 1945, MUC-BG-261 and 296, Box 2 of 5, MUC Files, Box 66X, RG 326, ANL, NARA.

92. Hewlett and Anderson, *The New World*, pp. 324–35; Fermi et al. to Compton, Nov. 18, 1944, MUC-RSM-234, copy 24, File V-23, Drawer 1, loose material, ANL, TIC; Wigner to Allison, July 13, 1944, MUC-EPW-93, Box 3 of 5, MUC Files, Box 66X, RG 326, ANL, NARA.

93. Hewlett and Anderson, *The New World*, p. 325.

94. Ibid.

95. Compton to Fermi, July 17, 1944, MUC-AC-1082, Box 1 of 5, MUC Files, Box 66X, RG 326, ANL, NARA.

96. Ibid.

97. Compton and Robert M. Hutchins to Groves, July 17, 1944, draft for discussion by advisory committee, MUC-AC-1083, and Compton to Hutchins, July 17, 1944, MUC-AC-1084, Box 1 of 5, MUC Files, Box 66X, RG 326, ANL, NARA; Minutes of the

Advisory Board meeting, August 21 at 7:30 P.M., Folder " August 1944," Box 2 of 4, Met Lab Compton Chron. Files 12/43–8/44, Box 34X, RG 326, ANL, NARA.

98. Compton, *Atomic Quest,* 196–97.

99. Nichols to Compton, March 17, 1945, Folder "March–May 1945," Box 2 of 4, Chron. File "Misc. Docs. June 1944–1949," Box 2X, RG 326, ANL, NARA; Compton to Nichols, Jan. 25, 1945, MUC-AC-1272, MUC Files, Box 66X, RG 326, ANL, NARA; Walter Bartky, minutes of policy meeting, Jan. 17, 1945, Report no. CS-2706, Box 3 of 4, Met Lab C-Series Reports, Box 54X, RG 326, ANL, NARA.

100. Fermi and Zinn to Compton, January 15, 1945, Folder "January 1–20, 1945," Box 3 of 4, Met Lab Compton Chron. Files, Box 34X, RG 326, ANL, NARA.

101. Compton to Nichols, Feb. 1, 1945, Series 5, 600.12 Metallurgical Project, MED Decimal Files, RG 77, NARA.

102. Compton and W. Bartky, special meeting of Policy Council, March 20, 1945, Report no. CS-2863, and Walter Bartky, minutes of policy meeting, March 21, 1945, Report no. CS-2862, Box 4 of 4, Met Lab C-Series Reports, Box 55X, RG 326, ANL, NARA.

103. Compton to Groves for attention of the secretary of war, March 5, 1945, Folder "March 1945," Box 4 of 4, Met Lab Compton Chron. File 3/45–1946, Box 34X, RG 326, ANL, NARA.

104. Walter Bartky, minutes of policy meeting, April 18, 1945, Declassified, Report no. CS-2970, "Minutes of Policy Meeting 4/18/45," Box 3 of 4, Met Lab C-Series Reports, Box 56X, RG 326, ANL, NARA.

105. John P. Howe, Metallurgical Laboratory Report for June 1945, Folder "HILSPEC 55084," TIC boxed material, ANL, TIC.

106. Zinn to Walter Bartky, June 15, 1945, Folder "Miscellaneous," undated & misc., Box 4 of 4, Met Lab Compton Chron. File 3/45–1946, Box 34X, RG 326, ANL, NARA.

107. Compton to W. W. Watson, July 5, 1945, MUC-AC-2751, Box 1 of 5, MUC Files, Box 66X, RG 326, ANL, NARA; Compton to Karl T. Compton, July 5, 1945, MUC-AC-2754, Box 1 of 5, MUC Files, Box 66X, RG 326, ANL, NARA.

108. A. H. Jaffey to Seaborg, Jan. 20, 1945, MUC-GTS-1303, Box 2 of 4, MUC Files, Box 67X, RG 326, ANL, NARA.

109. Grover C. Thompson to John Lansdale, Jr., July 11, 1945, "Atomic Weapons, Use of," Vertical File, HSTL; Hewlett and Anderson, *The New World,* pp. 398–99. At Compton's direction, between July 11 and 13, Daniels polled the Met Lab scientists with a secret ballot on five options. Of the 278 academic employees, the 150 who voted split. The largest number, 69 (46 percent), preferred "a military demonstration in Japan, to be followed by a renewed opportunity for surrender before full use of the weapons is employed." In second place, 26 percent voted for an experimental demonstration in the United States with Japanese representatives present; after that Japan would be given a chance to surrender before "full use" of the weapon. Twenty-three, or 15 percent, favored "the manner that from the military point of view" would most effectively encourage Japan's surrender with a minimum human cost to the Allies. A public experimental demonstration combined with withholding military use drew 11 percent of the votes,

and only three, or 2 percent, wanted to keep secret the new weapons developments and did not want the devices used in this war. See Compton to Daniels, July 11, 1945, MUC-AC-2761, Box 1 of 5, MUC Files, Box 66X, RG 326, ANL, NARA; Daniels to Compton, July 13, 1945,"Atomic Weapons, Use of," 233 Vertical File, HSTL; Alice Kimball Smith, *A Peril and a Hope. The Scientists' Movement in America: 1945–47* (Chicago: The University of Chicago Press, 1965), pp. 57–59, Metallurgical Laboratory Report for July 1945, Folder "HILSPEC 55083," TIC boxed material, ANL, TIC. Taken from the original documents found at the Truman Library and at the National Archives, the information on this poll as presented here differs somewhat from the reports in other secondary literature. It is not possible to determine whether Truman saw Daniels's report in July or August 1945.

110. Smith, *A Peril and a Hope*, p. 85; Meeting of Aug. 7 of Political and Social Implications, MUC-Atomic Scientists of Chicago, File V-74, Drawer 2, loose material, ANL, TIC; Arthur Jaffey to members of the Atomic Scientists of Chicago, Nov. 27, 1945, Folder 2, Box 77, John A. Simpson Papers, UCRL SC; Proposal for Reorganization of the Federation of American Scientists made by representatives from the Chicago, Clinton, and Los Alamos Groups to the Council of the FAS, untitled folder, Drawer 2, Cabinet 29, ANL, TIC; first council meeting of the Federation of American Scientists, Jan. 5–6, 1946, John A. Simpson Papers, UCRL SC; Executive Committee to members of the Atomic Scientists of Chicago, January 5–6, 1946, untitled folder, Drawer 2, Cabinet 29, ANL, TIC. Alice Kimball Smith, in *A Peril and a Hope,* has ably covered the nuclear scientists' movement from 1945 to 1947.

111. Hilberry, policy meeting, Aug. 22, 1945, Report no. CS-3135, "Policy Meeting 8/22/45," Box 1 of 4, Met Lab C-Series Reports, Box 58X, RG 326, ANL, NARA.

112. Allison to Seaborg, Aug. 15, 1945, Folder "Laboratories Argonne Administrative Future Program," Box 1 of 4, Met. Lab. Subj. Files 4–11, Box 32X, RG 326, ANL, NARA; Compton to Groves, Aug. 23, 1945, 334 Postwar Policy, MED Decimal Files, RG 77, NARA; Project Council policy meeting, Clinton Laboratories, Sept. 17, 1945, untitled folder, Box 4 of 4, Zinn Reading Files 1946–1955 & Misc., Box 32X, RG 326, ANL, NARA.

113. Seaborg, *History of Met Lab Section C-1: May 1945–May 1946,* p. 238; L. O. Morgan to Seaborg, Oct. 25, 1945, MUC-GTS-2042, Box 3 of 4, MUC Files, Box 67X, RG 326, ANL, NARA.

114. Seaborg, *History of Met Lab Section C-1: May 1945–May 1946,* pp. 153–54, and passim.

115. Hutchins to Groves, Oct. 6, 1945, Box 36, Groves, L. R., JRO, LC MSS.

116. L. C. Furney to members of the Laboratory Council, Nov. 6, 1945, Folder "Oct. 1945–April 1946," Lab. Council, Met Lab Mtgs., Box 2 of 4, Met Lab & ANL Mtgs. 1943–1948, Box 32X, RG 326, ANL, NARA; Seaborg, *History of Met Lab Section C-1: May 1945–May 1946,* p. 340.

117. Nichols to Compton, Nov. 19, 1945, 334 (Committees), Series 5, MED Decimal Files, RG 77, NARA; Hilberry to John T. Tate, Nov. 27, 1945, pp. 8–9, attached to Nichols to Compton, Nov. 19, 1945, untitled folder, Box 4 of 4, Zinn Reading Files 1946–1955 & Misc., Box 32X, RG 326, ANL, NARA.

118. Compton et al. to Col. K. D. Nichols, Dec. 5, 1945, Attachment no. 2 to "Agenda, Argonne Advisory Council Meeting, April 5–6, 1946," untitled folder, Box 4 of 4, Zinn Reading Files 1946–1955 & Misc., Box 32X, RG 326, ANL, NARA; Seaborg, *History of Met Lab Section C-1: May 1945–May 1946*, pp. 317–18.

119. "Recommendation from Advisory Committee on Argonne Operation," MUC-NH-2826, Box 3 of 4, MUC Files, Box 68X, RG 326, ANL, NARA; L. C. Furney to members of the Laboratory Council, Dec. 31, 1945, Folder "Oct. 1945–April 1946," Lab. Council, Met Lab Mtgs., Box 2 of 4, ANL, Met Lab & ANL Mtgs. 1943–1948, Box 32X, RG 326, ANL, NARA; Metallurgical Laboratory Report for December 1945, MUC-FD-L-131, Progress Report Met Lab Contract W-7401-ENG-37, signed contract copies, Book no. 3, 7/45–12/45, Box 4 of 5, Pettitt & Foote, Box 40, RG 326, ANL, NARA; Dewey M. Stowers to Groves, Feb. 1, 1946, 600.12 Metallurgical Project, Series 5, MED Decimal Files, RG 77, NARA.

120. Metallurgical Laboratory Report for Dec. 1945, Contract W-7401-ENG-37, MUC-FD-L-131, Met Lab Progress Reports 5.43–6.46, Drawer 2, Cabinet 29, ANL, TIC; Nichols to Groves, Jan. 22, 1946, 334 (1942–48) (Advisory Committee on R & D), Series 5, MED Decimal Files, RG 77, NARA.

121. L. C. Furney to members of the Laboratory Council, Jan. 30 and Feb. 26, 1946, and Daniels to Compton, Feb. 7, 1946, Folder "January 1946 and February 1946," Box 5 of 5, Met Lab Zinn Chron. File 11/45–4/46, Box 70X-71X, RG 326, ANL, NARA.

122. L. C. Furney to members of the Laboratory Council, Jan. 30, and Feb. 9, 1946, Folder "Oct. 1945–April 1946," Lab. Council, Met Lab Mtgs., Box 2 of 4, Met Lab & ANL, Mtgs. 1943–1948, Box 32X, RG 326, ANL, NARA; L. C. Furney to members of the Laboratory Council, Feb. 28, 1946, Folder "February 1946," Box 5 of 5, Met Lab Zinn Chron. File 11/45–4/46, Box 70X-71X, RG 326, ANL, NARA.

123. Hewlett and Anderson, *The New World*, p. 633; "Broad Policy on National Laboratories Recommended by General Groves' Advisory Committee on Research and Development," n.d., Attachment no. 1 to "Agenda, Argonne Advisory Council Meeting, April 5–6, 1946," untitled folder, Box 4 of 4, Zinn Reading Files 1946–1955 & Misc., Box 32X, RG 326, ANL, NARA.

124. Daniels to file, April 5 and 6, 1946, Folder "April 1946," Box 5 of 5, Met Lab Zinn Chron. File 11/45–4/46, Box 70X-71X, RG 326, ANL, NARA.

125. Nichols to General Groves, March 14, 1946, 334 (Advisory Committee on R & D), MED Decimal Files (1942–48), RG 77, NARA; L. C. Furney to members of the Laboratory Council, March 19, 1946, Folder "Oct. 1945–April 1946," Lab Council, Met Lab Mtgs., Met Lab & ANL Mtgs. 1943–1948, Box 2 of 4, Box 32X, RG 326, ANL, NARA.

126. R. E. Lapp to Daniels, March 31, 1946, Folder "1946" Box 4 of 4, undated & misc., Compton Chron. File 3/45–2/46, Box 34X, RG 326, ANL, NARA; Daniels to file, April 10, 1946, and L. C. Furney to members of the Laboratory Council, April 11, 1946, Folder "April 1946," Box 5 of 5, Met Lab Zinn Chron. File 11/45–4/46, Box 70X-71X, RG 326, ANL, NARA.

127. Daniels to file, April 10, 1946, and Hilberry, Minutes of the meeting of the Council of Representatives of Participating Institutions, Argonne National Laboratory,

Chicago, April 5–6, 1946, April 19, 1946, Folder "April 1946," Box 5 of 5, Met Lab Zinn Chron. File 11/45–4/46, Box 70X-71X, RG 326, ANL, NARA.

128. Daniels to file, April 5 and 6, 1946, Folder "April 1946," Box 5 of 5, Met Lab Zinn Chron. File 11/45–4/46, Box 70X-71X, RG 326, ANL, NARA.

129. Daniels to file, April 10, 1946, Folder "April 1946," Zinn Chron. File 11/45–4/46, Box 70X-71X, RG 326, ANL, NARA; Hilberry, Minutes of the meeting of the Board of Governors, Argonne National Laboratory, May 6, 1946, Folder "May 1946," Box 1 of 3, Met Lab Zinn Chron. File 5/46–3/47, Box 71X, RG 326, ANL, NARA.

130. Hilberry, Minutes of the meeting of the Board of Governors, Argonne National Laboratory, May 6, 1946, Folder "May 1946," Box 1 of 3, Met Lab Zinn Chron. File 5/46–3/47, Box 71X, RG 326, ANL, NARA.

131. L. C. Furney to members of the Laboratory Council, May 29, 1946, Folder "May 1946," Box 1 of 3, Met Lab Zinn Chron. File 5/46–3/47, Box 71X, RG 326, ANL, NARA.

132. Hewlett and Anderson, *The New World,* pp. 410–11.

133. H.R. 4280, 79th Cong., 1st sess., in James D. Nuse, comp., *Legislative History of the Atomic Energy Act of 1946 (Public Law 585, 79th Congress)* (Washington, D.C.: AEC, 1965), Vol. 1, pp. 933–61, Nuclear Regulatory Commission Law Library, Washington, D.C.; Hewlett and Anderson, *The New World,* pp. 411, 430.

134. Smith, *A Peril and a Hope,* pp. 144–47, 149, 157 (quotation). On p. 209 Smith reports that congressmen were impressed that 95 percent of the Manhattan Project scientists opposed the bill.

135. Ibid., pp. 194–200; L. C. Furney to members of the Laboratory Council, Feb. 28, 1946, Box 5 of 5, Met Lab Zinn Chron. File 11/45–4/46, Box 70X-71X, RG 326, ANL, NARA; Hewlett and Anderson, *The New World,* pp. 438–39, 513, 530, 714–22.

136. Stone, *Industrial Medicine,* xix; Hilberry, Minutes of the meeting of the Council of Representatives of Participating Institutions, Argonne National Laboratory, Chicago, April 5 and 6, 1946, Folder "April 1946," Box 5 of 5, Met Lab Zinn Chron. File 11/45–4/46, Box 70X-71X, RG 326, ANL, NARA; A. V. Peterson to Zinn, May 29, 1946, Folder "May 1946," Box 1 of 3, Met Lab Zinn Chron. File 5/46–3/47, Box 71X, RG 326, ANL, NARA.

137. Met Lab Report for December 1945, MUC-FD-L-131, Progress Report Met. Lab. Contract W-7401-ENG-37 signed contract copies, Book no. 3, 7/45–12/45, Box 4 of 5, Pettitt & Foote, Box 40, RG 326, ANL, NARA; Daniels to file, April 10, 1946, Folder "April 1946," Box 5 of 5, Met Lab Zinn Chron. File 1/45–4/46, Box 70X-71X, RG 326, ANL, NARA; R. E. Lapp and Daniels to A. H. Frye, March 21, 1946, and Daniels to Frye, March 24, 1946, Folder "Instrument Div. 3/45–6/48," Box 1 of 4, all in W. Zinn, Box 11, RG 326, ANL, NARA.

138. Daniels to J. J. Nickson, May 17, 1946, Health Division 3/45–5/49, Box 4 of 4, all in Zinn, Box 11, RG 326, ANL, NARA.

139. Daniels to file, April 10, 1946, Folder "April 1946," Box 5 of 5, Met Lab Zinn Chron. File, Box 70X-71X, RG 326, ANL, NARA.

140. L. C. Furney to members of the Laboratory Council, April 11, 1946, Folder "Oct. 1945–April 1946," Lab. Council, Met Lab Mtgs., Box 2 of 4, Met Lab & ANL Mtgs.

1943–1948, Box 32X, RG 326, ANL, NARA; Daniels to division directors, May 20, 1946, Hist. File–6, Drawer 2, Cabinet V23, ANL, TIC.

Chapter 2

1. "Operational History Argonne National Laboratory Fiscal Year 1946–1947," p. 19, RG 326, OPA, ANL, NARA-GL; Arthur H. Frye, Jr., to the district E engineer, Manhattan District, June 28, 1946, History Notes, Vol. 1, Folder 33, Box 6, Job 6263, RG 326, DOE; Report to the Joint Committee on Atomic Energy on the Acquisition of the Du Page County Site for the Argonne National Laboratory, Gen. Corres., Box 6, ANL–Du Page Files, JCAE, RG 128, ANL, NARA.

2. Hilberry to W. B. Harrell, Sept. 17, 1951, UCVP ANL, ANL History: Ten Year Summary Report, p. 18, ANL; N. Hilberry, Minutes of the Meeting of the Board of Governors, Argonne National Laboratory, Oct. 7, 1946, Hist. File–1946, Drawer 1, Cabinet V23, ANL, TIC; Operational History, Argonne National Laboratory, Fiscal Year 1946–1947, Program Correspondence Files, ANL unprocessed files, ANL Information Services, Box D 367106, CP-1 Commemoration, RG 326, NARA-GL.

3. Hewlett and Anderson, *The New World,* pp. 630–32, 636–37.

4. "The Atomic Energy Act of 1946," *United States Statutes at Large* 60:755–75; Richard G. Hewlett and Francis Duncan, *Atomic Shield,* Vol. 2 of *History of the United States Atomic Energy Commission* (University Park: Pennsylvania State University Press, 1969), p. 15.

5. Original ANL contract, November 22, 1946, Article II, 1(d), UCVP ANL, untitled folder, ANL.

6. L. C. Furney to E. F. Frolik, June 11, 1947, Folder "L. C. Furney, Staff Asst., 10/45–9/49," Box 4 of 4, W. Zinn, Box 21, RG 326, ANL, NARA; "Tentative Program for the Scientific and Technical Operations of the Argonne National Laboratory for the Year 1947–48," Folder "In Date Order," Box 3 of 4, Chron. File "Misc. Docs. 6/44–1949," Box 2X, RG 326, ANL, NARA.

7. Furney to division directors, July 13, 1946, Folder "OK for declass–ENP, Date Only & Non-ANL," Box 3 of 5, Hilberry Special and Important Historical E. N. Pettitt, Box 37, RG 326, ANL, NARA; Zinn to McCorkle et al., July 19, 1946, Folder "July thru August 1946," Box 1 of 3, Met Lab Zinn Chron. File 5/46–3/47, Box 71X, RG 326, ANL, NARA; "The History of Reactor Operations," p. 19, No. 228, Norbert Golchert Papers, ANL.

8. Furney to division directors, July 12, 1946, Folder "OK for declass.," Box 3 of 5, Hilberry Special and Important Historical E. N. Pettitt, Box 37, RG 326, ANL, NARA; "Design of a Fast Breeding Pile," Argonne National Laboratory Council Meeting, August 6, 1946, Folder "In Date Order," Box 3 of 4, Chron. File "Misc. Docs. 6/44–1949," Box 2X, RG 326, ANL, NARA.

9. Furney to division directors, July 12, 1946, Folder "OK for declass.," Box 3 of 5, Hilberry Special and Important Historical E. N. Pettitt, Box 37, RG 326, ANL, NARA; Edward E. Brown to Pres. Clayton F. Smith and members of the Board of Forest Preserve Commissioners, Aug. 9, 1946, and Robert P. Patterson to Clayton F. Smith, July 2,

1946, Folder 7, Box 28, VPSP, UCRL SC; Office of the Acting General Counsel to the Commissioners, January 21, 1947, Footnotes 1945–1963 (Chapters 1–3), Box 1007, Footnote Files, Vol. 2, AEC History, RG 326, AEC; Fred B. Rhodes to G. Lyle Belsley, May 13, 1947, AEC 601 ANL (12–3–46), Box 11233, Job 4327, AEC Secretariat, RG 326, DOE.

10. Daniels to Frank Spedding, Aug. 20, 1946, Correspondence Uncl., Box 2 of 6, F. Daniels, Box 6, RG 326, ANL; Hilberry to L. R. Hafstad, Nov. 2, 1949, p. 25, Folder "In Date Order," Box 3 of 4, Chron. File "Misc. Docs. 6/44–1949," Box 2X, RG 326, ANL, NARA.

11. Chas. G. Sauers to Compton, July 12, 1946; George B. McKibbin to Compton, Aug. 1, 1946; Compton to secretary of war, Aug. 9, 1946; and Kenneth C. Royall to Compton, Aug. 26, 1946, Box 7, Series 2, Argonne Lab, AHC.

12. Zinn to Col. A. H. Frye, Jr., Aug. 30, 1946, Folder "8/30/46," Box 2 of 4, ANL Reading File, Box 11X, RG 326, ANL, NARA.

13. Minutes of meeting of Laboratory Executive Committee, Sept. 20, 1946, Folder "9/20/46," Box 2 of 4, ANL Reading File, Box 11X, RG 326, ANL, NARA; David E. Lilienthal to Bourke B. Hickenlooper, Jan. 6, 1948, p. 7 of attached "Report to the Joint Committee on Atomic Energy on the Acquisition of the Du Page County Site for the Argonne National Laboratory," Gen. Corres., Box 5, ANL–Du Page Files, JCAE, RG 128, ANL, NARA.

14. O. C. Simpson to Zinn, Sept. 16, 1946, "Chemistry Div.–O. C. Simpson 12/44–12/49," Box 2 of 4, W. Zinn, Box 11, RG 326, ANL, NARA; Compton to Daniels, Sept. 18, 1946, and F. H. Spedding to Daniels, Sept. 18, 1946, "Folder 9/18/46," Box 2 of 4, ANL Reading File, Box 11X, RG 326, ANL, NARA; J. J. Nickson to Zinn, Sept. 30, 1946, and N. Hilberry, Minutes of meeting of Laboratory Executive Committee, Oct. 9, 1946, Folder "9/30/46," Box 3 of 4, ANL Reading File, Box 11X, RG 326, ANL, NARA; R. G. Gustavson to Daniels, Sept. 19, 1946, Hist. File–1946, Drawer 1, Cabinet V23, ANL, TIC; Zinn to Groves, Oct. 22, 1946, Folder 6, Box 3 of 5, W. Zinn, Box 19, RG 326, ANL, NARA.

15. N. Hilberry, Minutes of the Oct. 7, 1946, Argonne National Laboratory Board of Governors meeting, Hist. File–1946, Drawer 1, Cabinet V23, ANL, TIC; "Report to the Joint Committee on Atomic Energy on Acquisition of the Du Page County Site for the Argonne National Laboratory," p. 19, Gen. Corres., Box 6, ANL–Du Page Files, JCAE, RG 128, ANL, NARA.

16. Daniels to Groves, October 18, 1946, Folder "September thru October 1946," Box 1 of 3, Met Lab Zinn Chron. File 5/46–3/47, Box 7X, RG 326, ANL, NARA; Groves to Gen. Brereton, Nov. 3, 1947, 601 files, Series 5, RG 77, MED; Arthur H. Frye, Jr., to Gen. Groves, Oct. 24, 1946, Folder 6, Box 3 of 5, W. Zinn, Box 19, RG 326, ANL, NARA.

17. Statement by the president, Oct. 28, 1946, Folder "AEC Wash.–Commissioners 1946–1951," Box 4 of 4, ANL Zinn Reading Files 1946–1955 & Misc., Box 32X, RG 326, ANL, NARA; Daniels to David E. Lilienthal, Oct. 30, 1946, Personal FD, Box 2 of 3, Farrington Daniels, Box 8, RG 326, ANL, NARA; Daniels to Compton, Box 7, Series 2, Argonne Lab, AHC.

18. Donald F. Wood to director, Argonne National Laboratory, Dec. 12, 1946, Box

4 of 5, W. Zinn, AEC Wash.–Misc. 1949–1953, Box 18, RG 326, ANL, NARA; Original ANL contract, W-31-109-ENG-38, Nov. 22, 1946, untitled folder, UCVP ANL, ANL.

19. Hewlett and Anderson, *The New World*, p. 638.

20. David E. Lilienthal, *The Atomic Energy Years 1945–1950*, Vol. 2 of *The Journal of David E. Lilienthal* (New York: Harper & Row), p. 114; Report to the Joint Committee on Atomic Energy on the Acquisition of the Du Page County Site for the Argonne National Laboratory, p. 7, Gen. Corres., Box 6, ANL–Du Page Files, JCAE, RG 128, NARA; Argonne National Laboratory Report for November 1946, Part 1, Hist. File–1946, Drawer 1, Cabinet V23, ANL, TIC; Neil H. Jacoby, interview with the Atomic Energy Commission, Nov. 19, 1946, Folder 1, Box 29, Presidents' Papers 1951–60, UCRL SC; Memorandum of meeting of the commission, Nov. 21, 1946, History Notes, Vol. 1, Folder 27, Box 6, Job 6263, AEC Secretariat, RG 326, DOE.

21. Office of the Acting General Counsel to the Commissioners, January 21, 1947, Footnotes 1945–1963 (Chapters 1–3), Box 1007, Footnote Files, Vol. 2, AEC History, RG 326, AEC, DOE; Atomic Energy Commission Minutes of Meeting No. 24, January 23, 1947, AEC Minutes, RG 326, AEC, DOE.

22. C. A. Benz to Hoylande Young, July 22, 1947, Box D-913:83, ANL History File, Program Correspondence Files, unprocessed ANL records, Information Services, RG 326, NARA-GL. *The Argonne Bulletin*, Feb. 20, 1948, p. 2; Leonard Greenbaum, *A Special Interest* (Ann Arbor: University of Michigan Press, 1971), pp. 23–24.

23. E. E. Huddleson, Jr., to Carroll L. Wilson, Feb. 7, 1947, with notations by Wilson, AEC 601 ANL (12–3–46), Box 11233, Job 4327, AEC Secretariat, RG 326, DOE; *Chicago Daily News*, Feb. 21, 1947; *Chicago Journal of Commerce*, Feb. 24, 1947.

24. N. Hilberry, Notes on meeting of Laboratory Executive Committee of February 18, 1947, Folder "2/18/47," Box 3 of 6, ANL Reading File, Box 12X, RG 326, ANL, NARA; Area Engineer's Office press release for Feb. 20, 1947, Gen. Corres., Box 5, ANL–Du Page Files, JCAE, RG 128, ANL, NARA; E. E. Huddleson, Jr., to Carroll L. Wilson, Feb. 7, 1947, AEC 601 ANL (12–3–46), Box 11233, Job 4327, RG 326, AEC Secretariat, DOE; ANL Report for Feb. 1947, ANL-WHZ-74, March 7, 1947, Footnotes 1945–1963 (Chapters 1–3), Box 1007, Footnote Files, Vol. 2, AEC History, RG 326, DOE; *Chicago Daily News*, Feb. 21, 1947; Charles H. Elston et al., Report of the Subcommittee on Du Page County Site to the Joint Committee on Atomic Energy, June 17, 1948, p. 8, Du Page Laboratory Site, Entry 128, McMahon File, JCAE, RG 128, NARA.

25. *Chicago Daily News*, Feb. 21, 1947; *Chicago Journal of Commerce*, Feb. 24, 1947; Erwin O. Freund to Argonne National Laboratory, April 2, 1947, Folder 4, Box 221, President's Papers 1951–60, UCRL SC; Argonne National Laboratory Building Plans, Report 2, July 22, 1947, untitled folder, Box 3 of 5, W. Zinn, Box 19, RG 326, ANL, NARA.

26. Argonne National Laboratory Building Plans, Report 2, July 22, 1947, untitled folder, Box 3 of 5, W. Zinn, Box 19, RG 326, ANL, NARA.

27. "Acquisition of Permanent Site of Argonne National Laboratory," draft, April 7, 1947, and C. Wayland Brooks and Chauncey W. Reed to David E. Lilienthal, May 5, 1947, both in Gen. Corres., Box 5, ANL–Du Page Files, JCAE, RG 128, NARA; Minutes of AEC Meeting no. 45, May 6, 1947, RG 326, AEC Minutes, DOE; David E. Lilienthal

to Sen. Brooks and Rep. Reed, May 7, 1947, Gen. Corres., Box 5, ANL–Du Page Files, JCAE, RG 128, NARA.

28. C. A. Benz to Hoylande Young, July 22, 1947, Box D-913:83, ANL History File, Information Services, Program Correspondence Files, ANL unprocessed records, RG 326, NARA-GL; N. Hilberry, Minutes of the meeting of the Board of Governors, Argonne National Laboratory, December 1, 1947, Folder "OK for declass.–ENP, date only & non-ANL," Box 3 of 5, Hilberry Special and Important Historical E. N. Pettitt, Box 37, RG 326, ANL, NARA; N. Hilberry, Minutes of Laboratory Executive Committee meeting, May 20, 1947, Lab Council Minutes, OPA, ANL.

29. B. B. Hickenlooper to Hon. C. Wayland Brooks and Hon. Arthur H. Vandenberg, May 16, 1947, Gen. Corres., Box 5, ANL–Du Page Files, JCAE, RG 128, NARA; Col. Arthur H. Frye, Jr., to Jack Derry, May 27, 1947, and attached E. R. Fleury to Zinn, May 26, 1947, AEC 601 ANL (12–3–46), Box 11233, Job 4327, AEC Secretariat, RG 326, DOE; AT-40-1-Gen-42 Letter of Intent, UCVP ANL, Contracts AT-40-1-GEN-42, AEC-ANL Construction (Du Page), ANL; Univ. Chicago Press Release, June 20, 1947, Folder "April through July 1947," Box 2 of 3, Zinn Chron. File, Box 71X, RG 326, ANL, NARA; Report of the Subcommittee on Du Page County to the Joint Committee on Atomic Energy, pp. 11–12, Gen. Corres., Box 5, ANL–Du Page Files, JCAE, RG 128, NARA.

30. E. R. Fleury to Zinn, April 14, 1947, Folder "April thru July 1947," Box 2 of 3, Zinn Chron. File, Box 71X, RG 326, ANL, NARA; Zinn to Tammaro, July 14, 1927, Folder "Radiologic Physics Division 7/45–9/48," Box 4 of 4, W. Zinn, Box 14, RG 326, ANL, NARA; D. E. Wallace to Zinn, July 23, 1948, Folder "Radiologic Physics Division 7/45–9/48," Box 4 of 4, W. Zinn, RG 326, ANL, Box 14, NARA; Report to the Congress by the United States Atomic Energy Commission, Box 172, AEC Correspondence 7/46–1/47, JRO, LC MSS; The Argonne National Laboratory Report for January 1947, p. 3, Hist. File–1947, Drawer 1, Cabinet V23, ANL, TIC.

31. N. Hilberry, Minutes of Laboratory Executive Committee meeting, Feb. 11, 1947, Folder "2/11/47," Box 3 of 6, ANL Reading File, Box 12X, RG 326, ANL, NARA; J. J. Nickson, Waste Disposal and Proposal, March 21, 1947, Folder "3/21/47," Box 4 of 6, ANL Reading File, Box 12X, RG 326, ANL, NARA; J. E. Rose to Zinn, July 10, 1947, Folder "Radiologic Physics Division 7/45–9/48," Box 4 of 4, W. Zinn, Box 14, RG 326, ANL, NARA.

32. Hewlett and Duncan, *Atomic Shield*, pp. 27–28.

33. J. W. Hinkley et al., "Report of the Advisory Board on Relationships of the Atomic Energy Commission with Its Contractors," June 30, 1947, Folder 1, Box 33, VPSP, UCRL SC; T. S. Chapman, Minutes of Laboratory Executive Committee meeting, July 27, 1948, PD, ANL. The quotations are from the minutes of the Laboratory Executive Committee meeting.

34. Atomic Energy Commission Instruction GM-42, Organization and Management, Office of Chicago Directed Operations, Folder "COO-General 10/45–12/50," Box 1 of 5, AEC, W. Zinn, Box 18, RG 326, ANL, NARA; *The Argonne Bulletin*, Jan. 2, 1948, p. 1.

35. AEC, "Atomic Energy and the Physical Sciences," Seventh Semiannual Report (Washington, D.C.: GPO, 1950), pp. 182–84.

36. United States Atomic Energy Commission, "Letter from the Chairman and Members of the United States Atomic Energy Commission," in U.S. Sen., 80th Cong., 1st Sess., Doc. 96, pp. 5, 9, 11.

37. N. Hilberry, Minutes of Laboratory Executive Committee meeting of January 23, 1947, Footnotes 1945–1963 (Chapters 1–3), Box 1007, Footnote Files, Vol. 2, AEC History, RG 326, DOE.

38. Hewlett and Duncan, *Atomic Shield,* p. 29.

39. Ibid., pp. 29–30.

40. Report of the General Advisory Committee "Atomic Energy Commission Support of Basic Science," October 5, 1947, Footnote 29 Chapter 3 to Footnote 32 Chapter 5, Footnote 1947, Box 1008, Footnote Files, Vol. 2, AEC History, RG 326, DOE.

41. Proposal for Reactor Development Committee, October 1, 1947, AEC-Reactor Dev. Div. Reports, Box 175, JRO, LC MSS; Atomic Energy Commission Reactor Development Committee Report by the Division of Research, Footnote 19 Chapter 3 to Footnote 32 Chapter 5, Footnote 1947, Box 1008, Footnote Files, Vol. 2, AEC History, RG 326, DOE.

42. Atomic Energy Commission, Authorization of Fast Reactor—Argonne National Laboratory (Revision), Nov. 24, 1947, unique document no. SAA2000B69A001, Folder 6, Box 11233, Job 4327, AEC Secretariat, RG 326, AEC, DOE; N. Hilberry, Minutes of Executive Laboratory Committee meeting of Nov. 18, 1947, PD, ANL; AEC Minutes of Meeting no. 124, Nov. 19, 1947, AEC minutes, RG 326, DOE.

43. Hewlett and Anderson, *The New World,* p. 480; Hewlett and Duncan, *Atomic Shield,* 89, pp. 333–34.

44. N. Hilberry, Minutes of Laboratory Executive Committee meeting of Nov. 25, 1947, PD, ANL.

45. Atomic Energy Commission Instruction GM-42, I.1., p. 1, Folder "COO-General 10/45–12/50," Box 1 of 5, W. Zinn, Box 17, AEC Secretariat, RG 326, ANL, NARA.

46. Hoylande D. Young to H. C. Baldwin, Nov. 28, 1947, Folder "11/24–28/47," Box 5 of 6, ANL Reading File, Box 12X, RG 326, ANL, NARA; T. S. Chapman to David Saxe, Aug. 9, 1948, Folder "T. S. Chapman, Asst. Director, 10/47–10/48," Box 3 of 4, W. Zinn, Box 21, RG 326, ANL, NARA.

47. J. C. Franklin to all Clinton Laboratory employees, December 31, 1947, untitled folder, Box 4 of 4, Zinn Reading Files 1946–1955 and Misc., Box 32X, RG 326, ANL, NARA; AEC Press Release no. 80, Dec. 31, 1947, Footnotes 3 through 46 of Chapter 6, Footnote Files, Vol. 2, AEC History, RG 326, DOE; Oppenheimer to David E. Lilienthal, Jan. 2, 1948, Lilienthal, David E.—From J. R. Oppenheimer, Box 46, JRO, LC MSS; Oppenheimer to members of the General Advisory Committee, Jan. 3, 1948, Footnotes 3 through 46 of Chapter 6, Footnote Files, Vol. 2, AEC History, RG 326, DOE; J. C. Franklin to all Clinton Laboratory employees, Dec. 31, 1947, Zinn Reading Files 1946–1955, Box 32X, RG 326, ANL, NARA.

48. Atomic Energy Commission, Washington, D.C., Press Release no. 80, Dec. 31, 1947, for release on Jan. 1, 1948, Footnote 3 through 46 of Chapter 6, Footnote Files, Vol. 2, AEC History, RG 326, DOE.

49. H. Etherington to H. E. Metcalf, Jan. 5, 1948, untitled folder, Box 4 of 4, Zinn Reading Files 1946–1955 & Misc., Box 32X, RG 326, ANL, NARA; N. Hilberry, Minutes of Laboratory Executive meeting, Feb. 10, 1948, PD, ANL; Hewlett and Duncan, *Atomic Shield*, pp. 121–26.

50. Zinn to W. B. Harrell, Feb. 4, 1948, Folder "T. S. Chapman, Assistant Director, 10/47–10/48," Box 3 of 4, W. Zinn, Box 21, RG 326, NARA; Zinn to Harrell, May 1, 1948, Hist. File–1948, Drawer 1, Cabinet V23, ANL, TIC; N. Hilberry, Minutes of Argonne National Laboratory Board of Governors meeting, May 2, 1948, Hist. File–1948, Drawer 1, Cabinet V23, ANL, TIC.

51. Hilberry, *Elements of Basic Management Philosophy*, Sept. 22, 1953, p. 3, Folder "J. T. Bobbitt, Asst. Director, 6/50–7/52," Box 3 of 4, W. Zinn, Box 21, RG 326, ANL, NARA; Zinn to Harrell, Jan. 23, 1948, Appendix "A" of Argonne National Laboratory Facilities, Du Page County, Illinois, Folder 8, Box 11233, Job 4327, AEC 97, AEC Secretariat, RG 326, DOE.

52. Hewlett and Duncan, *Atomic Shield*, pp. 29–30; Zinn to J. B. Fisk, July 23, 1948, loose material in box, Box 3 of 6, F. Daniels, Box 6, RG 326, ANL, NARA; Zinn to George L. Weil, April 22, 1948, p. 15, Folder "W. H. Zinn (1) 2/26/46–3/17/50," Box 1 of 4, N. Hilberry & F. Daniels, Box 5, RG 326, ANL, NARA.

53. Zinn to Tammaro, Feb. 25, 1948, and Carroll L. Wilson to Tammaro, March 5, 1948, Folder "Washington-General Manager," AEC, Box 3 of 5, W. Zinn, Box 18, RG 326, ANL, NARA; Tammaro to Fisk, Feb. 26, 1948, Folder "Wash.-DN. of Research 5/47–10/48," AEC, Box 2 of 5, W. Zinn, Box 18, RG 326, ANL, NARA.

54. N. Hilberry, Minutes of Laboratory Executive Committee meeting, Feb. 17, 1948, PD, ANL.

55. "Report of the Subcommittee on Du Page County to the Joint Committee on Atomic Energy," pp. 11–12, Gen. Corres., Box 5, ANL–Du Page Files, JCAE, RG 128, NARA; N. Hilberry, Minutes of Laboratory Executive Committee meeting, March 8, 1948, PD, ANL; Edward B. Hayes to Sen. John W. Bricker, Nov. 4, 1947, and Report of the Subcommittee on Du Page County to the Joint Committee on Atomic Energy, both in Gen. Corres., Box 5, ANL–Du Page Files, JCAE, RG 128, NARA.

56. Argonne National Laboratory Facilities Du Page County Illinois, Folder 8, Box 11233, Job 4327, AEC 97, AEC Secretariat, RG 326, NARA; Minutes of AEC Meeting no. 173, May 12, 1948, AEC Minutes, RG 326, DOE; N. Hilberry, Minutes of the Laboratory Executive Committee meeting of May 18, 1948, PD, ANL.

57. Report of the Subcommittee on Du Page County to the Joint Committee on Atomic Energy, pp. 1, 11, 17, 19–21, Gen. Corres., Box 5, ANL–Du Page Files, JCAE, RG 128, NARA; Digest Chicago Newspapers, no. 8, June 18, 1948, untitled file, Drawer 2, Cabinet 29, ANL, TIC.

58. 60 Stat. 771 (See Sec. 12 (a) (7); 68 Stat. 95 3 (Sec. 172)).

59. N. Hilberry, Minutes of Laboratory Executive Committee meeting of June 22, 1948, and T. S. Chapman, Minutes of Laboratory Executive Committee meeting of July 27, 1948, PD, ANL; Thomas J. Page, Atomic Agriculture for Broadcast 081248, Aug. 12, 1948, untitled folder, Box 1 of 2, Seaborg, Box 9, RG 326, ANL, NARA. *Argonne Bulletin,*

Sept. 10, 1948, p. 1; H. K. Stephenson, Minutes of meeting of Argonne National Laboratory Board of Governors, Dec. 6, 1948, p.3, Hist. File–1948, Drawer 1, Cabinet V23, ANL, TIC; Hilberry to Hafstad, Nov. 28, 1949, p. 26, Folder 9, Box 28, VPSP, UCRL SC.

60. Hewlett and Duncan, *Atomic Shield,* pp. 186–88.
61. Ibid.
62. Ibid., p. 197.
63. Breeder Pile Discussion, June 19–20, 1945, Met Lab Report CF-3199; W. H. Zinn, Design of a Fast Breeder Neutron Pile (Argonne Laboratory, Metallurgical Laboratory, Jan. 25, 1946), Report CF-3414.
64. W. H. Zinn, Feasibility Report—Fast Neutron Test of Conversion (Argonne National Laboratory, Aug. 11, 1947), Report ANL-WHZ-112.
65. Richard G. Hewlett and Francis Duncan, *Nuclear Navy, 1946–1962* (Chicago: University of Chicago Press, 1974), pp. 9–12.
66. Ibid., pp. 35–38.
67. H. G. Rickover, Proposed Bureau of Ships Nuclear Power Program, March 22, 1948, ANL: Hewlett and Duncan, *Atomic Shield,* p. 190; "Nuclear Navy," *Argonne News,* March 6, 1952, p. 3; Paul W. Ager to Carroll L. Wilson, Dec. 11, 1946, Nuclear Prop. of Ships (STR & SIR), Vol. 1, Folder "458.12(12-11-46)," Box 11227, Job 9327, AEC Secretariat, RG 326, DOE; Zinn to H. Etherington, June 17, 1948, ORNL no. 48, Box 2 of 5, W. Zinn, Box 20, RG 326, ANL, NARA; "Nuclear Propulsion of Ships—Status of the Bureau of Ships," April 16, 1948, Nuclear Prop. of Ships (STR & SIR), Vol. 1, AEC 47/1, Folder "458.12(12-11-46)" Box 112227, Job 4327, RG 326, AEC Secretariat, DOE.
68. T. A. Solberg, Report of conference held at Argonne National Laboratory, May 4, 1948, AEC; Hewlett and Duncan, *Atomic Shield,* pp. 191–92.
69. Hewlett and Duncan, *Nuclear Navy,* pp. 68–69.
70. Ibid., pp. 79–82.
71. Tammaro to Westinghouse, letter contract, Dec. 10, 1948, AEC.
72. Neil H. Jacoby, interview with the Atomic Energy Commission, Nov. 19, 1946, Folder 1, Box 29, Presidents' Papers 1951–60, UCRL SC; N. Hilberry, Minutes of Argonne National Laboratory Board of Governors meeting, Dec. 2, 1946, Box 7, Series 2, Argonne Lab, AHC; Argonne National Laboratory Building Plans, Report 2, untitled folder, Box 3 of 5, W. Zinn, Box 19, RG 326, ANL, NARA; notes of A. L. Hughes on meeting of representatives of institutions participating in the Argonne National Laboratory, May 8, 1947, Box 6, Series 3, AHC.
73. Hoylande D. Young to Edward R. Trapnell, Nov. 26, 1947, Corridor 3-12, Drawer 1, ANL Historical Material, ANL-Univ. Interactions (1946), ANL, TIC.
74. N. Hilberry, Minutes of Argonne National Laboratory Board of Governors meeting, March 13, 1948, pp. 13–15, Hist. File–1948, Drawer 1, Cabinet V23, ANL, TIC; Information for the General Advisory Committee meeting April 5–6, 1949, Box 6, William B. Harrell Papers, UCRL SC.
75. J. C. Boyce to Zinn, Aug. 8, 1951, Folder "AEC Wash.–Commissioners 1946–1951," Box 4 of 4, Zinn Reading Files 1946–1955 & Misc. ANL, Box 32X, RG 326, ANL, NARA; Hewlett and Duncan, *Atomic Shield,* pp. 283–86.

76. N. Hilberry, Minutes of Laboratory Executive Committee meeting, May 5, 1948, p. 2, PD, ANL.

77. Zinn to all members of the scientific staff, July 23, 1948, Folder "Radiologic Physics Division 7/45–9/48," Box 4 of 4, W. Zinn, Box 14, RG 326, ANL, NARA.

78. N. Hilberry, Minutes of Laboratory Executive Committee meetings, March 30, 1948, p. 2, and May 5, 1948, p. 2, PD, ANL.

79. Mazuzan and Walker, *Controlling the Atom,* pp. 40–41.

80. Hilberry to Shields Warren, April 12, 1949, April 11–15, 1949, Box 2 of 4, Met Lab Reading Files, Box 13X, RG 326, ANL, NARA; Dr. Robert Hasterlik, interview by Donald Wood, Feb. 5, 1980, pp. 3, 5, ANL; Austin M. Brues to Paul B. Pearson, Dec. 9, 1949, Bio. & Med. Rsch. Div. Jan. 49–Jun. 50, Box 2 of 5, W. Zinn, Box 22, RG 326, ANL, NARA; Lewis H. Reed to Carroll Wilson, Feb. 25, 1947, Footnotes 1 thru 67, Chapter 8, Footnote Files, Vol. 2, AEC History, DOE; *Announced United States Nuclear Tests July 1945 through December 1983,* NVO-209 (Rev. 4) (Las Vegas: Office of Public Affairs, DOE, Nevada Operations Office, 1984), p. 2. See monthly reports of the laboratory radiological physics division, such as the one from Jan. 19, 1953–Feb. 15, 1953, Hist. File–1954, Drawer 4, Cabinet V23, ANL, TIC; *Argonne Bulletin,* July 30, 1948, p. 1; AEC, *Atomic Energy and the Life Sciences,* Sixth Semiannual Report (Washington, D.C.: GPO, 1949), p. 181.

81. J. E. Rose, Division of Health Physics, Report of activities, Oct. 1948 and Nov. 1948, Folder "Radiologic Physics Division 10/48–6/50," Box 4 of 4, W. Zinn, Box 14, RG 326, ANL, NARA.

82. Zinn to Brues et al., April 19, 1948, and Norbert J. Scully and Brues to Charles N. Rice, July 21, 1948, Bio. & Med. Rsch. Div. Jan. 46–Dec. 48, Box 2 of 5, W. Zinn, Box 22, RG 326, ANL, NARA; John W. White, "Strangest Garden in the World," *Science Digest,* January 1950, pp. 11–15, condensed from *Collier's;* E. L. Powers, Minutes of Division of Biological and Medical Research meeting of January 5, 1951, Bio. & Med. Rsch. Div., Box 2 of 5, W. Zinn, Box 22, RG 326, ANL, NARA.

83. Argonne Cancer Research Hospital, AEC 26/1, July 1, 1948, Footnotes 1–67 Chapter 8, Footnote Files, AEC History, Vol. 2, DOE; Minutes of meeting of Argonne National Laboratory Board of Governors, pp. 15–16, Oct. 4, 1948, Drawer 1, Cabinet V23, Hist. File–1948, ANL, TIC; Dr. Robert Hasterlik, interview by Donald Wood, Feb. 5, 1980, p. 2, OPA, ANL.

84. Shields Warren to Lowell T. Coggeshall, Feb. 19, 1948, Addenda 89-34, 4 (ACC), President's Papers, UCRL SC; For additional information on Maria Goeppert-Mayer and her Nobel Prize see Chapter 7.

85. F. C. Hoyt to Zinn, December 20, 1948, Dec. 20–24, 1948, Box 1 of 4, Met Lab Reading Files, Box 13X, RG 326, ANL, NARA; Information for the General Advisory Committee meeting, April 5–6, 1949, p. 14, Box 6, William B. Harrell Papers, UCRL SC; AEC Film Footage Program, Argonne National Laboratory, no. 43, Film Footage Script Notes 1951–1952 B, Box 1 of 4, N. Hilberry & F. Daniels, Box 5, RG 326, ANL, NARA; AEC, "Assuring Public Safety in Continental Weapons Tests," Thirteenth Semiannual Report (Washington, D.C.: GPO, 1953), p. 37.

86. Minutes of the Executive Committee meeting of the Atomic Scientists of Chicago, Jan. 14, 1947, Folder 2, Box 7, John A. Simpson Papers, UCRL SC; *Argonne Bulletin,* May 14, 1948, p. 2, April 29, 1949, p. 100, and Sept. 3, 1948, p. 194; *Phys. Rev.* 74(August 1, 1948): 235 and 76(July 1, 1959): 185; Louis Haber, *Women Pioneers of Science* (New York: Harcourt Brace Jovanovich, 1979), p. 93; *Argonne News,* Nov. 1963, p. 3; AEC, *Fifth Semiannual Report of the Atomic Energy Commission* (Washington, D.C.: GPO, 1949), p. 50.

87. Maria Mayer to Farrington Daniels, n.d. and not sent, PD, ANL; Robert G. Sachs, "Maria Goeppert-Mayer—Two-Fold Pioneer," *Physics Today,* February 1981, pp. 46–47; Joan Dash, *The Triumph of Discovery. Women Scientists Who Won the Nobel Prize* (New Jersey: Julian Messner, 1991), p. 7; Nichols to Groves, Sept. 5, 1945, RG 77, MED, Entry 5, 600.12, NARA; Maria Mayer to Daniels, n.d. and not sent, PD, ANL; Daniels to J. H. Mahoney, Feb. 18, 1946, PD, ANL; Argonne National Laboratory organization chart, Oct. 1, 1946, War Time Files, Series 2, Box 7, AHC.

88. Sachs, "Maria Goeppert-Mayer—Two-Fold Pioneer."

89. Information received by author on March 3, 1993, from Roger Mead, archivist, Los Alamos National Laboratory; Rodney P. Carlisle and August W. Giebelhaus, *Bartlesville Energy Center. The Federal Government in Petroleum Research 1918–1983,* DOE/BC/10126-1 (DOE, 1985), pp. 49–50.

90. *Argonne Bulletin,* Aug. 12, 1949, p. 144; Contract W-31-109-ENG-38, original ANL contract, Article XXI–Antidiscrimination, p. 18, untitled folder, UCVP ANL, ANL; Los Alamos Scientific Laboratory contract, Article XXIII–Antidiscrimination, Los Alamos National Laboratory Archive, A-83-033, Los Alamos National Laboratory; *Ebony,* Sept. 1949, pp. 24–26.

91. "Exhibit 'C' Biographical Data of Key Administrative and Technical Staff Members, Argonne National Laboratory," pp. 61–62, Biograph. Data of Key Adm. and Tech. Staff Members, Exhibit C, Box 2A, Massey Subj. Files, ANL; Zinn to James F. Corwin, Dec. 29, 1952, Information Division Nov. 1952–July 1953, Box 4 of 4, W. Zinn, Box 10, RG 326, ANL, NARA.

92. Margaret Nickson, Report no. CH-3831, "A Study of the Hands of Radiologists," June 1, 1947, Box 3 of 4, Met Lab C-Series Reports, Box 64X, RG 326, ANL, NARA.

93. *American Men and Women of Science,* 16th ed., 1986, OPA, ANL.

94. *Congressional Record,* 80th Cong., 2nd sess., 5166, 8953, 9034, 9067, 9070, 9073; Lilienthal, *Journals,* pp. 334, 337, 359–60.

Chapter 3

1. Hewlett and Duncan, *Atomic Shield,* pp. 326–32, 356.
2. Ibid., p. 357.
3. Ibid.
4. W. M. Manning and R. P. Rogers to T. S. Chapman, Nov. 3, 1948, Folder "Nov. 1–5, 1948," Box 1 of 4, Met Lab Reading Files, Box 13X, RG 326, ANL, NARA; *The Argonne Bulletin,* November 14, 1947, p. 2.
5. David E. Lilienthal to Sen. Brien McMahon, May 25, 1949, AEC Wash., Box 4 of

5, W. Zinn, Box 18, RG 326, ANL, NARA; AEC Meeting no. 264, May 4, 1949, AEC Minutes, RG 326, DOE; press release from the office of Sen. Brien McMahon, Press Releases JCAE-1949, Gen. Corres., Box 516, JCAE, RG 128, NARA; tentative draft of Argonne National Laboratory statement for press, June 1, 1949, Folder "Uranium, Missing and Associated Corr.," Box 2 of 4, N. Hilberry & F. Daniels, Box 5, RG 326, ANL, NARA.

6. N. Hilberry, Minutes of Laboratory Executive Committee meetings, May 25 and June 14, 1949, PD, ANL; AEC Meeting no. 264, May 4, 1949, RG 326, AEC minutes, DOE; press release from the office of Sen. Brien McMahon, Press Releases JCAE-1949, Gen. Corres., Box 516, JCAE, RG 128, NARA; AEC Meeting no. 298, August 3, 1949, AEC Minutes, RG 326, DOE.

7. Paul Fields, interview by Ruth R. Harris, Dec. 7, 1992, Argonne National Laboratory, p. 15.

8. N. Hilberry, Minutes of Laboratory Executive Committee meetings, May 25 and June 14, 1949, PD, ANL; Lilienthal to McMahon, June 15, 1949, Gen. Corres., Box 4, JCAE, RG 128, ANL, NARA.

9. Ernest W. Thiele, "Report to the Joint Congressional Committee on Atomic Energy on the Disappearance of Enriched Uranium at the Argonne National Laboratory," July 1, 1949, Folder "Ernest W. Thiele's Investigation for AEC of Missing Uranium from ANL," Box 2 of 4, N. Hilberry and F. Daniels, Box 5, RG 326, ANL, NARA; press release from the offices of Sen. Brien McMahon, July 11, 1949, Press Releases JCAE-1949, Gen. Corres., Box 516, JCAE, RG 128, NARA.

10. N. Hilberry, Minutes of Laboratory Executive Committee meetings, Oct. 28, 1947, and May 25 and June 14, 1949, PD, ANL; AEC Meeting no. 264, May 4, 1949, AEC Minutes, RG 326, DOE; press release from the office of Sen. Brien McMahon, Press Releases JCAE-1949, Gen. Corres., Box 516, JCAE, RG 128, NARA; AEC Meeting no. 298, August 3, 1949, AEC Minutes, RG 326, DOE; J. T. Bobbitt to Zinn, July 27, 1949, Folder "July 25-29, 1949," Box 3 of 4, Met Lab Reading Files, Box 13X, RG 326, ANL, NARA; "Exhibit 'C' Biographical Data of Key Administrative and Technical Staff Members Argonne National Laboratory," p. 11, Folder "Exhibit C, Biograph. Data of Key Adm. and Tech. Staff members ANL," Walter Massey Subj. Files, Box 2A, ANL.

11. N. Hilberry, Minutes of Laboratory Executive Committee meeting, May 10, 1949, p. 2, PD, ANL.

12. Argonne National Laboratory Facilities, Du Page County, Illinois, Feb. 7, 1949, Staff Paper no. AEC 97/2, AEC Report, W. Zinn, Box 19, RG 326, ANL, NARA; Argonne National Laboratory Facilities Du Page County, Illinois–AEC 97 Series, Status Report Extract, 3/1–3/15/49, and Carroll L. Wilson to Zinn, Feb. 24, 1949, both in Folder 8, Box 11233, Job 4327, AEC Secretariat, RG 326, DOE.

13. Information for the General Advisory Committee Meeting, April 5–6, 1949, Box 6, William B. Harrell Papers, UCRL SC.

14. N. Hilberry, Minutes of Laboratory Executive Committee meeting, April 19, 1949, PD, ANL; T. S. Chapman to Zinn, April 15, 1949, Folder "T. S. Chapman, Asst. Director, 11/48–11/49," Box 4 of 4, W. Zinn, Box 21, RG 326, ANL, NARA.

15. Tammaro to Zinn, May 6, 1949, Folder "Budget Division 1/49–12/49," Box 4 of 4, W. Zinn, Box 16, RG 326, ANL, NARA; AEC Meeting no. 269, May 12, 1949, p. 178, AEC Minutes, RG 326, DOE.

16. N. Hilberry, Minutes of Laboratory Executive Committee meeting, July 5, 1949, PD, ANL; "Service Irradiations," Folder "J. C. Boyce-Associate Director 11/52–8/53," Box 4 of 4, Zinn Reading Files 1946–1955 & Misc., Box 32X, RG 326, ANL, NARA.

17. Press Release no. 152, Jan. 16, 1949, Folder "AEC Wash.–Div. of Reactor Dev. 1949," Box 2 of 5, W. Zinn, Box 18, RG 326, ANL, NARA; N. Hilberry, Minutes of Argonne National Laboratory Executive Committee meeting, Jan. 25, 1949, PD, ANL; Hewlett and Duncan, *Atomic Shield,* pp. 209–10.

18. Hewlett and Duncan, *Atomic Shield,* p. 210; Location of Experimental Fast Breeder Reactor, Staff Paper no. AEC 79/2, Sept. 12, 1949, 412.13 ANL (11-13-47), Folder 6, Box 11233, Job 4327, AEC Secretariat, RG 326, DOE; Statement of Conditions for the Experimental Breeder Reactor Project, Oct. 13, 1949, Idaho Ops. Office, Box 1 of 5, Box 5 of 5 to Box 19, W. Zinn, Box 18, RG 326, ANL, NARA; Status of Reactor Projects, Oct. 17, 1949, Staff Paper no. AEC 152/4, 412.13 ANL (11-13-47), Folder 6, Box 11233, Job 4327, RG 326, AEC Secretariat, DOE.

19. *Argonne Bulletin,* March 25, 1949, p. 3; Acquisition of Reactor Testing Station, Aug. 1, 1949, Staff Paper no. AEC 142/11, Footnotes Chapter 7 Footnotes 23–86 Jan.–Dec. 1949, March–Dec. 1948, Box 1011, Footnote Files, AEC History, Vol. 2, RG 326, DOE; J. R. Huffman to Zinn, June 21, 1949, Folder "Materials Testing Reactor Project, May–June 1949," Box 1 of 4, Zinn Reading Files, Box 73X, RG 326, ANL, NARA; Stuart McLain to L. E. Johnston, July 29, 1949, Folder "Materials Testing Reactor Project, 7/49–10/49," Box 1 of 4, Zinn Reading Files, Box 73X, RG 326, ANL, NARA.

20. Zinn to Lawrence R. Hafstad, Feb. 26, 1949, Folder "AEC Wash.–Div. of Reactor Dev. 1949," Box 2 of 5, W. Zinn, Box 18, RG 326, ANL, NARA; Zinn to Tammaro, Aug. 24, 1949, Folder "CP-4 (EBR-I) Questionnaire, Corres.: Declass. General," Box 4 of 4, Reactor Hist. Files, Box 20X, RG 326, ANL, NARA.

21. Hafstad to Zinn, Oct. 13, 1949 draft, and Hafstad to L. E. Johnston, Nov. 2, 1949, both in Folder "AEC Wash.–Div. of Reactor Dev. 1949," Box 2 of 5, W. Zinn, Box 18, RG 326, ANL, NARA; Hafstad to Johnston, Nov. 22, 1949, 412.13 ANL(11-13-47), Folder 6, Box 11233, Job 4327, AEC Secretariat, RG 326, DOE.

22. Zinn to Hafstad, rough draft, n.d., confirming telephone conversation of Jan. 27, 1950, Folder "AEC Wash.–Division of Reactor Development 1950," Box 1 of 5, W. Zinn, Box 5 of 5, W. Zinn, Box 17, Box 18, RG 326, ANL, NARA; Stuart McLain to Zinn, May 12, 1950, May 8–12, 1950, Box 1 of 4, Met Lab Reading Files, Box 14X, RG 326, ANL, NARA; H. M. Mott-Smith to William L. Borden, June 27, 1950, NRTS-Idaho 1949–1951, Gen. Corres., Box 444, JCAE, RG 128, NARA; AEC Press Release no. 209, Sept. 23, 1949, DOE Historian's Office Archives; Location of Experimental Fast Breeder Reactor, Sept. 12, 1949, Staff paper no. AEC 79/2, 412.13 ANL (11-13-47), Folder 6, Box 11233, Job 4327, AEC Secretariat, RG 326, DOE; Statement of Conditions for the Experimental Breeder Reactor Project, Oct. 13, 1949, Folder "Idaho Ops. Office," Box 1 of 5, Box 5 of 5 to Box 19, W. Zinn, Box 18, RG 326, ANL, NARA; Status of Reactor Projects,

Oct. 17, 1949, Staff Paper no. AEC 152/4, 412.13 ANL (11-13-47), Folder 6, Box 11233, Job 4327, AEC Secretariat, RG 326, DOE; S. McLain to Zinn, Nov. 23, 1949, Folder "AEC Wash.–Div. of Reactor Dev. 1949," Box 2 of 5, W. Zinn, Box 18, RG 326, ANL, AEC; Stuart McLain, Minutes of Materials Testing Reactor Committee, Dec. 22, 1949, Folder "Dec. 27–30, 1949," Box 4 of 4, Met Lab Reading File, Box 13X, RG 326, ANL, NARA.

23. Albert Wattenberg, interview by Ruth Harris, p. 49, April 24, 1992, Washington, D.C.; Zinn, Material Submitted for Seventh Semiannual Report, 1949, Folder "AEC Wash.–Seventh Semiannual Report to Congress," Box 1 of 5, W. Zinn, Box 18 to 19, RG 326, ANL, NARA.

24. Information for the General Advisory Committee meeting April 5–6, 1949, pp. 16–17, Box 6, William B. Harrell Papers, UCRL SC; N. Hilberry, Minutes of Laboratory Executive Committee meeting, May 10, 1949, pp. 6–8, PD, ANL.

25. Hewlett and Duncan, *Nuclear Navy*, pp. 106–7; Zinn to Harrell, Dec. 19, 1949, Folder "Boiling Reactor Experiments, CP-5 6/48–9/50," Box 3 of 4, Zinn Chron. Files, Box 74X, RG 326, ANL, NARA; Hafstad to Zinn, October 25, 1949, Folder 101 "AEC Wash.–Div. of Reactor Dev., 1949," Box 2 of 4, W. Zinn, Box 10, RG 326, ANL, NARA; Zinn to Hafstad, Oct. 13, 1949, Staff Paper no. AEC 152/5, Oct. 26, 1949, Folder 6, Box 11233, Job 4327, AEC Secretariat, RG 326, DOE; Zinn to Tammaro, Dec. 20, 1949, Box 3 of 4, N. Hilberry, Folder 506, Box 3, RG 326, ANL, NARA. See also Hewlett and Duncan, *Nuclear Navy*, pp. 71–72.

26. W. P. Bigler to J. H. McKinley, Dec. 3, 1948, Folder "Naval Reactor Div. 4/48–7/49," Box 2 of 4, W. Zinn, Box 14, RG 326, ANL, NARA; *Argonne Bulletin*, Oct. 8, 1948, p. 214; H. Etherington, "Program for Development of a Nuclear Reactor for Submarine Propulsion," Feb. 17, 1949, Folder "Feb. 14–18, 1949," Box 2 of 4, Met Lab Reading Files, Box 13X, RG 326, ANL, NARA; H. Etherington to Zinn, Aug. 16, 1949, Folder ANL-HE-190, ANL-HE-170 to ANL-HE-199, ANL-HE-249, Box 1 of 5, ANL-HE-136 to Box 30, Box 5 of 5, Etherington Files, Box 29, RG 326, ANL, NARA; "Progress on Reactor Projects," Sept. 20, 1949, 458.12 (12-11-46), Nuclear Prop. of Ships (STR & SIR), Vol. 1, Box 11227, Job 4327, AEC Secretariat, RG 326, DOE; Hewlett and Duncan, *Nuclear Navy*, pp. 144–45.

27. Hewlett and Duncan, *Atomic Shield*, pp. 216–17; Richard G. Hewlett, "The Experimental Breeder Reactor No. 1: The Life Story of a Nuclear Reactor," unpublished ms., pp. 65–70, DOE.

28. "Report of Program of Chemistry Division" and attached "Report of Program of Biology Division," Nuclear Chemistry Section, ANL-HKS-3, 1949, Hist. File–1949, Drawer 2, File V-23, ANL, TIC; Austin M. Brues to Paul B. Pearson, Dec. 9, 1949, Folder 10, Box 28, VPSP, UCRL SC; Tammaro to Hafstad, Oct. 25, 1949, Staff Paper no. AEC 97/5, November 4, 1949, Folder 8, Box 11233, Job 4327, AEC Secretariat, RG 326, DOE; Hilberry to Hafstad, Nov. 28, 1949, p. 28, Folder 9, Box 28, VPSP, RG 326, UCRL SC; Zinn, Seventh Semiannual Report, Argonne National Laboratory, Folder "AEC Wash.–Seventh Semiannual Report to Congress," Box 1 of 5, W. Zinn, Boxes 18–19, RG 326, ANL, NARA.

29. *Atomic Energy and the Life Sciences,* Sixth Semiannual Report (Washington,

D.C.: GPO, 1949), pp. 90–91; AEC, "Major Activities in the Atomic Energy Programs. Jan.–June 1953." (Washington, D.C.: GPO, 1953), pp. 37–39; Dr. Robert Hasterlik, interview by Donald Wood, Feb. 5, 1979, p. 2, OPA, ANL.

30. Argonne National Laboratory, *Review of Non-Technical Activities for February 1950,* March 15, 1950, p. 3, and *Review of Non-Technical Activities for May 1950,* June 15, 1950, p. 1, ANL A. M. Brues-Misc., loose material, Box 2 of 4, Reading File, Box 36, RG 326, ANL, NARA.

31. Information for the General Advisory Committee Meeting April 5–6, 1949, Box 6, William B. Harrell Papers, UCRL SC; Hilberry to Hafstad, Nov. 26, 1949, p. 30, Folder 9, Box 28, VPSP, UCRL SC.

32. Minutes of Special Meeting of Board of Governors, Nov. 7, 1949, pp. 7, 10, Footnotes 1–62, Chapter 13, Box 1017, Footnote Files, Vol. 2, AEC History, RG 326, DOE; Hilberry to Hafstad, Nov. 26, 1949, p. 30, Folder 9, Box 28, VPSP, UCRL SC.

33. Minutes of Special Meeting of Board of Governors, Nov. 7, 1949, pp. 7, 10, Footnotes 1–62, Chapter 13, Box 1017, Footnote Files, Vol. 2, AEC History, RG 326, DOE; Hilberry to Hafstad, Nov. 26, 1949, p. 30, Box 28, Folder 9, VPSP, UCRL SC.

34. D. B. Langmuir, notes on discussion of national laboratory problems, Nov. 18, 1949, Op. & Mgmt. of Natl. Labs, 635.123 (9-26-49), Box 11230, Job 4327, Executive Secretariat, RG 326, DOE; Harrell to R. W. Harrison, Nov. 15, 1949, ANL History, 1946–59, UCVP ANL, ANL.

35. Hilberry to Hafstad, Nov. 28, 1949, pp. 37–39, Folder 9, Box 28, VPSP, UCRL SC.

36. Greenbaum, *A Special Interest,* pp. 40–43.

37. Announcement of Jan. 27, 1950, Folder "J. C. Boyce–Associate Director 8/49–10/52," Box 4 of 4, Zinn Reading Files, 1946–1955 & Misc., Box 32X, RG 326, ANL; "Participating Institution Program," Folder "AEC Wash.–GAC Mtg. at ANL, March 1951," Box 4 of 5, W. Zinn, Box 18, RG 326, ANL, NARA.

38. AEC Meeting no. 346, Dec. 21, 1949, AEC Minutes, RG 326, DOE; Federation of American Scientists resolution, Feb. 4, 1950, Folder 8, Box 28, President's Papers 1951–60, UCRL SC. The legislation upholding sweeping clearance requirements was the Independent Offices Appropriation Act of 1950.

39. Harrell to Robert M. Hutchins, Nov. 15, 1949, Folder "ANL History 1946–59," UCVP ANL, ANL; Zinn to Carleton Shugg, Dec. 3, 1948, Folder "Westinghouse no. 53," Box 2 of 5, W. Zinn, Box 20, RG 326, ANL, NARA.

40. AEC Meeting no. 346, Dec. 21, 1949, AEC Minutes, RG 326, DOE; Federation of American Scientists resolution, Feb. 4, 1950, Folder 8, Box 28, President's Papers 1951–60, UCRL SC.

41. David Saxe to Zinn, Dec. 19, 1949, Folder "Budget Division 1/49–12/49," Box 4 of 4, Box 16, W. Zinn, RG 326, ANL, NARA; McDaniel to Zinn, April 5, 1951, Folder "AEC Wash.–Div of Research 1951–53," Box 2 of 5, W. Zinn, Box 18, RG 326, ANL, NARA.

42. Howard Brown to Medford Evans, July 31, 1950, Folder "Medicine Health & Safety ANL 6/50–12/50," Box 3357, Job 1132, RG 326, DOE; Harrell to J. A. Cunningham, Nov. 2, 1950, W-31-109-ENG-38, Supp. no. 10, UCVP ANL, ANL.

43. Notes on Discussion of National Laboratory Problems, December 19, 1949, Footnotes 1–62 Chapter 13, Box 1017, Footnote Files, Vol. 2, AEC History, RG 326, DOE; Howard Brown and staff, "A Report on AEC Management Practices in Dealing with National Laboratories," March 10, 1950, Box 11230, Job 4327, AEC Secretariat, RG 326, DOE; Hewlett and Duncan, *Atomic Shield,* pp. 433–34.

44. Hafstad quoted in "Operating Policy of the Argonne National Laboratory," semi-processed file no. 17, Policy Advisory Board Meetings, 43.4 15 Oct 195B, RG 326, NARA-GL; AEC Meeting no. 416, June 1, 1950, AEC Minutes, RG 326, DOE. See also AEC Meeting no. 324, DOE.

45. Hewlett and Duncan, *Atomic Shield,* pp. 362–409, 466–67.

46. Ibid., pp. 430–31, 531.

47. Ibid., pp. 428–30, 531; Rodney Carlisle and Joan Zenzen, "Supplying the Nuclear Arsenal: Production Reactor Technology, Management, and Policy 1942–1992," chapter 4, p. 24, unpublished ms. draft.

48. Hilberry, Minutes of Laboratory Executive Committee meetings, July 18 and Aug. 15, 1950, PD, ANL; "Reply to G. L. Weil's questions of July 3, 1950," July 5, 1950, Folder "N. Hilberry & F. Daniels," Box 1 of 4, W. H. Zinn (11) 4/8/50–12/18/52, Box 5, RG 326, ANL, NARA.

49. Zinn to Tammaro, Oct. 16, 1950, Folder "Naval Reactor Div., 7/50–11/50," Box 2 of 4, W. Zinn, Box 14, RG 326, ANL; N. Hilberry, *Elements of Basic Management Philosophy,* Sept. 22, 1953, p. 4, Folder "J. T. Bobbitt, Asst. Director, 6/50–7/52," Box 3 of 4, W. Zinn, Box 21, RG 326, ANL, NARA.

50. N. Hilberry, Minutes of Laboratory Executive Committee meeting, Aug. 15, 1950, PD, ANL.

51. "Report of Trip to Argonne National Laboratory by W. J. Bergin and W. J. Sheehy, August 14–24, 1950," Gen. Corres., Box 4, JCAE, RG 128, NARA; Merle W. Griffith to gentlemen of the press, Feb. 13, 1951, Folder 5, Box 31, VPSP, UCRL SC; *Chicago Daily Tribune,* March 20, 1951, Part 1, p. 10F; Harrell to Tammaro, Sept. 14, 1953, and Tammaro to Harrell, Sept. 16, 1953, Folder 5, Box 31, VPSP, UCRL SC; Ramey quoted in Executive Session of the Subcommittee on Legislation, June 20, 1956, pp. 19–22, 6/20/56, unclassified executive hearings 1956, Box 4, JCAE, RG 128, NARA.

52. "Statement by the Director of Argonne National Laboratory, Feb. 2, 1951, Folder "10/49–12/51, L. C. Furney, Staff Asst.," Box 4 of 4, W. Zinn, Box 21, RG 326, ANL, NARA.

53. N. Hilberry, Minutes of Laboratory Executive Committee meeting, Aug. 15, 1950, PD, ANL; J. C. Boyce, "Courses in Reactor Technology at Site D," Sept. 26, 1950, Folder "J. C. Boyce–Associate Director 8/49–10/52," Box 4 of 4, Zinn Reading Files 1946–1955 & Misc., Box 32X, RG 326, ANL, NARA; Chicago Operations Office Program Assumptions for Fiscal Year 1953 Budget, April 6, 1951, untitled folder, Box 1 of 4, A. M. Brues-Misc. Reading File, Box 36, RG 326, ANL, NARA; M. H. Wahl, Planning Committee Meeting, April 4–5, 1951, April 10, 1951, 4/9–13/51, Box 4 of 4, Met Lab Reading Files, Box 14X, RG 326, ANL, NARA; Hewlett and Duncan, *Atomic Shield,* p. 531; *Argonne National Laboratory Review of Non-Technical Activities December 1950,* Jan. 15,

1951, p.1, Box 2 of 4, A. M. Brues-Misc. Reading File, Box 36, RG 326, ANL, NARA; Warren C. Johnson to R. W. Harrison, Nov. 7, 1952, Folder 8, Box 28, President's Papers 1951–60, UCRL SC; Stuart McLain, "Interim Report on Educational Programs at Argonne National Laboratory," Dec. 11, 1956, ANL History 1946–59, UCVP ANL, ANL.

54. Hewlett and Duncan, *Atomic Shield*, pp. 552–53; N. Hilberry, *Elements of Basic Management Philosophy*, p. 4, Folder "J. T. Bobbitt, Asst. Director, 6/50–7/52," Box 3 of 4, W. Zinn, Box 21, RG 326, ANL, NARA.

55. N. Hilberry, Minutes of Laboratory Executive Committee meeting, Feb. 27, 1951, Folder 504, Emergency Directory to Folder 23, Box 3 of 4, N. Hilberry, Box 3, RG 326, ANL, NARA; S. H. Paine, Jr., to Frank Foote, April 23, 1951, Folder "Metallurgy Division 1/51–7/52," Box 1 of 4, W. Zinn, Box 14, RG 326, ANL, NARA; Hilberry to Harrell, Sept. 17, 1951, untitled folder, Zinn Reading Files 1946–1955 & Misc., Box 4 of 4, Box 32X, RG 326, ANL, NARA.

56. Hewlett and Duncan, *Atomic Shield*, p. 418; Hewlett and Duncan, *Nuclear Navy*, pp. 142–47, 149–50, 231.

57. Hewlett and Duncan, *Atomic Shield*, p. 419.

58. Ibid., pp. 495–98; Hewlett, "The Experimental Breeder Reactor No. 1," pp. 65–80.

59. Hewlett and Duncan, *Atomic Shield*, p. 515; Hewlett and Duncan, *Nuclear Navy*, pp. 182–86.

60. W. B. Allen et al., "Geology and Ground-Water Hydrology of the Argonne National Laboratory Area, Du Page County, Illinois, 1949," File no. 41962 "Radiological Physics," Box 1 of 4, N. Hilberry & F. Daniels, Box 5, RG 326, ANL, NARA.

61. Argonne National Laboratory, *Review of Non-Technical Activities for March 1950*, April 15, 1950, p. 2, loose material, Box 2 of 4, A. M. Brues-Misc. Reading File, Box 36, RG 326, ANL, NARA.

62. Information for the General Advisory Committee meeting, April 5–6, 1949, p. 14, William B. Harrell Papers, UCRL SC.

63. W. M. Manning to Zinn, April 25, 1950, April 24–28, 1950, Box 3 of 4, LDO Reading File, Box 33, RG 326, ANL, NARA.

64. E. L. Powers, Minutes of the Argonne National Laboratory Division of Biological and Medical Research Program Committee meeting, Nov. 17, 1950, Folder "Bio. & Med. Rsch. Div.," Box 2 of 5, W. Zinn, Box 22, RG 326, ANL, NARA; Hasterlik and Brues to Harrell, July 16, 1951, Folder "Biological & Medical Research Division 7/50–1/53," Box 3 of 5, W. Zinn, Box 22, RG 326, ANL, NARA; Dr. Robert E. Rowland, interview by Ruth Harris, Dec. 7, 1992, Argonne National Laboratory.

65. Robert E. Rowland, interview by Ruth Harris, p. 7, Dec. 7, 1992, Argonne National Laboratory; Hasterlik and Brues to Harrell, July 16, 1951, Folder "Medicine, Health & Safety ANL-1951," Box 3357, Job 1132, RG 326, DOE.

66. Dr. Robert Hasterlik, interview by Donald Wood, Feb. 5, 1980, p. 5, OPA, ANL; AEC, "Major Activities in the Atomic Energy Programs. January–June 1952," Twelfth Semiannual Report (Washington, D.C.: GPO, 1952), p. 31.

67. *Argonne Bulletin*, Dec. 30, 1949, pp. 26–27, and July 7, 1950, p. 127.

68. AEC, Seventh Semiannual Report, Jan. 1950, p. 108; AEC, Tenth Semiannual Report, July 1951, p. 31.

69. AEC, *Atomic Energy and the Physical Sciences* (Washington, D.C.: GPO, 1950), p. 121.

70. W. D. Wilkinson to F. G. Foote, June 27, 1951, Folder "Metallurgy Division 1/51–7/52," Box 1 of 4, W. Zinn, Box 14, RG 326, ANL, NARA; Zinn and F. G. Foote to S. McLain et al., March 23, 1954, and A. B. Shuck to Program 1.5.3 file, March 29, 1954, 10-15-52-8-3-54, Met Program 1.5.3, Box 6, 1943–64 (FRCs 38–43) Acces. no. 75AZ55, AEC ANL Metallurgy Program Files, RG 326, NAGL.

71. Organization and By-laws of the Council of Participating Institutions of Argonne National Laboratory, adopted Jan. 8, 1951, AMU History Project, Box 1, William B. Harrell Papers, UCRL SC; W. Harrell, Argonne National Laboratory Operating Contract Negotiations, July 24, 1951, memorandum, ANL Contract W-31-109-ENG-38, Proposed Continuation beyond 6/30/52, UCVP ANL, ANL; Boyce to Zinn, Aug. 1, 1951, and Zinn to Gordon Dean, Aug. 13, 1951, Folder "AEC Wash.–Commissioners 1946–1951," Box 4 of 4, Zinn Reading Files 1946–1955 & Misc., Box 32X, RG 326, ANL, NARA; K. S. Pitzer to Tammaro, Jan. 26, 1951, Folder "AEC Wash–Div. of Reactor Dev., 1951," Box 1 of 5, W. Zinn, Box 5 of 5 to Box 18, W. Zinn, Box 17, RG 326, ANL, NARA.

72. Boyce to Zinn et al., Oct. 31, 1951, and Zinn to Boyce, Dec. 3, 1951, Folder "J. C. Boyce-Associate Director 8/49–10/52," Box 4 of 4, Zinn Reading Files 1946–1955 & Misc., Box 32X, RG 326, ANL, NARA.

73. W. Harrell, Argonne National Laboratory Operating Contract Negotiations, July 24, 1951 memorandum, ANL Contract W-31-109-ENG-38, Proposed Continuation beyond 6/30/52, UCVP ANL, ANL; "Years in the Life of a University," *University of Chicago Magazine,* October 1991.

74. AEC, "Assuring Public Safety in Continental Weapons Test" (Washington, D.C.: GPO, 1953), p. 39; Argonne National Laboratory, *Review of Non-Technical Activities June 1951,* July 15, 1951, pp. 1, 4, *Review of Non-Technical Activities July 1951,* Aug. 15, 1951, p. 3, and *Review of Non-Technical Activities August 1951,* Sept. 17, 1951, p. 2, all three in loose material, Box 3 of 4, A. M. Brues-Misc. Reading File, Box 36, RG 326, ANL, NARA.

75. *Argonne Bulletin,* Dec. 1, 1950, p. 13, March 21, 1952, p. 55, and June 6, 1952, p. 87.

76. Stuart McLain, "Interim Report on Educational Programs at Argonne National Laboratory," Dec. 11, 1956, pp. 4–7, ANL History 1946–59, UCVP ANL, ANL.

77. Ibid., pp. 6–8.

78. Ibid., pp. 12–16.

Chapter 4

1. Original versions of sections relating to Eisenhower Atoms for Peace program, the AEC's Five-Year Program in 1954, and the development of the AEC's peaceful nuclear reactor efforts were first published in Jack M. Holl, Roger M. Anders, and Alice L. Buck, *The United States Civilian Nuclear Power Policy, 1954: A Summary History* (Washington, D.C.: DOE, 1986).

2. Spencer Weart, *Nuclear Fear: A History of Images* (Cambridge, Mass.: Harvard University Press, 1988), pp. 155–69; Mazuzan and Walker, *Controlling the Atom*, pp. 2, 77; Irvin C. Bupp and Jean-Claude Derian, *Light Water: How the Nuclear Dream Dissolved* (New York: Basic Books, 1978), p. 4.

3. Richard G. Hewlett and Jack M. Holl, *Atoms for Peace and War, 1953–1961: Eisenhower and the Atomic Energy Commission* (Berkeley: University of California Press, 1989), pp. 5–16.

4. "Ordinary Bulbs—With a Historic Difference," *Argonne News*, Special Issue, Dec. 21, 1961, p. 3; Hewlett and Holl, *Atoms for Peace and War*, pp. 7–8.

5. Hewlett and Holl, *Atoms for Peace and War*, pp. 7–8.

6. Ibid.

7. Weart, *Nuclear Fear*, pp. 170–79.

8. Greenbaum, *A Special Interest*, p. 56.

9. Walter Zinn, "Conducted Tour of Argonne's Reactor Facilities," March 20, 1954, Folder "W. P. Bigler," W. Zinn Files, Box 4 of 4, RG 77, ANL, NARA; B. I. Spinrad, "The Reactor Engineering Division," *Argonne News*, June 1958, p. 4.

10. Hilberry to J. J. Flaherty, Manager Chicago Operations Office, April 25, 1955, Folder "Norman Hilberry," Walter Zinn Files, Box 3 of 4, RG 77, ANL, NARA; "Subcritical Reactors Use Materials from West Stands Pile," *Argonne News*, May 1956, p. 9; "Last Rites for Two Veteran Reactors," *Argonne News*, Dec. 1956, p. 3.

11. J. R. Huffman, "The Materials Testing Reactor Design," Oct. 1, 1953, Phillips Petroleum Co., Atomic Energy Division, IDO-16121-PPCo., Hist. File 1954, Drawer 4, ANL, TIC.

12. AEC, "Materials Testing Reactor" in *Thumbnail Sketch—National Reactor Testing Station*, Folder "NRTS," Box D-367105, Program Correspondence Files, unprocessed ANL records, RG 326, GLFRC; B. I Spinrad, "The Reactor Engineering Division," *Argonne News*, June 1958, pp. 4–5.

13. H. V. Lichtenberger, "The Experimental Breeder Reactor," n.d., Folder "EBR-I 1/54–1/55," Box 73X, Zinn Reading Files, RG 77, ANL, NARA; Warner E. Unbehaun, "History and Status of the EBR," April 15, 1953, Report no. AECD-3712, Hist. File 1953, Drawer 4, ANL, TIC.

14. "EBR Successfully Demonstrates Process of Breeding Atomic Fuel," *Argonne News*, July 1, 1953, p. 3; "Argonne in Idaho," *Argonne News*, Nov. 1959, p 6. See also, Norman Hilberry, "Why Celebrate the Ten Years of EBR-I," and Leonard J. Koch, "EBR-I: Its Development and Construction," in *Argonne News*, Dec. 21, 1961, Special Issue, pp. 4–9.

15. Spinrad, "The Reactor Engineering Division," pp. 4–5.

16. S. Untermyer, "Plans for the New Argonne Research Reactor, CP-5," *Nucleonics*, Jan. 1954, pp. 12–15; ANL, "Research Reactor," Folder "Unclassified Reports on CP-5," Box 4 of Box 74X, W. Zinn Chron. Files, RG 326, ANL, NARA; Zinn to J. C. Boyce, The Argonne Nuclear Reactor, Folder "J. C. Boyce," Box 4 of Box 32X, Zinn Reading Files, RG 326, ANL, NARA.

17. ANL, "Argonne Reactors—CP-5 and EBWR," n.d., OPA, ANL.

18. Ibid.

19. Stuart McLain, "Argonne National Laboratory," ms. history of the laboratory (ultimately published in the *Journal of the Central Association of Science and Math Teachers*, March 1958, and in *Atoms Industry*, April 1958), unprocessed OPA files, RG 326, ANL, GLFRC.

20. Spinrad, "The Reactor Engineering Division," p. 5.

21. Zinn to Hafstad, Feb. 17, 1954, Folder "Feb. 15–19, 1954," Zinn Reading Files, Box 1 of Box 18X, RG 326, ANL, NARA; Zinn to files, Folder "Boiling Reactor Experiment," ANL; Zinn Files, Box 2 of Box 74X, RG 326, ANL, NARA.

22. Oral history interview of Joseph Dietrich as transcribed by Marie Miyasaki, Oct. 23, 1979, ANL.

23. Ibid.; Zinn to files, July 23, 1954, Folder "Boiling Reactor Experiment," ANL, NARA.

24. Zinn to files, July 23, 1954, Folder "Boiling Reactor Experiment," Box 2 of Box 74X, Zinn Files, RG 326, ANL, NARA; Samuel Untermyer to B. R. Prentice, "Visit to Borax and ANL during Week of July 26, 1954," Aug. 3, 1954, File V-23, Hist. File 1954, Drawer 4, ANL, TIC.

25. Untermyer to Prentice, Aug. 3, 1954, File V-23, Hist. File 1954, Drawer 4, ANL, TIC; "Statement by Dr. Walter H. Zinn Concerning an Experiment Using the Boiling Reactor Principle," March 5, 1954, in JCAE, "Report of the Subcommittee on Research and Development on the Five-Year Reactor Development Program Proposed by the Atomic Energy Commission, March 1954," (Washington, D.C.: GPO, 1954), pp. 15–16; "Major Activities in the US Atomic Energy Program: The Reactor Program," *Bulletin of the Atomic Scientists* 10 (Nov. 1954): 360; "A Reactor Runs Away—As Planned," *Argonne News*, Aug. 1955, pp. 8–9.

26. Weart, *Nuclear Fear*, p. 169; *Our Friend the Atom* (Walt Disney, 1956), film.

27. W. F. Miller to A. V. Crewe, Past Contributions of AEC to Economic Development of the Country, April 3, 1963, Folder 3 "JCAE," Box 18, 434-89-0024, GLFRC.

28. "Lab Announces Completion of Electronic Brain," *Argonne News*, Feb. 4, 1953, pp. 1, 5, 10; Jean F. Hall, "Argonne's Version of the Institute's Digital Automatic Computer," *Argonne News*, May 1956, pp. 3–5.

29. "Scientists Complete and Operate World's Fastest Electronic Brain," *Argonne News*, Sept. 2, 1953, pp. 1, 3.

30. AEC, Commission Meeting No. 1126, Sept. 22, 1955, AEC Minutes, RG 326 USAEC, DOE.

31. "The Applied Mathematics Division," *Argonne News*, Nov. 1956, pp. 3–5; ANL, *Argonne's High-Speed Computers*, TPD-62-3, OPA, ANL; Press release, ANL-PIO-56, Dec. 17, 1957, OPA, ANL; Press release, ANL-PIO-424, March 14, 1963, OPA, ANL.

32. W. F. Miller, "Computers in the Nuclear Sciences, Past, Present, Future," address given to the meeting of the Armed Forces Communication and Electronics Association, Wheaton, Ill., Feb. 22, 1962, Folder 3, Box 18, 434-89-0024, JCAE, GLFRC.

33. "ANL Electronic Manipulator Provides 'Sense of Feel,'" *Argonne News*, Nov. 1957, p. 21.

34. Stuart McLain, "Five Year Program," Folder "McLain Program Coordinator," W. Zinn Files, Carton 22, Box 1, RG 326, ANL, NARA.

35. D. W. Eisenhower, "Atoms for Peace" speech, *Public Papers of the Presidents of the United States: Dwight D. Eisenhower: 1953* (Washington, D.C.: GPO, 1960), pp. 813–22; Jack M. Holl and Roger M. Anders, Introduction to *Atoms for Peace: Dwight D. Eisenhower's Address to the United Nations* (Washington, DC: NARA, 1990), p. 1.

36. Holl, Anders, and Buck, *United States Civilian Nuclear Power Policy,* pp. 1–2; JCAE, *Atomic Power and Private Industry* (Washington, D.C.: GPO, 1952).

37. Holl, Anders, and Buck, *United States Civilian Nuclear Power Policy,* pp. 1–2; Jack M. Holl, "Eisenhower's Peaceful Atomic Diplomacy: Atoms for Peace and the Western Alliance," *Materials and Society,* 1983, 7 (3/4): 365.

38. Francis Duncan and Jack M. Holl, *Shippingport: The Nation's First Atomic Power Station* (Washington, D.C.: DOE, 1983), pp. 4–9; Hewlett and Duncan, *Nuclear Navy,* pp. 225–34.

39. "Statement of Walter H. Zinn, Director, Argonne National Laboratory, to the Joint Committee on Atomic Energy, Monday, July 6, 1953," Folder "AEC Wash.–Joint Committee on Atomic Energy," W. Zinn, Box 18, RG 326, ANL, NARA.

40. JCAE, *Five-Year-Power Reactor Demonstration Program* (Washington, D.C.: GPO, 1954) pp. 2–4; Duncan and Holl, *Shippingport,* p. 9; Hewlett and Duncan, *Nuclear Navy,* pp. 236–39; Holl, Anders, and Buck, *United States Civilian Nuclear Power Policy,* p. 3.

41. Hewlett and Holl, *Atoms for Peace and War,* pp. 136–42; Richard G. Hewlett and Bruce J. Dierenfeld, *The Federal Role and Activities in Energy Research and Development, 1946–1980: An Historical Summary* (Oak Ridge, Tenn.: DOE, 1983), pp. 25–29.

42. W. Zinn, "Elementary Review of Basic Problems in Power Generation with Nuclear Reactors," *Argonne News,* Nov. 5, 1952, p. 3; W. Zinn, "Statement to the Joint Committee on Atomic Energy," July 6, 1953.

43. W. Zinn, "Statement to the Joint Committee," July 6, 1953.

44. W. Zinn, "Remarks prepared for delivery at the Atomic Industrial Forum, Washington, D.C., May 24, 1954," L. C. Furney, Staff Asst., W. Zinn Files, Box 1, Carton 22, RG 326, ANL, NARA.

45. Ibid., quotations taken from this source; also, Zinn, "Five Year Program," and "The Argonne Reactor Program," Nov. 2, 1953, Stuart McLain folder, W. Zinn Files, Carton 22, Box 1, RG 326, ANL, NARA.

46. ANL, "Program for the Development of a Fast Power Breeder Reactor," Feb. 5, 1953, Folder "EBR-II," W. Zinn Reading Files, Carton 73X, Box 2, RG 77, ANL, NARA.

47. Zinn to Arthur H. Barnes, Dec. 29, 1953, Folder "Reactor Engr. Div, 4–5/54," W. Zinn Files, Box 3, RG 326, ANL, NARA; R. K. Winkleblack to R. H. Layse, Feb. 5, 1954, Reading File 11/16/53–3/26/54, Box 1, RG 326, ANL, NARA.

48. Hewlett and Duncan, *Nuclear Navy,* p. 231; John Bell, "The Argonne National Laboratory's Role in the Development of Nuclear Propulsion for Underwater Naval Craft," Sept. 8, 1958, Box D367109, ANL Information Services, unprocessed files, RG 326, ANL, GLFRC; *Proposed ANL Naval Reactor Design Studies,* n.d.; J. B. Anderson to

S. McLain, Navy Reactor Program for Remainder of FY 1954 (Budget Activity No. 4510-09), and A. Amorosi to A. H. Barnes, Navy Program, May 11, 1954, all in Folder D230 "Schedules and Programs," Box 39, RG 326, ANL, NARA.

49. *General Reactor Design and Development, 4500 Program, Review of Present Programs,* Dec. 23, 1953, ANL-SM-1235, Folder "Stuart McLain Program Coordinator 9/53–9/56," W. Zinn Files, Carton 22, Box 1, RG 326, ANL, NARA.

50. Zinn to Barnes, June 1, 1954, Folder "April–June 1954, Reactor Engr. Div.," Box 4 of Box 15, W. Zinn Files, RG 326, NARA.

51. Petition of Argonne Scientists to the chairman and members of the U.S. Atomic Energy Commission, June 17, 1954, Folder 5, Box 31, VPSP, UCRL SC.

52. Hewlett and Holl, *Atoms for Peace and War,* pp. 110–12; *Washington Post and Time Herald,* May 24, 1954.

53. Minutes of the Laboratory Executive Committee meeting, May 6, 1958, Folder 3, Box 6, 434-89-0024, GLFRC.

54. Minutes of the Laboratory Executive Committee meeting, Oct. 14, 1955, Folder "W. H. Zinn," Box 1, William B. Harrell Papers, UCRL SC.

55. Board of Trustees notes, Sept. 9, 1954; Board of Trustees notes no. 7—committee on budget, Dec. 20, 1954, both ANL History, 1946–59, UCVP, ANL; W. Kenneth Davis to John A. Hall, Assignment of Foreign Scientists to AEC National Laboratories, n.d., Folder "AEC Wash.–Div of Reactor Development," Box 2 of Box 18, W. Zinn Files, RG 326, ANL, NARA; "The International School," *Argonne News,* March 1959, pp. 8–9.

56. AEC press release, "Enrollment Completed for New AEC Reactor School: 19 Nations Represented," March 5, 1955; "President's Greetings to 31 Students of Reactor Training School," March 2, 1955, both in White House Central Files, DDE.

57. Willard Libby to A. J. Goodpaster, Jan. 27, 1956; Bryce N. Harlow to Rep. Frank W. Boykin, Jan. 30, 1956, White House Central Files, DDE.

58. AEC, Commission Meeting No. 1102, July 15, 1955, AEC Minutes, RG 326, AEC, DOE Archives.

59. ANL, Press release on the International School of Nuclear Science and Engineering at Argonne National Laboratory, Jan. 30, 1958, ANL-PIO-67, OPA, ANL; Zinn to A. H. Barnes, Dec. 23, 1955, Folder "8/55–6/56, Reactor Engineering Div.," Box 4 of Box 15, Zinn Files, RG 326, ANL, NARA.

60. "The International School," *Argonne News,* March 1959, pp. 8–9.

61. Minutes of the Laboratory Executive Committee, Sept. 4, 1957, Folder 3 "Lab. Exec. Comm. Minutes," Box 6, 434-89-0024, GLFRC.

62. Hewlett and Holl, *Atoms for Peace and War,* pp. 232–35; "Roundup from Geneva, August 8 to 20," *Nucleonics,* Sept. 1955.

63. Minutes of the Laboratory Executive Committee, Oct. 14, 1955, Folder "W. H. Zinn," Box 1, William B. Harrell Papers, UCRL SC; W. H. Zinn for distribution, Aug. 19, 1954, Folder "KAPL&GE no. 30," Box 22: Box 5 of 5 to Box 23: Box 1 of 5, W. Zinn, RG 326, ANL, NARA.

64. W. P. Bigler to Scientific Division Directors, ANL contributions to International Conference, April 26, 1955, Folder "Peaceful Uses Conf, 1955," Box D-367116, ANL

Information Services Program correspondence, unprocessed records, RG 326, ANL, GLFRC; W. H. Zinn, "Design and Description of the Argonne Research Reactor CP-5," Paper 861; J. R. Dietrich, H.V. Lichtenberger, and W. H. Zinn, "Design and Operating Experience of a Prototype Boiling Water Power Reactor," Paper 851; W. H. Zinn, "Review of Fast Power Reactors," Paper 814, all in International Conference on Peaceful Uses of Atomic Energy, Aug. 8–20, 1955, File V-24, Drawer 1 "CP-5," ANL, TIC.

65. U. M. Staebler, "Power Reactor Development Program," April 9, 1955, Folder "AEC Wash.–Div of Reactor Develop. 1/55–7/55," W. Zinn Files, Box 2 of Box 18, RG 326, ANL, NARA; Press release, "Hilberry Reviews Argonne's Idaho Expansion Plans," May 16, 1957, ANL-4, OPA, ANL.

66. "Experimental Boiling Water Reactor Will Pursue Goal of Commercial Nuclear Power," *Argonne News*, Nov. 3, 1954, p. 3.

67. W. H. Zinn, "Review of Fast Power Reactors," Paper 814, International Conference on Peaceful Uses of Atomic Energy, Aug. 13, 1955, File V-24, Drawer 1 "CP-5," ANL, TIC; "Argonne at Geneva," *Argonne News*, Aug. 1955, pp. 3–5; Minutes of the History Committee meeting, Dec. 6, 1994, p. 27.

68. "The Geneva Conference—A Special Report," *Nucleonics*, Sept. 1955, passim.

69. "A Report on Geneva by Argonne Men Who Were There," *Argonne News*, Nov. 1955, p. 5.

70. The following section has appeared, in part, in Jack M. Holl, "The National Reactor Testing Station: Establishing the Atomic Energy Commission's Nuclear Reactor Research and Development Facility, 1949–1962," *Pacific Northwest Quarterly*, Jan. 1994, pp. 15–24.

71. Quotation from the testimony of Desblonde Deboisblanc, Reactor Physics and Engineering, Atomic Energy Division, Phillips Petroleum Company, at Idaho Falls, Idaho, Folder "Safeguard Information, 4/1/59–8/8/62," FRC 100375, RG 326, NAPNW; J. H. Kittle to F. G. Foote, EBR-I meltdown paper for ANS meeting, May 22, 1957, Folder "Metallurgy Program, Disassembly of EBR-I Core II," ANL Metallurgy Program Files, 1943–1964, Box 7, RG 326, GLFRC; Walter Zinn, "A Letter on EBR-I Fuel Meltdown," *Nucleonics*, June 1956, p. 33.

72. Zinn to distribution, Jan. 14, 1955, Folder "Metallurgy Program—Disassembly of EBR-I Core II," Box 7, ANL Metallurgy Program Files, 1943–1964, RG 326, ANL, GLFRC.

73. Mazuzan and Walker, *Controlling the Atom*, pp. 126–28; Zinn, "A Letter on EBR-I Fuel Meltdown," *Nucleonics*, June 1956, p. 33.

74. Ibid.

75. Mazuzan and Walker, *Controlling the Atom*, pp. 127–82; Zinn, "A Letter on EBR-I Fuel Meltdown"; J. H. Kittel, M. Novick, and R. F. Buchanan, "The EBR-I Meltdown—Physical and Metallurgical Changes in the Core," May 20, 1957 (presentation at the Third Annual Meeting of the American Nuclear Society, Pittsburgh, Pa., June 10–12, 1957).

76. "Twentieth Semiannual Report of the Atomic Energy Commission" (Washington, D.C.: GPO, 1956), pp. 45–46.

77. Zinn, "A Letter on EBR-I Fuel Meltdown."

78. "Nuclear Accidents are Everybody's Business," *Nucleonics*, May 1956, p. 39.

79. Quotation from "Atom's Biggest Worry—Getting Itself Insured," *Business Week*, March 31, 1956, pp. 119–22; *New York Times*, April 6, 1956, p. 12; "AEC Tells How Reactor Failed," *Business Week*, April 14, 1956, p. 34; "The Atom: Undercover Accident," *Time*, April 16, 1956, p. 25; "Idaho Reactor Damaged," *Science*, 1223 (April 27, 1956): 718.

80. Zinn, "A Letter on EBR-I Fuel Meltdown."

81. "Reactor Runs Amuck, No One Hurt," *Science Digest*, July 1956, p. 67; Brian Balogh, *Chain Reaction: Expert Debate and Public Participation in American Commercial Nuclear Power, 1945–1975* (Cambridge: Cambridge University Press, 1991), p. 34.

82. Weart, *Nuclear Fear*, pp. 176–82; Balogh, *Chain Reaction*, p. 137; Hewlett and Holl, *Atoms for Peace and War*, pp. 109–12; Mazuzan and Walker, *Controlling the Atom*, pp. 126–28.

83. Lawrence A. Kimpton, Text of speech at the dedication ceremony of the Experimental Boiling Water Reactor, Feb. 9, 1957, Folder "EBWR-ANL," Gen. Corres., Box 566, JCAE, RG 128, NARA.

84. Margaret Parker, "Argonne Laboratory Switches to Atomic Power," *The Daily Journal* (Du Page County, Ill.), Feb. 11, 1957, p. 1; Weart, *Nuclear Fear*, p. 280–88.

85. R. H. Armstrong and C. N. Kelber, "Argonaut—Argonne's Reactor for University Training," *Nucleonics*, March 1957, pp. 62–65; George C. Baldwin, Walter E. Carey, Ayhan Cilesiz, William R. Kimel, G. Stanley Klaiber, Gerard S. Pawlicki, and Frederick G. Prohammer, "Pedagogical Applications of an Argonaut Reactor," Aug. 1958, International School of Nuclear Science and Engineering, ANL; "Argonne Launches the Argonaut," ANL Press Release no. 168, April 4, 1957, OPA, ANL; "The Argonaut: Argonne's Nuclear Assembly for University Training," *Argonne News*, March 1957, pp. 6–7.

86. Hewlett and Holl, *Atoms for Peace and War*, pp. 341–45, 359–61.

87. Ibid., p. 410; Mazuzan and Walker, *Controlling the Atom*, pp. 93–121.

88. "AEC Authorizes Argonne National Laboratory to Construct Experimental Breeder Reactor No. 2," ANL press release, Feb. 27, 1958, ANL-PIO-78, OPA, ANL.

89. "ZPR-III," ANL press release, May 16, 1957, ANL-4, "Build Three New Reactor Facilities at ANL," ANL press release, Nov. 24, 1959, both OPA, ANL.

90. "TREAT," Argonne press release, Feb. 24, 1958, ANL-PIO-176; "TREAT," Argonne press release, June 4, 1958, ANL-PIO-114; "Niels Bohr Visits Argonne," *Argonne News*, Feb. 1958, p. 2.

91. Hilberry to Frank Pittman, Dec. 8, 1958, Status of Reactor Development Project, ANL Director Hilberry and Misc. Chron. Files, Box 28, RG 326, NARA; S. A. Bernsen and C. K. Soppet, "Boiling Reactor Experimentation Program at Argonne National Laboratory," 1957, File V-24, Drawer 2, loose material, ANL, TIC; "AEC Reactor in Illinois Assigned New Role Following Operation at Record Power Level," AEC Press Release no. E-422, Nov. 16, 1962, Folder "EBWR-ANL," Gen. Corres., Box 566, JCAE, RG 128, NARA; JCAE Request for Information for 202 Hearings, AEC 496/54, Feb. 6, 1959, Folder 2 "Organization and Management 7," Box 1389, Secretariat Records, JCAE, RG 326, AEC, DOE.

92. "AEC Authorizes Work on Low Power Military Package Reactor," ANL Press Release no. 136, OPA, ANL; "ALPR (Argonne Low Power Reactor): A Power Heat Package," *Nuclear Energy Engineer,* Jan. 1959; John McCone to Carl T. Durham, Aug. 12, 1958, Folder "Argonne Low Power Reactor," Gen. Corres. 1946–1977, Box 541, JCAE, RG 128, NARA.

93. Spinrad to Hilberry, "Future Reactor Program of Argonne National Laboratory," Feb. 21, 1958, Folder "Arbor," Box 4, Lab Director's Project Files, unprocessed files, RG 326, ANL, GLFRC; "Progress in the Peaceful Uses of Atomic Energy—a 3-Year Summary," Feb. 4, 1958, ANL-PIO-70, ANL; Norman Hilberry, "U.S. Explores Wide Front for Power Reactor Knowledge," Sept. 4, 1958, ANL-PIO-138-1, ANL.

94. Hewlett and Holl, *Atoms for Peace and War,* pp. 413–14.

Chapter 5

1. L. G. Ratner, "The Life and Death of a Particle Accelerator," in *Symposium on the History of the ZGS: AIP Conference Proceedings No. 60,* eds. Joanne S. Day, Alan D. Krisch, and Lazarus G. Ratner (New York: American Institute of Physics, 1980), pp. 397–400.

2. E. N. da C. Andrade, *Rutherford and the Nature of the Atom* (Garden City, New York: Anchor Books, 1964), p. 111; Michael N. Keas, "Ernest Rutherford," in *Nobel Laureates in Chemistry, 1901–1992,* ed. Laylin K. James (American Chemical Society and the Chemical Heritage Foundation, 1993), pp. 49–60.

3. Ratner, "The Life and Death of a Particle Accelerator," p. 399; Daniel S. Greenberg, *The Politics of Pure Science* (New York: The New American Library, 1969), pp. 210–11.

4. Hewlett and Anderson, *The New World,* p. 12; Steven Weinberg, *Dreams of a Final Theory: The Search for the Fundamental Laws of Nature* (New York: Pantheon Books, 1992), p. 266.

5. Hewlett and Duncan, *Atomic Shield,* pp. 228–37.

6. Ibid., pp. 80–81.

7. Ratner, "The Life and Death of a Particle Accelerator," p. 401; Greenberg, *The Politics of Pure Science,* p. 212–15; Hewlett and Duncan, *The Atomic Shield,* pp. 223–26.

8. Hewlett and Anderson, *The New World,* pp. 222–23, 229.

9. E.L. Goldwasser, "The Universities' Role," in *Symposium on the History of the ZGS: AIP Conference Proceedings No. 60,* eds. Joanne S. Day, Alan D. Krisch, and Lazarus G. Ratner (New York: American Institute of Physics, 1980), pp. 25–26; Greenberg, *Politics of Pure Science,* pp. 213–14.

10. Minutes of the Laboratory Council, April 11, 1947, File "Laboratory Council Minutes," OPA, ANL.

11. Boyce, Notes on the Midwest Cosmotron Project, April 29, 1953, Folder "Accelerator Corres. 1953," Box 3 of Box 32X, ANL Accelerator Corres. 1953–1956, RG 77, ANL, NARA; Allison, Fermi, Herb, Kruger, Livingood, Mitchell, Turner, Williams, and Zinn to T. H. Johnson, director, division of research, AEC, Jan. 30, 1953, Folder "AEC Wash.–Div of Research 1951–53," Box 2 of Box 18, W. Zinn, RG 326, ANL, NARA; Boyce to members of the Council of Participating Institutions and chairman of the physics

departments at these institutions, Feb. 6, 1953, Folder 9, Box 11, Enrico Fermi Papers, UCRL SC.

12. Johnson to Zinn, Feb. 16, 1953, Folder "Accelerator Corres. 1953," Box 3 of Box 32X, ANL Accelerator Corres., 1953–1956, RG 77, ANL, NARA.

13. F. Wheeler Loomis to Boyce, Feb. 23, 1953, Folder "Accelerator Corres. 1953," Box 3 of Box 32X, ANL Accelerator Corres. 1953–1956, RG 77, ANL, NARA.

14. Greenbaum, *A Special Interest,* pp. 24–25, 45–55.

15. A. Tammaro, manager, Chicago Operations Office to T. H. Johnson, April 3, 1953, "Security Interest in Midwest Cosmotron Project"; Boyce to Zinn, Nov. 20, 1953; Boyce, "Notes on the Midwest Cosmotron Project," April 29, 1953, all in Folder "Accelerator Corres. 1953," Box 3 of Box 32X, ANL Accelerator Corres. 1953–1956, RG 77, ANL, NARA; Boyce to Loomis, March 6, 1953.

16. Boyce, Notes on the Midwest Cosmotron Project, April 29, 1953.

17. Laboratory Council Executive Board minutes, June 1, 1953, as cited in Boyce to Zinn, July 20, 1953, Folder: Accelerator Corres. 1953, Box 3 of Box 32X, ANL Accelerator Corres. 1953–1956, RG 77, ANL, NARA.

18. Ibid.

19. Ibid.; Greenbaum, *A Special Interest,* pp. 63–64.

20. Boyce to Zinn, Midwest Cosmotron Project, July 20, 1953, July 30, 1953, and Sept. 16, 1953, all in Folder "Accelerator Corres. 1953," Box 3 of Box 32X, ANL Accelerator Corres. 1953–1956, RG 77, ANL, NARA.

21. Libby to Zinn, Sept. 23, 1953, Folder "Accelerator Corres. 1953," Box 3 of Box 32X, ANL Accelerator Corres. 1953–1956, RG 77, ANL, NARA.

22. Zinn to T. H. Johnson, Jan. 27, 1954, Folder "Accelerator Corres.," Box 3 of Box 32X, Accelerator Corres. 1953–1956, RG 77, ANL, NARA.

23. Ibid.

24. Zinn to file, Feb. 3, 1954 and Feb. 12, 1954; Johnson to Tammaro, manager, Chicago Operations Office, March 2, 1954; Johnson to Zinn, March 8, 1954, all in Folder "Accelerator Corres. 1954," Box 3 of Box 32X, ANL Accelerator Corres. 1953–1956, RG 77, ANL, NARA. See also Walter Zinn, Box 719, JCAE General Corres., JCAE, RG 128, NARA.

25. Zinn to Johnson, March 22, 1954 and Johnson to Zinn, March 24, 1954, Folder "Accelerator Corres. 1954," Box 3 of Box 32X, ANL Accelerator Corres. 1953–1956, RG 77, ANL, NARA.

26. AEC 603/17, Argonne Accelerator Project, June 3, 1954, Folder "Accelerator Corres. 1954," Box 3 of Box 32X, ANL Accelerator Corres. 1953–1956, RG 77, ANL, NARA.

27. Ibid.

28. White Paper and related documents, Midwestern Universities Research Association, received by the director, ANL, June 28, 1954; K. D. Nichols to the AEC, Meeting with Midwestern Research Association regarding support for a Midwest accelerator, 6/11/54; David Saxe to Zinn, July 2, 1954; Zinn to L. A. Turner, July 7, 1954, all in Folder "Accelerator Corres. 1954," Box 3 of Box 32X Accelerator Corres. 1953–1956, RG 77, ANL, NARA.

29. Council Executive Board minutes, July 8, 1954; Minutes of the meeting of the Organization Committee of MURA with members of the Argonne National Laboratory and the University of Chicago, July 16, 1954; A. L. Hughes, Professor of Physics, Washington Univ. to W. F. Libby, July 20, 1954, all in Folder "Accelerator Corres. 1954," Box 3 of Box 32X, ANL Accelerator Corres. 1953–1956, RG 77, ANL, NARA.

30. Norris to Anderson, Feb. 21, 1956, Folder "W. Zinn 1956," Box 720, Gen. Corres. 1946–1977, JCAE, RG 128, NARA.

31. "Need for a High Energy Laboratory in the Middle West," and "Desirable Characteristics of Such a Laboratory," both attached to Kruger to Johnson, May 6, 1955, Folder "Accelerator Corres. 1955," Box 3 of Box 32X, ANL Accelerator Corres. 1953–1956, RG 77, ANL, NARA. See also "Proposal for Cooperative Research in High Energy Physics through the Establishment of a Laboratory in the Middle West which Will Serve Best the Eduction and Scientific Needs of That Area," Folder "Plants, Labs, Buildings, and Land, Midwestern Universities," Vol. 4, Box 1288, AEC Secretariat, RG 326, AEC, DOE.

32. Kruger to Johnson, May 6, 1955; J. J. Livingood to Zinn, Proposals for Accelerator, March 2, 1955; Minutes of the Accelerator Conference Held at Argonne National Laboratory, April 23, 1955, all in Folder "Accelerator Corres. 1955," Box 3 of 32X, ANL Accelerator Corres. 1953–1956, RG 77, ANL, NARA.

33. Johnson to Kruger, May 18, 1955 in AEC 827/1, Midwestern Universities Research Association Proposal, May 26, 1955, Folder "Plants, Labs, Buildings, & Land," Vol. 1, Box 1288, AEC Secretariat, RG 326, AEC, DOE; Hatcher to Strauss, June 9, 1955 and Strauss to Hatcher, June 20, 1955, Folder "Accelerator Corres. 1955," Box 3 of 32X, ANL Accelerator Corres. 1953–1956, RG 77, ANL, NARA.

34. "Proposal for a High Current 25 Bev Proton Accelerator at Argonne National Laboratory," July 25, 1955, in Zinn to Johnson, July 28, 1955, Folder "W. Zinn 1955," Box 720, General Corres. 1946–1977, JCAE, RG 128, NARA; John J. Livingood, "The Beginning," in *Symposium on the History of the ZGS, AIP Conference Proceedings No. 60*, eds. Joanne S. Day, Alan D. Krisch, and Lazarus G. Ratner (New York: American Institute of Physics, 1980), pp. 4–5.

35. Hewlett and Holl, *Atoms for Peace and War*, p. 241.

36. W. B. Harrell to Zinn, Oct. 12, 1955, Folder "W. Zinn," Box 720, Gen. Corres. 1946–1977, JCAE, RG 128, NARA.

37. Notes on meeting of the Atomic Energy Commission with representatives of midwestern universities on the question of AEC support of a basic research center in the Midwest, August 3, 1955, Folder "Plants, Labs, Buildings, & Land Midwestern Universities," Vol. 1, Box 1288, AEC Secretariat, RG 326, AEC, DOE; Greenbaum, *A Special Interest*, p. 79.

38. Harrell to Kimpton, Aug. 30, 1955, and Harrell to Warren Johnson, Aug. 31, 1955, Folder "High Energy Accelerator (4)," Box 4, William B. Harrell Papers; Minutes of the Board of Trustees, Oct. 13, 1955, Box 45, Univ. Chicago Board of Trustees, all in UCRL SC; Greenbaum, *A Special Interest*, pp. 78–79.

39. Notes on meeting with members of the research and reactor subcommittee of

the General Advisory Committee, Oct. 31, 1955, Folder "Plants, Labs, Buildings, & Land, Midwestern Universities," Vol. 1, Box 1288, AEC Secretariat, RG 326, AEC, DOE.

40. Ibid.

41. W. B. McCool, "Memorandum for the Commissioners and General Manager: Discussion of Midwestern Research Center at Meeting 1144," Nov. 1, 1955, Folder "Plants, Labs, Buildings, & Labs, Midwestern Universities," Vol. 1, Box 1288, AEC Secretariat, RG 326, AEC, DOE.

42. Notes on meeting of the Atomic Energy Commission with representatives of the midwestern universities on the question of AEC support of a basic research center in the Midwest, Nov. 8, 1955; AEC 827/11, Notes on meeting with Midwest university presidents and MURA made by the secretary pro tem, Dec. 5, 1955 (see also AEC 827/6), all in Folder "Plants, Labs, Buildings, & Land, Midwestern Universities," Vol. 1, Box 1288, AEC Secretariat, RG 326, AEC, DOE.

43. AEC 827/7, Midwest Accelerator Program, Nov. 21, 1955; Quotation from AEC 827/8, Notes on meeting with representatives of ANL and University of Chicago, Nov. 22, 1955, both in Folder "Plants, Labs, Buildings & Land, Midwestern Universities," Vol. 1, Box 1288, AEC Secretariat, RG 326, AEC, DOE.

44. Norris to Anderson, Feb. 17, 1956, Folder "W. Zinn 1956," Box 720, General Corres. 1946–1977, JCAE, RG 128, NARA; Quotation from Zinn, Summary of Communication on Accelerator Situation (from Oct. 7, 1955), Nov. 11, 1956, Folder "Accelerator Corres. 1955," Box 3 of Box 32X, ANL Accelerator Corres. 1953–1956, RG 77, ANL, NARA.

45. Quotations from L. A. Turner to Zinn, Jan. 23, 1956; J. J. Livingood and M. Foss to Zinn, Jan. 25, 1956, both in Folder "Accelerator Corres. 1956," Box 3 of Box 32X, ANL Accelerator Corres. 1953–1956, RG 77, ANL, NARA.

46. Turner to Zinn, conversations with E. P. Wigner, Jan. 16, 1956; Livingood to Zinn, cocktail party for G.A.C.—Jan. 12, 1956, Jan. 17, 1956, both in Folder "Accelerator Corres. 1956," Box 3 of Box 32X, ANL Accelerator Corres. 1953–1956, RG 77, ANL, NARA.

47. L. C. Teng, "The ZGS—Conception to Turn-On," in *Symposium on the History of the ZGS: AIP Conference Proceedings No. 60,* eds. Joanne S. Day, Alan D. Krisch, and Lazarus G. Ratner (New York: American Institute of Physics, 1980), p. 11.

48. AEC, Minutes of 1164th Meeting, Jan. 18, 1956, DOE; AEC 827/15, Argonne Accelerator Program, Feb. 7, 1956, Folder "Plants, Labs, Buildings, & Land, Midwestern Universities," Vol. 1, Box 1288, AEC Secretariat, RG 326, AEC, DOE.

49. Quotation from "Ten Years of Leadership Draw to a Close," and J. J. Livingood, "Multi-Billion Volt Accelerator for Argonne," *Argonne News,* March 1956, pp. 3–5; Ratner, "Life and Death of a Particle Accelerator," pp. 403–4.

50. Zinn to Strauss, Jan. 6, 1956, Folder "1956," Box 720, General Corres., W. Zinn, JCAE, RG 128, NARA; W. B. Harrell to files, Jan. 23, 1956 and Harrell to files, Jan. 26, 1956, Folder "W. H. Zinn," Box 1, William B. Harrell Papers, UCRL SC; Kimpton to Zinn, Feb. 8, 1956, ANL Directorship, 1956–1957, Box 6, W. Massey Subj. Files, ANL.

51. Norris to Anderson, Feb. 21, 1956, Folder "1956," Box 720, Gen. Corres., W. Zinn, JCAE, RG 128, NARA.

52. Harrell to Kimpton, March 20, 1956; Harrell to Kimpton, March 26, 1956; Univ. of Chicago Board of Trustees Minutes, April 12, all in Folder "ANL Directorship 1956–1957," Box 6, W. Massey Subj. Files, ANL; AEC, Minutes of Meeting 1192, April 12, 1956, RG 326, AEC, DOE.

53. AEC, Minutes of Executive Session 1206, June 4, 1956; AEC, Minutes of Meeting 1210, June 26, 1956; AEC, Minutes of Executive Session 1220, Aug. 1, 1956.

Chapter 6

1. Seitz to Kimpton, Feb. 18, 1957, and Kimpton to Seitz, Feb. 22, 1957, both in Folder "ANL Directorship 1956–1957," Box 6, W. Massey Subj. Files, ANL.

2. Corbin Allardice and Edward R. Trapnell, "The First Reactor," (Washington, D.C.: DOE, 1992), pp. 14–15; Rhodes, *The Making of the Atomic Bomb*, pp. 438–39.

3. *Chicago Sun-Times*, March 3, 1957, p. 34.

4. "Dr. Norman Hilberry . . . Argonne's New Director," *Argonne News*, March 1957, pp. 2, 12.

5. Darleane C. Hoffman, "The Heaviest Elements," *Chemical and Engineering News*, May 2, 1994, pp. 24–34; Rhodes, *The Making of the Atomic Bomb*, pp. 211–15; "The Plutonium Story," *Argonne News*, December 1962, p. 9.

6. Glenn T. Seaborg, "Transuranium Elements," in *The Encyclopedia of Physics*, 3d ed., ed. Robert M. Besangon (New York: Van Nostrand Reinhold Company, 1985), pp. 1258–66.

7. Paul R. Fields, "Discovery and History of 252Cf," *Argonne Reviews*, October 1969, pp. 106–8.

8. Ibid.; Paul R. Fields, interview by Ruth R. Harris, Dec. 7, 1992.

9. "New Element, Number 100, Produced by Argonne Scientists," *Argonne News*, April 14, 1954, p. 3. The discovery of elements 99 and 100 and discussion of the nuclides found in the fallout debris were published in *Physical Review*, 99 (1955): 1048, 102 (1956): 108, and 119 (1960): 2000.

10. Hoffman, "The Heaviest Elements," p. 24; Seaborg, "The Transuranium Elements," p. 1263.

11. Paul Fields, comments of Argonne History draft, Meeting Minutes, Dec. 6, 1994, ANL History Committee, p. 44, ANL.

12. *Time*, July 22, 1957.

13. Seaborg, "The Transuranium Elements," pp. 1263–64; "Three Nation Team Creates Element 102," *Argonne News*, Aug. 1957, pp. 4–5.

14. Hoffman, "The Heaviest Elements," p. 25; Seaborg, "The Transuranium Elements," p. 1263.

15. Paul Fields, comments on Argonne History draft, Meeting Minutes, Dec. 6, 1994, ANL History Committee, p. 44, ANL.

16. "The National Laboratories and the Atomic Energy Commission, 1961," in Folder "History, A Review of Operations, Cannon," Box 6, W. Massey Subj. Files, ANL.

17. Quotations in this and next paragraph from Hilberry, *Elements of Basic Management Philosophy*, pp. 1–17.

18. Ibid., p. 22.

19. ANL, "Report on the Effect of AEC Policies and Practices on Management Policies and Practices of Argonne National Laboratory," Nov. 10, 1953; Zinn to Harrell, Feb. 19, 1954; Harrell to Tammaro, Feb. 26, 1954, all in Folder "Policy and Practice Guide," Box 21, W. Massey Subj. Files, ANL.

20. Hilberry, *Elements of Basic Management Philosophy*, pp. 39–42.

21. Ibid., p. 57 and passim.

22. J. H. Hansen, chairman of the executive board, to the members of the Council of Participating Institutions of Argonne National Laboratory, April 19, 1957, Folder "AUA Contract," Box 2, William B. Harrell Papers, UCRL, SC.

23. E. J. O'Donnell, S. J., to J. T. Rettaliata, Aug. 29, 1956; James A. McCain to Rettaliata, June 26, 1956; E. B. Fred to Rettaliata, June 11, 1956, all in Folder "Ad hoc File, Chemistry Div.," ANL Information Service, RG 326, ANL, NARA-GL.

24. Résumé of discussion, meeting of May 10 of Committee on Argonne National Laboratory, R. W. Harrison, W. E. Zinn, Enrico Fermi, and W. C. Johnson, May 10, 1954, Folder 2, Box 11, Enrico Fermi Papers, UCRL SC. For the entire package of comments solicited by Rettaliata, see Seitz to Simpson, Dec. 19, 1956, Folder "Ad hoc File, Chemistry Div.," ANL Information Service, RG 326, ANL, NARA-GL.

25. "Report of ad hoc Committee on Relationship between Argonne National Laboratory and the University Communities," as forwarded from Rettaliata to Kimpton, Feb. 8, 1957, Folder "AMU History Project," Box 1, William B. Harrell Papers, UCRL SC; Greenbaum, *A Special Interest*, pp. 84–92.

26. Quotation from Minutes of the Council of Participating Institutions, May 20, 1957, and July 17, 1957, Policy Advisory Board Meetings, 43.4, RG 326, ANL, NARA-GL; Harrell to Kimpton, May 20, 1957, and ANL press release, ANL-PIO-8, May 20, 1957, in Rettaliata Comm., Box 27A, W. Massey Subj. Files, ANL.

27. "Dr. Myers to Fulfill Argonne Post," *Argonne News*, June 1958, p. 3; "New Appointments Announced," *Argonne News*, Oct. 1958, p. 3.

28. "New University Organization Formed," *Argonne News*, June 1958, p. 7; Meeting of the Council of Participating Institutions and Associated Midwest Universities, June 10, 1958, Participating Institutions, Folder 43.2, Lab. Director's Project Files, 1955–1970, RG 326, ANL, NARA-GL.

29. Strauss to secretary of state, Dec. 4, 1958, as cited in Hewlett and Holl, *Atoms for Peace and War*, p. 445.

30. "Argonne at Geneva—1958," *Argonne News*, August 1958, pp. 4–5; Hewlett and Holl, *Atoms for Peace and War*, pp. 446–47; D. H. Lennox to R. W. Seidensticker, Nov. 11, 1994.

31. AEC, *The Future Role of the Atomic Energy Commission Laboratories: A Report to the Joint Committee on Atomic Energy* (Washington, D.C.: AEC, 1960), pp. v–viii; also AEC, *The Future Role*, Part 2, *The Individual Laboratories in the Next Decade*, Section 5, "Multiprogram Laboratories, Argonne National Laboratory," Folder "Med, Health, Safety, Role of the AEC Labs," Box 3357, Job 1132, AEC.

32. Hilberry, *Laboratory Objectives and Philosophy: Long Range Program for Argonne*

National Laboratory, 1959, Box 2, W. Massey Subj. Files, ANL; see also, ANL, *Long-Range Program,* Vol. 1 (ANL: Lemont, Ill., March 1959), pp. 4–5.

33. H. A. Stanwood, Jr., to C. L. Dunham, director, Division of Biology and Medicine, Nov. 30, 1959, Folder "Medical, Health, and Safety, Role of the AEC Labs," Box 3357, Job 1132, DOE; A. Tammaro to A. R. Luedecke, "Laboratory Directors Meeting," Chicago, Jan. 31–Feb. 1, 1959, Folder 20 "Organization and Management 6 1959," Box 1, Division of Biology and Medicine, Job 1132, DOE.

34. W. B. McCool, memorandum on meeting of laboratory directors, Chicago, Jan. 31–Feb. 1, 1959, AEC 956/2, Feb. 27, 1959, Folder 13 "Plants, Labs, Bldgs, & Land, Laboratories," Vol. 1, Box 1403, AEC Secretariat, DOE; Hilberry, *Laboratory Objectives and Philosophy,* p. 7.

35. McCune to McCone, June 25, 1959, and AIF, Inc., *Discussion on the Future Role of AEC-Owned Research and Development Installations Held Sept. 10, 1959,* Folder 1, Box 1, Job 11174, Collec. no. 434-89-0024, GLFRC.

36. Quotation from McCool, memorandum on meeting of laboratory directors, Chicago, Jan. 31–Feb. 1, 1959, pp. 5–6; Hilberry, *Laboratory Objectives and Philosophy,* p. 7.

37. Stanwood to Dunham, Nov. 30, 1959.

38. AEC, *The Future Role of the Atomic Energy Commission Laboratories,* pp. 63–66.

39. AEC, *The Future Role of the Atomic Energy Commission Laboratories,* pp. 65–66; Holl, Anders, and Buck, *U.S. Civilian Nuclear Power Policy,* pp. 7–8.

40. "Janus," *Argonne News,* March 1960, p. 8.

41. Quotation from AEC, minutes discussing AEC 956/6, Meeting 1528, July 15, 1959, Folder "1959 USAEC Minutes of Mtgs," Vol. 23, Box 3725, RG 326, AEC, DOE; AEC, *The Future Role of the Atomic Energy Commission Laboratories,* pp. 66–67.

42. AEC, *The Future Role of the Atomic Energy Commission Laboratories,* p. 69.

43. AEC press release, "12.5 Billion Electron Volt Accelerator Authorized for Argonne National Laboratory," Dec. 12, 1957. DOE.

44. "A Distinct Odor of Fish from Atomic Energy Commission," *Milwaukee Journal,* Dec. 18, 1957.

45. Hildebrand to John H. Williams, Aug. 15, 1956, Folder 1, Box 4, Samuel K. Allison Papers, UCRL SC; Harrell to Hilberry, April 25, 1958 and Harrell to Kimpton, Aug. 15, 1958, ANL History, UCVP ANL, ANL.

46. Ratner, "Life and Death of a Particle Accelerator," pp. 413–14.

47. Greenberg, *The Politics of Pure Science,* pp. 231–33.

48. Ratner, "Life and Death of a Particle Accelerator"; see also, Teng, "The ZGS—Conception to Turn-On," and E. L. Goldwasser, "The Universities' Role."

49. "ZGS," *Argonne News,* Jan. 1961, p. 4

50. ANL, *High Energy Physics at Argonne National Laboratory,* p. 44, TPD-63-37, ANL.

51. Ratner, "Life and Death of a Particle Accelerator," p. 415.

52. W. D. Walker, "The MURA-ANL-FERMILAB 30" Bubble Chamber," in *Symposium on the History of the ZGS: AIP Conference Proceedings No. 60,* eds. Joanne S. Day, Alan D. Krisch, and Lazarus G. Ratner (New York: American Institute of Physics, 1980), pp. 127–28.

53. Greenbaum, *A Special Interest,* pp. 119–20; Ratner, "Life and Death of a Particle Accelerator," pp. 416–17.

54. Hildebrand to Symon as quoted in Greenbaum, *A Special Interest,* p. 130.

55. The information in this and the next two paragraphs are from the following sources: "How Argonne National Laboratory Serves the University Community," 1959, Folder 1, Box 1, Job 11174, Collec. no. 434-89-0024, GLFRC; ANL, Long-Range Program, Vol. 1, March 1979, Proposals, Box 7, W. Massey Subj. Files, ANL; Greenbaum, *A Special Interest,* pp. 122–23.

56. Minutes of Policy Advisory Board, Argonne National Laboratory, Sept. 13, 1961, 43.4, Box 17, RG 326, ANL semi-processed records, NARA-GL.

57. ANL, *High Energy Physics at Argonne National Laboratory,* pp. 45–46, TPD-63-37, ANL; R. H. Hildebrand, "Scheduling Policy for the ZGS," Argonne Accelerator Users Group meeting held at ANL, May 11–12, 1962, TPD-62-49, pp. 5–6.

58. Minutes of Policy Advisory Board, Argonne National Laboratory, Sept. 13, 1961, 43.4, Box 17, RG 326, ANL semi-processed records, NARA-GL.

59. A. R. Luedecke, general manager, to Chairman McCone, 12/6/59, Folder 1 "Research and Development (General), Particle Accelerators," Vol. 1, Box 1424, RG 326, AEC Secretariat, DOE.

60. George Beadle to Seaborg, Oct 4., 1961, and A. V. Crewe to Beadle, Sept. 28, 1961, "Funding of the Zero Gradient Synchrotron Project," AEC 603/80, Oct. 13, 1961, Folder 2 "Research and Development, Particle Accelerators," Vol. 2, Box 1424, RG 326, AEC Secretariat, DOE.

61. Portions of this section have already appeared in Jack M. Holl, "The National Reactor Testing Station," *Pacific Northwest Quarterly,* Jan. 1994, p. 23; see also Mazuzan and J. Walker, *Controlling the Atom,* p. 341.

62. "Seaborg is New Chairman of AEC," *Argonne News,* Feb. 1961, p. 2; Holl, Anders, and Buck, *U.S. Civilian Nuclear Power Policy,* p. 9.

63. R. W. Harrison to Hilberry, Feb. 27, 1961; Ad hoc Advisory Committee on Director of the Argonne National Laboratory, Recommendations in Order of Preference, March 30, 1961; Hilberry to Staff, April 14, 1961; Warren C. Johnson to members of the Ad hoc Committee on Director for Argonne National Laboratory, May 3, 1961, all in ANL.

Chapter 7

1. Samuel Glasstone, ed., *The Effects of Atomic Weapons* (Los Alamos, N.Mex.: Los Alamos Scientific Laboratory, 1950), p. 342, table 11.28, as quoted in Barton C. Hacker, *Elements of Controversy: The Atomic Energy Commission and Radiation Safety in Nuclear Weapons Testing, 1947–1974* (Berkeley: University of California Press, 1994), p. 2.

2. "The Hypothetical Bomb and the Civil Defense Drill," *Argonne News,* June 1961, p. 6.

3. "Fallout Meter for under 20 Cents," *Argonne News,* Nov. 1962, pp. 4–6.

4. Hildebrand quoted in "New Laboratory Director Appointed," *Argonne News,* Oct.

1961, pp. 10–11; ANL press release, Oct. 1961, Folder "ANL," Vol. 1, Box 4, Gen. Corres., RG 128, JCAE, NARA.

5. Albert V. Crewe, interview by author, Aug. 11, 1992, University of Chicago; ANL press release, Oct. 11 and 16, 1961, Folder "ANL," Vol. 1, Box 4, Gen. Corres., RG 128, JCAE, NARA; Warren C. Johnson to ad hoc committee for Argonne National Laboratory Directorship, July 6 and Aug. 1961, ANL-Directorship, Box 6, W. Massey Subj. Files, ANL.

6. Teng, "The ZGS—Conception to Turn On," pp. 13–14; "Zero Gradient Synchrotron," and "The ZGS Is Working," *Argonne News*, Oct. 1963, pp. 5–8; ANL press release, Dec. 21, 1961, ANL-PIO-353, ANL.

7. Daniel J. Kevles, *The Physicists: The History of a Scientific Community in Modern America* (Cambridge: Harvard University Press, 1987), pp. 387–92; ANL press release, Dec. 21, 1961, ANL-PIO-353, ANL.

8. Quoted material from Teng, "The ZGS—Conception to Turn On," p.16; "Zero Gradient Synchrotron," and Lee C. Teng, "The ZGS is Working," in *Argonne News*, Oct. 1963, pp. 4–8; ANL press release, Dec. 21, 1961, ANL-PIO-353, ANL.

9. Goldwasser, "The Universities' Role," pp. 25–31; Minutes of the Policy Advisory Board, Sept. 23, 1961, Box 16, W. Massey Subj. Files, ANL; A. V. Crewe, interview by author, p. 8, Aug. 11, 1992, Chicago.

10. A. V. Crewe, "The Construction of the ZGS," in *Symposium on the History of the ZGS: AIP Conference Proceedings No. 60*, eds. Joanne S. Day, Alan D. Krisch, and Lazarus G. Ratner (New York: American Institute of Physics, 1980), pp. 18–24; A. V. Crewe, interview by author, Aug. 11, 1992; Edward J. Bauser, AEC, to James T. Ramey, JCAE, Oct. 3, 1961, Folder "ANL," Vol. 1, Box 4, Gen. Corres., JCAE, RG 128, NARA; Beadle to Crewe, Oct. 31, 1961, ANL-Director, Box 6, W. Massey Subj. Files, ANL.

11. Greenbaum, *A Special Interest*, pp. 130–36.

12. AMU Board of Directors, special meeting, Allerton Estate, Univ. of Illinois, Jan. 6–7, 1962, Policy Advisory Board Meetings, Folder 43.4, RG 326, NARA-GL.

13. Crewe, remarks to AMU-PAB joint meeting, May 7, 1962, in Policy Advisory Board Meetings, Folder 43.4, RG 326, NARA-GL.

14. AMU, Minutes of the Policy Advisory Board, Oct. 17, 1962, Policy Advisory Board meetings, Folder 43.4, RG 326, NARA-GL.

15. Greenbaum, *A Special Interest*, p. 141.

16. Ibid., pp. 144–45.

17. AMU, Minutes of the Policy Advisory Board, Oct. 17, 1962, Folder 43.4, RG 326, NARA-GL.

18. Goldwasser, "The Universities' Role," p. 27; AMU, Minutes of the Policy Advisory Board, July 17, 1963, Folder 43.4, RG 326, NARA-GL; Greenbaum, *A Special Interest*, pp. 150–51.

19. AMU, Minutes of the Policy Advisory Board, July 17, 1963, Folder 43.3, RG 326, NARA-GL.

20. AMU, Minutes of the Policy Advisory Board, Oct. 16, 1963, Folder 43.4, RG 326, NARA-GL.

21. Goldwasser, "The Universities' Role," pp. 27–28.

22. Quoted material from AEC, "Disapproval of MURA Proposal," AEC 603/93, Jan. 17, 1964, Folder 3 "R&D/Particle Accelerators," Vol. 3, Box 1424, RG 326, AEC Secretariat, DOE; Goldwasser, "The Universities' Role," p. 28.

23. Greenbaum, *A Special Interest*, p. 159.

24. AMU, Minutes of the Policy Advisory Board, Jan. 16, 1964, Folder 43.4, RG 326, NARA-GL; Greenbaum, *A Special Interest*, pp. 159–61.

25. Goldwasser, "The Universities' Role," p. 29.

26. Greenbaum, *A Special Interest*, pp. 165–67.

27. William Harrell to A. W. Peterson and E. L. Goldwasser, July 23, 1964 and Gerald F. Tape to Beadle and Williams, Aug. 28, 1964, ANL Negotiations on Future Operations and Management, 1964, W. Massey Subj. Files, ANL; Quoted material from Greenbaum, *A Special Interest*, pp. 170–73.

28. A. V. Crewe, interview by author, Aug. 11, 1992, Chicago.

29. Crewe to all employees, Oct. 20, 1964, Management Plan, Box 1, W. Massey Subj. Files, ANL; Comparison of Two Proposals for the Plan of Operation of Argonne National Laboratory under Tripartite Agreement, n.d., and Confidential Memorandum—Argonne Matter Proposed Strategy—Nov. 14, 1968, in Folder "ANL Negotiations on Future of Operations and Management," Box 14, W. Massey Subj. Files, ANL.

30. W. B. Harrell to the board of trustees, "New Five-Year Contract for Operation of Argonne National Laboratory," Nov. 10, 1966, ANL Contracts and Budgets, 1966–67, Box 27B, W. Massey Subj. Files, ANL.

31. Statement by Dr. A. V. Crewe, director, Argonne National Laboratory, Oct. 21, 1964, ANL press releases "News from ANL," and "Management and Organization," Box 6, W. Massey Subj. Files, ANL.

32. "Twelve Universities form Central States Universities to Promote Science Education," *Argonne News*, Aug. 1965, pp. 4–5.

33. "Two New Offices Established at Argonne to Expedite University Cooperation," *Argonne News*, Aug. 1965, pp. 4–5.

34. A. V. Crewe, "Scientific Research: An Investment for the Future," *Argonne National Laboratory News-Bulletin*, March 1963, pp. 13–15; Quotation from A. V. Crewe, "Public Needs to Understand Why and How of Basic Research," *Argonne News*, Dec. 1963, p. 3.

35. Robert G. Sachs, interview by Jack M. Holl, Dec. 4, 1992, University of Chicago.

36. Ibid.; "R. G. Sachs Named Associate Laboratory Director," *Argonne News*, Dec. 1963, p. 2.

37. Robert G. Sachs, interview by author, Dec. 4, 1992, University of Chicago.

38. Ibid.; W. D. Walker, "The MURA-ANL-Fermilab 30" Bubble Chamber," pp. 53–56.

39. Lolita Woodard, "How Xenon Tetrafluoride Was Discovered," *Argonne Laboratory News Bulletin*, June 1963, pp. 14–15.

40. ANL, *Annual Report: 1963*, p. 23.

41. "Maria Goeppert-Mayer: Nobel Prize—Physics 1963," *Argonne News*, Nov. 1963, pp. 3, 5.

42. Robert G. Sachs, interview by author, Dec. 4, 1992, University of Chicago. Paul Fields has a somewhat different recollection of these events in which he recalls hearing Fermi comment on Goeppert-Mayer's shell model at a seminar (Paul Fields, Comments on Argonne History manuscript, January 1995).

43. Minutes of the Executive Committee meeting of the Atomic Scientists of Chicago, Jan. 14, 1947, John A. Simpson Papers, Folder 2, Box 7, UCRL SC; *Physical Review* 74 (Aug. 1, 1948): 235, and 76 (July 1, 1959): 185; Louis Haber, *Women Pioneers of Science* (New York: Harcourt Brace Jovanovich, 1979), p. 93; "Maria Goeppert-Mayer," *Argonne News,* Nov. 1963, pp. 3, 5.

44. "The EBR-I is Designated National Historic Landmark," *Argonne News,* Sept.–Oct. 1966, pp. 6–9.

45. Holl, Anders, and Buck, *U.S. Civilian Nuclear Power Policy* (Washington, D.C.: DOE, 1986); AEC, *Civilian Nuclear Power . . . A Report to the President—1962* (Oak Ridge, Tenn.: AEC, 1962); AEC, *Annual Report, 1964,* pp. 98–101; AEC, *Annual Report, 1965,* pp. 125–28.

46. Holl, Anders, and Buck, *U.S. Civilian Nuclear Power Policy,* pp. 12–13; Richard G. Hewlett and Bruce J. Dierenfeld, *The Federal Role and Activities in Energy Research and Development, 1946–1980: An Historical Summary* (Oak Ridge, Tenn.: AEC, 1983), p. 49.

47. Albert Crewe, "The Role of a National Laboratory in the Power Development Field," address to the American Power Conference, April 28, 1965, AEC 956/20, May 6, 1965, AEC Secretariat, DOE.

48. Milton Shaw to Stephen Lawroski, "Cancellation of FARET," n.d., Folder "LMFBR," W. Massey Subj. Files, ANL; John T. Conway to JCAE, "AEC Fast Reactor Program Changes," Nov. 26, 1965 and Seaborg to Chet Holifield, Nov. 24, 1965, Gen. Corres., JCAE, RG 128, NARA; AEC, LMFBR Program, ANL press release, Nov. 26, 1965, Folder "LMFBR Prog. '65," W. Massey Subj. Files, ANL.

49. AEC, Minutes of Meeting 2252, Jan. 16, 1967; Meeting 2254; Meeting 2255; AEC 588/37, Fast Flux Test Facility (FFTF); AEC 588/39, Fast Breeder Power Reactor Development Program—Special Analytical Study; AEC 588/40—Supplement to AEC 588/37, AEC Secretariat, DOE; *Chicago Tribune,* Dec. 2, 1965.

50. Shaw to Crewe, Nov. 26, 1965, Folder "LMFBR Prog. '65," W. Massey Subj. Files, ANL.

51. Ibid.; The Role of ANL in the LMFBR Program, Nov. 24, 1965; Outline of Major RDT Needs, Nov. 1965; Functions of ANL LMFBR Program Office in the LMFBR Program, Nov. 1965, all in Folder "LMFBR Prog. '65," W. Massey Subj. Files, ANL.

52. "The Program Office," *Argonne News,* April 1968, pp. 6–9; Paul F. Gast, "LMFBR Problems—Lecture Notes," 1968 AMU-NEEC Faculty Student Conference, ANL, Aug. 26–30, 1968, File TPD-68-41, ANL, TIC; "The Overall Plan—LMFBR Program Plan," File TPD-68-49, ANL, TIC.

53. "The Role of ANL in Support of the Fast Flux Test Reactor Project," Nov. 1965, in Folder "LMFBR Prog. '65," W. Massey Subj. File, ANL.

54. "The Role of ANL in the LMFBR Program," Nov. 24, 1965, Folder "LMFBR Prog. '65," W. Massey Subj. File, ANL.

55. Argonne National Laboratory Reactor Engineering Division Review Commit-

tee meeting, December 6, 1966, AUA Review Committees—Reactor Engineering, Collec. no. 434-89-0024, GLFRC.

56. Minutes of the Ninth Meeting of the Board of Trustees of Argonne Universities Association, April 13, 1967, AUA Board of Trustees Meetings, Vol. 1,and AUA Board Comments—Reactor Development, both in Collec. no. 434-89-0024, GLFRC.

57. Quotations from AUA Education Committee meeting, Aug. 17, 1967, AUA Board Comments on Education, Box 89, Collec. no. 434-89-0024, GLFRC; Plan of Merger, AUA/AMU, October 9/25, 1967, AUA Admin., Box 83, Collec. no. 434-89-0024, GLFRC; Robert B. Duffield to all staff, "Argonne Center for Educational Affairs," Sept. 20, 1968, Educational Affairs, Box 27, Collec. no. 434-91-0014, GLFRC; Greenbaum, *A Special Interest,* pp. 184–85.

58. Argonne Advanced Research Reactor, excerpts from transcripts of hearings before the Subcommittee on Legislation of the Joint Committee on Atomic Energy, Folder "LMFBR Prog. '65," W. Massey Subj. Files, ANL.

59. "Argonne's Need for a High Flux Reactor," Lab Director's Project Files, RG 326, NARA-GL; Proposal for an Advanced Research Reactor for the Argonne National Laboratory, July 15, 1961; Leonard E. Link, "History of High Flux Reactor Development at Argonne Starting with Events Leading First to Mighty Mouse and Eventually to AARR," Jan. 4, 1963; ANL Research Reactors, n.d., all in Folder "LMFBR Prog. '65," W. Massey Subj. Files, ANL.

60. Joint Committee on Atomic Energy, "The Future Role of the Atomic Energy Commission Laboratories," January 1960, p. 65.

61. Seaborg to Crewe, Dec. 20, 1962 and Crewe to George W. Beadle, Jan. 7, 1963, W. Massey Subj. File, ANL.

62. Edward J. Bauser to files, Feb. 6,l963, Gen. Corres., RG 128, NARA; C. N. Kelber, "Comparison of AARR with other High Flux Reactors or under Construction," Jan. 4, 1963, Folder "LMFBR Prog. '65," W. Massey Subj. Files, ANL; Comptroller General of the United States, "Review of Selected Construction Projects-Atomic Energy Commission," Report to Joint Committee on Atomic Energy, Feb. 19, 1968, RG 128, NARA.

63. Argonne Advanced Research Reactor (AARR) meeting, AEC Headquarters, "H" Street Offices, March 2, 1966, and McDaniel to Crewe, March 16, 1966, Lab Director's Proj. Files, AARR, RG 326, NARA-GL.

64. Crewe to McDaniel, May 19, 1966; see also "Justification of Need—AARR," both in Lab Director's Proj. Files, AARR, RG 326, NARA-GL.

65. Crewe to Dunbar, October 4, 1966, Lab Director's Proj. Files, AARR, RG 326, NARA-GL.

66. I. Charak to distribution, trip report—AEC Division of Reactor Licensing, Feb. 3, 1967; Crewe to Kenneth A. Dunbar, March 10, 1967; Shaw to Dunbar, March 28, 1967; Crewe to Dunbar, April 6, 1967; Crewe to Dunbar, May 15, 1967, all in Lab Director's Proj. Files, AARR, RG 326, NARA-GL.

67. Shaw to P.A. Morris, director, Division of Reactor Licensing, May 12, 1967; Crewe to Dunbar, May 15, 1967; Shaw to Dunbar, May 15, 1967; Dunbar to Crewe, May 22, 1967, all in Lab Director's Proj. Files, AARR, RG 326, NARA-GL.

68. S. Lawroski to Dunbar, May 23, 1967; Crewe to Dunbar, May 24, 1967; Shaw to Lawroski, May 25, 1967; Crewe to Dunbar, May 26, 1967; M. Levenson to Lawroski, "AARR Pressure Vessel," all in Lab Director's Proj. Files, AARR, RG 326, NARA-GL.

69. *Argonne News,* July 1967, pp. 8–11.

70. Shaw to Dunbar, Aug. 24, 1967; Shaw to Dunbar, Sept. 14, 1967; Levenson to Dunbar, Sept. 14, 1967, all in Lab Director's Proj. Files, AARR, RG 326, NARA-GL; N. J. Palladino to Seaborg, Oct. 12, 1967; Harold L. Price to John T. Conway, Oct. 16, 1967, Gen. Corres., JCAE, RG 128, NARA.

71. *Chicago Daily News,* Dec. 7, 1966; *Chicago Sun-Times,* Dec. 12, 1966; *Argonne News,* Dec. 1966, p.10; University of Chicago press release, June 2, 1967, UC.

72. George Beadle to Duffield, Aug. 7, 1967; Harrell to Beadle, Aug. 9, 1967; Duffield to Beadle, Aug. 28, 1967, all in Folder "ANL Directorship," W. Massey Subj. Files, ANL; Howard C. Brown to John T. Conway, Aug. 16, 1967; AEC press release, Aug. 18, 1967, both in Gen. Corres., ANL, RG 128, JCAE, NARA; *Argonne News,* Sept. 1967, p. 2.

Chapter 8

1. Philip Powers, "A Challenge to Midwestern Universities, A Report by the President of Argonne Universities Association," Argonne Universities Association, Oct. 1967.

2. Ibid.

3. "Response to AEC Questions—Background on Formation of [Argonne Laboratory] Senate," Senior Scientists Group, Collec. no. 434-89-0024, GLFRC.

4. Duffield to Spofford G. English, Aug. 6, 1968; Responses to AEC Questions: Background on Formation of Senate, Aug. 6, 1968; Argonne National Laboratory Senate: Report of Membership, Bulletin no. 6, Sept. 30, 1968, all in Folder "Senior Scientists Group," Collec. no. 434-89-0024, GLFRC; "Argonne Senate Reviews Its First Year, Tells Plans," *Argonne News,* Nov. 1968, pp. 4–5.

5. Robert B. Duffield, "Some Remarks on Argonne National Laboratory Matters," AUA Board of Trustees Meeting, July 1, 1968, AUA Board of Trustees Meetings, Vol. 1, Collec. no. 434-84-0024, GLFRC.

6. Ibid.

7. Chet Holifield to Elmer B. Staats, controller general, Nov. 15, 1966, AUA Admin., Box 83, Collec. no. 434-89-0024, GLFRC; ANL, Report of the Study Group on Environmental Pollution, Feb. 27, 1967, TPD-67-14, ANL, TIC.

8. ANL, Report of the Study Group on Environmental Pollution, Feb. 27, 1967, TPD-67-14, ANL, TIC; AUA, Minutes of the Exec. Comm. for Argonne National Laboratory Affairs, Dec. 16, 1966, Folder "AUA Exec. Comm.," Vol. 1, Box 89, Collec. no. 434-89-0024, GLFRC.

9. E. J. Croke and B. Hoglund to B. I. Spinrad, Nov. 30, 1966, Folder "AUA Board Committee on Environ. Pollution," Box 89, Collec. no. 434-89-0024, GLFRC.

10. Rev. 9:2, 18 (Revised Standard Version); ANL, "Report of the Study Group on Environmental Pollution," Feb. 27, 1967, File TPD-67-14, ANL, TIC.

11. W. M. Manning to John T. Middleton, director, NCAPC, Sept. 6, 1967; S. G.

English to Middleton, Sept. 7, 1967; R. E. Hollingsworth to John T. Conway, JCAE, Oct. 25, 1967, all in JCAE, Gen. Corres., Air Pollution, Box 511, RG 128, NARA.

12. Argonne Position on Environmental Studies, June 21, 1968, AUA Board of Trustees Meetings, Box 88, Collec. no. 434-84-0024, GLFRC; ANL CES, "Federal Laboratories As Centers of Excellence in the Environmental Sciences, A Case Study," File TPD-72-74, ANL, TIC; ANL, *Annual Report: 1968*, p. 17; A. A. Jonke, "Reduction of Atmospheric Pollution by the Application of Fluidized Bed Combustion," Oct. 1968, File TPD-274, ANL, TIC.

13. ANL, *Annual Report: 1968*; ANL press release, "News from Argonne National Laboratory" (69-9), April 7, 1969, OPA, ANL; ANL, "Development of the Argonne Braille Machine," Dec. 1972, File TPD-72-90, ANL, TIC; AEC press release, "AEC Laboratory Developing New, Small Artificial Kidney for NIH," July 8, 1969, JCAE, Gen. Corres., Box 4, RG 128, NARA; ANL press release, "News From Argonne National Laboratory" (67-14), Aug. 14, 1967, OPA, ANL; "Development of High-Energy Batteries for Electric Vehicles," July 1970, File TPD-70-20, ANL, TIC.

14. D. C. White, "Report to Argonne Universities Association on Its Role in Environmental Pollution," April 4, 1967, Folder "AUA Board Committee Environ. Pollution," and AUA, Staff Paper No. 1 on the Salt Creek Project, Nov. 27, 1967, Folder "AUA Exec. Comm.," both in Box 89, Collec. no. 434-89-0024, GLFRC.

15. AUA, Staff Paper no. 1, Nov. 27, 1967; AUA, Minutes of the Executive Committee, Nov. 27, 1967, Folder "AUA Exec. Comm."; Minutes of the Second Meeting of the Environmental Pollution Committee, Folder "AUA Board Committee Environ. Pollution," all in Box 89, Collec. no. 434-89-0024, GLFRC.

16. AUA, Minutes of the Education Committee Meeting, Dec. 7, 1967, Folder "AUA Board Committee Education"; Philip Powers to J. Boyd Page, Sept. 4, 1968, Folder "AUA Exec. Comm.," both in Box 89, Collec. no. 434-89-0024, GLFRC.

17. AUA, Minutes of the Fourteenth Meeting of the Board of Trustees, Oct. 14, 1968, AUA Board of Trustees Meetings, Vol. 2, Box 88; Report of the AUA ad hoc Committee on Socio-Technological Research Organization, Folder "AUA Board Committee Environ. Pollution," Box 89; Minutes of the AUA Board Committee on Environ. Studies, Jan. 8, 1969, Folder "AUA Board Committee Environ. Pollution," Box 89, all in Collec. no. 434-89-0024, GLFRC.

18. L. E. Link, "Argonne Position on Environmental Studies—Additional Considerations," Jan. 3, 1969; Link to Duffield, "Argonne's Role with Environmental Studies," Jan. 7, 1969, Folder "AUA Board Committee Environ. Pollution," Box 89, Collec. no. 434-89-0024, GLFRC.

19. AUA, Minutes of the Sixteenth Meeting of the Board of Trustees, March 31, 1969, AUA Board of Trustees Meetings, Vol. 2, Box 88, Collec. no. 434-89-0024, GLFRC.

20. "The Problem of Man's Environment: Summary of Conference," *Argonne News*, Sept. 1969, pp. 6–9; Frederick D. Rossini, Summary Comments on the Problem of Man's Environment, July 29, 1969, Folder "AUA Board Committee Environ. Pollution," Box 89, Collec. no. 434-89-0024, GLFRC.

21. ANL, "A Proposed Study of the Effects of Heated Discharges in the Great Lakes,"

Sept. 3, 1968, File TPD-69-29, ANL, TIC; Robert D. O'Neill to Edward J. Bauser, JCAE, Gen. Corres., Pollution, Box 511, RG 128, NARA; ANL press release, "News from Argonne National Laboratory (70-12 and 70-42)," OPA, ANL; ANL, *Annual Report: 1969*, File TPD-69-5, ANL, TIC; ANL, *Annual Report: 1970*, File TPD-70-3, ANL, TIC.

22. Frederick D. Rossini, "Report to the ad hoc Committee on Environmental Programs," Oct. 20, 1969; AUA, Report of the AUA Special Committee on Environmental Programs, Jan. 20, 1970, both in Folder "AUA Board Committee Environ. Pollution," Box 89, Collec. no. 434-89-0024, GLFRC.

23. Duffield to staff employees, Dec. 1, 1969, AUA Administrative, Box 83, Collec. no. 434-89-0024, GLFRC; *Argonne News*, Dec. 1969, p. 2.

24. Résumé of meeting, AUA Board Committee Environ. Pollution, May 27, 1970; Recommendation to the AUA Board of Trustees by the Board Committee Environ. Studies, June 22, 1970; "A Condensation of R. Rossini's Report to the Annual Meeting of the AUA on October 12, 1970, Showing the Evolvement of AUA Policy Relative to Environmental Research," all in Folder "AUA Board Committee Environ. Pollution," Box 89, Collec. no. 434-89-0024, GLFRC.

25. AUA, "Midwest Regional Environmental Systems Program (MRESP)"—A proposal to the National Science Foundation, Sept. 1971, Box 28, Collec. no. 434-91-0014, GLFRC; "Historical Development of the Proposed Midwest Regional Environmental Systems Program," n.d. and AUA staff paper on the proposed Midwest Regional Environmental Systems Program, Jan. 22, 1971, in Folder "AUA Board Committee Environ. Pollution," Box 89, Collec. no. 434-89-0024, GLFRC; "Outline and Summary of the Midwest Regional Environmental Systems Program," Jan. 9, 1971, and Lynn E. Weaver to members of the AUA Board of Trustees, Feb. 16, 1971, in Folder "AUA Board Committee Environ. Pollution," Box 28, Collec. no. 434-91-0014, GLFRC.

26. W. C. Redman to M. V. Nevitt, March 19, 1971; Nevitt to A. C. Upstrom, April 26, 1971; E. J. Croke to L. E. Link, May 7, 1971, all in Folder "AUA Board Committee Environ. Pollution," Box 28, Collec. no. 434-91-0014, GLFRC; W. B. Harrell to W. B. Cannon, July 28, 1971, William B. Harrell Papers, Box 2, UCRL SC; ANL press release, "News from Argonne National Laboratory (71-20)," OPA, ANL.

27. Duffield to Spofford G. English, Nov. 2, 1971; English to Duffield, Nov. 12, 1971, Folder "AUA Board Committee Environ. Pollution," Box 28, Collec. no. 434-91-0014, GLFRC.

28. Daniel Alpert to Philip Powers, Jan. 10, 1971; Minutes of the meeting of the Board Committee Environ. Studies, Jan. 28, 1971; Alpert to presidents of the member universities of the Argonne Universities Association, Oct. 29, 1971; Alpert to Duffield, Oct. 31, 1971; Edwin L. Goldwasser to Norman Hackerman, Jan. 26, 1971, all in Folder "AUA Board Committee Environ. Pollution," Box 28, Collec. no. 434-91-0014, GLFRC.

29. Powers to members of the AUA Board of Trustees, Nov. 3, 1971; Norman Hackerman to the presidents, Nov. 3, 1971; Dean Alpert, "The Argonne Universities Association and the Support of Environmental Studies," Nov. 16, 1971; Resolution, Nov. 16, 1971, all in Folder "AUA Board Committee Environ. Pollution," Box 28, Collec. no. 434-91-0014, GLFRC; AUA, Minutes of the Twenty-Seventh Meeting of the Board of Trust-

ees, Nov. 16, 1971, AUA Board of Trustees Meetings, Box 88, Collec. no. 434-89-0024, GLFRC.

30. ANL CES, "Argonne National Laboratory Environmental Research Program," Sept. 1972, File TPD-72-76, ANL, TIC; ANL CES, "Measurements of Air Pollutant Concentrations at O'Hare International Airport, Chicago and Orange County Airport, California," File TPD-72-54, ANL, TIC; Duffield to C. L. Henderson, "Office of Regulation, U.S. AEC," May 3, 1971, Environmental Statement Project, Box 22, Collec. no. 434-89-0024, GLFRC.

31. Levenson to Duffield, Nov. 29, 1967; J. H. McKinley to Dunbar, Dec. 15, 1967; Arthur Schoenkaut, deputy controller, GAO, to Dunbar, "GAO Review of Selected AEC Construction Projects," Dec. 8, 1967, all in Lab Director's Project Files, AARR, RG 326, NARA-GL.

32. Schoenkaut to Dunbar, GAO Review of Selected AEC Construction Projects, Dec. 8, 1967, Lab Director's Proj. Files, AARR, RG 326, NARA-GL.

33. Dunbar to Shaw, "Construction Project Starts and Overruns—AARR," Jan. 9, 1968; Duffield to Dunbar, Jan. 16, 1968; Dunbar to McDaniel, "Argonne Advanced Research Reactor—Revised Cost Estimate," Jan. 24, 1968, all in Lab Director's Project Files, AARR, RG 326, NARA-GL.

34. Powers to McDaniel, Feb. 16, 1968, Lab Director's Project Files, AARR, RG 326, NARA-GL.

35. D. Connor to W. M. Manning, "Highlights of March 2 Meeting with Messrs. Bauser and Radcliffe," JCAE, March 15, 1968, Lab Director's Proj. Files, AARR, RG 326, NARA-GL.

36. McDaniel and Shaw to Dunbar, April 2, 1968; Dunbar to Duffield, April 2, 1968; telephone dictation from McDaniel, April 4, 1968, all in Lab Director's Project Files, AARR, RG 326, NARA-GL; *Argonne News,* April 1968, p. 3.

37. Otto Kerner to Crewe, Dec. 22, 1966, Fermi Natl. Accel. Lab, Collec. no. 434-89-0024, GLFRC.

38. Board Committee on High Energy Physics, Thirteenth Meeting of the AUA Board of Trustees, July 1–2, 1968, AUA Board of Trustees Meetings, RG 326, NARA-GL.

39. Ibid.

40. Norman Ramsey to Crewe, Feb. 14, 1967; Crewe to Wilson, March 13, 1967; Wilson to Manning, July 31, 1967, all in Fermi Natl. Accel. Lab., Collec. no. 434-89-0024, GLFRC.

41. Wilson to Duffield, Nov. 29, 1967; Lawroski to Shaw, Dec. 7, 1967; Shaw to Lawroski, Jan. 16, 1968; Duffield to Wilson, Jan. 24, 1968, Fermi Natl. Accel. Lab, Collec. no. 434-89-0024, GLFRC.

42. Robert B. Duffield interview, May 24, 1992; AEC Meeting 2343, Sept. 10, 1968, AEC Minutes, DOE; Duffield to Wilson, March 12, 1968; Wilson to Duffield, March 15, 1968; Lee Teng to Wallace Givens, director, Applied Mathematics Division, April 22, 1968; Wilson to Duffield, June 10, 1968; McKinley to Poillon, June 21, 1968, all in Fermi Nat. Accel. Lab., Collec. no. 434-89-0024, GLFRC.

43. Victor Weisskopf to Duffield and Wilson, Aug. 12, 1968, Folder "AUA Exec. Comm.," Collec. no. 434-89-0024, GLFRC.

44. Wilson to Weisskopf, Aug. 22, 1968, Folder AUA Exec. Comm.," Collec. no. 434-89-0024, GLFRC.

45. AEC, "Policy for National Action in the Field of High Energy Physics," Jan. 24, 1965, Folder "AUA Exec. Comm.," JCAE, Collec. no. 434-89-0024, GLFRC; Duffield to Weisskopf, Aug. 26, 1968, and Wilson to Weisskopf, Aug. 22, 1968, in Folder "AUA Exec. Comm.," Collec. no. 434-89-0024, GLFRC.

46. Goldwasser et al. to Powers, Oct. 6, 1967, and Powers to Goldwasser, Jan. 9, 1965, in Folder "AUA Board Committee High Energy Physics," Collec. no. 434-91-0014, GLFRC; ANL press release, "News from Argonne National Laboratory," July 1, 1968, ANL.

47. See High Energy Physics Advisory Panel, "Report on High Energy Physics," June 1969, File TPD-69-30, ANL, TIC; ANL press release, "News from Argonne National Laboratory," June 15, 1967, Jan. 22, 1969 and Oct. 13, 1969, all ANL-OPA.

48. Goldwasser to Duffield, Jan. 20, 1969; Duffield to Wilson, July 21, 1969; Cork, notes on meeting, Friday, Sept. 26, 1969 at ANL, all in Fermi Nat. Accel. Lab., 67–74, Collec. no. 434-89-0024, GLFRC.

49. George Casarett et al., "Report on the Workshop on Long Range Planning for the Argonne National Laboratory," Chicago, June 15–17, 1970, and Bruce Cork to AUA Long Range Planning Committee, June 4, 1970, in AUA Long Range Plans, Collec. no. 434-89-0024, GLFRC.

50. McDaniel to Powers, March 12, 1969; Board Committee on High Energy Physics to AUA Executive Board, June 23, 1972, both in Folder 59.6B "AUA Board Committee High Energy Physics," Collec. no. 434-91-0014, GLFRC.

51. Homer A. Neal to Powers, October 5, 1972, Folder 59.6B "AUA Board Committee High Energy Physics," Collec. no. 434-91-0014, GLFRC.

52. Ibid.

53. Report of Review Committee, Reactor Engineering Division, Argonne National Laboratory, July 12–13, 1967, Folder "AUA Review Committee Reactor Engineering," Box 86, Collec. no. 434-89-0024, GLFRC; Minutes of the Twelfth Meeting of the Board of Trustees of Argonne Universities Association, January 19, 1968, Folder "AUA Board of Trustees Meetings," Vol. 1, Collec. no. 434-89-0024, GLFRC.

54. Albert V. Crewe, "The Role of a National Laboratory in the Power Development Field," speech to the American Power Conference, April 28, 1965, AEC 956/20, May 6, 1965, AEC Secretariat, DOE.

55. Minutes of the meeting of the Executive Committee of the Argonne Universities Association, May 16, 1967, Folder "AUA Exec. Comm.," Box 89, Collec. no. 434-89-0024, GLFRC; Minutes of the Tenth Meeting of the Board of Trustees of Argonne Universities Association, June 29, 1967, AUA Board of Trustees, Vol. 1, Collec. no. 434-89-0024, GLFRC.

56. "Sodium Release and Fire in the EBR-II Sodium Boiler Plant Building Control Room on February 9, 1958," March 5, 1968, File TPD-68-12, ANL, TIC; H. O Monson to Murray Joslin, May 22, 1968; Ad hoc Committee for EBR-II and "Visit to EBR-II," April 11, 12, 13, 1968, in Folder "AUA Review Committee Reactor Engineering," Vol. 1, Collec. no. 434-89-0024, GLFRC.

57. H. O. Monson to Murray Joslin, May 22, 1968, Folder "AUA Review Committee Reactor Engineering," Vol. 1, Collec. no. 434-89-0024, GLFRC.

58. Shaw to Duffield, June 6, 1968, Director's Files, Box 6, Collec. no. 434-91-0015, GLFRC.

59. Ibid.

60. William B. Cannon to file, re Robert B. Duffield, Nov. 11, 1972, Folder "Review of Opers. of ANL by W. B. Cannon," Box 8, W. Massey Subj. Files, ANL.

61. Duffield to all staff, "Reactor Development Program," Sept. 6, 1968, Director's Files, Box 6, Collec. no. 434-91-0015, GLFRC; Minutes of meeting of Board Committee for Reactor Development, Oct. 28, 1968, Folder "AUA Board Committee Reactor Development," Vol. 1, Collec. no. 434-89-0024, GLFRC; Minutes of the meeting of the Exec. Comm. of Argonne Universities Association, Sept. 19, 1968, Folder "AUA Exec. Comm.," Box 89, Collec. no. 434-89-0024, GLFRC; ANL press release, "News from Argonne National Laboratory," Sept. 10, 1968, OPA, ANL.

62. EBR-II Review Committee Report, Aug. 12–13, 1969; EBR-II Review Committee Report, July 21–22, 1971; EBR-II Review Report Outline, July 19–20, 1972, all in Folder "AUA Review Committee EBR-II," Box 87, Collec. no. 434-89-0024, GLFRC; ANL, *Annual Report: 1973,* TPD-659, ANL, TIC.

63. ANL press releases, "News from Argonne National Laboratory" (69-12, 69-12T), April 18, 1969; ANL press release, "News from Argonne National Laboratory" (69-42), Nov. 20, 1969; ANL press release, "News from Argonne National Laboratory" (70-4), Jan. 13, 1970, all OPA, ANL.

64. Shaw to R. V. Laney, Feb. 4, 1972; Laney to R. K. Winkleblack, Feb. 11, 1972, Files, Folder "LMFBR Prog. '65," Box 6, W. Massey Subj. ANL; ANL press releases, "News from Argonne National Laboratory" (69-26), July 31, 1969, and "News from Argonne National Laboratory" (72-25), June 30, 1972, OPA, ANL; ANL, *Annual Report: 1972,* ANL, TIC.

65. William Kerr, "Reactor Development Committee Report to AUA Board of Trustees," Oct. 8, 1969, Folder "AUA Board Committee Reactor Development," Box 59, Collec. no. 434-89-0024, GLFRC.

66. Cannon to files, re Robert B. Duffield, Nov. 13, 1972, Folder "Review of Opers. of ANL," Box 8, W. Massey Subj. Files, ANL; Melvin Price to Cannon, Nov. 28, 1969, Folder "LMFBR," Box 6, W. Massey Subj. Files, ANL.

67. Cannon to files, re Robert Duffield, Nov. 13, 1972, Folder "Review of Opers. of ANL," Box 8, W. Massey Subj. Files, ANL.

68. ANL press release, "News from Argonne National Laboratory" (69-31), OPA, ANL; ANL, "Affirmative Action Compliance Program," 1971; William B. Harrell Papers, Box 2, UCRL SC; ANL, *Annual Report: 1970,* File TPD-70-3, ANL, TIC.

69. ANL press release, "News from Argonne National Laboratory" (70-6), Jan. 19, 1970, OPA, ANL; ANL, "LMFBR-Liquid Metal Fast Breeder Reactor," File TPD-70-21, ANL, TIC; *Argonne News,* March 1970, pp. 4–5.

70. Cannon to files, re Robert B. Duffield, Nov. 13, 1972, Folder "Review of Opers. of ANL," Box 8, W. Massey Subj. Files, ANL; Duffield to Milton Shaw, Feb. 12, 1970, Div. of Reactor Dev. and Tech., Box 17, Collec. no. 434-89-0024, GLFRC.

71. *Argonne News,* April 1970, p. 3; AUA, Minutes of the Twenty-Second Meeting of the Executive Committee, April 13, 1971, Folder "AUA Exec. Comm.," Vol. 2, Box 90, Collec. no. 434-89-0024, GLFRC.

72. Duffield to all employees, April 17, 1970, ANL Personnel Robert V. Laney, Box 20, W. Massey Subj. Files, ANL; Cannon and Power to Seaborg, draft, Feb 10, 1970, Folder "LMFBR," Box 6, W. Massey Subj. Files, ANL; Hewlett and Duncan, *Nuclear Navy.*

73. "Special Message to the Congress on Energy Resources," June 4, 1971, *Public Papers of the Presidents of the United States, Richard M. Nixon, 1971* (Washington, D.C.: GPO, 1972), pp. 706–7.

74. Minutes of meeting in R. V. Laney's office, "ANL's Role in the LMBBR Program," Box 3, William B. Harrel Papers, UCRL SC; Cannon to E. H. Levi and J. T. Wilson, "Management Improvements at the Argonne Laboratory," Nov. 18, 1972, Folder "Review of Opers. of ANL," Box 8, W. Massey Subj. Files, ANL; Cannon to Levi, Aug. 6, 1970, Folder "LMFBR Prog. '65," Box 6, W. Massey Subj. Files, ANL.

75. Duffield to associate directors, division directors, and key administrative officials, Dec. 17, 1971; Duffield to distribution, Jan. 31, 1972; Cannon to E. H. Levi and J. T. Wilson, "Management Improvements at the Argonne Laboratory," Nov. 18, 1972, all in Folder "Review of Opers. of ANL," Box 8, W. Massey Subj. Files, ANL; "The Argonne Laboratory: Relevance Closes In," *Science and Government Report,* Nov. 1, 1972.

76. Cannon to Powers, May 10, 1972, ANL Personnel Director Robert Duffield, Box 20, W. Massey Subj. Files, ANL.

77. S. A. Miller to Duffield, Laney, and Nevitt, April 14, 1972; AUA, Minutes of the Twenty-Ninth Meeting of the Board of Trustees, April 18, 1972 and Minutes of the Thirtieth Meeting of the Board of Trustees, in AUA Board of Trustees Meetings, Box 88, Collec. no. 434-89-0024, GLFRC.

78. Wilson to Edward H. Levi, June 2, 1972, Robert B. Duffield, Box 6, W. Massey Subj. Files, ANL.

79. Cannon to Levi, June 28, 1972, June 30, 1972, Sept. 5, 1972, Robert B. Duffield, Box 6, W. Massey Subj. Files, ANL.

80. AUA, Minutes of the Thirty-Second Meeting of the Board of Trustees, Oct. 9–10, 1972, AUA Board of Trustees Meetings, Box 88, Collec. no. 434-89-0024, GLFRC; "The Argonne Lab: Universities Seek Greater Influence," *Science & Government Report,* Dec. 1, 1972.

81. Duffield to Cannon, Nov. 19, 1972, Oct. 30, 1972; Cannon to files, Oct. 16, 1972; Duffield to all employees, all in Robert B. Duffield, Box 20, W. Massey Subj. Files, ANL.

82. "The Argonne Lab: Relevance Closes In," *Science and Government Report,* Nov. 1, 1972; James Quinn, "Duffield Forced Out? Argonne Director Quits," *Chicago Tribune,* Oct. 30, 1972.

83. Cannon to Levi, Nov. 16, 1972; S. D. Golden to Cannon, Appointment of Acting Director of Argonne National Laboratory, Nov. 29, 1972; Powers to the AUA Board of Trustees, Dec. 5, 1971, all in ANL.

84. Cannon to Levi and Wilson, Acting Director Issue, Dec. 8, 1972, W. Masey Subj. Files, Robert V. Laney, Box 20, ANL.

Chapter 9

1. Minutes of the 4125th Meeting of the GAC, July 9–11, 1973, Folder "GAC Mtg. July 9–11, 1973," GAC Minutes and Reports of Meetings, RG 326, NARA.

2. George J. Poli to Cannon, Nov. 17, 1972, Mgmt. Reviews 1971–1972, Box 1, W. Massey Subj. Files, ANL.

3. Robert Sachs, interview by the author, Dec. 10, 1991, Univ. of Chicago.

4. Minutes of the GAC, July 9–11, 1973, RG 326, NARA.

5. Cannon to Robert H. Bauer, March 12, 1974, William B. Harrell Papers, Box 2, UCRL, SC.

6. Robert V. Laney, "The Role of the National Laboratory," Nov. 28, 1972, in ANL Personnel Robt. V. Laney, Box 20, W. Massey Subj. Files, ANL. Also presented as speech to the Chicago chapter, American Nuclear Society, ANL, Nov. 30, 1972, and cited in *Argonne News*, Jan.–Feb. 1973, pp. 3, 6.

7. Ibid.

8. Ken Hechler, *Toward The Endless Frontier: History of the Committee on Science and Technology, 1959–1979* (Washington, D.C.: GPO, 1980), p. 615.

9. FPC, *Northeast Power Failure, November 9 and 10, 1965: A Report to the President, Dec. 6, 1965* (Washington, D.C.: GPO, 1965).

10. Ibid.

11. FPC, *El Paso Power Failure, December 2, 1965, A Report to the Committee on Commerce, United States Senate,* April 11, 1966; FPC, *Gulf States Utilities Company Power Failure, December 6, 1965, A Report to the Committee on Commerce, United States Senate,* April 28, 1966; Working Group Western Operations Subcommittee of the North American Power Systems Interconnection Committee, *Report on Disturbance in Western Interconnection, June 7, 1966*; San Francisco Regional Office, FPC, *Report on the Southern California Edison Company Power Interruption, July 19, 1966,* dated March 1967, all in IE242, Electric Outage Data, 1942–1965, Norton Savage Files, DOE.

12. FPC, *Power Interruption: Pennsylvania–New Jersey–Maryland Interconnection, June 5, 1967,* dated March 1968, IE242, Electric Outage Data, 1967–1968, Norton Savage Files, DOE.

13. Statement by Senator Edward M. Kennedy on electric power blackouts, submitted to Massachusetts Department of Public Utilities, Jan. 23, 1969; FPC, *Power Service Interruption, Cape and Vineyard Electric Company, Cape Cod Area of Massachusetts, July 3, 1967,* October 1967; FPC, *Report on the Boston Edison Company Service Interruption, December 28–30, 1968,* Jan 27, 1969, all in IE242, Electric Outage Data, 1967–1968, Norton Savage Files, DOE.

14. Holl, Anders, and Buck, *U.S. Civilian Nuclear Power Policy,* p. 15; Neil De Marchi, "Energy Policy under Nixon: Mainly Putting Out Fires," in *Energy Policy in Perspective: Today's Problems, Yesterday's Solutions,* ed. Craufurd D. Goodwin (Washington, D.C.: 1982), pp. 398, 406.

15. Hechler, *Toward the Endless Frontier,* pp. 655–69.

16. Jack Holl and Terrence R. Fehner, *Securing America's Energy Future: The United States Department of Energy, 1977–1989* (Washington, D.C.: GPO, 1990), pp. 4–5;

Elliot L. Richardson and Frank G. Zarb, "Perspective on Energy Policy," Energy Resources Council, Dec. 1976, DOE; Rodney P. Carlisle and Jack M. Holl, "The Significance of Energy Abundance in Recent American History," DOE Historians Office, April 1980, DOE; De Marchi, "Energy Policy under Nixon," pp. 447–52.

17. Hechler, *Toward the Endless Frontier*, pp. 668–69.

18. ANL Policy Committee Meeting, Jan. 22, 1973, Box 9, Collec. no. 434-89-0024, GLFRC.

19. AEC Division of Research, *Fusion* (Oak Ridge, Tenn.: AEC, n.d.).

20. ANL press release, "Energy Crisis," 1973, OPA, ANL; Argonne National Laboratory, *Annual Report: 1973,* File TPD-73-3, ANL; for an excellent history see, Joan L. Bromberg, *Fusion: Science, Politics, and the Invention of a New Energy Source* (Cambridge, Mass.: 1982).

21. Bromberg, *Fusion*, 8, 56–57, 261; ANL press release, "Energy Crisis," 1973, OPA, ANL; Argonne National Laboratory, *Annual Report: 1973,* File TPD-73-3, ANL.

22. ANL press release, "Energy Crisis," 1973, OPA, ANL; GAC, "Long-Range Research Review for the Atomic Energy Commission," March 1974, Box 15, Collec. no. 434-89-0024, GLFRC; Albert H. Teich and Mark E. Rushefsky, "Diversification at Argonne Laboratory," May 1976, Participating Institutions, Box 6, W. Massey Subj. Files, ANL.

23. ANL press release, "Energy Crisis," 1973, OPA, ANL; AEC, *Annual Report to Congress, 1974,* Jan. 17, 1975, pp. 36–37; Teich and Rushefsky, "Diversification at Argonne National Laboratory," May 1976, all in Participating Institutions, Box 6, W. Massey Subj. Files, ANL.

24. Cannon to Hollingsworth, Jan. 17, 1973; Cannon to ANL Faculty Advisory Committee, Jan. 22, 1973; Cedric Chernick to Cannon, Feb. 23, 1973, all in The Crisis and Reorganization of ANL, Box 9, W. Massey Subj. Files, ANL.

25. Cannon to E. H. Levi, Feb. 2, 1973; Laney to Cannon, Feb. 8, 1973; Laney to Shaw, Feb. 5, 1973; Robert Jackson, Jr., to division directors and department heads, Feb. 14, 1973; Laney to all employees, Feb. 13, 1973, all in Folder "ANL Budget Misc.," Box 18, W. Massey Subj. Files, ANL.

26. ANL press release, Feb. 14, 1973; Laney to Cannon, Feb. 8, 1973; Laney to all employees, Feb. 13, 1973, all in Folder "ANL Budget Misc.," Box 18, W. Massey Subj. Files, ANL.

27. Hollingsworth to Rep. John N. Erlenborn, March 12, 1973, and Martin R. Hoffman, general counsel to Sen. Charles H. Percy, March 23, 1973, both in Folder 18, "Procurements and Contracts," Univ. of Chicago, Box 7989, AEC Secretariat, DOE.

28. "New AEC Commissioner Pays Us a Visit," *Argonne News,* Dec. 1972, p. 5; "Dr. Ray Named Chairman, USAEC," *Argonne News,* March 1973, p. 2; "46,000 ANL, AEC-CH Documents Declassified," *Argonne News,* November 1973, p. 8; Robert Sachs, interview by author, Dec. 4, 1992, High Energy Physics Building, University of Chicago.

29. Louis Rosen to Cannon, Jan. 2, 1973, Folder "ANL Directors," Box 6, W. Massey Subj. Files, ANL; Cannon to Levi, Jan. 19, 1973, and Levi to Cannon, Jan. 24, 1973, Information Control on Release, Box 12, W. Massey Subj. Files, ANL.

30. "Dr. Robert G. Sachs: A Visit with Our New Director," *Argonne News,* April 1973,

pp. 1, 6–7; The NAS report was published in *Physics in Perspective*, Vol. 2, Part A, "The Core Subfields of Physics."

31. Levi to John T. Wilson, February 27, 1973; Cannon to Levi, March 1, 1973; Levi to all employees, March 13, 1973; Levi to Sachs, March 29, 1973; Robert Sachs, interview conducted by author, Dec. 4, 1992, High Energy Physics Building, University of Chicago, all in ANL.

32. Robert Sachs quoted in Cannon to Levi, Feb. 14, 1973, ANL 1969–1973, The Crisis and Reorganization of ANL, Box 6, W. Massey Subj. Files, ANL.

33. Robert Sachs quoted in "Dr. Robert G. Sachs: A Visit with Our New Director," *Argonne News*, April 1973, p. 1.

34. Robert Sachs, interview by author, Dec. 4, 1992, High Energy Physics Building, University of Chicago.

35. Robert Sachs, "The Director's Corner," *Argonne News*, July–Aug. 1973, 3.

36. *Atomic Energy Commission Research and Development Laboratories: A National Resource*, Sept. 1973, TID-26400, AEC.

37. Ibid.

38. "Report of the AEC-NSF Study of the Future Role of the Zero Gradient Synchrotron (ZGS)," Sept. 9, 1974, Folder "Zero Gradient Synchrotron," Box 12, W. Massey Subj. Files, ANL.

39. *Atomic Energy Commission Research and Development Laboratories: A National Resource*, Sept. 1973, TID-26400, AEC.

40. Ibid.; "J. J. Katz elected to NAS," *Argonne News*, May 1973, p. 2.

41. Robert Sachs, interview by author, Dec. 4, 1992, High Energy Physics Building, University of Chicago. Shaw's dialogue was re-created from Sachs's recollection.

42. Ibid.; Sachs, interview by Jack M. Holl and Ruth Harris, Dec. 10, 1991, High Energy Physics Building, University of Chicago.

43. Robert Sachs, interview by author, Dec. 4, 1992, High Energy Physics Building, University of Chicago.

44. AUA, Minutes of the Thirty-Sixth Meeting of the Board of Trustees, April 19, 1973, Vol. 4, Box 88, Collec. no. 434-89-0024, GLFRC.

45. Robert Gillette, "AEC Shakes Up Nuclear Safety Research," *Science*, vol. 180, June 1, 1973; Gillette, "Anders to the AEC," *Science*, vol. 181, August 3, 1973.

46. Milton Shaw to R. E. Hollingsworth, June 8, 1973, Folder 26, Box 7934, AEC Secretariat, DOE; AEC, Minutes of Commissioners' Exec. Sess. no. 21, June 14, 1973, Commissioners' Exec. Sess., Vol. 1, Box 3738, AEC Secretariat, DOE.

47. The White House Press Secretary, *President's Statement on Energy Summary Outline—Fact Sheet*, June 29, 1973.

48. AEC, Minutes of the Commissioners' Exec. Sess. no. 21, Commissioners' Exec. Sess., Vol. 1, Box 3738, AEC Secretariat, DOE.

49. Sachs to all employees, July 3, 1973, ERDA, Folder 4, Box 19, Collec. no. 434-89-0024, GLFRC; AUA, Minutes of the Thirty-Ninth Meeting of the Board of Trustees, Oct. 8–9, 1973, Vol. 4, Box 88, Collec. no. 434-89-0024, GLFRC.

50. Dixy Lee Ray, "The Nation's Energy Future: A Report to Richard M. Nixon,

President of the United States," Dec. 1, 1973, AEC WASH-1281; Quoted material from Robert Sachs, interview by author, Dec. 4, 1992, High Energy Physics Building, University of Chicago.

51. Jack M. Holl, "The Nixon Administration and the 1973 Energy Crisis: A New Departure in Federal Energy Policy," in *Energy and Transport: Historical Perspectives on Policy Issues,* eds. George H. Daniels and Mark H. Rose (Beverly Hills, Calif.: Sage Publications, 1982), pp. 149–56.

52. Robert G. Sachs, "National Energy Policy—A View from the Underside," in *National Energy Issues—How Do We Decide? Plutonium as a Test Case,* ed. Robert G. Sachs, Proceedings of a symposium of the American Academy of Arts and Sciences at Argonne National Laboratory, Sept. 29–30, 1978; Holl and Fehner, *Securing America's Energy Future,* pp. 5–8.

53. Robert G. Sachs, "Statement on H. R. 11510," Nov. 29, 1973, Folder 4, Box 19, Collec. no. 434-89-0024, GLFRC.

54. "The Energy Problem: How We at Argonne Will Meet It," *Argonne News,* Dec. 1973–Jan. 1974, p. 4.

55. AUA, Minutes of the Thirty-Eighth Meeting of the Board of Trustees, Sept. 7, 1973, and Paul W. McDaniel to the AUA Board of Trustees, Oct. 9, 1973, in AUA Board of Trustees Mtgs., Vol. 4, Box 88, Collec. no. 434-89-0024, GLFRC.

56. Excerpt from Minutes of the Thirty-First Meeting of the Exec. Comm., Nov. 13–14, 1973, "ANL Tripartite Contract Renewal '74," Box 23, W. Massey Subj. Files, ANL.

57. Cannon to Levi, Dec. 4, 1973, ANL "1969–73 Crisis and Reorganization," Box 9, W. Massey Subj. Files, ANL.

58. Paul W. McDaniel to members of the AUA Board of Trustees, Dec. 12, 1973, AUA Board of Trustees Mtgs., Vol. 4, Box 88, Collec. no. 434-89-0024, GLFRC; Bauer to McDaniel, Jan. 11, 1974, and McDaniel to the AUA Board of Trustees, Feb. 8, 1974, both in Folder "ANL Tripartite Contract Renewal '74," Box 23, W. Massey Subj. Files, ANL.

59. McDaniel to the AUA Board of Trustees, Feb. 8, 1974, Folder "ANL Tripartite Contract Renewal '74," Box 23, W. Massey Subj. Files, ANL.

60. AUA, excerpt from Minutes of the Thirty-Second Meeting of the Exec. Committee, Jan. 25, 1974; McDaniel to Levi, Jan. 27, 1974, mailgram, Folder "ANL Tripartite Contract Renewal '74," Box 23, W. Massey Subj. Files, ANL.

61. Levi to AUA presidents, Jan. 29, 1974; Levi to McDaniel, Jan. 29, 1974; John H. Roberson to AUA Board of Trustees, Jan. 30, 1974, all in Folder "ANL Tripartite Contract Renewal '74," Box 23, W. Massey Subj. Files, ANL.

62. Cannon to Levi and John T. Wilson, May 9, 1974, ANL Personnel, Director Robert Sachs, Box 20, W. Massey Subj. Files, ANL.

63. Bauer to John H. Roberson, Sec. AUA, Feb. 5, 1974; William Cannon to Edward Levi, Feb. 1, 1974; Lucille Schott to Cannon, Feb. 13, 1974; AUA, excerpt from Minutes of the Forty-First Meeting of the Board of Trustees, Feb. 20, 1974, all in Folder "ANL Tripartite Contract Renewal '74," Box 23, W. Massey Subj. Files, ANL; Cannon, memo to file, Feb. 4, 1974, ANL "1963–74 Crisis and Reorganization," Box 9, W. Massey Subj. Files, ANL.

64. Cannon to Levi, Feb 1. 1974; Bauer to McDaniel, March 15, 1974; McDaniel to Bauer, March 25, 1974; Thorne to McDaniel, June 12, 1974, all in Folder "ANL Tripartite Contract Renewal '74," Box 23, W. Massey Subj. Files, ANL; Cannon, memo to file, ANL 1969–73 Crisis and Reorganization, Box 9, W. Massey Subj. Files, ANL.

65. AUA, Minutes of the Special Meeting of the Members of the Argonne Universities Association, June 25, 1974; handwritten resolution passed at the AUA Board Meeting, June 25, 1974; McDaniel to presidents of AUA member universities, "Proposed Amendment of AUA's Founders Agreement," June 26, 1974, Folder "ANL Tripartite Contract Renewal '74," Box 23, W. Massey Subj. Files, ANL.

66. McDaniel to Bauer, July 19, 1974; John T. Wilson to Robert D. Thorne, July 9, 1974; McDaniel to the AUA Board of Trustees, July 12, 1974; Bauer to Cannon and McDaniel, July 16, 1974; Cannon to Bauer, July 19, 1974; Sachs to Levi, July 8, 1974, Folder "ANL Tripartite Contract Renewal '74," Box 23, W. Massey Subj. Files, ANL.

67. AEC Policy Session item, Extension of Contract no. W-31-109-ENG-38 with Argonne Universities Association and the University of Chicago for Operation of Argonne National Laboratory, SECY-75-416, Folder 18, Box 7989, AEC Secretariat, DOE.

68. "The Director's Corner," *Argonne News,* Feb. 1975, p. 2; Sen. Charles H. Percy to Sachs, Dec. 12, 1974; Statement of Dr. Robert C. Seamans, Jr., before the Joint Committee on Atomic Energy and the Senate Interior and Insular Affairs Committee, all in ERDA, Folder 5, Box 19, Collec. no. 434-89-0024, GLFRC; Robert Sachs, interview by author, Dec. 4, 1992, High Energy Physics Building, University of Chicago.

69. Sachs to division directors, department heads, and program managers, June 14, 1974; Sachs to Robert Thorne, July 16, 1974; Sachs to Thorne, July 25, 1974, all in ERDA, Folder 4, Box 19, Collec. no. 434-89-0024, GLFRC.

70. Sachs to Seamans, Feb. 11, 1975, ERDA, Folder 5, Box 19, Collec. no. 434-89-0024, GLFRC; Teller to Seamans, March 31, 1975, Folder "Lab and Field Coordination," Box 1604, Job 1211, RG 430, WNRC.

71. Robert C. Porter, "The Battery Program: 'It's Time to Bring Industry In,'" Seamans," *Argonne News,* May–June 1975, pp. 1–2.

72. Seamans to Yarymovych, July 2, 1975, and Yarymovych to asst. admins. et al., July 3, 1975, in Folder "Organization and Management—Field and Lab Utilization," Box 17, 430-79-1/Job 1216, RG 326, DOE; Robert Sachs, interview by author, Dec. 4, 1992, High Energy Physics Building, University of Chicago.

73. Report of the Field & Laboratory Utilization Study Group, Dec. 1975, ERDA-100; ERDA Field Organization: Capsule Summaries of Plants and Laboratories FY1975, Aug. 1976, ERDA-76-73; Preliminary Plan Chicago ERDA Center, April 1976, and Proposed Plan for ERDA Centers, May 1976, all in Folder "Organization and Management 7—ERDA Center Concept," Box 22, 430-80-5/Job 1220, RG. 326, DOE.

74. Robert C. Gunness to Seamans, Nov. 10, 1975; Seamans to Gunness, Nov. 17, 1975, Folder "Organization and Management 7—Field and Lab Utilization," Box 17, 430-79-1/Job 1216, RG 326, DOE; Report of the Field & Laboratory Utilization Study Group, Dec. 1975, ERDA-100, Folder "Organization and Management 7—ERDA Center Concept," Box 22, 430-80-5/Job 1220, RG. 326, DOE.

75. News from Argonne National Laboratory, Sept. 21, 1976, OPA, ANL; Sachs to Yarymovych, Sept. 27, 1976, Folder "Organization and Management 12-4—Contractor 1976," Box 24, 430-80-5/Job 1220, RG 326, DOE.

76. Robert Sachs, interview by author, Dec. 4, 1992, High Energy Physics Building, University of Chicago.

77. Ibid.

78. "CPC: A Viable Answer to Solar Energy Concentration," *Argonne News,* Jan. 1975, pp. 1,4; "The Role of Argonne National Laboratory in Implementing National Energy Policy," in Sachs to Bauer, Sept. 11, 1975, Folder 6, Box 19, Collec. no. 434-89-0024, GLFRC; ANL, "Advanced Energy Systems—Solar Energy Research," *Argonne 76,* File TPD-76-3.

79. "The Role of Argonne National Laboratory in Implementing National Energy Policy," in Sachs to Bauer, Sept. 11, 1975, Folder 6, Box 19, Collec. no. 434-89-0024, GLFRC.

80. "Special to *Joliet Herald-News,*" Jan. 19, 1975, OPA, ANL; ANL, "Basic Statistics," June 30, 1975, Folder 6, Box 19, Collec. no. 434-89-0024, GLFRC; "Budget Outlays by Multiprogram Laboratories by Direct Energy R&D Programs," in Yarymovych to Hollister H. Cantus, director, Office of Congressional Relations, April 2, 1976, Folder "Organization and Management 7—Field & Lab Utilization 1976," Box 22, 430-80-5/Job 1220, RG 326, DOE.

81. Sachs is quoted in P. Failla, asst. to the lab. director, program planning, to the staff, July 12, 1977, and Failla to the staff, ERDA Facilities Report, Aug. 19, 1977, in Vol. 7, ERDA, Folder 10, Box 19, Collec. no. 434-89-0024, GLFRC; ANL, Revised Long Range Plan, 1975–1981, Jan. 24, 1975, Box 1620, Job 1212, RG 430, WNRC; ANL, "Report on Capabilities, Activities and Resources," March 1975, ANL Information Services, RG 326, NARA-GL.

82. Melvin Price to Robert Rowland, ANL, April 14, 1972; Edward Bauser, JCAE, to Robert E. Hollingsworth, AEC, April 18, 1972; Hollingsworth to Bauser, May 2, 1972, all in "6–Plants, Labs, Buildings and Land—48 Labs," Vol. 3, Box 7838, RG 326, AEC Secretariat, DOE.

83. AEC, Division of Biology and Medicine, "AEC Program in Human Radiobiology," n.d., c. 1968, AUA Board of Trustees, Gen. Corres., Vol. 1, Box 88, Collec. no. 434-89-0024, GLFRC.

84. Robley D. Evans, *Comments on a National Center of Human Radiobiology,* Dec. 1967; Report of the Subcommittee of the Advisory Committee for Biology and Medicine U.S. Atomic Energy Commission to Study the Proposal for a National Center of Human Radiobiology, May 11, 1968, both in AUA Board of Trustees, Gen. Corres., Vol. 1, Box 88, Collec. no. 434-89-0024, GLFRC; Hacker, *The Dragon's Tail,* pp. 20–25.

85. AEC, Division of Biology and Medicine, "AEC Program in Human Radiobiology," n.d., c. 1968, AUA Board of Trustees, Gen. Corres., Vol. 1, Box 88, Collec. no. 434-89-0024, GLFRC.

86. Report of the Subcommittee of the Advisory Committee for Biology and Medicine U. S. Atomic Energy Commission to study the proposal for a National Center of

Human Radiobiology, 1968, AUA Board of Trustees, Gen. Corres., Vol. 1, Box 88, Collec. no. 434-89-0024, GLFRC.

87. Minutes of Review Committee, October 26–27, 1967, Radiological Physics Division, ANL, Folder "AUA Review Committee Radiological and Environmental Research," Vol. 1, Box 86, Collec. no. 434-89-0024, GLFRC.

88. Minutes of Review Committee, October 15–17, 1968, and October 15–17, 1969, Radiological Physics Division, ANL, and Report of the Subcommittee of the Advisory Committee for Biology and Medicine U.S. Atomic Energy Commission to Study the Proposal for a National Center of Human Radiobiology, 1968, all in AUA Review Committee Radiological and Environmental Research, Vol. 1, Box 86, Collec. no. 434-89-0024, GLFRC.

89. Melvin Price to James R. Schlesinger, Dec. 2, 1971, "6–Plants, Labs, Buildings, and Land—48 Labs," Dec. 4, 1993, Vol. 3, Box 7838, RG 326, AEC Secretariat, DOE; "Concise History of the Formation of the Center for Human Radiobiology," Appendix B, Minutes of the Advisory Committee for the Center for Human Radiobiology, Nov. 16–17, 1972, Box 8, W. Massey Subj. Files, ANL; Robert E. Rowland, interview by R. Harris, Dec. 7, 1992, Argonne National Laboratory.

90. Minutes of the Review Committee, Nov. 11, 12, 13, 1970, Radiological Physics Division, ANL, AUA Review Committee Radiological and Environmental Research, Vol. 1, Box 86, Collec. no. 434-89-0024, GLFRC.

91. Robert E. Rowland, interview by Ruth R. Harris, Dec. 7, 1992, Argonne National Laboratory.

92. "Human Radiobiology Center," *Argonne News*, March 1974, pp. 1–2; "Review of the Activities of the Radiological Physics Division . . . conducted on Nov. 3–5, 1971," and "Review for 1972 of the Radiological and Environmental Research Division, ANL," both in AUA Review Committee Radiological and Environmental Research, Vol. 1, Box 86, Collec. no. 434-89-0024, GLFRC; Minutes of the Advisory Committee for the Center of Human Radiobiology, Nov. 16–17, 1972, and R. E. Rowland, "Comments on the Report of the Advisory Committee for the Center for Human Radiobiology," March 25, 1973, both in Center for Human Radiobiology, Box 8, W. Massey Subj. Files, ANL.

93. W. K. Sinclair to Sachs, Aug. 9, 1973, "Comments on DBER Review of Research Programs, April 23–26, 1973," and Review of the Research Program of the Division of Biological and Medical Research at Argonne National Laboratory, June 26, 1973, both in Box 18, Collec. no. 434-89-0024, GLFRC; Remarks by Professor Henry Koffler, April 8, 1970, Folder 59.6, Box 28, Collec. no. 434-91-0014, GLFRC.

94. Manson Benedict, "Recommendations on Research Programs for Nuclear Energy Development," Sept. 20, 1973; ANL Biomedical and Environmental Research Program for ERDA, Jan. 10, 1975; W. K. Sinclair to James L. Liverman, Oct. 16, 1974, all in Folder 1, Box 19, Collec. no. 434-89-0024, GLFRC; ERDA, Balanced Program Plan for the Division of Biomedical and Environmental Research, July 1975, Box 8, Collec. no. 434-91-0015, GLFRC; ANL, Draft Plan 1976–1982, Nov. 1976, Box 2, Collec. no. 434-91-0014, GLFRC.

95. Hacker, *The Dragon's Tail*, pp. 66–68.

96. Ibid., pp. 67–68; AEC, "Response to Query on Plutonium Injection Study-Manhattan District," May 8, 1974, 3-Commissioner's Exec. Sess., Vol. 2, Box 3738, AEC Secretariat, DOE. Of the 18 plutonium patients, there were 13 males and 5 females and 15 whites and 3 blacks; 13 were 45 years or older, 2 were between 35 and 45, 2 were between 18 and 34, and 1 was 4 years old. Five of the patients died within 6 months, 2 died within the first year. One survived more than a year, 2 survived more than two years. One patient lived 14 years after injection, and another lived more than 20 years. The fates of two patients were unknown. Still another 16-year-old cancer patient was injected with americium in 1947. He died within 11 months. See W. H. Langham, S. H. Bassett, P. S. Harris, and R. E. Carter, "Distribution of Excretion of Plutonium Administered Intravenously to Man," Sept. 20, 1950, LA-1151, Los Alamos National Laboratory; Patricia W. Durbin, "Plutonium in Man: A New Look at the Old Data," *Radiobiology of Plutonium,* 1972; Giles L. Lofton, AEC Division of Inspection Report 44-2-326, Aug. 14, 1974, DOE.

97. "Patients Injected with Plutonium (Draft Report of 5–24–74)," 3-Commissioners Exec. Sess., Vol. 2, Box 3738, AEC Secretariat, DOE.

98. R. E. Rowland, "Plutonium Studies at the Center for Human Radiobiology (CHR)," Nov. 8, 1973, Folder "Center for Human Radiobiology," Box 8, W. Massey Subj. Files, ANL.

99. Robert E. Rowland, interview by Ruth R. Harris, Dec. 7, 1992, Argonne National Laboratory.

100. AEC, Minutes of Executive Session 74-28, May 13, 1974, 3-Commissioners Exec. Sess., Vol. 2, Box 3738, AEC Secretariat, DOE; R. E. Rowland, "Plutonium Studies at the Center for Human Radiobiology (CHR)," Nov. 8, 1973, Folder "Center for Human Radiobiology," Box 8, W. Massey Subj. Files, ANL.

101. James L. Liverman to Robert H. Bauer, manager, Chicago Operations Office, April 4, 1974, Center for Human Radiobiology, Box 24, Collec. no. 434-91-0015, GLFRC; AEC, Minutes of Executive Session 74-27, April 11, 1974, 3-Commissioner's Exec. Sess., Vol. 2, Box 3738, AEC Secretariat, DOE.

102. AEC, Minutes of Limited Attendance Session 74-68, April 15, 1974, 12-Organization and Management 6-Limited Attendance Session, Vol. 2, Box 7937; Minutes of Executive Session 74-24A, April 26, 1974; Minutes of Executive Session 74-28, 3-Commissioner's Exec. Sess., Vol 2, Box 3738, all in AEC Secretariat, DOE.

103. Draft report of ad hoc Committee convened on May 22, 1974, to provide recommendations on the scientific merit of studying the fate of plutonium in deceased persons who had received injections of plutonium years earlier, May 28, 1974, 3-Commissioners Exec. Sess., Vol. 2, Box 3738, AEC Secretariat, DOE; Patricia J. Lindop, review of *Radiobiology of Plutonium,* Stover and Jee, eds., *Environmental Pollution,* 5 (1973): 153.

104. AEC, Minutes of Commissioner's Executive Session 74-33, July 5, 1974, 3-Commissioners Exec. Sess., Vol. 2, Box 3738, AEC Secretariat, DOE.

105. Argonne OPA, May 11, 1975; "Argonne Loses Plutonium Vial," *Chicago Daily News,* May 12, 1975.

106. *Chicago Tribune,* May 12, 1975.
107. *Chicago Sun-Times,* May 13, 1975.
108. Robert G. Sachs, letter to the editor, *Chicago Sun-Times,* May 21, 1975.
109. "Investigation Report on Pu Loss Discovered on May 9, 1977, Sections 1 and 2," Folder 75-9-A, 430-88-5, RG 430, ERDA, WNRC; John McCarron, "How Do You Ship a 600-yard Dump?" *Chicago Tribune,* May 19, 1975.
110. "Argonne Mufs It," *Chicago Sun-Times,* May 13, 1975.
111. "Investigation Report on Pu Loss Discovered on May 9, 1975," Section 1, Folder 75-9-A, 430-88-5, RG 430, ERDA, WNRC.
112. Bauer to Dana R. Dixon, director, security division, May 14, 1975, Folder 79-9-A, 430-88-5, RG 430, ERDA, WNRC.
113. See Weart, *Nuclear Fear.*
114. Kameshwar C. Wali, "What Is High Energy Physics About?" in *Symposium on the History of the ZGS: AIP Conference Proceedings No. 60,* eds. Joanne S. Day, Alan D. Krisch, and Lazarus G. Ratner (New York: American Institute of Physics, 1980), pp. 82–83.
115. Ibid., pp. 83–84; DOE, *High Energy Physics: The Ultimate Structure of Matter,* p. 5.
116. "High Energy Physics," in GAC, *Long Range Research Review for the Atomic Energy Commission,* March 1974, pp. 24–25.
117. Ibid., pp. 26–28.
118. Ratner, "Life and Death of a Particle Accelerator," pp. 402–5.
119. John A. Erlewine, "Establishment of Two Subcommittees to Assist in Study of Future of the ZGS Accelerator," Consent Calendar Item, May 14, 1974, SECY-74-636, and John M. Teem to the commission, "Outline of Proposed Staff Paper—Underutilization of Present Facilities and New Facility Needs in High Energy Physics," June 4, 1974, both in Folder 8, Box 16, 326-75-7, RG 326, DOE.
120. "Report of the AEC-NSF Study of the Future Role of the Zero Gradient Synchrotron (ZGS)"; John A Erlewine, "AEC-NSF Study of the Zero Gradient Synchrotron (ZGS)," Commission Action, Aug. 8, 1974, SECY-75-119, Box 15, 326-75-7 RG 326, DOE.
121. "Report of the AEC-NSF Study of the Future Role of the Zero Gradient Synchrotron (ZGS)," Sept. 9, 1974, pp. 12-16, Box 12, W. Massey Subj. Files, ANL.
122. D. S. Greenberg, "Budget Cutter Sees Lots of Plump Targets in R&D Funds," *Science and Government Report,* Oct. 15, 1974.
123. Report of the subpanel on new facilities of the High Energy Physics Advisory Panel to the Atomic Energy Commission, July 30, 1974, Vol. 3, HEPAC, Box 37, Collec. no. 434-91-0015, GLFRC.
124. John Deutch to Sachs, Nov. 18, 1977, and James S. Kane to Robert H. Bauer, Nov. 25, 1977, both in ZGS Facility, Box 22, Collec. no. 434-91-0015, GLFRC; T. H. Fields, for Seamans meeting, 4/14/76, Folder 7, Box 19, Collec. no. 434-89-0024, GLFRC; DOE, *High Energy Physics: The Ultimate Structure of Matter and Energy,* 17; E. F. Parker, "History of the Polarized Beam," in *Symposium on the History of the ZGS: AIP Conference*

Proceedings No. 60, eds. Joanne S. Day, Alan D. Krisch, and Lazarus G. Ratner (New York: American Institute of Physics, 1980), pp. 143–57; J. B. Roberts, Jr., "The ZGS Polarized Beam Program," in *Symposium on the History of the ZGS: AIP Conference Proceedings No. 60,* eds. Joanne S. Day, Alan D. Krisch, and Lazarus G. Ratner (New York: American Institute of Physics, 1980), pp. 158–68.

125. GAO, "The Multiprogram Laboratories: A National Resource for Nonnuclear Energy Research, Development, and Demonstration," May 22, 1978, p. 19, EMD-78-62; Robert Sachs, interview by author, Dec. 4, 1992, High Energy Physics Building, University of Chicago.

Chapter 10

1. See Holl and Fehner, *Securing America's Energy Future* and Holl, Anders, and Buck, *U.S. Civilian Nuclear Power Policy,* in which significant portions of the introduction of this chapter originally appeared; ERDA-76-1, "A National Plan for Energy Research, Development and Demonstration: Creating Energy Choices for the Future," Vol. 1, April 15, 1976.

2. Holl, Anders, and Buck, *U.S. Civilian Nuclear Power Policy,* pp. 18–19; ANL, *Argonne 76,* File TPD-76-3, p. 4.

3. Carter quoted in Burton I. Kaufman, *The Presidency of James Earl Carter, Jr.* (Lawrence: University Press of Kansas, 1993), pp. 5–7; Sidney Kraus, ed., *The Great Debates: Carter vs. Ford, 1976* (Bloomington: Indiana University Press, 1979), pp. 461, 464–65.

4. Holl and Fehner, *Securing America's Energy Future,* pp. 14–15; Kaufman, *James Earl Carter, Jr.,* pp. 16–17; Kraus, *The Great Debates,* pp. 57–58.

5. Holl and Fehner, *Securing America's Energy Future,* pp. 15–16; Kaufman, *James Earl Carter, Jr.,* pp. 32–33.

6. Holl and Fehner, *Securing America's Energy Future,* p. 17; James E. Carter, "Address to the Nation on the Energy Problem," in Senate Committee on Energy and Natural Resources (SCEN), *Executive Energy Documents* (Washington, D.C.: GPO, 1978), pp. 371–77.

7. Kaufman, *James Earl Carter, Jr.,* p. 15.

8. Briefing by James Schlesinger at the White House, Sept. 13, 1977, and Schlesinger, "Remarks at the Energy Seminar of Georgia Business and Industry Association and the Georgia Office of Energy," Sept. 16, 1977, Atlanta, Georgia, DOE.

9. Holl and Fehner, *Security America's Energy Future,* pp. 17–18.

10. Robert Sachs, *Argonne 77,* File TPD-77-2, and *Argonne 78,* File TPD-78-2, ANL.

11. "25 Years Ago: EBR-I Produced First Electricity from the Atom," *Argonne News,* Jan. 1977, pp. 1, 8.

12. "*Today* Comes to Argonne," *Argonne News,* May–June 1977, p. 11.

13. Laney, "The Director's Corner," *Argonne News,* March 1977, p. 2.

14. "Controversial Energy Issues: Weinberg Outlines Acceptable Nuclear Future," *Argonne News,* April 1977, p. 6.

15. "Laney, Comey Debate: The Nuclear Controversy," *Argonne News,* April 1977, pp. 8–9.

16. Ibid.

17. "Percy Visits Argonne," *Argonne News*, May–June, 1977, p. 7.

18. Holl, Anders, and Buck, *U.S. Civilian Nuclear Power Policy*, p. 20; Spurgeon M. Keeny, Jr., ed., *Report of the Nuclear Energy Policy Study Group: Nuclear Power—Issues and Choices* (Cambridge, Mass.: 1977), pp. 271–99.

19. Sachs, "The Director's Corner—Carter's Energy Policy," *Argonne News*, May–June 1977, p. 2.

20. William C. Boesman, "The Role of the National Energy Laboratories in ERDA and Department of Energy Operations: Retrospect and Prospect," Dec. 13, 1977, Congressional Research Service, Library of Congress, pp. 47–48, 76–77.

21. GAO, *The Multiprogram Laboratories: A National Resource for Nonnuclear Energy Research, Development, and Demonstration*, Report of the Congress of the United States by the Comptroller General, EMD-78–62, May 22, 1978, p. 1.

22. Boesman, "The Role of the National Energy Laboratories," p. 79.

23. Albert Teich, "Bureaucracy and Politics in Big Science: Relations between Headquarters and the National Laboratories in AEC and ERDA," 1977 Annual Meeting of the American Political Science Association, Sept. 1–4, 1977, Washington, D.C.

24. Carter quoted in Boesman, "The Role of the National Energy Laboratories," pp. 89–91.

25. ANL, *1977 Institutional Plan (FY 1977–FY 1983): Executive Summary, Nov. 1977* (Washington, D.C.: DOE, 1977).

26. Teich, "Bureaucracy and Politics in Big Science," pp. 35–36.

27. William D. Metz "National Laboratories: Focused Goals and Field Work Hinted Under DOE," *Science*, Dec. 2, 1977, p. 198; *The Multiprogram Laboratories*, p. 58.

28. *The Multiprogram Laboratories*, p. 9.

29. Ibid., pp. 61–62; ANL, "Institutional Plan (FY 1979–FY 1984)," Jan. 1979, pp. 75–77, Box 2B, W. Massey Subj. Files, ANL.

30. *The Multiprogram Laboratories*, pp. 37–58.

31. Ibid., p. 35.

32. Henry V. Bohm, president, to AUA Board of Trustees, May 25, 1978, AUA Board of Trustees, Box 22, W. Massey Subj. Files, ANL.

33. Deutch to Sachs, July 11, 1978, AUA Board of Trustees, Box 22, W. Massey Subj. Files; ANL, "Institutional Plan (FY 1979–FY 1984)," Jan. 1979, Box 2B, W. Massey Subj. Files, ANL.

34. Robert Sachs, interview by author, Dec. 4, 1992, High Energy Physics Building, University of Chicago.

35. John T. Wilson to Faculty Committee on Argonne National Laboratory, March 15, 1978, and Cedric L. Chernick to Johnson, Sugarman, and Cannon, April 12, 1978, in ANL Directorship Search 1978, Box 20, W. Massey Subj. Files, ANL.

36. Chernick to Cannon, D. G. Johnson, and N. Sugarman, May 2, 1978, ANL Director's Search, Box 20, W. Massey Subj. Files, ANL.

37. Ibid.

38. Edward J. Croke to Cannon, Sept. 21, 1978, ANL Director's Search, Box 20, W. Massey Subj. Files, ANL.

39. Chernick to ANL Director Search Committee, May 30, 1978, ANL Director's Search, Box 20, W. Massey Subj. Files, ANL.

40. Chernick to John Schiffer and Martin Steindler, May 26, 1978, ANL Director's Search, Box 20, W. Massey Subj. Files, ANL.

41. Stephanie Derejko to Cannon, June 20, 1978, "Notes of ANL Director Search Committee Meeting, June 19, 1978," and Cannon, "Notes on Meeting with Deutch, July 18, 1978," Folder ANL Short List, Box 5, W. Massey Subj. Files, ANL.

42. Chernick, "Notes—ANL Search Committee," Sept. 13, 1978; D. Gale Johnson to Faculty Advisory Committee on Argonne National Laboratory, Sept. 20, 1978; John J. Roberts to Cannon, and Edward J. Croke to Cannon, both Sept. 21, 1978, ANL Director's Search, Box 20, W. Massey Subj. Files, ANL.

43. Cannon to Hanna H. Gray, Oct. 4, 1978, ANL Short List, Box 5, W. Massey Subj. Files, ANL.

44. Bohm to Provost D. Gale Johnson, Nov. 7, 1978, ANL Director's Search, Box 20, and Chernick to Cannon, Jan. 11, 1979, Phase 3, Box 5, W. Massey Subj. Files, ANL.

45. Chernick, "Notes," Dec. 7, 1978, ANL Director's Search, Box 20, W. Massey Subj. Files, ANL.

46. Peter A. Carruthers to Cannon, Jan. 30, 1978, ANL Director's Search, Box 20, W. Massey Subj. Files, ANL.

47. Cannon to John Deutch (draft), Feb. 5, 1979, and Cannon to the search committee, Feb. 5, 1979, ANL Director's Search, Box 20, W. Massey Subj. Files, ANL; Cannon to Deutch, April 13, 1979, and Deutch to Cannon, May 9, 1979, both in Massey Appt., Box 5, W. Massey Subj. Files, ANL.

48. "A Visit with Our New Director, Walter E. Massey," *Argonne News,* June 1979.

49. Ibid.; Lynn Norment, "Dr. Walter E. Massey: In Search of Solutions to the Energy Crunch," *Ebony,* Nov. 1979, pp. 89–94.

50. Cannon to Hanna H. Gray and D. Gale Johnson, April 13, 1979, Massey Appt., Box 5, W. Massey Subj. Files, ANL.

51. GAO, "U.S. Fast Breeder Reactor Program Needs Direction," Sept. 22, 1980, EMD-80-81, p. i.

52. Ibid., p. 11; "Alternative Breeding Cycles for Nuclear Power: An Analysis," Congressional Research Service, Library of Congress, Oct. 1978; "Nuclear Proliferation and Safeguards," Office of Technology Assessment, June 30, 1977; GAO, "Nuclear Reactor Options to Reduce the Risk of Proliferation and to Succeed Current Light Water Reactor Technology," U.S. General Accounting Office, May 23, 1979; "Nuclear Power Issues and Choices," Ford Foundation, March 1977.

53. GAO, "U.S. Fast Breeder Reactor Program Needs Direction," pp. 25–26.

54. "LMFBR Instrumentation—A Quiet but Important Story," pp. 76–48, OPA, ANL.

55. John I. Sackett, interview by author, August 18, 1993, Argonne-West; J. H. Kittel, L. A. Neimark, L. G. Walters, R. E. Einziger, and D. E. Mahagin, "Status of LMFBR Fuels and Materials Development," in *Fast Breeder Reactor Studies,* C. E. Till, et al., ANL-80-40, pp. 109ff.

56. "Report of the EBR-II Review Committee," Sept. 8–9, 1977, Univ. of Chicago Spec. Comm. for Reactor Dev., Box 37, 434-91-0015, GLFRC; J. A. Kyger to Sachs, Dec. 28, 1977, Response to EBR-II Review Committee Report, AUA EBR-II Rev. Comm., Box 87, 434-89-0024, GLFRC.

57. Robert F. Redmond to AUA Special Committee for Reactor Development, Sept. 1, 1977, Univ. of Chicago Spec. Comm. for Reactor Dev., Box 37, 434-91-0015, GLFRC.

58. Minutes of the AUA Special Committee for Reactor Development, Sept. 29, 1977, Univ. of Chicago Spec. Comm. for Reactor Dev., Box 37, 434-91-0015, GLFRC.

59. Charles H. Percy to Sachs, July 18, 1977, and Robert Laney to Percy, Aug. 29, 1977, Congressional Inquiries, Box 42, 434-91-0015, GLFRC.

60. Laney to Cannon, Aug. 21, 1978; Schlesinger to Frank Church, Aug. 24, 1978; ANL press release, James A. McClure, n.d., all in Folder "LMFBR Prog.," Box 6, W. Massey Subj. Files, ANL; Laney to Cannon, Sept. 20, 1978, ANL Short List, Box 5, W. Massey Subj. Files, ANL.

61. GAO, "U.S. Fast Breeder Reactor Program Needs Direction," pp. 33–34; "Report of the EBR-II Review Committee Meeting," Sept. 8–9, 1977, Univ. of Chicago Spec. Comm. for Reactor Dev., Box 37, 434-91-0015, GLFRC.

62. Philip L. Cantelon and Robert C. Williams, *Crisis Contained: The Department of Energy at Three Mile Island* (Carbondale: Southern Illinois University Press, 1982), pp. 1–7.

63. Ibid.

64. "Radiological Assistance Team—Argonne Responds to Three Mile Island Crisis," *Argonne News*, May 1979, p. 4.

65. Cantelon and Williams, *Crisis Contained*, pp. 167–79.

66. "Radiological Assistance Team," *Argonne News*, May 1979, p. 12.

67. "Three Mile Island—Argonne Assesses Environmental Impact of Cleanup," *Argonne News*, March 1980, p. 3.

68. R. Avery, et al., "An Assessment of the 'Report of the President's Commission on the Accident at Three Mile Island,'" ANL, Nov. 26, 1979, AUA Board of Trustees Mtgs., Box 88, 434-89-0024, GLFRC; *The Report of the President's Commission on the Accident at Three Mile Island: The Need for a Change: The Legacy of Three Mile Island* (Washington, D.C.: GPO, 1979).

69. Ibid.

70. See especially, Robert Gillette, "Nuclear Reactor Safety: A New Dilemma for the AEC," *Science*, Vol. 173, July 9, 1971; Gillette, "Nuclear Reactor Safety: At the AEC the Way of the Dissenter is Hard," *Science*, Vol. 174, May 5, 1972; Gillette, "Nuclear Safety (II): The Years of Delay," *Science*, Vol. 177, Sept. 8, 1972; Gillette, "Nuclear Safety (III): Critics Charge Conflicts of Interest," *Science*, Vol. 177, Sept. 15, 1972; Gillette, "Nuclear Safety (IV): Barriers to Communication," *Science*, Vol. 177, Sept. 22, 1972; Gillette, "Nuclear Safety: AEC Report Makes the Best of It," *Science*, Vol. 179, Jan. 26, 1973.

71. Holl, Anders, and Buck, *U.S. Civilian Nuclear Power Policy*, pp. 21–22; WASH-1000, "An Assessment of Accident Risks in U.S. Commercial Nuclear Power Plants" (Washington, D.C.: GPO, August 1974, October 1975).

72. Massey to H. V. Bohm, Nov. 27, 1979; R. Avery, et al., "An Assessment of the 'Report of the President's Commission on the Accident at Three Mile Island,'" AUA Board of Trustees Mtgs., Box 88, 434-89-0024, GLFRC.

73. ANL, "Institutional Plan FY 1984–FY 1989," Nov. 1983, ANL-3.4, 3.5, DOE.

74. ANL, "Institutional Plan FY 1980–FY 1985," December 1979, DOE; "TREAT—The First 20 Years Are Just the Beginning," *Argonne News*, January 1979, pp. 8–9, 13; ANL, "1974 Research Highlights," File TPD-74-1, pp. 4–5; *Argonne 76*, File TPD-76-3, pp. 6–7; "Argonne National Laboratory 1981: 35 Years of Leadership," File TPD-81-3, last three in Box 13, W. Massey, Subj. Files, ANL.

75. "EBR-II Is a Certified Cogeneration Facility," *Argonne News*, Jan.–Feb. 1981, p. 8.

76. "Report of the EBR-II Review Committee," Sept. 20–21, 1979, and J. A. Kyger to Massey, "Response to the 1979 Report of the AUA Review Committee for EBR-II," AUA-EBR-II Rev. Comm., Box 87, 434-89-0024, GLFRC; "Three Mile Island—It Cannot Happen in a Breeder," *Argonne News*, Sept. 1980, p. 3.

77. "Report of the EBR-II Review Committee," July 23–24, 1981, and "AUA Report of EBR-II Review, July 23–24, 1981, and ANL Responses," Univ. of Chicago Rev. Comm. Box 35, 434-91-0015, GLFRC.

78. "Report of the EBR-II Review Committee," Nov. 23–25, 1980, Univ. of Chicago Rev. Comm., Box 35, 434-91-0015, GLFRC.

79. "Power Run Up and Down—EBR-II Performs First 'Transient' Tests," *Argonne News*, May–June 1981, p. 10.

80. Ed Hahn, "EBR-II Earns an 'A' in Latest Tests," *Argonne News*, Nov.–Dec. 1982, p. 11.

81. Toni Grayson Joseph to John Deutch, May 1, 1979, and Joseph to Robert Bauer, May 24, 1979, Office of Energy Research, DOE.

82. Ling Committee, "Final Report on Management Review of Argonne National Laboratory," July 9, 1979, p. 15, DOE Review of ANL for Contract Renewal, Box 23, W. Massey Subj. Files, ANL.

83. Ibid.

84. Massey to N. Douglas Pewitt, Nov. 27, 1979, Mgmt. Rev., 1979, Box 1, W. Massey Subj. Files, ANL.

85. Ibid.

86. "Laney, Deputy Director, Retires," *Argonne News*, Dec. 1979, p. 10.

87. "News from Argonne National Laboratory," March 28, 1980, OPA 80-12; "Beckjord Named Deputy Director for Science and Technology," *Argonne News*, April 1980, p. 3.

88. William Massey, interview by author, June 1, 1992, ANL.

89. Massey to Deutch, Sept. 28, 1979; Ling Committee, "Final Report on Management Review of Argonne National Laboratory."

90. Massey to Deutch, Sept. 28, 1979.

91. Cannon's notes to Hanna Gray and Deutch to Massey, November 1, 1979, ANL Gen. Corres., Box 5, W. Massey Subj. Files, ANL.

92. Massey to Gray, Dec. 11, 1979, ANL Gen. Corres., Box 5, W. Massey Subj. Files, ANL.

93. Cannon to Massey, January 3, 1980, ANL Gen. Corres. Box 5, W. Massey Subj. Files, ANL.

94. "Massey Outlines 'State of the Laboratory,'" *Argonne News*, March 1980, p. 10; "The FY 1981 Outlook for Argonne," *Argonne News*, May 1980, p. 2; Minutes of the Executive Session, Seventy-First Meeting of the Board of Trustees, AUA, March 27, 1980, Vol. 7, Box 88, 434-89-0024, GLFRC.

95. "Massey is Guest at White House," *Argonne News*, May 1980, p. 4.

96. Massey, "Blueprint for the Future," *Argonne News*, June–July, 1980, p. 2.

97. Ibid.; Minutes of the Executive Session, Seventy-Second Meeting of the Board of Trustees, AUA, July 15, 1980, Vol. 7, Box 88, 434-89-0024, GLFRC; Minutes of the Forty-Eighth Meeting of the Executive Committee of the Board of Trustees, AUA, June 23, 1980, Box 20B, W. Massey Subj. Files, ANL.

98. "CP-5 Reactor Retires from Service," *Argonne News*, Sept. 1979, p. 8.

99. ANL, "Institutional Plan FY 1980–FY 1985," Dec. 1979.

100. Massey, "American Science and Technology: Sick or Healthy," address given at the Commonwealth Edison Lake Lawn Engineering Conference, Sept. 11, 1979, in Personnel, Misc., Box 21, W. Massey Subj. Files, ANL.

101. "Moving Up—Charles E. Till" *Argonne News*, Sept. 1980, p. 7; Till, interview by author, August 18, 1993, Argonne-West.

102. Minutes of the Seventy-Fourth Meeting of the Board of Trustees, AUA, Jan. 15, 1981, Box 27, 434-91-0014, GLFRC.

Chapter 11

1. "Massey Graduates . . . 25 Years Later," *Argonne News*, May–June 1981, p. 3.

2. William Massey, "Science Education: Role of the National Laboratories," *Argonne News*, July 1982, pp. 2, 4–5.

3. "Opening the Door to Equal Opportunity," *Argonne News*, May–June 1981, pp. 8–9.

4. "Handicapped Collegians Spend Week at Argonne," and "Exchange Students Build Bridges for the Future," *Argonne News*, Oct. 1981, pp. 12–13; "Employees Active in Their Communities," *Argonne News*, May 1983, p. 4.

5. "Cardiss Collins Congratulates Massey [for minority business award]," *Argonne News*, Oct. 1981, p. 12.

6. Henry V. Bohm to AUA Board of Trustees, May 8, 1979, University of Chicago Gen. Corres., Vol. 3, 43.19a, UCRL SC.

7. Ibid.

8. Chernick to file, re. Executive Session, AUA Board of Trustees Meeting, Aug. 1, 1979, ANL Gen. Corres., Box 5, W. Massey Subj. Files, ANL.

9. "The Role of the AUA in the Tripartite Contract: Resolution Adopted by the AUA Planning Committee and AUA Assessment and Budget Committee," May 12, 1960, ANL Contract, Box 5; AUA, Minutes of the Forty-Eighth Meeting of the Executive Committee of the Board of Trustees, June 23, 1980, Box 20B; Bohm to the chief executive officers of member universities and the Board of Trustees of AUA, Alternative Structure, Box 19, all in W. Massey Subj. Files, ANL.

10. Ibid.; AUA, Minutes of the Executive Session, Seventy-Second Meeting of the Board of Trustees, July 15, 1980, Box 88, Collec. no. 434-89-0024, GLFRC.

11. Cannon to Hanna Gray, Kenneth W. Dam, and Lewis Nosonow, Jan. 22, 1981, ANL Gen. Corres., Box 5, and Options for the University of Chicago—Confidential Limited Distribution—President's Office Only, n.d., Alternative Structure, Box 19, W. Massey Subj. Files, ANL.

12. GAO, *The Multiprogram Laboratories,* EMD-78-62, May 22, 1978.

13. GAO, "Draft Report: Argonne National Laboratory, A Case Study of the Department of Energy's Methods for Selecting National Laboratory Operating Contractor," March 1981, in ANL Contract 1981, Box 5, W. Massey Subj. Files, ANL.

14. GAO, "The Department of Energy Needs Better Procedures for Selecting a Contractor to Operate Argonne National Laboratory," June 8, 1981, EMD-81-66; "GAO Raps Argonne, Recommends $400 Million Fund Freeze," *Joliet Herald-News,* June 24, 1981.

15. William Massey, "State, City Technology Task Forces Adopt Complementary Approaches," *Argonne News,* Jan.–Feb. 1982, p. 2.

16. William Massey, "Director's Corner: DOE Laboratory Directors Meet," *Argonne News,* Nov. 1979, p. 2; Massey, "Director's Corner: Task Force Finds Obstacles to Interaction," *Argonne News,* Nov. 1981, p. 2; "'Spotlight on Argonne,' Industry Executives Learn about Research," *Argonne News,* Nov.–Dec. 1982, p. 7.

17. Bueche quoted in John Walch, "DOE Laboratories in the Spotlight," *Science* 213 (Aug. 14, 1981): 744.

18. Keyworth quoted in ibid.

19. Bromley quoted in ibid.; William Massey, "The Director's Corner: Congress Focuses on National Laboratories," *Argonne News,* Aug.–Sept. 1981, p. 2.

20. William Massey, statement before the Committee on Science and Technology, U.S. House of Representatives, July 29, 1981, W. Massey Subj. Files, Box 21, ANL.

21. William Massey, *State of the Laboratory: Moves Toward Excellence,* Director's Report to the AUA, Sixteenth Annual Meeting, Oct. 25–27, 1981, AUA-59.8, Annual Member's Meeting, Box 28, Collec. no. 434-91-0014, GLFRC.

22. William Massey, Statement before the Subcommittee on Energy Development of the Committee on Science and Technology, July 29, 1981, ANL.

23. Holl and Fehner, *Securing America's Energy Future,* p. 37.

24. James B. Edwards, secretary of energy, statement before the Senate Committee on Energy and Natural Resources, Feb. 23, 1981, Office of Executive Secretary, DOE; Holl and Fehner, *Securing America's Energy Future,* pp. 37–38.

25. Nathaniel Sheppard, Jr., "Concern Grows Over Policy on National Labs," *New York Times,* Dec. 8, 1981.

26. William Massey, "The Director's Corner: Reagan Administration—Not Anti-Science," *Argonne News,* Jan.–Feb. 1981, p. 2; "The Director's Corner: The Impact of the DOE Budget," *Argonne News,* March 1981, p. 2.

27. Massey to Robert H. Bauer, April 9, 1981, ANL Gen. Corres., Box 5, W. Massey Subj. Files, ANL.

28. Massey to file, Executive Session, Office of Energy Research—Laboratory Directors Meeting, Sept. 16, 1981, Lab Reorganization, Box 19, W. Massey Subj. Files, ANL; Massey to Gray, Oct. 1, 1981, Folder 1120.3, ANL General (1981–82), Lab Management Division, DOE.

29. Till, interview by author, Aug. 18, 1981, Argonne-West.

30. Massey, note to file, Executive Session, Office of Energy Research—Laboratory Directors meeting, Sept. 16, 1981, Lab Reorganization, Box 19, W. Massey Subj. Files, ANL.

31. Massey to Gray, Oct. 1, 1981; Walter Massey, interview by author, June 1, 1992, NSF, Washington, D.C.

32. AUA, Minutes of the Seventy-Seventh Meeting of the Board of Trustees, Sept. 14, 1981, Box 27, Collec. no. 434-91-0014, GLFRC.

33. Massey, interview by author, June 1, 1992; Lewis H. Nosanow and Eric S. Beckjord to Hanna H. Gray, "Suggestions on Your Action Concerning the Argonne Contract Question," July 27, 1981, Alternative Structure, Box 19, W. Massey Subj. Files, ANL.

34. Lewis H. Nosanow to Hanna Gray and Kenneth W. Dam, Aug. 18, 1981; Massey, note to file, Exec. Sess., Office of Energy Research Laboratory Directors Meeting, Sept. 11, 1981, Lab Reorganization, W. Massey Subj. Files, ANL; AUA, Minutes of the Seventy-Seventh Meeting of the Board of Trustees, Sept. 14, 1981, Box 27, Collec. no. 434-91-0014, GLFRC.

35. Options for the University of Chicago—Confidential Limited Distribution—President's Office Only, n.d., Alternative Structure, Box 19, and "A Revised Governance Structure for the Argonne National Laboratory," a proposal by the University of Chicago, July 1981, Gallagher-Laney Report, Box 18, W. Massey Subj. Files, ANL.

36. Gray quote from Gray to Russell, Oct. 6, 1981, ANL Contract, Box 5, W. Massey Subj. Files, ANL; Gray to Russell, Nov. 2, 1981, and The University of Chicago Plan for the Management of Argonne National Laboratory, Nov. 2, 1981, in Vol. 3, Univ. of Chicago, Gen. Corres., UC; The University of Chicago Plan for the Management of Argonne National Laboratory, Nov. 25, 1981, and Gray to Russell, Jan. 5, 1982, in AUA Board of Trustees, Box 27, Collec. no. 434-91-0014, GLFRC.

37. Gray to Russell, Oct. 6, 1981, ANL Contract, Box 5, W. Massey Subj. Files, ANL.

38. The University of Chicago Plan for Management of Argonne National Laboratory, Nov. 25, 1981, Alternative Structure, Box 19, W. Massey Subj. Files, ANL; AUA, Minutes of the Eighty-First Meeting of the Board of Trustees, Jan. 18, 1982; Russell to Gray, Dec. 22, 1981; Gray to Russell, Jan. 5, 1982; Clarifications addendum to the University of Chicago Plan for the Management of Argonne National Laboratory, December 22, 1981, all in AUA Board of Trustees, Box 27, Collec. no. 434-91-0014, GLFRC.

39. Gray to Russell, Oct. 1981, ANL Contract, Box 5, and The University of Chicago Plan for Management of Argonne National Laboratory, Nov. 25, 1981, Alternative Structure, Box 19, in W. Massey Subj. Files, ANL.

40. Henry V. Bohm to members of the AUA Board of Trustees, Dec. 23, 1981, Univ. of Chicago, Gen. Corres., Vol. 3, UC; AUA, Minutes of the Eighty-First Meeting of the Board of Trustees, Jan. 18, 1982, AUA Board of Trustees, Box 27, Collec. no. 434-91-0014,

GLFRC; John Walsh, "For Argonne, Criticism and a Comeback," *Science* 218 (Oct. 22, 1982): 354–57.

41. Robert H. Bauer to Alvin W. Trivelpiece, "Recommendation to Modify and Extend the Operating Contract for Argonne National Laboratory (ANL) W-31-109-ENG-38," Feb. 12, 1982; Bauer to Trivelpiece, "Subject: Extend/Compete Field Recommendation for Argonne National Laboratory," March 30, 1982.

42. Louis C. Ianiello to David M. Richman, "Contract to Run ANL," April 6, 1982; William S. Heffelfinger to the under secretary, "Action: Decision to Extend or Compete University of Chicago–Argonne Universities Association Contract W-31-109-ENG-38 for Operation of Argonne National Laboratory," July 1, 1982, all in DOE Office of Field Operations Management, DOE; William Massey, "Argonne Can Look Forward to a More Secure Future," *Argonne News*, Nov.–Dec. 1982, p. 2.

43. Trivelpiece to Rep. John E. Porter, Jan. 29, 1982, Folder "1120.3 ANL-General," Lab Management Division, DOE.

44. Edwards quoted in *Business Week*, July 19, 1982, p. 169; White House, "President Introduces Cost Control Survey Team," July 15, 1982, and "Fact Sheet on the President's Private Sector Survey on Cost Control in the Federal Government," July 15, 1982, both in Grace Report, Box 9, Collec. no. 434-91-0015, GLFRC.

45. "Critical Need for Energy Research and Development: The Role of the Midwest Research Labs," Hearing before the Subcommittee on Energy, Nuclear Proliferation, and Government Processes of the Committee on Governmental Affairs, U.S. Senate, March 22, 1982, Argonne, Ill., pp. 1–13.

46. Ibid.; "Says Sen. Percy at Argonne Dedication: Dry Scrubber a Boon to Illinois Coal," *Argonne News*, Mar.–Apr. 1982, pp. 3, 10.

47. Percy to Edwards, March 19, 1982, and Edwards to Percy as cited in "Critical Need for Energy Research and Development: The Role of the Midwest Research Labs," p. 5; Percy et al. to Edwards, May 11, 1982, and Edwards to Percy, June 22, 1982, in Folder "1120.3 ANL General (1981–1982)," Lab Management Div., DOE.

48. "Critical Need for Energy Research and Development: The Role of the Midwest Research Labs," pp. 1–26.

49. As cited in ANL, "Argonne National Laboratory: A Resource for the Nation," a document prepared for the Multiprogram Laboratory Panel of the Energy Research Advisory Board, Feb. 22, 1982, p. 3, TPD-82-7, ANL, TIC.

50. Ibid., p. 18.

51. Ibid., p. 13; Massey to Hanna Gray, Aug. 9, 1982, UC General, Box 1, Folder 1, Job 11174, ANL.

52. ANL, "Argonne National Laboratory: A Resource for the Nation," p. 13.

53. John Walsh, "For Argonne, Criticism and a Comeback," *Science* 218 (Oct 22, 1982): 354.

54. George A. Keyworth to Solomon J. Buchsbaum, June 3, 1982, and N. D. Pewitt to W. Kenneth Davis, June 3, 1982, in Packard Report no. 2, Office of Field Operations Management, DOE.

55. Keyworth to Buchsbaum, June 3, 1982, Packard Report no. 2, Office of Field Operations Management, DOE.

56. DOE, "Final Report of the Multilaboratory Panel of the Energy Research Advisory Board," Sept. 1982; OSTP, "Report of the White House Science Council: Federal Laboratory Review Panel" (Packard Report), July 15, 1983; "President's Private Sector Survey on Cost Control" (Grace Commission Survey); N. D. Pewitt to W. Kenneth Davis, June 3, 1982, "Draft: Talking Points for Deputy Secretary Davis in Discussions with OSTP Laboratory Study Panel," Oct. 7, 1982, all in Packard Report no. 2, Office of Field Operations Management, DOE.

57. *Chicago Sun-Times,* July 16, 1981, p. 25; R. Stephen Berry, "The Federal Laboratories," *Bulletin of the Atomic Scientists* (March 1984): 21–25; G. B. L., "What Will Be the Future Role of National Laboratories?" *Physics Today,* Jan. 1982, p. 53.

58. DOE, "Final Report of the Multiprogram Laboratory Panel of the Energy Research Advisory Board," Sept. 1982, p. 55.

59. Massey to the Board of Governors, Dec. 21, 1983, "Background on Multi-Year Funding for Research and Development at National Laboratories," p. 2, in Univ. of Chicago Board of Governors Exec. Comm., Box 1, Job 11174, ANL.

60. *Chicago Sun-Times,* July 16, 1982, p. 35; Philip M. Boffey, "Experts Criticize U.S. Laboratories: White House Science Advisers Cite 'Serious Deficiencies,'" *New York Times,* July 16, 1983; Berry, "The Federal Laboratories," pp. 21–25.

61. Boffey, "Experts Criticize U.S. Laboratories."

62. N. D. Pewitt to Guy W. Fiske, "Interagency Working Group on Federal Laboratories," Feb. 18, 1982; Trivelpiece to DOE assistant secretaries, "Report of the White House Council: Federal Laboratory Review Panel," Nov. 3, 1982, Packard Panel no. 2, Office of Field Operations Management, DOE.

63. Roger E. Batzel to William R. Frazer, senior vice president, academic affairs, Univ. of Calif., Dec. 5, 1983, Packard Report no. 2, Office of Field Operations Management, DOE.

64. Donald P. Hodel, "Reagan Administration Energy Policy: Are We Better Off Today Than Were in 1980?" speech before the National Press Club, June 14, 1984, Washington, D.C., *DOE/News:* N-84-032; Holl and Fehner, *Securing America's Energy Future,* pp. 38–40.

65. Massey to members of the Executive Committee, Board of Governors of Argonne National Laboratory, Sept. 16, 1983, and Memorandum of discussion, Fourth Meeting Executive Committee, Board of Governors of Argonne National Laboratory, Aug. 24, 1983, Folder 5, Box 3, Job 11174, ANL; Massey, responses to multiyear funding paper, April 26, 1984, Board of Governors Exec. Comm., Box 3, Folder 9, Job 11174, ANL.

66. Rauch replaced Bauer on Oct. 2, 1883; Memorandum of discussion, Fourth Meeting Executive Committee, Board of Governors of Argonne National Laboratory, Aug. 24, 1983, Board of Governors Executive Committee, Box 3, Folder 5, Job 11174, ANL; Massey to Gray, Aug. 9, 1982, UC General, Box 1, Folder 1, Job 11174, ANL.

67. Memorandum of discussion, First Meeting of the Board of Governors of the Argonne National Laboratory, Dec. 13, 1982; Memorandum of discussion, First Meeting of the Executive Committee, Board of Governors of Argonne National Laboratory, Jan. 28, 1983, Board of Governors meeting, UC General 2, Box 1, Job 11174, ANL.

68. "New Building Is Architectural First," *Argonne News,* Sept. 1981, p. 9.
69. Ibid.
70. Lewis Mumford, *The City in History* (New York: Harcourt, Brace & World, 1961).
71. "Argonne Awarded Grant to Commission Outdoor Sculpture," *Argonne News,* Aug.–Sept. 1981, pp. 8–9.
72. "A Picture-Perfect Sculpture Dedication," *Argonne News,* Oct. 1982, p. 4.
73. Ibid., pp. 4–5.
74. "Argonne, 201 Praised at Dedication," *Argonne News,* Oct. 1982, pp. 3, 13.
75. Minutes of the Fourth Meeting of the University of Chicago Board of Governors of Argonne National Laboratory, July 19, 1983, Board of Governors Meetings, Box 1, Job 11174, ANL.
76. Minutes of the First Meeting of the Scientific and Technical Advisory Committee of the University of Chicago Board of Governors of Argonne National Laboratory, April 15, 1983, University of Chicago Board of Governors, Box 2, Job 11174, ANL.
77. Massey to Trivelpiece and Shelby Brewer, Feb. 2, 1983, and "Materials Research and Development at Argonne," in *Descriptions of Major Facilities at Argonne National Laboratory,* April 6, 1984, University of Chicago Board of Governors, in Box 1, Folder 1, Job 11174, ANL.
78. "Basic Research Improves Future Materials, Technology," *Argonne National Laboratory, 1983–1984* (annual report), pp. 11–12; "Electron Microscopy Center for Materials Research," in *Descriptions of Major Facilities at Argonne National Laboratory,* April 6, 1984, University of Chicago Board of Governors, Box 1, Folder 1, Job 11174, ANL.
79. "Basic Research Improves Future Materials, Technology," *Argonne National Laboratory, 1983–1984 Annual Report,* pp. 11–12.
80. "HVEM–Tandem Facility: Searching for Damage at the Molecular Level," *Argonne News,* Nov. 1981, p. 3.
81. "IPNS-I," report to the Board of Trustees, University of Chicago, Nov. 4, 1982, University of Chicago Board of Governors, Box 1, Job 11174, ANL.
82. Walter Massey, "The IPNS Initiative and the Draft Report of the Review Panel on Neutron Scattering: A Position Paper for the Planning Committee of the Argonne Universities Association," Dec. 9, 1980, IPNS–1980, Box 28, W. Massey Subj. Files, ANL.
83. "Position Adopted by the Board of Trustees of the Argonne Universities Association at Its Meeting on Jan. 15, 1981," Appendix A, Seventy-Fourth Meeting of the Board of Trustees, Jan. 15, 1981, Box 27, Collec. no. 434-91-0014, GLFRC.
84. "Major Facility Proposed for ANL: $62 Million Intense Pulsed Neutron Source," *Argonne News,* March 1976, pp. 3–4; J. M. Carpenter and David L. Price, "An Intense Pulsed Neutron Source for Argonne National Laboratory," *IEEE Transactions on Nuclear Science* NS-22 (June 1975): 1768–71; Gerard H. Lander, "IPNS: Pointing the Way to New Frontiers in Materials Research," *logos,* TPD-LOGOS-1.
85. Massey, "The IPNS Initiative," p. 5.
86. Arthur L. Robinson, "Will U.S. Skip Neutron Scattering Derby?" *Science* 211 (Jan. 16, 1981): 259–63.

87. Massey to James S. Kane, June 11, 1980, IPNS-1980, Box 28, W. Massey Subj. Files, ANL.

88. Robinson, "Will U.S. Skip Neutron Scattering Derby?" Science 211 (Jan. 16, 1981): 259–63; Massey, "Report from the Director of Argonne National Laboratory on IPNS," Minutes of the Seventy-Fourth Meeting of the Board of Trustees, AUA, Jan. 15, 1981, Box 27, Collec. no. 434-91-0014, GLFRC.

89. Massey, "The IPNS initiative," p. 2.

90. Ibid., p. 7.

91. Walter Massey, "The Director's Corner: A Green Light for IPNS-I," *Argonne News*, May–June 1981, pp. 2, 13; Arthur L. Robinson, "Argonne's Pulsed Neutron Source Turned On," *Science* 214 (Sept. 4, 1981): 1097–99.

92. Massey, "Report from the Director of Argonne National Laboratory on IPNS," and "Report from the Director of Argonne National Laboratory," Minutes of the Seventy-Fifth Meeting of the Board of Governors, AUA, April 30, 1981, in Box 27, Collec. no. 434-91-0014, GLFRC; "IPNS-I," Report to the Board of Trustees, University of Chicago, Nov. 4, 1982, University of Chicago Board of Governors, Job 11174, ANL; Arthur Robinson, "Argonne's Pulsed Neutron Source Turned On," *Science* 213 (Sept. 4, 1981): 1097–99.

93. John Walsh, "For Argonne, Criticism and a Comeback," *Science* 218 (Oct. 21, 982): 354–57.

94. Lowell M. Bollinger, "ATLAS: The World's First Superconducting, Heavy-Ion Accelerator is Born," *logos*, vol. 1 no. 3, pp. 2–11.

95. Ibid.; "The linac—A Decade in the Making," *Argonne News*, July 1981, p. 9; "ATLAS Accelerator Off to a Flying Start," *Argonne News*, June–July 1985, p. 3.

96. Bollinger, "ATLAS, the World's First Superconducting, Heavy-Ion Accelerator is Born," *logos*, Vol. 1, No. 3, pp. 2–11.

97. "Superconducting linac is Formally Activated on June 26," *Argonne News*, July 1981, pp. 8–9.

98. Paul Fields comments, Minutes of the History Committee meeting, ANL, July 26, 1994, pp. 13–14.

99. Dennis Byrne, "Argonne Seeks Big Aid to Study Tiny Quark," *Chicago Sun-Times*, March 13, 1983; Ronald Kotulak, "Argonne Bids for 'Gentle Giant' to Bare Quirks of Quark," *Chicago Tribune*, Feb. 27, 1983.

100. K. L. Kliewer to file, Nov. 13, 1981, Spec. Board Comm. on the Electron Accel., Box 37, Collec. no. 434-91-0015, GLFRC; "Battle for an Electron Accelerator: Argonne Takes On SURA," *Physics Today*, July 1983, pp. 57–60.

101. Hanna Holborn Gray to University of Chicago Board of Trustees, March 10, 1983, ANL Gen. Corres., Box 5, W. Massey Subj. Files, ANL.

102. Dennis Byrne, "Argonne Seeks Big Aid to Study Tiny Quark," *Chicago Sun-Times*, March 13, 1983; Ronald Kotulak, "Argonne Bids for 'Gentle Giant' to Bare Quirks of Quark," *Chicago Tribune*, Feb. 27, 1983.

103. Massey, note to file, Nov. 11, 1981, and K. L. Kliewar to file, Nov. 13, 1981, in Spec. Board Comm. on the Electron Accel., Box 37, Collec. no. 434-91-0015, GLFRC.

104. Eliot Marshall, "Illinois and Virginia Scrap Over Accelerator," *Science* 220 (May 27, 1983): 929–32; J. P. Schiffer, memorandum to file, May 21, 1982, Spec. Board Comm. on the Electron Accel., and Massey, note to file, in Box 37, Collec. no. 434-91-0015, GLFRC; Massey, note to file, Feb. 2, 1983, GEM Project, Box 24, Collec. no. 434-91-0015, GLFRC.

105. DOE-NSF Nuclear Science Advisory Committee, "Report of the Panel on Electron Accelerator Facilities," GEM Project, Box 24, Collec. no. 434-91-0015, GLFRC.

106. Stephen Budiansky, "US Nuclear Physics: Argonne's Loss," *Nature* 412 (April 28, 1983): 743; "Argonne and Power Politics," *Chicago Tribune*, April 27, 1983.

107. ANL, "Comparative Economic Analysis of Alternative Sites for an Electron Accelerator Facility," May 1983; Gray to Donald P. Hodel, April 28, 1983; J. H. Norem to Massey, May 16, 1983; Nancy M. O'Fallon to University of Chicago Board of Governors, June 13, 1983; Great Lakes governors to Donald P. Hodel, all in Gem Project, Box 24, Collec. no. 434-91-0015, GLFRC; "Arguing for Argonne," *Chicago Sun-Times*, May 29, 1983.

108. Peter David, "Keyworth Answers the Critics," *Nature* 303 (June 9, 1983): 465; Eliot Marshall, "Illinois and Virginia Scrap over Accelerator," *Science* 220 (May 27, 1983): 929.

109. Massey, note to file, May 23, 1983, GEM Project, Box 24, Collec. no. 434-91-0015, GLFRC; Kim McDonald, "Argonne, 23 Universities Battle over Big Accelerator Contract," *The Chronicle of Higher Education*, June 8, 1983, p. 13; Eliot Marshall, "Argonne Puts in a Bid for Virginia's Accelerator," *Science* (June 10, 1983): 1133; Stephen Budiansky, "Designers Dispute over Machine," *Nature* 303 (July 7, 1983).

110. J. P. Schiffer to Massey, "For His Eyes Only," April 27, 1983, GEM Project, Box 24, Collec. no. 434-91-0015, GLFRC.

111. Massey to Trivelpiece, July 19, 1983, and Massey to Percy et al., July 19, 1983, in Board of Governors Membership-1982, Box 27A, W. Massey Subj. Files, ANL; "Argonne Surrenders Accelerator to SURA," *Physics Today*, Sept. 1983, pp. 41–42; Dennis Byrne, "Argonne Drops out of Race for Electron Accelerator," *Chicago Sun-Times*, July 20, 1983.

112. Alan Schriesheim, interview by author, July 12, 1993, Argonne National Laboratory; "Schriesheim Named Senior Deputy Laboratory Director," *Argonne News*, Aug. 1983, pp. 2,6; Paul Fields to Cynthia Wilkinson, October 24, 1995, OPA, ANL.

113. Alan Schriesheim, interview by author, July 12, 1993, Argonne National Laboratory; Walter Massey, interview by author, June 1, 1992, National Science Foundation, Washington, D.C.

114. Alan Schriesheim, interview by author, July 12, 1993, Argonne National Laboratory; "New Senior Deputy Director is a Builder: An Interview with Alan Schriesheim," *Argonne News*, Oct. 1983, pp. 8–10.

115. Holl, Anders, and Buck, *U.S. Civilian Nuclear Power Policy*, p. 24.

116. Charles Till, "The CRBR Termination and Its Effect on Argonne," *Argonne News*, Nov.–Dec. 1983, p. 2.

117. Massey to the Board of Governors, "The Future of Breeder Reactor Research

and Development at Argonne National Laboratory," Jan. 19, 1983, University of Chicago Board of Governors, Box 2, Job 11174, ANL.

118. "Laboratory Director's Report," Minutes of the Sixth Meeting of the University of Chicago Board of Governors of Argonne National Laboratory, Jan. 26, 1984, University of Chicago Board of Governors Meetings, Box 7, Job 11174, ANL.

119. Energy Research and Advisory Board, Multiprogram Laboratory Panel, September 1982; President's Private Sector Survey on Cost Control in the Federal Government, April 1983; White House Science Commission, Federal Laboratory Review Panel, May 1983, all in Box 27A, W. Massey Subj. Files, ANL.

120. The University of Chicago Board of Governors of Argonne National Laboratory, "Response to the Report of the White House Science Council Federal Laboratory Review Panel," May 1984, Board of Governors, Box 27A, W. Massey Subj. Files, ANL.

121. The University of Chicago Board of Governors of Argonne National Laboratory, "Report of the ad hoc Committee on the Future of Argonne," July 1984, Box 37, Collec. no. 434-91-0015, GLFRC.

122. Ibid.

123. W. Dale Compton and Stuart A. Rice to Hanna Gray, April 23, 1984, University of Chicago Board of Governors, Box 3, Job 11174, ANL; The University of Chicago Board of Governors of Argonne National Laboratory, "Report of the ad hoc Committee on the Future of Argonne," July 1984, Box 37, Collec. no. 434-91-0015, GLFRC.

124. Walter Massey and Alan Schriesheim, "State of the Laboratory Address," Nov. 16, 1983, University of Chicago Board of Governors Executive Committee, Box 3, Job 11174, ANL.

Chapter 12

1. Sandy Illian, "Rare Deer Add Charm to Argonne Lab Grounds," *The Doings*, Feb. 3, 1989; "Study Finds More Than 100 Fallow Deer at ANL," *Argonne News*, April–May 1992, pp. 6–7.

2. A. Steven Messenger, Walter R. Suter, and John A. Wagner, "Ecological Survey of Argonne National Laboratory," March 1969, ANL-7559; "Environmental Assessment: Proposed 7-GeV Advanced Photon Source," U.S. Department of Energy, Feb. 1990, DOE/EA-0389.

3. "Wildlife at the INEL," a brochure prepared by the Environmental Affairs Subcommittee of the Idaho Falls Section of the American Nuclear Society, n.d.

4. Richard Rhodes, *Nuclear Renewal: Common Sense About Energy* (New York: Viking Penguin, 1993), pp. 96–99.

5. Nancy L. Buschhaus, Kirk E. LaGory, Douglas H. Taylor, "Behavior in an Introduced Population of Fallow Deer during the Rut," *American Midland Naturalist*: 124 (1990): 318–29.

6. George A. Keyworth, "The Reagan Science Policy: Where We've Been, Where We're Going," remarks to the Council for the Advancement of Science Writing, San Francisco, Calif., Oct. 31, 1984.

7. Ibid.

8. Ibid.

9. Alan Schriesheim, "Keynote speaker at the Midwest Universities Energy Consortium Inc. Conference," Chicago, noted in press release, "News from Argonne National Laboratory 84-21," April 4, 1984.

10. Alan Schriesheim, interview by author, July 12, 1993, Argonne National Laboratory.

11. Ibid.

12. Ibid.

13. DOE Appraisal of Argonne, "Summary of Programmatic Performance," and "Summary of Support Function Performance," ANL (FY 1984); Director's comments, Executive Committee, Board of Governors; Minutes of the Tenth Meeting of the Executive Committee, The University of Chicago Board of Governors of Argonne National Laboratory, Jan. 18, 1985, all in University of Chicago Board of Governors Meeting Jan. 18, 1985, Folders 11 and 12, Box 3, Job 11174, ANL.

14. See, for example, J. G. Asbury and R. O. Mueller, "Solar Energy and Electric Utilities: Should They Be Interfaced?" *Science,* Feb. 4, 1977; J. G. Asbury and S. B. Webb, "Centralizing or Decentralizing: The Impact of Decentralized Electric Generation," *Public Utilities Fortnightly,* Sept. 25, 1980; T. G. Alston and J. G. Asbury, "Grain Alcohol: The Right Product in the Wrong Market?" Argonne National Laboratory Report ANL/EES-TM-79, March 16, 1980, ANL, TIC.

15. Director's comments, Executive Committee, Board of Governors, Jan. 18, 1985, University of Chicago Board of Governors Meeting Jan. 18, 1985, Folder 11, Box 3, Job 11174, ANL.

16. Laboratory Director's Report to the Board of Governors, Jan. 18, 1985, University of Chicago Board of Governors, Folder 2, Box 2, Job 11174, ANL; Minutes of the Tenth Meeting of the University of Chicago Board of Governors of Argonne National Laboratory, Jan. 18, 1985, University of Chicago Board of Governors, Box 3, Folder 12, Job 11174, ANL.

17. Mary L. Parramore to the Implementation Group, Implementation Group Activities, April 15, 1985, Job 11174, ANL Board of Governors Meeting April 26, 1985, Folder 12, Box 3, Job 11174, ANL.

18. Ibid.; Schriesheim to Trivelpiece, April 18, 1985, University of Chicago Board of Governors, Folder 7, Box 2, Job 11174, ANL.

19. Notes for Board and Executive Committee, Laboratory Director's Update, April 24, 1985, University of Chicago Board of Governors, Director's Report, Folder 2, Box 2, Job 11174, ANL.

20. Trivelpiece quoted in Mary L. Parramore to the Implementation Group, Implementation Group Activities, April 15, 1985, University of Chicago Board of Governors meeting April 26, 1985, Folder 12, Box 3, Job 11174, ANL.

21. Laboratory Director Report to Board of Governors, Jan. 17, 1985, University of Chicago Board of Governors, Folder 11, Box 3, ANL; "Herrington, Hodel confirmed for Energy, Interior," *DOE This Month,* Feb. 1985, p. 1; portions of this section appeared previously in Holl and Fehner, *Securing America's Energy Future,* pp. 51–52.

22. Holl and Fehner, *Securing America's Energy Future*, pp. 47–48; David Rogers and Paul Blustein, "Reagan Gives Cabinet Drastic Cuts in Many Widely Popular Programs," *Wall Street Journal*, Dec. 6, 1984.

23. Harris W. Fawell to the Illinois Congressional Delegation, May 9, 1985, University of Chicago Board of Governors, Folder 14, Box 3, ANL; Oak Ridge National Laboratory, "Multiprogram National Laboratories: Trends and Projections, FY1979–1990," Aug. 1985, Lab. Director's Meeting File, 9/30/85, Box 11, 434-91-0015, GLFRC.

24. U.S. House of Representatives, Committee on Appropriations, Hearings before the Subcommittee on the Department of the Interior and Related Agencies, 99th Cong., 1st Sess., March 29, 1985, pp. 758–60.

25. Ibid.

26. John S. Herrington to Harris W. Fawell, May 17, 1985, University of Chicago Board of Governors, Folder 14, Box 3, ANL; Herrington, Address to the Atomic Industrial Forum/American Nuclear Society, Nov. 13, 1985, San Francisco, Calif., Vol. 25, Box 9, Department of Energy, 434-91-0015, GLFRC.

27. ANL, "New Strategic Thrusts," April 1985, University of Chicago Board of Governors, Folder 2, Box 2, ANL.

28. Ibid.

29. Bruce S. Brown and John M. Carpenter, "Cold Neutrons, Enriched Target Open New Research for IPNS," *logos*, Summer 1985, pp. 11–15; Gerard H. Lander, "IPNS: Pointing the Way to New Frontiers in Materials Research," *logos*, 1982, pp. 2–7.

30. "ATLAS Accelerator Off to a Flying Start," *Argonne News*, June–July 1985, p. 3; "ATLAS Completed on Time, within Budget," Argonne National Laboratory: Research Highlights, 1985–1986, pp. 12–13; Alan Schriesheim, "State of the Laboratory," *Argonne Bulletin*, Dec. 16, 1985.

31. "ANL young scientists among top 100 in U.S.," *Argonne News*, Dec.–Jan 1984–1985, p. 4; "Top Innovators," *Argonne News*, Feb.–March 1986, p. 10. See also *Argonne News*, Oct.–Nov. 1984 and Oct.–Nov. 1985.

32. Charles Till, interview by author, May 23, 1994, Argonne-East.

33. Ibid.

34. Jerry E. Bishop, "New Type of Uranium Fuel Sets Record for Efficient Burning in Nuclear Reactor," *Wall Street Journal*, Feb. 3, 1989, p. B3; "The Soul of a New Machine," *Energy Daily*, March 10, 1989, pp. 5–6; Ron Winslow, "New Breeder Reactor May Operate More Safely, Produce Less Waste," *Wall Street Journal*, Dec. 9, 1988.

35. Conference Strategy Summary, Energy and Water Appropriations for Fiscal Year 1985, Folder "June 1984," Box 1, Management Council, Office of the Director, ANL; ANL, "New Strategic Thrusts—April 1985," Folder "April 1985," Box 2, Management Council, Office of the Director, ANL; Pat Moonier to Management Council, "Highlights of Management Council Meeting," May 20, 1985, Folder "Strategic Planning Board," Box 11, Management Council, Office of the Director, ANL; Ed Hahn, "New Reactor Concept Introduced at EBR-II 20-Year Anniversary at Argonne-West," *Argonne News*, Sept. 1984, p. 5.

36. Mitchell Rogovin, director, *Three Mile Island: A Report to the Commissioners and*

to the Public, Vol. 2, Part 1, Nuclear Regulatory Commission; Peter Miller, "A Comeback for Nuclear Power? Our Electric Future," National Geographic, Aug. 1991, pp. 62–87; Joseph G. Morone and Edward J. Woodhouse, The Demise of Nuclear Energy? Lessons for Democratic Control of Technology (New Haven: Yale University Press, 1989), pp. 85–89; John L. Campbell, Collapse of an Industry: Nuclear Power and the Contradictions of U.S. Policy (Ithaca: Cornell University Press, 1988), pp. 65–67; William Rankin, Barbara Melber, Thomas Overcase, and Stanley Nealey, Nuclear Power and the Public: An Update of Collected Survey Research on Nuclear Power, 1981, PNL-4048, report prepared for the Department of Energy by Battelle Human Affairs Research Center, Seattle, Washington.

37. Charles Till, Introduction to Stanley H. Fistedis, The Experimental Breeder Reactor-II: Inherent Safety Demonstration (Amsterdam: North-Holland, 1987), pp. 1–2; Rhodes, Nuclear Renewal, p. 105; "EBR-II Hits 20-Year Mark," Argonne News, July–Aug. 1984, pp. 6–7.

38. Fistedis, The Experimental Breeder Reactor-II, passim; "EBR-II Tests Prove Inherent Safety of Liquid-Metal Cooled Reactors," Argonne News, April–May, 1986, pp. 6–7.

39. Miller, "A Comeback for Nuclear Power?" pp. 72–73.

40. Joseph Palca, "US Keeps Fast Reactors Alive," Nature 320 (March 1986).

41. Charles Till, interview by author, May 23, 1994, Argonne-East; Miller, "A Comeback for Nuclear Power?"

42. Moonier to the Management Council, "Highlights of Management Council Meetings," April 1, 11, 15, and Dec. 16, 1985, Jan. 6 and Feb. 10, 1986; Folder "Strategic Planning Board," Box 11, Management Council, Office of the Director, ANL; Nancy M. O'Fallon to file, May 8, 1986, Folder "May 1986," Box 3, Management Council, Office of the Director, ANL.

43. Moonier to the Management Council, "Highlights of Management Council Meetings," March 18, 1985, August 21, 1985, and March 3, 1986, Folder "Strategic Planning Board," Box 11, Management Council, Office of the Director, ANL; "Top Technology, IFR Best in World," Argonne News, Aug.–Sept. 1985, p. 8.

44. Yoon I. Chang, "The Total Nuclear Power Solution," The World and I, April 1991, pp. 288–95; Jon Van, "Design Could Spur Nuclear Comeback," Chicago Tribune, March 10, 1991.

45. Chang, "The Total Nuclear Power Solution"; Scott Pendleton, "Nuclear Enters a New Era," Christian Science Monitor, March 6, 1991; Charles Till and Yoon I. Chang, "Integrating the Fuel Cycle at IFR," Nuclear Engineering International, Nov. 1992. See also, Charles E. Till, "Plutonium, Nonproliferation, and the Future of Nuclear Energy," talk presented at the workshop "The Role of Civilian Plutonium at Stanford University," March 29–30, 1994, Stanford, Calif., Folder "April 1994," Management Council, Office of the Director, ANL.

46. Minutes of the Fifth Meeting of the Scientific and Technical Advisory Committee of the University of Chicago Board of Governors of Argonne National Laboratory, March 27, 1985, Management Council, Office of the Director, ANL.

47. Rocky Barker, "DOE Urged to Improve Safety at 2 INEL Reactors," Idaho Falls Post-Register, Aug. 11, 1988.

48. Dave Jacqué, "Argonne Helps Reduce Risk from Soviet Reactors," *Argonne News*, June–July 1993, pp. 4–5.

49. "Team of Soviet Scientists Visits INEL," *Pocatello Idaho State Journal*, Nov. 4, 1990.

50. Ibid.; "Argonne to Lend Countries Nuclear Expertise," *Naperville Sun*, June 30, 1993.

51. The Public Citizen, "A Decade in Decline," as quoted in John T. Suchy, "Nuclear Power's Controversial Future," *The Rotarian*, Feb. 1991, pp. 14–16; Wendy M. Koch, "Argonne: Fast Reactor a Revolution in Power," *Kankakee (Illinois) Daily Journal*, Feb. 8, 1989.

52. Till, interview by author, May 23, 1994, Argonne-East.

53. Thomas W. Lippman, "Disputed Nuclear Program Reborn," *Washington Post*, April 13, 1993.

54. Jon Healey, "Budget-Cutting Blaze Could End Quest for a Perfect Reactor," *CQ*, May 29, 1993, pp. 1350–55.

55. Alan Schriesheim, "Reactor that Recycles Nuclear Waste Is on Clinton Cut List," *Portland Oregonian*, March 31, 1993; Schriesheim, "Unlikely "Green Machine' Is Environmentalists' Friend," *Palos Heights (Illinois) Weekly*, Oct. 28, 1993; National Research Council of the National Academy of Sciences, *Nuclear Power: Technical and Institutional Options for the Future* (Washington, D.C.: National Academy Press, 1992), p. 12.

56. Schriesheim, "Unlikely 'Green Machine' Is Environmentalists' Friend," *Palos Heights (Illinois) Weekly*, Oct. 28, 1993.

57. Reps. Sam Coopersmith and Martin R. Hoke, "Taxpayers Have Spent Too Much on Nuclear Reactor Program," *Traverse City (Michigan) Record Eagle*, June 17, 1993; Jon Healey, "Budget-Cutting Blaze Could End Quest for a Perfect Reactor," *CQ*, May 29, 1993, pp. 1350–55.

58. Lancelot quoted in Tim Jackson, "Politicians Gear Up to Fight about Reactor Research Money," *Pocatello Idaho State Journal*, May 30, 1993; Kevin Richert, "Environmentalists Push IFR Closure," *Post Register*, June 1, 1993; Ed Lane, "Groups Seek to Kill Liquid Metal Reactor," *Energy Daily*, June 11, 1993; Tim Jackson, "Lobbyists Push Clinton to Kill IFR in Recision Package," *Pocatello Idaho State Journal*, Oct. 31, 1993.

59. Denman quoted in Jon Healey, "Reactor Technology Faces Meltdown," *Washington Times*, June 3, 1993; Ed Lane, "IFR Debate Part of Larger 'Energy Future' Battle," *Energy Daily*, Aug. 5, 1993.

60. Schriesheim, "Reactor That Recycles Nuclear Waste Is on Clinton Cut List," *Portland Oregonian*, March 31, 1993.

61. Till, interview by author, May 23, 1994, Argonne-East; Jon Healey, "Budget-Cutting Blaze Could End Quest for a Perfect Reactor," *CQ*, May 9, 1993, pp. 1350–55.

62. Schriesheim to file, May 16, 1994, Folder "May 1994," Management Council, Office of the Director, ANL.

63. Andrew Gottesman, "Argonne Research Work Might Be Saved," *Chicago Tribune*, Sept. 20, 1994; Basil Talbott, "Plan to Save Argonne Jobs Goes to White House for OK," *Chicago Sun-Times*, Sept. 20, 1994.

64. Jim Ritter and Basil Talbott, "Argonne Nuclear Recycling Survives," *Chicago Sun-Times,* March 28, 1993; Jon Healey, "Environmentalists Oppose Rod Recycling," *Washington Times,* June 3, 1993; "Does Banana Peel of History Await IFR?" *R&D Magazine,* June 1993; Anne Hazard, "Argonne Reactor Unnecessary for Plutonium Disposal, Report Says," *Chicago Tribune,* Jan. 29, 1994; Committee on International Security and Arms Control, National Academy of Sciences, *Management and Disposition of Excess Weapons Plutonium* (Washington, D.C.: National Academy Press, 1994). See also, "Liquid Metal Reactors," hearing before the Subcommittee on Energy and Power of the Committee on Energy and Commerce, U.S. House of Representatives, June 9, 1993.

65. "Reactors and Compromise: President Bill Clinton's Budget Compromises Are Sending Conflicting Signals about His Seriousness," *Nature* 362 (April 22, 1993).

66. "An Unscientific Defeat for Argonne," *Chicago Tribune,* Aug. 9, 1994.

Chapter 13

1. Alan Schriesheim, "The New Uncertainty Principle: Our National Labs in the 21st Century," remarks delivered to Sigma Xi, General Motors Research Center, Nov. 9, 1994, Warren, Michigan, in OPA, ANL.

2. Hewlett and Holl, *Atoms for Peace,* p. 209.

3. The source for this and the next paragraph is Schriesheim, "The New Uncertainty Principle."

4. Partial transcript of Clinton's December 21, 1992, press conference in *Inside Energy* (Dec. 23, 1992), p. 5; excerpts from statement and testimony of O'Leary confirmation hearing, Jan. 19, 1993, in *DOE This Month,* Jan. 1993, pp. 2–3; U.S. Department of Energy, *Fueling a Competitive Economy: Strategic Plan* (Washington, D.C.: DOE, 1994), DOE/-S-0108; Terence Fehner and Jack M. Holl, *Department of Energy 1977–1994* (Washington, D.C.: DOE, 1994), pp. 77–101.

5. "Report of the HEPAP Subpanel on the Future of the High Energy Physics Program at Argonne," n.d., Vol. 4, HEPAP, Box 37, Collec. no. 434-91-0015, GLFRC.

6. Frank Fradin, interview by author, May 24, 1994, Argonne-East.

7. M. Mitchell Waldrop, "Gambling on the Supercollider," *Science* 221 (Sept. 9, 1983): 1038–40.

8. Ibid.

9. Hanna Gray to Donald P. Hodel, July 19, 1983, Board of Governors, Box 27A, W. Massey Subj. Files, ANL; E. L. Berger, "Summary of Discussions at High Energy Physics Advisory Panel Meeting," Feb. 7–8, 1983, Vol. 4, HEPAP, Box 37, Collec. no. 434-91-0015, GLFRC; K. L. Kliewer to A. Schriesheim, April 30, 1984; Summary of Management Council meeting, May 29, 1984, Folder "Strategic Planning Board," Box 11, Management Council, Office of the Director, ANL; ANL, proposal that Argonne National Laboratory serve as the host laboratory for R&D on the Superconducting Super Collider, n.d.; ANL, High Energy Physics Program Budget, April 1984; all but first two in Vol. 5, Misc. Corres., Box 38, Collec. no. 434-91-0015, GLFRC.

10. "Report of the High Energy Physics Division Review Committee—Argonne National Laboratory," May 15–17, 1984, and May 8–10, 1985, Vol. 3, University of Chi-

cago Review Committees, Box 34, Collec. no. 434-91-0015, GLFRC. See also R. Diebold to Massey, Schriesheim, and Kliewer, "SSC at Argonne," May 4, 1984, and T. H. Fields to Schriesheim and Kliewer, "Critical Planning Issues Regarding Accelerator R&D at Argonne," Vols. 5 and 6, Misc. Corres., Box 38, Collec. no. 434-91-0015, GLFRC.

11. "Panel Puts Aside Differences, List Major Materials Facilities," *Physics Today*, Sept. 1984, pp. 57–59; Arthur L. Robinson, "Major Materials Facilities Ranked," *Science* 225 (Aug. 3, 1984): 704; William Sweet, "Key Decision Nears on Proposed Argonne Photon Source," *Physics Today*, Jan. 1987, p. 71.

12. Minutes of the Ninth Meeting of the Executive Committee of the University of Chicago Board of Governors of Argonne National Laboratory, Oct. 1, 1984, University of Chicago Board of Governors, Folder 11, Box 3, ANL; Summary of Management Council Meeting, Aug. 13, 1984, Folder "Strategic Planning Board," Box 11, Management Council, Office of the Director, ANL.

13. Laboratory Director's report for the Board of Governors, July 25, 1984, University of Chicago Board of Governors, Folder 2, Box 7, ANL; J. B. Cohen to Massey, Nov. 9, 1984; Nancy M. O'Fallon to file, "6 GeV Synchrotron Special Committee Meeting Nov. 8, 1984 (Nov. 16, 1984), and March 2, 1985 (March 4, 1985)"; K. L. Kliewer to file, comments from the first meeting of the 6 GeV Special Committee, all in 43.31, Box 37, Collec. no. 434-91-0015, GLFRC; Schriesheim to Gray and Massey, Dec. 9, 1985, Box 3, Management Council, Office of the Director, ANL.

14. Cohen to Massey, Nov. 9, 1984, 43.31, Box 37, Collec. no. 434-91-0015, GLFRC; notes for Board and Executive Committee, Laboratory Director's update, April 24, 1985, University of Chicago Board of Governors, Folder 2, Box 2, ANL; Pat Moonier to Management Council, "Highlights of Management Council Meeting," March 18, 1985, Folder "Strategic Planning Board," Box 11, Management Council, Office of the Director, ANL.

15. Schriesheim to staff, March 14, 1985, 43.31, Box 37, Collec. no. 434-91-0015, GLFRC.

16. Laboratory Director's report to the Executive Committee, Oct. 5, 1984, University of Chicago Board of Governors, Box 1, ANL; Minutes of the Ninth Meeting of the University of Chicago Board of Governors for Argonne National Laboratory, Oct. 11, 1984, and Director's comments, Executive Committee of the Board of Governors, Jan. 17, 1985, in University of Chicago Board of Governors, Folder 11, Box 3, ANL.

17. Director's items, Summary of Management Council Meetings, Nov.–Dec., 1985, Folder "Strategic Planning Board," Box 11, and Irving Shain to Alvin Trivelpiece, July 21, 1986, Box 4, in Management Council, Office of the Director, ANL.

18. Pat Moonier to the Management Council, "Highlights of Management Council Meeting," Jan. 27, 1986 and June 23, 1986, Folder "Strategic Planning Board," Box 11, Management Council, Office of the Director, ANL; Alan Schriesheim, interview by author, July 12, 1993, Argonne National Laboratory.

19. "An Argonne National Laboratory/Midwestern University Proposal for a 6-GeV Synchrotron Source," n.d., 43.31, Box 37, Special Committee for the 6 GeV Synchrotron Light Source, Collec. no. 434-91-0015, GLFRC; "Argonne to Get 6 GeV Funding: Initial Design Underway," *Argonne News*, Feb.–March 1986, pp. 4–5.

20. William Sweet, "Key Decision Nears on Proposed Argonne Photon Source," *Physics Today,* Jan. 1987, p. 71; Kenneth L. Kliewer, "Most Powerful Light Brightens Future of Materials Science," *logos,* Spring 1986, pp. 2–7.

21. Alan Schriesheim, "7-GeV Advanced Photon Source," testimony before the Subcommittee on Energy Research and Development of the House Committee on Science, Space, and Technology, U.S. House of Representatives, March 23, 1988, Folder "March 1988," Management Council, Office of the Director, ANL.

22. Kliewer, "Most Powerful Light Brightens Future of Materials Science," *logos,* Spring 1986, pp. 2–7; Jon Van, "Argonne's Bright Light Draws Researchers," *Chicago Tribune,* April 19, 1992; Cheryl Wade, "Brilliant! Powerful X-Ray Tool to Look inside Molecules," *Midland Daily News,* March 21, 1993; Jim Ritter, "Argonne X-Ray Machine Has a Positive Image," *Chicago Sun-Times,* March 21, 1993.

23. David Baurac, "APS to Produce Most Brilliant X-Ray Beams for Research," *Argonne News,* March–April 1988, pp. 4–5; "Argonne Researchers Develop Key Component for APS," *Argonne News,* Nov.–Dec. 1988, p. 3.

24. K. L. Kliewer, "Argonne Perspective on 6-GeV Synchrotron Source," n.d., 43.31, Box 37, Special Committee for the 6 GeV Synchrotron Light Source, Collec. no. 434-91-0015, GLFRC; Jon Van, "Unraveling Mysteries of Nylon: Teams to Take Uncommon Look at Common Products, *Chicago Tribune,* Dec. 12, 1993; "Brightest X-ray Research Planned: Dow, Du Pont Share with Northwestern," *American Metal Market,* Jan. 12, 1994.

25. Alan Schriesheim, interviews by author, July 12, 1993 and May 25, 1994, Argonne National Laboratory; Pat Moonier to Management Council, "Highlights of Management Council Meeting," June 23, 1986, Folder "Strategic Planning Board," Box 11, Management Council, Office of the Director, ANL.

26. "Atomic Energy Work Wins Lawrence Award for Moncton," *Argonne News,* June–July 1987, p. 7; "Moncton Appointed Associate Laboratory Director for APS," *Argonne News,* Jan.–Feb. 1989, p. 5.

27. David Moncton, interview by author, May 24, 1994, Argonne-East.

28. Ibid.

29. Hilary J. Rauch, Manager Chicago Operations Office, to Robert O. Hunter, director, Office of Energy Research, DOE, Argonne National Laboratory (ANL) Institutional Planning On-Site Review—July 12, 1988, Folder "ANL," Office of Field Operations Management, DOE; Moncton to colleagues, May 20, 1988, Harris W. Fawell, Newsletter, June 8, 1988.

30. Dan Rostenkowski to Tom Bevill, April 28, 1989, Folder "May 1989," Box 10, and Hazel O'Leary to Illinois governor Jim Edgar, Jan. 4, 1994, Folder "Jan. 1994," in Management Council, Office of the Director, ANL; Harris W. Fawell, "New X-Ray Research Deemed Promising," *Downer's Grove Reporter,* Jan. 23, 1992.

31. Frank Fradin, interview by author, May 24, 1994, Argonne-East; Marcia Clemmit, "Physicists Play a Hands-On-Role in Super Facilities Construction," *The Scientist,* May 27, 1991.

32. "Ground Broken for Advanced Photon Source," *Argonne News,* Aug.–Sept. 1990, p. 3; Moncton, "APS Unique in Breadth of Mission," *Argonne News,* Aug.–Sept. 1990, p. 2.

33. ANL, *Research Highlights, 1993–94* (annual report), p. 11.

34. ANL, "Structural Biology Center's Beamlines Open for X-Ray Crystallography Studies," *Research Highlights, 1992–93* (annual report), p. 40; Jon Van, "Argonne Sees Light at End of New Project," *Chicago Tribune*, June 5, 1990; Jeffrey Bils, "Argonne to Get Biology Center," *Chicago Tribune*, March 18, 1994.

35. Rongguang Zhang and Edwin Westbrook, "Structural Biology Center Reveals Key to Cholera Toxin," *logos*, Spring 1993, pp. 8–13.

36. Ibid.; Harvey Drucker, interview by author, May 24, 1994, Argonne-East.

37. Harvey Drucker, interview by author, May 24, 1994, Argonne-East.

38. Ivan Labat, "Decoding the Human Genome Is as Fast and Easy as ACGT," *logos*, Spring 1994, pp. 2–6.

39. "Superconductivity Agreement Scores 'First' for Laboratory," *Argonne News*, Nov.–Dec. 1988, p. 13.

40. Mark M. Banaszak Holl, interview by author, September 20, 1994, Providence, Rhode Island.

41. "Argonne Named Superconductivity Research Center," *Argonne News*, Aug.–Sept. 1987, p. 3.

42. Frank J. Adrian and Dwaine O. Cowan, "The New Superconductors," *Chemical and Engineering News*, Dec. 21, 1992, p. 24; "Superconductivity: Evolution Becomes a Revolution," *logos*, Autumn 1987, pp. 2–3.

43. John Holmes, "A Quantum Leap in Record Speed," *Insight*, May 25, 1987, pp. 8–13; Alan M Wolsky, Robert F. Giese, and Edwards J. Daniels, "The New Superconductors: Prospects for Applications," *Scientific American*, Feb. 1989, pp. 61–69.

44. John N. Maclean, "Superconductor Motor Built," *Chicago Tribune*, Jan. 6, 1988, p. 3.

45. "Argonne Scientists Achieve First in Superconductivity Research," and Richard W. Weeks, "Industry Has Many Applications for New Superconductors," *logos*, Autumn 1987, pp. 4–9.

46. Roger B. Poeppel, "Ceramic Engineering May Solve Problems with Superconductors," *logos*, Autumn 1987, pp. 12–16; "Argonne Makes History in Test of New Wire," *Argonne News*, April–May 1987, p. 3.

47. "Fradin Coordinates Superconductivity Programs," *Argonne News*, Aug.–Sept. 1987, p. 3.

48. Pat Moonier to the Management Council, "Highlights of Management Council Meeting," May 4, 1986, Folder "Strategic Planning Board," Box 11, Management Council, Office of the Director, ANL; Alan M. Wolsky, Edward J. Daniels, and Robert F. Giese, "New Superconductors May Have Significant Economic Impact," *logos*, Autumn 1987, pp. 22–26.

49. "Current Superconductivity: Records Set at Argonne," *Research Highlights, 1993–94* (annual report), pp. 14–15.

50. Ken Gray, " Argonne Solves Mystery of 'Unpredictable' Behavior in Magnetic Fields," *logos*, Winter 1992, pp. 16–18.; "X-Ray Imaging Improves Wire Production Methods," *Research Highlights, 1993–94* (annual report), p. 16; "Joint Project Targets Practical Superconducting Wire," *Argonne News*, June–July 1992, p. 14.

51. David Baurac, "'Super Bearing' May Carry a Heavy Load," *Argonne News*, Oct.–Nov. 1993, pp. 5, 15; "Levitating Bearings are Product of Meissner Effect," *logos*, Winter 1992, p. 19.

52. David Baurac, "'Super Bearing' May Carry a Heavy Load," *Argonne News*, Oct.–Nov. 1993, pp. 5, 15; "Levitating Bearings are Product of Meissner Effect," *logos*, Winter 1992, p. 19. See also the entire special edition,"Superconductivity: 5 Years' Progress," *logos*, Winter 1992; ANL, *Frontiers: Research Highlights, 1994–95*, pp. 29–30.

53. "A Plan for the University of Chicago/Argonne Laboratory Development Corporation," Folder "1120.3 ANL General (1985)," Laboratory Management Division, DOE; Brian Frost to Schriesheim, "Patent Awards Policy at Argonne," July 27, 1987, Folder "July 1987," Box 5, Management Council, Office of the Director, ANL; Claudia M. Caruana, "Technology Transfer at Argonne National Laboratory," *Chemical Engineering Progress*, Aug. 1987, pp. 94–98.

54. Harvey Drucker, "Argonne Enters 'Age of CRADAs,'" *Argonne News*, June–July 1993, pp. 2,10.

55. Caruana, "Technology Transfer at Argonne National Laboratory," *Chemical Engineering Progress*, p. 95; "Technology Transfer Efforts Paying Off," *Argonne News*, June–July 1990, pp. 6–7.

56. Brian Frost to the ANL Management Council, "Relationships between ANL and ARCH," April 3, 1987, Folder "April 1987," Box 5, Management Council, Office of the Director, ANL.

57. Mary Kunzweiler, "Science Meets the Marketplace: Argonne Entrepreneurs Flourish," *Argonne News*, June–July, 1986, pp. 10–11; Caruana, "Technology Transfer at Argonne National Laboratory," *Chemical Engineering Progress*, Aug. 1987.

58. David Young, "Firms Test the Waters in Argonne Spinoff Program," *Chicago Tribune*, March 15, 1991.

59. ANL, *Institutional Plan, FY 1994–FY 1999*, Sept. 1993, pp. 51–52; ANL, "Training Mission Linked with Protection of Natural Resources," *Spacial Analysis for Energy and Environmental Resource Management*, 1994, p. 11.

60. John Hudson, *History of Chemistry* (New York: Chapman and Hall, 1992), p. 281.

61. Mark Fernau and David South, "Global Climate Change Challenge: Numbers, Numbers Everywhere, but Not Enough to Make Policy Decisions," *logos*, Spring 1992, pp. 14–19; John Henry Dreyfuss, "Argonne Monitors Worldwide Climate Changes," *Argonne News*, Aug.–Sept. 1993, p. 4.

62. ANL, *Institutional Plan, FY 1994–FY 1999*, Contract W-31-109-ENG-38, Sept. 1993, pp. 87–91.

63. Mele Feazell, "Contaminated Government Site to Be Restored after Fifty Years," *Waste Tech News*, Dec. 13, 1993.

64. Alex Rodriguez, "Environmental Hazards Lurking at Argonne Lab," *Arlington Heights Daily Herald*, Jan. 27, 1989; Rodriguez, "Argonne Pollution Questions Swirl," *Arlington Heights Daily Herald*, December 19, 1989; Rodriquez, "Group: Argonne Lab Contaminating Creek," *Arlington Heights Daily Herald*, Nov. 14, 1990.

65. "Estimated Cost of Cleaning up DOE's Labs," *The Scientist,* March 6, 1989; Alex Rodriguez, "Argonne to Clean Up Its Act, Environmentally," *Arlington Heights Daily Herald,* May 30, 1991.

66. Bob Secter, "Picnickers, A-Bomb Waste Sharing Refuge in Illinois," *Los Angeles Times,* June 24, 1990.

67. Ibid.

68. James Hill and Janita Poe, "Activists Warn of Forests' Nuclear 'Hot Spots,'" *Chicago Tribune,* Aug. 31, 1992; "Phelan Wants Nuke Cleanup," *Southtown Economist,* Dec. 2, 1992; Rachel Melcer, "Set Removal of 12 'Hot Spots,'" *Chicago Southwest Courier,* April 27, 1993.

69. Robert Bergsvik, "Phelan Wants Nuke Cleanup," *Southtown Economist,* Dec. 2, 1992; Barbara Dargis, "Greenpeace Renews Call for Site Cleanup," *Chicago Heights Star,* Dec. 3, 1992; Tom Houlihan, "A Look at Red Gate Woods, or Is That Really a Giant Salamander," *Chicago Heights Star,* Dec. 6, 1992.

70. Bernie Biernacki, "Dept. of Energy Resumes Its Study of Red Gate Woods," *Downers Grove Reporter,* March 10, 1993; David Kalisz, "Red Gate Survey/Clean-Up Expected to Be Completed Next Year," *Desplaines Valley News,* Nov. 25, 1993.

71. Dermot Connolly, "Officials Lay Waste To Idea of Woods as a Dumping Ground," *Chicago Southwest Courier,* April 26, 1994; Ken O'Brien, "Red Gate Woods on the Rebound," *Chicago Tribune,* Oct. 27, 1994.

72. David A. Martin, "Labor of Love: Environmental Cleanup," *about . . . time,* Mar. 1993, pp. 26–27; Harvey Drucker, interview by author, May 24, 1994, Argonne-East.

73. As quoted in Martin, "Labor of Love," *about . . . time,* March 1993, p. 26; David Poyer, "Blacks, Hispanics Hit Harder by Non-Electric Energy Rate Hikes," *logos,* Fall 1993, pp. 7–11.

74. Martin, "Labor of Love," *about . . . time,* March 1993, p. 26; ANL, "Argonne Expertise Gives New Life to an Old Industrial Chicago Neighborhood," *Research Highlights, 1993–94* (annual report), p. 32.

75. Harvey Drucker, interview by author, May 24, 1994, Argonne-East.

76. Martin, "Labor of Love," *about . . . time,* March 1993, p. 26.

77. "Argonne/Bethel Partnership for Sustainable Urban Development Community," *New Life,* Oct. 1994, p. 1.

78. Schriesheim, "Explorers Program Makes Its Mark," *Argonne News,* Feb.–March 1991, p. 2; ANL, "Pioneering Science Explorers Education Program Expands across the Nation," *Research Highlights, 1991–92* (annual report), pp. 54–55.

79. ANL, "Succeeding With Science: Scientific Education Programs at Argonne National Laboratory," 1991, as conveyed from Schriesheim to James D. Watkins, June 25, 1991, Executive Secretariat's Office, ES-91-012578, DOE.

80. Maryka Bhattacharyya, "WPI Initiator's Position Focuses on Current, Future Women Scientists," *Argonne News,* Dec. 1991–Jan. 1992, pp. 2, 12.

81. Marion Thurnauer, "Progress of Women at Argonne: Sustaining the Positive Momentum," *Argonne News,* Oct.–Nov. 1993, pp. 2, 10; "Women at Argonne," n.d., Folder "August 1993," Management Council, Office of the Director, ANL.

82. Jan K. Bohren to the Management Council, March 9, 1993, Folder "March 1993," Management Council, Office of the Director, ANL.

83. Ibid.; "Women at Argonne," n.d., Folder "August 1933," Management Council, Office of the Director, ANL.

84. Marion Thurnauer, "Progress of Women at Argonne; Sustaining the Positive Momentum," *Argonne News,* Oct.–Nov. 1993, p. 10; "Girls Need Professional Women as Role Models, Organization Says," *Chicago Tribune,* April 18, 1994.

85. "Women's Science Careers Search," *Palos Citizen,* May 17, 1990; "Program Features Science Careers for Women," *Naperville Sun,* May 3, 1991; "Women in Science Program," *Downers Grove Reporter,* April 15, 1994.

86. D. K. Hanson, N. K. Meshkow, M. C. Thurnauer, "Graduate School and Beyond," a panel discussion from the conference "Science Careers in Search of Women," April 6–7, 1989, Folder "Argonne National Laboratory," Office of Field Management, DOE; "Poster Session," *Argonne News,* Oct.–Nov. 1994, p. 7.

87. Ronald E. Yates, "Refitting Cold War Science," *Chicago Tribune,* Oct. 26, 1992; Joseph Haggin, "Technology Transfer Challenges National Labs," *Chemical and Engineering News,* Sept. 13, 1993, p. 32; "Technology Transfer Efforts Paying Off," *Argonne News,* June–July 1990, pp. 6–7.

88. Harvey Drucker, "Argonne Enters the 'Age of CRADAs,'" *Argonne News,* Oct.–Nov. 1992, pp. 2, 10; Fehner and Holl, *Department of Energy,* p. 89.

89. ANL, "Technology Transfer: Enhancing U.S. Competitiveness," in *Meeting the Manufacturing Challenge,* n.d. (c. 1994), p. 25; "Lab Teams with Baxter Healthcare to Fight AIDS," *Argonne News,* April–May 1991, p. 5.

90. ANL, "Battery Analysis and Diagnostic Laboratory for Industrial Users," in *Argonne National Laboratory and the Transportation Industry: Partners in Innovation,* n.d. (c. 1994), p. 30; Edwin Darby, "A Powerful Drive for Electric Cars," *Chicago Sun-Times,* Nov. 12, 1992.

91. "Argonne Developing Electric Car Battery," *Chicago Heights Star,* Nov. 22, 1992; "Argonne Hopes to Plug-In Electric Battery," *Chicago Heights Star,* Nov. 8, 1992; ANL, "Inventing a High-Powered Lithium/Iron-Disulfide Battery for Electric Vehicles," *Argonne National Laboratory and the Transportation Industry: Partners in Innovation,* n.d. (c. 1994), p. 28.

92. Dan McCosh, "Emerging Technologies for the Supercar," *Popular Science,* June 1994, p. 95; Dawn Stover, "One-Stop Shopping at the National Labs," *Popular Science,* June 1994, p. 96; Alan Schriesheim, "The New Uncertainty Principle."

93. As quoted in Schriesheim, "The New Uncertainty Principle."

94. Matthew L. Wald, "More Join the Quest for Cleaner Cars," *New York Times,* Dec. 16, 1993; Harvey Drucker, interview by author, May 24, 1994, Argonne-East.

95. "Forum to Aid Small Auto Suppliers," *Herald-News,* July 24, 1994.

96. "Lab Works on Future Cars," *Darien (Illinois)* and *Lemont (Illinois) Metropolitan,* July 21, 1994; "How to Please 1,700 Scientists and Engineers: Argonne National Labs (sic) Cranks Up the World's First High-Performance 128 Processor Supercomputer," *Lan Computing and Intech,* Jan. 1994.

97. Rick Stevens, "Parallel-Processing Computers Solve Complex Problems Fast," *logos,* Summer 1994, pp. 2–7.

98. Ibid.

99. Ibid.; Tim Studt, "Promise of Rich Payoffs Drives Computer-Aided Chemistry," *R&D Magazine,* Sept. 1993, pp. 28–30; Schriesheim quoted in Mele Feazell, "Industry, Academia and Government Gain Easy Access to IBM's Supercomputer," *Waste Tech News,* Nov. 30, 1993.

100. DOE Strategic Plan, April 1994, in Folder "April 1994," Management Council, Office of the Director, ANL; ANL, "Draft Institutional Plan, FY 1995–FY 2000," May 1994, p. 125; Harvey Drucker, interview by author, May 24, 1994, Argonne-East.

101. Alan Schriesheim, "The Changing Role of the National Laboratories," *CHEMTECH,* Feb. 1994, pp. 33–34.

102. M. Granger Morgan and Robert M. White, "A Design for New National Laboratories," *Issues in Science and Technology,* Winter 1993–94, pp. 29–32; "The National Labs and American Competitiveness," *Harvard Business School Bulletin,* Feb.–March 1994, pp. 37–38.

103. "The GOCO Concept in the New DOE: Laboratory Directors' Consensus View," DOE Laboratory Directors' meeting, March 25, 1992, Washington, D.C., Office of Field Operations Management, DOE.

104. Dennis Byrne, "Big Science Critics Accomplish Little," *Chicago Sun-Times,* Dec. 9, 1993; "Who You Gonna Call? Porkbusters!?" *Chicago Tribune,* March 18, 1993.

105. Lewis M. Branscomb, "National Laboratories in the Nineties," Director's Special Colloquium, April 1984, Argonne National Laboratory, Lab Director's meeting, Box 11, Collec. no. 434-91-0015, GLFRC.

106. Ronald E. Yates, "Refitting Cold War Science—New Role for National Labs; Make U.S. Competitive," *Chicago Tribune,* Oct. 26, 1992.

107. Alan Schriesheim, "Toward a Golden Age for Technology Transfer," *Issues in Science and Technology,* Winter 1990–1991, pp. 52–58; "United States National Labs: How Does Their Research Measure Up?" *The Scientist,* June 14, 1993, p. 14; Will Lepkowski, "National Laboratories Enter New Era of Hope Mixed with Uncertainty," *Chemical and Engineering News,* Aug. 14, 1995, 19–24.

108. Schriesheim, "The New Uncertainty Principle."

Index

Italicized page numbers refer to figures and maps.

AARR. *See* Argonne Advanced Research Reactor (AARR, A²R²)
Abbadessa, John, 301–2
Abbott Corporation, 465
Abboud, Robert, 480
Abelson, Philip H., 177, 216n
accelerators: colliding beam, 330, 462; construction of, 76, 160–61, 163–64, 168–69, 171; development of, 153–54, 166, 170–72, 209, 217–18, 326–27; evaluation of, 329–31; funds for, 155, 160, 166, 169; location and control of, 158–70; principle of, 152–53; public attitudes toward, 326; pulsed-neutron beams produced by, 412; Soviet operation of, 169, 171–72, 328; support for, 157–58, 461–62; upgrades for, 416–18; utilization of, 155–58, 328–29, 419–20. *See also* bubble chambers; cyclotrons; National Accelerator Laboratory; synchrotrons; *names of specific accelerator projects*
Access to Argonne campaign, 386
accidents: explosions, 118–21; fires, 266, 271, 446. *See also* meltdowns
Acheson-Lilienthal committee, 54
actinides, recycling of, 456–57. *See also* transuranic elements
Adams, Henry, 504
adrenocorticotropic hormone (ACTH), 102
Advanced Computing Research Facility, 438
Advanced Light Source, 472
Advanced Photon Source (APS) project: administration of, 470–73, 499; benefits of, 469–72, 494; construction of, 472–73; design of, 467–68, *468;* environmental impact of, 484; experiments using, 474–75; funds for, 472–73; planning for, 463–67; support for, 438, 461

Advisory Committee for Biology and Medicine, 75
Advisory Committee on Uranium, 3–5
AEC. *See* Atomic Energy Commission (AEC)
affirmative action: evaluation of, 309; leadership in, 271, 349–50, 378; and women's status, 491–93
AFSR (Argonne fast-source reactor), 194
AGS (Alternating Gradient Synchrotron), 169, 171, 196, 211, 295, 328
AIDS virus, research on, 474, 495
air, monitoring of, 97, 101. *See also* air pollution
aircraft: proposals for nuclear-powered, 109, 127, 128; used in nuclear tests, 178
air pollution: and battery-powered automobiles, 288–89; effects of, 484; identification of, 482–83; legislation on, 249, 333; sources of, 455; study of, 257; system model for, 249–50
Alabama, fire at plant in, 446
alkali battery, 250
Allerton estate, meeting at, 213–14
AlliedSignal (company), 498
Allison, Samuel K.: appointment for, 25, 29; and pile experiments, 9, 15; recruitment of, 5; role of, 8, 10, 12–13, 25
alloy 49. *See* plutonium
Alpert, Daniel, 256
alpha emitters, 101
alpha particles, 153
ALPR (Argonne Low-Power Reactor), 150, 205
Alternating Gradient Synchrotron (AGS), 169, 171, 196, 211, 295, 328
Alvarez, Luis W., 208
American Association of Physics Teachers, 338, 353
American Chemical Society, 297, 429
American Council on Education, 353
American Iron and Steel Institute, 439

American Nuclear Society, 378
American Physical Society, 133, 160, 417, 478
American Power Conferences, 230–31, 310, 311
americium, 177, 582n96
Ames Laboratory (Iowa): administration of, 221, 343; budget of, 395; collaboration with, 26; and superconductivity, 478
Amoco Oil Company, 383, 442, 490, 494
Amorosi, Alfred, 232–33, 271
AMU. *See* Associated Midwest Universities (AMU)
Anderson, Clinton, 148, 170, 204–5
Anderson, Herbert L., 16, 18–19, 173–74, 209
Anderson, N. Leigh, 443
Anker-Johnson, Betsy, 492
ANL. *See* Argonne National Laboratory
ANLAER (Assembly on Equal Rights), 349–50
anticancer drugs, 312
anti-intellectualism, 144
Anti-Jet-Lag Diet, 482
antinuclear groups: attitudes of, 452–53; and environmental contamination, 486–88; and IFR decision, 457
APS. *See* Advanced Photon Source (APS) project
ARBOR (Argonne boiling reactor), 150–51
ARCH (Argonne–University of Chicago Development Corporation), 481–83, 502
Arco (Idaho). *See* Idaho
Argonaut reactor (Argonne Nuclear Assembly for University Training), 147, 190, 193, 234, 516
Argonne Advanced Research Reactor (AARR, A^2R^2): benefits of, 235, 259; cancellation of, 257–59, 268, 274; controversy over, 235–36; development of, 235–41; funds for, 236–38
Argonne boiling reactor (ARBOR), 150–51
Argonne Cancer Research Hospital, 75, 90, 316
Argonne fast-source reactor (AFSR), 194
Argonne Forest, 13–14, 20
Argonne graduate center, 214–16
Argonne Laboratory senate, 245–46
Argonne Low-Power Reactor (ALPR), 150, 205
Argonne National Laboratory: accelerators at, 157–58, 162–72, 196–97, 201, 209; administration of, 40–42, 48–51, 53, 79–80, 90–91, 93–94, 105–6, 172–73, 183–86, 235, 239, 305, 343, 381–82; AMU partnership with, 200–201; analogy for, 442; animosity toward, 214; challenges to, 243–48, 282, 501–4; chronology of, 509–13; construction of, 15, 66, 84, 89; criticism of, 81–82, 96–97, 348, 371; electricity for, 149–50; employees at, 29–31, 297, 314, 331, 408, 505–7; environmental contamination of, 485–88; and ERDA priorities, 313–15; establishment of, 1, 3, 22–23, 41, 47; evaluations of, 308–13, 345–46, 368–70, 374, 380–82, 388–89, 394; future of, 28–29, 32, 38–42, 229, 340, 372–76, 426–28, 437–40, 502–4; goals of, 39, 45–46, 62–63, 87, 104, 151, 157, 163, 183, 223, 246–48, 274, 281, 341, 370, 397–400, 438; identity of, 90–93; leadership for, 47–50, 173–75, 183–84, 206, 208–10, 347–51, 507; legislation for, 42–44, 46, 48; maps of, *517–19;* name of, 22; and national agenda, 341–42, 427, 459–60, 500–501; policy advisory board on, 188–89; political support for, 389–91, 395–97, 439, 441–42, 448, 453–56; and postwar planning, 33–34; protests by, 133–34; radiological assistance teams from, 360–61; reorganization of, 102, 132–33, 268–70, 438; role of, 38–42, 44–46, 93, 203, 231–34, 243–45; sites for, 13–14, 40–41, 51–53, *52,* 55–57, *56,* 64–66, 89; status of, 31, 188, 294–300, 442–43; urban rehabilitation by, 488–90; wildlife at, 430, *431, 432–33. See also* Idaho National Engineering Laboratory; National Reactor Testing Station (NRTS); *names of specific programs and reactors*
Argonne Nuclear Assembly for University Training, 147, 190, 193, 234, 516
Argonne Tandem-Linear Accelerator System (ATLAS): design of, 416–18, *417,* 478; funds for, 418; operation of, 443; significance of, 398, 416; support for, 422; upgrade for, 438
Argonne Universities Association (AUA): access to Argonne for, 386, 391; agreement evaluated by, 379–81; and Argonne administration, 270–71, 274–77, 280–81, 299, 346–51, 386, 392–93; and basic research, 247; and energy issues, 345–46; and environmental studies, 250–57; establishment of, 221; expectations of, 242; fees for, 276; goals of, 265; headquarters of, 304; and high-energy physics, 260, 264; and human radiobiology, 317–18; key personnel at, 508; members of, 221–22, 307–8; policy manifesto by, 243–48; and reactor develop-

ment, 259, 267, 270; role of, 235, 246, 306, 353–54, 369, 372, 378–82, 391; and ZGS, 330–31. *See also* Tripartite Agreement
Argonne–University of Chicago Development Corporation (ARCH), 481–83, 502
Argonne-West. *See* Idaho National Engineering Laboratory; National Reactor Testing Station (NRTS)
arms race. *See* nuclear proliferation
Arrhenius, Svante August, 483
arts, at ANL, 406–7, 436–37
Asbury, Joseph, 438
Assembly on Equal Rights (ANLAER), 349–50
Associated Midwest Universities (AMU): and Argonne graduate center, 215–16; Argonne partnership with, 200–201; board members of, 213–14; demise of, 221–22; establishment of, 189; role of, 199–200, 209, 220, 235; and Tripartite Agreement, 220–21; on university-laboratory relations, 213–14
Associated Universities, Inc., 160, 219, 220–21, 391
Astrom, Bjorn, 180n
ATLAS. *See* Argonne Tandem-Linear Accelerator System (ATLAS)
atom: computer model for, 498; nuclear structure of, 227–29, 419–20; radiation's effect on, 410; research on, 152–54; shell structure of, 227–29; x rays for study of, 469
Atomic Age, symbols of, 122
atomic bomb: feasibility of, 17; governance of, 36–37; possibilities of, 5–6; and public attitudes, 110; public information on, 36–38; race for, 1, 16, 27, 145–46; Soviet detonation of, 81, 88–89; success of, 35–36. *See also* nuclear weapons; thermonuclear weapons
Atomic Bomb Casualty Commission, 315
atomic energy: control of, 42–44; military applications of, 6; support for, 334. *See also* Atomic Energy Commission (AEC); atomic piles; nuclear energy
Atomic Energy Act (1946): amendments to, 126–29; ANL established under, 42–44, 46, 48; components of, 54, 66, 79; impact of, 49; loophole in, 97; restrictions in, 110, 135
Atomic Energy Act (1954), 129, 135, 250, 255
Atomic Energy Commission (AEC): abolition of, 278–81; and accelerators, 157–58, 162–67, 169, 172, 196–97, 327–28; administration of, 79–80, 91–92, 175, 192–94, 221, 233–34, 247, 306; and Argonne administration, 270–77, 293, 308–9; and bubble chamber, 199; budget of, 127–28, 155–56, 191–93, 279, 290, 294–95, 329–30; and cancer research, 75, 90; changes in, 204–6; and commercialization, 194–95; and computer development, 122–25; and contractor relations, 92–93, 193; criticism of, 148; and EBR-I meltdown, 141–44; and EBR-II construction, 149; and energy plan, 301; and environmental research, 100–101, 248, 256–57; establishment of, 48–49; evaluation by, 308–9; five-year program of, 128–29; focus of, 104–5; funds from, 203–4, 249; and fusion research, 286–87; goals of, 59, 60–63, 87–88, 108–10, 127, 168, 195, 262; and H-bomb development, 94–95, 286; and high-energy physics, 260, 329–31; human radiobiology studies sponsored by, 316–18; and institutional cooperation, 90–91; and international cooperation, 72–73; key personnel at, 507; and laboratory senate, 246; members of, 53–54, 81, 94–96, 104, 162, 167, 205, 208, 270, 290; and nuclear bombs, 88, 128, 178–79; and nuclear waste, 73–74; and Oppenheimer's security clearance, 133–34; organization of, 49, 58–59; and plutonium patients, 321–23; power of, 49–50; public attitudes toward, 290–91; and radiation protection, 74; and reactor projects, 62–64, 66–68, 71–73, 85–87, 150–51, 230–34, 258–59; reorganization of, 303; review of, 50; role of, 209, 244; and security issues, 81–84, 133–34; and site selection, 55–56, 65–66; and training issues, 136–37; and university-laboratory relations, 214, 219–20; and waste disposal, 73–74. *See also* General Advisory Committee (GAC); Joint Committee on Atomic Energy (JCAE); U.S. Department of Energy (DOE)
"atomic farm," demonstration of, 190–91
Atomic Industrial Forum, 193
atomic piles: approach to, 24; construction of, 7, 28; development of, 15, 28, 35, 40; instrumentation for, 23; intermediate, 10; locations of, 9, 13–14; poisoning of, 31; research on, 51; types of, 15–16. *See also* chain reaction; nuclear reactors
Atomic Scientists of Chicago, 36, 43
atomic shell structure, 77, 227–29
Atoms for Peace program: context of, 126–28; expositions for, 125; goals of, 413, 459; implementation of, 129; international compo-

nent of, 135–38, 192; and security issues, 134; support for, 144, 146, 173. *See also* International Conference on the Peaceful Uses of Atomic Energy
AT&T Bell laboratories, 499
Atterling, Hugo, 180n
Audubon Society, and Integral Fast Reactor (IFR), 448
Aurandt, Paul (Paul Harvey), 96–97
Austin Company, 66, 87, 89
automobiles: industrial-laboratory cooperation on, 495–97; testing breaks for, 498. *See also* batteries
Avery, Robert, 268, 362–64, 444, 450–51
AVIDAC (Argonne's Version of the Institute's Digital Automatic Computer), 123
avionics, ergonomics in, 367–68

Bacher, Robert F., 54, 73, 156
Baker, Howard, 390
Baldridge, Malcolm, 395
barium, 1–2, 479
Barnes, Arthur H., 132–33
Bartlesville (Okla.), energy technology labs in, 336
basic research: vs. applied research, 61–62, 109, 244–45, 499–500; budget for, 223; choices in, 247–48; criticism of, 348, 369–70; vs. demonstration reactor, 265, 299; funds for, 282, 314, 328, 345–46, 373, 433–34; and IPNS, 414–15; vs. national agenda, 341–42; rationale for, 303, 327–28, 342; reinvigoration of, 281; support for, 223, 226, 235, 275, 309. *See also* materials science
Bates Linear Accelerator, 420. *See also* Massachusetts Institute of Technology (MIT): and accelerator projects
batteries: lithium-sulfur, 288–89, 310, 311, 337, 495–97; research on, 250, 480
Batzel, Roger, 402
Bauer, Robert, 305, 307, 394, 404
Bauser, Edward J., 259
Baxter Healthcare Corporation, 495
Bayh-Dole Act (1984), 481, 482
Beadle, Alan, 180n
Beadle, George: and Argonne directors, 208–10, 240; and Argonne graduate center, 214–16; contract negotiations by, 214; and high-flux reactor, 236; and MURA accelerator, 217; and university-laboratory relations, 218–20; and ZGS, 204
Beckjord, Eric S., 371, 373

Behnke, Wallace B., Jr., 438
Benedict, Manson, 233
berkelium, 177
Berkner, Lloyd V., 160–61
Berlin crisis, 208
Berry, R. Stephen, 401
beryllium: exposure to, 74; self-sustaining fission in, 5; as shield, 114; toxicity of, 102
betatron, 155
Bethe, Hans, 206, 291, 433
Bethel Lutheran Church, 488
Bethel New Life, Inc., 488–90
Bettis Atomic Power Laboratory, radiological assistance teams from, 360–61
Bettis Field. *See* Westinghouse Company
Bevatron: benefits of, 163; construction of, 169, 171; development of, 154–55; role of, 211; support for, 328
Bevill, Tom, 472
Beyea, Jan, 448
Bhattacharyya, Maryka, 491
Big Rock Nuclear Power Plant (Charlevoix, Mich.), 254
Bikini Atoll, 50, 178–79
Billings Hospital, 75, 320, 529n89
biochemistry, and advanced photon source, 469
biology research: demonstrations in, 190; focus of, 74, 101–2; initiatives in, 318–20; patents in, 482; reactor for, 195; review of, 409. *See also* biomedical research; Structural Biology Center
biomedical research: and advanced photon source, 469; on AIDS virus, 474, 495; on cancer, 75, 102, 319, 323; on disease, 473–75, 495; on drugs, 191, 312, 473–74; focus of, 46, 74, 101–2; on radiation, 101–2, 190–91; in radiobiology, 315–23; review of, 409
biotechnology, 473–75
bismuth phosphate method, 26–27
Blaw-Knox Construction Company, 87
Bloch, Erich, 476, 501
Bloch, Maj. E. J., 39
BNL. *See* Brookhaven National Laboratory (BNL)
board of governors (ANL): administration of, 404; agenda of, 426–28, 437–40; character of, 424; establishment of, 391–93; IFR evaluated by, 450–51; organization by, 408–9
Bobbitt, John T., 84, 206–7
Boeing Corporation, 494
Bohm, Henry V.: on Argonne directors, 351; and

AUA withdrawal, 379–82; role of, 345–46; on university-laboratory relations, 386, 389
Bohr, Niels, 2, 149
Boiling Reactor Experiment-I. *See* boiling-water reactors; BORAX (reactors)
boiling-water reactors: benefits of, 130; commercialization of, 150–51, 194–95; development of, 95, 149–51; explosion of, 118–21; fuel for, 129–30; model of, 139; shortcomings of, 131. *See also* BORAX (reactors); Experimental Boiling-Water Reactor (EBWR)
Bollinger, Lowell, 416–18, 443
bone marrow transplants, 102
BORAX (reactors): construction of, 139; design of, 118–19, *119*, 515–16; research with, 149; safety of, 130. *See also* Experimental Boiling-Water Reactor (EBWR)
Borden, William L., 133
Borg-Warner Corporation, 383
Born, Max, 228
Borst, Lyle B., 523n32
Boyce, Joseph C.: and accelerators, 160–61, 164; appointment for, 189; role of, 157–59; and security issues, 91–92; and university-laboratory relations, 187
Boyer, Marion W., 94, 104
braille-reading machine, 250
breach tests, 367
breeder reactors: administration of, 265–68; AEC project for, 230–34; commercialization of, 273, 357; construction of, 99–100; development of, 60–62, 371; environmental impact of, 297; funds for, 314, 346, 373, 440; future of, 298–99, 339–40, 426; and nuclear proliferation, 338–40, 444–45; problems with programs for, 290–91, 398; reorganized programs for, 268–70, 354–58; safety research on, 363–65; uncertainties about, 280. *See also* fast-breeder reactors; *names of specific reactor projects*
Breit, Gregory, 8
Bricker, John, 65–66
Briggs, Lyman, 3, 5
Brinkman, William F., 413–15
Brobeck, William, 154
Brokaw, Tom, 337–40
Broken Arrow (group), 486–88
Bromley, D. Allan: and accelerator project, 420–22; and Argonne directorship, 276; evaluation by, 384–86
Brookhaven National Laboratory (BNL): academics at, 386; accelerators at, 76, 155, 160–61, 163, 166, 169, 171, 196–97, 211, 217, 295, 327–28, 330, 462; administration of, 167, 219, 220–21, 276, 343, 392; bubble chamber designed by, 263; budget of, 160, 289, 294–95, 441, 506; competition from, 466; computer at, 124; employees at, 297, 506; environmental contamination at, 486; goals of, 104, 214; heavy-ion collider at, 467, 472; omega minus particle found by, 224–25; political support for, 390; radiobiology at, 316; radiological assistance teams from, 360–61; reactors at, 110–11, 236, 270, 411–12, 414; and risk reduction after Chernobyl, 452; site for, 49; status of, 156–57; and superconductivity, 478; synchrotron at, 468–69, 473; training at, 200–201; university relations with, 72
Brooks, Charles W., 56–57
Brown, George E., 454–56
Brown, Howard C., Jr., 92–93
Brown's Ferry (Ala.), fire at plant in, 446
Brown University: Argonne director from, 352–53; disease research at, 474
Brues, Austin M., 57, 102
bubble chambers: benefits of, 224–25, 263, 329; description of, 198, 263; politics of, 199–200; role of, 211
Buckley, John L., 253
budget (ANL): and advanced photon source, 472–73; by board of governors, 408; Carter's cuts in, 373, 375, 440; Clinton's cuts in, 453–56, 499; for construction, 204; and CRBR cancellation, 425–26; division of, 289–91, 293–95, 401; and energy initiatives, 314, 331–32; evaluation of, 384–85; for general maintenance, 192; limitations on, 64–65; and long-range planning, 398–99; Nixon's cuts in, 272; politics of, 81, 84–85, 87, 279–80, 330; Reagan's cuts in, 388, 395–96, 438–42, 448, 465; and research priorities, 50, 223, 381; yearly amounts of, 505–6
Bueche, Arthur M., 384
building 201, construction of, 404–8, *405*
Bulletin of the Atomic Scientists, establishment of, 36
Bureau of the Budget, investigation by, 92
Bush, George, 453, 495
Bush, Vannevar: administration of, 9, 33; on postwar research, 43; role of, 4–6, 8, 12, 25; on university-laboratory relations, 33; uranium committee organized by, 4

616 ■ Index

Butler, Margaret, *124*
Byrne, Dennis, 420
Byrne, Jane, 382–83, 389, 435

Caldecott, Richard, 255–56
California, electrical short-circuiting in plant in, 446
californium, 177–79
Cambridge Electron Accelerator. *See* Massachusetts Institute of Technology (MIT): and accelerator projects
Camp Upton (Long Island, N.Y.). *See* Brookhaven National Laboratory (BNL)
Canada: cooperation with, 73; nuclear accident in, 143; pile projects in, 27–28; security breach in, 62
Canadian–United States Eastern Interconnection (CANUSE), 283–84
cancer research, 75, 102, 319, 323
Cannon, William B.: animosity toward, 242; and Argonne administration, 270–77, 289, 293, 306–7, 347, 350–51, 373; and environmental research, 255–56; and Tripartite Agreement, 305, 380–81
Cantlon, John, 254
Cantril, Simeon, 12
CANUSE (Canadian–United States Eastern Interconnection), 283–84
Carnegie Foundation, 4
Carnegie Institute, 2, 42
Carpenter, Jack, 412, 443
Carruthers, Peter A., 351
Carson, Rachel: *Silent Spring*, 460
Carter, James Earl (Jimmy): appointees by, 353; background of, 333–34; and breeder reactors, 354–58, 453; budget of, 373, 375, 440; election of, 334; energy policies of, 280–81, 334–41, 343–45, 373–74; nuclear policy of, 334, 338–40; and Three Mile Island, 362
Castle-Bravo thermonuclear test, 128, 179
Caterpillar Corporation, 494
CATS (collaborative access teams), 473
CBA (colliding beam accelerator or ISABELLE), 330, 462
Center for Educational Affairs, 375
Center for Human Radiobiology: administration of, 319; establishment of, 315–18; and plutonium patients, 320–23
Central States Universities, Inc. (CSU), 222
ceramics: for fuels, 445; superconducting, 476–80

Cerenkov radiation, collection of, 313
CERN (Switzerland): acelerators at, 166, 169, 196, 211; W and Z particles discovered at, 462
cesium, 486
Chadwick, James, 15, 209
Chain, Bobby, 377
chain reaction: celebration of first, *19,* 19–20; centralized studies of, 7; first sustained, 1, *18,* 18–20, 175, 514; possibility of, 2–3, 10; for producing plutonium, 16
Chalk River plant (Canada), 25, 143
Chamberlain, Owen, 216n
Chang, Yoon, 439, 444, 457
Change (magazine), recognition from, 353
Chapman, Thomas S., 82, 84
chemistry: and advanced photon source, 469–70; advances in, 75–77; demonstrations in, 190; "hot atom," 77; initiatives in, 440; of noble gases, 225–27, *226;* research on, 102–3; and science agendas, 460
Chernick, Cedric, 225
Chernobyl accident: analysis of, 451–52; impact of, 446–48, 460
Chicago: and Argonne's political support, 390–91; economy of, 382–83, 488–90; institutions in, 490; science education in, 377–78; and site location, 51–53
Chicago Air Pollution Systems Analysis Program, 249–50
Chicago Department of Air Pollution Control, 249
Chicago Pile 1 (CP-1): accounts of, 522n24; construction of, 16–17; dismantling of, 22–23; first chain reaction at, 1, *18,* 18–20, 175, 514; operation of, 16, 18–19, 51, 175
Chicago Pile 2 (CP-2): description of, 111, *111;* radiation shield for, 23; role of, 76; shutdown of, 113; size of, 23, *23*
Chicago Pile 3 (CP-3): construction of, 27–28; operation of, 51; role of, 76, 103; success of, 31
Chicago Pile 3' (CP-3'), 111, *112,* 113
Chicago Pile 4. *See* Experimental Breeder Reactor Number 1 (EBR-I)
Chicago Pile 5 (CP-5): benefits of, 137, 201; decontamination of, 485; description of, *116,* 116–17, 515; experiments using, 201, 411–12; location of, 111; model of, 139; retirement of, 374–75, 411; role of, 236, 258, 296; safety of, 117–18
Chicago Science Explorers, 490

Chicago Symphony Orchestra, 407
Chicago *Tribune,* on reactor research, 458
China Syndrome, 360
chlorophyll, research on, 297
Cho, Yanglai, 465
cholera toxin, research on, 474
chopper, development of, 76
Chrysler Corporation, 495
Chu, J. C., 122–23
Cisco (sailboat), 254
Citizens for a Better Environment, 338
civil defense: mock exercises for, 207–8; and power outages, 283; and Three Mile Island, 360. *See also* safety
Classen, Howard, 225
Clean Air Act (1967), 249, 333
climate, changes studied in, 483–84
Clinch River Breeder Reactor (CRBR; Tenn.): authorization of, 333, 340; cancellation of, 425–26, 441, 444, 453–54; debate over, 354–58, 424; future of, 343, 354; legacy of, 454; safety testing for, 364
Clinton, William Jefferson (Bill): administration of, 494, 496; appointees of, 453; budget of, 453–56, 499; and fuel reprocessing, 456–58; nuclear power opposed by, 453–56, 460
Clinton Laboratories (Tenn.): administration of, 33, 63–64; personnel of, 28, 46, 523n32; pile projects at, 21, 26, 40, 63; research at, 24, 40; role of, 21–22; university relations with, 25. *See also* Oak Ridge National Laboratory
cloud chambers, use of, 198
coal: increased production of, 334; scrubbers for, 396; supplies of, 287–88
Coal and Shale Processing and Combustion panel, 301
coal technology: research on, 302, 396, 469; and safety, 433; technology transfer of, 311–12
Coggeshall, Lowell T., 75
Cohen, Jerry B., 464–66, 482
Cold War: and accelerator projects, 168–69, 328; Argonne's role in, 192; and computer development, 122; and conferences, 138; effects of, 109–10, 135, 383–84, 455, 500; end of, 459; environmental impact of, 484; and high-flux reactor, 237; and training program, 136–38. *See also* communism; Soviet Union
Cole, Frank, 465
Cole, Kenneth D., 12

collaborative access teams (CATS), 473
colliding beam accelerator (CBA or ISABELLE), 330, 462
Collins, Cardiss, 378
Colorado State University, radiobiology at, 316
Columbia University: nuclear fission confirmed at, 2; and personnel issues, 33; pile experiments at, 3–4, 9
Comey, David D., 338–39
Commission on Industrial Competitiveness, 439
Committee on Science and Technology (U.S. House), 383–84, 390, 397
Commonwealth Edison, 150–51, 480
communism: crusade against, 81–83, 126, 134; and nuclear challenge, 137–38; and security issues, 92, 96–97
community development corporation, cooperation with, 489–90
compound parabolic concentrator (CPC), 312–14, 337
Compton, Arthur Holly, 4; administration of, 8–9, 11–12, 20–21; appointment for, 34; and Argonne's future, 29–31; and Argonne's origins, 3, 22, 41; background of, 3–4; on bomb's use, 36; and chain reaction, 7, 18–20; characteristics of, 24–25, 28–30, 32–33; family of, 7; on Hanford design, 26; and health concerns, 10–12, 31; and Manhattan Project, 27; Met Lab formed by, 7–9; and Met Lab goals, 27–28; on nuclear science's future, 24; and personnel issues, 32–33; and pile projects, 14, 27–28; postwar planning by, 33–34, 38–40; reorganization by, 12–13, 21; role of, 3–6, 8, 10, 17, 32–33, 35, 176; on scientists' role, 37; and site selection, 13–14, 21, 23–24, 52
Compton, Karl, 7
Compton, W. Dale, 437
computers: development of, 76–77, 122–25, 201; and human radiobiology center, 317; simulations by, 190, 363–64; supercomputing, 408, 497–99; virus analysis with, 474
Conant, James B.: appointment of, 59; and chain reaction, 17, 20; recommendations by, 12; role of, 5–6, 8, 25
concentration processes, 27
Congressional Research Service, 341
constant-gradient synchrotron, 171
Continuous Electron Beam Accelerator Facility (Va.), 472

contracts: and AEC, 92–93, 193; for ANL, 219–20, 394–95, 408; cooperative, 494–99; with EPRI, 344; with Exxon Nuclear, 344; with government departments, 344; investigation of, 92–93; issues for, 193; with Japan, 344; for minorities, 378; with NASA, 344; for national laboratories, 506; negotiations for, 214; with NRC, 344; with OSRD, 521n19. *See also* university-laboratory relations
Control Data Corporation, 465
controlled thermonuclear research (CTR). *See* thermonuclear energy (fusion energy)
Cook County Forest Preserve Commission, 14, 51–52, 55, 113, 487
coolants: accidents involving, 360, 365–68; experiments with, 109, 149, 364; light-water, 114; measurement of, 312; role of, 69; and safety measures, 118; shutting off, 142; sodium as, 131, 151, 364–68, 447–48, 516; for submarine propulsion system, 70, 88–89; towers for, 95; water vs. graphite and heavy water as, 67
Coolidge, William David, 468
Cooper, Charles M., 13, 26
Cooperative Research and Development Agreements (CRADAs), 494–99
Coppersmith, Sam, 454
Cork, Bruce, 262–64
Corps of Engineers. *See* U.S. Army Corps of Engineers
cosmic rays, 175–76
cosmotrons, 155, 157, 159–62, 327
Council of Participating Institutions, 186–89, 220. *See also* Associated Midwest Universities (AMU)
Council on Economic Competitiveness, 501–2
Courant, Richard, 105
CP-1. *See* Chicago Pile 1 (CP-1)
CP-2. *See* Chicago Pile 2 (CP-2)
CP-3. *See* Chicago Pile 3 (CP-3)
CP-3'. *See* Chicago Pile 3' (CP-3')
CP-5. *See* Chicago Pile 5 (CP-5)
CP-6. *See* plutonium production reactors (CP-6); Savannah River Laboratory: reactors at
CPC (compound parabolic concentrator), 312–14, 337
CRADAs (Cooperative Research and Development Agreements), 494–99
Crawford, John W. (Jack), 371
CRBR. *See* Clinch River Breeder Reactor (CRBR; Tenn.)

Crewe, Albert V., *210, 253;* and accelerator projects, 217–18; administration of, 223, 232, 235, 306, 314; appointments for, 189, 208, 212; and Argonne graduate center, 214–16; background of, 208–10; and basic research, 223, 226, 235; and educational programs, 222; and reactor projects, 230–31, 236–39; retirement of, 240; and synchrocyclotron, 166, 196; and university-laboratory relations, 213–14, 218–20, 222–23; and ZGS, 210–12
Crkvenjakov, Radomir, 475
Croke, E. J., 248, 256
crystallography, 474
crystal spectrometer, 76
CSU (Central States Universities, Inc.), 222
CTR (controlled thermonuclear research). *See* thermonuclear energy (fusion energy)
Cunningham, Burris B., 37
Curie, Marie, 176
Curie, Pierre, 176
curium, 177–78, 180–81
cyclotrons: availability of, 103; development of, 154; research on, 46, 156, 201; transuranic elements discovered with, 177–81

Daniels, Farrington, *41;* appointment for, 35; and Argonne's establishment, 38–42; on Argonne's role, 44–46; on Manhattan Project records, 44; and postwar planning, 38–42; and reactor research, 40, 61; scientists polled by, 532–33n109; and site selection, 53
Davis, Stanley A., 271
Davis, W. Kenneth, 394, 456
DBER (Division of Biological and Environmental Research), 321
Dean, Gordon E., 94, 96, 108–9, 115–16
deer, at ANL, 430, *431,* 433
Deere and Company, 383
Dehmer, Patricia, 443
Delaware, blackout in, 283–84
Denman, Scott, 454–55
Des Plaines River, 485
detectors, construction of, 438, 461–62
Detroit (Mich.). *See* Enrico Fermi Fast-Breeder Reactor (Detroit)
Deutch, John M.: administration of, 330–31, 379; appointment for, 344; associates of, 424; and director appointment, 350–54, 369; evaluations by, 345–46, 368–70, 382; Massey's confrontation with, 371–72; support for, 373

deuterium, 286
dial painters, radium exposure of, 102, 315–18
Dietrich, Joseph R., 139–40
Dirksen, Everett, 236
discrimination, 78
disease, research on, 473–75, 495. *See also* AIDS virus, research on; cancer research
Disney, Walt, 122
Ditmars, Jack, 489
Division of Biological and Environmental Research (DBER), 321
Division of Reactor Development and Technology (RDT), 232, 240, 265–66, 300. *See also* Shaw, Milton
Division of Reactor Licensing, 238–39
Division of Reactor Safety Research, 300
DNA, research on, 475
Doan, Richard L., 8, 10, 25
Donets, Evgeny C., 181
dosimeters, 146
Doub, William O., 300, 322
Dow Chemical Company, 383, 470
Downers Grove (Ill.), and Argonne site, 56
Drmanac, Radoje, 475
Drucker, Harvey: and AIDS research, 474; and cooperative research, 494–95; program of, 498–99; and urban rehabilitation, 489; on USCAR, 497
drugs: anticancer, 312; APS experiments on, 473–74; radioactive tracers for, 191
Dubos, Réne Jules, 252–53
DuBridge, Lee, 7
Duffield, Priscilla, 261
Duffield, Robert B., *243, 253;* and accelerator project, 260–63; administration of, 242, 252, 270–74, 279, 293, 306; appointment for, 240; and AUA's challenge, 246–48; background of, 240–41; criticism of, 267; departure of, 274, 275, 281; and environmental studies, 250–52, 254–56; and Hot-Fuel Examination Facility, 269–70; and laboratory senate, 245–46; and reactor projects, 257–59, 265–70; reorganization by, 268–70
Dukakis, Michael, 453
Dulles, John Foster, 190
Dunbar, Kenneth, 239
Du Page County (Ill.): acquisition of site in, 55–57, 64–66; Civil Defense Control Center in, 207; construction in, 79; facilities in, *67;* and nuclear waste disposal, 57–58, 73–74; and radiation exposure, 67, 74; reactor in, 111, 116–18, 145–47; surveys of, 52–53, 101. *See also* Palos Park site
DuPont Corporation: and advanced photon source, 470; collaboration with, 24, 26, 442; and facility locations, 21; and H-bomb development, 95, 97–98; role of, 15, 17, 20–22
Duquesne Light Company (Pittsburgh), 128
Durham, Carl T., 145
Dvorak, Anthony, 484–85

Eastern Europe, U.S. assistance for, 451–52
Eastman, Dean, 464
EBR-I. *See* Experimental Breeder Reactor Number 1 (EBR-I)
EBR-II. *See* Experimental Breeder Reactor Number 2 (EBR-II)
EBWR. *See* Experimental Boiling-Water Reactor (EBWR)
educational programs: and academics, 386; administration of, 235; benefits of, 103–5, 137, 200–201; for community development, 490; evaluation of, 187; expansion of, 271, 377–78, 428; fellowships in, 91–92; graduate center for, 214–16; leadership in, 353, 375–76; limitations on, 201; and science explorers, 490, *491;* university support for, 222. *See also* International School of Nuclear Science and Engineering
EDVAC (computer), 122
Edwards, James B., 387–88, 395–97, 456
Ehret, Charles F., 482
E. I. du Pont de Nemours and Company (DuPont). *See* DuPont Corporation
Einstein, Albert, 3
einsteinium, 179–80
Eisenberger, Peter, 465–67
Eisenhower, Dwight D.: AEC briefing for, 108–10; budget of, 160; election of, 106; and insurance issues, 148; and international cooperation, 135–36, 138; Oppenheimer's security clearance suspended by, 133–34. *See also* Atoms for Peace program
electrical system, short-circuiting in, 446
electric power: alternative sources of, 288–89, 300–303, 339–40; for cars, 495–96; costs of, 128, 328, 489; and economic concerns, 454–55; possibilities for, 107, 126; reactors for, 60; shortages of, 282–84, 334–35; successful nuclear generation of, 60, 99–100, *108,* 108–9, 115, 139, 336–37, 515; transmission and production of, 476–80

Electric Power Research Institute (EPRI), 344, 452, 495
electrochemical pyroprocessing, 445–46, 449–50
electron microscopes, 398, 409–11, *410*
element 93 (neptunium), 177
element 94. *See* plutonium
element 95 (americium), 177, 582n96
element 96 (curium), 177–78, 180–81
element 97 (berkelium), 177
element 98 (californium), 177–79
element 99 (einsteinium), 179–80
element 100 (fermium), 179–81
element 101 (mendelevium), 179
element 102 (nobelium), 176–82
elementary particle research program, 263
Elgin State Hospital, 102
Eli Lilly and Company, 465, 471
Elston, Charles H., 65–66
Emergency Control Center (ANL), 207–8
energy: changing policies on, 373–74; commercialization of technology for, 311–12; consumption of, 301, 459; dependence on, 283–84; government controls on, 335–36, 343; public attitudes toward, 285; shortages in, 273, 279, 284; socioeconomic issues in, 345; sources of, 301, 460. *See also* electric power; nuclear power plants; solar energy
Energy and Water Appropriations Bill (1989), 472
energy conservation: Carter's plan for, 280–81, 334–41, 343–45, 373–74; Ford's plan for, 280, 333–36; funds for, 314, 319; impact of, 434; renewed emphasis on, 460
energy crisis: and accelerator project, 328–29; Carter's plan for, 280–81, 334–41, 343–45, 373–74; impact of, 282; legacy of, 433–34; Nixon's plan for, 273, 278–80, 285, 300–304, 336; oil embargo in, 302, 460; power failures in, 282–84; and research funding, 384; response to, 303–4, 334–35; responsibility for, 309, 403
Energy Independence Authority, 333
Energy Policy Office, 301
Energy Reorganization Act (1974), 310
Energy Research Advisory Board (ERAB): and accelerator project, 421; and advanced photon source, 465–66; and board of governors, 426–27; evaluations by, 389–90, 397, 400–401, 404; on government-laboratory relations, 500; members of, 424; on multiprogram labs, 397, 408; role of, 399

energy research and development: Argonne's initiatives in, 286–89, 303–4, 309–10, 312, 336, 342, 346, 489–90; and budget cuts, 290; Carter's plan for, 335–36, 340; commercialization of, 345, 383, 387–88; Ford's plan for, 333–34; funds for, 302–3, 384–86; impetus for, 341, 399–400, 403; leadership in, 455; materials science in, 411; priorities in, 285, 338–39; Reagan's plan for, 433–35
Energy Research and Development Administration (ERDA): administration of, 310–12, 331, 342; Argonne's role in, 303; and biological initiatives, 319–20; cooperation with, 495–96; establishment of, 280–81, 310; goals of, 301, 344; key personnel of, 507; logo of, 312; solar initiatives of, 313–14, 336
English, Spofford G., 37, 246
ENIAC (computer), 122
Enrico Fermi Fast-Breeder Reactor (Detroit): construction of, 141–42, 149; partial meltdown of, 234, 446; status of, 230
Enrico Fermi Institute of Nuclear Studies (Chicago): and accelerator project, 196–97; staff of, 209, 240, 289, 291, 293
environment: and Manhattan Project's hazards, 45; monitoring of, 97, 100–101; and safety measures, 118; and science's role, 282
environmental studies: authorization of, 250–52; conference on, 252–54; establishment of, 242, 247–49, 280; funds for, 250, 252, 255–57, 302, 344; initiatives in, 254–57, 319–20, 483–85; on lab grounds, 430, 432–33; local impact of, 378; on pollutants, 249–50, 319–20, 498; significance of, 251–53; on sites for nuclear power plants, 257
E. O. Lawrence Memorial Award, 471
ergonomics, and safety issues, 367–68
Erlenborn, John N., 290
Erlewine, John, 274, 276, 307
Eshbach, O. W., 38–39
Esso Research and Engineering Company, 423–24
Etherington, Harold, 70, 88, 98
ethics, and atomic bomb aftermath, 36
ETR, testing materials from, 269
EURATOM, establishment of, 126
Europe: high-energy physics in, 462; reactor research in, 237
Evans, Robley D., 315–17
Experimental Boiling-Water Reactor (EBWR): construction of, 139–40, 516; dedication of, *145*, 145–47; model of, 190; role of, 148,

Index ■ 621

260, 516; success of, 149–50. *See also* BORAX (reactors)
Experimental Breeder Reactor Number 1 (EBR-I): development of, 113–14, 227; electricity from, *108,* 108–9, 115, 336–37, 515; experiments with, 130; fuel meltdown of, 141–44; as landmark, 229, 337; new core for, 149; role of, 114–16; safety of, 131; status of, 111; view of, *115*
Experimental Breeder Reactor Number 2 (EBR-II): accidents of, 266; administration of, 268–69, 272, 298; construction of, 140, 194; design of, 140, 148–49, 365–68, *366,* 516; development of, 131; evaluation of, 356–57; experiments on, 356, 447–48; fuel for, 131, 439, 449–50; future of, 426; model of, 139; operation of, 234, 265–69, 296; reorganization of programs for, 268–70; role of, 233, 265–66, 374, 426; and safety, 357–58, 364–68, 444, 446–48, 451; shutdown of, 457–58; status of, 151, 230, 356–57; success of, 355, 365, 516; testing materials in, 269, 333, 340, 355–56, 367–68. *See also* Integral Fast Reactor (IFR)
Exxon Nuclear, 344
Exxon Research and Engineering Company, 423–24, 442, 466, 470–71

facilities: architecture of, 405; and Argonne goals, 382; and arts, 406–7, 436–37; building 201 (administration), 404–7, *405;* condition of, 374–75, 424, 436; criteria for, 52, 65, 238; decontamination of abandoned, 57–58, 485–88; grounds of, 430, 432–33, 436; main administration buildings, *186;* permanent location for, 51–53; for reactors, 296; review of, 427–28; role of, 185; scattered nature of, 89–90; temporary, 48, 66, *67*
fallout: home meter for measuring, 208; from nuclear testing, 97, 178–79; research on, 254; from Three Mile Island, 361
fallout shelters, and mock exercises, 207
FARET (Fast Reactor Test Facility), 230–31, 233
fast-breeder reactors: approval of, 62; design of, 68–69; development of, 131; need for, 61; safety of, 131, 364; site for, 86–87. *See also names of specific reactor projects*
fast-fission pile, 40, 46
Fast Flux Test Facility (FFTF): administration of, 231–32, 272; budget of, 291; construction of, 231, 367; mock-ups for, 269; program for, 231–33, 289, 356–57; safety testing for, 364
Fast Reactor Test Facility (FARET), 230–31, 233
fatalities, from nuclear reactor explosion, 205
Fawell, Harris W., 442, 455, 472, 501
FBI. *See* Federal Bureau of Investigation
FCF (Fuel Cycle Facility), *445,* 445–46, 449, 516
Feder, Harold M., 248–49
Federal Aviation Commission, 257
Federal Bureau of Investigation (FBI), 82–83, 324
Federal Energy Agency, 303
Federal Power Commission, 283–84
Federal Technology Transfer Act (1986), 494
Federation of American Scientists, 36, 92, 133
Federation of Atomic Scientists, 36
Fee, Darrell C., 443
Fermi, Enrico, *30;* on accelerators, 156–57; appointments for, 8, 22, 29, 33, 35, 59; English of, 17; first controlled chain reaction by, 1, 18–20, 175; focus of, 15–16; and nuclear fission, 2; and pile projects, 9, 10, 15, 23, 28, 31, 524n45; and postwar planning, 32, 34; role of, 3, 13, 35, 173–74, 227–29, 491, 514; security clearance for, 9; and transuranic elements, 176–77
Fermilab. *See* National Accelerator Laboratory (later called Fermilab)
fermium, 179–81
Fernau, Mark, 484
Fett, Gilbert H., 233
Feynman, Richard, 327
FFAG (fixed-field alternating-gradient) accelerator, 165–66
FFTF. *See* Fast Flux Test Facility (FFTF)
Fields, Paul R.: on accelerators, 418; associates of, 205, 424; and security issues, 83; on shell model, 566n42; on transplutonium elements, 177–78, 180–82
Filbey, E. T., 8
Finkel, Miriam, 79, 223
Finucane, Dan, 233
fires: at Alabama plant, 446; in EBR-II, 266; at plutonium fabrication plant, 271
fish, research on, 254
Fisk, James B.: and accelerators, 155, 168; appointment for, 58–59; influence by, 85, 156; and reactor projects, 61, 68
fission research: location of, 23; products of,

27, 449; techniques in, 286, 411–12. *See also* atomic energy; nuclear energy
fixed-field alternating-gradient (FFAG) accelerator, 165–66
Flaherty, J. J., 135
Flanders, D. A., 122
Flerov, Georgii N., 181–82
Floberg, John F., 193
fluidized-bed combustor, 288, 337
flywheels, and superconductivity, 480
Ford, Bacon & Davis, Inc., 57
Ford, Gerald R.: energy policies of, 280, 333–36; on plutonium reprocessing, 338
Ford Foundation, 162, 340
Ford Motor Company, 494, 495
Forsling, Wilhelm, 180n
Foss, Martyn (Magnet Man), 172
fossil fuels research, 314, 319, 448, 455. *See also* coal; natural gas; oil supplies
Fradin, Frank, 461, 465, 478–79
France: breeder technology in, 375; neutron-scattering research in, 415; synchrotrons in, 472
Franck, James, 4, 32, 223
Frautschi, Steven C., 463
Fred, E. B., 187
Freed, Sherman, 178
Freeman, A., 259
Freund, Erwin O., 55–56, 430
Freund Lodge, 74, 405, 436
Friedman, Arnold M., 180
Frisch, Otto R., 2
Froman, Darol K., 12
Frost, Brian, 481
Frye, Col. Arthur H., 40–41, 52–53
Ft. Riley (Kans.), environmental impact statement on, 483
Fuel Cycle Facility (FCF), *445,* 445–46, 449, 516
fuels: burnup efficiency of, 356; ceramic, 445; closed cycle for, 444–45; experiments with, 149; meltdown of, 141–44; metal, 356, 439, 444–45, 447–50; oxide, 356; pins for, 367; proliferation-resistant, 357; reprocessing of, 445–46, 449–50, 456–58; rods for, 99; and security issues, 140; tested in transient bursts, 368; thorium-uranium, 340, 357; for training reactor, 147; zirconium, 444–45, 447–48. *See also* plutonium; uranium
fusion research: funds for, 319; review of, 409; techniques in, 286–89. *See also* magnetohydrodynamics (MHD); thermonuclear weapons

Galvin Task Force, 501n
gamma rays, 3, 117, 190
gas-cooled reactors, 61, 70
gaseous diffusion separation plant, 67
gasoline, rationing of, 17
Gast, Paul, 232, 271
Geiger, Hans, 153
Gell-Mann, Murray, 216n, 327
GEM (GeV Electron Microtron): description of, 419–20; postmortem on, 422–24, 461, 464; rejection of, 420–21
General Accounting Office (GAO): Argonne's management evaluated by, 381–82, 388–89, 391; and breeder program restructuring, 355, 358; high-flux reactor reviewed by, 258; Tripartite Agreement evaluated by, 380–81
General Advisory Committee (GAC): and accelerators, 162, 168, 197, 328; and energy research, 287–88; establishment of, 49; goals of, 60–62, 280; and H-bomb development, 94; and high-energy priorities, 216–18; members of, 59–60; role of, 66; and university-laboratory relations, 215–16
General Dynamics Corporation, 241, 267, 272
General Electric Company: collaboration with, 28; navy contracts with, 70, 334; reactors for, 61, 150–51; research at, 132
General Motors Corporation, 495–96
genome project, 475
GEORGE (computer), 124–25
Georgetown University, radium toxicity studies at, 315–16
geothermal energy, 319
Germany: neutron-scattering research in, 415; nuclear fission discovered in, 1–2; research in, 462
Getz, G. T., 196–97
Ghiorso, Albert, 37, 177, 179, 181
Glennan, T. Keith, 104
global warming, studies on, 483–84
gluons, 419
GOCO facilities, 49, 385, 500. *See also* national laboratories
Goeppert-Mayer, Maria, *228;* appointment for, 77–78, 227; role of, 223–24, 491; on shell theory, 227–29
Goertz, Ray C., 125, 140
Golden (Colo.), solar institute in, 336
Goldhaber, Maurice, 276
Goldwasser, E. L. (Ned): on Argonne's role, 196–97; on AUA, 256; and high-energy pri-

orities, 196, 216; and MURA accelerator, 218; and university-laboratory relations, 156, 219; and ZGS, 212
Gordon, Kermit, 217–18
Gould, Inc., 383
Grace, J. Peter, commission headed by, 395, 401, 427
Graham, Dan, 407
Graham, Lawrence, 489
Graham, William, 475–76
Gramm-Rudman Act, 448
graphite: bricks of, 16–17; recycling of, 149; role of, 111, 117; stability of, 26
graphite plants, security issues for, 528n80
Gray, Hanna Holborn: and Argonne board of governors, 391–93; and Argonne's political support, 389–91; and Argonne's renewal, 403–4; election of, 346; and Massey's appointment, 352, 353, 372; and SSC, 463; testimony by, 396–97; and Tripartite Agreement, 380
Great Britain: accelerators in, 153, 166; bomb casualties in, 17; cooperation with, 5, 72–73; neutron research in, 209, 415; reactors in, 28, 180, 364, 375
Great Lakes, research on, 254, 257, 319–20
Greece, communists in, 62
Greenberg, Daniel, 197, 330
Greenewalt, Crawford, 17–19
greenhouse gas effect, 484
green machine. *See* Integral Fast Reactor (IFR)
Greenpeace, 487
groundwater, contamination of, 101
Groves, Gen. Leslie: and Argonne's future, 39–40; and budget, 50; and cyclotron research, 154; and facility locations, 21; on Met lab's goals, 27–29; philosophy of, 15–16; and postwar planning, 34, 38–40; role of, 14, 17, 22, 24–25, 49; and scientists' role, 37; security measures of, 25, 30; and site selection, 51–53, 55
Gunness, Robert, 311–12
Gustavson, R. A., 38–39

Hackerman, Norman, 256–57
hadron, research on, 329
Hafstad, Lawrence R.: and accelerator location, 160; on administration, 90–91, 93; and BORAX-I, 119; and environmental conference, 253; operating policy by, 93–94; and reactor projects, 85–88
Hagstrom, Ray, 443

Hahn, Otto, 1–2, 135, 177
halogen poisoning, 529n89
Hamermesh, Morton, 162
handicapped, educational programs for, 378
Hanford Engineering Development Laboratory: administration of, 343; assistance for, 23, 29, 31; budget of, 289, 294–95, 506; design of, 26; employees at, 29, 297, 506; environmental contamination at, 486, 488; pile projects at, 26–27, 63; reactors at, 110, 176, 231–33; responsibility for, 22; separation plant at, 21; wildlife at, 430. *See also* Fast Flux Test Facility (FFTF)
Harrell, William B.: and accelerator location, 161; administration of, 93, 175; and construction costs, 89; and educational program, 72; and environmental studies, 256; and high-flux reactor, 237, 239; and Met Lab transition, 40; and permanent laboratory, 57; and personnel decisions, 210, 527n70; recommendations by, 92, 104
Harrisburg (Pa.). *See* Three Mile Island (TMI)
Harvard University: and accelerator project, 196; competition from, 466
Harvey, Paul, 96–97
Harwell Atomic Energy Research Establishment (Great Britain), 180
Hasterlik, Robert, 74, 102
Hattiesburg (Miss.), scientist from, 352, 377
Hawkins, H. G., Jr., 30–31
hazardous-gas monitor, 482–83
hazardous waste, cleanup of, 485. *See also* nuclear waste
H-bomb. *See* thermonuclear weapons
health issues: center for, 315–18; concerns over, 10–12; cooperation in, 495; divisions for, 12; increase in, 31–32; for plutonium use, 36; and reactor sites, 67–68. *See also* biomedical research
heat transfer system, 99
heavy-ion linear accelerator (HILAC), 181
heavy water: in CP-5, 116–17; as neutron moderator, 95
heavy-water reactors: description of, 98, 100–111, *112*, 113; research on, 27–28; security issues for, 528n80. *See also* Chicago Pile 3 (CP-3)
Heisenberg, Werner, 3–4, 502–3
helium, 103, 353
hemodialyzer (artificial kidney), 250
HEPAP. *See* High-Energy Physics Advisory Panel (HEPAP)

HERA-ZEUS detector, 438, 462
herbicides, production of, 484–85
Herrington, John S., 440–42, 449, 476
hextron project, 419
HFBR (High-Flux Beam Reactor), 412
HFEF (Hot-Fuel Examination Facility), 269–70, 296, *296*, 446
Hickenlooper, Bourke B., 57, 79–82, 96
high-energy physics: AEC policies on, 262; description of, 326–27; evaluation of, 329–30, 408; focus on, 152; funds for, 155–56, 160, 191–92, 203–4; importance of, 327–28; leadership in, 462–63; priorities in, 216–18, 460–62; public attitudes toward, 326; role of, 195–96, 211; Soviet competition in, 169, 171–72; support center for, 461–62. *See also* accelerators
High-Energy Physics Advisory Committee, 399, 408
High-Energy Physics Advisory Panel (HEPAP), 260, 261, 330, 461–62
High Energy Physics Building, 211
High-Flux Beam Reactor (HFBR), 412
High-Flux Isotope Reactor, 412
high-flux reactors: development of, 61, 71; future of, 236; for isotope production (HFIR), 236; site for, 63–64, 67. *See also* Argonne Advanced Research Reactor (AARR, A^2R^2); Chicago Pile 5 (CP-5); materials-testing reactors (MTR)
high-sulfur dry scrubber, 396
high-temperature reactors, 40, 63
high-voltage accelerators, 40
High Voltage Electron Microscope (HVEM)–Tandem Accelerator, 398, 409–11, *410*
HILAC (heavy-ion linear accelerator), 181
Hilberry, Norman, *185, 253;* and accelerators, 164; administration of, 25, 65, 75, 90–91, 151, 172, 173; appointment for, 9, 42, 51, 135, 175; and Argonne's future, 39; and Argonne's goals, 88; background of, 175–76; and budget, 85, 201; and lab's responsibility, 21; and national laboratory system, 192–94; philosophy of, 183–86; and reactor projects, 113, 140; retirement of, 205–6, 208; and Rettaliata report, 189; and security issues, 134; and site selection, 13–14, 51–52; and training issues, 135–38, 200–201
Hildebrand, Roger: and accelerator project, 196–98, 204, 209, 211–12; appointment for, 189; and bubble chamber, 199; characteristics of, 200; influence by, 224; and university relations, 200, 213

Hirohito (emperor of Japan), 36
Hirschfelder, Joseph C., 216
Hitler, Adolf, 2–3, 27
Hodel, Donald P.: and accelerator project, 421; appointment for, 402–3; and multiprogram labs, 428; recommendations to, 437–38; replacement for, 440
Hoffman, Darleane C., 176, 182
Hoglund, B., 248
Hogness, T. R., 28, 32, 39
Hoke, Martin, 454
Holifield, Chet: and Argonne administration, 270; and energy plan, 303; and environmental issues, 248, 253; and reactor projects, 204–5, 237; and research focus, 300
Hollingsworth, Robert E.: and Argonne administration, 270, 274, 276, 293; and breeder reactor project, 265; complaints to, 300; and environmental pollution studies, 248
Holm, Lennart, 180n, 181
Homogeneous Reactor Experiment (HRE-I), 128
Honeywell Corporation, 465
Hornig, Donald, 351
Horwitz, Philip, 443
Hosmer, Craig, 270
"hot atom" chemistry, 77
Hot-Fuel Examination Facility (HFEF), 269–70, 296, *296*, 446
Houlihan, Tom, 487
Hovde, Frederick L., 168
Hoyt, Frank C., 76–77
HRE-I (Homogeneous Reactor Experiment), 128
Huberman, Eliezer, 475
Huffman, John R., 99
Hull, Harvard L., 51
Hull, John, 480
Human Genome Project, 475
humans: injected with plutonium, 320–23; radiation's effects on, 315–18. *See also* biomedical research; Center for Human Radiobiology; health issues; safety
Humphrey, Hubert, 218
Humphrey, Paul, 118
Hunter, Meredith, 485
Hutchins, Robert M.: and Argonne administration, 92; and facility locations, 21; and postwar planning, 33, 37, 40; role of, 3, 8, 16
HVEM (High Voltage Electron Microscope)–Tandem Accelerator, 398, 409–11, *410*

hydrogen-3. *See* tritium
hydrojet propulsion, 132

Ianniello, Louis, 394, 467
IBM. *See* International Business Machines
ICBMs (intercontinental ballistic missiles), 132
Idaho: BORAX-I test in, 118–21; reactor proving ground in, 86–88; reactors constructed in, 98–100. *See also* Idaho National Engineering Laboratory; National Reactor Testing Station (NRTS)
Idaho National Engineering Laboratory: administration of, 343; closure of facilities at, 457; safety research at, 358; wildlife at, 430, 432, *432*. *See also* National Reactor Testing Station (NRTS); *names of specific reactor projects*
IFR. *See* Integral Fast Reactor (IFR)
Illinois: environmental standards in, 485; funds from, 257, 472, 481; lab site in, 52
Illinois Department of Nuclear Safety, 486
Illinois Environmental Protection Agency, 249–50
Illinois Institute of Environmental Quality, 257, 295
Illinois Institute of Technology, cooperation with, 498
Illinois Superconducting Corporation, 483
Illinois Warning Center, 207
in-core testing (INCOT), 355–56
Independent Offices Appropriation Act (1950), 546n38
Indiana, lab site in, 52
indium foils, 113
Industrial Photography, awards from, 475–76, 476
Industrial Research, awards from, 312, 482
Industrial Research Institute, 383
industrial sites, cleanup of, 489–90
industry: and advanced photon source, 465, 467, 470; Argonne's relation with, 72–73, 244, 349; assistance for, 439–40; boiling-water reactors for, 150–51; and contract issues, 193, 494–99; cooperation with, 72–73, 428, 434–35, 494; and economic development, 383–84; energy research by, 387–88; and environmental studies, 250–51; funds from, 439; and government research funds, 384–85; IFR funds from, 454; rapid changes in, 499–500; reactors for, 110, 126–28, 195; role of, 273; vs. scientific research, 183. *See also* technology transfer

Inglis, David, 77, 228
Inland Seas (ship), 254
inner city, economic development of, 488–90
Institute for Advanced Study, 77, 122–23
Institute for Nuclear Studies: accelerator summit at, 164–65; seminar at, 160; staff of, 173–74; status of, 156–57
Institute of Nuclear Science and Engineering, 222
Institut Laue-Langevin (ILL; France), 415
instrumentation: for cars, 497; complexity of, 362; construction of, 13, 44–45; development of, 23, 76, 355–56; and ergonomics, 367–68; malfunctions in, 359, 362; Pooh names for, 17; radiation detecting, 31; shortcomings of, 76
Integral Fast Reactor (IFR): cancellation of, 456–58, 498; concept of, *445*, 445–48; debate over, 452–57; development of, 439, 516; evaluation of, 450–51; fuel recycling by, 449–50; significance of, 443–44; support for, 437–38, 448–50
Intense Pulsed Neutron Source (IPNS): challenge to, 413–15; construction of, 412–13; description of, 411; funds for, 412–16, 442; operational level of, 428; review of, 408–9; significance of, 398, 409; status of, 396; success of, 394, 442–43; support for, 415–16
intercontinental ballistic missiles (ICBMs), 132
Intermagnetics General Corporation, 479–80
internal emitters, 79, 102
International Atomic Energy Agency, 126, 137
International Business Machines (IBM), 125, 497–99
International Conference on the Peaceful Uses of Atomic Energy: in 1955, 138–41, 148, 167, 168, 201; in 1958, 148, 190–91
International School of Nuclear Science and Engineering: establishment of, 135–37; housing for, 200; staff of, 147; students of, 201; support for, 173
International Union of Pure and Applied Chemistry (IUPAC), 182
International Union of Pure and Applied Physics (IUPAP), 181–82
Iowa State College: AEC programs at, 58; collaboration with, 26
IPNS. *See* Intense Pulsed Neutron Source (IPNS)
Iran, hostages in, 373
irradiation, experiments on, 114

isotopes: barium, 1–2; conference on, 187; helium, 103; separation of, 3; in transuranic elements research, 26, 180. *See also* radioisotopes

Jackson, Harold E., 419–20
Jackson, Henry, 234
Jackson, Max, 233
Jacobson, L. O., 10
Jacobson, Norman, 232–33
James, Ralph A., 177
James, William, 335
JANUS reactor, 195, 297
Japan: bombs dropped on, 36, 315; contracts with, 344; IFR supported by, 454–55; nuclear technology in, 237, 375, 501; synchrotrons in, 472
JCAE. *See* Joint Committee on Atomic Energy (JCAE)
Jeffries, Zay, 28, 32, 34
Jensen, J. Hans, 77, 188–89, 227, 229
Jesse, William P., 30
John Deere and Company, 383
John Jay Hopkins Laboratory (San Diego), 240–41
Johns Hopkins University: fission research at, 2; policies of, 223–24
Johnson, D. Gale, 347
Johnson, Edwin C., 43
Johnson, Lyndon B.: and accelerator project, 218; and environmental issues, 252; and nuclear power, 229, 337; staff of, 351
Johnson, Thomas H.: and accelerators, 160–66, 168–70; role of, 157–58
Johnson, Warren, 217, 219–20
Johnston, J. Bennett, 456, 472
Joint Committee on Atomic Energy (JCAE): and accelerators, 167, 170, 197; and AEC appointments, 54; and Argonne administration, 60, 270, 274; and atomic energy legislation, 127; and BORAX-I explosion, 121; and budget, 234; and civilian research and development, 128, 130; and energy research, 286; and environmental studies, 250; establishment of, 50; and H-bomb, 94; and human radiobiology, 315–18; and international competition, 191; members of, 81; and political change, 204–5; and reactor projects, 87, 148, 236–37, 258–59, 273; role of, 96, 244, 247, 280; and security issues, 82–84, 96–97; and site selection, 57, 65–66; and Soviet capabilities, 89

Joint Institutes of Nuclear Research (Soviet Union), 181
Joliot-Curie, Frederic, 2
Jones, Haydn, 13
Jones, Roger W., 273
Jonke, Albert A., 287–88
Jordy, Sam, 489
Josephson, Linda, 487
Joslin, Murray, 250–51
Joyce, James, 327

Kaiser Wilhelm Institute for Chemistry, 1
Kaper, Hans, 436–37
KAPL (Knolls Atomic Power Laboratory), 147
Karpinski, Gene, 454
Karraker, Bill, 208
Katz, Joseph J., 297
Kaufman, Burton, 335
Kemeny commission, 361–62, 365
Kennedy, Edward, 284
Kennedy, John F.: and accelerator project, 217–18; appointments by, 205; and civil defense, 208; election of, 204; and high-flux reactor, 236
Kennedy, Joseph W., 11
Kepler, Johannes, 497
Kerner, Otto, 237, 259
Kerr, William, 270
Kerst, Donald W., 155, 162, 166
Keyworth, George: and advanced photon source, 464; on Argonne initiatives, 439–40; on energy crisis, 433–34; on industrial competitiveness, 377; and peer review, 421; on Reagan's energy goals, 433; on science funding, 384, 397, 400, 423; on technology transfer, 434–35
kidney, artificial (hemodialyzer), 250
Kilgore, Harley M., 43
Killian, James R., Jr., 197
Kimpton, Lawrence: and accelerators, 167–68, 170; and Argonne administration, 104, 173; and EBWR, 145; and university-Argonne relations, 186–87, 189
King, Martin Luther, Jr., 488
Kissinger, Henry A., 293
Klein, Rudolph, *124*
Kliewer, Kenneth, 463, 464–65, 470
Knapp, Gordon, 465
Knolls Atomic Power Laboratory (KAPL), 147
Knuth, August, 18
Koch, Leonard J., 148
Koffler, Henry, 319

Korea, communists in, 81
Korean War, 95–98
Kouts, Herbert J. C., 300
Kruger, P. Gerald, 157, 161–62, 164–65
Kurath, Dieter, 77, 228
Kurchatov Institute of Atomic Energy (Soviet Union), 181
Kurtis, Bill, 490
Kuznetsov, Ivan, 452
Kyle, Martin, 301

Labat, Ivan, 475
laboratory's evening academic program (LEAP), 271
Lake Michigan, research on, 254
Lancelot, Jill, 454–55
Laney, Robert V.: administration of, 273, 276, 281–82; appointment for, 272–73, 277; background of, 272; and budget cuts, 289–90; and energy plans, 286, 337–40; opposition to, 350–51; and plutonium MUF, 323–24; retirement of, 371
Lapp, Ralph, 40–41
Laramie (Wyo.), energy technology labs in, 336
LARC I (computer), 124
Larsen, Clarence, 300
Latimer, Wendell, 12
Lawrence, Ernest O., *30;* on cyclotrons, 153–55; research by, 4; role of, 6–7, 35
Lawrence Berkeley Laboratory: academics at, 386; administration of, 221, 276, 343; Advanced Light Source at, 472; budget of, 294–95, 506; cooperation with, 498; employees at, 297, 506; environmental contamination at, 486; funds for, 203–4, 289; political support for, 396; and SSC, 463; synchrotron at, 466; training at, 215; university relations of, 372
Lawrence Livermore Laboratory: academics at, 386; administration of, 221, 343; budget of, 289, 294–95, 441, 506; computer at, 124; employees at, 297, 506; environmental contamination at, 486; goals of, 214; and plutonium recycling, 457; and Reagan's policies, 402
Lawroski, Harry, 268
Lawroski, Stephen, 73, 233
Lazarus, Steven, 482–83
League of Conservation Voters, 454–55
LEAP (laboratory's evening academic program), 271

Lederman, Leon, 466
Lee, Malcolm H., 271
Lee, T. D., 216n
legislation: on air pollution, 249, 333; on Argonne's establishment, 42–44; on NSF, 44; on patents, 481–82; on technology transfer, 481–82, 494. *See also* Atomic Energy Act (1946); Atomic Energy Act (1954); Atomic Energy Commission (AEC)
Lennox, David H., 147, 190
leptons, 327
Levenson, Milton, 235, 239, 268, 271–72
Levi, Edward H., 43, 274, 293, 306–7
levitation, of magnets, 475–76, *476,* 480
Lewis, Warren K., committee headed by, 17–20
Libby, Willard: and accelerators, 162, 163, 166–72; and Argonne's school, 136–37; and radiobiology, 74; and reactor projects, 113; on university-laboratory relations, 174
Lichtenberger, Harold V.: and BORAX-I test, 120; and reactor construction, 99, 139; research by, 89, 132; role of, 68–69
light-water reactors: competitiveness of, 129–30; fuel for, 357; future of, 426; recycling fuel from, 450; replacement for, 338; research on, 300; safety of, 362–64, 366–67, 433; uses of, 128
Lilienthal, David E.: appointment for, 53–54, 79–80; and Argonne site, 55, 57, 65; attack on, 81; H-bomb opposed by, 94; resignation of, 94, 96; and security issues, 82
Lindop, Patricia A., 323
lines of assurance (LOA), 364
Ling, James, 368–69, 371–72
Link, Leonard E., 255
liposome encapsulation, 312
Lippman, Thomas, 453
liquid metal, research on, 69, 70, 89, 114, 132. *See also* sodium
Liquid Metal Fast Breeder Reactor (LMFBR) program: administration of, 267–68, 270–72, 274, 291; commercialization of, 273, 354; debate over, 354–55; demise of, 281; funds for, 264, 273, 279–80, 289–90, 358, 373; instrumentation for, 355–56; model for, 230; as priority, 265–68, 296, 354; program for, 231–34; restructuring of, 354–58; safety of, 354, 363–65; support for, 269, 291, 333, 336. *See also* Clinch River Breeder Reactor (CRBR; Tenn.); Experimental Breeder Reactor Number 2 (EBR-II)

Lisbon Nuclear Safety Initiative, 452
lithium-sulfur battery, 288–89, 310, 311, 337, 495
Liverman, James, 321, 322
Livingood, John J., 162, 166, 171–72, 189, 210
Lloyd, Marilyn, 354
LMFBR. *See* Liquid Metal Fast Breeder Reactor (LMFBR) program
Long Range Nuclear Option workshop, 301
Longworth, Richard C., 382
Loofbourow, John R., 58
Loomis, F. Wheeler, 38–39, 59, 158–59
loop-design breeder, 426
Los Alamos National Laboratory: administration of, 49, 221, 343; Argonne's relations with, 84; budget of, 294–95, 506; computer at, 123–24; employees at, 29–30, *30*, 78, 240, 297, 506; environmental contamination at, 486; establishment of, 22; fusion research at, 287; goals of, 214; political support for, 396; protests by, 133–34; pulsed-neutron source for, 413–14, 416; radiobiology at, 316; reactor at, 111; research at, 24, 26; training at, 216; transuranic elements research at, 178–79
Love Canal, effects of, 460
Lucas, Scott, 51
Lynch, David, 466

magnetic bearing, 480
magnetic fusion, 386
magnetohydrodynamics (MHD), 287, 310
magnets: construction of, 478; and fusion research, 287; levitation of, 475–76, *476,* 480
Magnuson, Warren G., 43
Maine, nuclear power referendum in, 446
Mallinson, George G., 222
Malm, John G., 225
Mancuso, Thomas F., 316
Manhattan Engineer District (MED): administration of, 48, 58–59; leadership for, 15; and pile development, 35; reports for, 17; site for, 14. *See also* Manhattan Project; Metallurgical Laboratory
Manhattan Project: army control of, 21–22, 25; compartmentalization of, 25–26, 30–31, 135; environmental hazards of, 45; facilities in, 14, 40; gun vs. implosion method for, 26; instruments for, 44–45; legacy of, 385; Met Lab's status in, 21; nuclear waste buried by, 57–58; personnel for, 24–25, 48; plutonium experiments by, 320–23; and postwar planning, 32; records of, 44; and research priorities, 154; security of, 25; Soviet spying on, 97; support for, 15–16. *See also* atomic bomb
MANIAC I (computer), 123
MANIAC II (computer), 123–24
Manning, Winston M.: and laboratory senate, 245; role of, 46, 89, 178, 180, 239; and security issues, 134
Mares, Jan, 408
Mark I/Mark II cores, 98
Mark III core, 149
Marquette University, Argonne's relations with, 187
Marsden, Ernest, 153
Marshall, Brig. Gen. James C., 14
Marshall, W. R., 215
Marshall Islands, nuclear testing in, 128
Maryland, blackout in, 283–84
Massachusetts: blackout in, 283–84; nuclear power referendum in, 446
Massachusetts Institute of Technology (MIT): and accelerator projects, 163–64, 196, 211, 327, 419–23; competition from, 466; and plutonium patients, 321–22; Radiation Lab at, 7; radium toxicity studies at, 315–17
Massey, Walter E., *352;* administration of, 353–54, 368–72, 375–76, 408, 427, 481; appointments for, 352, 404, 428, 435; and Argonne's future, 397–98, 403–4, 437–40; and Argonne's political support, 389–91; and Argonne's revitalization, 369–70; background of, 351–53; and board of governors, 393–94; and budget, 373–74, 388–89, 401; and building 201, 404–7; and GEM, 420; goals of, 374; and IPNS, 411, 413–15; and national laboratories, 382–83, 385–87; and new director, 424–25, 429; and science education, 377–78; testimony by, 396–97; and university relations, 372–73, 379–82
materials science: cooperation in, 409; funds for, 416; history of, 398; new initiatives in, 408; support for, 438. *See also* Advanced Photon Source (APS) project; High Voltage Electron Microscope (HVEM)–Tandem Accelerator; Intense Pulsed Neutron Source (IPNS)
materials-testing reactors (MTR): accidents with, 141; construction of, 87; criticality of, 100; design of, 97; development of, 113–14; role of, 98–99, 114; site for, 86–87; transuranic element research with, 178–79

May, Alan Nunn, 62
May, Andrew Jackson, 43
Mayer, Joseph, 78, 223–24, 227
May-Johnson bill, 43–44
McCarthy, Joseph, 126, 134. *See also* communism: crusade against
McClure, James, 357–58, 446
McCone, John A., 193–94
McCormack, Mike, 285–86
McCune, Francis K., 193
McDaniel, Paul W.: and accelerator project, 237–38, 258–59; background of, 304; on laboratories vs. universities, 261; and Tripartite Agreement, 304–9; and ZGS, 264
McKellar, Kenneth D., 54
McKinley, John, 184
McLain, Stuart, 86–87, 97–99, 105
McMahon, Brien, 43–44, 81, 94, 96
McMillan, Edwin M., 154, 177, 276
McTague, John, 438
MED. *See* Manhattan Engineer District (MED)
media, and plutonium MUF, 323–25
medical ethics, and plutonium patients, 322
medical research. *See* biomedical research
medium-energy electron accelerator (MEEA), 409
Meissner effect, 475–76, *476*, 480. *See also* superconductivity
Meitner, Lise, 1–2
meltdowns: of EBR-I, 141–44; of Fermi Fast-Breeder Reactor, 234; prevention of, 365–68; studies of, 363–64; of Three Mile Island, 358–63
mendelevium, 179
MERLIN (computer), 124
mesothorium, 315–16
Metallurgical Laboratory: administration of, 8–10, 12, 25, 182; and Argonne site, 13–14; changing attitudes at, 23–24; collaborations of, 26–28; employees at, 24–25, 28–31; establishment of, 7–9; future of, 32–34, 37–40; goals of, 9–10, 13, 15, 20, 26–27, 35; health concerns at, 10–12, 31–32; legacy of, 44–46; mission of, 26–29; pile projects at, 10, 13; plutonium isolated at, 14; and postwar planning, 33–34; reorganization of, 12–13, 21–22; security issues at, 30–31; status of, 21–22; support for, 15–18; university relations with, 8, 32–33, 37–38
metallurgy: of fuel elements, 97–98; research on, 103

meteorology, monitoring of, 101
Metropolitan Edison (Pa.). *See* Three Mile Island (TMI)
Meyers, John, 471
MHD (magnetohydrodynamics), 287, 310
MHTGR (modular high-temperature gas-cooled reactor), 453
Michelson, Albert, *4*
Michigan. *See* Enrico Fermi Fast-Breeder Reactor (Detroit); University of Michigan
microtron. *See* GEM (GeV Electron Microtron)
Midwest: accelerator for, 161–64, 212, 217–18; Argonne's role in, 243–45; and high-flux reactor, 258–59; research center for, 157–58, 161; research funding in, 155–56. *See also* Associated Midwest Universities (AMU); Midwest Universities Research Association (MURA)
Midwest Regional Environmental Systems Program (MRESP), 255–56
Midwest Universities Research Association (MURA): and accelerator projects, 164–65, 169–70, 196–97, 217–18, 224; attitudes of, 218, 224; and bubble chamber, 199; competition for, 197; influence by, 171, 173–75, 186–87, 203, 210; members of, 165, 170; opposition by, 165–68, 197–98; and Tripartite Agreement, 220–22. *See also* university-laboratory relations
MIKE, test of, 178–79
military: and atomic liaison committee, 49; control by, 43; influence by, 88–89, 95–96, 107; postwar role of, 35, 40; projects for, 434; reactors for, 97–100. *See also* U.S. Army; U.S. Department of Defense; U.S. Navy
Military Policy Committee, 25, 32
Miller, W. F., 124–25
Mills, Adm. Earle W., 85–86
Milne, A. A., 17
Milsted, John, 180
minorities: and affirmative action, 349–50; contracts for, 378; educational programs for, 377–78; outreach programs for, 490; role of, 77–79
mission statements: context of, 427; criticism of, 400–401; governors' suggestions for, 427–28; need for, 381–82; and refitting Cold War science, 501–2
MIT. *See* Massachusetts Institute of Technology (MIT)
modular high-temperature gas-cooled reactor (MHTGR), 453

630 ■ Index

Moncton, David E., 470–73, 479
Mondale, Walter, 357
Monsanto Chemical Company, 42, 110
Monson, Harry O., 266–67
Moore, T. V., 13
Morgantown (W.Va.), energy technology labs in, 336
Morrison, Philip, 27
Motorola Corporation, 465, 494
MRESP (Midwest Regional Environmental Systems Program), 255–56
MTR. *See* materials-testing reactors (MTR)
MUF (material unaccounted for), plutonium, 323–25
Mulliken, Robert S., 32
multitevatron particle accelerator. *See* Superconducting Super Collider (SSC)
mu meson particle, 211–12
Mumford, Lewis, 406
MURA. *See* Midwest Universities Research Association (MURA)
Murphy/Jahn Associates (architects), 405
Museum of Science and Industry (Chicago), 48, 53, 72
Myers, Frank E., 189, 222

NACA (National Advisory Committee for Aeronautics), 5
Nader, Ralph, 452
NAPCA (National Air Pollution Control Administration), 288
NASA (National Aeronautics and Space Administration), 344, 498
National Academy of Sciences, 5, 297, 315, 397, 451, 454, 457. *See also* National Research Council
National Accelerator Laboratory (later called Fermilab): accelerators at, 295, 462; administration of, 343; architecture of, 406; and Argonne's high-energy physics program, 461–62; assistance for, 261–62; budget of, 330, 395; establishment of, 259–60; funds for, 260–61, 327–28; influence by, 263–64, 274–75; scientists for, 261–62; staff recruited from, 465; support for, 327–28
National Advisory Committee for Aeronautics (NACA), 5
National Aeronautics and Space Administration (NASA), 344, 498
National Air Pollution Control Administration (NAPCA), 288

National Bureau of Standards, 3, 323, 412, 419–20, 466
National Cancer Institute, 75, 102
National Carbon Company, 26
National Center for Air Pollution Control, 249–50
National Committee on Radiation Protection, 74, 101–2
National Competitiveness Technology Transfer Act (1989), 494
National Defense Research Committee (NDRC), 4–6
National Electron Accelerator Laboratory (NEAL), 420
National Endowment for the Arts, 407
National Environmental Policy Act (1969), 252
National Environmental Research Park, Argonne-West as, 432
National Geographic, on Integral Fast Reactor (IFR), 448
National Heart and Lung Institute (NHLI), 288
National Institutes of Health (NIH), 388
national laboratories: administration of, 58–60, 92–93, 192–94, 220–21, 270, 274–75, 342–43; benefits of, 182–83, 434–35; budgets of, 193, 289–90, 294, 343–44, 384–85, 388, 399; challenges for, 44–46, 499–504; contractors for, 506; and economic development, 383–84; and energy policy, 301–3, 341–42; environmental cleanup of, 485–88; environmental studies at, 248–50; establishment of, 49; evaluations of, 281–82, 310–13, 400; future of, 340, 397–400, 455; goals of, 59, 156, 195, 280, 282, 381–84; identities of, 460; model for, 194–95; multiprogram type of, 281, 294, 342–45, 428, 459, 506, 517; new priorities developed for, 399–400; operating policy for, 93–94; overhaul ordered for, 400–403; possible closing of, 389–90, 400; postwar planning for, 33–35; proposals for, 6–7, 35, 40; "reinvention" of, 460; role of, 243–45, 383–86; sites for, 51–53; and social change, 378; status of, 191–96, 204–6, 294, 336; training at, 215–16; university relations with, 72–73, 195–96; and waste disposal, 73–74. *See also* weapons laboratories; *names of specific laboratories*
National Laboratory–Industrial Research Institute Task Force, 383
National Nuclear Energy Series, 44, 78

National Reactor Testing Station (NRTS): accidents at, 118–21, 141, 205; administration of, 265–66, 268–69, 294; budget of, 289, 425; LBJ at, 229, 337; reactors at, 111, 118, 140, 149–50, 269; research at, 113–14; students at, 137. *See also* Idaho National Engineering Laboratory; Transient Reactor Test Facility (TREAT); *names of specific reactor projects*

National Research Council, 105, 412, 464

National Science Board, 353

National Science Foundation (NSF): and accelerators, 419–21; budget of, 388; funds from, 155, 163, 203, 257, 295; legislation for, 44; and lithium-sulfur battery, 288–89; policy for, 353; proposals to, 255–57; and science education, 377–78; and ZGS review, 329

National Security Council, 95, 127

national synchrotron light source (NSLS), 466, 473

National Taxpayer's Union, 454–55

National Warning System, 207, 283

natural gas: deregulation of, 333, 357; shortages of, 334–35

Natural Resources Defense Council, 485

Nature (periodical): on EBR-II tests, 448; on IFR cancellation, 457

Nautilus (submarine), 514–15

Naval Research Laboratory (NRL), 2–3

NDRC (National Defense Research Committee), 4–6

Neal, Homer A., 264–65

NEAL (National Electron Accelerator Laboratory), 420

neptunium, 177

neutrino studies, 263, 329

neutron hodoscope, 364

neutron research: discoveries in, 15, 209; focus of, 34, 117; funds for, 413–16; organization of, 6–7; review of, 413–15; on scattering, 408, 412–15, 443; support for, 2–6. *See also* high-flux reactors; Intense Pulsed Neutron Source (IPNS); nucleons

neutron spallation, 412

Nevada: nuclear testing in, 97; nuclear waste site in, 488

Nevitt, Michael V., 272, 276

New England Deaconess Hospital Cancer Research Institute, radiobiology at, 316

New Explorers (program), 490

New Generation Vehicle Initiative, 496

New Jersey, blackout in, 283–84

New Jersey College of Medicine, radium toxicity studies at, 315–16

New York City, blackout in, 283

New York Health and Safety Laboratory, radiobiology at, 316

New York University, radium toxicity studies at, 315–16

NHLI (National Heart and Lung Institute), 288

Nichols, Lt. Col. Kenneth D.: and accelerators, 163–64; and Argonne's future, 38–41; and Argonne-university relations, 72; and governance, 48; and personnel, 28–29; and plutonium patients, 322; and postwar planning, 33–34, 38–40; role of, 14–15, 49; and security issues, 134; and site selection, 13, 52

Nickson, James J., 12, 31, 45, 57–58

Nickson, Margaret, 79

NIH (National Institutes of Health), 388

niobium resonator, 417–18

nitrogen, and superconductivity, 477–78

Nixon, Richard M.: appointments by, 290, 300; budget by, 272; election of, 284; energy policies of, 229, 273, 278–80, 285, 300–304, 336; foreign policy of, 293

Nobel, Alfred, 182

Nobel Institute of Physics (Sweden), 180

nobelium, 176–82

Nobel Prize. *See* Fermi, Enrico; Goeppert-Mayer, Maria; Seaborg, Glenn T.

noble gases, 225–27, *226*

Norris, George, 167

North American Aviation, Inc., 128

North Carolina State College: reactor at, 110; training at, 137, 173

Northwestern University: and advanced photon source, 470; and air pollution studies, 249; and bubble chamber, 199; students of, 105

Novick, Meyer (Mick), 115, 268

NRL (Naval Research Laboratory), 2–3

NSAC (Nuclear Science Advisory Committee), 419–21

NSF. *See* National Science Foundation (NSF)

NSLS (national synchrotron light source), 466, 473

nuclear energy: Clinton's opposition to, 453; conferences on, 138–41; electricity generated by, 99–100, *108*, 108–9, 115, 336–37, 515; government's role in, 385, 388–89;

peaceful uses for, 107, 126–28, 138–41, 190–91; possibilities for, 107, 110; public attitudes toward, 278–79, 281–82, 325–26, 337, 353, 432–33, 446, 452; role of, 334, 339–40, 354–58, 444; for submarines, 69–71, 86, 88–89, 96

Nuclear Engineering Congress (Ann Arbor, Mich.), 125

nuclear fission, 1–6

nuclear power plants: commercialization of, 194–95, 339–40; development of, 131, 149–51, 333; disappointments in, 147–48; environmental impact statements for, 257; fuel for, 114, 129–31, 444–45; funds for, 128–31, 191–92; insurance for, 143, 148; international number of, 337–38; and nuclear proliferation, 337–40, 354–55, 357, 444, 528n80; and ownership issues, 129; pilot for, 139–40; and reactor dedication, 145

nuclear proliferation: concerns over, 337–40, 354–55, 357, 444, 528n80; fear of, 88, 94, 122; and fuel development, 357; IFR's solution to, 449–50. *See also* Cold War

nuclear reactors: administration of, 230–34; civilian development of, 128, 130, 132–33, 139–40, 294; construction of, 89; debate over, 367–68; for defense purposes, 97–100; development of, 40, 46, 408; drive mechanisms for, 98; environmental impact of, 484–85; explosion of, 120–21; fuel for, 103, 114–15, 129–31, 444–45; funds for, 294, 314, 438, 442; and H-bomb development, 94–95; increased focus on, 59–62, 86–87; locus of research on, 60, 62–64, 66–68; meltdown of, 141–44; private development of, 126–28; reorganized programs for, 268–70; safeguards for, 100–101; status of, 110–11; summary of, 514–16, *515;* testing materials for, 89, 410–11. *See also* atomic piles; coolants; fuels; *names of specific types of reactors*

Nuclear Regulatory Commission (NRC): contracts with, 344; establishment of, 301; on safety, 432–33; and Three Mile Island, 359, 362

nuclear rocket reactor, 516

nuclear science: advances in, 75–77; basic research vs. applied research in, 61–62, 109, 244–45; characteristics of, 4, 176; competition in, 109–10; control of, 42–44, 63–64; future of, 24, 375; international competition in, 191; military applications of, 40;

postwar planning for, 32–35, 42–44; public information on, 36–38; and remote control development, 125–26; role of, 279–81. *See also* nuclear energy

Nuclear Science Advisory Committee (NSAC), 419–21

nuclear structure, 227–29

nuclear testing, 97–98, 101, 178–79

nuclear war, 88, 94, 122

nuclear waste: amount of, 130; contamination by, 485–88; disposal of, 45, 57–58, 101, 449; environmental cleanup of, 485; IFR's solution to, 449–50; and security issues, 82–83, 96, 444; and site selection, 53, 487–88; studies of, 73–74, 101; types of, 449

nuclear weapons: development of, 484–85; fuel for, 338–40, 444; fuel recycled from, 450; mock exercises, 207–8; tests of, 50. *See also* atomic bomb; nuclear proliferation; thermonuclear weapons; weapons laboratories

nucleons, 419–20

nuclides, 178–79

Oak Ridge Automatic Computer Logical Engine (ORACLE), 123, *124*

Oak Ridge National Laboratory: academics at, 386; administration of, 63–64, 107, 220, 343; budget of, 289, 294–95, 506; cancer research at, 75; computer at, 123; contracts with, 344; employees at, 297, 506; environmental contamination at, 486; facilities at, 67–68; fusion research at, 287; future of, 340; goals of, 63–64, 214; and nuclear-powered aircraft, 109; and nuclear-powered submarines, 70; political support for, 390; radiobiology at, 316; radioisotope research at, 85; radiological assistance teams from, 361; reactors at, 67–68, 99, 110–11, 113, 128, 138, 176, 236, 238, 270, 412, 414, 467; site for, 49; training at, 137, 271; university relations for, 187; waste shipped to, 57. *See also* Clinch River Breeder Reactor (CRBR; Tenn.)

O'Fallon, Nancy M., 448

Office of Coal Research, 288

Office of Emergency Planning, 283

Office of Energy Research, 344–46, 382

Office of Health and Environmental Research, 473

Office of Naval Research, 155, 203, 295

Office of Nuclear Energy, 456

Office of Price Administration, 17
Office of Science and Technology Policy (OSTP), 389–90
Office of Scientific Research and Development (OSRD): contracts of, 521n19; establishment of, 5; plan by, 6; reports for, 17; role of, 8, 12–14. *See also* S-1 committee
O'Hare Conference on the Environment, 252–54
O'Hare International Airport, air pollution at, 257
oil supplies: embargo on, 302, 460; level of, 287, 334; research on, 301
Okhawa (physicist), 166
O'Leary, Hazel, 456–57, 460, 494–95, 499, 501n
Olsen, Bob, 125
omega minus particle, 224–25
Onnes, Heike Kamerlingh, 476–77
OPEC (Organization of Petroleum Exporting Countries), 302, 434, 460
Operational Reliability Testing program (ORTP), 367–68
Operation Crossroads, 50
Operation Ranger, 97
Oppenheimer, J. Robert: appointment for, 59; and Argonne's establishment, 43; H-bomb opposed by, 94; on peaceful uses, 107; role of, 22, 29, 35; security clearance for, 133–34, 144
ORACLE (Oak Ridge Automatic Computer Logical Engine), 123, *124*
Orange County Airport (Calif.), air pollution at, 257
Oregon, nuclear power referendum in, 446
Organization of Petroleum Exporting Countries (OPEC), 302, 434, 460
Orlemann, Edwin F., 37
ORTP (Operational Reliability Testing program), 367–68
oscillators, 76, 154
OSRD. *See* Office of Scientific Research and Development (OSRD)
OSTP (Office of Science and Technology Policy), 389–90
Ottawa (Ill.), watchmakers in, 102, 315–18
oxygen, and superconductivity materials, 479
ozone hole, 460, 483–84

PACE (Precision Analog Computing Equipment), 125
Pacific Northwest Laboratory: administration of, 343; budget and staff of, 506; contracts with, 344; cooperation with, 498; and risk reduction after Chernobyl, 452. *See also* Hanford Engineering Development Laboratory
Pacific Proving Grounds, MIKE tested at, 178–79
Pacific Science Center, 298
Packard, David, report by, 400–403, 427, 428
Packer Engineering Corporation, 494
Page, J. Boyd, 251–52
Palfry, John, 219–20
Palladino, Nunzio J., 347
Palos Park site: acquisition of, 55–57; contamination of, 485–88; facilities at, 51–53, *52, 56;* returned to forest preserve, 113. *See also* Chicago Pile 2 (CP-2); Chicago Pile 3 (CP-3); Du Page County (Ill.)
Panelon Electron Accelerator Facilities, 420
Panofsky, Wolfgang K. H., 197, 216n
parallel supercomputers, 497–99
Parker, Margaret, 145–47
particle accelerators. *See* accelerators
Partnership for a New Generation Vehicle Initiative, 496–97
patents, 481–83
Patterson, Robert L., 51–52
Paulini, Joseph W., 475
Pavilion/Sculpture for Argonne (Graham), 406–7
PBF (Power-Burst Facility), 269
Peach Bottom (Pa.) nuclear power plant, 241, 267
peer review, 413–15, 421
Pennsylvania: blackout in, 283–84; nuclear accident in, 358–63
Pennsylvania State University, training at, 137, 173
Percy, Charles H.: and accelerator project, 421; and Argonne's political support, 290, 357, 395–97; and proliferation risk, 339
Perlman, Isadore, 9, 37
personnel: administration of, 271; board of governors' role in, 392–93; and breeder program restructuring, 358; changes in, 29–32; community leadership by, 378; cuts in, 272, 279, 289–90, 441; evaluation of, 273–74, 370–71, 493; "graying" of, 278–79; increase in, 84; lack of, 28; opportunities for, 297; and plutonium MUF, 325; policies for, 59, 78; and postwar planning, 38–40, 48; precautions for, 74, 102, 146; recycling of, 24–

25; and security issues, 62, 82; training for, 271; wages of, 48, 173, 401, 527n70. *See also* affirmative action; scientists; security

Peterson, A. W., 219

Peterson, Maj. A. V., 22, 37

Pewitt, Douglas, 368, 415

Pewitt, E. Gayle, 435

PFR (prototype fast reactor), 364

Phelan, Richard, 487

Phillips Petroleum Company, 8, 114

Physically Handicapped in Science Program, 378

physics: advances in, 75–77; research on, 102–3; and university-Argonne relations, 187. *See also* high-energy physics

Piore, Emanuel R., 197

Piore panel, 197–98, 203

Pittsburgh (Pa.), energy technology labs in, 336

Planchon, Peter, 447–48

plants, radiation demonstrations with, 190–91

plutonium: for bomb, 20; contamination by, 486; dangers of, 10, 36; discovery of, 177; in fuel rods, 444–45, 447–48; implosion technique for, 26; isolation of, 14–16; laboratory dedicated to, 103; missing (MUF), 323–25; patients injected with, 320–23; production of, 9–11, 16, 20, 31; public information on, 37; purification of, 27; recovery of, 131, 140; reprocessing and recycling of, 337–40; security issues for, 82–83, 337–40, 354–55, 357, 444, 528n80; separation of, 9, 26; studies of, 7; toxicity of, 31, 324; weapon-grade vs. power, 130. *See also* Redox process, development of

plutonium patients, 320–23

plutonium production reactors (CP-6), 12, 15, 97–98, 515. *See also* Savannah River Laboratory: reactors at

Plutonium Project Record (PPR), 38, 44

plutonium-239: creation of, 114–16; missing (MUF), 323–25; research on, 177–78

plutonium-240, 26

plutonium-244, 178

plutonium-245, 178

plutonium-246, 178

Poeppel, Roger, 478

polarized-beam facility. *See* Zero Gradient Synchrotron (ZGS)

politics: and accelerator construction, 168–72; and accelerator location and control, 158–60; and advanced photon source, 471–72; and AEC appointees, 79–80; and AEC changes, 204–6; and Argonne administration, 350; and Argonne's possible closing, 389–91; context of, 81, 107, 242; and energy policy, 285, 333–36, 373–75; and national labs' future, 502; at peaceful uses conference, 190; and peer review, 421; and reactor projects, 148, 299–300, 448, 452–56; and research priorities, 247, 337–38; and science education, 389; and security, 82–84; and site selection, 65–66. *See also* Cold War; science politics

Pollard, Robert, 452

Power-Burst Facility (PBF), 269

Power Reactor Demonstration Program, 148, 195

Power Reactor Development Company (PRDC), 141–42, 149, 230

power reactor innovative small module (PRISM), 453

Powers, Philip: appointment for, 235; and Argonne administration, 270–71, 274, 275; and environmental studies, 251, 254–55; opposition to, 256; and personnel decisions, 262; and policy direction, 243–47; and reactor projects, 239, 258; resignation of, 275; and Tripartite Agreement, 240; and ZGS funds, 264

Poyer, David, 489

PPR (Plutonium Project Record), 38, 44

PRDC (Power Reactor Development Company), 141–42, 149, 230

Precision Analog Computing Equipment (PACE), 125

President's Science Advisory Committee (PSAC), 197

Press, Frank, 377

pressurized-water reactors, 70, 114, 514. *See also* U.S. Navy, reactor program of

Price, David, 412

Price, Melvin: and environmental conference, 253; and FARET cancellation, 231; and high-flux reactor, 236–37; and human radiobiology, 315–18; and reactor insurance, 148; and site selection, 65–66

Price-Anderson Act (1957), 147–48

Princeton University: and accelerator project, 196, 204, 327; and computer development, 77, 122–23; fusion research at, 287, 343; pile experiments at, 9

PRISM (power reactor innovative small module), 453

Project Independence, 302
property: acquisition of, 49–50, 55–57, 64–66; negotiations over, 52–53; search for, 51–53. *See also* facilities; Palos Park site
protein crystallography, 474
protons, in accelerators, 153–54
prototype fast reactor (PFR), 364
PSAC (President's Science Advisory Committee), 197
public attitudes: in Sweden, 446; toward accelerators, 326; toward AEC, 290–91; toward atomic bomb, 110; toward energy, 285; toward energy crisis, 334–35; toward high-energy physics, 326; toward intellectualism, 144; toward nuclear energy, 278–79, 281–82, 325–26, 337, 353, 432–33, 446, 452; toward science, 144, 290
Public Citizen, 455
public relations: and Argonne's image, 349; and declassification, 290–91; and EBR-I meltdown, 143–44; and plutonium MUF, 323–25; and plutonium patients, 320–22; and reactor dedication, 145–47; and science education, 389
pulsed-spallation facility, 412
Purcell, E. M., 216n
Putnam, Frank, 409

quality assurance issue, 371–72
quarks, theory of, 327, 419

Rabi, Isidor Isaac, 30; on accelerators, 168–71; and personnel issues, 33; research by, 2
Rabinowitch, Eugene, 36
radiation: alarms for, 360; checkpoints for, 146; collection of, 313; controls for, 16, 18–19; effects of, 79, 89, 101–2, 190, 297; exposure to, 10–12, 74, 529n89; and fuel costs, 130; monitoring of, 31, 97; and public attitudes, 110; recording damage from, 409–11; and remote control development, 125–26; research on, 109, 117; safety measures for, 57–58; toxicity of, 101–2, 324; and waste disposal, 45. *See also* Center for Human Radiobiology; fallout
radiation illness, 318
Radiation Lab (Mass.), 7
radioactivity, discovery of, 153
radioisotopes: bone-seeking, 318; cancer treatment with, 90; locus of research on, 85; medical research on, 75; patients injected with, 320–23; production of, 60. *See also* internal emitters; isotopes

radiological assistance teams, duties of, 360
Radiological Society of North America, 79
radium: discovery of, 176; exposure to, 74, 102, 315–18; health hazards of, 10, 36; toxicity of, 101–2
Ramey, James T.: and Argonne administration, 270; at conference, 253; and Harvey's attempted security breach, 97; and reactor projects, 230, 234, 265; and research focus, 300
Ramsey, Norman, 206, 216–18
Rancho Seco reactor, electrical short-circuiting in, 446
Rasmussen, Norman, 339, 363
Rasmussen reactor safety study, 339, 363
Ratner, Larry, 152–53, 198, 199
Rauch, Hilary, 404, 437
Ray, Dixy Lee, *299;* appointment for, 290; and Argonne administration, 293, 298, 307; and energy plan, 300–304; goals of, 290–91, 298; leadership of, 299–300, 305; and plutonium patients, 322
RDT. *See* Division of Reactor Development and Technology (RDT)
REACH (educational program), 271
Reactor Safeguards Committee, 67, 101, 120, 141
Reactor Safety Review Committee, 144
Reagan, Ronald: administration of, 394–95, 413, 440, 475, *477;* and breeder reactor, 425–26; budget of, 388, 395–96, 438–42, 448, 465; election of, 376, 380; energy policies of, 385, 387–89, 398, 433–35; and government-industry cooperation, 383; labs' overhaul ordered by, 400–404; on non-weapon research, 394; and research-and-development expenditures, 399–400; science advice for, 377, 384
Reck, Ruth, 483
recycling, research on, 489
Red Gate Woods Preserve, 486–88
Redox process, development of, 63, 77
Reed, Chauncey W., 56–57
Reed, F. W., *210*
regenerative reactors, 130, 130n
Regenstein Library (Chicago), 406
Relativistic Heavy Ion Collider, 472
Remote Control Engineering Laboratory, 125–26
reprocessing technology: and nuclear weapons proliferation, 338–40; for plutonium, 337–40. *See also* breeder reactors

Research and Development, awards from, 443
Rettaliata, John T., committee headed by, 187–89
Revelation (Bible), and environmental pollution, 249
Rhodes, Richard, 447
Rickover, Capt. Hyman G.: and Argonne's role, 132; and civilian projects, 128; management philosophy of, 230–31, 232, 267, 270, 298, 300; and reactor research, 70–71, 88–89, 98, 100, 114; staff of, 230, 272
Ridenour, Louis J., 155
Roberson, John H., 199, 215
Robert Taylor high-rise housing project, 489
Robinson, Ben, 489
robots, 126, 146
Rocky Glen Forest Preserve, 55
Roddis, Louis, 397
Roentgen, Wilhelm Conrad, 468, 473
Roosevelt, Franklin D., 3–5, 8, 34, 335
Rose, John E., 57
Rosen, Louis, 276, 291
Rosenberg, Ethel, 97
Rosenberg, Julius, 97
Rosner, Carl H., 480
Rossini, Frederick J., 250–51, 255
Rostenkowski, Daniel, 439, 448
Rothchild, Maurice L., 430
Rowe, Hartley, 59
Rowen, Henry S., 253
Rowland, Robert E., 317–18, 321–22, 397–99
Royall, Kenneth C., 52
Ruggiero, Alessandro, 465
Russell, George A., 379
Rutherford, Ernest, 153
Rutherford laboratory (U.K.), 415

Sachs, Robert G. (Bob), *292, 299;* administration of, 280, 293–94, 305–8, 311, 312, 346; appointments for, 223, 289, 291, 299; background of, 223–24, 276, 291–93, 297–98, 330; and Carter's nuclear policy, 339–40; and energy plans, 301–3, 309–10, 337–38, 342; and ERDA priorities, 314–15; goals of, 298–300; and high-energy physics, 329; and human radiobiology, 318; legacy of, 331–32; and national laboratories, 336, 344; and plutonium MUF, 324–25; resignation of, 262, 346–47; and Shaw's attitude, 297–98; and solar energy, 312–14, 337; and ZGS, 224–25
Safe Energy Communication Council, 454

safety: antinuclear criticism of, 452–53; and BORAX-I, 118–21; and breeder reactors, 363–65; changing standards for, 486–87; checkpoints for, 146; components of, 364; and CP-5, 117–18; debate over, 338–40, 362–63, 432–33; and environmental research, 100–101; and ergonomics, 367–68; fuel and coolant issues in, 444–45, 447–48; funds for, 344; and high-flux reactor, 238; importance of, 271; and management techniques, 230–31; mock exercises for, 207–8; as priority, 300; and public relations issues, 143–44; and radiation research, 57–58, 117–18; and radiobiology studies, 315–18; reassessment of, 205; and remote control development, 125–26; responsibility for, 127; in routine operations, 365; standards for, 323; at Three Mile Island, 362. *See also* accidents; civil defense
safety research and engineering facilities (SAREF), 358, 364
Saft America Inc., 495
Sandia Laboratory: administration of, 343; budget and staff of, 289, 506; contracts with, 344
Savannah River Laboratory: administration of, 343; reactors at, 95, 97–98, 100, 110, 149
Sawmill Creek, and nuclear waste, 485
Schenectady (N.Y.). *See* General Electric Company
Schiffer, J. P., 422
Schlesinger, James R.: administration of, 335, 336, 344, 347, 350; appointments for, 270, 290; and Argonne's director, 276; and Argonne's status, 274; and breeder reactor program, 357–58; and radium exposure research, 317
Schriesheim, Alan, *435;* administration of, 427, 436–37, 442–43, 481, 490–93; and advanced photon source, 464–68, 470–72; appointments for, 423–24, 428–29; and Argonne's future, 502–3; background of, 423–24; and budget, 438–39, 441–42, 448; and Clinton administration, 453–56; and Cold War's end, 459; and cooperative research, 498–99; and directorship, 424–25, 429; and economy, 455; and integral fast reactor, 443–46, 448–51, 453–54; retirement of, 503; and science agendas, 460; and SSC, 463; and strategic planning, 437–40; and superconductivity, 475–76, *477,* 478; and technology transfer, 435

Schriesheim, Bea, 436
Schultz, George, 476
science: criticism of, 501; definition of, 348–49; education in, 377–78; future of, 375; postwar context of, 182–83; public attitudes toward, 290, 459–60; role of management in, 183–86. *See also* nuclear science; science politics; scientists
Science Careers for Women conference, 492, 493
Science Digest, awards from, 443
science explorers, program for, 490, *491*
science politics: context of, 242, 278; and IFR cancellation, 458; leadership in, 289; and Tripartite Agreement, 246
Science (periodical), on high-energy physics, 462
Scientific and Technical Advisory Committee (STAC), 408–9
scientists: academic appointments for, 428; and accelerator location, 158–64; on bomb's use, 36–37, 532–33n109; compartmentalization of, 25–26, 30–31; competition among, 176; in Congress, 285–86; cooperation among, 5–6, 26–28, 50, 90–91, 135–37, 180–81, 184, 528n80; evaluations of, 273–74, 386–87; and facility shortcomings, 89; freedom of, 59, 184; health concerns of, 10–12; influence by, 278, 348–49, 375–76; opportunities for, 279; patents for, 481–83; public confidence in, 144; recognition for, 184–85; recruitment of, 184, 402; and research organization, 6–8, 42–44, 202–4; responsibility of, 375–76; reunion of CP-1, *45;* role of, 37; and security issues, 9, 50, 62, 90–92, 133–34, 158–61; senate for, 245–46; training for, 91–92, 104–5, 135–38, 147, 173
scrubbers (coal), 396
sculpture, at ANL, 406–7
SDI (Strategic Defense Initiative), 441
Seaborg, Glenn T., *38;* appointments for, 9, 37, 59, 204, 205; and Argonne director's appointment, 208; colleagues of, 11; departure of, 39; and first controlled chain reaction, 20; health of, 31; and high-flux reactor, 236–37; lectures by, 105; and Lewis committee, 17–18; plutonium research by, 9, 14, 177, 179–82; proposals by, 35; role of, 12, 46, 218, 220, 491; Soviet Union visited by, 237; and transuranic isotopes, 26; and university-laboratory relations, 218

Seamans, Robert, Jr.: administration of, 310–12, 314, 342, 346; appointment for, 309
Searle Corporation, 465
Seawolf (submarine), 334
Secretary of Energy Advisory Board (SEAB), 500, 501n
security: and AEC, 81–84, 133–34; and Atoms for Peace program, 134; and bomb's use, 36–38; and BORAX-I test, 120; and classified research, 87; and compartmentalization, 25–26, 30–31; criticism of, 92, 133–34; double-edged nature of, 459; effects of, 30; of fissionable material, 82–84; and foreign-born scientists, 9; frustrations with, 50, 90–92, 158–59, 161; industry's concerns with, 140; and instrument construction, 44–45; and Korean War, 96–97; loosened restrictions on, 129; and plutonium MUF, 323–25; programs under, 135; and reactor research, 62; and scientists' collaboration, 90–91, 528n80; and university relations, 103–4. *See also* nuclear proliferation
SEFOR (Southwest Experimental Oxide Reactor), 230
Seitz, Frederick: and accelerators, 217–18; and advanced photon source, 464; and Argonne directorship, 42, 175; and cosmotron, 159; and high-energy priorities, 216; and university-Argonne relations, 187
Selig, Henry, 225
semi-works research, 21, 97–98
separation: of isotopes, 3; of plutonium, 9, 26; processes of, 27; of uranium, 78
SERI (Solar Energy Research Institute), 336
Sesonsk, Alexander, 233
Setti, Riccardo Levi, 313
Shain, Irving, 466–67
Shaw, Milton: and accelerator, 260; administration of, 270–77, 291, 311, 401; appointment for, 230; attitudes of, 297–98; and breeder reactor project, 230–33, 265–68, 270; and budget, 289; conflict with, 233–34, 265; departure of, 299–300; goals of, 270; and high-flux reactor, 237, 239–40, 258–59
Shenoy, Gopal, 465
Shepard, Kenneth, 417–18
Shippingport (Pa.): nuclear power station at, 128; students at, 137
Sierra Club, 454–55, 486
signage, priorities in, 311
Silent Spring (Carson), 460
Simmelman, J. G., *210*

Simpson, Oliver C., 89, 187
SIP (Sam's Infernal Pile). *See* Chicago Pile 5 (CP-5)
SLAC. *See* Stanford Linear Accelerator Center (SLAC)
SL-I (Stationary Low-Power Reactor), 150, 205
Smith, Harold D., 43
Smyth, Henry D.: and accelerator location, 160, 164; appointment for, 25; and Argonne's role, 28, 104; history by, 36; and postwar planning, 32; and research organization, 7; role of, 6, 96
Snapp, Roy, 108
Snow, C. P., 407
social change, 242, 378
social science: and energy issues, 345; research in, 349, 489–90
Socio-Technological Research Organization (STRO), 251–54
sodium: as coolant, 131, 151, 447–48, 516; properties of, 365–68, 447; in safety testing, 364
sodium-potassium alloy, 69
Sodium Reactor Experiment (SRE), 128
solar energy: and architecture, 405–6; collectors for, 312–14, 337; funds for, 314, 319; research on, 302, 310, 336; support for, 448
Solar Energy Research Institute (SERI), 336
solid methan moderator, 443
S-1 committee, 5–8, 21, 25
Sonnenschein, Hugo, 456
South, David, 484
South Carolina, reactors built in, 95, 97
Southeast Universities Research Association (SURA), and accelerator project, 419–22
Southern Governor's Conference, 248
Southwest Experimental Oxide Reactor (SEFOR), 230
Soviet Union: and accelerators, 169, 171–72, 328; atomic test by, 81, 88–89; competition with, 126, 137–38, 455, 459, 500–501; and conference participation, 138–41; cooperation with, 478; and high-energy physics, 169, 171–72; monitoring of debris from, 101; nuclear accident in, 446–48, 451, 460; reactor research in, 237; research focus of, 194, 223; satellites of, 190; and security issues, 62; transuranic elements research in, 181–82; troop movements of, 14, 17; U.S. assistance for, 451–52
spark chambers, use of, 198

SPEAKEASY (computer language), 482
Spedding, Frank, 12, 16–18, 38–39
Spencer, Bruce, 451–52, 457
spin-orbit coupling, 228–29
Spinrad, B. I., 115, 147, 149–51
split-ring resonator, 417–18
SP1 parallel supercomputer, 497–99
Sproul, Robert G., 37
Sputnik, 190
SRE (Sodium Reactor Experiment), 128
SSRL (Stanford synchrotron radiation laboratory), 466
STAC (Scientific and Technical Advisory Committee), 408–9
Stagg Field, *11;* first controlled reaction at, 1; football abolished from, 3; pile experiments at, 7, 10, 16. *See also* Chicago Pile 1 (CP-1)
Stahr, Elvis J., 218, 219, 252–53
Stalin, Joseph, 62
Stanford Linear Accelerator Center (SLAC): administration of, 343; architecture of, 406; budget of, 289; competition from, 211, 466; components sent to, 375; funds for, 203, 261; proposal for, 163–64, 197; support for, 328, 330; upgrade for, 416–17
Stanford synchrotron radiation laboratory (SSRL), 466
Star Wars (Strategic Defense Initiative or SDI), 441
Stationary Low-Power Reactor (SL-I), 150, 205
St. Charles Countians Against Hazardous Waste, 485
Stearner, S. Phyllis, 79
Stearns, Joyce C., 30, 34
steel, initiative on, 439–40
Steinbach, Edward S., 13
Stephenson, H. Kirk, 72
Stetter, Joseph, 482–83
Stevens, Rick, 497
Stevenson-Wydler Technology Innovation Act (1980), 481
Stewart, Irvin, 521n19
Stimson, Henry L., 25, 34
Stockman, David, 456
Stone, Robert S., 12–13, 32, 35, 37, 39
Stone & Webster (engineering firm), 13–15
STR (Submarine Thermal Reactor). *See* submarines, propulsion project for
Strassman, Fritz, 1–2, 135, 177
Strategic Defense Initiative (SDI), 441
Strauss, Adm. Lewis L.: and accelerators, 165–

67, 170, 174; and cancer research, 75; criticism of, 148; and H-bomb development, 94; and nuclear energy, 107, 145, 151; on peaceful uses conference, 190; and Soviet competition, 169; and Zinn's resignation, 168, 172–73
strip-mined land, reclamation of, 320
STRO (Socio-Technological Research Organization), 251–54
Strong Memorial Hospital, and plutonium patients, 321
strontium 90, contamination by, 486
Structural Biology Center, 438, 473–74
Studier, Martin, 177–78
subatomic physics, 76–77, 198–200
subcritical reactors, 113
submarines, propulsion project for, 69–71, 86, 88–89, 96, 98, 100, 109, 514–15
Submarine Thermal Reactor (STR). *See* submarines, propulsion project for
Sugarman, Nathan, 347
"Supercar," 496
supercomputing, initiatives in, 408, 497–99
superconducting linac, 417–18. *See also* Argonne Tandem-Linear Accelerator System (ATLAS)
Superconducting Products (company), 479
Superconducting Super Collider (SSC): cancellation of, 460, 473; competition for, 471–72; design for, 408, 462–63
superconductivity: center for applications of, 476, 502; commercialization of, 482–83; description of, 476–77; development of, 417–18, 478–80; and levitation of magnets, 475–76, 476
Superconductivity Research Center for Applications, 476, 502
SURA (Southeast Universities Research Association), and accelerator project, 419–22
Sweden: cyclotron in, 180–81; public attitudes in, 446
swimming-pool research reactor, 138
Switzerland. *See* CERN (Switzerland)
Symon, Keith, 199–200
synchrotrons, 154–55, 166, 209. *See also* Advanced Photon Source (APS) project; Zero Gradient Synchrotron (ZGS)
Szilard, Leo, 2–3, 9–10, 18, 36

Taecker, Rollin G., 147, 222
Tammaro, Alfonso, 58–59, 65, 170
tandem accelerator, 172
tandem-linear accelerator. *See* Argonne Tandem-Linear Accelerator System (ATLAS)
Tape, Gerald F.: and Argonne management, 391–93; and breeder reactor project, 265; and national laboratory system, 192; role of, 258; Soviet Union visited by, 237
Task Force on Energy (U.S. House), 285
TAT (training and technology), 271
Tate, John T., 38–39, 42
technetium, 486
Technical Advisory Panel, 260
technology transfer: agenda for, 403; agreements for, 494–97; center for, 428, 435, 481–82, 502; of coal technology, 311–12; components of, 494; definition of, 434–35; difficulties of, 480–81; legislation on, 481–82, 494; mechanism for, 502; in steel industry, 439–40; and superconductivity, 482–83
Technology Transfer Center, 428, 435, 481–82, 502
Teem, John, 301
television, in remote control use, 125–26, 146
Teller, Edward, 2–3, 224, 227, 229, 310
temperature coefficients, 118, 141–42
Teng, Lee C.: and accelerator project, 171–72; and advanced photon source, 465; recruitment of, 166; and ZGS, 198, 210–11
Tennessee Eastman Corporation, 51
Tennessee Valley Authority (TVA), 54
Teunis, Ronald, 447
tevatron, 462
Texas, blackout in, 283–84
Thalgott, F. W., 132
thermonuclear energy (fusion energy), 286–89. *See also* fusion research
thermonuclear weapons: components needed for, 95; debate over, 94; development of, 88; funds for, 191–92; secrecy of, 179; tests of, 128, 178
Thiele, Ernest W., 83
Thomas, Charles A., 32, 110
Thompson, James R., 396
Thompson, R. W., 198–99
Thompson, S. G., 177
Thompson, William M., 125
thorium, 34, 114–15, 484–85
thorium-uranium fuel, 340, 357
Thornburgh, Richard, 361
Thorne, Robert D., 307, 340, 350–51
3M Corporation, 465
Three Mile Island (TMI): aftermath of, 364–65, 368, 374, 444, 446, 460; assessment and

640 ■ Index

cleanup of, 362–63; design of, 366–67; nuclear accident at, 358–63
Thurnauer, Marion, 493
Till, Charles E.: appointment for, 376; and Argonne's future, 389, 439; background of, 376; on CRBR cancellation, 425–26, 444; and integral fast reactor, 444–47, 449–54, 457–58
T. J. Watson Research Center, 499
Today (television program), 336–40
Tolman, Richard, 5, 25, 32
torpedoes, nuclear power for, 132
toxic gases, identification of, 482–83
Transducer Research, Inc., 483
Transient Reactor Test Facility (TREAT), 149, 269, 296, 364, 446
transuranic elements: and bubble chambers, 198–200; discovery of, 176–82; production of, 449; research on, 179
TRIGA nuclear reactor, 241
Tripartite Agreement: and AEC, 307; and Argonne administration, 276, 373; AUA denunciation of, 306–8, 379–82; basis for, 331; criticism of, 351, 391; description of, 220–21; end of, 382, 393–94; evaluation of, 380–82; implementation of, 235, 242; implications of, 245–46, 255–56, 305; operation of, 220–23, 379–80; renewals of, 273, 275, 304–6, 308–9, 368, 378
tritium: contamination by, 486, 488; and fusion research, 287; meter for, 312; production of, 95, 103; research on, 191
Trivelpiece, Alvin: and advanced photon source, 464–65, 467; and Argonne contract, 394; and Argonne's future, 389, 396–97, 437–38, 440–42; associates of, 424; and budget, 448; and GEM project, 423; and national laboratory system, 400; and SSC, 463
TRUEX radioactive waste reduction process, 443
Truman, Harry S: AEC appointments by, 53–54, 79–80, 104; and Argonne's establishment, 42–44; and bomb's use, 36–37; H-bomb supported by, 94–95; and Korean War, 95; and postwar planning, 34–35; reelection of, 81; and Soviet bomb, 88
Tuck, James, 166
Tulgey Woods, 56
Turner, Louis A., 102, 171

Uncertainty Principle, 502–3

Union Carbide Corporation, 63, 220
United Nations, 126, 138–41
United States: blackouts in, 283–84; energy consumed by, 301; technology shortcomings in, 282. *See also* public attitudes; *specific U.S. agencies and government offices*
United Way projects, 488–90
universities: and accelerators, 154–70; funds for, 155–56, 203, 384–85; as model, 184–85; reactors for, 147; role of, 35, 384–87; and training issues, 136–37. *See also* university-laboratory relations
university-laboratory relations: and accelerator projects, 158–62; and advanced photon source, 464–67; Allerton meeting on, 213–14; Argonne's role in, 72–73; and breeder-reactor management, 233–34; changes in, 235, 372–73; commitment to cooperative, 195–96; competition in, 386–87, 419–20; criticism of, 386; and environmental studies, 251, 255–57; focus on, 434–35; and funding for projects, 422; governors' review of, 428; and high-energy priorities, 216–18; and identity, 90–92; and laboratory senate, 245; LBJ on, 218; of Met Lab and Chicago, 37–41, 48–49; and Rettaliata plan, 186–89; Williams committee on, 218–20. *See also* Argonne Universities Association (AUA); Associated Midwest Universities (AMU); educational programs; Midwest Universities Research Association (MURA)
University of California at Berkeley: accelerator at, 76, 154–55, 169, 211, 217, 328, 467; AEC programs at, 58, 214; collaboration with, 26; employees at, 37; facilities operated by, 4, 37, 49, 92; fission research at, 2; fusion research at, 287; and training, 216; transuranic elements research at, 177–79, 181
University of Chicago: and accelerators, 161, 164–65; and air pollution studies, 249; animosity toward, 213–14, 242, 391; architecture of, 406; and Argonne administration, 37–41, 48–49, 72–73, 92, 104, 188, 214–21, 347–51; and Argonne board of governors, 391–93; and Argonne graduate center, 214–16; Argonne's contract with, 394–95, 408; and AUA evaluation, 380–81; and Clinton Engineer Works, 25; cooperation with, 498; employees at, 507; expeditions from, 176; fees for, 276; funds for, 203; Met Lab's relations with, 8, 32–33, 37–38; NDRC contract

Index ■ 641

with, 5, 8; and personnel issues, 25; physics at, 155; and plutonium patients, 320; policy advisory board for, 188–89; role of, 219–20, 353–54, 369, 372–73, 391–92; and site selection, 21, 53, 55; support from, 200. *See also* Argonne–University of Chicago Development Corporation (ARCH); Institute for Nuclear Studies; Stagg Field; Tripartite Agreement
University of Detroit, students from, 105
University of Illinois: and accelerator project, 419–20, 422; and Allerton meeting, 213–14; Argonne's relations with, 187–88; and bubble chamber, 199; funds for, 203; physics at, 155; staff of, 241
University of Illinois at Chicago, cooperation with, 498
University of Iowa. *See* Ames Laboratory (Iowa)
University of Kansas, Argonne's relations with, 187
University of Maryland, cooperation with, 498
University of Michigan: and bubble chamber, 199; collaboration with, 331; competition from, 466; environmental research by, 254; labs at, 137
University of Minnesota, research at, 4
University of Missouri, reactor at, 412
University of Pennsylvania: and accelerator project, 196, 204, 327; and computer development, 122
University of Pittsburgh School of Public Health, radiobiology at, 316
University of Rochester, and plutonium patients, 321
University of Washington School of Medicine, radiobiology at, 316
University of Wisconsin: and accelerator project, 161, 164; Aladdin machine at, 466; Argonne's relations with, 187; and bubble chamber, 199
Untermyer, Sam, 116, 118, 120–21
uranium: availability of, 14, 60, 114, 357, 426, 445; contamination by, 486; development of, 484–85; enriched vs. natural, 67; in fuel rods, 444–45, 447–48; graphite bricks implanted with, 16–17; gun method for, 26; health hazards of, 10; for IPNS, 442–43; long burnup of, 129–30; miners of, 316; natural vs. enriched, 113; private enrichment of, 333; production of, 103; recovery of, 77, 140; regenerative fission in, 5; separation of, 78; splitting atoms of, 1–3, 177
uranium-233, 20, 114–16
uranium-235, 5, 6, 99, 114, 116, 131
uranium-238, 6, 20, 34, 114–15, 179
uranium-239, 177
uranium-253, 179
uranium-255, 179
Uranium Beam Facility, 438
uranium oxide, 82–84
uranium-plutonium-zirconium alloy, for fuel, 444–45, 447–48
urban rehabilitation, 488–90
Urey, Harold C., 5–6, 78
U.S. Advanced Battery Consortium (USABC), 495–97
U.S. Army: Bureau of Ordnance of, 2–3; compartmentalization in, 25; control by, 37; environmental damage by, 484–85; environmental impact statements for, 483; Intelligence and Security Division of, 30–31; and lithium-sulfur battery, 288; portable reactor for, 150, 205; project records for, 44. *See also* U.S. Army Corps of Engineers
U.S. Army Corps of Engineers: and Argonne's establishment, 41; control by, 21–22, 25; and plant construction, 12–13; and plutonium patients, 322; role of, 12–14, 20; and site selection, 21, 23–24, 51–53, 55–57. *See also* Manhattan Engineer District (MED); Manhattan Project
U.S. Bureau of Mines, personnel for, 78
U.S. Coast Guard, 482
U.S. Council for Automotive Research (USCAR), 496–97
U.S. Department of Commerce, nonweapons research in, 394
U.S. Department of Defense: budget of, 388–89, 440–42, 463; contracts with, 344; funds from, 439, 448; and power outages, 283; requirements of, 466; research and development in, 388. *See also* military
U.S. Department of Energy (DOE): and accelerator project, 419–21, 461; administration of, 340–43, 346–47, 374, 383, 401, 456; and advanced photon source, 471; and Argonne's future, 371, 440; and breeder reactor project, 355–57, 425–26; budget of, 343–44, 351, 383–84, 388–90, 415–16, 425, 467, 473, 481; and commercialization, 345, 383; criticism of, 365, 381–82, 391, 400–403; and director appointment, 350–51;

and environmental cleanup, 485–88; establishment of, 280–81, 335–36; evaluation by, 348, 368–70, 374, 394, 413–14; goals of, 341–42, 403, 427, 460; industrial cooperation with, 494–97; key personnel at, 508; opposition to, 380; and Reagan's election, 376, 395–97; restructuring/attempted abolishment of, 387, 395–97, 400, 402–3, 456; and superconductivity, 478–79; transition to, 331–32; and ZGS review, 330
U.S. Department of Energy and Natural Resources, organization of, 279
U.S. Department of Health, Education, and Welfare, contracts with, 249, 344
U.S. Department of Labor, Women's Bureau of, 493
U.S. Department of the Interior: contracts with, 344; energy technology labs in, 336
U.S. Environmental Protection Agency, 257, 288, 295, 485
users groups: for advanced photon source, 471, 473; experiments by, 386; funds for, 204; guidelines of, 202–3; and high-energy physics research support, 461–62; interests of, 264; model for, 212; and politics, 197–201; role of, 202–4, 260, 428; for synchrotrons, 471; for ZGS, 212, 217–18, 260, 264
U.S. Geological Survey, survey by, 101
U.S. International Nuclear Fuel Cycle, 371
U.S. Navy: nuclear bomb tests for, 50; and nuclear-powered submarines, 69–71, 88–89, 96, 272, 334, 514; petroleum reserves of, 333; reactor program of, 98, 100, 109, 114, 128, 132, 343, 451; reactor site for, 86
U.S. Public Interest Research Group, 454
U.S. Steel Corporation, 383
U.S. War Department: and Argonne site, 52; and Manhattan Project, 15
U.S. Weather Bureau, 74
Utah Power and Light Company, 516

Van de Graaff generators, 76, 156, 201, 416–17
Vibrio cholerae (cholera bacterium), 474
Vietnam War, context of, 260–61, 279
Vineyard, George H., 342
visitors, access for, 91–92
von Neumann, John, 122–23, 169–70
Voorhees, Walker, Foley & Smith (contractors), 66

Wade, David, 451

Waldman, Bernard, 219
Wali, Kameshwar, 327
Walker, Robert L., 329–30
Walker, W. D., 199
Walkowicz, Don, 497
Wallace, George, 125
Walsh, John, 393, 399, 416
Walters, Leon, 445
Warner, J. C., 168
Warren, Shields, 74
Washington Conference on Theoretical Physics, 2
waste management, environmental research on, 254. *See also* hazardous waste, cleanup of; nuclear waste
Waste Management Corporation, 490
watchmakers, radium exposure of, 102, 315–18
Waterfall Glen Forest Preserve, 485
water reactors, development of, 132. *See also* boiling-water reactors; heavy-water reactors; light-water reactors
watersheds, environmental research on, 254–55
Watkins, James D., 490, 501n
Watson, Cecil J., 31
Wattenberg, Albert, 16, 87
Watts, Richard, 18, 19
weapons laboratories: and automobile technology, 496; budget for, 192–93, 294, 441. *See also* Lawrence Livermore Laboratory; Los Alamos National Laboratory; nuclear weapons; Sandia Laboratory
Weapons Neutron Research facility (WNR-PSR), 413–14
Weart, Spencer, 144
Weaver, Lynn, 255
Weil, George, 18
Weinberg, Alvin, 67–68, 98, 107, 215, 338
Weinberger, Caspar, 475–76
Weisskopf, Victor, 206, 261–62
Weldon Springs (Mo.), cleanup of, 484–85
Westbrook, Edwin, 473
West Garfield Park (Ill.): economic development of, 488–90; science explorers of, *491*
Westinghouse Company, 70–71, 88–89, 98, 100, 132
Weston (Ill.). *See* National Accelerator Laboratory (later called Fermilab)
West Stands. *See* Stagg Field
Wheeler, John A., 2, 31
Whitaker, Martin D., 10

White, David C., 250
White, Milton G., 6–7, 20, 25, 32
White House Science Council, 400–402, 404
Wigner, Eugene: appointment for, 6, 8; and first controlled chain reaction, 19; and Hanford design, 26; and health concerns, 11; Nobel Prize for, 227; and nuclear fission discovery, 2–3; and pile experiments, 9, 15; proposals by, 35; role of, 10, 13, 24, 68
wildlife, at ANL, 430, *431*, 432–33
Wilkinson, W. D., 103
Williams, John H., 166, 216–20
Williams, Roger, 24
Willow Springs (Ill.), as nuclear free, 487
Wilson, Carroll L.: administration of, 58–59, 64, 72, 79; appointment for, 54; influence by, 85; and navy contracts, 71; resignation of, 94; and site selection, 55, 64–65; and university relations, 91
Wilson, John T., 290, 347, 350
Wilson, Robert R., 206, 260–63, 274–75
Wilson, Volney C., 5–6, 16
Winston, Roland, 312–13
Wisconsin, lab site in, 52. *See also* University of Wisconsin
WNR-PSR (Weapons Neutron Research facility), 413–14
Wojcicki, Stanley, 462
Wollan, Ernest O., 12
women: and director position, 351; educational programs for, 378; outreach programs for, 490; role of, 77–79; status of, 491–93
Women in Science program, 490–93
Woods, Leona, 18
Woods Hole committee, on high-energy physics, 462–63
World War II: and Argonne's origins, 1; and Met Lab's future, 33–34; and nuclear race, 2–3, 6; troop movements in, 14, 17
W particle, 462
WTTW (television station), 490

xenon fluorides, 225–26, *226*
xenon-135, 31
Xerox PARC, 499
x rays, 3, 10, 467–69
x-ray synchrotron. *See* Advanced Photon Source (APS)

Yale University, disease research at, 474
Yanders, Armon, 304

Yarymovych, Michael I., 310–12
Yates, Ronald, 501
Yates, Sidney, 441, 448, 455
YBC 1-2-3, 479
Yom Kippur War, and oil supplies, 302
Young, Hoylande D., 78
yttrium, 479

Zager, B. A., 181
Zaluzec, Nestor J., 443
Zero Gradient Synchrotron (ZGS), *326*; benefits of, 198, 200–201, 224–25, 261–62, 329–31; construction of, 189, 196, 210–11; dedication of, 223, 326; design of, 196–97, 217, 478; evaluation of, 329–30; facility for, 211; funds for, 203–4, 264; future of, 262–65; guidelines for experiments on, 202–3, 219; legacy of, 412, 416, 419; operation of, 328; politics of, 199–200, 212, 328; power of, 211–12, 327; public interest in, 337; requirements of, 197–98; shutdown of, 329–31, 346, 375, 411, 461; status of, 295; users committee for, 212, 217–18, 260, 264
Zero Power Assembly, 96
Zero Power Plutonium Reactor (ZPPR), 268–69, 296, 333, 446
Zero Power Reactor Building, 90
Zero Power Reactor Number 1 (ZPR-I), 149
Zero Power Reactor Number 2 (ZPR-II), 149
Zero Power Reactor Number 3 (ZPR-III), 149, 269
Zero Power Reactor Number 6 (ZPR-VI), 149, 269, 296
Zero Power Reactor Number 9 (ZPR-IX), 149, 269, 296
zero-power reactors: construction of, 149, 268; description of, 269, 515; status of, 296–97
ZEUS detector, 438, 462
ZGS. *See* Zero Gradient Synchrotron (ZGS)
ZING, 414
ZING-P, 412, 414
Zinn, Walter H., *41*, *253*; and accelerators, 156–59, 162–64, 166–68, 170–72; administration of, 48–51, 60, 78–79, 81, 84–85, 90–94, 105–6; and AEC appointees, 54, 59–62; appointment for, 30–31, 42, 46; and Argonne's construction, 89; and Argonne's future, 28; and Argonne's goals, 87–88, 132–33, 152; background of, 47; and cancer research, 75; and EBR-I meltdown, 141–44; and electricity generation, 99–100; and en-

ergy plan, 301; and fallout from nuclear tests, 97; and H-bomb development, 95; and instrumentation, 23; and Korean War, 95–96; and nuclear fuels, 129–31; and nuclear waste, 73–74; and pile projects, 31, 40, 46, 175; and postwar planning, 34–35; proposals by, 35, 40; and reactor projects, 62–64, 66–69, 86–88, 97–99, 114, 118–21, 128, 130–33, 139–40, 151; resignation of, 165, 168, 171–74; role of, 10, 16, 35, 138–41, 514; and security issues, 82–84, 134; and site selection, 51–53, 55–57, 65–66; and submarine project, 70–71, 88–89, 96; and training program, 135–37; and university relations, 72–73, 103–4

ZIP (safety rod), 16, 18–19, 175

ZIP (Zinn's Infernal Pile). *See* Experimental Breeder Reactor Number 1 (EBR-I)

zirconium: in fuel rods, 444–45, 447–48; production of, 98; research on, 77

Z particle, 462

ZPPR (Zero Power Plutonium Reactor), 268–69, 296, 333, 446

Zweig, George, 327

JACK M. HOLL, professor of history at Kansas State University, Manhattan, was principal investigator and author for the Argonne history project. From 1980 to 1988 he served as chief historian of the U.S. Department of Energy. He is the coauthor, with Richard G. Hewlett, of *Atoms for Peace and War, 1953–1961: Eisenhower and the Atomic Energy Commission.* He has also coauthored histories of the Department of Energy and of U.S. civilian nuclear power policy, 1954–84.

Richard G. Hewlett is a senior partner and chairman of the board of History Associates, Inc. He is the former chief historian of the Atomic Energy Commission, the Energy Research and Development Administration, and the Department of Energy and the coauthor of a three-volume history of the Atomic Energy Commission.

Ruth R. Harris, retired director of research at History Associates, Inc., coordinated research on the Argonne history project. She is the author of *Dental Science in a New Age,* written for the National Institute of Dental Research.

Alan Schriesheim served as the director of Argonne National Laboratory from 1984 to 1996.